Maths for Economics

W9-AEX-816

Maths for Economics

Third Edition

Geoff Renshaw
with contributions from Norman Ireland

OXFORD
UNIVERSITY PRESS

OXFORD

UNIVERSITY PRESS

Great Clarendon Street, Oxford OX2 6DP

Oxford University Press is a department of the University of Oxford.
It furthers the University's objective of excellence in research, scholarship,
and education by publishing worldwide in

Oxford New York

Auckland Cape Town Dar es Salaam Hong Kong Karachi
Kuala Lumpur Madrid Melbourne Mexico City Nairobi
New Delhi Shanghai Taipei Toronto

With offices in

Argentina Austria Brazil Chile Czech Republic France Greece
Guatemala Hungary Italy Japan Poland Portugal Singapore
South Korea Switzerland Thailand Turkey Ukraine Vietnam

Oxford is a registered trade mark of Oxford University Press
in the UK and in certain other countries

Published in the United States
by Oxford University Press Inc., New York

© Geoff Renshaw 2012

The moral rights of the author have been asserted
Database right Oxford University Press (maker)

First published 2005
Second edition 2009

All rights reserved. No part of this publication may be reproduced,
stored in a retrieval system, or transmitted, in any form or by any means,
without the prior permission in writing of Oxford University Press,
or as expressly permitted by law, or under terms agreed with the appropriate
reprographics rights organization. Enquiries concerning reproduction
outside the scope of the above should be sent to the Rights Department,
Oxford University Press, at the address above

You must not circulate this book in any other binding or cover
and you must impose the same condition on any acquirer

British Library Cataloguing in Publication Data

Data available

Library of Congress Cataloging in Publication Data

Data available

Typeset by Graphicraft Limited, Hong Kong
Printed in Italy on acid-free paper by L.E.G.O. S.p.A. – Lavis TN

ISBN 978–0–19–960212–4

3 5 7 9 10 8 6 4 2

To my wife, Irene, for her unstinting moral and practical support;
and to my mother and father, to whom I owe everything.

Contents

Part One Foundations

Part Two Optimization with one independent variable

Part Three Mathematics of finance and growth

Part Four Optimization with two or more independent variables

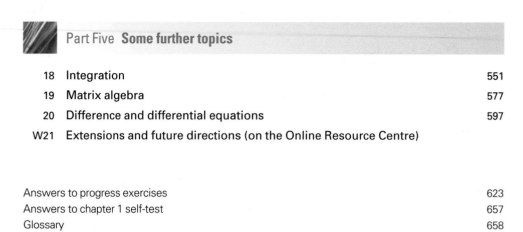

Part Five **Some further topics**

Detailed contents

Part Two **Optimization with one
          independent variable**

Part Five **Some further topics**

About the author

Geoff Renshaw was formerly a lecturer and is now an associate fellow in the Economics Department at Warwick University. He has lectured mainly in the areas of international economics, national and international economic policy, and political economy, but also taught maths to economists for more than thirty years. His teaching philosophy has always been to remember his first encounters with new ideas and techniques, and keep in mind how difficult they seemed then, even though they may seem obvious now. Geoff has always endeavoured to keep things simple and down to earth and infect his students with his own enthusiasm for economics.

Geoff was educated at Oxford and the London School of Economics. Before becoming an academic he worked in the research department of the Trades Union Congress. Most of his career has been spent at Warwick University, but he has also taught at Washington University, St Louis, and at Birmingham University.

He has also been a consultant to the International Labour Organization (a UN agency) and spent two years in Geneva working on international trade and economic relations between industrialized and developing countries, in addition to a year in Budapest, where he headed a project on the Hungarian labour market. Geoff has also consulted for the United Nations Industrial Development Organization, and has spent time in Vienna and Warsaw working on the Polish economy.

Geoff has published several books on industrial adjustment, north–south trade and development, and multinational corporations.

Outside of economics and politics Geoff enjoys studying the English language, practising DIY on houses and cars, and thinking up new inventions—none successful as yet. He is married with three children.

About the contributor

Norman Ireland has been a professor of economics at the University of Warwick since 1990, and was Chair of the Department of Economics from 1994 to 1999. He was joint Managing Editor of the *International Journal of Industrial Organization* from 1986 to 1992, and is currently a member of the editorial board of the *Journal of Comparative Economics*. He has published two books in the field of industrial organization and a number of articles across several fields of economics, but particularly in industrial economics, public economics, and economic theory. He has always been involved in teaching mathematics for economists, or mathematical economics, at various levels from first year undergraduate to Master's level.

About the book

This book is intended for courses in maths for economics taken in the first year, or in some cases in the second year, of undergraduate degree programmes in economics whether they be single honours or combined honours courses. It has its origins in lectures that I gave for many years to first year economics students at Warwick University.

Students arriving to study economics at British universities are highly diverse in their prior exposure to both maths and economics. Some have studied maths to the age of 17 or 18 (GCE AS and A2 level in the UK) and arrive at university with some degree of competence and confidence in maths. Others have studied maths only to the age of 16 (GCSE level) and many of these have forgotten, or perhaps never fully understood, basic mathematical techniques. There are also many students from abroad, whose backgrounds in maths are highly varied. Moreover, some students beginning economics at university have previously studied economics or business studies at school, while others have not.

The degree courses taken at university by students of economics are also highly diverse. In some courses, economics is the sole or main subject. Consequently maths for economics is prioritized and sufficient space is created in the curriculum to allow it to be explored in some depth. In other degree courses economics forms only a part, and sometimes a small part, of a combined-subject programme that includes subjects such as business studies, philosophy, politics, and international studies. Then, the crowded curriculum often leaves little space for studying maths for economics.

This diversity, in both students' prior knowledge of maths and economics and their course requirements, creates a challenge for anyone attempting to write a textbook that will meet the needs of as many students and their courses as possible. This book seeks to respond to this challenge and thereby enable every reader, whether mathematically challenged or mathematically gifted, and whether they are specializing in economics or not, to realize their true potential in maths for economics, and thereby develop the tools to study economics more effectively and more rewardingly.

More specifically, in responding to this challenge I have attempted to give the book four core structural characteristics:

1 Confidence building

Recognizing that many economics students found maths difficult and unrewarding at school, and have often forgotten much of what they once knew, part 1 of the book is devoted entirely to revision and consolidation of basic skills in arithmetic, algebraic manipulation, solving equations and curve sketching. Part 1 starts at the most elementary level and terminates at GCSE level or a little above. It should be possible for every student to find a starting point in part 1 that matches his or her individual needs, while more advanced students can of course proceed directly to part 2. More guidance on finding the appropriate starting point is given in the chapter map on pages xviii–xix.

2 Steady learning gradient

Many textbooks in this area develop their subject matter at a rapid pace, thus imposing a steep learning curve on their readers. This often leaves students with a weak maths background feeling lost, while even students who are relatively strong in maths sometimes fail to grasp concepts fully and to understand the economic analysis behind the various techniques and applications they are learning. To avoid these pitfalls, I have tried to give this book a carefully calibrated learning gradient that starts from the most basic level but gradually increases in mathematical sophistication as the book progresses. Consequently, no reader need be lost or left behind, and hopefully will go beyond a rote-learning approach to mathematics to achieve (perhaps for the first time) true understanding.

3 Comprehensive explanation

Many textbooks skim briefly over a wide range of mathematical techniques and their economic applications, leaving students able to solve problems in a mechanical way but feeling frustrated by their lack of real understanding.

In this book I explain concepts and techniques in a relatively leisurely and detailed way, using an informal style, trying to anticipate the misconceptions and misunderstandings that the reader can so easily fall victim to, and avoiding jumps in the chain of reasoning, however small. Wherever possible, every step is illustrated by means of a graph or diagram, based on the adage that 'one picture is worth a thousand words'. Many of the explanations are by means of worked examples, which most students find easier to understand than formal theoretical explanations. There is extensive cross-referencing both within and between chapters, making it easy for the reader to quickly refresh their understanding of earlier concepts and rules when they are re-introduced later. There is also a glossary which defines all of the key terms in maths and economics used in the book.

4 Economic applications and progress exercises

As soon as it is introduced, every core mathematical technique is immmediately applied to an economic problem, but in a way that requires no prior knowledge of economics. While this is challenging to the reader because it requires grappling with mathematics and economics simultaneously, I feel that it is essential to renew and reinforce the reader's motivation. Additionally, progress exercises have been strategically positioned in every chapter. I regard these as an integral part of the book, *not* an optional extra. Their answers are at the end of the book, while much supplementary material can be found at the book's Online Resource Centre (www.oxfordtextbooks.co.uk/orc/renshaw3e/). For more ambitious readers and lengthier courses, the final part of the book contains some relatively advanced topics, and there is a further supplementary chapter W21 at the Online Resource Centre.

How to use the book

To the student

Of course, the way in which you use this book will be primarily dictated by the requirements of your course and the instructions of your lecturer or tutor. However, at university you are expected to undertake a significant amount of independent study, much of which needs to be self-directed. The chapter map is intended to help you with this. You will see that the book caters for three levels of prior maths knowledge, labelled A, B, and C in the map. Even if you don't feel you fit neatly into any of these categories, studying the flow chart should help you to choose your own personal route through the book.

Although much effort has gone into making this book as user-friendly as possible, studying maths and its application to economics can never be light reading. In a single study session of 1–2 hours you should not expect to get through more than a few sections of a single chapter. To achieve a full understanding you may need to re-read some sections, and even whole chapters. It is usually better to re-read something that you don't fully understand, rather than pressing on in the hope that enlightenment will dawn later. You should always take notes as you read. It also greatly helps understanding if you work through with pencil and paper all the steps in any chain of mathematical reasoning. Tedious, but worth the effort. 'No pain, no gain' is just as true of mental exercise as it is of physical exercise.

Above all, it is essential that you attempt the progress exercises at the end of each section, as this is the only reliable way of testing your understanding. Worked answers to most of the questions are at the end of the book, with further answers on the book's website, or Online Resource Centre, which is at

www.oxfordtextbooks.co.uk/orc/renshaw3e/

There you will also find more exercises with answers, and a wide range of additional material such as how to use Excel® to plot graphs.

To the lecturer or tutor

At first sight this book may appear excessively long for many courses in maths for economics, which, in today's crowded syllabuses, are often quite short. However, this length is deceptive, for two reasons. First, explanations are quite detailed, facilitating independent study and thereby economizing on teaching time. For example, at Warwick University those who need to study part 1 of the book do so as an intensive revision programme, most of which is independent study, in the first two weeks of term.

Second, the range of material covered between the first and last chapters is so wide that it is extremely unlikely that any course would find the whole book appropriate for study. Rather, there are at least three overlapping study programmes within the book, as outlined in the chapter map. If none of these three suggested study programmes is suitable, the map may help you to design a path through the book that matches your syllabus requirements and the characteristics of your students.

Please note too that the book's website, or Online Resource Centre (see address above), contains much useful supplementary material, including a bank of exercises and answers reserved (by means of a password) for lecturers which can therefore be used for setting tests and examinations. There is also an additional chapter, W21, of more advanced material written by Professor Norman Ireland (see main contents pages).

Chapter map: *alternative routes through the book*

Choose A, B, or C as your starting point, then follow the arrows

(A) You have forgotten almost all of the maths you ever knew and want to make a completely fresh start.

(B) You have passed GCSE maths or an equivalent exam taken at age 16+, but you have done no maths since and now feel the need for some revision.

(C) You have passed AS/A2 maths or equivalent exams taken at age 17+ and 18+ and are fairly confident in your maths knowledge at this level.

Part one **Foundations**

Chapter 1. This starts from the lowest possible level and aims to rebuild basic knowledge and self-confidence. Be sure to complete the progress exercises and the **self-test** at the end of the chapter.

Take the **self-test** at the end of chapter 1; answers are at the end of the book. If you struggle with this, read chapter 1 and complete the progress exercises before going on.

Chapters 2–5. These revise the algebra component of GCSE maths or equivalent maths exam taken at age 16+. Chapters 3–5 contain in addition some economic applications.

In chapter 5, sections 5.5–5.9 go a little beyond GCSE maths and you can skip them if you wish, but be sure to study sections 5.10–5.12 on inequalities as these are important in economics.

Part two **Optimization with one independent variable**

Chapters 6 and 7. These introduce differentiation, a powerful mathematical technique widely used in economics. You may find these chapters a little difficult initially, but hard work at this stage will pay off later in your studies.

Chapters 8 and 9. These apply to economics the techniques of differentiation learned in chapters 6 and 7. Chapter 8 is concerned with a firm's costs, the demand for its product and its profit-seeking behaviour. Chapter 9 is devoted to the concept of elasticity.

If you are joining the book at this point because you have passed AS/A2 maths or equivalent exams, you will find that you are already familiar with all the pure maths used in these chapters. However, you may feel the need to browse chapters 6 and 7 for revision purposes. You should also study the economic applications in chapters 3–5 (see detailed contents pages). This will also help you to tune in to the book's notation and style.

Part three Mathematics of finance and growth

Chapter 10. This important chapter introduces the key concept of present discounted value, and also how to calculate growth rates, effective interest rates and repayments of a loan. The maths is fairly simple and mostly covered in the GCSE syllabus, though its economic application will of course be new.

This chapter is not closely linked to any other chapters and can be read at any time.

Chapters 11–13. These chapters explain the maths of logarithmic and exponential functions, which are used widely in economics. These concepts are covered in AS/A2 maths, though less fully. If you find you know the maths already, skip to the economic applications in sections 11.8, 12.9–12.11, and 13.8–13.10.

Part four Optimization with two or more independent variables

Chapters 14–17. These four chapters are, in a sense, the core of the book. The maths in these chapters will be new to all students, but is a natural extension of part 3 and you should find it no more difficult than earlier chapters.

Chapters 14 and 15 introduce functions with two or more independent variables, their derivatives, and maximum/minimum values. This material, although new to all students, is a natural extension of chapters 6 and 7 (and earlier chapters).

Chapter 16 explains the Lagrange multiplier, an optimization technique with many important uses in economics. Chapter 17 introduces some new but quite simple mathematical concepts and techniques: homogeneous functions, Euler's theorem, and the proportionate differential.

The economic applications—to cost minimization, profit maximization, and consumer choice among others—take up about one-half of chapters 14–16, and most of chapter 17.

Part five Some further topics

There are four chapters in this part, each of which can be studied independently of one another and of the rest of the book. All chapters contain economic applications.

Chapter 18 introduces the mathematical technique of integration, with some applications to economics. The maths will be familiar if you have taken AS/A2 maths, but will also be well within the capacity of any student who has progressed this far in the book.

Chapter 19 is concerned with matrix algebra, which some students of AS/A2 maths will have met before, but which again will be fairly readily understood by any sufficiently motivated student.

Chapter 20 introduces difference and differential equations, which will be new to all students but which are in part merely an extension of work in chapter 13.

Finally, **chapter 21** develops three relatively advanced topics as a taster for students who want to carry their study of mathematical economics further. Owing to space constraints this chapter is located on the book's Online Resource Centre **www.oxfordtextbooks.co.uk/orc/renshaw3e/**.

Guided tour of the textbook features

Maths can seem like a daunting topic if you have not studied it for a while, and you may be somewhat surprised to find how much maths there seems to be in university economics courses. However, once you have overcome your initial fears you will find that the maths techniques used in mainstream economics are quite straightforward and that using them can even be enjoyable! This guided tour shows you how best to utilize this textbook and get the most out of your study, whether or not you have studied maths at A-level.

Objectives

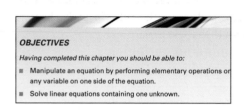

Each chapter begins with a bulleted list of learning objectives outlining the main concepts and ideas you will encounter in the chapter. These serve as helpful signposts for learning and revision.

Progress exercises

At the end of each main section of each chapter you will have the opportunity to complete a progress exercise, designed to test your understanding of key concepts before moving on. You are strongly recommended to complete these exercises to help reinforce your understanding and identify any areas requiring further revision. Solutions to the progress exercises are at the end of the book, with further materials at the Online Resource Centre at www.oxfordtextbooks.co.uk/orc/renshaw3e/.

Examples

You understand the theory, but how is it used in practice? Examples play a key role in the book, from short illustrative examples that demonstrate a formula in use to more involved worked examples that show step by step how an individual problem is solved.

Graphs and diagrams

There is an old saying: one picture is worth a thousand words. Reflecting this, verbal explanations are reinforced by numerous graphs and diagrams that will help you understand both the maths techniques and the economic applications.

EXAMPLE 10.18

If I deposit €100 in a bank that pays interest at a nominal rate of 1 will I get back after 5 years if interest is added (a) annually; (b) m

Answer:

(a) Using rule 10.4, $y = a(1 + r)^x$ with $a = 100$, $r = 0.1$ (because 10

$y = a(1 + r)^x = 100(1 + 0.1)^5 = 161.05$

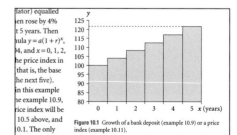

Figure 10.1 Growth of a bank deposit (example 10.9) or a price index (example 10.11).

Economic applications

Each key mathematical idea in the text is applied to an economic situation, so that you can immediately see the usefulness and relevance of maths in solving economic problems, and its significance in economic methodology.

4.13 Economic application 1: supply an

At the beginning of this chapter we said that we needed to develop s
in order to be able to analyse non-linear relationships in econom
equipped for this analysis, as we first need to look at some other n
next chapter. However, to conclude this chapter on quadratic equati
briefly indicate some economic relationships that are likely to be qua

Rules

Each chapter highlights the most important rules (key definitions and relationships) that you need to know to complete the maths that underpins the economics. You should memorize and revise the rules in each chapter before moving on to the next topic.

RULE 10.4 The compound growth formula

The formula is: $y = a(1 + r)^x$

where

a = 'principal' (initial sum invested, or initial value of the v
r = annual proportionate interest rate (for example, if the i
    proportionate interest rate is $r = 10 \div 100 = 0.1$)
x = number of years
y = future compounded value (= value of initial sum + cun

Hints

Hint boxes have been included throughout the text to alert you to common mistakes and misunderstandings, so that you can proceed with your studies with confidence.

Hint Beware of a mistake that is often made in handling gr
The mistake is to assume that if a variable grows at 4% per
is 4% × 5 = 20%. This would be correct if growth followed
section 10.4); but, as explained above, growth almost invar
growth formula (rule 10.4).

Summaries

The central points and concepts covered in each chapter are distilled into summaries at the end of chapters. These provide a mechanism for you to reinforce your understanding and can be used as a revision tool.

Summary of sections 3.1–3.7

In sections 3.1–3.4 we showed how any equation could be r
mentary operations (adding, multiplying and so on) to bo
the key distinction between variables and parameters in any
one unknown can be solved (rule 3.2).

In sections 3.5 and 3.6 we introduced the idea of a functio
For any linear function $y = ax + b$ the graph of this functio
referred to as a linear function. The slope is given by the cor

End-of-chapter checklists

The topics in each chapter are presented in checklist form at the end of every chapter to allow you to reflect on your learning and 'tick' each topic as you master it, before moving on if you wish to the further exercises on the Online Resource Centre.

Checklist

Be sure to test your understanding of this chapter by
attempting the progress exercises (answers are at the
end of the book). The Online Resource Centre contains
further exercises and materials relevant to this chapter
www.oxfordtextbooks.co.uk/orc/renshaw3e/.

The overall objective of this chapter was to refresh
and renew your understanding of the basic rules
that govern the manipulation of numbers. Specific-

✔ Fraction
  verting f
  and vice
  by a gi
  changes

✔ Index n
  pressing

✔ Powers

Guided tour of the Online Resource Centre

www.oxfordtextbooks.co.uk/orc/renshaw3e/

The Online Resource Centre that accompanies this book contains a further chapter, chapter W21, which due to space limitations had to be omitted from the book itself. The chapter, written by Norman Ireland of Warwick University, provides an introduction to some more advanced topics which should help undergraduate students intending to take further modules in mathematical economics in their second or later years of study, as well as postgraduate students.

The Online Resource Centre also provides students and lecturers with ready-to-use teaching and learning resources. These are free of charge and are designed to maximize the learning experience. Below is a brief outline of what you will find.

For students

Solutions to progress exercises

Once you've attempted the progress exercises in the text you can check the solutions at the end of the book. Some of the exercises also have expanded solutions at the Online Resource Centre to enhance your understanding.

Further exercises

The best way to master a topic area is through practice, practice, and more practice! A bank of questions, with answers, additional to the progress exercises in the book itself, has been provided for each chapter in the book to allow you to further test your understanding of the topics.

'Ask the author' forum

If you are struggling with a particular problem, or just cannot seem to get your head around a specific technique or idea, then you can submit your question to the author via the interactive online forum created for this text. As well as replying directly to you by email, Geoff will post his responses to all questions and comments from both students and lecturers on this site.

Instructions on how to use Excel® and Maple

An introduction to the use of Excel® and Maple software for graph plotting and solving equations has been created for both students and instructors, and includes demonstrations of how these software programs can assist in the use of maths for economics.

For adopting lecturers

Test exercises for instructors

One of the greatest burdens facing instructors is the need to continually prepare fresh assessment material. To aid this task, a suite of additional exercises, with answers, has been created. As these are password protected and hence not available to students, they are suitable for use by instructors in assignments and examinations. Lecturers who have adopted the book are assigned a password.

Graphs from the text

Again for instructors only, all graphs from the book have been provided in high-resolution format for downloading into presentation software or for use in assignments and exam material.

Acknowledgements

In preparing the third edition of this book I am again greatly indebted to the OUP editing and production team for their limitless encouragement and advice and their unfailing enthusiasm for the project. Specifically I warmly thank Kirsty Reade, my commissioning editor; my production editor, Joanna Hardern; Charlotte Dobbs, text designer; and Gemma Barber who designed the book's cover. I would also like to thank Peter Hooper, Kirsten Shankland, and Helen Tyas for their parts in the development of this new edition. For their immensely hard work and relentless attention to detail I am very grateful to the copy-editor, Mike Nugent, and the proofreader, Paul Beverley. I also thank June Morrison for compiling a very comprehensive and well-structured index. For their past and, I hope, future management of the book's Online Resource Centre, I am grateful to Fiona Loveday, Fiona Goodall, and Sarah Brett. Most of those named above also worked on the second edition of the book, for which I take this opportunity to thank them again. However, I should also like to repeat my thanks to the many others who have been involved in various ways and at various times in the production of this book. In particular I thank Tim Page and Jane Clayton, two former OUP staff without whom this book would almost certainly never have seen the light of day.

Amongst my colleagues at Warwick University and elsewhere, I am especially grateful to Norman Ireland who, having regrettably declined to become a co-author, agreed to write a lengthy and extremely valuable chapter, as well as setting numerous exercises and offering much general encouragement and support. I also owe a huge debt to Peter Law, whose meticulous checking and painstaking comments on many of the chapters saved me from a large number of small errors and a small number of large errors. Jeff and Ann Round also gave me valuable and very patient advice. Peter Hammond, despite having co-authored a book with which this one attempts to compete, was also very patient and helpful on a number of points. I am also grateful to those users of previous editions who have taken the trouble to email me, sometimes in praise and sometimes to point out errors. Both types of communication are very welcome.

As ever I am profoundly grateful to my wife Irene, who has been unfailingly patient and supportive throughout the three editions of this book. The late Mary Pearson greatly encouraged my labours, as did Lavinia McPherson – her 100 years notwithstanding. As always, the remaining shortcomings of this book are entirely my responsibility.

Part One
Foundations

- Arithmetic
- Algebra
- Linear equations
- Quadratic equations
- Some further equations and techniques

Chapter 1
Arithmetic

OBJECTIVES

Having completed this chapter you should be able to:

- Add, subtract, multiply, and divide with positive and negative integers.
- Use brackets to find a common factor and a common denominator.
- Add, subtract, multiply, and divide fractions and decimal numbers.
- Convert decimal numbers into fractions and vice versa.
- Convert fractions into proportions and percentages and vice versa.
- Increase or decrease any number by a given percentage, and calculate percentage changes.
- Express time series data in index number form.
- Understand and evaluate powers and roots.
- Carry out all of the above tasks both with and without a calculator.
- Round numbers to any number of significant figures or decimal places.
- Convert numbers to and from scientific notation.

If you are not sure whether you need to study this chapter, take the test at the end of the chapter, then grade your performance using the answers at the end of the book.

1.1 Introduction

In this chapter we revise the basic concepts of numbers and operations with them. We look at positive and negative numbers, fractions, decimals, and percentages. We also review powers, such as squares and cubes, and square roots and cube roots. We examine the basic operations of addition, subtraction, multiplication, and division, as well as finding powers and roots. We review how these operations are performed with pen and ink only, and with the aid of a calculator.

You may think that all this is surely unnecessary in the twenty-first century when we have incredibly sophisticated, powerful, and fast calculators and computers on hand to do all these tasks for us. It is certainly true that the ability to perform large numbers of complex calculations quickly and accurately entirely 'by hand', which was a skill much in demand half a century ago, is now—thankfully—no longer needed. But although we all have powerful machines at our elbow to take the drudgery out of calculation, the problem known as 'garbage in, garbage out' remains. That is, we must have the necessary maths skills to be able to identify the tasks that we want our calculator or computer to perform; to know how to instruct these machines to perform the tasks; and finally, to be able to understand and interpret the answers the machine spews out. To these we might add a fourth very useful skill: the ability to look at an answer generated by a calculator or computer and assess whether it is broadly correct, or whether it is garbage due to some foolish error in data input or choice of formula that we have made.

This chapter is an essential preparation for the study of algebra, which begins in chapter 2. We can define algebra as the study of general relationships between numbers that are unspecified and therefore identified by symbols—usually, letters of the alphabet. To prepare ourselves for the study of these general relationships, we need to understand the properties of *specific* numbers and the rules for manipulating them. This chapter aims to provide you with this understanding.

1.2 Addition and subtraction with positive and negative numbers

When we are working with positive and negative numbers, the key point to understand is that the '+' and '−' signs serve two distinct purposes:

(1) We place a '+' or '−' sign in front of a number to indicate whether that number is positive or negative. Thus (+5) is a positive number and (−3) is a negative number. The brackets are not strictly necessary, but we have added them to make it absolutely clear that the '+' and '−' signs are *attached* to the numbers.

(2) We place a '+' or '−' sign *between* two numbers to indicate whether the operation of addition or subtraction is to be performed.

This results in four possible operations:

Case (a) Adding a positive number.

Case (b) Adding a negative number.

Case (c) Subtracting a positive number.

Case (d) Subtracting a negative number.

Let us examine these four cases one by one. To make our thinking more concrete, we consider the bank accounts of two people, Ann and John. We suppose that Ann has a balance of (+5) euros in her bank account, and John has a balance of (+3) euros in his.

For case (a), adding a positive number, suppose Ann and John are planning to marry, and therefore add John's bank balance to Ann's to arrive at their combined wealth. The necessary addition is clearly (+5) + (+3) = (+8) euros.

We show this diagrammatically in figure 1.1(a). Bank balances are measured along the horizontal lines. To the right of the zero point, balances are positive (a credit balance). To the left of the zero point, balances are negative ('in the red', or a debit balance). Ann's balance is (+5) and John's is (+3). Their combined balance is found by aligning the zero of John's balance with the (+5) mark on Ann's balance, as shown, giving their combined balance of (+8).

For case (b), adding a negative number, we suppose Ann and John are again adding John's bank balance to Ann's to arrive at their combined wealth. This time, though, we assume John is 'in the red'; he is overdrawn at the bank, having a balance of (−3). The necessary addition is therefore (+5) + (−3) = (+2) euros.

We show this diagrammatically in figure 1.1(b). Ann's balance is (+5) as before, but John's is now (−3) and is therefore placed to the left of the zero mark. Their combined balance is found

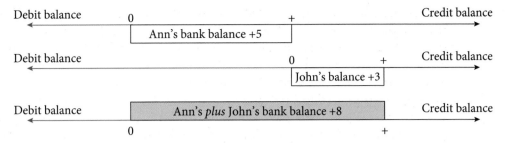

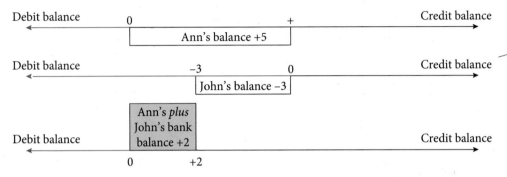

Figure 1.1(a) Adding a positive number.

Figure 1.1(b) Adding a negative number.

as before by aligning the zero of John's balance with the (+5) mark on Ann's balance. This time, though, because John's negative balance is measured to the left, their combined balance is only (+2). In words, Ann's credit balance of (+5) is partly cancelled out by John's debit balance of (−3), leaving a net balance of only (+2).

Moving on to case (c), subtracting a positive number, we suppose now that Ann and John have abandoned their marriage plans and have drifted into an argument about which of them is the wealthier. They decide to settle this by subtracting John's bank balance from Ann's. We suppose that Ann has a balance of (+5) euros in her bank account, and John has a balance of (+3) euros in his. The necessary subtraction is therefore (+5) − (+3) = (+2). Clearly, Ann's balance is greater than John's by (+2) euros.

We show this diagrammatically in figure 1.1(c). Ann's balance is (+5) and John's is (+3). We subtract John's balance from Ann's by aligning the (+3) mark on John's balance with the (+5) mark on Ann's. The difference between the two balances is then (+2).

Finally we consider case (d), subtracting a negative number. Let us suppose that Ann and John are continuing to argue about who is the wealthier, to be settled by subtracting John's bank balance from Ann's. This time, though, we assume John is overdrawn at the bank, having a balance of (−3). The necessary subtraction is therefore (+5) − (−3) = (+8) euros.

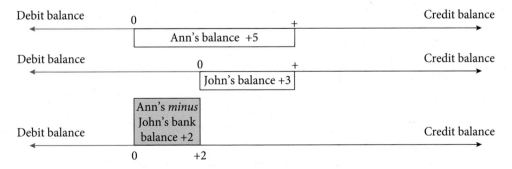

Figure 1.1(c) Subtracting a positive number.

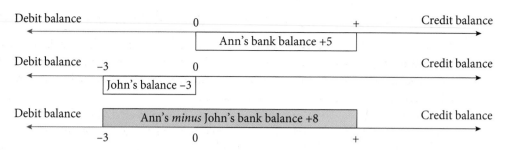

Figure 1.1(d) Subtracting a negative number.

We show this diagrammatically in figure 1.1(d). Ann's balance is (+5) and John's is (−3). We subtract John's balance from Ann's by aligning the zero mark on John's balance with the zero mark on Ann's. The difference between the two balances is then (+8). This may seem puzzling at first sight, but hopefully becomes clear when we realize that John would need to pay (+8) euros into his account in order to have the same balance as Ann. Thus Ann is (+8) euros better off than John.

Let us now collect together the results of the four cases we have just examined:

Case (a) Adding a positive number.   We found that $(+5) + (+3) = (+8)$.

Case (b) Adding a negative number.   We found that $(+5) + (−3) = (+2)$.

Case (c) Subtracting a positive number.   We found that $(+5) − (+3) = (+2)$.

Case (d) Subtracting a negative number.   We found that $(+5) − (−3) = (+8)$.

From the four cases we can derive the following rules for addition and subtraction:

RULE 1.1 Adding and subtracting positive and negative numbers
Rule 1.1a
From cases (a) and (d) we see that the rule is:

If two numbers are separated by two plus signs (case (a)) or by two minus signs (case (d)), we must add the two numbers together.

Rule 1.1b
From cases (b) and (c) we see that the rule is:

If two numbers are separated by either a plus sign followed by a minus sign, or a minus sign followed by a plus sign, we must *subtract* the second number from the first.

In practice, of course, we don't usually bother to place a '+' sign in front of positive numbers; nor do we bother with the brackets for positive numbers. So cases (a)–(d) above would actually be written as:

(a) $5 + 3 = 8$ The result, 8, is Ann and John's combined bank balance, when they both have positive balances.

(b) $5 + (−3) = 2$ The result, 2, is their combined balances, when John is in the red.

(c) $5 − 3 = 2$ The result, 2, is the difference between Ann's balance and John's, when they both have positive balances.

(d) $5 − (−3) = 8$ The result, 8, is the difference between Ann's balance and John's, when John is in the red.

As a matter of language, or terminology, note that when two numbers are added together, the result is called a sum. When one number is subtracted from another, the result is called a difference.

Adding and subtracting positive and negative numbers on your calculator

To follow this book you will need what is usually described as a 'scientific' calculator. This need not cost more than about £8. It is worth spending some time learning how your calculator works. As there are small differences between different makes and models of calculator, the advice given here is necessarily somewhat general.

On your calculator you will find a key marked (−) or +/− (depending on the make and model of your calculator) which you should press to tell the calculator that the number is negative. On some calculators you press this before keying in the number; on others you can also press it after. Using this key you can evaluate, say, (−5) + (−3) with the key strokes

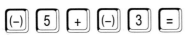

and the answer, −8, should appear in the display screen. Once you are confident that you know the rules of signs (rule 1.1), you will probably evaluate (−5) + (−3) with one key stroke fewer by keying in

which of course gives the same answer, −8.

Progress exercise 1.1

Without using your calculator, calculate:

(a) $14 + (-3) - (-9)$ (b) $52 - (-7) + (-6)$ (c) $(-3) + 6 - (-7) + (-6)$

(d) $(-8) - 4 - (-6) + (-2)$ (e) $(-15) - (-9) - (+8)$ (f) $(-2) + 4 - (+2) + (-2)$

Then use your calculator to check your answers.

1.3 Multiplication and division with positive and negative numbers

Multiplication

We can explain the rules for multiplication of positive and negative numbers in the following way. If a number is multiplied by +1, the number is left unchanged. If a number is multiplied by −1, the number is left unchanged in absolute magnitude but its sign is reversed: that is, if it was previously positive, it becomes negative and vice versa. Thus:

$(+5) \times (+1) = (+5)$ Multiplication by (+1) has no effect.

$(+5) \times (-1) = (-5)$ Multiplication by (−1) causes (+5) to become (−5).

$(-5) \times (+1) = (-5)$ Multiplication by (+1) has no effect.

$(-5) \times (-1) = (+5)$ Multiplication by (−1) causes (−5) to become (+5).

However, in order to be consistent, these rules must hold for multiplication by *any* number, not just −1. This implies that:

(a) $(+5) \times (+3) = (+15)$

(b) $(+5) \times (-3) = (-15)$

(c) $(-5) \times (+3) = (-15)$

(d) $(-5) \times (-3) = (+15)$

From these cases we can deduce the rule for multiplication.

RULE 1.2 **Multiplying positive and negative numbers**

When multiplying two numbers together:

If the two numbers are *both positive* (case (a)) or *both negative* (case (d)), the result is *positive*.

If *one* of the numbers is *positive* and the *other negative* (cases (b) and (c)), the result is *negative*.

Since we don't usually bother to place a '+' sign in front of positive numbers, nor do we bother with the brackets for positive numbers, cases (a)–(d) above would actually be written as:

(a) $5 \times 3 = 15$

(b) $5 \times (-3) = (-15)$

(c) $(-5) \times 3 = (-15)$

(d) $(-5) \times (-3) = 15$

Multiplication by 0

Note that any number $\times\, 0 = 0$. To understand why this is so, it may help to think as follows. We can think of 3×1 as meaning 'take 1 box containing 3 objects', which will give us 3 objects. Similarly, we can think of 3×0 as meaning 'take 0 boxes containing 3 objects', which will give us zero objects.

Division

Division simply reverses multiplication. Therefore it must obey exactly the same sign rules as multiplication, otherwise there would be a danger that when we multiplied and then divided by the same number, we would not get back to the number we started with.

Therefore the rule for division is as follows:

RULE 1.3 **Dividing positive and negative numbers**

When dividing one number by another:

If the two numbers are *both positive* (case (e) below) or *both negative* (case (f)), the result is *positive*.

If *one* of the numbers is *positive* and the other *negative* (cases (g) and (h) below), the result is *negative*.

Thus for example:

(e) $(+15) \div (+3) = (+5)$

(f) $(-15) \div (-3) = (+5)$

(g) $(-15) \div (+3) = (-5)$

(h) $(+15) \div (-3) = (-5)$

Note an important difference between multiplication and division: while 15×3 is the same thing as 3×15, this is not true of division; $15 \div 3$ is *not* the same thing as $3 \div 15$.

Division of a number by itself

A special case of division which we will need later is that when a number is divided by itself, the result is 1. For example, $15 \div 15 = 1$. If you have difficulty seeing why this is true, think of dividing ('sharing') 15 sweets between 15 children; the result is 1 sweet per child.

This also holds when the number in question is negative. This follows from rule 1.3 above, which tells us that $(-15) \div (-5) = (+3)$. Applying this rule when the numerator and denominator are equal gives us $(-15) \div (-15) = (+1)$. The only exception to this is when the number in question is 0 (see below).

Division by 0

Division of any number by 0 is said to be undefined; that is, it has no meaning. You may find this puzzling, but it is not logically necessary that every mathematical expression must have a meaning. We can write down a meaningless mathematical expression, just as we can write down a meaningless word! So if x is any number, $\frac{x}{0}$ is undefined or, in other words, meaningless. This is also true, of course, if x itself is 0. We explain this a little further in section 2.8.

Other ways of writing the division operation

Above, we indicated the division operation by using the traditional '÷' sign. There are two other ways of indicating a division operation. The first is by writing a fraction in which the first number appears in the top (called the numerator) and the second in the bottom (the denominator). The second is by putting a forward slash ('/') between the two numbers.

For example:

(i) $15 \div 3$ may also be written as $\frac{15}{3}$ or 15/3

(j) $(-15) \div (-3)$ may also be written as $\frac{-15}{-3}$ or $(-15)/(-3)$

(k) $(-15) \div 3$ may also be written as $\frac{-15}{3}$ or $(-15)/3$

(l) $15 \div (-3)$ may also be written as $\frac{15}{-3}$ or $15/(-3)$.

Of these alternative notations, the traditional '÷' sign is not much used, possibly because it is easily mistaken for a '+' sign. In this book we shall mostly use the fraction notation ($\frac{15}{3}$) to indicate division. However, we will also use the forward slash notation (15/3), even though you may find it less clear, because this notation takes up less space on the page and doesn't threaten to disturb the line spacing in a word-processed or printed document.

In most of this book, apart from this chapter and the next, we shall follow the normal convention and omit the '+' sign and the brackets when writing positive numbers. We shall also omit the brackets from negative expressions except where such omission would result in ambiguity.

Some more points of terminology are worth noting. First, a **fraction** is also often called a **ratio** or a quotient. Second, when two numbers are multiplied together, the result is called a **product**. The two numbers are called **factors** of the product. When one number is divided by another, the result is called a **quotient**.

Multiplying and dividing positive and negative numbers on your calculator

As we saw above for addition and subtraction, a scientific calculator can handle multiplication or division of negative numbers, provided you press the key marked (−) or its equivalent, to tell the calculator that the number is negative. On some calculators you must press this key before keying in the number; on others, after. Using this key you can evaluate, say, $(-5) \times (-3)$ with the key strokes

and the answer, 15, should appear in the display screen. On some calculators you may need to key in the brackets too.

Progress exercise 1.2

Without using your calculator, calculate:

(a) $3 \times (-7)$ (b) $(-4) \div (-2)$ (c) $(-6) \times (-3)$

(d) $(+6) \div (-3)$ (e) $(-5) \times (+2)$ (f) $(-18) \div (+6)$

1.4 Brackets and when we need them

'Mixed' operations

When addition and subtraction are *mixed* with multiplication and division, you get different answers depending on which part of the calculation you do first. For example,

$$6 + 8 \div 2$$

is ambiguous. If you do the addition first, you get $14 \div 2 = 7$; but if you do the division first, you get $6 + 4 = 10$. There appears to be no way of knowing which answer the writer of this expression intended.

To avoid this ambiguity, mathematicians have adopted the convention or customary rule which says that in a long mathematical expression we should carry out operations in the order **D-M-A-S**: that is, first, any Division operations; then any Multiplication; then Addition; and finally Subtraction. Applying these conventions to the expression above tells us to do the division first, so the correct answer is 10.

A further convention is that in any expression involving a power, also known as an exponent, such as 3^2 (which means 3×3), the exponent should be evaluated first. (Exponents are examined in section 1.14 below.) This means that we must carry out operations in the order **E-D-M-A-S**, where E denotes any exponent (= power).

If we want to override the E-D-M-A-S rule, we do this by using brackets. The rule for brackets is that any mathematical operation inside brackets must be done first, before applying the E-D-M-A-S rule. Taking this on board, the rule is as follows.

RULE 1.4 Order of operations

When working out any long mathematical expression, the various operations must be done in the order **B-E-D-M-A-S**: that is, first, anything inside Brackets; then any Exponent; then Division operations; then any Multiplication; then Addition; and finally Subtraction.

Therefore in the expression $(6 + 8) \div 2$, the brackets tell us that we must do the addition first, giving us the answer 7. Thus, to summarize, B-E-D-M-A-S means that

$$6 + 8 \div 2 = 6 + 4 = 10$$

but

$$(6 + 8) \div 2 = 14 \div 2 = 7$$

However, because as noted in section 1.3 above the division sign ($\div$) is not often used, these two expressions would not be written in this way. Instead, using the fraction notation, we would write

$$6 + 8 \div 2 \qquad \text{as} \qquad 6 + \frac{8}{2}$$

but

$$(6 + 8) \div 2 \qquad \textit{as} \qquad \frac{6 + 8}{2}$$

This last is particularly important to note. It means that $\frac{6+8}{2}$ should be read as $\frac{(6+8)}{2}$. Thus the addition must be done before the division.

Similarly, something such as $\frac{6+3}{2+8}$ should be read as $\frac{(6+3)}{(2+8)}$. It is essential to keep this in mind when applying the B-E-D-M-A-S rule.

Hint Forgetting to follow this rule is the most common source of error in basic maths.

Expanding (multiplying out) brackets

EXAMPLE 1.1

Consider: $3 \times (4 + 5)$

Applying the B-E-D-M-A-S rule, we evaluate this as $3 \times 9 = 27$. However, we also get this answer if we evaluate $3 \times (4 + 5)$ as

$$3 \times (4 + 5) = (3 \times 4) + (3 \times 5) = 12 + 15 = 27$$

Thus in order to remove the brackets we must take *each* of the terms inside the brackets (the 4 and the 5) and multiply it by the multiplicative term in front of the first bracket (the 3), then add the results together.

This diagram may help you remember the procedure:

$$3 \times (4 + 5) = (3 \times 4) + (3 \times 5)$$

That is, the 3 multiplies both the 4 and the 5, and the results are added.

This process is called multiplying out, or expanding, the expression we started with.

EXAMPLE 1.2

As a second example, consider:

$$4 \times (3 + 5 - 2)$$

Applying the B-E-D-M-A-S rule, we evaluate this as $4 \times 6 = 24$. Alternatively, by multiplying out we get

$$\begin{aligned} 4 \times (3 + 5 - 2) &= (4 \times 3) + (4 \times 5) + [4 \times (-2)] \\ &= 12 + 20 - 8 \\ &= 24 \quad \text{(as before)} \end{aligned}$$

As in example 1.1, we see that when multiplying out, the 4 outside the brackets multiplies each of the terms (3, 5, and −2) inside the brackets, giving us the three components 4×3, 4×5, and $4 \times (-2)$. We then add these three components to get the answer, 24. Note that we must be careful with signs; the last component is $4 \times (-2)$.

Again, a diagram may help you to see what we have done:

$$4 \times (3 + 5 - 2) = (4 \times 3) + (4 \times 5) + [4 \times (-2)]$$

EXAMPLE 1.3

As a third example, we consider a case where the multiplicative term outside the brackets is negative. In such a case, some care with rules on signs is necessary.

Consider: $(-3) \times (4 - 5 + 6)$

Applying the B-E-D-M-A-S rule, we evaluate this as $(-3) \times 5 = -15$. Alternatively, multiplying out gives

$$(-3) \times (4 - 5 + 6) = \underbrace{[(-3) \times 4]}_{\text{neg.}} + \underbrace{[(-3) \times (-5)]}_{\text{pos.}} + \underbrace{[(-3) \times 6]}_{\text{neg.}}$$

$$= -12 + 15 - 18 = -15$$

This is the correct answer, since $(-3) \times (4 - 5 + 6)$ and $-12 + 15 - 18$ are both equal to −15.

Generalization

Generalizing from examples 1.1–1.3 gives us the following rule:

> **RULE 1.5** **Multiplying out brackets**
>
> To multiply out an expression such as $4 \times (3 + 5 - 2)$, multiply each of the numbers inside the brackets (3, 5, and −2 in this case) by the multiplicative term in front of the first bracket (4, in this case), and add the results. Care is necessary with signs.

EXAMPLE 1.4

A special but important case of rule 1.5 arises when the multiplicative term is (−1). Consider:

$$-(4 + 5)$$

Applying B-E-D-M-A-S, this becomes −(9), or simply −9.

What happens when we try to remove the brackets by multiplying out? To do this, we must realize that −(4 + 5) is actually a shortened form of

$$(-1) \times (4 + 5)$$

So when we multiply out using rule 1.5, we get

$$(-1) \times (4 + 5) = [(-1) \times 4] + [(-1) \times 5] = (-4) + (-5) = -9$$

Thus in the expression −(4 + 5), the minus sign in front of the first bracket causes the 4 and the 5 inside the brackets to change sign to −4 and −5 after multiplying out. As we shall meet cases like this quite frequently, it's worth looking at another example:

EXAMPLE 1.5

Consider: $-(-4 + 5 - 3)$

As in example 1.4, we treat this as a shortened form of

$$(-1) \times (-4 + 5 - 3)$$

so on multiplying out it becomes

$$[(-1) \times (-4)] + [(-1) \times 5] + [(-1) \times (-3)] \quad \text{which equals } 4 - 5 + 3$$

Thus we have

$$-(-4 + 5 - 3) = 4 - 5 + 3$$

So removal of the brackets by multiplying out has in this case caused a reversal of *all* the signs inside the brackets. Forgetting this is a very common mistake.

'Nested' brackets

You'll notice that in examples 1.2–1.5 we met expressions such as

$$[(4 + 21) \div (7 - 2)] \div 5$$

In cases like this, where we have brackets inside brackets (which mathematicians rather charmingly call 'nesting'), the rule is that we work out the value of the expression by starting with the innermost brackets and working outwards. In this example, the answer is 1. Note that the outer ('square') brackets are necessary to tell us which of the two division operations to carry out first. Without the square brackets, the answer could be either 1 or 25. (Check this for yourself!)

1.5 Factorization

This rather forbidding term simply means *reversing* the process of multiplying out that we examined in the previous section.

EXAMPLE 1.6

Consider: $12 + 15$

First, note that $12 = 3 \times 4$, and for this reason we say that 3 and 4 are **factors** of 12. We also say that 3 is a **cofactor** of 4, and vice versa, because they are jointly factors of 12.

In the same way, $15 = 3 \times 5$, and we say that 3 and 5 are factors of 15.

So obviously we can write:

$$12 + 15 = (3 \times 4) + (3 \times 5)$$

Because 3 is a factor of both 12 and 15, it is called a **common factor**.

The common factor, 3, can now be taken outside of a single pair of brackets, to give

$$(3 \times 4) + (3 \times 5) = 3 \times (4 + 5)$$

What we have just done is called **factorization**. We can check that what we have done is correct by multiplying out the brackets in $3 \times (4 + 5)$. By rule 1.5:

$$3 \times (4 + 5) = (3 \times 4) + (3 \times 5) = 12 + 15 \quad \text{(what we started with)}$$

Thus factorization reverses rule 1.5.

A diagram may help you visualize the process of taking the common factor outside of a pair of brackets:

$$12 + 15 = (3 \times 4) + (3 \times 5) = 3 \times (4 + 5)$$

We shall use factorization extensively throughout this book. Therefore it is important that you understand the process and feel comfortable with it. Don't forget that you can always easily check that you have performed the process of factorization correctly by multiplying out the result (using rule 1.5) and confirming that you get back to where you started from.

 Hint You can only factorize in this way when the numbers are separated by a + or a − sign. The method is not valid for something such as $12 \div 15$ or 12×15.

Next we look at two more examples of factorization.

EXAMPLE 1.7

Consider: $12 + 20 - 8$

Here $12 = 4 \times 3$; $20 = 4 \times 5$; and $-8 = 4 \times (-2)$

Thus 4 is a common factor of 12, 20, and −8, because they are all divisible by 4 without remainder. (Another common factor is 2, but 4 is the *highest* common factor.) Therefore we can write

$$12 + 20 - 8 = [4 \times 3] + [4 \times 5] + [4 \times (-2)]$$
$$= 4 \times (3 + 5 - 2)$$

In words, we take the common factor, 4, outside of, and multiplying, a pair of brackets with the cofactors, 3, 5, and −2, inside. (Check for yourself that $12 + 20 - 8$ and $4 \times (3 + 5 - 2)$ both equal 24.)

EXAMPLE 1.8

In this case, we consider a negative common factor. Consider:

$$-4 - 8 + 2$$

Clearly +2 is a common factor of these three numbers. Therefore we can write:

$$-4 - 8 + 2 = [2 \times (-2)] + [2 \times (-4)] + [2 \times 1]$$

Then we can take the 2 outside of a single pair of brackets and get:

$$-4 - 8 + 2 = 2 \times (-2 - 4 + 1)$$

Equally, though, we could use −2 as the common factor. Then we get:

$$-4 - 8 + 2 = [(-2) \times 2] + [(-2) \times 4] + [(-2) \times (-1)]$$
$$= (-2) \times (2 + 4 - 1)$$

It is important to remember, when multiplying out this last expression, that all the terms inside the bracket have to be multiplied by a negative number, −2, and this will change their signs. Check this for yourself, by multiplying out $(-2) \times (2 + 4 - 1)$ in order to get back to the starting point, $-4 - 8 + 2$.

Using brackets on your calculator

A scientific calculator is programmed to obey the B-E-D-M-A-S convention in mathematics: that is, to deal with brackets first, then exponents, then to perform division and multiplication operations *before* addition and subtraction. Thus if you want to evaluate $(3 \times 4) + (10 \div 2)$, you key in:

and the correct answer, 17, should appear on the display. It is not necessary to key in any brackets because the calculator will automatically obey B-E-D-M-A-S and do the multiplication and division before the addition.

But suppose instead you want to evaluate $3 \times (4 + 10) \div 2$; that is, you want the addition to be done first. Then you must use the bracket keys provided on the calculator's keyboard. Thus you key in:

The calculator, obeying B-E-D-M-A-S, then does the addition first because it is inside brackets, and therefore the correct answer, 21, should appear on the display. The order in which the multiplication and division are done has no effect on the answer (check this for yourself).

? ***Progress exercise 1.3***

Factorize:

(a) $18 + 6 - 36$ (b) $150 + 220$ (c) $-12 - 3 + 6$

Multiply out:

(d) $5 \times (15 + 12)$ (e) $-6 \times (2 + 5)$ (f) $-3 \times (-2 + 1)$

1.6 Fractions

In arithmetic the idea of fractions such as $\frac{1}{4}$, $\frac{1}{2}$, and $\frac{3}{8}$ is hopefully familiar. As noted earlier, a fraction such as $\frac{2}{3}$ is just another way of writing $2 \div 3$. Thus a fraction (also called a ratio or quotient) results whenever we divide one number by another. A fraction such as $\frac{2}{3}$ is often written as 2/3 or $^2/_3$. Remember that in any fraction the top part is called the **numerator** and the bottom part is called the **denominator**. Thus in $\frac{3}{8}$ the numerator is 3 and the denominator 8.

We will now look at the rules for manipulating fractions.

Simplifying fractions

A key truth about fractions is that any fraction is left unchanged in its value when its numerator (top) and denominator (bottom) are both divided by the same number. Thus, given

$$\frac{5}{10}$$

we can divide the numerator and denominator by 5 and get

$$\frac{5}{10} = \frac{(5 \div 5)}{(10 \div 5)} = \frac{1}{2}$$

Almost without thinking, we see that this is correct because 5 is half of 10 and 1 is half of 2.

We chose to divide the numerator and denominator by 5 because we saw straight away that 5 is a *common factor* of both 5 and 10. In a more difficult case we have to find the factors of the numerator and denominator first, in order to identify any common factors. For example, given

$$\frac{12}{15}$$

we first see that 4 and 3 are factors of 12, because $4 \times 3 = 12$; and 5 and 3 are factors of 15, because $5 \times 3 = 15$. So we have

$$\frac{12}{15} = \frac{4 \times 3}{5 \times 3}$$

Now we see that 3 is a common factor of both the numerator and denominator. So if we now divide numerator and denominator by 3, we get

$$\frac{12}{15} = \frac{4 \times 3}{5 \times 3} = \frac{4 \times 3 \div 3}{5 \times 3 \div 3}$$

and since division reverses multiplication, in the numerator we have $4 \times 3 \div 3 = 4$; and in the denominator we have $5 \times 3 \div 3 = 5$. So finally we have

$$\frac{12}{15} = \frac{4 \times 3}{5 \times 3} = \frac{4 \times 3 \div 3}{5 \times 3 \div 3} = \frac{4}{5}$$

This process is called simplification, because $\frac{4}{5}$ is simpler than $\frac{12}{15}$.

The process is also often called **cancelling**, because when we divide top and bottom by 3 we are effectively cancelling, or reversing, the multiplication of top and bottom by 3. When you are working with paper and pen, a quick way of carrying out this step is to draw a line through the multiplying 3s, like this:

$$\frac{12}{15} = \frac{4 \times 3}{5 \times 3} = \frac{4 \times \cancel{3}}{5 \times \cancel{3}} = \frac{4}{5}$$

Warning: don't try to do this to something with + or − signs in the numerator and/or denominator (see rule 1.4 above, and the hint that follows it). For example, the following is *false*:

$$\frac{7}{8} = \frac{4 + 3}{5 + 3} = \frac{4 + \cancel{3}}{5 + \cancel{3}} = \frac{4}{5}$$

The above is wrong because the numerator and denominator of $\frac{7}{8}$ don't have a common factor. A fraction won't simplify unless such a common factor exists. For example, in $\frac{15}{24}$ the numerator and denominator both have 3 as a factor, so it simplifies to $\frac{5}{8}$.

The process of cancelling can also be reversed: that is, we can *multiply* both numerator and denominator by the same number. For example,

$$\frac{3}{8} = \frac{3 \times 4}{8 \times 4} = \frac{12}{32}$$

This operation is permissible because multiplying both numerator and denominator by the same number leaves a fraction unchanged in its value.

Using your calculator to simplify fractions

A scientific calculator can handle fractions. On most calculators the relevant key is marked: a^b/c. To key in, say, $\frac{3}{8}$ requires the key strokes

Here the a^b/c key tells the calculator that the 3 and the 8 are numerator and denominator respectively of the fraction.

To simplify a fraction, such as $\frac{18}{72}$, we key in

and the result appears on the display as $1 \rfloor 4$. The $\rfloor$ symbol (or something similar depending on the make of your calculator) indicates a fraction, which in this example we read of course as $\frac{1}{4}$.

Progress exercise 1.4

Simplify, where possible, using your calculator if necessary:

(a) $\frac{8}{40}$ (b) $\frac{12}{132}$ (c) $-\frac{6}{39}$ (d) $\frac{15}{55}$

1.7 Addition and subtraction of fractions

Addition of fractions

EXAMPLE 1.9

Consider: $\frac{1}{2} + \frac{1}{4}$

The way we tackle this is to see, intuitively, that a half is the same thing as two quarters, so we add two quarters and one quarter to get the answer, three quarters. In symbols:

$$\frac{1}{2} + \frac{1}{4} = \frac{2}{4} + \frac{1}{4} = \frac{3}{4}$$

What we have done here is to rearrange things so that both fractions have the same denominator: in this case, 4. This is called a **common denominator**. A common denominator is any number that the two denominators, 2 and 4, will divide into without remainder. Having done this, we simply add the numerators (2 and 1) to get the answer, $\frac{3}{4}$.

EXAMPLE 1.10

Consider: $\dfrac{1}{5} + \dfrac{1}{2}$

Again we see, after a little thought, that one fifth is the same thing as two tenths, and a half is the same as five tenths, so we add these together to get the answer, seven-tenths. In symbols:

$$\frac{1}{5} + \frac{1}{2} = \frac{2}{10} + \frac{5}{10} = \frac{7}{10}$$

Again we have rearranged both fractions so that they have a common denominator, 10. Having done this, we simply add the numerators (2 and 5) to get the answer, $\frac{7}{10}$.

A systematic method

We solved the two examples above by intuition. We can make this intuitive process systematic, as follows. Consider again the previous example:

$$\frac{1}{5} + \frac{1}{2}$$

To rearrange things so that both fractions have a common denominator, we proceed in two steps. First, we take the first fraction, $\frac{1}{5}$, and multiply both its numerator and denominator by the denominator, 2, of the second fraction. This gives us

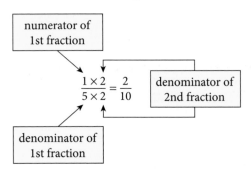

This is a permissible step because we know from section 1.6 above that multiplying the numerator and denominator of a fraction by the same number leaves the fraction unchanged in its value.

Second, we take the second fraction, $\frac{1}{2}$, and multiply both its numerator and denominator by the denominator, 5, of the first fraction. This gives us

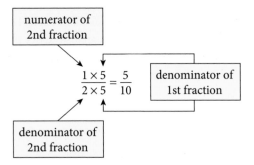

Finally we add these two together, which gives us our answer:

$$\frac{1}{5} + \frac{1}{2} = \frac{2}{10} + \frac{5}{10} = \frac{7}{10}$$

Let's now test out this method on two more examples.

EXAMPLE 1.11

Consider: $\dfrac{1}{8} + \dfrac{1}{4}$

First, we take the first fraction, $\frac{1}{8}$, and multiply its numerator and denominator by 4, the denominator of the second fraction. This gives us $\frac{1 \times 4}{8 \times 4} = \frac{4}{32}$. Then we multiply the numerator and denominator of the second fraction by the denominator of the first. This gives us $\frac{1 \times 8}{4 \times 8} = \frac{8}{32}$. Finally we add the two results, so our answer is

$$\frac{1}{8} + \frac{1}{4} = \frac{4}{32} + \frac{8}{32} = \frac{12}{32}$$

Although this answer is correct, we haven't quite finished because $\frac{12}{32}$ can be simplified further. The numerator and denominator share a common factor, 4 (that is, both 12 and 32 are divisible by 4 without remainder). So, by dividing its numerator and denominator by 4 we can simplify our answer further to

$$\frac{12}{32} = \frac{12 \div 4}{32 \div 4} = \frac{3}{8}$$

In this example the need for further simplification arose because our method led us to use 32 as the common denominator of the two fractions. If we had used 8 instead, the need for further simplification would not have arisen. This is because 8 is the *lowest* common denominator; it is the smallest number divisible by both 8 and 4 without remainder. However, the advantage of our method is that we don't need to find the lowest common denominator—a task that is not always easy.

EXAMPLE 1.12

Consider: $\dfrac{1}{7} + \dfrac{1}{11}$

Applying the same method as before, we take the first fraction, $\frac{1}{7}$, and multiply both its numerator and denominator by 11, the denominator of the second fraction. This gives us $\frac{1}{7} = \frac{1 \times 11}{7 \times 11} = \frac{11}{77}$. Then we take the second fraction, $\frac{1}{11}$, and multiply both its numerator and denominator by 7, the denominator of the first fraction. This gives us $\frac{1}{11} = \frac{1 \times 7}{11 \times 7} = \frac{7}{77}$.

Finally we add the two results to get the answer:

$$\frac{1}{7} + \frac{1}{11} = \frac{11}{77} + \frac{7}{77} = \frac{18}{77}$$

Unlike example 1.11, this answer cannot be simplified because 18 and 77 have no common factor.

Subtraction of fractions

The method for subtraction is exactly the same as the method for addition. That is, we have to find a common denominator before carrying out the subtraction of one numerator from the other. The only difference is that we now have a minus sign instead of a plus sign between the two fractions.

EXAMPLE 1.13

Consider: $\dfrac{1}{2} - \dfrac{2}{5}$

First we multiply the numerator and denominator of the first fraction by the denominator of the second fraction. We get $\frac{1 \times 5}{2 \times 5} = \frac{5}{10}$.

Then we multiply the numerator and denominator of the second fraction by the denominator of the first fraction. We get $\frac{2 \times 2}{5 \times 2} = \frac{4}{10}$.

Finally we subtract the second from the first, and get

$$\frac{1}{2} - \frac{2}{5} = \frac{5}{10} - \frac{4}{10} = \frac{1}{10}$$

Adding or subtracting three or more fractions

This complication is fairly easily handled, as the following example shows.

EXAMPLE 1.14

Consider: $\dfrac{1}{8} + \dfrac{1}{4} + \dfrac{1}{5}$

There are two methods.

Method 1: we add in two steps. Step 1: we add the $\frac{1}{8}$ and the $\frac{1}{4}$, using our method from above. So $\frac{1}{8}$ becomes $\frac{1 \times 4}{8 \times 4} = \frac{4}{32}$ and $\frac{1}{4}$ becomes $\frac{1 \times 8}{4 \times 8} = \frac{8}{32}$. Adding these gives $\frac{1}{8} + \frac{1}{4} = \frac{4}{32} + \frac{8}{32} = \frac{12}{32}$ which simplifies by cancelling to $\frac{3}{8}$

Step 2: we add this to the $\frac{1}{5}$ in the question, giving

$$\frac{3}{8} + \frac{1}{5} = \frac{3 \times 5}{8 \times 5} + \frac{1 \times 8}{5 \times 8} = \frac{15}{40} + \frac{8}{40} = \frac{23}{40}$$

Method 2: we multiply numerator and denominator of each fraction by *both* (*all*) of the other two (or more) denominators, and add the results. This gives

$$\frac{1}{8} + \frac{1}{4} + \frac{1}{5} = \frac{1 \times 4 \times 5}{8 \times 4 \times 5} + \frac{1 \times 8 \times 5}{4 \times 8 \times 5} + \frac{1 \times 8 \times 4}{5 \times 8 \times 4} = \frac{20}{160} + \frac{40}{160} + \frac{32}{160} = \frac{92}{160} = \frac{23}{40}$$

You should use whichever of these two methods you find easier to understand and remember.

In practice you will probably use your calculator for most calculations of this kind. But it's important to do a few using pen and paper alone, so that you understand what is going on inside your calculator!

Positive and negative fractions

All of the fractions we have considered up to this point have been positive. However, fractions can be negative just as whole numbers can, and they then obey all the rules of signs as specified in rules 1.1–1.3 in sections 1.2 and 1.3 above. For example, $\frac{1}{2} - (-\frac{2}{5}) = \frac{1}{2} + \frac{2}{5}$ (from rule 1.1); $(-\frac{1}{2}) \times (-\frac{1}{5}) = +\frac{1}{10}$ (from rule 1.2); and so on. In particular, note that

$$\frac{(-2)}{(-5)} = (-2) \div (-5) = 2 \div 5 = \frac{2}{5} \text{ (from rule 1.3)}$$

Also, $\dfrac{(-2)}{5} = -\dfrac{2}{5}$; and $\dfrac{2}{(-5)} = -\dfrac{2}{5}$ (also from rule 1.3)

This last simply says that if either the numerator or the denominator of a fraction is negative, then the fraction as a whole is negative.

Expressions containing both whole numbers and fractions

When confronted with something such as $3\frac{1}{4} - 5\frac{1}{8}$, there are two ways to tackle it.

Method 1: We write $3\frac{1}{4}$ as $3 + \frac{1}{4}$, and similarly write $5\frac{1}{8}$ as $5 + \frac{1}{8}$. Therefore:

$$3\frac{1}{4} - 5\frac{1}{8} = (3 + \frac{1}{4}) - (5 + \frac{1}{8}) = 3 + \frac{1}{4} - 5 - \frac{1}{8}$$

(remembering the rule of signs when removing brackets; see rule 1.5 and example 1.4 in section 1.4 above). This can be re-ordered as

$$(3 - 5) + (\tfrac{1}{4} - \tfrac{1}{8}), \quad \text{which equals } (-2) + (+\tfrac{1}{8}), \quad \text{which equals } -1\tfrac{7}{8}$$

Using your calculator to add or subtract fractions

You will find you can use this method quite easily on your calculator. That is, you simply key in: $3 - 5 + \frac{1}{4} - \frac{1}{8}$, using the a^b/c key for the fractions (see section 1.6 above). When you press the 'equals' key, the display will show $-1 \rfloor 7 \rfloor 8$, which we read as $-1\frac{7}{8}$.

Method 2: We can write $3\frac{1}{4}$ as $3 + \frac{1}{4}$, which equals $\frac{12}{4} + \frac{1}{4}$, which equals $\frac{13}{4}$. Similarly, $5\frac{1}{8}$ can be written as $\frac{41}{8}$. Therefore:

$$3\frac{1}{4} - 5\frac{1}{8} = \frac{13}{4} - \frac{41}{8}, \text{ which equals } \frac{26}{8} - \frac{41}{8} = -\frac{15}{8} = -1\frac{7}{8} \text{ as before.}$$

Progress exercise 1.5

Calculate, without using your calculator:

(a) $\frac{2}{5} + \frac{1}{2}$ (b) $\frac{1}{2} - \frac{1}{2}$ (c) $3\frac{1}{4} - \frac{1}{2}$ (d) $1\frac{1}{4} + 5\frac{5}{16}$ (e) $\frac{1}{12} - \frac{1}{4} + \frac{1}{3}$ (f) $\frac{5}{8} + \frac{4}{7}$

1.8 Multiplication and division of fractions

Multiplication of fractions

The rule for multiplying one fraction by another is very simple.

RULE 1.6 Multiplying fractions

To multiply one fraction by another, simply multiply the two numerators together to obtain the numerator of the answer, and multiply the two denominators together to obtain the denominator of the answer. Then simplify the result, if possible.

EXAMPLE 1.15

$$\frac{3}{8} \times \frac{2}{5} = \frac{3 \times 2}{8 \times 5} = \frac{6}{40} = \frac{3}{20}$$

This generalizes to any number of terms, for example 3:

$$\frac{3}{8} \times \frac{2}{5} \times \frac{3}{4} = \frac{3 \times 2 \times 3}{8 \times 5 \times 4} = \frac{18}{160} = \frac{9}{80}$$

Rule 1.6 also holds when a whole number is multiplied by a fraction. For example:

$$15 \times \frac{2}{5}$$

Here we can make use of the fact that 15 is the same thing as $\frac{15}{1}$. So the problem becomes

$$\frac{15}{1} \times \frac{2}{5} = \frac{15 \times 2}{1 \times 5} = \frac{30}{5} = 6$$

Reciprocals

From the multiplication rule, we can get

$$4 \times \frac{1}{4} = \frac{4}{1} \times \frac{1}{4} = \frac{4 \times 1}{1 \times 4} = \frac{4}{4} = 1$$

Here, 4 and $\frac{1}{4}$ are said to be **reciprocals**. By definition, their product is unity (in other words, when you multiply one by the other the result is 1).

Equivalence of 'of' and multiplication

Sometimes we meet a question such as: what is half of 18? Obviously the answer is 9, but we need a procedure for answering this kind of question so that we don't have to rely on intuition, which can lead us astray.

The key insight is to see that 'half of 18' is found by *multiplying* 18 by $\frac{1}{2}$. That is,

$$\text{'half of 18'} = 18 \times \frac{1}{2} = \frac{18}{2} = 9$$

So 'of' just means 'multiplied by'. For example,

$$\frac{3}{4} \text{ of } \frac{1}{2} = \frac{3}{4} \times \frac{1}{2} = \frac{3 \times 1}{4 \times 2} = \frac{3}{8}$$

Division of fractions

From the multiplication rule (rule 1.6 above) we can get

$$10 \times \frac{1}{2} = \frac{10}{1} \times \frac{1}{2} = \frac{10 \times 1}{1 \times 2} = 5$$

But we also know that

$$10 \div 2 = 5$$

Combining these, it must therefore be true that

$$10 \div 2 = 10 \times \frac{1}{2} \text{ (since they both equal 5)}$$

So dividing by 2 is equivalent to multiplying by its reciprocal, $\frac{1}{2}$.

In this example we were dividing by a whole number, 2. But the rule must hold equally when dividing by a fraction. So, for example:

$$10 \div \frac{2}{5} \quad \text{must equal} \quad 10 \times \frac{5}{2}$$

This gives us the fundamental rule for dividing by a fraction:

RULE 1.7 Dividing by a fraction

To divide by a fraction, simply invert the fraction you are dividing by, then multiply instead of dividing.

EXAMPLE 1.16

$$1 \div \frac{2}{3}$$

The rule says that we first invert the fraction we are dividing by (that is, $\frac{2}{3}$ becomes $\frac{3}{2}$) and then *multiply* by it instead of dividing. Thus

$$1 \div \frac{2}{3} = 1 \times \frac{3}{2} = \frac{3}{2} = 1\frac{1}{2}$$

This answer may seem puzzling, but it may help to think in words. The question is: 'If a cake of size 1 is divided into slices, each of size two-thirds, how many slices will be obtained?' And the answer is: 'One and a half.' Figure 1.2 may help you to see this.

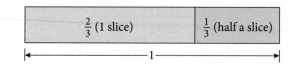

Figure 1.2

The rule holds equally when one fraction is divided by another. For example:

$$\frac{1}{8} \div \frac{1}{4} = \frac{1}{8} \times \frac{4}{1} = \frac{4}{8} = \frac{1}{2}$$

Here the question is: 'How many quarters are there in an eighth?' And the answer is: 'A half'. (In other words, an eighth is only half as large as a quarter.)

Factorizing fractions

Expressions containing several fractions can be factorized in exactly the same way as we factorized whole numbers in section 1.5 above. Consider for example:

$$\frac{1}{10} + \frac{3}{20}$$

Here we notice that $\frac{1}{10}$ can be written as $\frac{1 \times 1}{5 \times 2}$, which equals $\frac{1}{5} \times \frac{1}{2}$. Also $\frac{3}{20}$ can be written as $\frac{1 \times 3}{5 \times 4}$, which equals $\frac{1}{5} \times \frac{3}{4}$. Therefore

$$\frac{1}{10} + \frac{3}{20} = \left(\frac{1}{5} \times \frac{1}{2}\right) + \left(\frac{1}{5} \times \frac{3}{4}\right)$$

The two sets of brackets share a common factor, $\frac{1}{5}$. So we can take this outside a single pair of brackets and get

$$\left(\frac{1}{5} \times \frac{1}{2}\right) + \left(\frac{1}{5} \times \frac{3}{4}\right) = \frac{1}{5} \times \left(\frac{1}{2} + \frac{3}{4}\right)$$

Check for yourself that this is correct by multiplying out $\frac{1}{5} \times (\frac{1}{2} + \frac{3}{4})$. You should get back to what we started with: $\frac{1}{10} + \frac{3}{20}$.

The most difficult part of this operation is the first step: identifying a common factor. With practice you can become proficient at this, though it's never easy.

Expressions consisting of both whole numbers and fractions

When confronted with something such as $3\frac{1}{2} \div 2\frac{1}{3}$, we tackle it by converting both numbers into fractions. That is, $3\frac{1}{2} = 3 + \frac{1}{2} = \frac{6}{2} + \frac{1}{2} = \frac{7}{2}$, and similarly $2\frac{1}{3} = 2 + \frac{1}{3} = \frac{6}{3} + \frac{1}{3} = \frac{7}{3}$. Therefore

$$3\frac{1}{2} \div 2\frac{1}{3} = \frac{7}{2} \div \frac{7}{3} = \frac{7}{2} \times \frac{3}{7} = \frac{21}{14} = \frac{3}{2} = 1\frac{1}{2}$$

Note that if you use your calculator to simplify $\frac{21}{14}$, the display will show $1 \rfloor 1 \rfloor 2$, which must be read as $1\frac{1}{2}$.

Fractions involving addition or subtraction in numerator or denominator

When meeting something such as $\frac{3+2}{8}$ it is essential to remember that this may also be written as $(3 + 2) \div 8$ (see section 1.4 above). This way of writing it reminds us that the addition of the 3 and the 2 must be carried out *before* dividing by 8. Thus

$$\frac{3+2}{8} = \frac{5}{8}$$

An alternative way of expressing this same point is to say that in the expression $\frac{3+2}{8}$, both the 3 and the 2 must be divided by the 8. Thus

$$\frac{3+2}{8} = \frac{3}{8} + \frac{2}{8} = \frac{5}{8}$$

Yet another way of expressing the same point is to say that in any fraction such as $\frac{3+2}{8}$, it is permissible to divide or multiply the *whole* of the numerator and the *whole* of the denominator by the same number (see section 1.6). Thus if we divide the whole of the numerator and the whole of the denominator by 8, we get

$$\frac{3+2}{8} = \frac{\frac{3}{8} + \frac{2}{8}}{\frac{8}{8}} = \frac{\frac{3}{8} + \frac{2}{8}}{1} = \frac{3}{8} + \frac{2}{8} = \frac{5}{8}$$

If you forget this, you may be tempted to cancel the 2 in the numerator with the 8 in the denominator. Then you get

$$\frac{3+2}{8} = \frac{3 + \frac{2}{2}}{\frac{8}{2}} = \frac{3+1}{4} = 1$$

which of course is the wrong answer. It is wrong because the 3 in the numerator should also have been divided by 2. *Forgetting this is a very common mistake!*

Cancelling fractions when only multiplication or division is involved

In contrast to the above, it is quite OK to cancel freely between numerator and denominator when only multiplication or division signs appear in numerator and denominator. For example, consider $\frac{3 \times 2}{8}$. Compared to the previous case, $\frac{3+2}{8}$, we now have a multiplication sign instead of a plus sign in the numerator. In this case you *can* cancel the 2 with the 8 and get the right answer. Doing so gives

$$\frac{3 \times 2}{8} = \frac{3 \times \frac{2}{8}}{\frac{8}{8}} = 3 \times \frac{2}{8} = \frac{6}{8} = \frac{3}{4}$$

Summarizing, whenever you have something like $\frac{3+2}{8}$ or $\frac{16}{6-4}$ (that is, a fraction with addition or subtraction in the numerator or denominator), you must do the addition or subtraction first, before doing any cancelling. Alternatively, you can divide or multiply the *whole* of the numerator and the *whole* of the denominator by the same number.

But when you have something like $\frac{3 \times 2}{8}$ or $\frac{6}{8 \div 3}$ (that is, a fraction with *only* multiplication or division in the numerator and denominator), you can cancel individual elements in the numerator and denominator, if you wish. But you will also get the right answer if you choose instead to perform the multiplication or division in numerator or denominator *first*, and then do the cancelling.

Progress exercise 1.6

1. Calculate the following without using your calculator. Then check your answers using your calculator.

   (a) $\frac{1}{3} \times \frac{3}{5}$ (b) $\frac{4}{9} \div \frac{2}{3}$ (c) $\frac{1}{8}$ of $\frac{3}{4}$ (d) $2\frac{1}{4} \times 3\frac{1}{6}$ (e) $5\frac{1}{4} \div \frac{1}{2}$ (f) $\frac{3}{8} \div \frac{1}{3} \times 1\frac{1}{5}$

2. Factorize, if possible:

   (a) $\frac{1}{4} + \frac{3}{8}$ (b) $-\frac{2}{9} + \frac{3}{11}$ (c) $\frac{3}{16} + \frac{3}{8}$ (d) $\frac{5}{12} - \frac{1}{3}$ (e) $\frac{11}{16} + \frac{5}{12}$ (f) $\frac{3}{15} + \frac{5}{16}$

1.9 Decimal numbers

Decimal fractions

By 'decimal fraction' we mean a number in decimal format that is between zero and 1. A decimal fraction is simply another way of writing a fraction in which the denominator is 10, 100, 1000 and so on. Thus for example:

Decimal fraction	Equivalent fraction
0.1	$\frac{1}{10}$
0.01	$\frac{1}{100}$
0.001	$\frac{1}{1000}$
0.125	$\frac{125}{1000}$

and so on.

The pattern here is that the non-zero digits after the decimal point give us the *numerator* of the fraction, while the number of digits after the decimal point gives us the number of zeros in the *denominator* of the fraction. Thus for 0.001 the pattern is:

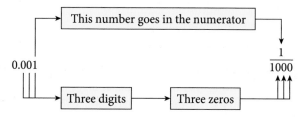

Thus for the decimal number 0.001, the equivalent fraction is $\frac{1}{1000}$. As a second example, consider 0.025. Applying the rule above, the 25 goes in the numerator, and because there are three digits after the point in 0.025, we put three zeros in the denominator. Thus $0.025 = \frac{25}{1000}$ (which simplifies to $\frac{1}{40}$).

But note that in making the conversion we don't count a zero or zeros at the *right-hand* end of the decimal number. For example, in 0.10 the right-hand zero doesn't count because 0.10 is the same thing as 0.1. Similarly $0.5400 = 0.54$, and so on.

Decimals with whole numbers

A whole number (also known as an **integer**) can be combined with a decimal fraction. For example, 3.125 consists of the whole number 3 together with the decimal fraction 0.125.

Multiplication and division of decimal numbers

One of the attractions of decimal numbers is that multiplication and division by 10 are very easy. The rules are as follows.

Multiplication

When a decimal number is multiplied by 10, its decimal point moves one place to the right. For example, $0.375 \times 10 = 3.75$. When multiplying by 100, the decimal point moves two places to the right. For example, $2.653 \times 100 = 265.3$. When multiplying by 1000, the decimal point moves three places to the right, and so on. When multiplying, we may have to insert one or more new zeros on the right-hand side. For example, $0.3 \times 100 = 30.0$. Here, in order to move the decimal point two places to the right, we had to insert an extra zero to the right of the 3. We also inserted an extra zero *after* the decimal point, though this was not strictly necessary, because 30.0 is simply 30. We put it there solely to keep track of where the decimal point is.

Division

When a decimal number is divided by 10, its decimal point moves one place to the left. For example, $37.5 \div 10 = 3.75$. When dividing by 100, the decimal point moves two places to the left. For example, $265.3 \div 100 = 2.653$. When dividing by 1000, the decimal point moves three places to the left, and so on. When dividing, we may have to insert one or more new zeros at the left-hand end of the number. For example, $4.64 \div 100 = 0.0464$. Here, in order to move the decimal point two places to the left, we had to insert a new zero before the 4. We also inserted a new zero *before* the decimal point; this is not strictly necessary but is put there to draw our attention to the decimal point.

Converting decimal numbers into fractions and vice versa

Converting a decimal number into a fraction

To do this, simply apply the definitions above, then simplify where possible. For example:

(1) $0.5 = \frac{5}{10}$, which after cancelling $= \frac{1}{2}$

(2) $0.05 = \frac{5}{100}$, which after cancelling $= \frac{1}{20}$

(3) $0.6 = \frac{6}{10}$, which after cancelling $= \frac{3}{5}$

(4) $0.625 = \frac{625}{1000}$, which after cancelling $= \frac{5}{8}$

(5) $0.7 = \frac{7}{10}$; no cancelling possible as 7 and 10 do not have a common factor.

In most cases you will probably want to use your calculator for the cancelling operations such as those above, using the method explained in section 1.6 above. Before reaching for your calculator it's a good idea to try to form some rough expectation of what the answer will be. Then if you make a mistake keying in, your suspicions will be aroused when the answer appears. For example, in $\frac{625}{1000}$ the numerator is a little more than half of the denominator, so we expect an answer of a little more than one half. Therefore the calculator's answer of $\frac{5}{8}$ seems correct, as 5 is a little more than half of 8.

Converting a fraction into a decimal number

You may find that you can convert some simple fractions into decimals intuitively—that is, without much conscious thought. For example:

$$\frac{1}{2} = 0.5; \qquad \frac{9}{10} = 0.9; \qquad \frac{1}{5} = 0.2$$

But you will probably use your calculator for most conversions of this type. Since calculators are designed to deal primarily with decimal numbers, they convert fractions into decimal numbers very easily. For example, to convert $\frac{3}{8}$ into decimal, simply key in

and the answer, 0.375, should appear on the display.

Rounding of numbers

When dealing with large numbers it's often convenient to work with an approximation which is easier to say, remember or manipulate. This is called **rounding**. For example, if on graduating you are offered a job with a salary of £19,780 per year, you might well report this to your friends and family as an offer of £20,000—not in order to exaggerate, but simply because £20,000 is easier to say and remember, and for most purposes is accurate enough. You have then 'rounded up' your salary to the nearest thousand pounds.

Similarly, if you were offered a Saturday job in a shop which paid £3.049 per hour (that is, three pounds and 4.9 pence) you might, when discussing with friends, say that the job paid £3 per hour. In this case you have 'rounded down' the hourly pay to the nearest pound. If you wanted a more accurate approximation you might say the job paid £3.05 per hour, in which case you have rounded to the nearest penny. (This is based on an actual case, incidentally.)

When rounding, the convention is that any final digit that is less than 5 is rounded down, while any final digit that is 5 or greater is rounded up. So, for example, 34 when rounded to the nearest 10 becomes 30, while 35 becomes 40.

Another rounding procedure is when we round to, say, two significant figures. Then, after rounding as above, we set all the digits, except the first two, equal to zero. For example, 153,965 becomes 150,000 when rounded to two significant figures. When rounded to three significant figures it becomes 154,000.

Rounding of decimal fractions

A strange feature of the number system is that some fractions have no exact decimal equivalent. For example, when you try to convert $\frac{2}{3}$ into its decimal equivalent, your calculator displays 0.66666666 with as many 6s as the display can accommodate. We can never write down the *exact* decimal equivalent of $\frac{2}{3}$, because we could never find a piece of paper large enough. We express this by writing $\frac{2}{3} = 0.66$ (recurring), meaning that the 6s go on forever.

If, however, we want to do any calculations, we must choose a manageable decimal number which approximately equals $\frac{2}{3}$. This is a form of rounding. For example, we might choose 0.666 (which is slightly less than $\frac{2}{3}$) or 0.667 (which is slightly more than $\frac{2}{3}$). If we made the first choice, we would write: $\frac{2}{3} = 0.666$ (rounded down to three decimal places). If we made the second choice, we would write: $\frac{2}{3} = 0.667$ (rounded up to three decimal places). Obviously the number of decimal places we include in our approximation depends on how accurate we want our answer to be.

1.10 Adding, subtracting, multiplying, and dividing decimal numbers

You will probably do most of these tasks on your calculator, but again it's important to understand the principles so that you can do a rough check on the answers that your calculator displays.

Adding and subtracting

When adding and subtracting decimal numbers, it's crucial to keep the decimal points of all of the numbers aligned as shown below. For example, we add 0.25 and 0.035 as follows. Note that 0.25 is the same thing as 0.250.

```
 0.25
 0.035 +
 0.285
```

Multiplying and dividing

When multiplying or dividing one decimal number by another, it's easy to get the decimal point in the wrong place in your answer. For example, $0.2 \times 0.3 = 0.06$. At first sight, this answer seems wrong because of the 'extra' zero in the answer, after the decimal point. That the answer is correct, however, is seen when we do the same calculation with the numbers in fractional form. We know that $0.2 = \frac{2}{10}$ and $0.3 = \frac{3}{10}$. So

$$0.2 \times 0.3 = \frac{2}{10} \times \frac{3}{10} = \frac{2 \times 3}{10 \times 10} = \frac{6}{100}; \text{ and finally } \frac{6}{100} = 0.06.$$

Thus we see that the 'extra' zero results from the fact that, in the fractional calculation, we have $10 \times 10 = 100$ in the denominator.

Another way of seeing why we get this answer is to note that $\frac{2}{10} \times \frac{3}{10}$ is the same thing as $\frac{2}{10}$ of $\frac{30}{100}$ (see section 1.8). So we are trying to find two tenths of 30 hundredths. Since one tenth of 30 hundredths is 3 hundredths, two tenths is 6 hundredths, or $\frac{6}{100}$.

Related problems can also arise when doing decimal division. Consider for example $0.5 \div 0.02$. The correct answer is 25, but this somehow seems too large. The best way of seeing that 25 is the correct answer is to write the calculation as $\frac{0.5}{0.02}$, then multiply numerator and denominator by 100 (which of course will leave the fraction unchanged). This gives

$$\frac{0.5}{0.02} = \frac{0.5 \times 100}{0.02 \times 100} = \frac{50}{2} = 25$$

Looking again at the problem, $0.5 \div 0.02$, we see that the answer (25) is large because 0.02 is very much smaller than 0.5 and therefore will divide into it many times (25 times, in fact!).

Progress exercise 1.7

1. Attempt these without using your calculator. Then check your answers using your calculator.

(a) 0.135×10 (b) 1.275×100 (c) $0.32 \div 10$ (d) $4.65 \div 100$

(e) 0.25×1.3 (f) $0.15 \div 0.05$ (g) 0.032×0.5 (h) $0.01 \div 0.2$

2. Convert the following into their equivalent fractions, cancelling as appropriate.

(a) 0.36 (b) 0.875 (c) 3.051

3. Convert the following into their equivalent decimals.

(a) $\frac{3}{5}$ (b) $\frac{3}{8}$ (c) $\frac{3}{20}$ (d) $\frac{2}{3}$

1.11 Fractions, proportions, and ratios

A **fraction**, also known as a **ratio**, is a way of comparing one number with another. For example, if we are told that 30 students have failed a certain exam, this does not in itself tell us very much (except perhaps that there are 30 rather unhappy students somewhere on the planet). But if we are also told that 150 students took the exam, we can now say that 30 out of 150 failed. A natural way of expressing this information is to write it as the fraction:

$$\frac{\text{number who failed}}{\text{number who took the exam}} = \frac{30}{150}$$

By dividing numerator and denominator by 30, this fraction simplifies to $\frac{1}{5}$. This tells us that one-fifth of the students, or one student in every five who took it, failed the exam. We say that, of students taking the exam, the **proportion** of failures was one-fifth.

Another slightly different way of presenting this information is to say that the ratio of failures to total number taking the exam was 30:150. Again this expression may be simplified to 1:5 by dividing both sides of the ratio by 30. This division is a valid operation because 30:150 is really just another way of writing $\frac{30}{150}$. Writing a ratio as 30:150 is a form of presentation not used very often, except when there are three or more numbers to be compared. Suppose for example we are told that 30 students failed an exam, 120 passed, and a further 20 were absent. Then we could say that passes, fails, and absentees were in the ratio 120:30:20, which simplifies to 12:3:2. Thus we might say that for every 12 passes there were 3 fails and 2 absentees. This form of presentation thus permits three (or more) numbers to be compared.

1.12 Percentages

What is a percentage?

The Latin *per centum* translates roughly as 'of every hundred'. Thus 40 *per centum*, which quickly became shortened to 40 per cent or 40%, means 40 in every 100. But '40 in every 100' can also be written as the fraction, or proportion, $\frac{40}{100}$. Thus in general a **percentage** (that is, a number written in per cent format) is the same thing as the numerator of a fraction in which the denominator is 100. Thus:

EXAMPLE 1.17

$10\% = \frac{10}{100}$ (which, after simplification, equals $\frac{1}{10}$).

Thus 10% is the same thing as 10 in every 100, or 1 in every 10.

EXAMPLE 1.18

$3.5\% = \frac{3.5}{100}$; but this mixes decimals and fractions in the same expression, so we multiply top and bottom by 10, and obtain $\frac{35}{1000}$ to avoid this mixing.

Thus 3.5% is the same thing as 3.5 in every one hundred, or 35 in every 1000.

EXAMPLE 1.19

$17.54\% = \frac{17.54}{100}$; but again, to avoid mixing decimals and fractions, we multiply top and bottom, this time by 100, and obtain $\frac{1754}{10000}$.

Thus 17.54% is the same thing as 17.54 in every one hundred, or 1754 in every ten thousand.

A percentage is an alternative to a proportion as a way of comparing one number with another. For example, in the previous section we considered an example in which 30 out of 150 students had failed an exam. We saw that failures as a proportion of the total number taking the exam were $\frac{30}{150}$, which simplifies to $\frac{1}{5}$. But $\frac{1}{5}$, after multiplying top and bottom by 20, is the same thing as $\frac{20}{100}$, which equals 20% (from the definition of a percentage, above). Thus to say that $\frac{1}{5}$ failed the exam, and that 20% failed, are equivalent statements.

Converting a fraction into a percentage and vice versa

The rules are very simple. First:

 RULE 1.8 Converting a fraction into a percentage

To convert any fraction into a percentage, simply multiply by 100.

EXAMPLE 1.20

Given: $\frac{1}{2}$. The equivalent percentage is found as $\frac{1}{2} \times 100 = \frac{100}{2} = 50$ (per cent).

EXAMPLE 1.21

Given: $\frac{1}{8}$. The equivalent percentage is $\frac{1}{8} \times 100 = \frac{100}{8} = 12.5$ (per cent).

EXAMPLE 1.22

Given: $\frac{3}{80}$. The equivalent percentage is $\frac{3}{80} \times 100 = \frac{300}{80} = 3.75$ (per cent).
To do this example on a calculator, we key in

[3] [÷] [8] [0] [x] [1] [0] [0] [=]

giving the answer 3.75. Alternatively, we could key in

[3] [a^b_c] [8] [0] [x] [1] [0] [0] [=]

and the answer displayed should be $3 \rfloor 3 \rfloor 4$, which we read as $3\frac{3}{4}$ (per cent).

The second rule is:

RULE 1.9 Converting a percentage into a fraction
To convert any percentage into a fraction, simply reverse rule 1.8: that is, divide by 100.

EXAMPLE 1.23

Given: 50%. This becomes $50 \div 100 = \frac{50}{100} = \frac{1}{2}$
To do this example on your calculator, key in

[5] [0] [a^b_c] [1] [0] [0] [=]

and the answer displayed should be $1 \rfloor 2$, which means $\frac{1}{2}$. If instead you key in

[5] [0] [÷] [1] [0] [0] [=]

you will get the answer 0.5, which of course is the decimal equivalent of $\frac{1}{2}$.

EXAMPLE 1.24

Given: 12.5%. This becomes $\frac{12.5}{100} = \frac{125}{1000} = \frac{1}{8}$
To do this example on your calculator, key in

[1] [2] [5] [a^b_c] [1] [0] [0] [0] [=]

and the answer displayed should be $1 \rfloor 8$, which means $\frac{1}{8}$. Note that if you key in

[1] [2] [.] [5] [a^b_c] [1] [0] [0] [=]

the 12.5 confuses the calculator, as it cannot convert a decimal number into a fraction. You
will get the answer 0.125, which of course is the decimal equivalent of $\frac{1}{8}$.

Converting a decimal number into a percentage and vice versa

Sometimes we may be given a proportion in the form of a decimal number rather than a fraction. For example, if 1 in 5 students have failed an exam we might be told that the proportion

of fails is 0.2. To convert such a decimal number into a percentage, or vice versa, the rules are exactly the same as for fractions.

 RULE 1.10 Converting a decimal number into a percentage

To convert any decimal number into a percentage, simply multiply by 100.

EXAMPLE 1.25

Given: 0.5. The equivalent percentage is $0.5 \times 100 = 50\%$.

EXAMPLE 1.26

Given: 0.125. The equivalent percentage is $0.125 \times 100 = 12.5\%$.

 RULE 1.11 Converting a percentage into a decimal number

To convert any percentage into a decimal number, simply reverse rule 1.10: that is, divide by 100.

EXAMPLE 1.27

Given: 12.5%. The equivalent decimal number is $12.5 \div 100 = \frac{12.5}{100} = \frac{125}{1000} = 0.125$.

EXAMPLE 1.28

Given: 2.5%. The equivalent decimal number is $2.5 \div 100 = \frac{2.5}{100} = \frac{25}{1000} = 0.025$.

You may wish to use your calculator for these, but multiplying and dividing by 100 is so easy that it hardly seems worth the effort!

Expressing one number as a percentage of another

To do this, divide the first number by the second, resulting in a fraction. Then simply follow the rule for fractions (rule 1.8 above): that is, multiply by 100.

EXAMPLE 1.29

Given: 80 students sit an exam and 36 fail. The percentage who failed is found as

$$\frac{36}{80} \times 100 = \frac{360}{8} = 45 \text{ (per cent)}.$$

Finding a given percentage of a number

EXAMPLE 1.30

To find, say, 5% of 27, recall that 'of' means 'multiply' (see section 1.8 above). Also, from the definition of a percentage, $5\% = \frac{5}{100}$. Therefore

$$5\% \text{ of } 27 = \frac{5}{100} \times 27 = \frac{135}{100} = 1.35 \text{ (per cent)}$$

It may help your understanding of this to see that $\frac{5}{100} \times 27$ is the same as $\frac{27}{100} \times 5$. Thus to find 5% of 27, we first divide 27 into 100 parts. Each of these is 1% of 27, so we must then multiply by 5 to arrive at 5% of 27.

Increasing a number by a given percentage

EXAMPLE 1.31

To increase 150 by 20%, we first find 20% of 150. Following example 1.30:

$$20\% \text{ of } 150 = 150 \times \frac{20}{100} = \frac{3000}{100} = 30.$$

So our answer is $150 + 30 = 180$.

Note that our answer is 100% of 150 (= 150), plus 20% of 150 (= 30). This means that we can get to our answer a little faster by increasing 150 by 120%, or equivalently finding 120% of 150. Following example 1.30:

$$120\% \text{ of } 150 = 150 \times \frac{120}{100} = \frac{18000}{100} = 180.$$

Generalizing from this example, we arrive at a rule:

RULE 1.12 Increasing a given number by a given percentage

The rule is:

$$\text{Answer} = (\text{Given number}) \times \left(\frac{100 + \text{Given percentage increase}}{100} \right).$$

A negative percentage change

Rule 1.12 can equally well handle cases where the given percentage increase is negative. Suppose Britain's exports were €220 billion last year but this year are 3.5% lower. Then from rule 1.12 this year's exports are given by:

$$\text{This year's exports} = (\text{Last year's exports}) \times \left(\frac{100 + \text{given percentage increase}}{100} \right).$$

In this case the percentage increase is *minus* 3.5%, so we have:

$$\text{This year's exports} = (\text{Last year's exports}) \times \left(\frac{100 - 3.5}{100} \right)$$

$$= 220 \times \left(\frac{96.5}{100} \right) = 212.3 \text{ (billion euros)}.$$

Calculating value added tax (VAT)

VAT exists in many countries and is calculated as a percentage of the seller's price before VAT has been added. Thus applying rule 1.12 we arrive at a rule for calculating the price including VAT:

RULE 1.13 Calculating price including VAT from price before VAT

$$\text{Price including VAT} = (\text{Price before VAT}) \times \left(\frac{100 + \text{percentage VAT rate}}{100} \right).$$

The price including VAT is, of course, the price paid by the buyer. The price before VAT is the price received by the seller. The difference between the two prices is the amount of VAT, which the seller transfers to the government.

EXAMPLE 1.32

Suppose the price of a mobile phone is €150 before VAT at 17.5% has been added. Then:

$$\text{Price including VAT} = (\text{€}150) \times \left(\frac{100 + 17.5}{100} \right) = \text{€}176.25 \text{ (using a calculator).}$$

The amount of VAT is €176.25 − €150 = €26.25. We can check that this is correct by finding 17.5% of €150, which equals $\frac{17.5}{100} \times 150 = \text{€}26.25$.

Reversing rule 1.13

Sometimes we know the price including VAT and we want to work backwards to find the price before VAT. In rule 1.13 we took the known price *before* VAT and multiplied it by $\frac{100 + \text{VAT rate}}{100}$ to arrive at the price including VAT. We now want to reverse this, and since multiplication is reversed by division it follows that we must now take the known price *including* VAT and divide it by $\frac{100 + \text{VAT rate}}{100}$ to arrive at the price before VAT. Thus the rule for reversing rule 1.13 is:

> **RULE 1.14** Calculating price before VAT from price including VAT
>
> $$\text{Price before VAT} = (\text{Price including VAT}) \div \left(\frac{100 + \text{percentage VAT rate}}{100} \right).$$

We can check that rule 1.14 is correct using example 1.32. If we are told that the price including VAT is 176.25, then using rule 1.14 we can calculate the price before VAT as

$$\text{Price before VAT} = 176.25 \div \left(\frac{100 + 17.5}{100} \right) = 150 \text{ (using calculator).}$$

Hint A very common mistake is made in this type of calculation. If we are told that the price including VAT is €176.25, then it is very tempting to think that if we deduct 17.5% from this we will get back to the price before VAT. So following example 1.30 we might calculate the VAT as 17.5% of 176.25, giving

$\frac{17.5}{100} \times 176.25 = 30.84$

and thus conclude that the price before VAT is 176.25 − 30.84 = 145.41.

But this is wrong, because we know from example 1.32 that the amount of VAT is 26.25, and the price before VAT is 150. What have we done wrong? Our mistake was to calculate the VAT as 17.5% of 176.25, the price *including* VAT, when the VAT is actually 17.5% of the price *before* VAT.

Generalization of rules 1.13 and 1.14

Obviously, rules 1.13 and 1.14 don't apply only to VAT calculations. We can use these rules above to increase *any* number by *any* given percentage, and to reverse that increase.

Using your calculator for percentages

For all but the simplest calculations involving percentages, you will need to use your calculator. Note, though, that if your calculator has a key marked %, it is probably best to ignore it. This key is put there mainly for use in the business world, where needs and methods differ from ours.

At the Online Resource Centre, we show how to calculate percentages using the Microsoft Excel®️ database and spreadsheet program. www.oxfordtextbooks.co.uk/orc/renshaw3e/

1.13 Index numbers

Data for a huge range of economic variables, such as industrial production, prices, exports, and unemployment, are collected and published in the UK by the Office of National Statistics (ONS) (www.statistics.gov.uk) or similar bodies at regular time intervals such as weekly, monthly, or annually. Each of these data sets is called a **time series**.

For example, column (2) of table 1.1 gives UK GDP (gross domestic product, a measure of the total value of goods and services produced in the economy) for the years 1995–2009, measured in millions of pounds (so the 1995 figure is £964,780 million or, rounded to the nearest billion, £965 billion). (1 billion = 1 thousand million.) The effects of inflation, which increases the *value* of goods and services produced even if their *quantity* is unchanged, have as far as possible been removed. For this reason the series in table 1.1 is referred to as real GDP, or volume GDP.

Column (2) of table 1.1 gives us the *level* of GDP in each year, but for some purposes we may be more interested in the *growth* of GDP over the 12 years. The growth of GDP is more easily seen if we reformulate column (2) in index number form. To do this we divide each year's GDP by the GDP of 1995: that is, by 964,780. The result of this is shown in column (3) of the table. Not surprisingly, the figure for 1995 is now 1, since we have divided the 1995 GDP by itself. All the other figures are now expressed as a proportion of the 1995 level. (See section 1.11 if you are unsure what a proportion is.) Thus 1996 GDP was 1.0289 of 1995 GDP, 1997 GDP was 1.0629 of 1995 GDP, and so on.

The numbers in column (3) are not very user friendly because they are close to 1 and involve four decimal places. It seems sensible, therefore, to multiply them all by 100, which as we know from section 1.12 has the effect of turning a proportion into a percentage. The results are shown in column (4) where the first number, 100, tells us that 1995 GDP was 100% of 1995 GDP. The figure for 1996, 102.89, tells us the GDP in 1996 was 102.89% of GDP in 1994. The figure for 2002 tells us that 2002 GDP was 123.87% of 1995 GDP, and so on.

Table 1.1 UK GDP, 1995–2009.

Year	GDP (£m)[a]	GDP relative to 1995 GDP	GDP index[b] (1995 = 100)	Year-to-year growth[c] (%)
(1)	(2)	(3)	(4)	(5)
1995	964,780	1.0000	100.00	–
1996	992,617	1.0289	102.89	2.89
1997	1,025,447	1.0629	106.29	3.31
1998	1,062,433	1.1012	110.12	3.61
1999	1,099,327	1.1395	113.95	3.47
2000	1,142,372	1.1841	118.41	3.92
2001	1,170,489	1.2132	121.32	2.46
2002	1,195,035	1.2387	123.87	2.10
2003	1,228,595	1.2734	127.34	2.81
2004	1,264,852	1.3110	131.10	2.95
2005	1,292,335	1.3395	133.95	2.17
2006	1,328,363	1.3769	137.69	2.79
2007	1,364,029	1.4138	141.38	2.68
2008	1,363,139	1.4129	141.29	−0.07
2009	1,296,390	1.3437	134.37	−4.90

[a] Source: Office of National Statistics.   [b] Calculated using rule 1.15.
[c] Calculated using rule 1.16. Rounded to two decimal places.

Column (4) is called an **index number series** for GDP with 1995 = 100. The year 1995 is called the **base year**. We could have chosen any year as the base year, but the first year in the time series is the most commonly chosen.

From this example we can derive the following rule:

> **RULE 1.15 Calculating an index number series**
>
> For any data in time series form, to convert it into an index number series with year 1 as the base year, divide every year's value by the year 1 value and multiply each by 100.

Measuring year-to-year growth

As is clear from column (4), an attraction of presenting time series data in index number form is that we can simply read off from the table the cumulative growth of the variable since the base year. For example, the cumulative growth between 1995 and 2009 was 34.37% (that is, $134.37 - 100$).

Note, though, that we cannot read off the year-to-year growth from column (4), except for the growth from year 0 (the first year, 1995) to year 1. Thus we can see that between 1995 and 1996 GDP rose from 100 to 102.89, an increase of 2.89 per cent. But for other years, to get the year-to-year growth we must perform a new set of calculations.

Suppose, for example, that in table 1.1 we wanted to calculate the growth of GDP between 1998 and 1999. To do this we first have to calculate the change in GDP between 1998 and 1999, which from column (2) is:

$$1999 \text{ GDP} - 1998 \text{ GDP} = 1{,}099{,}327 - 1{,}062{,}433 = 36{,}894$$

This is the absolute change in GDP, in millions of pounds. If we divide this by GDP in 1998, we get:

$$\frac{1999 \text{ GDP} - 1998 \text{ GDP}}{1998 \text{ GDP}} = \frac{1{,}099{,}327 - 1{,}062{,}433}{1{,}062{,}433} = \frac{36{,}894}{1{,}062{,}433} = 0.034726$$

which measures the change in GDP as a proportion of its initial level. Finally, as this is an inconveniently small number we multiply it by 100 to turn it into a percentage. So our final figure is $0.034726 \times 100 = 3.47\%$ (rounded to two decimal places).

The year-to-year growth rates, calculated in this way, are shown in column (5) of table 1.1. Check some of the calculations for yourself.

From this example we can derive a general rule:

> **RULE 1.16 Percentage growth in a time series**
>
> The percentage growth in a time series, Y, from year t to year $t + 1$ is calculated as:
>
> $$\left(\frac{\text{value of } Y \text{ in year } (t+1) - \text{value of } Y \text{ in year } t}{\text{value of } Y \text{ in year } t} \right) \times 100$$
>
> Note that this formula is also often written as:
>
> $$\left(\frac{\text{value of } Y \text{ in year } t+1}{\text{value of } Y \text{ in year } t} - \frac{\text{value of } Y \text{ in year } t}{\text{value of } Y \text{ in year } t} \right) \times 100$$
>
> which simplifies to:
>
> $$\left(\frac{\text{value of } Y \text{ in year } t+1}{\text{value of } Y \text{ in year } t} - 1 \right) \times 100$$

Finally, note that in rule 1.16 it is not necessary to use the actual values of GDP from column (2) of table 1.1. The index number values in column (4) will do the job equally well, because they are in the same ratio to one another as the original data. Thus we can also calculate the percentage growth of GDP between 1998 and 1999, using rule 1.16 with the data in index number form, as $\left(\frac{113.95 - 110.12}{110.12}\right) \times 100 = 3.47$ (rounded to 2 d.p.; the same as we calculated above using the actual GDP data).

At the Online Resource Centre, we show how the calculation of index numbers and growth rates can be speeded up enormously using Microsoft Excel®.
www.oxfordtextbooks.co.uk/orc/renshaw3e/

Progress exercise 1.8

1. Using your calculator if necessary, convert the following into percentages, rounded to two decimal places.

   (a) $\frac{1}{5}$ (b) $\frac{2}{9}$ (c) $\frac{80}{320}$ (d) $\frac{115}{92}$

   (e) 0.15   (f) 0.6   (g) 0.04   (h) 0.035

2. Find:

   (a) 45% of 952   (b) 17.5% of 2,903,100   (c) 0.4% of 9.5750

3. A shop offers 'buy two, get one free' on a certain item. What is the equivalent percentage price reduction?

4. In a sale, a shop reduces its prices by the amount of the VAT on each item. If VAT is 17.5%, what is the percentage price reduction? (Hint: the answer is *not* 17.5%.)

5. Car A uses 10 litres of fuel and car B uses 5 litres, for every 100 km travelled. By what percentage does car A's fuel consumption exceed car B's? What will be the percentage reduction in fuel used if I switch from car A to car B? Why are the two answers different?

1.14 Powers and roots

Squares and square roots

Squares

A 'square' means two equal numbers multiplied together. For example, 3 squared means 3×3, usually written as 3^2. Thus $3 \times 3 = 3^2 = 9$. In the expression 3^2, the 3 is called the **base** and the 2 is called the **power** or **exponent**.

If we were reading aloud, we could read 3^2 as 'three squared'. Historically, this terminology developed because $3^2 = 9$ is the area of a square with sides of length 3. When reading aloud we could also read 3^2 as '3 raised to the power 2'. Similarly, $2^2 = 2 \times 2$; $5^2 = 5 \times 5$; $10^2 = 10 \times 10$, and so on.

Square roots

The process that is the reverse of squaring a number is called taking (or finding) its **square root**. Thus 5 is the square root of 25 because $5^2 = 25$. We write the square root of 25 as $\sqrt{25}$. Thus $\sqrt{25} = 5$. We can also write $(\sqrt{25})^2 = 5^2 = 25$. This says that if we square the square root, we arrive back where we started. It is important to note that $\sqrt{25}$ is positive, by definition.

However, from rule 1.2 we know that $(-5)^2 = (-5) \times (-5) = 25$. Since $5 = \sqrt{25}$, $-5 = -\sqrt{25}$, and therefore $(-5)^2 = (-\sqrt{25})^2 = 25$. So there are two numbers which, when squared, equal 25. One is the square root of 25, $\sqrt{25}$, which equals 5. The other is the negative of the square root, $-\sqrt{25}$, which equals −5. We can write these two statements more compactly as $\pm\sqrt{25} = \pm 5$. (The symbol $\pm$ means 'plus or minus'.)

What is true of 25 is true of any positive number. Any given positive number has a square root, which is positive. There are two numbers which, when squared, equal the given number. One is the square root. The other is the negative of the square root.

Square roots of negative numbers

Suppose we are seeking the square root of -25. Obviously, $+5$ won't do, because $(+5)^2 = +25$, not -25 as required. Similarly, -5 won't do either, because $(-5)^2 = +25$, not -25 as required. Generalizing from this example, we conclude that the square root of a negative number does not exist.

(The previous sentence is slightly economical with the truth. In more advanced maths the concept of an *imaginary* number is introduced, and this number has the property that its square is negative. Thus to be completely precise we should not say that the square root of a negative number does not exist. It exists, but is not a *real* number. We touch briefly on this question at various points later, and examine it more closely in chapter W21, to be found at the Online Resource Centre (www.oxfordtextbooks.co.uk/orc/renshaw3e/).)

Cubes and cube roots

Cubes

A 'cube' means three equal numbers multiplied together. For example, 4 cubed means $4 \times 4 \times 4$, and is written as 4^3. Historically, the idea of a cube arose because $4 \times 4 \times 4$ gives the volume of a cube with sides of length 4. We also read 4^3 as '4 raised to the power 3'. Similarly, $2^3 = 2 \times 2 \times 2$; $5^3 = 5 \times 5 \times 5$; and so on.

Note that when a negative number is cubed, then, using rule 1.2, the result is negative. For example

$$(-2)^3 = (-2) \times (-2) \times (-2) = (+4) \times (-2) = -8$$

Cube roots

The process which reverses that of cubing a number is called taking (or finding) the cube root. Thus 4 is the cube root of 64 because $4^3 = 4 \times 4 \times 4 = 64$. We write the cube root of 64 as $\sqrt[3]{64}$. So therefore $\sqrt[3]{64} = 4$ because $4^3 = 64$. Generalizing from this example, we see that the cube root of any positive number is itself necessarily positive.

Cube roots of negative numbers

We saw above that a negative number has no square root. However, a negative number does have a cube root. For example, we saw immediately above that $(-2)^3 = -8$. Therefore, by definition, -2 is the cube root of -8. We write this statement as $\sqrt[3]{-8} = -2$. Generalizing from this example, we see that the cube root of any negative number is itself necessarily negative.

Fourth powers and roots

Following the pattern of the square and cube, a number raised to the fourth power means four equal numbers multiplied together. For example, $5^4 = 5 \times 5 \times 5 \times 5 = 625$. Reversing this means taking the fourth root. Thus 5 is the fourth root of 625. We write this as $\sqrt[4]{625} = 5$. We can also write $(\sqrt[4]{625})^4 = 5^4 = 625$. This says that if we take the fourth root of 625, then raise it to the power 4, we arrive back where we started, at 625.

The fourth root of any positive number is necessarily positive. We can explain this as follows. Taking the fourth root of a number is equivalent to taking a square root, twice. For example, if we take the square root of 625, we get $\sqrt{625} = 25$. If we then take the square root of 25 (that is, the square root of the square root of 625) we get $\sqrt{25} = 5$; that is, the fourth root of 625. Thus we can write: $\sqrt[4]{625} = \sqrt{\sqrt{625}} = 5$; or, in words, the square root of the square root equals the fourth root. However, the square root of any number is positive by definition (see under square roots, above); so each time we take a square root, we obtain a positive number. Therefore the fourth root of any positive number, since it is the square root of the square root of that number, must itself be positive.

However, it is also true that $(-5)^4 = (-5) \times (-5) \times (-5) \times (-5) = +25 \times +25 = +625$. Since $-5 = -\sqrt[4]{625}$, we have $(-5)^4 = (-\sqrt[4]{625})^4 = 625$. So there are two numbers which, when raised to the power 4, equal 625. One is the fourth root of 625, $\sqrt[4]{625}$, which equals 5. The other is the negative of the fourth root of 625, $-\sqrt[4]{625}$, which equals -5. We can write these two statements more compactly as $\pm\sqrt[4]{625} = \pm 5$.

Suppose we try to find the fourth root of a negative number. For example, if we were looking for $\sqrt[4]{-625}$, obviously $+5$ won't do, because $5^4 = +625$, not -625. Equally, -5 won't do, because $(-5)^4 = +625$, not -625. Generalizing from this example, we conclude that the fourth root of a negative number does not exist.

Higher powers and roots

Obviously we can raise any number to a higher and higher power. For example, taking 4 as the base we can have $4^5 = 4 \times 4 \times 4 \times 4 \times 4 = 1024$, and so on. And in each case we can reverse the process by taking the corresponding root; for example the fifth root of 1024 (written as $\sqrt[5]{1024}$) is 4.

Summary and generalization on roots

We looked at the square roots of positive numbers and saw that $\sqrt{25} = 5$ was positive by definition, but both $5^2 = 25$ and $(-5)^2 = 25$. In the case of the fourth root we saw that $\sqrt[4]{625} = 5$ was positive, but both $5^4 = 625$ and $(-5)^4 = 625$. This pattern holds for any *even* numbered root of any positive number. Any even numbered root (sixth, eighth, tenth, and so on) is positive, but raising either the root or the negative of the root to the relevant power takes us back to the number again. For example, $\sqrt[10]{1024} = 2$, but both $2^{10} = 1024$ and $(-2)^{10} = 1024$.

We also saw that the square root and the fourth root of a negative number do not exist. This too generalizes: for any negative number, no even numbered root exists. For example, $\sqrt[10]{-1024}$ does not exist.

When we looked at the cube root of a positive number, we saw that this was necessarily positive. This generalizes to any *odd* numbered root (fifth, seventh, and so on). For example, $\sqrt[7]{128} = 2$, because $2^7 = 128$. We also saw that the cube root of a negative number exists, and is negative. This too generalizes to any odd numbered root. For example, $\sqrt[5]{-32} = -2$.

Fractional powers and roots

In all of the examples above we considered only base numbers that were integers (whole numbers). However, we can also raise a fraction to some power, or find the root of a fraction. For example, taking a positive base of $\frac{1}{4}$ and squaring it, we get

$$\left(\frac{1}{4}\right)^2 = \frac{1}{4} \times \frac{1}{4} = \frac{1}{16}$$

Note that $\frac{1}{16}$ is less than $\frac{1}{4}$. Thus when we square $\frac{1}{4}$, or any other positive fraction, the result is a *smaller* number. The same is true of $(\frac{1}{4})^3 = \frac{1}{64}$, $(\frac{1}{4})^4 = \frac{1}{256}$, and so on. The higher is the power to which a positive fraction is raised, the smaller is the result.

We obtain the root of a positive fraction by reversing the power operation. For example, $\sqrt{\frac{1}{16}} = \frac{1}{4}$ (because $(\frac{1}{4})^2 = \frac{1}{16}$); $\sqrt[3]{\frac{1}{64}} = \frac{1}{4}$; and so on. Note that any root of a positive fraction is positive. This follows the pattern for the roots of integers, discussed above.

We can also raise a negative fraction to some power. For example, taking a base of $-\frac{1}{3}$ and squaring it, we get $(-\frac{1}{3})^2 = (-\frac{1}{3}) \times (-\frac{1}{3}) = \frac{1}{9}$. The result is positive because we are multiplying one negative number by another (see section 1.3 above). So when we square a negative fraction, the result is a *larger* number (that is, $\frac{1}{9}$ is greater than $-\frac{1}{3}$). This is true whenever we raise a negative fraction to an *even* numbered power.

In contrast, the cube of $-\frac{1}{3}$ is

$$(-\tfrac{1}{3})^3 = (-\tfrac{1}{3}) \times (-\tfrac{1}{3}) \times (-\tfrac{1}{3}) = \tfrac{1}{9} \times (-\tfrac{1}{3}) = -\tfrac{1}{27}$$

Here the result is negative because the last step in the calculation involves multiplying a positive number ($\frac{1}{9}$) by a negative number ($-\frac{1}{3}$). So when we cube a negative fraction, the result is a *larger* number (that is, $-\frac{1}{27}$ is greater than $-\frac{1}{3}$). This is true whenever we raise a negative fraction to an *odd* numbered power.

In all cases involving fractions, we obtain the root by reversing the power operation. For example, $\sqrt{\frac{1}{4}} = \frac{1}{2}$; $\sqrt[3]{-\frac{1}{27}} = -\frac{1}{3}$. The pattern of roots for fractions is exactly the same as the pattern for the roots of integers, discussed above. For a positive fraction, any root (square, cube, fourth, and so on) is positive. For a negative fraction, only odd numbered roots (cube, fifth, seventh, and so on) exist. These roots are negative.

Powers and roots on your calculator

Raising a number to any power

All but the cheapest calculators have a key for raising a number to any power. On most calculators the key is marked $\wedge$, but on some it is marked y^x. Here are some examples of use of the $\wedge$ key. (The $\wedge$ symbol is also used in mathematical computer programs such as Excel® and Maple.)

EXAMPLE 1.33

To find 3^2, key in: ⎡3⎤ ⎡∧⎤ ⎡2⎤ ⎡=⎤

and the answer, 9, should appear.

EXAMPLE 1.34

To find 3^4, key in: ⎡3⎤ ⎡∧⎤ ⎡4⎤ ⎡=⎤

and the answer, 81, should appear.

You can raise any positive number to any power in this way.

EXAMPLE 1.35

To raise a negative number to some power, on most calculators you must use brackets. Thus to find $(-4)^2$ you key in:

with the answer 16.

If you omit the brackets, some calculators will evaluate $-(4^2) = -16$. On some calculators you don't need brackets; they have a key marked $^+/-$. Pressing this key before keying in the number 4 gives one answer, and pressing it after the number gives the other answer.

Finding a root

Finding a root reverses finding a power, so we have to make the ∧ or y^x key work backwards. We do this by pressing the key marked 'SHIFT' or '2nd F' or 'INV' (depending on the make and model of your calculator), followed by y^x or ∧. Here is an example, from a calculator with a ∧ key and a 'SHIFT' key.

EXAMPLE 1.36

In example 1.34 we saw that to find 3^4 we key in:

| 3 | | ∧ | | 4 | | = | with answer 81

Suppose we want to reverse this; that is to find the fourth root of 81, written as $\sqrt[4]{81}$. (This is the number which, when raised to the power 4, equals 81.) We key in

| 4 | | Shift | | ∧ | | 8 | | 1 | | = | with the answer 3.

This reverse process can be quite confusing at first, but try to hang on to the key fact: to say that 3 is the fourth root of 81 is equivalent to saying that $3^4 = 81$.

Negative powers

Consider this series:

$$2^5 = 32$$
$$2^4 = 16 \quad (= 32 \div 2)$$
$$2^3 = 8 \quad (= 16 \div 2)$$
$$2^2 = 4 \quad (= 8 \div 2)$$

As we look down the list starting from the top, we see that on the left-hand side the power to which 2 is raised decreases by 1, and on the right-hand side we get from one number to the next by dividing by 2. Therefore it seems logical that we should be able to continue the list by dividing repeatedly by 2. We then get:

$$2^1 = 2 \quad (= 4 \div 2)$$
$$2^0 = 1 \quad (= 2 \div 2)$$
$$2^{-1} = \tfrac{1}{2} \quad (= 1 \div 2)$$
$$2^{-2} = \tfrac{1}{4} \quad (= \tfrac{1}{2} \div 2)$$
$$2^{-3} = \tfrac{1}{8} \quad (= \tfrac{1}{4} \div 2)$$

and so on. Thus we see that $2^0 = 1$, and negative powers denote fractions. In the same way, if we use a base of 3 instead of 2, we get:

$$3^5 = 243$$
$$3^4 = 81 \quad (= 243 \div 3)$$
$$3^3 = 27 \quad (= 81 \div 3)$$
$$3^2 = 9 \quad (= 27 \div 3)$$
$$3^1 = 3 \quad (= 9 \div 3)$$
$$3^0 = 1 \quad (= 3 \div 3)$$
$$3^{-1} = \tfrac{1}{3} \quad (= 1 \div 3)$$
$$3^{-2} = \tfrac{1}{9} \quad (= \tfrac{1}{3} \div 3)$$
$$3^{-3} = \tfrac{1}{27} \quad (= \tfrac{1}{9} \div 3)$$

and so on. Again, $3^0 = 1$, and negative powers denote fractions. The only difference is that, with base 3, we get from one number to the next on the right-hand side by dividing by 3 instead of 2.

There is a symmetry between the positive and negative powers which is worth noting. For example, $2^3 = 8$, while $2^{-3} = \frac{1}{8}$. So $2^{-3} = \frac{1}{2^3}$. Similarly, $3^2 = 9$, while $3^{-2} = \frac{1}{9}$. So $3^{-2} = \frac{1}{3^2}$. And so on.

To find, say, 2^{-3} on your calculator, the keystrokes are: 2; ^; (−); 3; =. The result should be 0.125 (which equals $\frac{1}{2^3} = \frac{1}{8}$). We will return to powers and roots in chapter 2, section 2.9.

Progress exercise 1.9

Find, preferably without using your calculator:

(a) 3^3 (b) 2^5 (c) $\sqrt{36}$ and $-\sqrt{36}$ (d) $\sqrt[4]{81}$

(e) 5^3 (f) $\sqrt[3]{27}$ (g) $\sqrt[2]{400}$ (h) 2^{-3}

1.15 Standard index form

Standard index form, also known as scientific notation, is a way of writing very large or very small numbers with less risk of error in writing or reading. In scientific notation, any number can be written as a number between 1 and 10, multiplied by 10 raised to some power. For example,

$$10 = 1 \times 10^1$$
$$1 = 1 \times 10^0$$
$$375 = 3.75 \times 10^2$$
$$\tfrac{1}{10} = 0.1 = 1 \times 10^{-1}$$
$$1000000 = 1 \times 10^6$$
$$\tfrac{1}{1000000} = 1 \times 10^{-6}$$

Confusion can arise when numbers that are less than 1 are given in decimal form, rather than as fractions. For example, the standard index form, 1×10^2 means 1 followed by two zeros; that is, 100. Similarly 1×10^{-2} means 1, divided by 1 followed by two zeros; that is, 1 divided by 100; that is, $\frac{1}{100}$. This is quite straightforward. However, $\frac{1}{100}$ in decimal form is 0.01, and if we are asked to express 0.01 in standard index form, we must remember that although in 0.01 there is only *one* zero after the decimal point, its standard index form is 1×10^{-2}, not 1×10^{-1} as we might be tempted to believe.

The above examples are trivial in themselves and are purely for illustrative purposes. Scientific notation really becomes useful when dealing with very large or very small numbers.

Hint Your calculator, and spreadsheet software such as Microsoft Excel®, switches automatically to scientific notation when dealing with very large and very small numbers. For example, my calculator can show a maximum of 10 digits on its display, so the largest number it can display in standard notation is 9,999,999,999, which is 1 short of 10 billion. So if I ask it to calculate 1 billion × 10 (which of course equals 10 billion), it switches to scientific notation and displays the answer as 1×10^{10}, which means a 1 followed by 10 zeros: that is, 10,000,000,000, which is 10 billion or 10,000 million. (In UK official statistics, most English speaking countries, and the economics profession, 1 billion equals one thousand million. The definition of 1 billion as one million million is now archaic.)

1.16 Some additional symbols

To conclude this chapter, here are some symbols that we will be using later and that may be new to you. Although new symbols are always a little off-putting, try to remember that they were invented to help us, and that they don't bite!

$\neq$ means 'is *not* equal to'

$\approx$ means 'is approximately equal to'

$\equiv$ means 'is identically equal to'. This conveys a stronger meaning than the = sign. The difference between $\equiv$ and = is important in economics and we shall return to it later.

$\pm$ means 'plus or minus'

∞ means 'infinity'. (Note that infinity is *not* 'a very large number'. Any number, however large, is *finite*; while the essence of infinity is that it is *not* finite.)

$-\infty$ means 'minus infinity'

$|x|$ means the **absolute value** of x: that is, ignoring its sign. For example, $|{-4}| = 4$.

$>$ means 'greater than' (for example, $4 > 3$)

$<$ means 'less than' (for example, $3 < 4$)

Finally, some symbols that you may find useful in setting out your work clearly are:

$\therefore$ means 'therefore'

$\Rightarrow$ means 'implies' or 'leads to'. For example, $x = 3 \Rightarrow x + 1 = 4$

Checklist

Be sure to test your understanding of this chapter by attempting the progress exercises (answers are at the end of the book). The Online Resource Centre contains further exercises and materials relevant to this chapter www.oxfordtextbooks.co.uk/orc/renshaw3e/.

The overall objective of this chapter was to refresh and renew your understanding of the basic rules that govern the manipulation of numbers. Specifically, we have reviewed:

✔ **Signs.** Rules governing signs when adding, subtracting, multiplying, and dividing positive and negative whole numbers.

✔ **Brackets.** When we need them and how to use them to take out common factors.

✔ **Fractions and decimals.** Adding, subtracting, multiplying, and dividing fractions and decimal numbers.

✔ **Decimal–fraction conversion.** Converting decimal numbers into fractions and vice versa.

✔ **Fractions, proportions, and percentages.** Converting fractions into proportions and percentages and vice versa. Increasing/decreasing a number by a given percentage; calculating percentage changes.

✔ **Index numbers.** What an index number is. Expressing time series data in index number form.

✔ **Powers and roots.** What they mean and how they are related. Manipulating and calculating powers and roots.

✔ **Using a calculator and making sense of the answers it gives.**

✔ **Rounding to significant figures or decimal places.**

✔ **Scientific notation.**

If you have worked carefully through the chapter, and especially the exercises, you should be feeling more confident of your understanding of basic concepts and manipulative rules of maths. If not, then you may need to re-read parts of this chapter before proceeding to chapter 2.

Self-test exercise (answers at end of the book)

1. Calculate, without using your calculator:

   (a) $6 \times 5 - 1$ (b) $4 \div -3 + 2$

   (c) $5 \times (3 + 4)$ (d) $(2 - 5 + 8)$

   (e) $\frac{3}{8} - \frac{1}{4} \times \frac{1}{2}$ (f) $\frac{2+3}{6-4}$

   (g) $\frac{1}{2}\left(\frac{3}{5} \times \frac{3}{10}\right)$ (h) $\frac{1}{4} + \frac{3}{16} \div \frac{1}{4}$

   (i) $2\frac{1}{8} - 1\frac{3}{16}$

2. Factorize, where possible:

   (a) $\frac{3}{8} - \frac{9}{11}$ (b) $\frac{5}{16} - \frac{15}{32}$ (c) $\frac{5}{9} - 1\frac{7}{18}$

3. What is 417.785 when rounded (a) to two decimal places; (b) to three significant figures?

4. What is the fractional equivalent of (a) 0.25; (b) 0.125?

5. What is the decimal equivalent of (a) $\frac{3}{8}$; (b) $\frac{1}{15}$ when rounded to three decimal places?

6. Convert the following into percentages, rounded to two decimal places: (a) $\frac{3}{8}$; (b) 0.175.

7. Calculate, without using a calculator: (a) 6^3; (b) $\sqrt[3]{125}$; (c) $\sqrt[4]{16}$.

8. (a) Express the following time series data in index number form with 1990 = 100.

Year	GDP (£m)
1990	750,674
1991	740,407
1992	741,860
1993	759,143
1994	792,717
1995	815,234

   (b) What was the percentage growth in GDP between (i) 1994 and 1995; (ii) 1990 and 1995?

9. If all of my income is taxed at a rate of 22% and I paid £5000 in tax last year, what was my income (a) before tax; (b) after tax?

Chapter 2
Algebra

OBJECTIVES

Having completed this chapter you should be able to:

- Add, subtract, multiply, and divide algebraic expressions.
- Use brackets to indicate the order of elementary operations, find a common factor and a common denominator.
- Add, subtract, multiply, and divide algebraic fractions.
- Understand powers and roots including fractional and negative powers.
- Understand and apply the rules for manipulating powers and roots.
- In logic, distinguish between necessary and sufficient conditions.

2.1 Introduction

In algebra we use symbols to denote unspecified or variable numbers. The symbols are either letters of the alphabet $(a, b, c, \ldots, x, y, z)$, or letters of the Greek alphabet $(\alpha, \beta, \gamma, \ldots, \chi, \psi, \omega)$. A list of letters of the Greek alphabet, together with a guide to pronunciation, is in the appendix at the end of this chapter. Mostly, but not invariably, we use lower-case symbols: that is, a, b, c rather than A, B, C. Because these letters stand for unspecified or variable numbers, we often refer to them as 'unknowns' or *variables*. In word-processed or printed works, we normally write them in *italic* font to make them stand out better.

Algebra is a form of shorthand which is used to express complex statements clearly and concisely. For example, suppose my car uses 8 litres of petrol on the journey to and from work, and that when I get there I have to pay €1.20 to park the car for the day. Suppose also that the price in euros of 1 litre of petrol is some unknown or unspecified amount that we can denote by the symbol z. Thus 8 litres of petrol will cost me $8 \times z$ euros. Then the daily cost, in euros, of my travel to and from work (consisting of fuel costs and parking costs) is given by:

$$8 \times z + 1.20$$

Now let us suppose that over a period of time I make an unspecified number, n, of these journeys. Then the total cost, in euros, of the n journeys will be:

$$n \times 8 \times z + n \times 1.20$$

This algebraic statement translates into words as: 'If I make a certain number of journeys to and from work, the total cost is the number of journeys multiplied by 8, multiplied by the price of a litre of petrol, plus the number of journeys multiplied by 1.20.' This is quite a long, complicated sentence and its meaning is not easy to grasp. Thus we can see that the algebraic method is a much more compact and precise way of saying something that is much more cumbersome when expressed in words. The downside of this, of course, is that we have to invest some time and effort (including, as always, lots of practice) before we can easily understand and manipulate algebraic expressions. But this effort soon brings a reward. Hopefully when you have got a little further into this book you will start to experience some of the payoff from your efforts.

2.2 Rules of algebra

Given that in algebra the symbols we use always denote numbers, it follows naturally that all the rules for manipulating numbers that we revised in chapter 1 apply equally to the manipulation of algebraic symbols. However, there are a few wrinkles in making the transition from arithmetic to algebra, which we'll now consider.

In algebra, as in arithmetic, the symbols $+$, $-$, $\times$, $\div$ mean 'add', 'subtract', 'multiply', and 'divide' respectively.

However, we saw in chapter 1 that the division sign ($\div$) is seldom used because it is easily confused with the $+$ sign. This is equally true in algebra. Instead we write $\frac{a}{b}$, or $^a/b$, or a/b to denote 'some number denoted by the symbol a divided by some other number denoted by the symbol b'. An expression of the form $\frac{a}{b}$ is called an algebraic fraction, or a ratio, or sometimes a quotient. As in arithmetic, the part above the line (a, in this case) is called the numerator while the part below the line (b, in this case) is called the denominator.

In arithmetic, we write 4×3 to denote '4 multiplied by 3'. In the same way, in algebra we can write $a \times b$ to denote 'a multiplied by b'. But because the multiplication sign can easily be mistaken for a number called x, it is not often used in algebra. Instead 'a multiplied by b' is written simply as 'ab' (or occasionally '$a.b$', where the dot means $\times$). This also has the attraction of being a more compact notation. (On some calculators, mobile phones and spreadsheet software such as Microsoft Excel®, the symbol * is used to denote multiplication.)

It is worth repeating that the rules of algebra are exactly the same as the rules of arithmetic that we revised in chapter 1. This means that if ever you are in doubt whether some operation you have performed is correct, you can always check by putting numbers in place of the algebraic symbols.

2.3 Addition and subtraction of algebraic expressions

When adding or subtracting, we must obey rules 1.1a and 1.1b in section 1.2. These were:

RULE 1.1 **Adding and subtracting positive and negative numbers**

Rule 1.1a
If two numbers are separated by two plus signs, or by two minus signs, we must *add* the two numbers together.

Rule 1.1b
If two numbers are separated either by a plus sign followed by a minus sign, or by a minus sign followed by a plus sign, we must *subtract* the second number from the first.

Thus for any two unspecified numbers denoted by a and b, the rules are:

(1) $a + (+b) = a + b$ (rule 1.1a)

(2) $a - (-b) = a + b$ (rule 1.1a)

(3) $a + (-b) = a - b$ (rule 1.1b)

(4) $a - (+b) = a - b$ (rule 1.1b)

Rules 1.1a and 1.1b may be summarized as:

- If the signs are the same, the result is addition (cases 1 and 2 above).

- If the signs differ, the result is subtraction (cases 3 and 4 above).

What we can add, and what we can't

If we are given, for example, $2 + 2 + 2$, we can easily see that this can be written more simply and compactly as 3×2. Similarly, if we are given $a + a + a$, this can also be written as $3 \times a$, which we write as $3a$ to avoid using the '$\times$' sign. Similarly, $b + b = 2b$, and so on. Here the 3 is called the **coefficient** of a, 2 the coefficient of b, and so on. We usually write $2b$ rather than $b2$, although both have the same meaning. That is, we put the coefficient in front of the variable rather than after it.

If we are given $a + a + a + b + b$, we can collect the as together as $3a$, and collect the bs together as $2b$. So

$$a + a + a + b + b = 3a + 2b$$

Note that we cannot add the 3 and the 2, because 3 refers to the as and 2 to the bs. If you are unsure on this point, try thinking of a as being a litre of wine, and b a kilo of bread. Can we add 3 litres of wine to 2 kilos of bread?

2.4 Multiplication and division of algebraic expressions

Multiplication

When multiplying, we must obey rule 1.2 in chapter 1. This was:

RULE 1.2 Multiplying positive and negative numbers

When multiplying two numbers together:

If the two numbers have the same sign, the result is positive.

If they have different signs, the result is negative.

Thus for any two unspecified numbers denoted by a and b, the rules are:

(1) $(+a) \times (+b) = +ab$ (same signs, result is positive)

(2) $(-a) \times (-b) = +ab$ (same signs, result is positive)

(3) $(+a) \times (-b) = -ab$ (different signs, result is negative)

(4) $(-a) \times (+b) = -ab$ (different signs, result is negative)

Division

When dividing, we must obey rule 1.3 in chapter 1. This was:

RULE 1.3 Dividing positive and negative numbers

When dividing one number by another:

If the two numbers have the same sign, the result is positive.

If the two numbers have different signs, the result is negative.

Thus for any two unspecified numbers denoted by a and b, the rules are:

(1) $(+a) \div (+b) = \frac{+a}{+b} = +\frac{a}{b}$ (same signs, result is positive)

(2) $(-a) \div (-b) = \frac{-a}{-b} = +\frac{a}{b}$ (same signs, result is positive)

(3) $(+a) \div (-b) = \frac{+a}{-b} = -\frac{a}{b}$ (different signs, result is negative)

(4) $(-a) \div (+b) = \frac{-a}{+b} = -\frac{a}{b}$ (different signs, result is negative)

From here on we shall follow the normal convention and omit the '+' sign in front of positive algebraic symbols. We shall also omit the brackets unless they are necessary to avoid ambiguity.

As noted above, we shall not often use the '×' sign for multiplication as it is too easily confused with an unspecified number named x. This has the following consequence for the way in which we write various multiplications:

Instead of:	we write:	which simplifies to:
$a \times b$	ab	no further simplification
$(-a) \times (-b)$	$(-a)(-b)$	ab (applying rule 1.2)
$(-a) \times b$	$(-a)b$	$-ab$ (applying rule 1.2)
$a \times (-b)$	$a(-b)$	$-ab$ (applying rule 1.2)

As was also discussed above, we will not use the '÷' sign. This has the following consequence for the way in which we write various divisions:

Instead of:	we write:	which simplifies to:
$a \div b$	$\frac{a}{b}$	no further simplification
$(-a) \div (-b)$	$\frac{-a}{-b}$	$\frac{a}{b}$ (applying rule 1.3)
$(-a) \div b$	$\frac{-a}{b}$	$-\frac{a}{b}$ (applying rule 1.3)
$a \div (-b)$	$\frac{a}{-b}$	$-\frac{a}{b}$ (applying rule 1.3)

Here are some worked examples to show rules 1.1 and 1.2, and the conventions regarding notation we have just discussed, in operation.

(1) $a + (-b)$ simplifies to $a - b$

(2) $3a + (-5b) - a - (-2)$ simplifies to $2a - 5b + 2$

(3) $10p + q + (-2q) - 3p + (-5p)$ simplifies to $2p - q$

(4) $-7 + 3x + (-4y) - (-5y) + 8x$ simplifies to $11x + y - 7$

(5) $5a \times 4b \div 2c$ simplifies to $\frac{10ab}{c}$

(6) $a \times 3b \div 2c$ is normally written as $\frac{3ab}{2c}$, which can't be simplified any further

(7) $a \times (-c) \div (-e)$ is normally written as $\frac{-ac}{-e}$, which simplifies to $\frac{ac}{e}$

(8) $x \div 2y \times (-3z)$ is normally written as $\frac{-3xz}{2y}$, which can't be simplified any further

Progress exercise 2.1

1. Write in their simplest forms:

(a) $-9a - 4a + 8a - a$ (b) $a + 2b - 3b + 6 - 5a - b$

(c) $14a - 5b - a + 6b - 5a + 2b$ (d) $5x + 2y - 3z + x - 2y - 2z + 2x + y - 5z$

(e) $5ab - 2ab + 3ba - 5ab$ (f) $6ax - 4bx + 2b - cb$

2. Rewrite the following without using the '×' and '÷' signs, and find the numerical value of each when $a = 2$, $b = 5$, and $c = -4$.

(a) $3a \times 4b$ (b) $a \div -ac$ (c) $-2a \times 3b \times -5c$ (d) $-2a \times -b \div -4c$

2.5 Brackets and when we need them

'Mixed' operations

As we saw in chapter 1, when addition and subtraction are mixed with multiplication and division, we get different answers depending on which part of the calculation we do first. For example, $6 + 8 \div 2$ is ambiguous. If we do the addition first, we get $14 \div 2 = 7$; but if we do the division first, we get $6 + 4 = 10$.

To avoid this ambiguity, rule 1.4 of chapter 1 tells us to carry out operations in the order **B-E-D-M-A-S**: that is, first carry out any operations inside Brackets, followed by any Exponent, then Division, then Multiplication, then Addition, and finally Subtraction.

EXAMPLE 2.1

$a + b \div c$

The B-E-D-M-A-S rule tells us that we must first *divide b* by *c*, and then *add a*.

However, as the division sign (÷) is not used in algebra, we do not write $a + b \div c$. Instead, we write $a + \frac{b}{c}$. Therefore in $a + \frac{b}{c}$ we must first *divide b* by *c*, and then *add a*.

EXAMPLE 2.2

$(a + b) \div c$

The B-E-D-M-A-S rule tells us that we must first *add a* to *b*, and then *divide* the result by *c*.

Again, because the division sign (÷) is not used in algebra, we do not write $(a + b) \div c$. Instead we write $\frac{a + b}{c}$. Therefore in $\frac{a + b}{c}$ we must treat the numerator as if it were in brackets; we must *first* add *a* and *b*, and then divide the result by *c*.

Note carefully the difference between examples 2.1 and 2.2. The key difference between the two is that in 2.1, only *b* is divided by *c*, while in 2.2 both *a* and *b* are divided by *c*.

EXAMPLE 2.3

Given, say, $4b^2$, the B-E-D-M-A-S rule tells us that we must deal with the exponent (the power, 2) before the multiplication. So $4b^2$ means $4 \times (b \times b)$, not $4 \times 4 \times b \times b$. If we want the power 2 to operate on the 4 as well as on *b*, we must write $(4b)^2$, not $4b^2$.

Expanding or multiplying out brackets

An example of multiplying out brackets from chapter 1 was

$3 \times (4 + 5)$

which we evaluate as

$$3 \times (4 + 5) = (3 \times 4) + (3 \times 5)$$

The arrows above help us to remember the procedure: that is, the 3 multiplies both the 4 and the 5, and the results are added. This process is called *multiplying out*, or *expanding*, the expression we started with.

In the same way, any equivalent algebraic expression can be multiplied out.

EXAMPLE 2.4

$a \times (b + c)$, which we write as $a(b + c)$ to avoid using the '$\times$' sign. Thus

$$a(b + c) = (a \times b) + (a \times c) \quad \text{which we write as } ab + ac.$$

EXAMPLE 2.5

$$a \times (b + c + d)$$

This becomes

$$a(b + c + d) = (a \times b) + (a \times c) + (a \times d) \quad \text{which we write as } ab + ac + ad.$$

EXAMPLE 2.6

Care is needed with signs when the multiplicative term outside the brackets is negative.

$$-a(b + c) = [(-a) \times (+b)] + [(-a) \times (+c)] = (-ab) + (-ac)$$

$$= -ab - ac$$

EXAMPLE 2.7

The same as example 2.6, but with $a = 1$

$-1(b + c)$, which we write simply as $-(b + c)$

Following exactly the pattern of example 2.6, we get

$$-1(b + c) = [(-1) \times (+b)] + [(-1) \times (+c)]$$

$$= [(-b)] + [(-c)]$$
$$= -b - c$$

From these examples we can derive the rule:

RULE 2.1 Multiplying out brackets

When brackets are removed by multiplying out, a minus sign outside the brackets means that the sign of each term previously inside the brackets is reversed.

Rule 2.1 is not really a new rule, but merely a restatement of rule 1.5 in chapter 1. If you are still uncertain about multiplying out, refer back to rule 1.5 and example 1.4 in chapter 1.

Factorization

Factorization is the reverse of multiplying out. In chapter 1 we factorized the expression

$$12 + 20 - 8$$

by noting that all three of these numbers were divisible by 4, without remainder. We separated out this common factor, as follows.

$$12 + 20 - 8 = (4 \times 3) + (4 \times 5) - (4 \times 2)$$
$$= 4 \times (3 + 5 - 2)$$

We can factorize algebraic expressions in the same way.

EXAMPLE 2.8

Suppose we are given: $ab + ac$

(Recall that ab means $a \times b$, and so on.) We see that both terms have a common factor, a. This common factor can be taken outside of a pair of brackets, as follows:

$$ab + ac = [a \times b] + [a \times c] = a \times (b + c), \quad \text{which we write as } a(b + c).$$

EXAMPLE 2.9

When we are factorizing, we must be careful with signs. Given: $-ab - ac$, we see that $(-a)$ is a common factor. Thus

$$-ab - ac = [(-a) \times b] + [(-a) \times c]$$
$$= (-a) \times (b + c), \quad \text{which we write as } -a(b + c)$$

Note the '+' sign inside the brackets. It is tempting to give the answer as $-a(b - c)$. Check for yourself, by multiplying out $-a(b - c)$, that this is not correct. A simple numerical example (such as $a = 2$, $b = 3$, $c = 4$) also shows it to be incorrect.

Progress exercise 2.2

1. Multiply out, and simplify:

(a) $a(b - c) + b(c - a)$ (b) $3(a - b + c) - 2(a + b - c)$

(c) $4z - (z + 3z - 3y)$ (d) $4a - (-4a) - 2ab$

2. Factorize the following (where possible):

(a) $4x + 8y$ (b) $5x - 10y$ (c) $-3x - 9xy$ (d) $-ax - 2ay$

(e) $-5ab + 15b$ (f) $4y - 12xy$ (g) $3ax - 9xy$ (h) $15xyz + 5yz - 10xy$

2.6 Fractions

In algebra, fractions such as $\frac{a}{b}$ appear frequently. This is because in algebra we avoid using the '÷' sign and therefore write $\frac{a}{b}$, or a/b instead of $a \div b$. Thus a fraction (also called a ratio or quotient) appears whenever we divide one number by another. Operations with algebraic fractions obey the same rules as arithmetical fractions (see chapter 1, sections 1.7 and 1.8).

Simplifying fractions

In chapter 1, section 1.8 we saw that with any arithmetic fraction we can multiply or divide the numerator and denominator by the same number without affecting the value of the fraction. In the fraction $\frac{3}{6}$, if we divide both numerator and denominator by 3 we get $\frac{3/3}{6/3}$. We can then cancel in both numerator and denominator. In the numerator, 3 divided by 3 equals 1. In the denominator, 6 divided by 3 equals 2. So we get

$$\frac{3}{6} = \frac{3/3}{6/3} = \frac{1}{2}$$

We thus converted $\frac{3}{6}$ into $\frac{1}{2}$, which is a simpler expression. This is a valid step because both numerator and denominator are divisible by 3 without remainder. It's intuitively fairly obvious that if you divide a cake into 6 equal slices and then eat 3 of them, then you have eaten half of the cake. We can do exactly the same thing with an algebraic fraction, as in the following example.

EXAMPLE 2.10

$$\frac{ac}{ab}$$

Here we see that both numerator and denominator can be divided by a, without remainder. Thus

$$\frac{ac}{ab} = \frac{\frac{ac}{a}}{\frac{ab}{a}}$$

In the numerator, c is both multiplied and divided by a, and since division reverses multiplication the two as cancel each other out, leaving us simply with c. Similarly the two as in the denominator cancel, leaving us simply with b. (Try a numerical example if you are uncertain on this point.) Thus

$$\frac{ac}{ab} = \frac{\frac{ac}{a}}{\frac{ab}{a}} = \frac{c}{b}$$

Again we have simplified by cancelling the a, which is common to both numerator and denominator.

The process of cancelling can also be reversed: that is, we can *multiply* numerator and denominator by any number. For example,

$$\frac{a}{b} = \frac{a \times c}{b \times c} = \frac{ac}{bc}$$

Here we have multiplied both numerator and denominator by c. This leaves the numerical value of the fraction unchanged. Check this for yourself, giving a, b, and c any values you wish.

2.7 Addition and subtraction of fractions

Addition

EXAMPLE 2.11

Consider:

$$\frac{a}{b} + \frac{c}{d}$$

In adding these two fractions we have to find a common denominator: that is, something that is divisible by both denominators without remainder (see chapter 1, section 1.7). The way we can do this is to multiply numerator and denominator of each fraction by the denominator of the other fraction.

Thus we first multiply top and bottom of $\frac{a}{b}$ (the first fraction) by d (the denominator of the second fraction). Thus

$$\frac{a}{b} \quad \text{becomes} \quad \frac{a}{b} \times \frac{d}{d} = \frac{ad}{bd}$$

Second, in the same way we then multiply top and bottom of $\frac{c}{d}$ (the second fraction) by b (the denominator of the first fraction). Thus

$$\frac{c}{d} \quad \text{becomes} \quad \frac{c}{d} \times \frac{b}{b} = \frac{cb}{bd}$$

Then we add the results, to get

$$\frac{a}{b} + \frac{c}{d} = \frac{ad}{bd} + \frac{cb}{bd} = \frac{ad + cb}{bd}$$

which is as far as we can take the calculation without knowing the values of a, b, c, and d.

If you had difficulty following this, try working through it again with specific numerical values, such as $a = 1$, $b = 2$, $c = 3$, and $d = 4$.

Subtraction

The method is exactly the same as the method for addition. To show this, example 2.12 below is the same as example 2.11, but with a minus sign between the fractions instead of a plus sign.

EXAMPLE 2.12

$$\frac{a}{b} - \frac{c}{d} = \frac{ad}{bd} - \frac{cb}{bd} \quad \text{(from example 2.11)}$$

$$= \frac{ad - cb}{bd}$$

Three or more fractions

Things are a little more tricky when there are three or more fractions to be added. Probably the easiest way of tackling this is to add the fractions two at a time.

EXAMPLE 2.13

$$\frac{a}{b} + \frac{c}{d} + \frac{e}{f}$$

We first add the first two fractions together, leaving the third fraction, $\frac{e}{f}$, aside for the moment. Following the method of example 2.11 above, we get

$$\frac{a}{b} + \frac{c}{d} = \frac{ad + cb}{bd}$$

We now have to add to this the third fraction, $\frac{e}{f}$. So our given problem becomes

$$\frac{ad + cb}{bd} + \frac{e}{f}$$

We now add these two fractions, again using the method of example 2.11. That is, we first multiply the top and bottom of the first fraction by f, the denominator of the second fraction. This gives

$$\frac{f(ad + cb)}{fbd}$$

Notice the use of brackets to indicate that the *whole* of the top of the first fraction, $(ad + cb)$, must be multiplied by f.

Second, we multiply the top and bottom of the second fraction by bd, the denominator of the first fraction. This gives

$$\frac{bde}{bdf}$$

Finally we add these two together. So our final answer is

$$\frac{a}{b} + \frac{c}{d} + \frac{e}{f} = \frac{f(ad + cb)}{fbd} + \frac{bde}{bdf} = \frac{f(ad + cb) + bde}{fbd}$$

If you had difficulty following this, try working through it again with specific numerical values, such as $a = 1, b = 2, c = 3, d = 4, e = 5, f = 6$. (The answer is $2\frac{1}{12}$.)

The above method does not always result in the lowest common denominator being found, but has the merit of being reliable and reasonably straightforward. Fortunately the need to do this does not arise very often in economics.

2.8 Multiplication and division of fractions

Multiplication

The rule for multiplying algebraic fractions is exactly the same as for arithmetical fractions (see chapter 1, section 1.8). We simply multiply the two numerators together to obtain the numerator of the answer, and multiply the two denominators together to obtain the denominator of the answer. Thus in general

$$\frac{a}{b} \times \frac{c}{d} = \frac{ac}{bd}$$

This generalizes to any number of terms. For example

$$\frac{a}{b} \times \frac{c}{d} \times \frac{e}{f} = \frac{ace}{bdf}$$

Reciprocals

As in arithmetic, for any number b (provided b is not zero):

$$b \times \frac{1}{b} = \frac{b}{1} \times \frac{1}{b} = \frac{b \times 1}{1 \times b} = \frac{b}{b} = 1$$

Thus b and $\frac{1}{b}$ are said to be *reciprocals* because when they are multiplied together, the result is 1.

Division

As in arithmetic, we invert the second fraction and multiply (see chapter 1, section 1.8, rule 1.7). Thus:

EXAMPLE 2.14

$$\frac{a}{b} \div \frac{c}{d} = \frac{a}{b} \times \frac{d}{c} = \frac{ad}{bc}$$

This method also works when the second term is not a fraction, as follows:

EXAMPLE 2.15

$$\frac{a}{b} \div c$$

Since c is the same thing as $\frac{c}{1}$, the problem becomes

$$\frac{a}{b} \div \frac{c}{1} = \frac{a}{b} \times \frac{1}{c} = \frac{a}{bc}$$

The method also works when the first term is not a fraction, as follows:

EXAMPLE 2.16

$$a \div \frac{c}{d} = \frac{a}{1} \times \frac{d}{c} = \frac{ad}{c}$$

Factorizing fractions

Algebraic fractions can be factorized in the same way as arithmetic fractions (see chapter 1, section 1.8).

EXAMPLE 2.17

$$\frac{a}{b} + \frac{ad}{c}$$

Here, a appears in the numerator of both fractions. In other words, a is a factor that is common to both numerators. So we can rearrange the expression as

$$\frac{a}{b} + \frac{ad}{c} = \left[\frac{a}{1} \times \frac{1}{b}\right] + \left[\frac{a}{1} \times \frac{d}{c}\right]$$

From this we see that $\frac{a}{1}$ (which of course is the same thing as a) is a common factor, so we can take it outside a pair of brackets and get

$$\frac{a}{1}\left(\frac{1}{b} + \frac{d}{c}\right) = a\left(\frac{1}{b} + \frac{d}{c}\right)$$

Check for yourself that this is correct by multiplying out.

EXAMPLE 2.18

$$\frac{ac}{bd} + \frac{abc}{de}$$

Here, $\frac{ac}{d}$ is common to both terms of the addition. So

$$\frac{ac}{bd} + \frac{abc}{de} = \left[\frac{ac}{d} \times \frac{1}{b}\right] + \left[\frac{ac}{d} \times \frac{b}{e}\right]$$

$$= \frac{ac}{d}\left(\frac{1}{b} + \frac{b}{e}\right)$$

Check for yourself that this is correct by multiplying out.

Cancelling between numerator and denominator

EXAMPLE 2.19

When meeting fractions that have addition or subtraction in numerator or denominator, such as $\frac{a+b}{bc}$, it is tempting to try to cancel the b in the numerator with the b in the denominator, and thereby get $\frac{a+1}{c}$ (see chapter 1, section 1.8). This is not correct; check for yourself by means of a simple numerical example that $\frac{a+b}{bc}$ does *not* equal $\frac{a+1}{c}$ (except when $b = 1$).

The mistake here is to forget that, when cancelling, the *whole* of the numerator and the *whole* of the denominator must be divided or multiplied by the same term.

Thus, given $\frac{a+b}{bc}$, we can divide the whole of the numerator and the whole of the denominator by b, and get

$$\frac{a+b}{bc} = \frac{\frac{a}{b} + \frac{b}{b}}{\frac{bc}{b}} \quad \text{which, after cancelling the } bs, \text{ becomes} \quad \frac{\frac{a}{b} + 1}{c}$$

However, this cancellation, although valid, is not very useful as the new expression is no simpler than the one we started with.

In contrast to the possible mistake in example 2.19, when meeting a fraction with *only* multiplication or division in the numerator and denominator, we *can* cancel individual elements in the numerator and denominator, if we wish, as the next two examples illustrate.

EXAMPLE 2.20

Given $\frac{ab}{bc}$, this has only multiplication in numerator and denominator, so we can divide the whole of the numerator and the denominator by b and get

$$\frac{ab}{bc} = \frac{\frac{ab}{b}}{\frac{bc}{b}} = \frac{a}{c}$$

EXAMPLE 2.21

Given:

$$\frac{\frac{a}{b}}{\frac{c}{b}}$$

this has only division in numerator and denominator. We can multiply the whole of the numerator and the denominator by b and get

$$\frac{\frac{a}{b}}{\frac{c}{b}} = \frac{\frac{a}{b} \times b}{\frac{c}{b} \times b}, \quad \text{which after cancelling becomes } \frac{a}{c}$$

Progress exercise 2.3

1. Express each of the following as a single algebraic fraction.

   (a) $\frac{3}{a} - \frac{6}{b}$ (b) $\frac{a}{b} - \frac{c}{d}$ (c) $\frac{5}{3a} - \frac{4}{2b}$ (d) $a - \frac{a}{b}$

   (e) $\frac{4b}{3a} - \frac{2c}{5d}$ (f) $\frac{5a}{a+b} + \frac{3c}{c+d}$ (g) $\frac{b}{a} - \frac{2c}{c+d}$ (h) $\frac{a}{a(b+c)} - \frac{d}{d+e}$

2. Express each of the following as a single algebraic fraction, and simplify where possible.

   (a) $\frac{a}{b} \times \frac{d}{c}$ (b) $\frac{a}{bc} \times \frac{bc}{a}$ (c) $ab \div \frac{b}{d}$ (d) $\frac{ab}{c} \div \frac{d}{c}$ (e) $\frac{2a}{x+b} \times \frac{x+b}{c}$ (f) $\frac{3}{y+1} \div \frac{6}{x+2}$

Some special cases

Before leaving fractions, let us look at some special cases.

(1) For any number x (provided x is not equal to zero),

$$\frac{x}{x} = 1$$

In words, any number divided by itself is equal to 1.

(2) If x is equal to zero, we have

$$\frac{x}{x} = \frac{0}{0}$$

Now, it is very tempting to say that $\frac{0}{0} = 0$ but this is not the case. In fact, $\frac{0}{0}$ is said to be *undefined*: that is, it has no meaning. If this is puzzling, remember that it is not a logical necessity that everything we can write down must have a meaning!

(3) Provided x is not equal to zero,

$$\frac{0}{x} = 0$$

Here the logic is that since division is repeated subtraction, the question here is: 'How many times can you subtract x from 0?' And the answer is: 'Zero times'.

(4) This is the most difficult special case. We have to decide what meaning to attach to

$$\frac{x}{0} \quad \text{(when } x \text{ is any number except zero)}$$

It is tempting to write

$$\frac{x}{0} = \infty \quad \text{(recall from chapter 1 that the symbol '}\infty\text{' denotes 'infinity').}$$

We could attempt to justify this by arguing that, since division is really nothing more than repeated subtraction, the logic here is that we are asking: 'How many times can we subtract zero from any positive number?' A plausible answer is: 'An infinite number of times'.

However, this answer runs into trouble because an essential feature of division is that it must be reversible by means of multiplication. For example, if $\frac{a}{b} = c$, then it must follow that $a = bc$. But when we try to apply this reversibility test to the expression $\frac{x}{0} = \infty$, we get $0 \times \infty = x$, which doesn't really make sense. So we are forced to conclude that $\frac{x}{0}$, like $\frac{0}{0}$, is undefined. Another way of putting this is to say that division by zero is not logically possible, since its outcome is undefined.

2.9 Powers and roots

We examined powers and roots in arithmetic in some detail in section 1.14. Here we will simply translate our findings there from arithmetic into algebra.

For any number, a, by definition we have:

$$a^2 = a \times a$$
$$a^3 = a \times a \times a$$

and in general

$$a^n = a \times a \times a \times a \times a \times \ldots \times \ldots \quad \text{(with } a \text{ appearing } n \text{ times, on the right).}$$

Reversing the process of raising a number, a, to the power n, is known as taking or finding the nth root. For example, if b is a positive number, and $a^2 = b$, then by definition a is the square root of b, written as $a = \sqrt{b}$. Therefore $a^2 = b$ implies $(\sqrt{b})^2 = b$. Note that a, the square root of b, is by definition positive. However, since $a^2 = b$, it must also be true that $(-a)^2 = b$ (because $(-a)^2 = a^2$; see rule 1.2). And since $a = \sqrt{b}$, this implies $(-\sqrt{b})^2 = b$. So both a (the square root of b) and $-a$ (the negative of the square root of b) have the property that, when squared, the result is b. We write the previous sentence in mathematical form as $(\pm a)^2 = (\pm\sqrt{b})^2 = b$.

In contrast, if b is negative its square root does not exist, because there is no (real) number which, when squared, is negative.

The above properties hold for any *even* numbered root (square root, fourth root, sixth root, and so on). For example, if b is a positive number, and $a^4 = b$, then a is the fourth root of b, written as $a = \sqrt[4]{b}$. We can show that a is positive (see section 1.14). But since $a^4 = b$, we also have $(-a)^4 = b$ (because $a^4 = (-a)^4$; see rule 1.2) and therefore $(\pm a)^4 = (\pm\sqrt[4]{b})^4 = b$. If b is negative, its fourth root does not exist as a real number.

Looking next at cubes and cube roots, if b is positive, and $a^3 = b$, then by definition a is the cube root of b, written as $a = \sqrt[3]{b}$. We can be sure that a is positive, since if a is negative, a^3 is also negative and therefore cannot equal b, which is assumed positive. Unlike the even numbered roots considered above, if b is negative its cube root exists, and must be negative. The above properties hold for any *odd* numbered root (cube root, fifth root, seventh root, and so on).

2.10 Extending the idea of powers

In this section we will derive some simple but important rules for manipulating powers, which we'll demonstrate by means of examples.

EXAMPLE 2.22

Consider $a^2 \times a^3$

By definition, $a^2 = a \times a$; and $a^3 = a \times a \times a$

Therefore $a^2 \times a^3 = a \times a \times a \times a \times a$

But, by definition $a \times a \times a \times a \times a = a^5$

Also, since $5 = 2 + 3$, a^5 can be written as a^{2+3}

So, combining these results we have $a^2 \times a^3 = a^{2+3} = a^5$

In words, we *add* the powers when the two numbers (a^2 and a^3) are *multiplied* together.

Generalizing from this example:

RULE 2.2 **Multiplication with powers**

$$a^n \times a^m = a^{n+m}$$

Note that this rule and those that follow are valid for any values of *a*, *m*, and *n*.

Hint We can only apply rule 2.2 where the two terms have the same base and are multiplied together. Thus we can't apply rule 2.2 to something like $a^2 + a^3$ because there is a '+' sign, not a multiplication sign, between a^2 and a^3. In other words, $a^2 + a^3$ does *not* equal a^5. (Check this for yourself, with $a = 2$, say.)

And we can't apply rule 2.2 to something like $a^2 \times b^3$ because the first term has base *a* and the second has base *b*.

EXAMPLE 2.23

Consider $\frac{a^5}{a^3}$

By definition, $a^5 = a \times a \times a \times a \times a$; and $a^3 = a \times a \times a$

Therefore $\frac{a^5}{a^3} = \frac{a \times a \times a \times a \times a}{a \times a \times a}$

which, after cancelling three of the *a*s in the numerator with the three *a*s in the denominator, becomes

$a \times a$

which by definition equals a^2. Also, since $2 = 5 - 3$, a^2 is the same thing as a^{5-3}

Collecting these results, we have found that $\frac{a^5}{a^3} = a^{5-3} = a^2$

In words, we *subtract* the powers when one number (a^5) is *divided* by another (a^3).

Generalizing from this example, we have:

RULE 2.3 **Division with powers**

$$\frac{a^m}{a^n} = a^{m-n}$$

EXAMPLE 2.24

Consider $(a^2)^3$

By definition, $(a^2)^3 = (a \times a)^3 = (a \times a) \times (a \times a) \times (a \times a) = a^6$

Thus we have found that $(a^2)^3 = a^{2 \times 3} = a^6$

Generalizing from this example, we have:

RULE 2.4 Raising a power to a power

$(a^m)^n = a^{m \times n} = a^{mn}$

EXAMPLE 2.25

Consider $(ab)^3$

By definition, $(ab)^3 = (a \times b)^3 = (a \times b) \times (a \times b) \times (a \times b)$

But, since the brackets on the right-hand side are unnecessary and numbers can be multiplied together in any order, we can rewrite $(a \times b) \times (a \times b) \times (a \times b)$ as

$a \times a \times a \times b \times b \times b = a^3 \times b^3 = a^3 b^3$

Thus we have found that $(ab)^3 = a^3 b^3$.

Generalizing from this example, we have:

RULE 2.5 Removing brackets with powers

$(ab)^n = a^n b^n$

Hint Don't confuse $(ab)^n$ with $(a+b)^n$. Rule 2.5 tells us that $(ab)^n = a^n b^n$, but $(a+b)^n$ does NOT equal $a^n + b^n$. If you're not sure why, try a numerical example; say, $a = 2$, $b = 3$, $n = 2$.

EXAMPLE 2.26

Consider $\frac{a^m}{a^m}$

Since any number divided by itself equals 1, we must have $\frac{a^m}{a^m} = 1$

But, from rule 2.3 above, we also have $\frac{a^m}{a^m} = a^{m-m} = a^0$

Combining these two results, we have $a^0 = 1$. (See also section 1.14.)

Generalizing from this example, we have:

RULE 2.6 Meaning of the power zero

$a^0 = 1$ (where a is any number except zero).

2.11 Negative and fractional powers

Negative powers

We met negative powers in section 1.14. Here we obtain the same interpretation by a more formal method.

EXAMPLE 2.27

The meaning of a^{-2}. Consider $a^{-2} \times a^2$.

By rule 2.2, $a^{-2} \times a^2 = a^{-2+2} = a^0$. By rule 2.6, $a^0 = 1$. Thus $a^{-2} \times a^2 = 1$. But this can be true only if a^{-2} and a^2 are reciprocals (see section 1.8). That is, $a^{-2} = \frac{1}{a^2}$.

Generalizing from this example, we have:

> **RULE 2.7 Meaning of a negative power**
>
> $a^{-n} = \dfrac{1}{a^n}$

This rule means that a^{-n} and a^n are reciprocals. When multiplied together, the result is 1. (See section 1.8.)

The nth root of a negative power

Given that rule 2.7 gives a meaning to a negative power, we must consider what is the corresponding negative root. If we are given two numbers a and b and a value of n such that $a^{-n} = b$, it is tempting to follow the pattern of section 2.9, and write $a = \sqrt[-n]{b}$. However, this is meaningless. The correct approach is as follows. Given $a^{-n} = b$, from rule 2.7 $a^{-n} = \frac{1}{a^n}$, therefore $\frac{1}{a^n} = b$, from which $a^n = \frac{1}{b}$. From this, following the pattern of section 2.9, $a = \sqrt[n]{\frac{1}{b}}$.

For example, given $a^{-n} = b$, with $a = {}^1/2$ and $n = 2$, we have from rule 2.7:

$$a^{-n} = \left(\frac{1}{2}\right)^{-2} = \frac{1}{({}^1/2)^2} = \frac{1}{{}^1/4} = 4$$

So $b = 4$. We can check that this is correct, as follows. With $b = 4$, and if $a = \sqrt[n]{\frac{1}{b}}$, we have: $a = \sqrt{\frac{1}{4}} = \frac{1}{2}$, which is correct.

Fractional powers

We have so far met positive and negative powers. Here we meet fractional powers for the first time.

EXAMPLE 2.28

Consider $a^{1/2}$. (Note that 1/2 means $\frac{1}{2}$; see section 1.6.)

Using rule 2.4, $(a^{1/2})^2 = a^{(1/2) \times 2} = a^1 = a$.

But, from the definition of a square root, $(\sqrt{a})^2 = a$ (that is, squaring and taking the square root cancel each other out).

Therefore, since $(a^{1/2})^2$ and $(\sqrt{a})^2$ both equal a, they must equal one another.

Therefore $(a^{1/2})^2 = (\sqrt{a})^2$

But this can only be true if $a^{1/2} = \sqrt{a}$. In words, a raised to the power one-half is the same thing as the square root of a.

From section 2.9 we recall that $\sqrt{a}$ exists only if a is positive, and if $\sqrt{a}$ exists it is positive by definition. Therefore the same is true of $a^{1/2}$.

Using the same method as in example 2.28, we can show that $a^{1/3} = \sqrt[3]{a}$. In words, a raised to the power one-third is the same thing as the cube root of a. From section 2.9 we recall that $\sqrt[3]{a}$ is positive if a is positive, and negative if a is negative. Therefore the same is true of $a^{1/3}$.

Similarly, we can show that $a^{1/4} = \sqrt[4]{a}$. In words, a raised to the power one-quarter is the same thing as the fourth root of a. In section 2.9 we stated (but did not prove) that $\sqrt[4]{a}$ has the same properties as $\sqrt{a}$. That is, $\sqrt[4]{a}$ exists only if a is positive, and if $\sqrt[4]{a}$ exists it is necessarily positive. Therefore the same is true of $a^{1/4}$.

From example 2.28 and the two paragraphs above we can derive the following generalization:

RULE 2.8 Fractional powers

$$a^{1/n} = \sqrt[n]{a}$$

Note that if a is negative and n is an even number, this root does not exist as a real number. However, rule 2.8 remains valid.

EXAMPLE 2.29

This is a small extension of example 2.28.

Consider $a^{2/3}$.   (Note that 2/3 means $\frac{2}{3}$; see section 1.6.)

Using rule 2.4, $a^{2/3} = (a^2)^{1/3}$. Using rule 2.8, $(a^2)^{1/3} = \sqrt[3]{a^2}$

Combining these, we have $a^{2/3} = \sqrt[3]{a^2}$.

The generalization of the previous example is:

RULE 2.9 Fractional powers

$$a^{m/n} = \sqrt[n]{a^m}$$ (Note that m/n means $\frac{m}{n}$)

Note that negative and fractional powers can be combined. Consider for example $a^{-1/2}$. By rule 2.7, $a^{-1/2} = \frac{1}{a^{1/2}}$. By rule 2.8, $a^{1/2} = \sqrt{a}$. Combining these, we get $a^{-1/2} = \frac{1}{\sqrt{a}}$.

2.12 The sign of a^n

In sections 2.9–2.11, and also in section 1.14, we have looked carefully at powers and roots. We are now ready to draw a conclusion that we will use quite frequently in later chapters. The conclusion is that if a is positive, then a^n is positive for any n. In words, this says that if any positive number is raised to any power, the result is positive.

We can demonstrate this as follows. Assuming a is positive, there are four cases:

(1) n is a positive integer ($n = 1, 2, 3, 4$ and so on). Then clearly a^n is positive.

(2) n is a negative integer ($n = -1, -2, -3, -4$ and so on). Then if, say, $n = -3$, from rule 2.7 we have $a^{-3} = \frac{1}{a^3}$. And since, from (1), a^3 is positive, then $\frac{1}{a^3}$ is clearly positive and so therefore is a^{-3}. This is obviously valid not only for $n = -3$ but for any value of n when n is a negative integer. Therefore in this case a^n is positive.

(3) n is a positive fraction ($n = \frac{1}{2}, \frac{1}{3}, \frac{1}{4}$, and so on). Then if, say, $n = \frac{1}{2}$, from rule 2.7 we have $a^{1/2} = \sqrt{a}$. If $n = \frac{1}{3}$, from rule 2.7 we have $a^{1/3} = \sqrt[3]{a}$. If $n = \frac{1}{4}$, from rule 2.7 we have $a^{1/4} = \sqrt[4]{a}$, and so on. But in section 1.14 (see also section 2.9) we showed that all roots of a positive number (square root, cube root, fourth root, and so on) were positive. Therefore in this case a^n is positive.

(4) n is a negative fraction ($n = -\frac{1}{2}, -\frac{1}{3}, -\frac{1}{4}$, and so on). This is merely a combination of cases (2) and (3). For example, if $n = -\frac{1}{3}$, from rules 2.7 and 2.9 we have $a^{-1/3} = \frac{1}{\sqrt[3]{a}}$, which is positive since we showed in (3) that $\sqrt[3]{a}$ is positive. Clearly this is valid whenever n is a negative fraction. Therefore in this case a^n is positive.

We therefore conclude:

 RULE 2.10 The sign of a^n

If a is positive, then a^n is positive for any value of n.

Note that rule 2.10 assumes that a is a real number (see section 1.14).

2.13 Necessary and sufficient conditions

The distinction in logic between a necessary and a sufficient condition is important in both maths and economics. You are probably already half-consciously aware of the concept but nevertheless we will review it explicitly here.

Suppose there is some event or state of the world that we will label X and another that we will label Y. Then there are four ways in which a logical or causal connection may run from X to Y.

(1) X may be a necessary condition, but not a sufficient condition, for Y. For example, if X is the state of being a male, and Y is the state of being a father, then X is a necessary but not a sufficient condition for Y. X is *necessary* for Y because you cannot be a father without (first) being a male. But X is *not sufficient* for Y, because you can be a male without being a father.

(2) X may be neither necessary nor sufficient for Y. For example, if X is the state of being a male, and Y is the state of being a parent, then X is neither a necessary nor a sufficient condition for Y. X is *not necessary* for Y, because you can be a parent without being male (by being a mother). X is *not sufficient* for Y, because you can be a male without being a parent.

(3) X may be sufficient, but not necessary, for Y. For example, if X is the state of being a father, and Y is the state of being a male, then X is a sufficient but not a necessary condition for Y. X is *sufficient* for Y because you cannot be a father without being male. But X is *not necessary* for Y, because you can be male without being a father.

(4) X may be both a necessary and a sufficient condition for Y. For example if X is the state of being married, and Y is the state of being a spouse, then X is *both necessary and sufficient* for Y. This is because being married and being a spouse have the same dictionary definition, so anyone who is married is, by definition, a spouse; and anyone who is a spouse is, by definition, married.

We can illustrate these four logical possibilities with diagrams (called Venn diagrams, after their inventor). Possibility 1 is shown in figure 2.1a. Box X contains all males, while box Y contains all fathers. Box Y lies entirely within box X, showing that all fathers are males. But box Y is smaller than box X, showing that it is possible to be male (thus in box X) but not a father (thus not in box Y).

Possibility 2 is shown in figure 2.1b. Box X contains all males, and box Y contains all parents. Box Z contains all females. Because box Y lies only partly in box X, we see that being male is not a necessary condition for being a parent; there exist parents who are not males. Because box Y covers only part of box X, we can see that being a male is not a sufficient condition for being a father; there exist males who are not parents.

Possibility 3, like possibility 1, is shown in figure 2.1a. Because box Y lies entirely inside box X, being a father is a sufficient condition for being a male, but it is not a necessary condition, as box Y is smaller than box X.

Possibility 4 is shown in figure 2.1c. Box X (those who are married) coincides exactly with box Y (those who are spouses). Thus we see that being married is both a necessary and a sufficient condition for being a spouse.

In addition, there is also the question of the reverse logical or causal relationship running from Y to X. In possibility 1, being in box Y (a father) is sufficient but not necessary for being in box X (a male). In possibility 2, being in box Y (a parent) is neither necessary nor sufficient for being in box X (a male). In possibility 3, being in box Y (a father) is sufficient but not necessary for being in box X (a male). In possibility 4, being in box Y (a spouse) is both necessary and

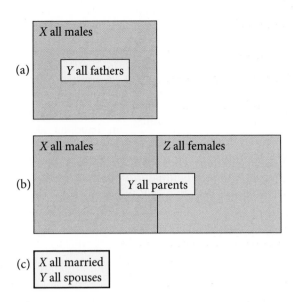

Figure 2.1 Examples of necessary and sufficient conditions.

sufficient for being in box X (married). Of course, the relationship running from Y to X is not independent of that running from X to Y, but we will not pursue the exact nature of this interdependence here. If you study the examples above carefully, you may be able to discover this interdependence for yourself.

Hopefully these examples serve to convey the key distinction between a necessary and a sufficient condition. This distinction is important in almost all reasoning and argument, so understanding it has a usefulness extending far beyond this book. However, it is also a distinction that we will use at several key points in later chapters.

Progress exercise 2.4

1. Simplify:

(a) $(-a)^2; (-a)^3; (-a)^4; (-a)^5$ (b) $(-2a)^2; (-2a)^3; (-2a)^4; (-2a)^5$

(c) $(-\frac{c}{3})^2; (-\frac{c}{3})^3; (-\frac{c}{3})^4; (-\frac{c}{3})^5$ (d) $8t^2 \div (-4t)$

(e) $(-4x^3)^3 \div (-2x^2)$ (f) $15x^3y^2 \div (-5x^2y^3)$

2. Remove the brackets from the following.

(a) $(xy)^2$ (b) $(x^3y)^{1/3}$ (c) $(100x)^{1/2}$ (d) $(\frac{x}{y})^2$

(e) $[\frac{x^{-1/2}}{y^2}]^2$ (f) $(x^2y^2)^{1/2}$ (g) $(x^2y^2x^3y^3)^{-1/6}$ (h) $(\frac{x^{-1}}{y^{-1/2}})^2$

Checklist

Be sure to test your understanding of this chapter by attempting the progress exercises (answers are at the end of the book). The Online Resource Centre contains further exercises and materials relevant to this chapter www.oxfordtextbooks.co.uk/orc/renshaw3e/.

In this chapter we have extended to algebra the manipulative rules that we reviewed in chapter 1.

The topics that we reviewed were:

✔ **Signs.** Rule of signs when adding, subtracting, multiplying, and dividing algebraic expressions.

✔ **Brackets and factorization.** Using brackets to indicate the order of algebraic operations and for factorization.

✔ **Fractions and common denominators.** Adding, subtracting, multiplying, and dividing algebraic fractions, including finding a common denominator when adding/subtracting fractions.

✔ **Powers and roots.** Meaning and rules for manipulating algebraic powers and roots including fractional and negative powers.

✔ **Necessary and sufficient conditions.** Four types of logical or causal relationship between event or state of the world X and event or state of the world Y.

If you have worked carefully through this chapter, and especially the exercises, your confidence in handling algebraic expressions should have increased. If so, you are ready to proceed to the slightly more advanced rules and techniques of algebra in chapter 3.

Appendix The Greek alphabet

Note that some Greek characters are visually difficult to distinguish from English language characters, and therefore are not used in mathematics.

Upper case	Lower case	Say as
A	α	alpha
B	β	beta
Γ	γ	gamma
Δ	δ	delta
E	ε	epsilon
Z	ζ	zeta
H	η	eta
Θ	θ	theta
I	ι	iota
K	κ	kappa
Λ	λ	lambda
M	μ	mu
N	ν	nu
Ξ	ξ	xi (pronounced 'ksi')
O	o	omicron
Π	π	pi
P	ρ	rho
Σ	σ	sigma
T	τ	tau
Y	υ	upsilon
Φ	ϕ (or φ)	phi
X	χ	chi (pronounced 'ki')
Ψ	ψ	psi
Ω	ω	omega

Chapter 3
Linear equations

OBJECTIVES

Having completed this chapter you should be able to:

- Manipulate an equation by performing elementary operations on it, so as to isolate any variable on one side of the equation.

- Solve linear equations containing one unknown.

- Recognize an explicit or implicit linear function, and identify its slope and intercept parameters.

- Plot the graph of a linear function and solve a linear equation by graphical methods.

- Recognize and solve linear simultaneous equations with two unknowns by both algebraic and graphical methods.

- Apply simultaneous equations methods to finding the equilibrium in linear supply and demand models and in the basic macroeconomic income–expenditure model.

- Carry out simple comparative static exercises with these models.

3.1 Introduction

We said in chapter 2 that algebra is a way of expressing complex relationships clearly and concisely. One of the key types of relationship in algebra is the equation. An **equation** is any mathematical expression that contains an 'equals' sign. It may help you when tackling equations to keep in mind that an equation is merely a *statement* in shorthand about the relationship between whatever is on the two sides of the 'equals' sign. Thus any equation can be translated into words. For example, the equation $c = 2a + 3b$ translates as 'There exist three numbers such that the first equals the sum of two times the second plus three times the third.' With even a simple equation such as this, the translation into words is quite lengthy and its meaning obscure. That is why algebra exists, and is increasingly essential in today's world. Even those responsible for drafting UK Acts of Parliament recently, with great reluctance, allowed an equation to appear in an Act.

Another useful way of looking at an equation is to see it as conveying information about one or more variable or unknown numbers. This information has the effect of restricting the possible value or values of the unknown(s). For example, if we denote Ann's income by the symbol x and John's income by y, this in itself gets us nowhere. But if someone gives us the information that their combined income is €60,000 per year, we can write the equation $x + y = 60,000$ (euros). This equation both conveys information about their incomes and restricts the possible values of their incomes. For example, assuming that negative income is impossible, we now know that neither earns more than €60,000. Similarly, if someone now tells us that Ann earns three times as much as John, we can write this information as $x = 3y$, which further restricts the possible values of x and y.

In this chapter we examine the rules governing the manipulation of equations and their application to a certain type of equation called a linear equation. We will introduce the idea of a function, which plays a central role in economic analysis, and show how to draw the graphs of linear functions. We will explain what is meant by a **solution** to an equation, and show how to solve linear equations and linear simultaneous equations. Finally, we will demonstrate some applications of linear functions and simultaneous equations in economics.

Equations and identities

Some equations express statements that are *conditionally* true: for example, $x^2 = 9$. In words, this equation makes the statement 'There exists a number, x, which has the property that, when squared, the result is 9.' This statement is true provided $x = +3$ or -3. If x does *not* equal $+3$ or -3, the statement is false. We say that the values $x = +3$ or $x = -3$ *satisfy* the equation $x^2 = 9$. The task of finding these values is called 'solving' the equation.

Other equations express statements that are *unconditionally* true: for example,

$$2 + 3 = 5 \quad \text{or} \quad a + a + b + b + b = 2a + 3b$$

The first of these statements is unconditionally true because its truth follows from the definitions of '2', '+', '3', '=', and '5'. The second statement is also unconditionally true because it is true for *any* values of the variables a and b. We call these equations **identities** and often use the symbol '≡' instead of '='. The symbol '≡' means 'is identically equal to'.

3.2 How we can manipulate equations

We manipulate equations in order to extract useful information from them. The key rule is as follows:

RULE 3.1 Operations on an equation

Any elementary operation may be performed on an equation, provided the operation is performed on the *whole* of both sides of the equation.

By 'elementary operation' we mean (1) adding, subtracting, multiplying, or dividing by a constant or an unknown; or (2) raising to any power (except the power zero).

Provided this rule is obeyed, an equation is left unchanged by these operations. By 'unchanged' we mean that the conditions upon which its truth depends remain the same. This is best seen in an example.

EXAMPLE 3.1

Consider: $x^2 = 9$

This statement is true when $x = +3$ or -3.

Now let us perform some elementary operation on both sides of the equation: say, adding 7. We then have

$$x^2 + 7 = 9 + 7$$

which is also true provided x equals either $+3$ or -3. (Check this for yourself.) In other words, the conditional truth expressed in the statement is left unchanged by adding 7 to both sides. That is why adding 7 to both sides is a legitimate step. In contrast, if we added 7 only to the right-hand side this would change the statement, as we would then have

$$x^2 = 9 + 7$$

which is true when $x = +4$ or -4.

Examples of manipulating equations

When manipulating equations, it's vital to remember that any operation must be performed on the *whole* of each side of the equation. Failure to do this is a very common mistake.

In deciding what operations to perform on an equation, it helps to keep in mind that we manipulate equations in order to extract useful information from them. Very often, but not invariably, our objective is to isolate the unknown, x, on one side of the equation. Here are some very simple examples of different types of operation to illustrate this.

EXAMPLE 3.2

This example illustrates adding or subtracting a constant on both sides of an equation.

If we are given $x - 4 = 9$

we can add 4 to both sides, giving

$$x - 4 + 4 = 9 + 4$$

Tidying this up, we are left with

$$x = 13$$

Thus the operation of adding 4 to both sides of the equation $x - 4 = 9$ has enabled us to extract the information that the equation is a true statement provided $x = 13$. We say that $x = 13$ is the value of x that *satisfies* the equation, or is the *solution* to it. We also say that $x = 13$ is the value of x that is *consistent with* the equation. Note that when the solution, $x = 13$, is substituted into the equation $x - 4 = 9$, the equation becomes $9 = 9$, an identity. This is always true: when the solution to any equation is substituted into it, the equation becomes an identity.

EXAMPLE 3.3

This example and the following three illustrate multiplying or dividing by a constant.

Given: $\frac{x}{3} = 15$

we can multiply both sides by 3, giving

$$3\left(\frac{x}{3}\right) = 45$$

Multiplying out the brackets, this becomes

$$\frac{3x}{3} = 45$$

and from this the 3s on the left-hand side can be cancelled to give

$$x = 45$$

Thus the operation of multiplying both sides of the equation $\frac{x}{3} = 15$ by 3 has enabled us to extract the information that the equation is a true statement provided $x = 45$. This value of x is the value that satisfies the equation, therefore is the solution to it. When $x = 45$ is substituted into the equation $\frac{x}{3} = 15$, the equation becomes $15 = 15$, an identity.

EXAMPLE 3.4

Given: $\frac{x}{4} = \frac{2x-1}{3}$

we can first multiply both sides by 4, giving

$$4\left(\frac{x}{4}\right) = 4\left(\frac{2x-1}{3}\right)$$

Cancelling the 4s on the left-hand side, and multiplying out the right-hand side, gives

$$x = \frac{8x-4}{3}$$

Then we can multiply both sides by 3 and get

$$3x = 8x - 4$$

from which, by adding 4 to both sides and subtracting $3x$ from both sides, we get

$$4 = 5x$$

Finally, we divide both sides by 5 and get

$$\frac{4}{5} = x$$

This is the value of x that satisfies the equation, or is the solution to it.

EXAMPLE 3.5

Consider: $\frac{x+20}{x} + 4 = 9$

This is slightly different from the previous example, because we need to multiply both sides by x, the variable or unknown number, instead of by a constant. Then we get

$$x\left(\frac{x+20}{x} + 4\right) = 9x$$

Notice the brackets on the left-hand side, which are necessary to ensure that the *whole* of that side is multiplied by x.

When we multiply out the brackets on the left-hand side, this becomes

$$\frac{x(x+20)}{x} + 4x = 9x \quad \text{(notice that the 4 becomes } 4x\text{)}$$

In the first term on the left-hand side, the xs cancel between numerator and denominator (refer back to chapter 2 if you're not sure on this point). This gives

$$x + 20 + 4x = 9x$$

Collecting together the xs on the left-hand side, we get

$$20 + 5x = 9x$$

Then we can subtract $5x$ from both sides, giving

$$20 + 5x - 5x = 9x - 5x$$

which simplifies to

$$20 = 4x$$

Finally, we divide both sides by 4 and get

$$5 = x$$

Thus by means of several **elementary operations** we have extracted the information that the statement $\frac{x+20}{x} + 4 = 9$ is true when $x = 5$. In other words, $x = 5$ satisfies, or is the solution to, this equation.

EXAMPLE 3.6

This example illustrates that we must take care when multiplying or dividing by an unknown, as information can be lost. Consider

$$x(x+1) = 2x$$

If we divide both sides by x, we get

$$\frac{x(x+1)}{x} = \frac{2x}{x}$$

and after cancelling the xs between numerator and denominator on both sides, this becomes

$$x+1 = 2$$

After subtracting 1 from both sides we get

$$x = 1$$

We can check that this solution is correct by substituting $x = 1$ into the first equation, which then becomes

$$1(1+1) = 2 \times 1 \quad \text{which is an identity}$$

However, by dividing by x we 'lost' another solution to the equation we were given, namely $x = 0$. (Check for yourself by substitution that this is indeed a second solution.) Because dividing by zero is not a legitimate step (see sections 1.3 and 2.8), when we divided by x we were implicitly assuming that x was not equal to zero. Given this assumption, our solution $x = 1$ was perfectly correct. But our unconscious assumption led us to overlook the alternative solution, $x = 0$.

EXAMPLE 3.7

This example illustrates raising both sides of an equation to the same power.

Given $(x-7)^2 = 9^2$

we can raise the whole of both sides to the power $\frac{1}{2}$ to give

$$[(x-7)^2]^{\frac{1}{2}} = [9^2]^{\frac{1}{2}}$$

(Incidentally, raising both sides to the power $\frac{1}{2}$ is, of course, the same as taking the positive square root of both sides—see chapter 2, rule 2.8.)

Using rule 2.4 from chapter 2, which says that $(a^n)^m = a^{nm}$, on the left-hand side we have

$$[(x-7)^2]^{\frac{1}{2}} = (x-7)^{2 \times \frac{1}{2}} = (x-7)^1 = x-7$$

Similarly, on the right-hand side

$$[9^2]^{\frac{1}{2}} = 9^{2 \times \frac{1}{2}} = 9^1 = 9$$

So therefore $[(x-7)^2]^{\frac{1}{2}} = [9^2]^{\frac{1}{2}}$ becomes $x-7 = 9$, from which $x = 16$.

 Hint Don't raise both sides of an equation to the power zero, since anything raised to the power zero equals 1 (see rule 2.6). So the equation will collapse to $1 = 1$.

The previous examples showed how to manipulate equations to obtain a solution. A slight variation on this is when the equation contains more than one variable or unknown and we want to isolate one of them on one side of the equation. This unknown is then called the 'subject' of the equation. By convention we arrange the equation so that the subject is on the left-hand side. This task is commonly called 'changing the subject' of an equation (and is also sometimes called transposition of formulae, or transformation of an equation). As usual we'll explain it by examples.

EXAMPLE 3.8

Consider $x + y = 1$

By subtracting x from both sides, we obtain $y = 1 - x$, with y the subject of the transformed equation. Equally, if we want to make x the subject, we can subtract y from both sides and get $x = 1 - y$.

EXAMPLE 3.9

Consider: $\frac{3x + 4b}{2c} = 1$

This equation contains three variables or unknowns: x, b, and c. Suppose we are told that $b = 3$ and $c = 9$, and are asked to find the value of x that results. In order to do this, it is necessary to isolate x on one side of the equation. We can do this by performing elementary operations. First, we multiply both sides by $2c$, giving

$$3x + 4b = 2c$$

Then we subtract $4b$ from both sides, and divide both sides by 3. This gives

$$x = \frac{2c - 4b}{3}$$

Now we are ready to substitute the given values, $b = 3$ and $c = 9$, into the equation. This gives

$$x = \frac{2(9) - 4(3)}{3} = \frac{18 - 12}{3} = 2$$

This technique, of isolating x on one side of the equation, is also known as 'expressing x in terms of b and c'. By the same method we could alternatively have isolated b or c on the left-hand side. (Try this for yourself.)

EXAMPLE 3.10

Given $a - b = x(c - nd)$, suppose we are asked to 'find n in terms of a, b, c, and x'. This means we are being asked to isolate n on one side of the equation, with everything else on the other side of the equals sign.

Since n is on the right-hand side, it seems sensible to attack the problem by trying to strip away other variables on the right-hand side. We can get rid of the x on the right-hand side by dividing both sides of the equation by x. This gives

$$\frac{a - b}{x} = \frac{x(c - nd)}{x}$$

which, after cancelling the xs on the right-hand side, becomes

$$\frac{a - b}{x} = c - nd$$

Next, we can subtract c from both sides and multiply both sides by $-\frac{1}{d}$, giving

$$\frac{-1}{d}\left(\frac{a - b}{x} - c\right) = \frac{-1}{d}(c - nd - c)$$

When we multiply out the brackets on the right-hand side we are left only with n, so the equation becomes

$$\frac{-1}{d}\left(\frac{a - b}{x} - c\right) = n$$

Finally, because the -1 on the left-hand side looks a little untidy, many mathematicians would want to rearrange the signs on the left-hand side to get rid of it, then swap the left- and right-hand sides to get

$$n = \frac{1}{d}\left(c - \frac{a-b}{x}\right)$$

Be sure to check each step of this for yourself.

3.3 Variables and parameters

An equation may contain one or more unknown values, which we denote by a letter of the alphabet. These unknown values in an equation are called **unknowns** or **variables**.

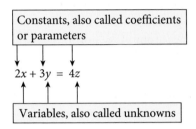

Figure 3.1

An equation may also contain one or more given or known values. The known values in an equation are called constants, **coefficients** or **parameters**. For example, see figure 3.1.

Sometimes when we are discussing an equation, we may want to look at it in a very general way, without specifying any particular values for the parameters. In such a case we assign letters instead of numbers to the parameters. By convention we choose letters from the beginning of the alphabet, in order to distinguish them from the variables or unknowns, which are normally assigned letters from the end of the alphabet. Letters that denote either variables or unspecified constants are also normally written in *italic* font. Thus, for example, in figure 3.2 the letters a, b, and c denote unspecified constants or parameters, and the letters x, y, and z denote variables or unknowns.

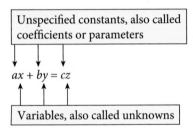

Figure 3.2

We sometimes use **subscripts** and **superscripts** as part of our labelling of variables or parameters. For example, later in this chapter we will write q^D and q^S, where the superscripts D and S distinguish quantity demanded and quantity supplied (respectively, of course). Similarly, if we are using the symbol t to denote a tax rate, then if the tax rate increases we might write t_0 as the initial tax rate and t_1 as the new tax rate, after the increase. On another occasion we might write t and $t^\star$ to distinguish them, or t and $\hat{t}$. You will meet many other notational devices like these in your study of economics. They can be a little off-putting at first, but try to remember that they are just ways of sticking labels on things for identification purposes.

3.4 Linear and non-linear equations

An equation is said to be linear if none of the variables that appear in it is raised to any power other than the power 1. For example, $x + 3 = 9$ is a linear equation because x (the variable, or unknown) is raised to the power 1. Similarly, $y = 3x - 5$ is linear because the two variables, x and y, are raised to the power 1. Conversely, $x^2 - 4 = 16$ is non-linear and so is $y = x^{1/2} + 5$.

 Hint An equation may *look* linear when in fact it isn't. For example, $x + 3 = \frac{1}{x}$. This looks at first sight to be linear because x appears to be raised to the power 1, but we must remember that $\frac{1}{x}$ is the same thing as x^{-1}. So the equation is not linear. This is also apparent when we multiply both sides by x, for then it becomes $x^2 + 3x = 1$, and the x^2 term tells us this is not a linear equation.

Solving linear equations with one unknown

If we are given a linear equation containing only one variable or unknown, we can solve it by using the manipulative techniques explained in section 3.2. Here are two examples of this.

EXAMPLE 3.11

Suppose we are given $\frac{1}{3}x + 4 = 7$

This is a linear equation with one variable or unknown (x) and three constants or parameters ($\frac{1}{3}$, 4, and 7).

To solve it, we have to isolate x on one side of the equation. To do this we can subtract 4 from both sides, then multiply both sides by 3. The equation then becomes

$$3\left(\frac{1}{3}x + 4 - 4\right) = 3(7 - 4) \quad \text{which simplifies to} \quad x = 9$$

Here $x = 9$ is the *solution* to the given equation in the sense that when $x = 9$, the equation is an unconditionally true statement (check this for yourself). We say that $x = 9$ is the value of x that *satisfies* the given equation, or that when $x = 9$, the given equation, $\frac{1}{3}x + 4 = 7$, becomes an *identity*. (If you are uncertain about the meanings of any of the words italicized in this paragraph, refer back to section 3.2.)

EXAMPLE 3.12

Suppose we are given $-4x + 6 = 8$

In this case we can isolate x on the left-hand side by subtracting 6 and dividing by -4. The equation then becomes

$$\frac{-4x + 6 - 6}{-4} = \frac{8 - 6}{-4} \quad \text{which simplifies to} \quad x = \frac{2}{-4} = -\frac{1}{2}$$

Thus $x = -\frac{1}{2}$ is the solution to the given equation.

Generalization

Can *every* linear equation with one unknown be solved in the same way as the two examples above? The answer is yes, and we can prove this as follows.

Every linear equation with one unknown has the *general form*

$$ax + b = c$$

where x is the variable or unknown and a, b, and c are *unspecified* parameters. If you find this difficult, think of a, b, and c as being given constants in any particular case, such as $a = -4$, $b = 6$, $c = 8$ (these were the parameters of example 3.12 above). At this point, we don't want to restrict ourselves to considering any specific case, so we haven't assigned any specific values to the parameters a, b, and c. The attraction of this general approach is that any conclusions we are able to draw will be valid for *any* specific values of a, b, and c (although we must restrict a to being non-zero, otherwise x disappears and our equation no longer has an unknown!).

We can find the solution to this general form by first subtracting b from both sides of the equation above, and then dividing both sides by the parameter a. The equation then becomes

$$\frac{ax + b - b}{a} = \frac{c - b}{a}$$

which after simplifying the left-hand side becomes

$$x = \frac{c - b}{a}$$

Notice that we can check that this solution is correct by substituting our solution for x into the given equation. Thus, in the general form $ax + b = c$, we put $\frac{c-b}{a}$ in place of x, which gives

$$a\left(\frac{c - b}{a}\right) + b = c$$

After cancelling the *a*s, this reduces to

$$c - b + b = c$$

that is, an identity. It is an identity because it is true whatever the values of *a*, *b*, and *c*. (Refer back to section 3.1 above if you are uncertain about identities.) This confirms that $x = \frac{c-b}{a}$ is the solution to the equation $ax + b = c$.

We can restate the above result as a rule:

> **RULE 3.2 The linear equation and its solution**
>
> Any linear equation with one unknown has the general form
>
> $$ax + b = c$$
>
> where *x* is the variable or unknown and *a*, *b*, and *c* are unspecified parameters (with $a \neq 0$). Its solution is
>
> $$x = \frac{c - b}{a}$$

Progress exercise 3.1

1. My electricity supplier's tariff (= payment or charging scheme) has three components. First, I pay £9.50 per quarter, irrespective of how much electricity I use. Second, I pay 6.22 pence for every kilowatt-hour (kWh) used between 8 am and midnight (the daytime rate). Third, I pay 2.45 pence for every kWh used between midnight and 8 am (the night or off-peak rate).

   (a) Calculate my bill if I use 500 kWh at the day rate and 200 at the night rate.

   (b) I estimate that my washing machine and dishwasher each use 50 kWh per quarter. At present, I use them only between 8 am and midnight. If I buy time-switches which permit me instead to use them only between midnight and 8 am, by how much will my electricity bill fall?

2. The bus fare from my home to the university is £1.75 in each direction. A monthly season ticket (allowing unlimited travel on this route) costs £28.

   (a) Write down two equations showing the monthly cost of my travel as a function of the number of journeys (i) when I pay for each trip and (ii) when I buy a season ticket.

   (b) What is the minimum number of trips such that the monthly season ticket is the cheaper mode of payment?

3. My mobile phone pay-as-you-go tariff charges 35 cents per minute for the first 3 minutes of calls per day and 12 cents per minute for the remaining minutes each day. Assuming I make at least 3 minutes of calls each day, write down an equation giving my total expenditure on calls per year (365 days) in terms of the average number of minutes of calls per day. What will my annual expenditure be if I average 15 minutes of calls per day?

4. For each of the following equations, find *x* in terms of the other parameters or variables. (Hint: this means you have to rearrange the equation so as to isolate *x* on the left-hand side.)

   (a) $ax + b = c$ (b) $a + b = k(c - dx)$ (c) $b = c - \frac{(9x)^2}{3a}$ (d) $\frac{x+a}{x-b+d} = c$

   (e) $a = \frac{bx^2 + c}{d}$ (f) $a = b\left(1 - \frac{c}{x}\right)$ (g) $\frac{ax}{b} + 1 = \frac{c}{b}$ (h) $\frac{a(x+b)}{c} = ad$

5. Solve the following equations for *x*.

   (a) $15x = 9$ (b) $2x - 5 = 7$ (c) $\frac{1 + 0.5x}{2} = 5$ (d) $\frac{5}{9}x + \frac{1}{3} = \frac{1}{5}$

3.5 Linear functions

An equation such as $y = 2x + 1$ is an example of a linear equation with two variables, x and y, and two parameters, 2 and 1. There are many pairs of values of x and y that satisfy this equation (that is, which are consistent with it). For example, $x = 1$ and $y = 3$. In fact, we can assign any value we wish to x, and always find a value for y that satisfies the equation. To illustrate this point, let us arbitrarily assign the values −10, 0, 5, and 50 to x. The corresponding values of y that are then necessary to satisfy the equation are calculated in table 3.1.

Table 3.1 Some values of x and y that satisfy $y = 2x + 1$.

Value of x	−10	0	5	50
Corresponding value of y, calculated from $y = 2x + 1$	$2(-10) + 1 = -19$	$2(0) + 1 = 1$	$2(5) + 1 = 11$	$2(50) + 1 = 101$

Equally, we could have assigned some arbitrary values to y and worked out the corresponding values of x necessary to satisfy the equation. (Try this yourself as an exercise.) The key point is that there is no *unique* solution, in the sense of a single pair of values of x and y that uniquely satisfy the equation $y = 2x + 1$. Instead, an infinite number of pairs of values of x and y satisfy the equation.

For this reason, an equation such as this, which contains two variables, is called a **function** (or a **relation**, though relation has a broader meaning than function). A function does not pin down the variables involved to any unique pair of values, but defines a relationship, or mutual dependence, between them. The relationship exists because, if the equation is to be satisfied, then as soon as a value is assigned to one of the variables the value of the other variable is determined. We use the concept of a function extensively in economics, and we will examine the concept more deeply in later chapters, including functions containing more than two unknowns.

In this chapter we will consider linear functions. However, sometimes we want to refer to a function relating x and y, but without specifying the precise form. In that case we write

$$y = f(x) \quad \text{or} \quad y = g(x) \quad \text{or perhaps} \quad y = \phi(x)$$

This notation is a little frightening at first sight, especially the Greek letter ϕ ('phi'—see appendix 2.1). But $f(x)$, which we say aloud as 'function of x' or 'f of x', simply means 'some mathematical expression that contains the variable x'; and similarly for $g(x)$ and $\phi(x)$. These are simply alternative shorthand ways of saying that two variables called x and y are related in some way that we are unwilling or unable to express more precisely. From this point onwards we will use the word 'function' as an alternative to 'equation' when referring to any relationship between two or more variables or unknowns.

Dependent and independent variables

In any equation (or function) involving two variables, such as $y = 2x + 1$, the variable that appears on the right-hand side of the equation (in this example, x) is by convention called the **independent variable**. We view this variable as being free to take any value we choose to assign to it.

The variable that stands alone on the left-hand side of the equation (in this example, y) is known as the **dependent variable**. (In section 3.2 above we called it the 'subject' of the equation.) We view y as being the variable that depends, for its value, on the value assigned to x. Because of this dependence, we say that 'y is a function of x'.

The inverse function

If we are given a function where y is the dependent variable, there is nothing to stop us re-arranging the equation so as to isolate x on the left-hand side. For example, given $y = 2x + 1$, by subtracting 1 from both sides and dividing by 2, we arrive at $x = \frac{1}{2}y - \frac{1}{2}$, where x is now the dependent variable. The functions $y = 2x + 1$ and $x = \frac{1}{2}y - \frac{1}{2}$ are said to be **inverse functions**. The relationship between x and y is the same in the two equations, in the sense that any pair of values of y and x that satisfies one equation will automatically satisfy the other. (For example, $x = 1$, $y = 3$; try some others for yourself.) The only difference between a function and its inverse is that in one, the dependent variable is y; and in the other, x.

Later in this book we will look more closely into the question of dependent versus independent variables. It is an important question in economic analysis.

3.6 Graphs of linear functions

Returning to the equation (or function):

$$y = 2x + 1$$

we can show the relationship between x and y in the form of a graph.

Preliminaries

First, some revision concerning the layout and terminology of graphs is probably helpful. We take any flat, or plane, surface—such as a sheet of graph paper—and divide it into four zones, or quadrants, by drawing a horizontal line and a vertical line across it. These lines are called axes. By convention the independent variable, x, is measured along the *horizontal axis*, and the dependent variable, y, along the *vertical axis* (see figure 3.3). The point at which the two axes cut is defined as the point where both x and y have the value zero, and this point is called the **origin**. Positive values of x are measured to the right of the origin and negative values to the left. Positive values of y are measured above the origin and negative values below it. The axes extend indefinitely far in all four directions.

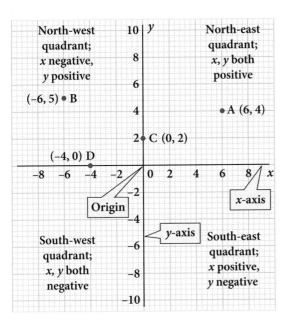

Figure 3.3 The four quadrants of a graph.

Coordinates

The surface of the graph paper is called the $0xy$ plane. (A plane is a flat surface.) Every point on the surface corresponds to a unique pair of values of x and y. These values are called the **coordinates** of the point. For example, at point A in figure 3.3 we can see that the value of x is 6, because a line dropped vertically from A passes through the x-axis at the point $x = 6$. Similarly, at A the value of y is 4, because a horizontal line from A cuts the y-axis at $y = 4$. We therefore say that the coordinates of point A are $x = 6$, $y = 4$, which we write for greater compactness as (6, 4). Note that by convention the first number inside the brackets is always the x-coordinate and the second number the y-coordinate.

As a second example, the coordinates of point B in figure 3.3 are (–6, 5). As a third example, consider point C in figure 3.3. Clearly $y = 2$ at this point. Slightly less obviously, the value of x at point C is $x = 0$. This is because a line dropped vertically from point C (and therefore lying on top of the y-axis) cuts the x-axis at $x = 0$. Thus the coordinates of point C are $(0, 2)$. By the same reasoning, *any* point lying on the y-axis has an x-coordinate of zero.

As a final example, consider point D in figure 3.3. Clearly $x = -4$ at this point. In this case, the value of y at point D is $y = 0$. This is because a horizontal line from point D (and therefore coinciding with the x-axis) cuts the y-axis at $y = 0$. Thus the coordinates of point D are $(-4, 0)$. By the same reasoning, *any* point lying on the x-axis has a y-coordinate of zero.

Note that point A is in the top right quadrant, or north-east quadrant, where both x and y are positive. Point B is in the top left or north-west quadrant, where y is positive but x negative. In the bottom left or south-west quadrant both x and y are negative, while in the bottom right or south-east quadrant x is positive and y negative. (The quadrants are sometimes referred to as the 1st, 2nd, 3rd, and 4th quadrants in the order just discussed.)

Plotting a graph

To plot the graph of $y = 2x + 1$, we begin by assigning a series of values to x. These can be any values we choose. Let us assign the values $-4, -3, -2, -1, 0, 1, 2, 3, 4$. Next, we calculate, for each of these values of x, the value of y necessary to satisfy the equation $y = 2x + 1$. These calculations are shown in table 3.2, called a **table of values**.

Table 3.2 Values for graph of $y = 2x + 1$ (see figure 3.4).

x	$y = 2x + 1$
-4	$y = 2(-4) + 1 = -7$
-3	$y = 2(-3) + 1 = -5$
-2	$y = 2(-2) + 1 = -3$
-1	$y = 2(-1) + 1 = -1$
0	$y = 2(0) + 1 = 1$
1	$y = 2(1) + 1 = 3$
2	$y = 2(2) + 1 = 5$
3	$y = 2(3) + 1 = 7$
4	$y = 2(4) + 1 = 9$

Each row of the table now gives us a pair of values of x and y that satisfy the equation $y = 2x + 1$. For example, the first row tells us that the pair $x = -4$, $y = -7$ satisfies the equation. We then transfer these pairs of values of x and y to the graph paper as coordinates of points (see figure 3.4). For example, the pair $x = -4$, $y = -7$ corresponds to point A on the graph with coordinates $(-4, -7)$. All of these points on the graph satisfy the equation. Finally, we join up the points. The resulting straight line is called the graph of $y = 2x + 1$ and gives us all the pairs of values of x and y (between $x = -4$ and $x = 4$) that satisfy the equation $y = 2x + 1$.

Of course, joining up the points on the graph to make a line is a valid step only if x and y are **continuous variables**: that is, they can take any numerical value. When this is not the case we have what is called a discontinuous function.

Discontinuous functions

One way in which this type of function can occur is when x, for whatever reason, can take on only the integer (whole number) values, $x = 1$, $x = 2$, $x = 3$, and so on. For example, suppose y and x are related by the linear function $y = 2x + 6$, with x restricted to positive integer values. To draw the graph we first draw up the table of values (table 3.3), where as usual we assign integer values to x. When we transfer these values to a graph we get only a series of disconnected points, as in figure 3.5a. However tempting it may be to connect up these points with a line, this is not legitimate as it would violate the

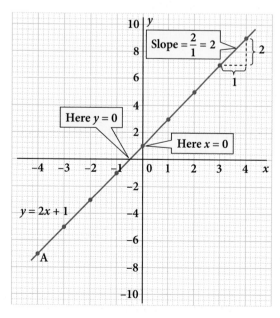

Figure 3.4 Graph of the linear function $y = 2x + 1$.

Table 3.3 Values for graph of $y = 2x + 6$.

x	0	1	2	3	4	5	6	7	8	9	10
y	6	8	10	12	14	16	18	20	22	24	26

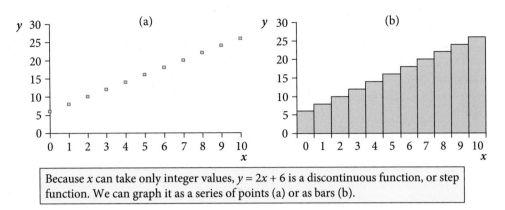

Because x can take only integer values, $y = 2x + 6$ is a discontinuous function, or step function. We can graph it as a series of points (a) or as bars (b).

Figure 3.5 Graph of the discontinuous function $y = 2x + 6$.

condition that x can assume only integer values. For example, y can take the value 11 only if x can take the value 2.5, which by assumption is impossible. Instead, we have a discontinuous function in which y jumps to a new value when x increases in a jump from one integer value to the next.

The disconnected points hanging in the air in figure 3.5(a) look rather lost and forlorn. Therefore it is quite common for the graph to be presented as a bar or column chart (see figure 3.5(b)), in which the height of the bar gives the value of y corresponding to the given integer value of x. A function such as this, where x can only vary in jumps, is often called a **step function**, and we can see the steps in figure 3.5(b).

Here we have used a linear function, $y = 2x + 6$, to explain the idea of a discontinuous or step function. However, there is no reason why a step function should be linear; the relationship between y and x could have any form. The defining characteristic is simply that x is not a continuous variable. We will meet step functions quite frequently later in this book.

3.7 The slope and intercept of a linear function

There are two important features of the graph of $y = 2x + 1$ in figure 3.4:

(1) *The slope, or gradient, of the curve.* (Note that in geometry a straight line is also classed as a curve, although this is a contradiction in everyday speech.) We can measure this slope by examining the increase in y that results when x increases by one unit. This is the definition of slope that we use in everyday life. For example, if we are walking up a hill, and 1 metre of horizontal movement takes us up by 0.5 metres, we say that the slope is 0.5 or 50%.

We can find the slope of the curve either by looking at the table of values or by studying the graph. Whichever you look at, you will quickly see that, when x increases by 1 unit, y increases by 2 units. For example, when x increases by 1 unit from $x = -4$ to $x = -3$, y increases by 2 units from -7 to -5. And when x increases by 1 unit from $x = 2$ to $x = 3$, y increases from 5 to 7. Thus the slope is 2, and because the graph is a straight line its slope is the same wherever we measure it. Because 2 is positive, an increase in y results in an increase in x, and we say that the graph is positively sloped. (In chapter 6 we will look at the analysis of slope when the graph is not a straight line.)

(2) *The position of the graph, relative to the axes.* Conventionally, we measure this by looking at the value of y at the point where the graph cuts the y-axis. Again we can find this either by looking at the table of values or by inspecting the graph itself. Either way you should find it easy to see that the curve cuts the y-axis at the point $y = 1$. This point is called the y-intercept of the graph. At this point, by definition, $x = 0$.

To summarize, we have found that the graph of

$$y = 2x + 1$$

is a straight line with a gradient, or slope, of 2 and an intercept on the y-axis of 1. This is no coincidence. The generalization of this result is:

RULE 3.3 The graph of $y = ax + b$

The graph of

$$y = ax + b \quad \text{(where } a \text{ and } b \text{ are any two constants)}$$

has a slope of a and intercepts (cuts) the y-axis at $y = b$. If a is positive, the graph slopes upwards from left to right. If a is negative, the graph slopes downwards from left to right.

Let us now look at some examples to see how this general rule works out in specific cases.

EXAMPLE 3.13

Plot the graph of $y = 3x - 4$ between $x = -2$ and $x = +4$.

First we construct the table of values, shown in table 3.4. Then we transfer the pairs of values of x and y from the table to the graph paper as coordinates of points. For example, the first point on the left has coordinates $(-2, -10)$. The resulting graph is seen in figure 3.6.

Table 3.4 Values for graph of $y = 3x - 4$.

x	-2	-1	0	1	2	3	4
$y = 3x - 4$	$3(-2) - 4$	$3(-1) - 4$	$3(0) - 4$	$3(1) - 4$	$3(2) - 4$	$3(3) - 4$	$3(4) - 4$
	$= -10$	$= -7$	$= -4$	$= -1$	$= 2$	$= 5$	$= 8$

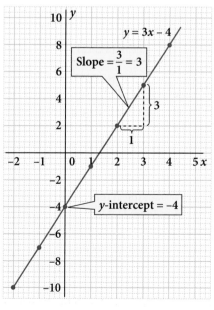

Figure 3.6 Graph of the linear function $y = 3x - 4$.

Let us check whether this example conforms to rule 3.3 above. First, consider the slope or gradient. By looking at either the table of values or the graph itself, we can see that when x increases by 1 unit, y increases by 3 units. For example, when x increases from 2 to 3, y increases from 2 to 5. This conforms to rule 3.3 above, which says that the 3 in the equation $y = 3x - 4$ gives us the slope of the graph.

Second, consider the intercept on the y-axis. As explained above, the y-intercept is where the graph cuts the y-axis, at which point $x = 0$. From the table of values or from looking at the graph (figure 3.6), we can see that when $x = 0$, $y = -4$. This conforms to rule 3.3 above, which says that the -4 in the equation $y = 3x - 4$ gives us the y-intercept of the graph.

EXAMPLE 3.14

Plot the graph of $y = -3x + 2$ between $x = -3$ and $x = +3$.

First we construct the table of values (table 3.5). Then we transfer the pairs of values of x and y from the table to the graph paper as coordinates of points. For example, the first point on the left has coordinates $(-3, 11)$. The resulting graph is seen in figure 3.7.

Table 3.5 Values for graph of $y = -3x + 2$.

x	-3	-2	-1	0	1	2	3
$y = -3x + 2$	$-3(-3) + 2$ $= 11$	$-3(-2) + 2$ $= 8$	$-3(-1) + 2$ $= 5$	$-3(0) + 2$ $= 2$	$-3(1) + 2$ $= -1$	$-3(2) + 2$ $= -4$	$-3(3) + 2$ $= -7$

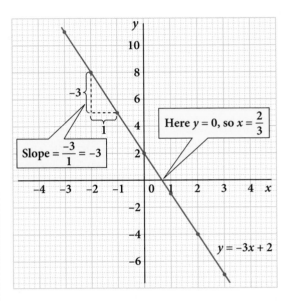

Figure 3.7 Graph of the linear function $y = -3x + 2$.

Again let us check whether this example conforms to rule 3.3 above. First, by looking at either the table of values or the graph itself, we can see that when x increases by 1 unit, y decreases by 3 units. For example, when x increases from 1 to 2, y decreases from -1 to -4. This conforms to rule 3.3 above, which says that the -3 in the equation $y = -3x + 2$ gives us the slope of the graph. Note that this is an example of a graph with a *negative slope* because an increase in x leads to a decrease in y. Thus the graph slopes downwards from left to right. The negative slope of the graph is a consequence of the fact that the constant that multiplies x is negative.

Second, examine the intercept on the y-axis. We know that this is where the graph cuts the y-axis, at which point $x = 0$. From the table of values or the graph (figure 3.7), we can see that this is where $y = 2$. This conforms to rule 3.3 above, which says that the 2 in the equation $y = -3x + 2$ gives us the y-intercept of the graph.

Finally, you have probably already noticed that when plotting the graph of a linear function, it is not necessary to construct a table of values with five or six pairs of values for x and y, as we did in this example and the previous one. Because we know the graph is going to be a straight line, we need only to find the coordinates of two points lying on the line. Then we can simply use a ruler to draw a line passing through the two points. However, it's a good idea to find three points rather than two, as an insurance against error in calculating the coordinates.

The intercept on the *x*-axis

There's another feature of the graph in figure 3.7 that is worth noting: the point at which the graph cuts the x-axis (known as the x-intercept). From figure 3.7 we can see that the graph cuts the x-axis somewhere between $x = \frac{1}{2}$ and $x = \frac{3}{4}$. No graph can ever be drawn with sufficient accuracy to give us the exact value of x. However, we can find this point precisely by algebraic methods. We know that at every point on the x-axis, $y = 0$ (see section 3.6 above if you are uncertain about this). And we also know that $y = -3x + 2$ at every point on the graph. Therefore, where the graph cuts the x-axis, we must have $y = 0$ and $y = -3x + 2$. The only way both can be true at the same time is if:

$$-3x + 2 = 0$$

Solving this equation gives $x = \frac{2}{3}$. Thus the algebra gives us the exact value of the x-intercept and confirms that our graph is drawn with reasonable accuracy.

Generalization

In the same way, we can find the x-intercept of *any* linear equation. Given the general form $y = ax + b$ (where a and b are any two constants), the x-intercept is found at the point on the graph where $y = 0$. Since $y = ax + b$, when $y = 0$ we must have:

$$ax + b = 0$$

By subtracting b from both sides and dividing both sides by a, we arrive at:

$$x = \frac{-b}{a}$$

This result is sufficiently important to be worth stating as a rule:

> **RULE 3.4** Solution of the linear equation $ax + b = 0$
>
> The linear equation $ax + b = 0$ (where a and b are given constants) intercepts (cuts) the x-axis at the point where $x = \frac{-b}{a}$.

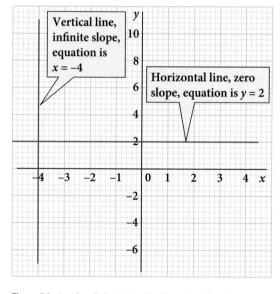

Figure 3.8 Graphs of a horizontal and a vertical function.

The case where the graph is a horizontal line

We have seen that in the general form $y = ax + b$, the graph slopes upwards from left to right if the parameter a is positive, and downwards from left to right if a is negative.

The case where the graph is a horizontal line can also be handled. If the coefficient a is zero, then ax is also zero, so the function $y = ax + b$ degenerates into $y = b$. Thus x disappears from the equation, and therefore variation in x has no effect on y. We are left with $y = b$, and since the parameter b gives the intercept on the y-axis, the graph is therefore a horizontal line at a distance b units from the x-axis (see figure 3.8).

The case where the graph is a vertical line

This case is slightly more difficult. We know that, in $y = ax + b$, if a is large and positive, the graph will slope upwards very steeply from left to right. Similarly, if a is large and negative the graph will slope downwards very steeply from left to right. However, since a must always be a finite number, then no matter how large in absolute value a becomes (whether positive or negative), its graph will never be truly vertical. Thus we conclude that the form $y = ax + b$ is not *completely* general, since it cannot include the case of a vertical graph. We return to this in the next section.

Implicit linear functions

We know that the graph of $y = ax + b$ is a straight line with the slope given by the parameter a and y-intercept given by the parameter b. Sometimes, however, a linear function is presented to us in a form such as

$$6x - 2y + 8 = 0$$

This is known as an **implicit function** because x and y both appear on the same side of the equation. Because both x and y are on the same side of the equation, we cannot say which is the independent variable and which the dependent. The dependency between x and y can be described as mutual, in the sense that as soon as a value is assigned to one variable, the value of the other is determined. For example, if $y = -5$, the equation becomes: $6x - 2(-5) + 8 = 0$, from which $x = -3$.

To turn the equation above into the form we are familiar with, we can rearrange it by subtracting $6x + 8$ from both sides, and then dividing by -2. This gives

$$y = 3x + 4$$

In this form we can see immediately that we have a linear function with a slope of 3 and an intercept on the y-axis of 4. We can also immediately visualize the shape of its graph.

Generalization

Given the implicit linear function

$$Ax + By + C = 0$$

where A, B, and C are parameters (constants).

By subtracting $Ax + C$ from both sides, and dividing both sides by B, we can transform the function into the form

$$y = \frac{-A}{B}x - \frac{C}{B}$$

This reformulation of the relationship between the two variables is called an **explicit function**. By isolating y on one side of the equation, it shows us explicitly how y depends on x.

This function when graphed will therefore have a slope of $\frac{-A}{B}$ and an intercept on the y-axis of $\frac{-C}{B}$. Whether the slope and the intercept are positive or negative will depend, of course, on whether A, B, and C are positive or negative.

Although the convention in maths is that y is the dependent variable and is therefore to be isolated on the left-hand side of the equation, this is by no means an absolute rule. There may well be occasions when we want to treat x as the dependent variable, and therefore wish to isolate x on the left-hand side instead. Given the implicit function $Ax + By + C = 0$, we can achieve this by subtracting $By + C$ from both sides and then dividing by A. The result is:

$$x = \frac{-B}{A}y - \frac{C}{A}$$

In this form we say that x is an explicit function of y.

We can now resolve the puzzle left at the end of the previous section concerning the equation of a vertical graph. In the equation above, if $B = 0$ we have $x = -\frac{C}{A}$, which when graphed is a vertical line at a distance $-\frac{C}{A}$ units away from the y-axis. Thus the form $Ax + By + C = 0$ is a slightly more general form of the linear function than the form $y = ax + b$, because the former can handle the case of a vertical line (see figure 3.8).

Summary of sections 3.1–3.7

In sections 3.1–3.4 we showed how any equation could be manipulated by performing the elementary operations (adding, multiplying, and so on) to both sides of the equation. We stated the key distinction between variables and parameters in any equation. Any linear equation with one unknown can be solved (rule 3.2).

In sections 3.5 and 3.6 we introduced the idea of a function, involving two or more variables. For any linear function $y = ax + b$ the graph of this function is a straight line, which is why it is referred to as a linear function. The slope is given by the constant, a. The intercept of the graph on the y-axis is given by the constant b.

If the parameter a is positive, the graph slopes upwards from left to right, and we say that the slope is positive. Conversely, if a is negative, the graph slopes downward from left to right, and we say that the slope is negative (because an increase in x results in a decrease in y). If $a = 0$, the graph is a horizontal line at a distance b units from the x-axis.

A linear function may also take the implicit form $Ax + By + C = 0$, where A, B, and C are parameters. Its graph is a horizontal line if $A = 0$, and a vertical line if $B = 0$. It can be rearranged in explicit form as $x = \frac{-B}{A}y - \frac{C}{A}$, provided $A \neq 0$.

3.8 Graphical solution of linear equations

Suppose we are asked to solve the linear equation

$$-3x + 2 = 0$$

We learned how to solve this by algebraic methods in section 3.7 above. Now we want to look at a way of solving it *graphically*. First, we refer back to example 3.14 above, where we plotted the graph of $y = -3x + 2$ (see figure 3.7).

There we saw that we find the x-intercept of the graph by making use of the fact that, at the point where the graph cuts the x-axis, $y = 0$ (since $y = 0$ everywhere on the x-axis). But if $y = 0$ at this point, it must also be true that $-3x + 2 = 0$. So by solving the equation $-3x + 2 = 0$, we find the value of x at which the graph cuts the x-axis. The solution in this example is $x = \frac{2}{3}$.

Thus the value of x that satisfies the equation $-3x + 2 = 0$ is also the value of x at which the graph of $y = 3x + 2$ cuts the x-axis.

Generalization

This approach is valid in general for any linear function $y = ax + b$. From rule 3.4 we know that the graph of $y = ax + b$ cuts the x-axis at the point where $x = -\frac{b}{a}$. At this point, $y = 0$; and since $y = ax + b$, it follows that $ax + b = 0$ also. We will state this as a rule:

RULE 3.5 Graphical solution of a linear equation

The solution to the linear equation $ax + b = 0$ is found graphically at the point where the graph of $y = ax + b$ cuts the x-axis. At this point, $x = -\frac{b}{a}$.

The only way this solution could fail would be if $a = 0$, for then the function collapses to $y = b$, which we know is a horizontal line b units above or below the x-axis. As x disappears from the equation, there is nothing to solve.

In fact, this conclusion generalizes to *any* function of x, whether linear or not. If we write $y = f(x)$, where $f(x)$ denotes a function of unspecified algebraic form, then at the point where its graph cuts the x-axis we have $y = 0$ and therefore $f(x) = 0$. The value of x at this point is therefore the solution to the equation $f(x) = 0$. We will see more of this idea in the next chapter.

Of course, the graphical method of solving an equation can never fully replace the algebraic method, because a graph can never be drawn with complete accuracy. But the graph serves as a useful check on the accuracy of our algebra. Moreover, as the saying goes, 'one picture is worth a thousand words'—the graphical approach can often give us a better understanding of the nature of the relationship between two or more variables, as we shall see later in this book.

 The Online Resource Centre shows how Excel can be used to plot the graphs of linear functions and solve linear equations. www.oxfordtextbooks.co.uk/orc/renshaw3e/

 Progress exercise 3.2

1. Write down the coordinates of the points (a) to (h) shown in the figure below.

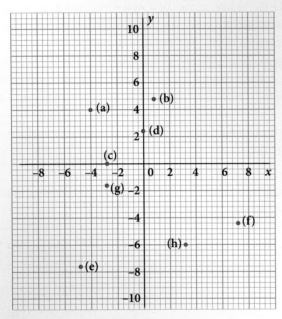

2. On a sheet of graph paper, draw in the x- and y-axes and choose an appropriate scale. Then mark the points in the $0xy$ plane with the following coordinates:

(a) $(3, 5)$ (b) $(5, -1/2)$

(c) $(0, 8)$ (d) $(4, 0)$

(e) $(-5, -3)$ (f) $(5, -8)$

(g) $(8, -2)$ (h) $(-5, 0.25)$

3. For each of the following implicit linear functions, express y as a function of x. Use this information to plot their graphs for the range of values of x stated.

(a) $2y + 4 = 6x$; for values of x between -4 and 5

(b) $3y + 12x = -24$; for values of x between -4 and 5

(c) $-5x = 10y + 50$; for values of x between 0 and 12

(d) $\frac{1}{6}x = y + \frac{1}{2}$; for values of x between -12 and 12

(e) $3x + 4y = 5$; for values of x between -10 and 10

4. Using your answers to 3(b) above, find approximate graphical solutions to the following linear equations. Then check your answers by algebraic methods.

(a) $3x - 2 = 0$ (b) $-4x - 8 = 0$ (c) $-\frac{1}{2}x - 5 = 0$ (d) $\frac{1}{6}x - \frac{1}{2} = 0$ (e) $-\frac{3}{4}x + \frac{5}{4} = 0$

5. For each of the following, write down the slope and the y-intercept, and calculate the x-intercept.

(a) $y = 2x + 5$ (b) $y = \frac{1}{3}x + 12$ (c) $y = \frac{1}{2}x + 20$ (d) $y = 0.25x + 0.5$ (e) $y = \frac{1}{5}x + 5$

3.9 Simultaneous linear equations

Simultaneous equations are not as difficult as they at first sound. As usual we'll develop the ideas by means of examples.

EXAMPLE 3.15

Suppose we are told that two variables x and y are related to one another by the linear function (or equation)

$$y = 3x \tag{3.1}$$

(Note that from now on we will number important equations, for easy reference.)

Further, we are told that the variables are also related by the linear function

$$y = x + 10 \tag{3.2}$$

In words, we are told that the two statements '$y = 3x$' and '$y = x + 10$' are both true at the same time. That is why they are called *simultaneous* equations.

Now, obviously there are many pairs of values of x and y that satisfy equation (3.1): for example, $x = 1$, $y = 3$; and $x = 2$, $y = 6$. And there are many pairs of values of x and y that satisfy equation (3.2): for example, $x = 1$, $y = 11$; and $x = 2$, $y = 12$. (Check these for yourself.) But none of these pairs of values satisfies *both* equations. Yet this is precisely our task here: to find the pair of values of x and y that satisfies *both* of these equations. This is what 'solving simultaneous equations' means.

The method of solution we will use is actually very simple. The key insight is that, if the two equations are both true simultaneously, the value of y in equation (3.1) will be the same as the value of y in equation (3.2). Therefore, the left-hand side of equation (3.1) and the left-hand side of equation (3.2) will be equal to one another. But, if the two left-hand sides are equal, then the two right-hand sides must be equal too. So, if the two equations are both true simultaneously, we will have

$$3x = x + 10$$

This is a linear equation with only one variable or unknown, x. By subtracting x from both sides and dividing both sides by 2, we quickly find that its solution is

$$x = 5$$

So now we know that when the ys in the two equations are equal to one another, $x = 5$. To complete our solution, we need to find what that value of y is. We can find this by substituting $x = 5$ into either equation (3.1) or equation (3.2). Substituting $x = 5$ into equation (3.1) gives

$$y = 3(5) = 15$$

So $x = 5$, $y = 15$ is the solution. This pair of values of x and y satisfies both equations simultaneously. Let's check that this is indeed true, by substituting these values into both equations. Then in equation (3.1) we get

$$15 = 3(5)$$

and in equation (3.2) we get

$$15 = 5 + 10$$

Both equations become *identities* when the solution values are substituted in. Thus both equations are satisfied by our solution values, which are therefore correct. Be sure to check all of the above working before reading on.

The solution method that we used in the example above is worth some further thought, as it is used frequently in both maths and economics. Essentially we *imposed the condition* that the ys in equations (3.1) and (3.2) were equal—something that is only true when x and y have taken on their solution values. This enabled us to derive the equation $3x = x + 10$. The solution to this equation gave us the value that x must have when the ys in equations (3.1) and (3.2) are equal; that is, in the solution. Thus we knew that we had found the solution value of x. From there, it was only a small step to finding the solution value of y.

EXAMPLE 3.16

Consider the simultaneous equations

$$8x + 4y = 12 \tag{3.3}$$
$$-2x + y = 9 \tag{3.4}$$

In this example both equations are written as implicit functions: that is, both x and y are on the same side of each equation. Therefore, before we can apply the method of the previous example, we must first use elementary operations to rearrange both of the equations in explicit form, in which either y or x is isolated on one side of both equations. We choose to isolate y. In equation (3.3) we divide both sides by 4 and then subtract $2x$ from both sides. This gives us

$$y = -2x + 3 \tag{3.3a}$$

In equation (3.4) we simply add $2x$ to both sides of the equation, giving

$$y = 2x + 9 \tag{3.4a}$$

Because we have performed only elementary operations, we know that whenever equation (3.3a) is satisfied, so also is equation (3.3). Similarly, whenever (3.4a) is satisfied, so is (3.4). (See section 3.2 if you are unsure on this point.) Therefore, if we solve (3.3a) and (3.4a) as simultaneous equations we will automatically find the solution to the simultaneous equations we were given, (3.3) and (3.4).

As in the previous example, we make use of the fact that, in the solution, the y in equation (3.3a) will equal the y in equation (3.4a), and therefore the right-hand side of (3.3a) will equal the right-hand side of (3.4a). Thus we will have

$$-2x + 3 = 2x + 9$$

This equation is solved by adding $2x$, subtracting 9, and dividing by 4. This gives

$$x = -\frac{3}{2}$$

We then find y by substituting $x = -\frac{3}{2}$ into either (3.3a) or (3.4a), giving $y = 6$.

As a final check we substitute our solution, $x = -\frac{3}{2}, y = 6$, into equations (3.3) and (3.4), as follows:

$$8\left(\frac{-3}{2}\right) + 4(6) = 12 \qquad \text{in (3.3)}$$
$$-2\left(\frac{-3}{2}\right) + 6 = 9 \qquad \text{in (3.4)}$$

As both equations then become identities (check this for yourself), our solution is correct.

EXAMPLE 3.17

The simultaneous equations

$$y = 2x - 10 \tag{3.5}$$
$$\frac{5 - y}{3} = x \tag{3.6}$$

Here we cannot immediately proceed as we did in example 3.15, because neither the left-hand sides nor the right-hand sides of the two equations are equal to one another. We could tackle this problem, as we did in example 3.16, by rearranging equation (3.6) to isolate y on one side. To do this, we multiply by 3, subtract $3x$, and add y, giving

$$y = -3x + 5 \tag{3.7}$$

Then we could solve (3.5) and (3.7) using the same method as in the previous two examples. Instead, however, we will use this example to illustrate a slight variation of that method. This again makes use of the key fact that, when both equations are satisfied, the value of y in equation (3.5) will be the same as the value of y in equation (3.6). This means that we can replace y in equation (3.6) with what y is equal to in equation (3.5): that is, $2x - 10$. When we make this replacement (or substitution), equation (3.6) becomes

$$\frac{5 - (2x - 10)}{3} = x \tag{3.8}$$

When we remove the brackets in the numerator of the left-hand side, and multiply by 3, this becomes

$$15 - 2x = 3x \quad \text{from which, by elementary operations:} \quad x = 3$$

From here we proceed exactly as in the previous two examples. To find y we substitute $x = 3$ into either equation (3.5), (3.6), or (3.7). Choosing equation (3.5) we get

$$y = 2x - 10 = 2(3) - 10 = -4$$

So our solution is $x = 3$, $y = -4$.

As a check that our solution is correct, we substitute the solution values for x and y into equations (3.5) and (3.6). This gives

$$-4 = 2(3) - 10 \hspace{8cm} \text{in (3.5)}$$

$$\frac{5 - (-4)}{3} = 3 \hspace{8cm} \text{in (3.6)}$$

As both of the above are identities, our solution is correct.

Summary on simultaneous equations

Solving a pair of simultaneous equations means finding a pair of values for x and y that satisfies both equations. The method of solution that we used in examples 3.15–3.17 made use of the fact that, in the solution, the value of y in one equation is the same as the value of y in the other. This means that we can replace y in one equation with what y is equal to in the other equation. This gives us an equation containing only x, which we can solve. (In some cases it may be more convenient to replace x rather than y, and thus obtain an equation containing only y which we can solve.) In example 3.16 some preliminary manipulation was necessary before we could do this. There are other techniques for solving simultaneous equations, some of which you may have learned in your previous studies of maths, and which you may prefer to stick with. These are simply variants of the basic method we have used.

3.10 Graphical solution of simultaneous linear equations

EXAMPLE 3.18

The first pair of simultaneous equations that we considered in the previous section (example 3.15) was

$$y = 3x \hspace{8cm} (3.9)$$

$$y = x + 10 \hspace{8cm} (3.10)$$

Let us plot the graphs of these two functions on the same axes. Table 3.6 gives some suitable values of x and y.

The graphs of the two functions are shown in figure 3.9. We see that the two graphs intersect at the point where $x = 5$, $y = 15$. (This can also be seen in the table of values.) These values were our solution to this pair of simultaneous equations. On reflection we should not find this surprising, for by definition the graph of $y = 3x$ consists of all those points with coordinates that satisfy the equation $y = 3x$. Similarly, the graph of $y = x + 10$ consists of all those points with coordinates that satisfy the equation $y = x + 10$. The point of intersection of the two graphs lies, by definition, on both. Therefore the coordinates of this point satisfy both equations simultaneously.

Another way of seeing this is to note that at the point of intersection the value of y from $y = 3x$ is the same as the value of y from $y = x + 10$. That these two ys were equal was exactly the assumption we made in section 3.9 in solving simultaneous equations algebraically.

Table 3.6 Values for graphs of $y = 3x$ and $y = x + 10$.

x	−5	−3	0	3	5	7	9
$y = 3x$	−15	−9	0	9	15	21	27
$y = x + 10$	5	7	10	13	15	17	19

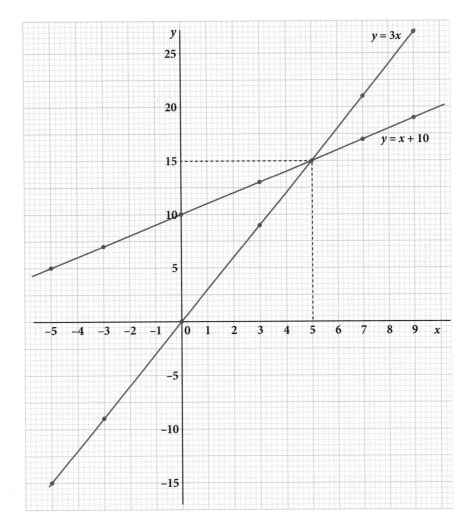

Figure 3.9 Solution to $y = 3x$; $y = x + 10$.

From example 3.18 we see that the rule is as follows:

RULE 3.6 Graphical solution of a pair of simultaneous linear equations

The graphical solution to a pair of simultaneous linear equations is given by the coordinates of the point of intersection of their graphs.

In chapter 4 we will see that this rule is also valid for simultaneous non-linear equations, and that with non-linear equations there is a possibility of more than one solution.

EXAMPLE 3.19

As a further illustration of this key point, let us repeat this exercise for the third example of simultaneous equations from the previous section (example 3.17). The equations were

$$y = 2x - 10 \qquad\qquad\qquad\qquad (3.5) \text{ repeated}$$

$$\frac{5-y}{3} = x \qquad \text{(3.6) repeated}$$

However, before we can plot the graph of equation (3.6) we must rearrange it so as to make y an explicit function of x, which we do by isolating y on the left-hand side. We did exactly this in example 3.17 above, and we got

$$y = -3x + 5 \qquad (3.7)$$

We will actually plot the graph of equation (3.7), but this is equivalent to plotting the graph of equation (3.6) because the two are identically equal. Because any pair of values of x and y that satisfies equation (3.7) automatically satisfies equation (3.6), their graphs have identical coordinates.

Table 3.7 gives some suitable values, and the resulting graphs are shown in figure 3.10. There we see that the two graphs intersect at the point $(3, -4)$. This is as expected, because we already know that $x = 3$, $y = -4$ is the solution to the simultaneous equations $y = 2x - 10$ and $y = 3x + 5$.

Table 3.7 Values for $y = 2x - 10$ and $y = -3x + 5$.

x	-4	-3	-2	-1	0	1	2	3	4	5	6	7
$y = 2x - 10$	-18	-16	-14	-12	-10	-8	-6	-4	-2	0	2	4
$y = -3x + 5$	17	14	11	8	5	2	-1	-4	-7	-10	-13	-16

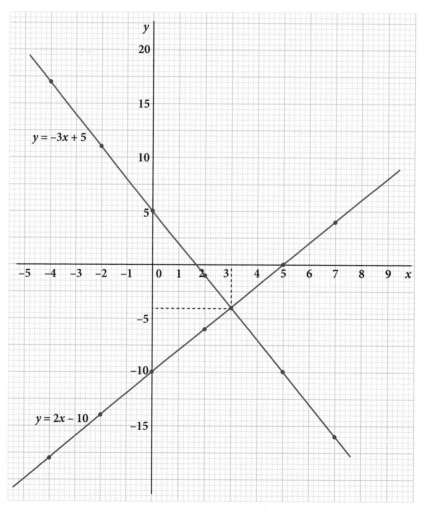

Figure 3.10 Solution to $y = 2x - 10$; $y = -3x + 5$.

3.11 Existence of a solution to a pair of linear simultaneous equations

An important question is whether we can be sure that any pair of linear simultaneous equations has a solution. Since the solution is where their graphs intersect, we can deduce that one way in which a solution might not exist is if the two graphs are parallel lines.

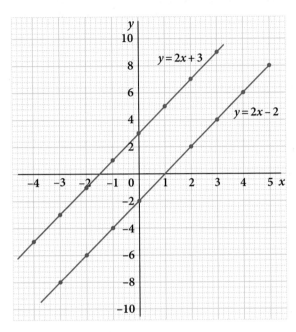

For example, in figure 3.11 we show the graphs of $y = 2x + 3$ and $y = 2x - 2$. The first function has a slope of 2 and a y-intercept of 3, while the second function has a slope of 2 and a y-intercept of -2. Since parallel lines never intersect, it is obvious that the simultaneous equations $y = 2x + 3$ and $y = 2x - 2$ do not have a solution. This is confirmed algebraically when we try to solve them by setting the two right-hand sides of these equations equal to one another. The result is

$$2x + 3 = 2x + 5$$

Subtracting $2x$ from both sides gives

$$3 = 5$$

which, of course, is a contradiction. So no solution exists. In such a case, we say that the two equations are *inconsistent*.

Figure 3.11 Inconsistent equations: $y = 2x + 3$; $y = 2x - 2$.

A second way in which the solution may fail is if the two graphs have not only the same slope but the same intercept, so that one lies on top of the other. The problem then is not that there is no solution, but that there are too many, since any point that lies on one graph automatically lies on the other. Therefore any pair of values that satisfies one equation automatically satisfies the other. For example, suppose we are given the simultaneous equations

$$y = 2x + 3 \tag{3.11}$$

$$x = \tfrac{1}{2}y - \tfrac{3}{2} \tag{3.12}$$

If we use the second equation to substitute for x in the first equation (see example 3.17), we get

$$y = 2\left(\tfrac{1}{2}y - \tfrac{3}{2}\right) + 3$$

which, by removing the brackets and tidying up, becomes

$$y = y$$

This is an identity, meaning that any value of y that satisfies one equation automatically satisfies the other. This means that if we pick an x completely at random, say $x = 1$, and then substitute this into equation (3.11), we get $y = 5$. And this pair of values, $x = 1$, $y = 5$ also satisfies equation (3.12) (check for yourself). So $x = 1$, $y = 5$ satisfies both equations, and this is true *whatever* value we assign to x. The reason for this is that equation (3.12) is not a new equation but simply a rearrangement of equation (3.11), as you can check for yourself by multiplying both sides of equation (3.12) by 2, then adding 3 to both sides. In such a case we say that the two equations are not *independent*.

Thus we conclude that a pair of linear simultaneous equations has a unique solution if, and only if, the equations are consistent and independent.

Three equations with two unknowns

An interesting case arises if we are given three linear equations with two unknowns and asked to solve them simultaneously. Each of the equations, when graphed, is a straight line. One possible configuration is shown in figure 3.12a. The three straight lines do not intersect at a unique point, and therefore there is no pair of values of x and y that can satisfy all three equations simultaneously. The set of three equations with two unknowns is said to be *over-determined*. The problem is that each equation imposes a restriction on the permissible combinations of x and y. With three equations and only two unknowns, there are too many restrictions for a solution to exist.

(The opposite case arises when there is only one equation and two unknowns; then we have the problem of *under-determinacy*, as there are too few restrictions on the permissible values of x and y for a unique solution to exist.)

However, another possible configuration of the three equations is shown in figure 3.12b. The three equations are

$$y = x + 10 \tag{3.13}$$

$$y = -2x + 25 \tag{3.14}$$

$$y = -\tfrac{1}{2}x + 17\tfrac{1}{2} \tag{3.15}$$

(a)

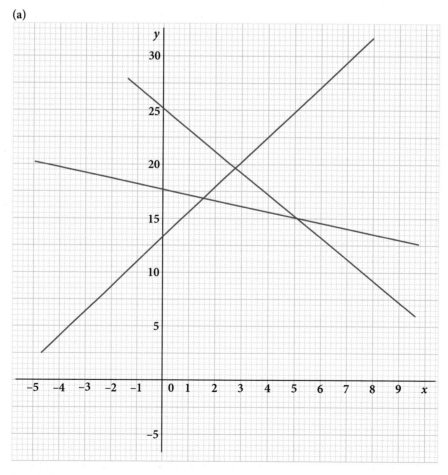

Figure 3.12(a) Three equations with two unknowns.

(b)

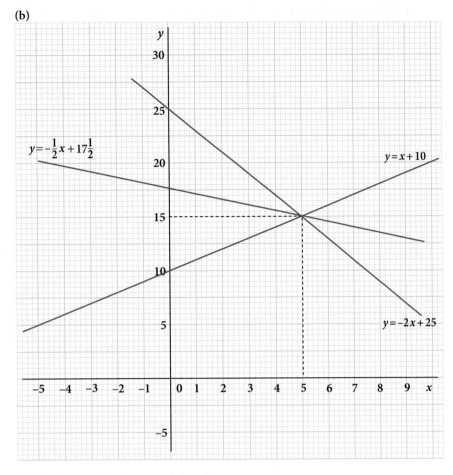

Figure 3.12(b) Three equations that are not independent.

Their graphs intersect at a single point, so the coordinates of this point, (5, 15), satisfy all three equations simultaneously. This point is therefore the solution. (Check this for yourself by substituting this pair of values into each of the three equations above.) However, this intersection at a common point has been achieved by a trick: the three equations are not independent of one another. Although it is by no means obvious, the third equation was constructed as the average of the other two. Using the rule that the average of any two numbers a and b is $\frac{a+b}{2}$, the average of the right-hand sides of equations (3.13) and (3.14) is

$$\frac{x+10+(-2x+25)}{2} = \frac{-x+35}{2} = \tfrac{1}{2}x+17\tfrac{1}{2}$$

which, of course, is the right-hand side of equation (3.15). Thus for any given value of x, the value of y in equation (3.15) is the average of the ys in equations (3.13) and (3.14). This guarantees that any pair of values of x and y that satisfies equations (3.13) and (3.14) automatically satisfies equation (3.15). So the latter equation is not independent, and is therefore redundant. Unlike figure 3.12a, this set of simultaneous equations is not over-determined because the number of *independent* equations does not exceed the number of unknowns.

It can be proved that whenever three linear equations with two unknowns intersect at a common point, this can only be because the three equations are not independent of one another. One equation is always obtainable as some combination (such as the average, in this example) of the other two.

3.12 Three linear equations with three unknowns

Up to this point we have considered only simultaneous linear equations with two unknowns. But the basic idea extends quite naturally to a set of three equations involving three unknowns. For example, suppose there are three variables or unknowns, which we can label x, y, and z. Suppose also we know that they are related to one another through three equations:

$$z = x + y + 4 \tag{3.16}$$

$$z = 2x - y \tag{3.17}$$

$$z = 3x - 4y \tag{3.18}$$

To solve this set of three simultaneous equations, we have to find a trio of values for x, y, and z that satisfies all three equations simultaneously. The method is a simple extension of the method we have already developed for two simultaneous equations with two unknowns. We take any two equations, say (3.16) and (3.17). Then we know that, in the solution, the z in equation (3.16) will equal the z in equation (3.17). This means that the left-hand sides of these two equations will equal one another, and so therefore will the two right-hand sides. Therefore we can set the two right-hand sides equal to one another, and get

$$x + y + 4 = 2x - y$$

which simplifies to

$$x = 2y + 4 \tag{3.19}$$

So now we know that, in the solution, x will equal $2y + 4$. Therefore we can replace x with $2y + 4$ in equations (3.17) and (3.18), giving

$$z = 2(2y + 4) - y = 3y + 8 \tag{3.17a}$$

$$z = 3(2y + 4) - 4y = 2y + 12 \tag{3.18a}$$

These are a pair of simultaneous equations with two unknowns, y and z. We can solve them in the usual way, by setting their right-hand sides equal to one another. This gives

$$3y + 8 = 2y + 12 \quad \text{with solution:} \quad y = 4$$

We then substitute $y = 4$ into (3.17a) or (3.18a) and find z as $z = 20$. We also substitute $y = 4$ into equation (3.19) and find $x = 12$.

So our solution is $x = 12$, $y = 4$, $z = 20$. We can check our solution by substituting these values into equations (3.16), (3.17), and (3.18). If all three equations become identities, our solution is correct. Check this for yourself.

The graphical representation of these three equations is difficult because we need three axes (one for each variable) at right angles to one another. This is tackled in part 4 of this book.

Four or more equations and unknowns

As the number of linear equations and unknowns increases, the method of solution that we have described above becomes increasingly laborious and error-prone. A more powerful technique for solving large sets of simultaneous linear equations is developed in chapter 19.

However many equations and unknowns we have to deal with, the conditions for the existence of a solution discussed in section 3.11 continue to hold. The number of independent equations must equal the number of unknowns, and the equations must be consistent with one another. The existence of a solution to non-linear simultaneous equations is discussed in section 4.12.

Summary of sections 3.8–3.12

In section 3.8 we saw that the solution to a linear equation (or any other equation) is found at the point where the graph of the corresponding function cuts the x-axis. In section 3.9 we showed, in examples 3.15–3.17, how to solve simultaneous linear equations. In section 3.10 we showed that the solution was at the point of intersection of their graphs. In section 3.11 and 3.12 we showed that a solution exists only if the number of independent equations equals the number of unknowns, and if the equations are consistent, including cases with more than two equations and unknowns.

The Online Resource Centre shows how Excel® can be used to plot the graphs of simultaneous equations and find their solution graphically. www.oxfordtextbooks.co.uk/orc/renshaw3e/

Progress exercise 3.3

1. Solve the following pairs of linear simultaneous equations by algebraic methods. In each case, sketch the graphs for values of x between -5 and $+10$ and thus verify your answers graphically.

   (a) $y = x + 2, y = 2x - 5$ (b) $y = 3x + 0.5, y = 0.25x - 5$ (c) $y = -2x + 3, y = 3x - 2$

   (d) $y = -5x + 10, y = 3x - 6$ (e) $y = 0.75x - 4, y = 2.5x - 11$

3.13 Economic applications

Now that we have revised quite extensively the basic rules of algebra, we are ready to see some applications to economics. In the remainder of this chapter we will look at two applications, one from microeconomics and the other from macroeconomics. You will not need to have studied economics previously in order to understand these examples.

3.14 Demand and supply for a good

Demand and supply are the basic building blocks of economic analysis. The concept of the **demand function** for a good or service is based on the theory of consumer behaviour. This postulates that an individual with a given income will divide his or her expenditure between the various goods and services available in the market place in such a way as to achieve the maximum possible utility or satisfaction. We will explore this more fully in later chapters, but for the moment all we need to know is that this theory suggests that an individual will normally wish to buy more of any good when its price falls (provided all other relevant factors remain constant).

By adding together the demands of all individuals for a good, we can derive the *market* demand function for a good as a relationship between the market price, p, and the total of all consumers' demands, q^D. For simplicity we will assume this function to be linear; thus we can write it as

$$q^D = -ap + b$$

where a and b are parameters (constants), both assumed positive. The minus sign in front of the parameter a, together with the assumption that a itself is positive, means that the market demand curve is negatively sloped. The quantity demanded is taken to be the dependent variable because we assume that, for each individual, the market price is given and it is simply up to him/her to decide how much to buy at that price.

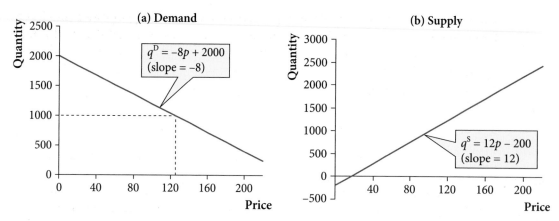

Figure 3.13 Demand and supply functions for apples.

As a specific example, if we are looking at the demand for apples and find that $a = 8$ and $b = 2000$, then our demand function is

$$q^D = -8p + 2000$$

which is graphed in figure 3.13(a). Note that q^D is a flow of purchases per unit of time, perhaps thousands of kilos per week in the case of apples, while p is the price per unit, for example 150 cents per kilo. The intercept of 2000 on the q^D axis tells us that if the price were zero, consumers would buy 2000 (thousand kilos per week); while the slope of –8 tells us that an increase (decrease) in price of 1 (cent per kilo) causes quantity demanded to fall (rise) by 8 (thousand kilos per week).

The concept of the **supply function** is based on the theory of the behaviour of the firm. We will explore this more fully in later chapters, but at this stage all we need to know is that theory suggests that under certain conditions firms in the pursuit of profit will choose to supply more of their output when its market price rises, provided everything else that influences the firm's behaviour remains constant. From this we obtain the market supply function (that is, the combined supply of all firms) as a relationship between the quantity supplied (offered for sale), q^S, and the market price, p. Assuming for simplicity that this relationship is linear, we can write it as

$$q^S = cp + d$$

where c and d are parameters. The parameter c must be positive, reflecting our assumption that quantity supplied increases when the price rises. The parameter d is likely to be zero or negative; this is explained below.

Again we take quantity supplied to be the dependent variable because we assume, in the simplest case at least, that the market price is given as far as any individual producer is concerned and each merely has to decide what quantity of apples to sell in the market at that price. Complications such as the fact that apples come in many qualities and varieties, and more importantly the possibility that individual producers may be able to influence the price by offering more or less for sale, are ignored here.

In the case of apples, we might find that the supply function is

$$q^S = 12p - 200$$

which is graphed in figure 3.13b. The slope of the supply function, 12, tells us that when the price increases (decreases) by 1 cent per kilo the quantity supplied by all firms in aggregate increases (decreases) by 12 (thousand kilos per week).

Note that the supply function has a negative intercept of –200 on the q^S axis, and an intercept of $16^2/_3$ on the p-axis. This implies that the quantity supplied will be negative when the price is less than $16^2/3$. This doesn't really make sense, as firms cannot supply negative quantities to the market (unless they become buyers, in which case they should be included in the demand function). So we assume simply that quantity supplied is zero when price is below $16^2/3$. We can interpret the price of $16^2/3$ (or lower) as being a price so low that it is unprofitable for any producer to supply any apples.

Market equilibrium

Now that we have a demand function and a supply function, it seems a logical next step to bring the two together. We do this by introducing the concept of market equilibrium. The key point is that the demand function tells us what quantity consumers of apples wish or are willing to buy at any given price, and similarly the supply function tells us what quantity sellers of apples wish to sell at any given price. We say that the market for apples is in equilibrium when the quantity consumers wish to buy equals the quantity sellers wish to sell, that is when

$$q^D = q^S$$

This equation is known as the market **equilibrium condition**. If it is not satisfied, then either $q^D > q^S$, implying that there are frustrated would-be buyers who are unable to find a seller; or $q^D < q^S$, which implies the opposite situation. Both imply disequilibrium in the market.

Let us collect together our equations for the market for apples. We have the demand function and the supply function

$$q^D = -8p + 2000 \tag{3.20}$$

$$q^S = 12p - 200 \tag{3.21}$$

together with the equilibrium condition

$$q^D = q^S \tag{3.22}$$

From a mathematical point of view this is a set of three simultaneous equations with three unknowns: q^D, q^S, and p. The solution is very simple. When equation (3.22) is satisfied (as it must be, in the solution) the left-hand side of equation (3.20) will equal the left-hand side of equation (3.21). But in that case the two right-hand sides of equations (3.20) and (3.21) must equal one another, and we will have

$$-8p + 2000 = 12p - 200$$

Solving this equation for p gives $p = 110$ (cents per kilo). This is the equilibrium price. We can then find the equilibrium quantity (which is both supplied and demanded at this price) by substituting $p = 110$ into either the supply function or the demand function (or both, if we want to be absolutely sure that we haven't made a mistake in our algebra). Substituting $p = 110$ into the demand function gives

$$q^D = -8(110) + 2000 = -880 + 2000 = 1120$$

and substituting into the supply function gives

$$q^S = 12(110) - 200 = 1320 - 200 = 1120$$

This solution is shown graphically in figure 3.14, where we have plotted the demand function and the supply function for apples on the same axes, so that we now measure both q^D and q^S on the vertical axis, depending on which curve we are looking at. The market equilibrium is found where the supply and demand functions intersect, at $p = 110$, $q = 1120$. At this price, $q^S = q^D$.

At any other price there is a disequilibrium in the market. For example, in figure 3.15, if $p = 60$, we can read off the quantity demanded from the demand function as $q^D = 1520$ and the quantity supplied from the supply function as $q^S = 520$. Thus at this price $q^D > q^S$ and we say

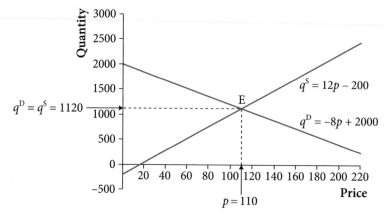

Figure 3.14 Equilibrium in the market for apples.

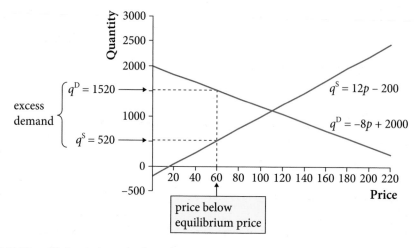

Figure 3.15 Disequilibrium in the market for apples.

that there is **excess demand**. This is because the price, $p = 60$, is below the equilibrium price of 110. Note that at this stage we are not attempting to analyse in any detail what would happen next. However, it seems reasonable to assume that when $q^D > q^S$, frustrated would-be buyers will offer a higher price, so the price will start moving up towards its equilibrium value. Similarly, when p is above its equilibrium value of 110, $q^D < q^S$, a situation described as **excess supply**. It then seems plausible that frustrated would-be sellers will offer to sell at a lower price and the price will start moving down towards its equilibrium value. Analysing these processes in more detail is the subject matter of *economic dynamics*, which we examine in chapter 20 and chapter W21, which is to be found at the Online Resource Centre.

3.15 The inverse demand and supply functions

If you have studied economics before reading this book, you may have been somewhat puzzled by figures 3.13, 3.14, and 3.15. This may be because they follow the convention in mathematics of putting the dependent variable, quantity demanded or supplied, on the vertical axis. Among economists, however, there is a long-established convention of putting price on the vertical axis when graphing demand or supply functions. We will now rework our analysis of the market for apples to conform to this convention.

Dealing first with the demand function, we have to rearrange the function so as to isolate p on the left-hand side of the equation. In the case of our demand function for apples, $q^D = -8p + 2000$,

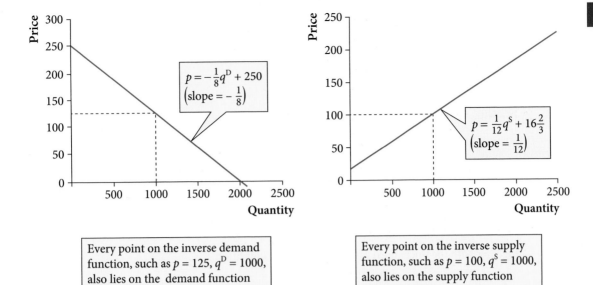

Every point on the inverse demand function, such as $p = 125$, $q^D = 1000$, also lies on the demand function (figure 3.13(a)).

Every point on the inverse supply function, such as $p = 100$, $q^S = 1000$, also lies on the supply function (figure 3.13(b)).

Figure 3.16 Inverse demand function for apples.

Figure 3.17 Inverse supply function for apples.

we can do this by subtracting 2000 from both sides and then dividing both sides by -8 (check this for yourself). This gives us

$$p = -\tfrac{1}{8}q^D + 250 \tag{3.23}$$

This is called the *inverse demand function*. Its graph is plotted in figure 3.16. The demand function (figure 3.13a) and its inverse (figure 3.16) express exactly the same mathematical relationship between p and q^D, in the sense that any pair of values of p and q^D that satisfies one equation will automatically satisfy the other (such as $q^D = 1000$, $p = 125$). We can view p in equation (3.23) as being the **demand price**: that is, the (maximum) price that buyers are willing to pay for the quantity q^D. Let us write p^D to denote the demand price, as defined in the previous sentence. Then, as a variant of equation (3.23), we can write

$$p^D = -\tfrac{1}{8}q^D + 250 \tag{3.23a}$$

In exactly the same way, to conform with the conventions of economics we have to derive the *inverse supply function*. We do this by taking the supply function $q^S = 12p - 200$, adding 200 to both sides, then dividing both sides by 12 (check this for yourself). This gives

$$p = \tfrac{1}{12}q^S + 16\tfrac{2}{3} \tag{3.24}$$

This inverse supply function is graphed in figure 3.17. Again, there is no difference in the information conveyed by the supply function (figure 3.13b) and its inverse (figure 3.17). We can view p in equation (3.24) as being the **supply price**: that is, the (minimum) price that sellers are willing to accept for the quantity q^S. If we denote the supply price by p^S, a variant of equation (3.24) is

$$p^S = \tfrac{1}{12}q^S + 16\tfrac{2}{3} \tag{3.24a}$$

Why do economists often prefer to work with the inverse supply and demand functions? Partly, it is merely convention, and conventions are not easily changed. Most microeconomics textbooks work with the inverse demand and supply functions and their graphs, and we will also do so later in this book. However, the decision to work with the inverse functions does not imply any shift in view as to whether it is quantity or price that is the independent variable.

In the market for apples we will continue to assume, as we did in section 3.14, that buyers and sellers take the market price as given and choose how many apples to demand or supply at that price. However, in other contexts we might take a different view. This is a matter for economic analysis which we will consider later in this book.

Market equilibrium

We can use the inverse demand and supply functions to find the market equilibrium. Our model carried over from section 3.14 now consists of the inverse demand and supply functions

$$p^D = -\tfrac{1}{8}q^D + 250 \tag{3.23a}$$

$$p^S = \tfrac{1}{12}q^S + 16\tfrac{2}{3} \tag{3.24a}$$

together with the unchanged market equilibrium condition

$$q^D = q^S \tag{3.22 repeated}$$

However, we now have only three equations but a total of four unknowns; p^D, p^S, q^D, and q^S. An additional unknown has appeared because we have introduced a distinction between the demand price and the supply price, a distinction that was absent in section 3.14. But there is now a new equilibrium condition, which is that in equilibrium the price paid by buyers must equal the price received by sellers; that is

$$p^D = p^S \tag{3.25}$$

Now we have a set of four simultaneous equations with four unknowns, p^D, p^S, q^D, and q^S. The method of solution is as follows. If equation (3.22) is satisfied (as it must be, in the solution), then q^D and q^S are identical. Let us write q to denote this common value of both q^D and q^S. Then we can replace both q^D in equation (3.23a) and q^S in equation (3.24a) with their common value, q.

Similarly, if equation (3.25) is satisfied, p^D and p^S are identical, with a common value that we will denote by p. Then we can replace both p^D in equation (3.23a) and p^S in equation (3.24a) with their common value, p. After these replacements or substitutions, equations (3.23a) and (3.24a) become

$$p = -\tfrac{1}{8}q + 250 \tag{3.23b}$$

$$p = \tfrac{1}{12}q + 16\tfrac{2}{3} \tag{3.24b}$$

Now we are on familiar territory, as we have two equations with two unknowns. In the usual way, we can say that, in the solution, the value of p in equation (3.23b) and the value of p in equation (3.24b) must be equal. Therefore the right-hand sides of the two equations must be equal too, which gives us

$$-\tfrac{1}{8}q + 250 = \tfrac{1}{12}q + 16\tfrac{2}{3}$$

The solution to this equation is $q = 1120$, from which $p = 110$. That is, the common value of q is $q = q^D = q^S = 1120$ and the common value of p is $p = p^D = p^S = 110$. These are, of course, the same answers that we obtained in section 3.14, using the supply and demand functions rather than their inverses. Market equilibrium using the inverse demand and supply functions is shown in figure 3.18. Note that we now measure both p^D and p^S on the vertical axis.

The pair of simultaneous equations (3.23b) and (3.24b) is an example of a **reduced form**, where we have reduced the number of equations in the model from four to two as a step towards solving it. In economics we frequently work with reduced forms.

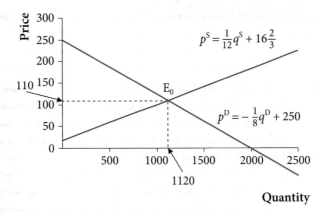

Figure 3.18 Equilibrium with inverse supply and demand functions.

Economic modelling

Our analysis of the market for apples in this section and the previous section is an example of **economic model** building, or simply modelling. A model of an economic process aims to identify in a simplified form the key features of that process and to draw conclusions that enlarge our understanding of the process and permit predictions about the effects of changes, for example in parameter values.

In the market for apples, the components of our model are, first, the demand and supply functions (or their inverses), which are called **behavioural relationships** because they describe how buyers and sellers behave when given the opportunity to buy or sell at various prices; and second, the market *equilibrium conditions*. Most economic models, however complex they may be, consist mainly of these two types of equation, though additional equations which are definitions, and therefore identities, may also appear. From a mathematical point of view, note also that in our two models, the first using the supply and demand functions and the second their inverses, the number of equations equalled the number of unknowns. This illustrates the conclusion reached at the end of section 3.11: in order for a set of simultaneous equations to have a solution, a necessary (but not sufficient) condition is that the number of equations exactly equals the number of unknowns. (Recall that it is not a sufficient condition because the solution nevertheless fails if the equations are either inconsistent or not independent.)

3.16 Comparative statics

As well as solving our economic model of the market for apples and thereby finding the equilibrium price and quantity, we can also use the model to analyse the effects of shifts in the supply or demand function due to change in buyers' preferences or in production technology, or government intervention such as the imposition of a tax. This analysis is called **comparative statics**, since we compare the initial or previous equilibrium with the new equilibrium that follows a shift or disturbance.

EXAMPLE 3.20

A standard example of comparative statics is the effect of a tax. Let us suppose that, in our model of the apple market from sections 3.14 and 3.15, the government imposes a tax of T cents per kilo on sellers. This means that for every kilo of apples that they sell, sellers must give the government I cents. This type of tax, where the tax bill depends on the *quantity* of apples sold (and not on their price) is called a **specific tax** or a per-unit tax. Tobacco and alcohol are among the goods subject to specific taxes in the UK.

We can analyse the effects of this tax using either the supply and demand functions from section 3.14 or their inverses from section 3.15. We will take the latter option because the effects of a tax are intuitively easier to understand when price is treated as the dependent variable.

The tax creates a new, tax-modified, inverse supply function. Before the tax, the minimum price that sellers were willing to accept for supplying quantity q^S was, by definition, p^S. But with the tax, sellers must recover an additional amount, T cents, on every kilo sold, so the minimum price they are now willing to accept for quantity q^S is $p^S + T$. If we denote this new minimum selling price by p^{S+}, then we have

$$p^{S+} = p^S + T \tag{3.26}$$

and from equation (3.24a) we have

$$p^S = \tfrac{1}{12}q^S + 16\tfrac{2}{3} \tag{3.24a repeated}$$

Using (3.24a) to eliminate p^S from (3.26), we have the tax-modified inverse supply function

$$p^{S+} = \tfrac{1}{12}q^S + 16\tfrac{2}{3} + T \tag{3.27}$$

Note that this does not replace the original inverse supply function, but exists alongside it. The original inverse supply function gives the *net* price received by sellers (after tax has been paid), and the tax-modified inverse supply function the *gross* price (including tax) at which sellers are willing to supply quantity q^S. Note too that (3.27) is not an *independent* equation, as it was obtained by using (3.24a) to substitute for p^S in equation (3.26). It does not give us any information additional to what is already contained in those two equations.

On the demand side, the tax has no direct effect. The maximum price that buyers are willing to pay for any given quantity continues to be given by the inverse demand function from section 3.15 above:

$$p^D = -\tfrac{1}{8}q^D + 250 \tag{3.23a repeated}$$

The equilibrium condition from section 3.14 above also remains valid:

$$q^D = q^S \tag{3.22 repeated}$$

However, the tax necessitates modifying our previous equilibrium condition, $p^D = p^S$ (equation (3.25) above). As discussed above, we now have

$$p^{S+} = p^S + T = p^D \tag{3.28}$$

This equilibrium condition says that the supply price, plus tax, must equal the demand price.

Let us pause for a moment and assess our model. We have six unknowns: p^D, p^S, q^D, q^D, p^{S+}, and T. We also appear to have six equations: (3.26), (3.24a), (3.27), (3.23a), (3.22), and (3.28). However, this is an illusion, as equation (3.27) is not independent, as already noted, and therefore gives us no new information. So we have only five *independent* equations, but six unknowns. However, if we are now told that the tax rate is, say, $T = 50$ (cents per kilo), then this gives us a sixth equation and our model has the same number of independent equations as unknowns, as is necessary for a solution (but not sufficient, as it remains possible that the equations are inconsistent; see section 3.11).

There are many ways of solving this set of simultaneous equations. We will tackle it in a way that is almost identical to that in section 3.15. First, if equation (3.22) is satisfied (as it must be, in the solution), then q^D and q^S are identical, with a common value that we can write as q. So we can replace both q^D in equation (3.23a) and q^S in equations (3.24a) and (3.27) with this common value, q.

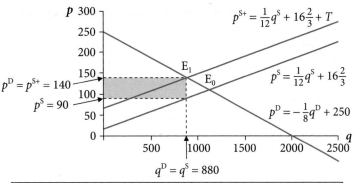

The tax shifts the supply function up by the amount of the tax, $T = 50$. The equilibrium moves from E_0 to E_1. Quantity falls from 1120 to 880. The market price rises from 110 to 140. The price received by sellers falls from 110 to 90. The area of the shaded rectangle measures the tax revenue.

Figure 3.19 Effects of a specific tax.

Second, equation (3.28) tells us that the left-hand sides of (3.23a) and (3.27) are equal in the equilibrium. Therefore we can set their right-hand sides equal, which gives

$$\tfrac{1}{12}q + 16\tfrac{2}{3} + T = -\tfrac{1}{8}q + 250$$

Finally, in this equation we replace T with 50, resulting in

$$\tfrac{1}{12}q + 16\tfrac{2}{3} + 50 = -\tfrac{1}{8}q + 250$$

This linear equation has only one variable, q; hence by elementary operations we can solve for q. We obtain $q = 880$. This of course is both q^D and q^S, as they are equal in equilibrium. We can then find p^D by substituting $q = 880$ into the inverse demand function (equation (3.23a)), resulting in $p^D = 140$. This also equals p^{S+}, from equation (3.28), so from (3.26) we then have $p^S = 140 - 50 = 90$.

(Be sure to check all this out for yourself, with pen and paper, before carrying on.)

The solution is shown graphically in figure 3.19. The original equilibrium is at E_0, where the inverse supply and demand curves cut (see also figure 3.18). The imposition of the tax shifts the equilibrium to E_1, where the tax-modified inverse supply curve cuts the inverse demand curve.

We see that the tax-modified inverse supply function lies 50 cents above the original inverse supply function; 50 cents per kilo is, of course, the specific tax rate. The common sense of this is that, for any given quantity, suppliers now require a price 50 cents higher, because the price before tax was already, by assumption, the minimum that they were willing to accept. This higher price causes buyers to reduce their demand, from 1120 to 880. The price paid by buyers rises from 110 to 140, while the price received by sellers falls from 110 to 90. Thus the 'burden' of the tax is shared between buyers and sellers. In this example buyers pay most of the tax (30 out of 50), but with other assumptions about the shapes of the supply and demand functions the distribution of the burden between buyers and sellers would be different.

From the graph we can read off $p^D = p^{S+} = 140$; $p^S = 90$, and $q^D = q^S = 880$. The tax revenue received by the government is given by the area of the shaded rectangle. This is because the (horizontal) length of the rectangle is quantity (880 thousand kilos) and its height is the tax per unit (50 cents); so its area is 880 thousand $\times$ 50 = 44,000 thousand cents = 440 thousand euros (per week).

EXAMPLE 3.21

As a second example of comparative statics we will consider the effect of an ad valorem tax on apples. An **ad valorem tax** is a tax in which the tax bill per kilo of apples sold is calculated as a proportion or percentage of the price received by the seller. It is also sometimes known as a proportionate tax. Value added tax (VAT) is an example of an ad valorem tax in the UK. (Ad valorem means 'according to value' in Latin.) As in the previous example we will assume that the tax is collected by sellers. Now, however, instead of the tax being a fixed amount, T cents, per unit sold, it is levied as a fixed proportion, t, of the supply price, p^S, at which the unit is sold. For example, if $t = 0.5$ and $p^S = 50$ cents, the seller has to pay $tp^S = 0.5 \times 50 = 25$ cents to the government per unit sold. If the supply price falls to, say, 40 cents, then the tax falls too, to $0.5 \times 40 = 20$ cents per unit sold. Thus, although the *rate* of tax is constant, the *amount* of the tax, per unit sold, varies with the supply price. (If you are happier with percentages than with proportions, a proportionate tax rate of $t = 0.5$ equals a percentage tax rate of $100t = 100(0.5) = 50$ per cent.)

As in the previous example, the tax creates a new, tax-modified, inverse supply function. Before the tax, the minimum price that sellers were willing to accept for supplying quantity q^S was, by definition, p^S. But with the tax, sellers must recover an additional amount, tp^S cents, on every kilo sold, so the minimum price they are now willing to accept for quantity q^S is $p^S + tp^S$. If we denote this new minimum selling price by p^{S+}, then we have

$$p^{S+} = p^S + tp^S = p^S(1 + t) \tag{3.29}$$

From equation (3.24a) we have

$$p^S = \tfrac{1}{12}q^S + 16\tfrac{2}{3} \tag{3.24a repeated}$$

Combining these, we have the tax-modified inverse supply function:

$$p^{S+} = \left(\tfrac{1}{12}q^S + 16\tfrac{2}{3}\right)(1 + t) \tag{3.30}$$

As in the previous example, this does not replace the original inverse supply function, but exists alongside it. As before, (3.30) is not an independent equation, as it was obtained by using (3.24a) to substitute for p^S in equation (3.29). It does not give us any information additional to what is already contained in those two equations.

The next two equations are carried over unchanged from the previous example. They are the inverse demand function:

$$p^D = -\tfrac{1}{8}q^D + 250 \tag{3.23a repeated}$$

and the equilibrium condition:

$$q^D = q^S \tag{3.22 repeated}$$

As in the previous example, the next equation of our model is the equilibrium condition, which says that the supply price, plus tax, must equal the demand price. Using equation (3.29), this condition is

$$p^{S+} = p^S(1 + t) = p^D \tag{3.31}$$

Finally, we need one more equation to make the number of independent equations equal to the number of unknowns. This last equation is the numerical value of the tax rate, t. We will assume that $t = 0.25$, or 25%. So we are now ready to solve the model.

Following the path of the previous example, we first note that if equation (3.22) is satisfied (as it must be, in the solution), then q^D and q^S are identical, with a common value that we can write as q. So we can replace both q^D in equation (3.23a) and q^S in equations (3.24a) and (3.30) with this common value, q.

Second, equilibrium condition (3.31) tells us that the left-hand sides of (3.23a) and (3.30) are equal in the equilibrium. Therefore we can set their right-hand sides equal, which gives

$$\left(\tfrac{1}{12}q + 16\tfrac{2}{3}\right)(1 + t) = -\tfrac{1}{8}q + 250$$

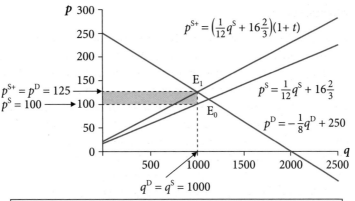

$p^{S+} = \left(\frac{1}{12}q^S + 16\frac{2}{3}\right)(1+t)$

$p^{S+} = p^D = 125$
$p^S = 100$

$p^S = \frac{1}{12}q^S + 16\frac{2}{3}$

$p^D = -\frac{1}{8}q^D + 250$

E_1
E_0

$q^D = q^S = 1000$

> Because the tax is a fixed proportion, $t = 0.25$, of the sellers' price, the tax-modified inverse supply curve p^{S+} is not parallel with the pre-tax inverse supply curve p^S. Quantity falls from 1120 to 1000. The market price rises from 110 to 125. The price received by sellers falls from 110 to 100. The shaded area measures the tax revenue.

Figure 3.20 Effects of an ad valorem (proportionate) tax.

After substituting $t = 0.25$ into this equation, we have

$$\left(\tfrac{1}{12}q + 16\tfrac{2}{3}\right)(1 + 0.25) = -\tfrac{1}{8}q + 250$$

Using elementary operations, we find the solution to this equation as $q = 1000$. Substituting this into the inverse demand function (3.23a), we find $p^D = 125$. From (3.31) this also equals p^{S+}. We can find p^S either from equations (3.31) (with $t = 0.25$) or (3.24a) as $p^S = 100$.

The solution is shown graphically in figure 3.20. The original equilibrium is at E_0, where the inverse supply and demand curves cut (see also figure 3.18). The imposition of the tax shifts the equilibrium to E_1, where the tax-modified inverse supply curve cuts the inverse demand curve. Note that the tax-modified supply curve is not parallel with the original supply curve (the supply curve without tax). This is because the tax is levied as a constant *proportion* of the supply price before tax. Therefore the *amount* of tax, which is the vertical distance between the two supply curves, increases as the supply price increases. This may be seen in equation (3.30), where the tax rate t affects p^{S+} *multiplicatively* rather than additively, unlike the previous example (equation 3.27) where the tax was a fixed amount, $T = 50$ cents per kilo, and p^{S+} lay 50 cents above p^S, a parallel shift.

As in the previous example, the shaded rectangle in figure 3.20 measures the tax revenue. We can calculate the revenue as $(p^D - p^S)q$. As q is measured in thousands of kilos per week, $(p^D - p^S)q = 25$ cents $\times$ 1000 thousand per week = 25,000 thousand cents per week = 250 thousand euros per week.

As in the previous example, the 'burden' of the tax is divided between buyers and sellers. Although the tax rate is 25%, the price paid by buyers rises from 110 to 125, an increase of only 13.6%. This is because the price received by sellers falls from 110 to 100, a reduction of 9.1%. The *amount* of the tax is 25 cents, which is 25% of the selling price, 100 cents.

> **»**   **Hint** This distinction between a parallel shift in a curve, as in example 3.20, and a proportionate shift, as in example 3.21, occurs very frequently in economic analysis. It is worth studying carefully.

When the tax is levied on buyers, not sellers

In examples 3.20 and 3.21 we assumed that the tax was levied on sellers (meaning that the tax bill was addressed to them). This resulted in the tax-modified supply curve and tax-inclusive supply price, p^{S+}. You may be wondering how the analysis would change if the tax were levied on buyers instead. The answer is that this would result in a tax-modified *demand* curve and a tax-inclusive demand price. The tax-modified demand curve would lie below the original or pre-tax demand curve, because the price that buyers would be willing to pay to the sellers for any given quantity would be reduced by the amount of the tax to be paid. However, in both cases (the specific tax in example 3.20 and the ad valorem tax in example 3.21), the with-tax equilibrium is unchanged. Whether the tax is levied on sellers or buyers makes no difference to its effect on market price and quantity sold. The demonstration of this is left to you as an exercise. (Hint: you will need to set up a tax-modified inverse *demand* function and a tax-inclusive demand price, p^{D-}, where the minus sign indicates that buyers are now willing to pay less to the sellers for any given quantity.)

3.17 Macroeconomic equilibrium

Macroeconomics is concerned with the working of the economy as a whole rather than with individual markets such as the market for apples. In this application of linear equations to economics, we examine how the equilibrium level of income for the whole economy is determined.

The key relationships of our model of the economy are as follows. Households considered in aggregate earn their incomes (whether in the form of wages, salaries or profits) by producing output. Therefore aggregate household income, Y, must equal the value of output, Q. Note that it is a fairly strong, though not universal, convention among economists to use capital (upper-case) letters to denote macroeconomic variables.

In turn, this output must be bought by somebody, otherwise it would not be produced. So the value of output, Q, must equal aggregate expenditure or demand, E. Thus we have two identities, $Y \equiv Q$ and $Q \equiv E$. Combining these we arrive at

$$Y \equiv E \tag{3.32}$$

This is the first equation, or more precisely identity, of our macroeconomic model. It says that aggregate income and aggregate expenditure are necessarily equal.

Next, aggregate expenditure consists of consumption expenditure by households, C, plus investment expenditure by firms, I. By investment we mean purchases of newly produced machinery, buildings, and other equipment. This gives us a second identity:

$$E \equiv C + I \tag{3.33}$$

The third equation in the model concerns the planned or desired consumption expenditure of households. As first proposed by the economist J.M. Keynes in the 1930s, we assume that planned household consumption is a function of household income. This relationship is called the **consumption function**. If we assume for simplicity that this function is linear, we can write it as $\hat{C} = aY + b$, where $\hat{C}$ is planned or desired consumption and a and b are both assumed positive. The parameter a is called the marginal propensity to consume (*MPC*). We restrict the *MPC* to being less than 1 (that is, $a < 1$) for reasons that will become clear shortly. As a specific example of this linear consumption function, we might have

$$\hat{C} = 0.5Y + 200 \tag{3.34}$$

which is graphed in figure 3.21. The slope of this function is 0.5, and this is the *MPC*. The *MPC* gives us the increase in planned consumption, $\hat{C}$, that occurs when income, Y, rises by one unit.

In passing, we can define planned or desired saving, $\hat{S}$, as $\hat{S} \equiv Y - \hat{C}$. This is an identity because in this simple model the decision not to spend €1 of income on consumption is necessarily a decision to save that €1. Substituting for $\hat{C}$ in this identity, it becomes $\hat{S} = Y - (0.5Y + 200) =$

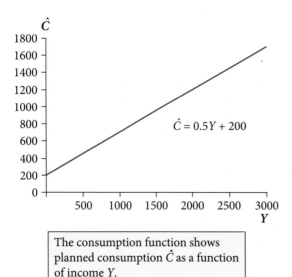

The consumption function shows planned consumption $\hat{C}$ as a function of income Y.

Figure 3.21 A linear consumption function.

$0.5Y - 200$. This is called the **savings function**, and 0.5 is the marginal propensity to save (*MPS*).

The distinction between C and $\hat{C}$ is crucial. C is actual output, and therefore actual sales, of consumption goods, while $\hat{C}$ is a subjective magnitude, planned or desired consumption. The two are not necessarily equal, and consumers will be frustrated whenever they differ from one another. Since this frustration will trigger changes in spending, which in turn will lead to changes in income, we can say that an equilibrium condition for this economy is

$$\hat{C} = C \tag{3.35}$$

Equations (3.32)–(3.35) above thus constitute our macroeconomic model. As discussed earlier in the context of the market for apples, the model consists of two identities, a behavioural relationship (the consumption function) and an equilibrium condition.

Our immediate problem is that we have only four equations but five unknowns, Y, E, I, C, and $\hat{C}$. So this set of simultaneous equations does not have a unique solution. In mathematical language, the set of simultaneous equations is under-determined. However, let's see how far we can get towards a solution. An obvious first step is to substitute equation (3.32) into equation (3.33), giving

$$Y \equiv C + I \tag{3.36}$$

This identity says that aggregate income must necessarily equal the value of consumption goods and investment goods produced (and sold). If we then use equation (3.35) to substitute for $\hat{C}$ in equation (3.34), we arrive at

$$C = 0.5Y + 200 \tag{3.37}$$

We can then substitute (3.37) into (3.36), and get

$$Y = 0.5Y + 200 + I$$

After some elementary operations this becomes

$$Y = \frac{200 + I}{0.5} \tag{3.38}$$

Equation (3.38) gives Y as a function of I, but as there are two variables or unknowns it does not give us a unique solution for Y. Equation (3.38) is said to be a *reduced form* of the model which consists of equations (3.32)–(3.35). This simply means that (3.38) was obtained by combining (3.32)–(3.35). Unlike (3.36), equation (3.38) is not an identity because it contains a behavioural relationship, the consumption function. The concept of a reduced form is important in economics (see also section 3.15).

Note also that in equation (3.38) the denominator on the right-hand side, 0.5, equals 1—*MPC*. Now we can see why the *MPC* is restricted to values less than 1. If the *MPC* were permitted to equal 1, the denominator of (3.38) would be zero, and we know from sections 1.3 and 2.8 that division by 0 is undefined. And if the *MPC* were permitted to be greater than 1, the denominator of the right-hand side of (3.38) would be negative, which would create horrendous problems for our model that we will not explore here. (Try to work out what they might be.)

Because equation (3.38) contains two variables and is therefore under-determined, we cannot determine the equilibrium level of income, Y, which is the main purpose of our model, unless we can come up with an additional equation telling us how I is determined. The simplest (but completely unimaginative) way to arrive at such an equation is to assume that the level of investment expenditure by firms, I, is equal to some fixed value, say

$$I = 800 \tag{3.39}$$

We then say that I is an **exogenous variable**, meaning that its value is determined or explained in some way that is outside of our model and which we can therefore simply take as given. A variable that is determined or explained inside our model, such as C and Y in this case, is called an **endogenous variable** ('exo' meaning without and 'endo' within).

Equation (3.39) gives us an additional equation, $I = 800$, without introducing any additional unknowns. The whole model now consists of equations (3.32)–(3.35), plus (3.39). So we now have a set of five simultaneous equations with five unknowns, $Y, E, C, \hat{C},$ and I. However, rather than solving the whole set of equations again, we can now find Y simply by substituting $I = 800$ into equation (3.38) above, which is the reduced form of equations (3.32)–(3.35). Equation (3.38) then becomes

$$Y = \frac{200 + I}{0.5} = \frac{200 + 800}{0.5} = 2000$$

This is the equilibrium level of Y in the sense that the plans or desires of households to spend on consumption goods $(\hat{C})$, and firms' spending on investment goods (I), are consistent with the economy's actual production of these goods $(C + I)$. A natural question that arises at this point concerns the dynamics of the model; that is, what happens when there is a disequilibrium. We will not address this here in our brief introductory treatment.

Graphical treatment

The model and its solution are shown graphically in figure 3.22, a very widely used figure known as 'the Keynesian cross', named for the economist J.M. Keynes who originated this model. Aggregate income, Y, is measured on the horizontal axis. On the vertical axis we measure both actual aggregate expenditure, E, and planned aggregate expenditure, $\hat{E}$. It may seem confusing at first to have two variables measured on the same axis, but this will become clear shortly.

First, we plot the graph of $E \equiv Y$ (equation (3.32)). Referring to E measured on the vertical axis, this is a linear function with an intercept of zero and a slope of 1. At every point on this line, expenditure equals income. Also, from equation (3.33), we have $Y \equiv C + I$ at every point on this line.

Second, we define planned aggregate expenditure, $\hat{E}$, as $\hat{E} = \hat{C} + I$. Thus planned aggregate expenditure, also known as aggregate demand, comprises planned consumption, $\hat{C}$, plus investment, I. If we then substitute equations (3.34) and (3.39) into this equation, it becomes:

$$\hat{E} = \hat{C} + I = 0.5Y + 200 + 800$$
$$= 0.5Y + 1000 \tag{3.40}$$

Referring now to $\hat{E}$ on the vertical axis, the graph of (3.40) in figure 3.22 is a straight line with a slope of 0.5 and an intercept of 1000.

Thus in figure 3.22 we have two curves. For any given level of Y, from the curve $E \equiv Y$ we can read off on the vertical axis the level of *actual* expenditure, E. Similarly from the curve $\hat{E} = 0.5Y + 1000$ we can read off on the vertical axis the level of *planned* expenditure, $\hat{E}$. At point J, where the two curves intersect, we have the level of Y at which $\hat{E} = E$.

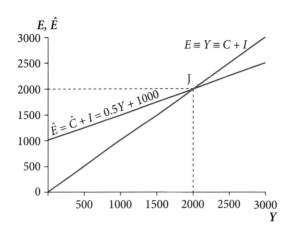

Equilibrium is at J, where planned expenditure $\hat{E} = \hat{C} + I$ and actual expenditure $E \equiv Y$ are equal. Equilibrium $Y = 2000$, of which $C = \hat{C} = 1200$ and $I = 800$.

Figure 3.22 The income–expenditure model.

This is the equilibrium point for this economy. The reason is that, since $\hat{E} = \hat{C} + I$ as defined above, and $E \equiv Y = C + I$ (from equations (3.32) and (3.36)), when $\hat{E} = E$ we have $\hat{C} + I = C + I$ and therefore $\hat{C} = C$. This is the equilibrium condition for our model (see equation (3.35)). Only at point J are planned and actual consumption equal.

Comparative statics

As we did with the model of the supply and demand for apples in section 3.16, we can carry out a comparative statics exercise to examine how the equilibrium level of Y changes when, say, investment increases by one unit. In the example above we had $I = 800$ and the equilibrium level of Y was $Y = 2000$. If for some reason investment now rises to $I = 801$, we can quickly find the new equilibrium level of income by substituting $I = 801$ into the reduced form, equation (3.38). This gives

$$Y = \frac{200 + I}{0.5} = \frac{200 + 801}{0.5} = 2002$$

Thus an increase in I of 1 results in an increase in Y of 2. In general it is easy to see from the reduced form, equation (3.38), that any given increase in investment is multiplied by 2 to arrive at the resulting increase in Y. We say that the **investment multiplier** is 2. In figure 3.22 an increase in investment is revealed as an upward shift in the aggregate demand function, $\hat{E} = \hat{C} + I$, which consequently cuts the $E \equiv Y$ line further to the right of J, at a higher level of Y (not shown in figure 3.22).

Generalization

Our macroeconomic model is easily generalized. Only two equations need modification. First, we replace the specific consumption function $\hat{C} = 0.5Y + 200$ with the general linear form $\hat{C} = aY + b$. Second, we assume a fixed but unspecified value, $I = \bar{I}$, for investment. We then have the following model:

$$Y \equiv E \tag{3.41}$$

$$E \equiv C + I \tag{3.42}$$

$$\hat{C} = aY + b \tag{3.43}$$

$$\hat{C} = C \tag{3.44}$$

$$I = \bar{I} \tag{3.45}$$

This can quickly be reduced to three equations. First, combine (3.41) and (3.42) to get $Y \equiv C + I$. Then combine (3.43) and (3.44) to get $C = aY + b$. We then have three equations:

$$Y \equiv C + I$$

$$C = aY + b$$

$$I = \bar{I}$$

Substituting the latter two equations into the first, and performing a few elementary operations, gives

$$Y = \frac{b + \bar{I}}{1 - a} = \frac{1}{1 - a}(b + \bar{I}) = \frac{1}{1 - MPC}(b + \bar{I}) \tag{3.46}$$

This is an important result. It shows that the equilibrium level of income is determined as the sum of investment, $\bar{I}$, plus the intercept term, b, of the consumption function, multiplied by $\frac{1}{1 - MPC}$. The term $\frac{1}{1 - MPC}$ is known as 'the multiplier'. The multiplier is greater than one provided the MPC is positive but less than one (check this for yourself).

Comparative statics in the general case

As before, we can examine how the equilibrium level of Y changes when investment increases. Suppose investment is initially at some level $\bar{I}_0$. Then the initial level of Y, which we'll label Y_0, is, from equation (3.46),

$$Y_0 = \frac{1}{1 - MPC}(b + \bar{I}_0)$$

If investment then rises to some new level, $\bar{I}_1$, then income consequently rises to some new level, Y_1, which is given by

$$Y_1 = \frac{1}{1 - MPC}(b + \bar{I}_1)$$

So we can calculate the *change* in Y as $Y_1 - Y_0$, which from the two equations above is

$$Y_1 - Y_0 = \frac{1}{1 - MPC}(b + \bar{I}_1) - \frac{1}{1 - MPC}(b + \bar{I}_0)$$

This rearranges as

$$Y_1 - Y_0 = \frac{1}{1 - MPC}[(b + \bar{I}_1) - (b + \bar{I}_0)] \quad \text{which simplifies to:}$$

$$Y_1 - Y_0 = \frac{1}{1 - MPC}[\bar{I}_1 - \bar{I}_0] \tag{3.47}$$

So in general the change in income, $Y_1 - Y_0$, equals the change in investment, $\bar{I}_1 - \bar{I}_0$, multiplied by the multiplier, $\frac{1}{1 - MPC}$.

Note that equation (3.47) becomes a more complicated expression in more complex models that allow for taxation and government spending and for exports and imports. The equation also becomes more complex if the consumption function is non-linear, for then the *MPC* varies with income.

Summary of sections 3.13–3.17

In sections 3.13–3.16 we explored in some depth a model of supply and demand for a good, assuming that the supply and demand functions were linear. The equilibrium price and quantity are found as the intersection of either the supply and demand functions (in which quantity is the dependent variable), or the inverse functions (with price as the dependent variable). We conducted a comparative statics exercise, analysing the effect of a tax on the equilibrium price and quantity. In section 3.17 we constructed a simple macroeconomic model of aggregate supply and demand, and found how a change in investment affected the level of income via the multiplier.

The Online Resource Centre gives more examples of economic modelling with linear functions and how they can be graphed using Excel®. www.oxfordtextbooks.co.uk/orc/renshaw3e/

Progress exercise 3.4

1. (a) Given the following supply and demand functions, find the equilibrium price and quantity. In each case, show the solution graphically.

(i) $q^D = -0.75p^D + 10$, $q^S = 2p^S - 1$; for values of p between 0 and 10

(ii) $q^D = -2p^D + 40$, $q^S = 3p^S - 15$; for values of p between 0 and 20

(b) Comment on the possibility of market equilibrium, and illustrate graphically, if the supply and demand functions are

$q^D = -3p^D + 14; q^S = 2p^S - 11$; for values of p between 0 and 10

2. Find the equilibrium price and quantity, given the inverse demand and supply functions:

$p^D = -3q^D + 30; p^S = 2q^S - 5$

3. Given the following supply and demand functions:

$q^D = -5p^D + 100; \quad q^S = 15p^S - 100$

(a) Find the equilibrium price and quantity.

(b) Find the inverse demand and supply functions, and verify that solving these simultaneously gives the same equilibrium price and quantity as in (a).

(c) Illustrate (a) and (b) graphically.

4. Given the following inverse supply and demand functions:

$p^D = -0.25q^D + 15; p^S = 0.1q^S + 8$

(a) Find the equilibrium price and quantity.

(b) Suppose the government imposes an ad valorem tax of 50% of the supply price. Find the new equilibrium price and quantity.

(c) Sketch the relevant graphs, showing the solutions.

5. (a) Given the following macroeconomic model, find the equilibrium levels of income and consumption, and illustrate diagrammatically.

$Y \equiv E$

$E \equiv C + 1$

$\hat{C} = 0.8Y + 100$

$I = 550$

$C = \hat{C}$

where Y = aggregate income, E = aggregate expenditure, C = consumption by households, I = investment by firms, $\hat{C}$ = planned or desired consumption.

(b) Suppose business leaders become more optimistic about the future demand for their products, and consequently increase their investment to 700. Find the new level of Y, the change in Y, the change in C, and the investment multiplier in this case.

(c) Suppose the government, alarmed by the rise in income because it fears that inflation might increase, tries to persuade households to reduce their marginal propensity to consume in order to restore income to its previous level. Calculate the size of the reduction in the MPC required to achieve this. (Hint: you need to derive an equation in which $1 - MPC$ is the dependent variable.)

6. Given the following macroeconomic model,

$Y \equiv E$

$E \equiv C + 1$

$\hat{C} = 0.85Y + 100$

$I = 650$

$C = \hat{C}$

where the variables are defined as in question 5.

(a) Find the equilibrium levels of income and consumption, and illustrate diagrammatically.

(b) If investment falls to 400, find the new equilibrium values of Y and C.

(c) Given this fall in investment, find the increase in the MPC required to restore Y to its original level.

Checklist

Be sure to test your understanding of this chapter by attempting the progress exercises (answers are at the end of the book). The Online Resource Centre contains further exercises and materials relevant to this chapter www.oxfordtextbooks.co.uk/orc/renshaw3e/.

In this chapter we progressed beyond the basic manipulative rules of chapter 2 and began putting algebra to work in solving equations and plotting graphs. We also introduced some simple economic applications. The key concepts and techniques were:

✔ **Equations and identities.** Distinction between an equation and an identity. Manipulation of an equation so as to isolate any variable on one side of the equation.

✔ **Variables and parameters.** Distinction between a variable and a parameter or constant.

✔ **Solving linear equations.** Solving linear equations containing one unknown.

✔ **Equations and functions; explicit and implicit functions.** Distinction between an equation and a function. Explicit and implicit linear functions. Identifying slope and intercept parameters of a linear function.

✔ **Graphical methods.** Plotting graphs of linear functions. Solving linear equations graphically.

✔ **Simultaneous equations.** Understanding linear simultaneous equations with two unknowns. Solving by both algebraic and graphical methods.

✔ **Economic applications:** equilibrium and comparative statics. Using simultaneous equation methods to solve linear supply and demand models and the macroeconomic income–expenditure model. Simple comparative static exercises in these models.

By now your confidence in your mathematical abilities should be strengthening and you can see the relevance of mathematical methods in economics. If so, you are ready to move on to chapter 4, where we consider our first non-linear relationships.

Chapter 4
Quadratic equations

OBJECTIVES

Having completed this chapter you should be able to:

- Recognize a quadratic expression, understand how it arises and how to factorize it.
- Solve a quadratic expression using trial-and-error methods and also by using the formula.
- Understand why some quadratic expressions cannot be factorized and why some have a single repeated factor.
- Recognize a quadratic function and sketch its graph.
- Understand the relationship between the algebraic and graphical solution of a quadratic equation.
- Recognize and solve simultaneous quadratic equations.
- Recognize and understand the role of quadratic functions in supply and demand models and as total cost or total revenue functions.

4.1 Introduction

In the previous chapter we examined linear equations and functions. We saw that a linear *equation* has the general form $ax + b = c$, where x is the variable or unknown and a, b, and c are constants or parameters. This equation is said to be linear because x is raised only to the power 1, and to no higher or lower power. A linear *function* has the general form $y = ax + b$ and specifies a relationship between the two variables or unknowns, x and y. This function is linear, in this case because x and y are both raised only to the power 1, and to no higher or lower power. We also saw that the graph of the function $y = ax + b$ is a straight line, which gives a concrete meaning to the term 'linear'.

As we saw in the previous chapter, the key feature of a linear relationship between two variables x and y is that the slope of the graph is constant, meaning that a 1-unit increase in x always increases or decreases y by the same amount. While some relationships in economics may be of this simple linear type, obviously we need to develop the ability to handle non-linear relationships too, where a given change in x does not always have the same effect on y. In this chapter we will examine quadratic equations and quadratic functions, which are the simplest type of non-linear relationship. We will show that a quadratic function, when graphed, produces a characteristically U-shaped curve. We will show how to solve quadratic equations, including simultaneous quadratic equations. At the end of the chapter we look briefly at two applications of quadratic functions in economics.

In chapter 5 we will briefly examine some other non-linear functional relationships. Fortunately, it turns out that knowledge of only a handful of function types is sufficient to carry us through to quite an advanced level of economic analysis. So, although you may find this chapter and the next somewhat challenging, you can take comfort from the fact that what you will learn here will, if understood and digested, carry you a long way in your study of economics.

4.2 Quadratic expressions

Expanding (multiplying out) pairs of brackets

Consider an expression such as

$$(a + b)(c + d)$$

where a, b, c, and d are unspecified constants.

This is a product: that is, it is $(c + d)$ *multiplied* by $(a + b)$. We can think of this product as measuring the area of a rectangle with width of $(a + b)$ and height of $(c + d)$, as in figure 4.1.

But we see from figure 4.2 that the total area can be broken down into two sub-rectangles. The rectangle on the left has width a and height $c + d$, so its area is $a(c + d)$. The rectangle on the right has width b and height $c + d$, so its area is $b(c + d)$. Therefore the total area is $a(c + d) + b(c + d)$.

Further, from figure 4.3 we can see that the total area can also be broken down into four sub-rectangles. The top left-hand sub-rectangle has a width of a and a height of c, so its area is ac. Similarly, the other three sub-rectangles have areas of ad, bc, and bd.

Therefore the area of the rectangle is

$(a + b)(c + d)$	(from figure 4.1)
$a(c + d) + b(c + d)$	(from figure 4.2)
$ac + ad + bc + bd$	(from figure 4.3)

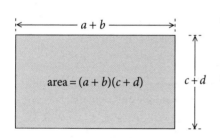

Figure 4.1

Therefore we have:

$$(a + b)(c + d) = a(c + d) + b(c + d) = ac + ad + bc + bd$$

This tells us how to 'expand' (multiply out) an expression such as $(a + b)(c + d)$. In step 1 we multiply $(c + d)$ first by a, and then by b, and add the results. This gives us figure 4.2: that is,

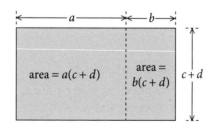

Figure 4.2

$$(a + b)(c + d) = a(c + d) + b(c + d)$$

In step 2 we simply multiply out the brackets on the right-hand side, in the way we explained in section 2.5. This gives us figure 4.3: that is,

$$a(c + d) + b(c + d) = ac + ad + bc + bd$$

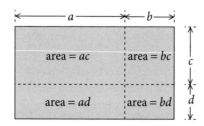

Figure 4.3

In words, each term in the second bracket (c and d) is multiplied by each term in the first bracket (a and b), and the results are added.

So the rule is:

RULE 4.1 Multiplying out brackets

$$(a+b)(c+d) \equiv a(c+d) + b(c+d)$$
$$\equiv ac + ad + bc + bd$$

Note that rule 4.1 holds for *any* values of a, b, c, and d. In other words, the relationships in rule 4.1 are not equations, but *identities*. That is why we have used the '$\equiv$' rather than the '$=$' sign. (Refer back to section 3.1 if you are uncertain about this distinction.)

Hint When applying this rule, be careful to obey the sign rules (see rule 2.1 in chapter 2). For example:

$$(a-b)(c+d) = a(c+d) - b(c+d)$$
$$= ac + ad - bc - bd$$

Note the change of sign when multiplying out $-b(c+d)$.

A special case of rule 4.1 arises when $a = c$ and $b = d$. Then we have $(a+b)(a+b) \equiv (a+b)^2$. This is shown in rule 4.2:

RULE 4.2 Multiplying out brackets: a special case

$$(a+b)(a+b) \equiv (a+b)^2$$
$$\equiv a(a+b) + b(a+b)$$
$$\equiv a^2 + ab + ba + b^2$$
$$\equiv a^2 + 2ab + b^2$$

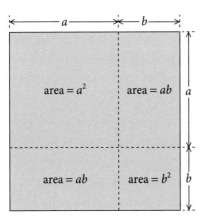

Figure 4.4

As we can see in figure 4.4, rule 4.2 gives us the area of a square with sides of length $a + b$.

In our discussion up to this point, we have left open the question whether a and b are variables or constants, as this question has no effect on the reasoning that leads to rules 4.1 and 4.2. However, we shall frequently meet expressions of the form

$$(x+a)(x+b)$$

where x is a variable or unknown, and a and b are constants or parameters. (Refer back to section 3.3 if you are unsure of the meaning of any of these words.)

Applying rule 4.1 to this expression, we get rule 4.3:

RULE 4.3 Multiplying out brackets: another special case

$$(x+a)(x+b) \equiv x(x+b) + a(x+b)$$
$$\equiv x^2 + xb + ax + ab$$
$$\equiv x^2 + (a+b)x + ab$$

It is worth taking careful note of the pattern in rule 4.3, as we shall meet it frequently. This rule tells us that

$$(x+a)(x+b) \equiv x^2 + (a+b)x + ab$$

The right-hand side, $x^2 + (a+b)x + ab$, is an example of what is called a **quadratic expression**.

A quadratic expression by definition contains an unknown, x, raised to the power 2 (that is, x^2). It *may* also contain a term in which x is raise to the power 1 (that is, x), but it must not contain any term in which x is raised to any higher or lower powers. It *may* also contain a constant, which is ab in the form above. Note especially the use of the '$\equiv$' sign, which reminds us that the above equation is an *identity*: that is, it is true irrespective of the values of x, a, and b. (As a historical note, this came to be called a quadratic expression because it gives us the area of a rectangle with one side of length $x + a$ and the other side of length $x + b$. A rectangle has four sides; and 'quad' means four.)

Progress exercise 4.1

1. Explain why $(a + b)^2$ does not equal $a^2 + b^2$.

2. Expand (= multiply out) the following:

(a) $(x + 3)(x + 2)$ (b) $(x + 5)(x - 4)$ (c) $(2x + 1)(x + \frac{1}{4})$

(d) $(x - 4)^2$ (e) $(x + 5)(x - 5)$

4.3 Factorizing quadratic expressions

Rule 4.3 tells us that if we *expand* (multiply out) the expression

$(x + a)(x + b)$

we get

$x^2 + (a + b)x + ab$

which is called a quadratic expression.

  Factorizing a quadratic expression is simply the reverse of this process. We start with

$x^2 + (a + b)x + ab$

and work back to find the factors of this expression, $(x + a)$ and $(x + b)$.

Hint If you are feeling a little lost, remember that 4 and 3 are factors of 12 because $4 \times 3 \equiv 12$. In the same way, from rule 4.3, $(x + a)$ and $(x + b)$ are the factors of $x^2 + (a + b)x + ab$ because

$(x + a)(x + b) \equiv (x + a) \times (x + b) \equiv x^2 + (a + b)x + ab$

Technique for factorizing a quadratic expression

EXAMPLE 4.1

Suppose we are asked to factorize

$x^2 + 9x + 20$

Let us assume that the factors are $(x + A)$ and $(x + B)$, where A and B are two numbers that we haven't yet found (but are hoping to find in the near future!).

Given this, then from the definition of factors we know that

$(x + A)(x + B) = x^2 + 9x + 20$ .... (4.1)

But we also know from rule 4.3 that

$(x + A)(x + B) = x^2 + (A + B)x + AB$ .... (4.2)

(Check for yourself that equation (4.2) is correct.)

Since the left-hand sides of equations (4.1) and (4.2) are identical, their right-hand sides must be identical too. Therefore

$$x^2 + 9x + 20 = x^2 + (A + B)x + AB$$

But this can only be true if $A + B = 9$ and $AB = 20$.

So we are looking for two numbers, A and B, whose sum is 9 and whose product is 20. After a little thought, it occurs to us that this will be true if $A = 5$ and $B = 4$, for then we have $A + B = 5 + 4 = 9$, and $AB = 5 \times 4 = 20$.

Therefore the factors are $(x + 5)$ and $(x + 4)$. We can check that this is correct by expanding

$(x + 5)(x + 4)$ which by rule 4.3 gives

$$(x + 5)(x + 4) = x^2 + 9x + 20$$

which is what we were asked to factorize.

EXAMPLE 4.2

$$x^2 - 5x + 6$$

Following the same procedure as in example 4.1, we get

$$(x + A)(x + B) = x^2 - 5x + 6 \tag{4.3}$$

and

$$(x + A)(x + B) = x^2 + (A + B)x + AB \tag{4.4}$$

Comparing the right-hand sides of the two equations, we see that we must have $A + B = -5$ and $AB = 6$. This example is a little trickier than example 4.1 because of the minus sign. But we can deduce that A and B must have the same sign (that is, both positive or both negative) because their product, AB, is positive.

We can list some possible values of A and B, given that they must have the same sign and that their product, AB, must equal 6:

AB	A	B	$A + B$
6	6	1	7
6	3	2	5
6	−6	−1	−7
6	−3	−2	−5

Clearly it is only the last of these that gives the correct values for both AB and $A + B$. So we conclude that $A = -3$ and $B = -2$. Therefore the factors are $(x - 3)$ and $(x - 2)$. We check this using rule 4.3:

$$(x - 3)(x - 2) = x^2 - 2x - 3x + 6$$
$$= x^2 - 5x + 6$$

which is what we were asked to factorize.

EXAMPLE 4.3

$$3x^2 - 6x - 72$$

In this case we first take the 3 outside brackets, giving us

$$3x^2 - 6x - 72 = 3(x^2 - 2x - 24)$$

On the right-hand side, we ignore the 3 for the moment and look for the factors of $x^2 - 2x - 24$. Following the same procedure as in example 4.1, we get

$$(x + A)(x + B) = x^2 - 2x - 24 \tag{4.5}$$

and

$$(x + A)(x + B) = x^2 + (A + B)x + AB \tag{4.6}$$

Comparing the right-hand sides of the two equations, we see that we must have $A + B = -2$ and $AB = -24$ This case is a little complicated because both $A + B$ and AB are negative. But we can deduce that A and B must have opposite signs (that is, one positive and the other negative) because their product, AB, is negative. So let us assume that A is negative and B positive. Also, if A is negative and B positive, then A must be greater in absolute magnitude than B because their sum, $A + B$, is negative.

Using these inferences, we can list some possible values of A and B, given that their product must be -24:

AB	A	B	$A + B$
-24	-24	1	-23
-24	-12	2	-10
-24	-8	3	-5
-24	-6	4	-2

We see that it is only the last of these that gives the right values for both AB and $A + B$. So we conclude that the factors of $x^2 - 2x - 24$ are $(x - 6)$ and $(x + 4)$. That is,

$$x^2 - 2x - 24 = (x - 6)(x + 4)$$

Multiplying both sides by 3, we get

$$3x^2 - 6x - 72 = 3(x - 6)(x + 4)$$

Thus the factors of $3x^2 - 6x - 72$ are 3, $(x - 6)$, and $(x + 4)$.

 Hint Not every quadratic expression can be factorized. Later we shall see exactly what conditions must be fulfilled in order that a quadratic expression may be factorized.

Clearly, the method of factorizing quadratic expressions used above is rather unsatisfactory because it relies on essentially 'trial and error' methods. Fortunately, there is a much more powerful method of factorizing quadratic expressions, which we examine shortly.

Meanwhile, you may be wondering *why* we should want to factorize quadratic expressions. This will become clear in the next section.

4.4 Quadratic equations

EXAMPLE 4.4

In the previous section we considered quadratic *expressions* such as $x^2 - 2x - 24$. These contained no '=' sign, and therefore were not equations. Suppose now we want to solve the *equation*

$$x^2 + 9x + 20 = 0 \tag{4.7}$$

This is an example of a *quadratic equation*. Equation (4.7) is quadratic because the left-hand side is a quadratic expression, as defined at the end of section 4.2 above. It is quadratic because x is raised only to the power 2 and the power 1 (because $9x \equiv 9x^1$), and to no other powers.

To solve it, we factorize the left-hand side. Using the 'trial and error' methods of the previous section, we find that the factors are $(x + 5)$ and $(x + 4)$. So we know that

$$x^2 + 9x + 20 = (x + 5)(x + 4)$$

Moreover, as noted earlier, this equation is an *identity*: that is, it is true for all values of x. This fact permits us to replace $x^2 + 9x + 20$ in equation (4.7) with something *identical*, namely the product $(x + 5)(x + 4)$. This gives

$$(x+5)(x+4) = 0 \tag{4.8}$$

The key point is that since equation (4.8) is *identical* to equation (4.7), it follows that any value of x that satisfies (4.8) will automatically satisfy (4.7). In other words, any value of x that is a solution to (4.8) is also a solution to (4.7). So if we can solve (4.8), we have also solved (4.7).

So now we want to solve equation (4.8). Why is (4.8) any easier to solve than (4.7)? The answer is: because the left-hand side of (4.8) is a *product*. We have something of the form

$$M \times N = 0$$

where $M = (x+5)$ and $N = (x+4)$.

Now, the product $M \times N$ equals zero whenever *either* $M = 0$ *or* $N = 0$ (because 'zero times anything is zero'—see section 1.3 if you are uncertain on this point).

Thus the product $(x+5)(x+4)$ equals zero whenever *either* $(x+5) = 0$ *or* $(x+4) = 0$. So

$$(x+5)(x+4) = 0 \tag{4.8 repeated}$$

is satisfied when

either $(x+5) = 0$ (which is true when $x = -5$)

or $(x+4) = 0$ (which is true when $x = -4$)

Therefore we conclude that $x = -5$ and $x = -4$ are the solutions (also known as **roots**) to equation (4.7).

We can check that these solutions are correct by substituting each of them in turn into equation (4.7). When $x = -5$, equation (4.7) becomes the identity

$$(-5)^2 + 9(-5) + 20 = 25 - 45 + 20 = 0$$

and when $x = -4$, equation (4.7) again becomes the identity

$$(-4)^2 + 9(-4) + 20 = 16 - 36 + 20 = 0$$

To summarize this example, to solve the quadratic equation $x^2 + 9x + 20 = 0$, we first factorized the left-hand side and obtained $(x+5)(x+4) = 0$. This equation is satisfied when $x = -5$ or $x = -4$. These values are therefore the solutions (or roots, as they are usually called) to the quadratic equation $x^2 + 9x + 20 = 0$.

Let's look at another example of solving a quadratic equation.

EXAMPLE 4.5

$$3x^2 - 6x - 72 = 0 \tag{4.9}$$

To solve equation (4.9), we factorize the left-hand side. From example 4.3, we know that the factors of $3x^2 - 6x - 72 = 0$ are 3, $(x-6)$ and $(x+4)$. So we know that

$$3x^2 - 6x - 72 \equiv 3(x-6)(x+4)$$

and moreover this equation is an *identity*—it is true for all values of x. Therefore, whenever $3(x-6)(x+4)$ equals zero, $3x^2 - 6x - 72$ also equals zero because the two are identically equal.

Now, $3(x-6)(x+4) = 0$ when *either* $x = 6$ *or* $x = -4$. (Check this for yourself.) Therefore we conclude that $x = 6$ and $x = -4$ are solutions to equation (4.9).

We can check that these solutions are correct by substituting each of them in turn into equation (4.9). When $x = 6$, equation (4.9) becomes the identity

$$3(6)2 - 6(6) - 72 = 108 - 36 - 72 = 0$$

When $x = -4$, equation (4.9) again becomes the identity:

$$3(-4)^2 - 6(-4) - 72 = 48 + 24 - 72 = 0$$

Generalization

To solve any quadratic equation of the form

$$ax^2 + bx + c = 0 \quad \text{(where } a, b, \text{ and } c \text{ are given constants)}$$

we first take the a outside brackets, giving us

$$a\left(x^2 + \frac{b}{a}x + \frac{c}{a}\right) = 0$$

Next, we look for the factors of $x^2 + \frac{b}{a}x + \frac{c}{a}$. Let us assume that we find the factors to be

$$(x + A) \text{ and } (x + B) \quad \text{(where } A \text{ and } B \text{ are two numbers)}$$

If these are the factors, then by definition we have

$$(x + A)(x + B) \equiv x^2 + \frac{b}{a}x + \frac{c}{a}$$

After multiplying both sides by a, this becomes

$$a(x + A)(x + B) \equiv ax^2 + bx + c$$

As this is an identity, when the left-hand side equals zero, the right-hand side must equal zero too. And the left-hand side equals zero when $x = -A$ or $x = -B$. These values are therefore the solutions to the quadratic equation $ax^2 + bx + c = 0$.

4.5 The formula for solving any quadratic equation

In the previous two sections we factorized quadratic expressions by 'trial and error' methods. This is not very satisfactory as it relies too much on guesswork and inspiration. Fortunately, a formula exists (which we will not prove here) for solving *any* quadratic equation. This takes us straight to the solutions (also known as roots), and also tells us whether a solution exists. The formula is as follows:

> **RULE 4.4 Solving a quadratic equation**
>
> Given any quadratic equation
>
> $$ax^2 + bx + c = 0 \quad \text{(where } a, b, \text{ and } c \text{ are given constants)}$$
>
> the solutions (roots) are given by the formula
>
> $$x = \frac{-b \pm \sqrt{b^2 - 4ac}}{2a}$$

Notice in the formula the symbol '$\pm$', meaning 'plus *or* minus'. Thus the formula generates the necessary two solutions, as we take first the 'plus' option, then the 'minus'. This will become clearer in the examples below.

This formula looks rather daunting at first sight, but if you say it aloud a few times you will soon find you have memorized it. And with a little practice you should have no difficulty in applying it correctly. Here are two examples of rule 4.4 in action:

EXAMPLE 4.6

$$x^2 + 5x + 6 = 0$$

Comparing this with the standard form

$$ax^2 + bx + c = 0$$

we see that $a = 1$, $b = 5$, and $c = 6$. Substituting these values into the formula (see rule 4.4), we get

$$x = \frac{-b \pm \sqrt{b^2 - 4ac}}{2a} = \frac{-5 \pm \sqrt{5^2 - 4(1)(6)}}{2(1)} = \frac{-5 \pm \sqrt{25 - 24}}{2}$$

$$= -\frac{5 \pm \sqrt{1}}{2} = either \frac{-5 + 1}{2} \text{ or } \frac{-5 - 1}{2} \quad \text{(note that the square root of 1 is 1)}$$

$$= either -2 \text{ or } -3$$

These solutions (roots) can be checked by substituting each in turn into the given equation. Be sure to check this for yourself.

Notice also that these solutions imply that the factors of $x^2 + 5x + 6$ are $(x + 2)$ and $(x + 3)$. We can confirm this by expanding $(x + 2)(x + 3)$, which gives

$$(x + 2)(x + 3) \equiv x^2 + 3x + 2x + 6 \equiv x^2 + 5x + 6 \quad \text{(as we expected).}$$

Note we have used the '$\equiv$' sign here to indicate an identity (see section 4.2 and especially rule 4.1 above).

EXAMPLE 4.7

$$x^2 + 4x - 12 = 0$$

Comparing this with the standard form

$$ax^2 + bx + c = 0$$

we see that $a = 1$, $b = 4$, and $c = -12$. Substituting these values into the formula in rule 4.4, we get

$$x = \frac{-b \pm \sqrt{b^2 - 4ac}}{2a} = \frac{-4 \pm \sqrt{4^2 - 4(1)(-12)}}{2(1)} = \frac{-4 \pm \sqrt{16 + 48}}{2}$$

$$= \frac{-4 \pm \sqrt{64}}{2} = either \frac{-4 + 8}{2} \text{ or } \frac{-4 - 8}{2} \quad \text{(note that } \sqrt{64} = 8\text{)}$$

$$= either \ 2 \text{ or } -6$$

These solutions (roots) can be checked by substituting each in turn into the given equation. Be sure to check this for yourself.

The solutions imply that the factors of $x^2 + 4x - 12$ are $x - 2$ and $x + 6$. We can confirm this by expanding $(x - 2)(x + 6)$, which gives

$$(x - 2)(x + 6) \equiv x^2 + 6x - 2x - 12 \equiv x^2 + 4x - 12$$

Note once more that the '$\equiv$' sign indicates an identity (true for all values of x).

4.6 Cases where a quadratic expression cannot be factorized

In section 4.3 above we said that not every quadratic expression could be factorized. Now we are ready to see why. Consider the following example:

EXAMPLE 4.8

$$x^2 + 2x + 2 = 0$$

Let us see what happens when we try to solve this equation using the formula. In this case we have $a = 1$, $b = 2$, and $c = 2$. Substituting these values into the formula (see rule 4.4) gives

$$x = \frac{-b \pm \sqrt{b^2 - 4ac}}{2a} = \frac{-2 \pm \sqrt{2^2 - 4(1)(2)}}{2(1)}$$

$$= \frac{-2 \pm \sqrt{4 - 8}}{2} = \frac{-2 \pm \sqrt{-4}}{2}$$

At this point we grind to a halt. This is because, in the numerator, we have $\sqrt{-4}$, so to go any further requires us to find the square root of *minus* 4, and we know from section 1.14 that a negative number has no square root. Thus we conclude that the equation $x^2 + 2x + 2 = 0$ has no solutions (roots).

Actually the conclusion of example 4.8 is not quite correct, as it stands. Mathematicians, who have very tidy minds, found it very unsatisfactory that some quadratic equations had solutions and some did not. They therefore defined a new type of number, called an imaginary number, which has the property that its square is negative. (We mentioned this in section 1.14.) We won't go into the details here, but this definition makes it possible to conclude that the equation above *does*, in fact, have roots, but these roots are not *real* numbers. Thus the formally correct conclusion is that the equation $x^2 + 2x + 2 = 0$ has no *real* roots. We examine this question more closely in chapter W21, which is available at the Online Resource Centre.

Generalization

The case above arose because, in the formula (rule 4.4), the expression under the square root sign was negative. When we look again at the formula

$$x = \frac{-b \pm \sqrt{b^2 - 4ac}}{2a}$$

it is clear that this problem will arise whenever b^2 is less than $4ac$, for then $b^2 - 4ac$, the expression under the square root sign, is negative. For example, when $a = 1$, $b = 4$, $c = 5$, then $b^2 = 16$ and $4ac = 20$, so $b^2 - 4ac = 16 - 20 = -4$.

So we conclude that the quadratic equation $ax^2 + bx + c = 0$ has no real solutions (roots) if the values of the parameters a, b, and c are such as to make b^2 less than $4ac$.

4.7 The case of the perfect square

EXAMPLE 4.9

$$x^2 + 6x + 9 = 0$$

Let's see what happens when we try to solve this, using the formula. In this case we have $a = 1$, $b = 6$, and $c = 9$. Substituting these values into the formula gives

$$x = \frac{-b \pm \sqrt{b^2 - 4ac}}{2a} = \frac{-6 \pm \sqrt{6^2 - 4(1)(9)}}{2(1)}$$

$$= \frac{-6 \pm \sqrt{36 - 36}}{2} = \frac{-6}{2} = -3$$

Thus the equation has only one solution (root), -3. This is because the expression under the square root sign happens, in this case, to equal zero (and the square root of zero is zero). It therefore disappears ('drops out') from the solution, leaving only one solution.

You might think this means that $x^2 + 6x + 9$ has only one factor, $(x + 3)$. But this can't be right, since we must multiply $(x + 3)$ by something in order to produce $x^2 + 6x + 9$. That *something* is $(x + 3)$. In other words, we have

$$(x + 3)(x + 3) = x^2 + 6x + 9$$

The expression $x^2 + 6x + 9$ is called a **perfect square** because it equals the area of a square with sides of length $(x + 3)$. The equation $x^2 + 6x + 9x = 0$ has two equal solutions, also described as a single, repeated solution.

Generalization

The case above arose because, in the formula, the expression under the square root sign was zero. When we look again at the formula

$$x = \frac{-b \pm \sqrt{b^2 - 4ac}}{2a}$$

it is clear that this problem will arise whenever the parameters a, b, and c of the equation $ax^2 + bx + c = 0$ are such as to make b^2 equal to $4ac$. The formula then collapses down to

$$x = \frac{-b}{2a}$$

For example, when $a = 2$, $b = 4$, $c = 2$, then $b^2 = 4ac = 16$, so $b^2 - 4ac = 16 - 16 = 0$.

Summary of sections 4.1–4.7

When $(x + A)(x + B)$ is expanded (multiplied out) it becomes the quadratic expression $x^2 + (A + B)x + AB$ (see rule 4.3). Then $(x + A)$ and $(x + B)$ are the factors of $x^2 + (A + B)x + AB$.

We factorize $x^2 + (A + B)x + AB$ by finding two numbers, A and B, such that $A + B$ is the coefficient of x and AB is the constant term. The solutions (roots) to the quadratic equation $x^2 + (A + B)x + AB = 0$ are therefore $x = -A$ and $x = -B$.

However, the formula in rule 4.4 gives us an easier method of solving quadratic equations of the form $ax^2 + bx + c = 0$. There are three classes of solution to this equation, depending on the values of the parameters, a, b, and c:

(1) If $b^2 > 4ac$, there are two distinct roots.

(2) If $b^2 = 4ac$, there is only one root, described as a repeated root.

(3) $b^2 < 4ac$, there are no roots (more precisely, there are no roots that are real numbers, but roots may be found by defining a new type of number, an imaginary number).

Progress exercise 4.2

1. By trial and error, solve the following quadratic equations (where possible). Then check your answers by solving them using the formula.

(a) $x^2 - 3x + 2 = 0$ (b) $x^2 + 5x + 4 = 0$ (c) $x^2 + 4x - 5 = 0$

(d) $x^2 + \frac{3}{4}x + \frac{1}{8} = 0$ (e) $x^2 - 16 = 0$

(f) $2x^2 + 8x + 6 = 0$ Hint: if $ax^2 + bx + c = 0$, then $x^2 + \frac{b}{a}x + \frac{c}{a} = 0$

(g) $-x^2 - 0.5x + 3 = 0$ (h) $x^2 + 8x + 16 = 0$ (i) $x^2 - x - 2 = 0$ (j) $x^2 + 3x + 3 = 0$

What makes the answers to (h) and (j) different from the others?

4.8 Quadratic functions

If we write, for example

$$y = x^2 + 5x + 6$$

we have a *function*: that is, a relationship between an independent variable, x, and a dependent variable, y. (See section 3.5 if you are unsure what a function is.) Because x is raised to the power 2 (and to no higher power), this is called a quadratic function. The general form of the quadratic function is

$$y = ax^2 + bx + c \quad \text{(where } a, b, \text{ and } c \text{ are parameters)}$$

The example above, $y = x^2 + 5x + 6$, is obtained from the general form by setting $a = 1$, $b = 5$, and $c = 6$.

Let's consider what the graph of a quadratic function will look like, by looking at some examples.

EXAMPLE 4.10

Starting from the general form, $y = ax^2 + bx + c$, let's take the case where $a = 1$, $b = 0$ and $c = 0$. This gives us the simplest case of a quadratic function:

$$y = x^2$$

If we draw up a table of values (table 4.1) for values of x between -4 and $+4$, and plot the resulting values as a graph, the result is seen in figure 4.5. Notice that y is positive when x is either positive or negative. When x is zero, y is also zero.

This curve is called a **parabola**. The U-shape arises from the fact that x^2 (and therefore y) is positive when x is either positive or negative. For example, $y = 4$ when $x = +2$ and also when $x = -2$. In general, y has the same value when $x = x_0$ as it does when $x = -x_0$. Therefore the curve has two arms or branches, lying symmetrically about the y-axis. If you fold the graph paper along the y-axis, the two branches of the graph meet.

Table 4.1 Values for graph of $y = x^2$.

x	-4	-3	-2	-1	0	1	2	3	4
$y = x^2$	$(-4)^2$	$(-3)^2$	$(-2)^2$	$(-1)^2$	$(0)^2$	$(1)^2$	$(2)^2$	$(3)^2$	$(4)^2$
	$= 16$	$= 9$	$= 4$	$= 1$	$= 0$	$= 1$	$= 4$	$= 9$	$= 16$

EXAMPLE 4.11

If we repeat example 4.10, but this time with $a = 3$ instead of $a = 1$, we obtain

$$y = 3x^2$$

Table 4.2 gives values of y for x between -4 and $+4$, and the resulting curve is also plotted in figure 4.5. From the table, and also in the graph, we can see that, for any given value of x, y is now three times as large as in example 4.10. For example, when $x = +2$ or -2, the curve $y = x^2$ gives $y = 4$, while the curve $y = 3x^2$ gives $y = 3 \times 4 = 12$.

Table 4.2 Values for graph of $y = 3x^2$.

x	-4	-3	-2	-1	0	1	2	3	4
$y = 3x^2$	$3(-4)^2$	$3(-3)^2$	$3(-2)^2$	$3(-1)^2$	$3(0)^2$	$3(1)^2$	$3(2)^2$	$3(3)^2$	$3(4)^2$
	$= 48$	$= 27$	$= 12$	$= 3$	$= 0$	$= 3$	$= 12$	$= 27$	$= 48$

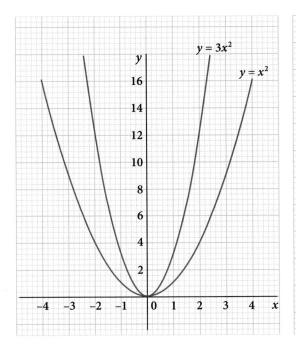

Figure 4.5 Graphs of $y = x^2$ and $y = 3x^2$.

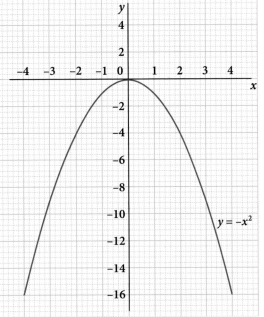

Figure 4.6 Graph of $y = -x^2$.

EXAMPLE 4.12

If we repeat example 4.10 but with $a = -1$ instead of $a = +1$, we get

$$y = -x^2$$

The effect is that, compared with example 4.10, the sign of y associated with any given value of x is reversed. For example, when $x = -3$, the function $y = x^2$ gives $y = (-3)^2 = +9$, while the function $y = -x^2$ gives $y = -[(-3)^2] = -9$. This inverts the graph (see figure 4.6). We haven't bothered to compile a new table of values, as it would be exactly the same as the table for $y = x^2$, but with every value of y negative instead of positive. At $x = 0$ the two curves coincide, of course.

EXAMPLE 4.13

This repeats example 4.10, but this time we set $c = 4$. This gives us

$$y = x^2 + 4$$

Compared with example 4.10, we have added a constant, 4, so it should be fairly obvious that this shifts the whole curve bodily upwards by 4 units (see figure 4.7). The effect of this is that the curve now cuts (intercepts) the y-axis at $y = 4$. Similarly, if we had set $c = -4$ instead of $+4$, this would have shifted the whole curve bodily downwards by 4 units. The curve would then intercept the y-axis at $y = -4$ (see also figure 4.7).

EXAMPLE 4.14

If we repeat example 4.10 but with $b = 3$, we have

$$y = x^2 + 3x$$

Let's consider where this curve will lie in comparison with $y = x^2$. When x is positive, so also is $3x$, and therefore the graph of $y = x^2 + 3x$ will lie *above* the graph of $y = x^2$. For example, when $x = 1$, $3x = 3$, and therefore when $x = 1$, the graph of $y = x^2 + 3x$ will lie 3 units above the graph of $y = x^2$ (see figure 4.8). Conversely, when x is negative, so also is $3x$,

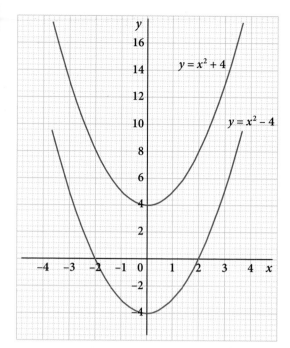

Figure 4.7 Graphs of $y = x^2 + 4$ and $y = x^2 - 4$.

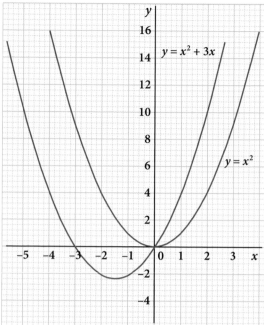

Figure 4.8 Graphs of $y = x^2$ and $y = x^2 + 3x$.

so the graph of $y = x^2 + 3x$ will lie *below* the graph of $y = x^2$. When $x = 0$, $3x$ also equals zero, so $y = 0$ and the two curves coincide.

Thus the addition of the $3x$ term has the effect of shifting the curve up when x is positive and down when x is negative. If we subtract $3x$ instead of adding it, the direction of shift is reversed: that is, the graph of $y = x^2 - 3x$ lies below the graph of $y = x^2$ when x is positive and above it when x is negative.

Summary

Reviewing the five previous examples, we can draw the following conclusions about the shape of the quadratic function $y = ax^2 + bx + c$.

(1) The x^2 term gives the graph an approximate U-shape, called a parabola. If the parameter a is positive, the graph looks like a U. If the parameter a is negative, the graph looks like an inverted U (that is, like this: $\cap$). The absolute magnitude of a determines how steeply the curve slopes up (or down).

(2) The constant term, c, determines the intercept of the curve on the y-axis.

(3) The x term shifts the parabola up and down. If the parameter b is positive, the curve shifts up when $x > 0$ and down when $x < 0$. If b is negative, this shift is reversed. The absolute magnitude of b determines the strength of the shift effect.

4.9 The inverse quadratic function

Here we consider a question which may appear to have little point now, but which is important later in this book. It concerns a function that is a close relative of the quadratic function $y = x^2$:

$$y = \sqrt{x} \quad \text{(which we can also write as } y = x^{1/2} \text{ or } y = x^{0.5}\text{)}$$

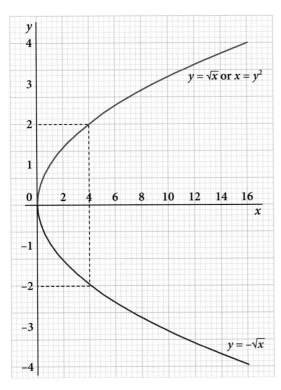

Figure 4.9 Graphs of $y = \sqrt{x}$ and $y = -\sqrt{x}$.

What does the graph of this function look like? To answer this, we first consider the function $x = y^2$. This is a quadratic function like $y = x^2$, except that it has x instead of y as the dependent variable. So its graph is the usual U-shaped curve but now with its two branches or arms oriented towards the x-axis rather than the y-axis, as in figure 4.9.

If we now take square roots on both sides of $x = y^2$, it becomes $y = \sqrt{x}$. Because the square root of any number is positive, by definition (see sections 1.14 and 2.9), y is positive. The graph of $y = \sqrt{x}$ is shown in figure 4.9. It is that part of the graph of $x = y^2$ lying above the x-axis; that is, where y is positive. This is an important function, for it gives us the relationship between any positive number, x, and its square root, y. The part of the graph of $x = y^2$ lying below the x-axis in figure 4.9 is the graph of another function, $y = -\sqrt{x}$.

Thus the function $x = y^2$ does not have a true inverse. The branch of the function $x = y^2$ where y is positive has $y = \sqrt{x}$ as its inverse; and the branch where y is negative has $y = -\sqrt{x}$ as its inverse.

You might be thinking that we could splice together the graphs of $y = \sqrt{x}$ and $y = -\sqrt{x}$ to make a combined graph, $y = \pm\sqrt{x}$. This would include both positive and negative values of y; for example, when $x = 4$, we would have $y = \pm 2$ (see dotted lines in figure 4.9). Thus perhaps $y = \pm\sqrt{x}$ could be considered to be the inverse function to $x = y^2$.

The problem with this is that $y = \pm\sqrt{x}$ is not a function, but merely a *relation*, a somewhat looser connection between two variables. This is because the definition of a function requires that there be no more than one value of the dependent variable (y in this case) associated with any given value of x. This condition is not satisfied in the case of $y = \pm\sqrt{x}$. For example, as we have just seen, when $x = 4$, y may be either $+2$ or -2. Thus there are two values of y associated with any positive value of x. So we must continue to view $y = \sqrt{x}$ and $y = -\sqrt{x}$ as distinct functions.

These ideas extend straightforwardly to other even-numbered roots, such as $y = \sqrt[4]{x}$, but we will not pursue this here. The problem of a non-existent inverse function does not arise with odd numbered roots. For example, as we shall see in the next chapter, the inverse of the function $x = y^3$ is the function $y = \sqrt[3]{x}$.

4.10 Graphical solution of quadratic equations

EXAMPLE 4.15

Let us consider the function $y = x^2 + x - 6$, which is graphed in figure 4.10.

We see that the graph cuts the x-axis at $x = +2$ and $x = -3$. Recall that, everywhere along the x-axis, $y = 0$ (see section 3.6 if you are uncertain on this point). Therefore $y = 0$ at these two points. But we know also that $y = x^2 + x - 6$ at these two points, because they lie on the curve. Thus two things are true at the two points, $x = +2$ and $x = -3$:

$$y = 0 \quad \text{and} \quad y = x^2 + x - 6$$

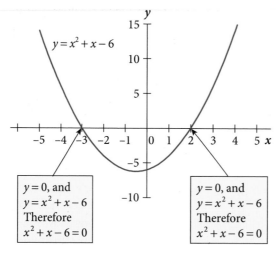

$y = x^2 + x - 6$

$y = 0$, and
$y = x^2 + x - 6$
Therefore
$x^2 + x - 6 = 0$

$y = 0$, and
$y = x^2 + x - 6$
Therefore
$x^2 + x - 6 = 0$

Figure 4.10 Graph of $y = x^2 + x - 6$.

Combining these two facts, we can conclude that $x^2 + x - 6 = 0$ at these points. Therefore $x = +2$ and -3 are the solutions to the quadratic equation $x^2 + x + 6 = 0$.

Thus the values of x at which the graph of the quadratic *function* $y = x^2 + x - 6$ cuts the x-axis give us the solutions (roots) of the quadratic *equation* $x^2 + x - 6 = 0$.

This exactly parallels the case of the linear function that we examined in section 3.8, except that we now have two solutions because the graph cuts the x-axis twice.

We can check this result algebraically. The equation $x^2 + x - 6 = 0$ factorizes easily as

$$(x - 2)(x + 3) = 0$$

which is true when either $x = +2$ or $x = -3$. Thus the algebra confirms the information we obtained from the graph.

EXAMPLE 4.16

Consider the function $y = -2x^2 + 3x + 5$, which is graphed in figure 4.11.

The graph cuts the x-axis at $x = -1$ and $x = 2\frac{1}{2}$. Therefore $y = 0$ at these two points, because they lie on the x-axis. But we know also that $y = -2x^2 + 3x + 5$ at these two points, because they lie on the curve. Combining these two facts, we can conclude that $-2x^2 + 3x + 5 = 0$ at these points.

Therefore $x = -1$ and $x = 2\frac{1}{2}$ are the solutions to the quadratic equation $-2x^2 + 3x + 5 = 0$.

Thus, as in example 4.15, the values of x at which the graph of the quadratic *function* $y = -2x^2 + 3x + 5$ cuts the x-axis give us the solutions (roots) of the quadratic *equation* $-2x^2 + 3x + 5 = 0$.

We can check this result algebraically. Using rule 4.4 with $a = -2$, $b = 3$, and $c = 5$, the solution to the equation $-2x^2 + 3x + 5 = 0$ is

$$x = \frac{-b \pm \sqrt{b^2 - 4ac}}{2a} = \frac{-3 \pm \sqrt{9 - 4(-2)(5)}}{-4} = \frac{-3 \pm \sqrt{49}}{-4} = -1 \text{ or } 2\frac{1}{2}$$

Thus the algebra and the graph corroborate one another.

EXAMPLE 4.17

The function $y = x^2 + 6x + 12$ is graphed in figure 4.12. As the graph does not cut the x-axis at any point, we can conclude immediately that the equation $x^2 + 6x + 12 = 0$ has no (real) solutions. This is confirmed when we apply rule 4.4 with $a = 1$, $b = 6$, and $c = 12$, for we then get

$$x = \frac{-b \pm \sqrt{b^2 - 4ac}}{2a} = \frac{-6 \pm \sqrt{36 - 4(1)(12)}}{2} = \frac{-6 \pm \sqrt{-12}}{2}$$

Since we have a negative number under the square root sign, this equation has no solutions that are real numbers.

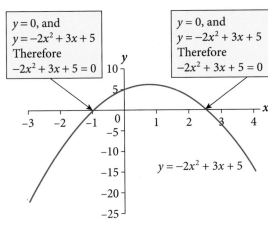

$y = 0$, and
$y = -2x^2 + 3x + 5$
Therefore
$-2x^2 + 3x + 5 = 0$

$y = 0$, and
$y = -2x^2 + 3x + 5$
Therefore
$-2x^2 + 3x + 5 = 0$

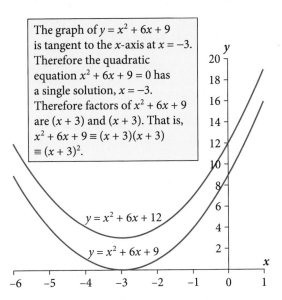

The graph of $y = x^2 + 6x + 9$
is tangent to the x-axis at $x = -3$.
Therefore the quadratic
equation $x^2 + 6x + 9 = 0$ has
a single solution, $x = -3$.
Therefore factors of $x^2 + 6x + 9$
are $(x + 3)$ and $(x + 3)$. That is,
$x^2 + 6x + 9 \equiv (x + 3)(x + 3)$
$\equiv (x + 3)^2$.

Figure 4.11 Graph of $y = -2x^2 + 3x + 5$.

Figure 4.12 Graphs of $y = x^2 + 6x + 12$ and $y = x^2 + 6x + 9$.

We can conclude from example 4.17 that, in general, when the graph of a quadratic function lies entirely above or below the x-axis, the corresponding quadratic equation has no real solutions. As mentioned in section 4.6, solutions for such equations can be found by defining an imaginary number which has the property that its square is negative. For further discussion, see chapter W21 section 4, to be found at the Online Resource Centre.

EXAMPLE 4.18

The function $y = x^2 + 6x + 9$ is also graphed in figure 4.12. We see that the graph is tangent to the x-axis at $x = -3$. This is the case of the 'perfect square' discussed in section 4.7 above. The corresponding quadratic equation is

$$x^2 + 6x + 9 = 0$$

Applying rule 4.4 with $a = 1$, $b = 6$, and $c = 9$, we get

$$x = \frac{-b \pm \sqrt{b^2 - 4ac}}{2a} = \frac{-6 \pm \sqrt{36 - 4(1)(9)}}{2} = \frac{-6}{2} = -3$$

Because $b^2 = 4ac$, the formula collapses to a single solution, -3, which we call a repeated solution.

We can also arrive at this conclusion by factorizing $x^2 + 6x + 9 = 0$. It factorizes as $(x + 3)(x + 3) = 0$, so we have a single (repeated) solution, $x = -3$.

It is easy to see that example 4.18 generalizes. Whenever the graph of a quadratic function $y = ax^2 + bx + c$ is tangent to the x-axis at a point, we know that the corresponding quadratic equation, $ax^2 + bx + c = 0$, is a perfect square.

We saw in section 4.7 that the quadratic expression $ax^2 + bx + c$ is a perfect square if $b^2 = 4ac$. The single repeated solution is then $x = -\frac{b}{2a}$. In example 4.18 we have $b^2 = 4ac = 36$.

At the Online Resource Centre we show how, using Excel®, quadratic functions can be graphed and quadratic equations solved. www.oxfordtextbooks.co.uk/orc/renshaw3e/

Progress exercise 4.3

1. Sketch the graphs of the following quadratic functions. (Hint: it *may* help if you factorize them.) Take values of x between -5 and $+5$ unless otherwise stated.

   (a) $y = x^2 + x - 6$ (b) $y = -2x^2 + 3x + 5$ (c) $y = x^2 - 2x - 3$

   (d) $y = 4x^2 + 8x + 4$ between $x = -8$ and $x = 2$ (e) $y = x^2 + 8$

2. Use your graphs from question 1 to find approximate solutions to the following quadratic equations, where possible. Then check your answers by factorization or by applying the formula.

   (a) $x^2 + x - 6 = 0$ (b) $-2x^2 + 3x + 5 = 0$ (c) $x^2 - 2x - 3 = 0$

   (d) $4x^2 + 8x + 4 = 0$ (e) $x^2 + 8 = 0$

   What makes (d) and (e) different from the others?

3. You are given the following information about a quadratic function $y = ax^2 + bx + c$. In each case, find a, b, and c and sketch the graph.

   (a) $y = 0$ when $x = 2$ or -3; $y = -12$ when $x = 0$.   (b) $a = -2$; $y = 0$ when $x = -4$ or -2.

4. A quadratic function $y = ax^2 + bx + c$ has the following properties: $a = -2$, $y = 0$ when $x = -3$ or -2. Find b and c and sketch the graph.

5. A quadratic function $y = ax^2 + bx + c$ has the following properties: $a = -3$; $y = 0$ when $x = -2$ or x_0; and $y = 6$ when $x = 0$. Find b, c, and x_0 and sketch the graph.

4.11 Simultaneous quadratic equations

EXAMPLE 4.19

Suppose we are asked to solve the simultaneous equations

$$y = 2x^2 + 3x + 2 \qquad\qquad (4.10)$$

$$y = x^2 + 2x + 8 \qquad\qquad (4.11)$$

At first sight this appears rather difficult, because quadratic equations are still somewhat unfamiliar to us, and here we have two of the beasts! However, there is no reason why we can't solve this pair of simultaneous equations using exactly the same techniques that we used to solve simultaneous linear equations in section 3.9.

As we did in section 3.9, we can make use of the fact that, if the two equations are both satisfied simultaneously by a pair of values of x and y, the value of y in equation (4.10) will be the same as the value of y in equation (4.11). Therefore, the left-hand side of equation (4.10) and the left-hand side of equation (4.11) will be equal to one another. But, if the two left-hand sides are equal, then the two right-hand sides must be equal to one another too. So we can simply *set* the right-hand sides of the two equations equal to one another, then solve the resulting equation. This will give us the value of x that satisfies both equations simultaneously. Doing this, we get

$$2x^2 + 3x + 2 = x^2 + 2x + 8$$

After subtracting the right-hand side from both sides, this becomes

$$2x^2 + 3x + 2 - [x^2 + 2x + 8] = 0 \quad \text{which simplifies to}$$

$$x^2 + x - 6 = 0$$

We can solve this quadratic equation either by using 'trial and error' methods to find factors, or by employing the formula (see rule 4.4 above). In this case, trial and error methods quickly factorize the equation as

$$x^2 + x - 6 \equiv (x+3)(x-2) = 0$$

so the solutions are $x = -3$ and $x = +2$. We can find the solution values for y by substituting these values into either equation (4.10) or equation (4.11). Using equation (4.10) we get

when $x = -3$, $\quad y = 2(-3)^2 + 3(-3) + 2 = 11$ $\quad$ and

when $x = +2$, $\quad y = 2(2)^2 + 3(2) + 2 = 16$

As a final check we should substitute $x = -3$, $y = 11$ and $x = 2$, $y = 16$ into both equations. If both equations then become identities, our solutions are correct. (Check this for yourself.)

4.12 Graphical solution of simultaneous quadratic equations

In the previous example we used algebraic methods to solve the simultaneous equations

$$y = 2x^2 + 3x + 2$$

$$y = x^2 + 2x + 8$$

and found the solutions to be $x = -3$, $y = 11$ and $x = 2$, $y = 16$.

Now let us look at the graph of the two equations (see figure 4.13). From the graph, we see that the two curves cut at $x = -3$, $y = 11$ and $x = 2$, $y = 16$. This is because any point where the two curves cut one another lies on both curves, and therefore the coordinates of that point must satisfy both equations simultaneously. This parallels exactly the reasoning of sections 3.9–3.10, where we examined linear simultaneous equations.

In general, therefore, the solutions to a pair of simultaneous quadratic equations are given by the coordinates of the points of intersection of the two curves.

Simultaneous quadratic equations with no (real) solutions

We have just seen that the solutions to a pair of simultaneous quadratic equations are found at the points where their graphs cut. If their graphs do not cut, the pair of simultaneous quadratic equations do not have a real solution. For example, consider the simultaneous quadratic equations

$$y = -2x^2 - 3x - 4$$

$$y = x^2 + 2x + 8$$

These two functions are graphed in figure 4.14, from which we see that they never intersect. So, as a pair of simultaneous quadratic equations they have no real solution. The two equations are *inconsistent*, as previously discussed for linear simultaneous equations (see section 3.11). We can confirm this algebraically by setting the two right-hand sides equal to one another. This gives

$$-2x^2 - 3x - 4 = x^2 + 2x + 8 \quad \text{which rearranges as}$$

$$3x^2 + 5x + 12 = 0$$

Using rule 4.4, we get

$$x = \frac{-b \pm \sqrt{b^2 - 4ac}}{2a} = \frac{-5 \pm \sqrt{25 - 4(3)(12)}}{6} = \frac{-5 \pm \sqrt{-119}}{6}$$

The negative number under the square root sign tells us that there are no real solutions.

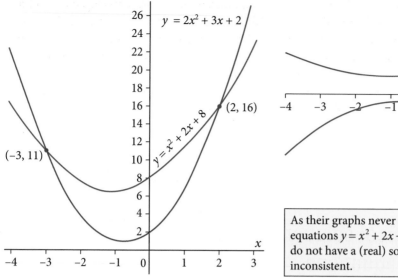

Figure 4.13 Solution to $y = 2x^2 + 3x + 2$; $y = x^2 + 2x + 8$.

As their graphs never intersect, the simultaneous equations $y = x^2 + 2x + 8$ and $y = -2x^2 - 3x - 4$ do not have a (real) solution. The equations are inconsistent.

Figure 4.14 Graphs of $y = -2x^2 - 3x - 4$ and $y = x^2 + 2x + 8$.

 The Online Resource Centre shows how simultaneous quadratic equations can be graphed and solved graphically using Excel®. www.oxfordtextbooks.co.uk/orc/renshaw3e/

 ### Summary of sections 4.8–4.12

The graph of the quadratic function $y = ax^2 + bx + c$ has a characteristic approximate U-shape (if $a > 0$) or inverted U-shape (if $a < 0$). The solutions to the quadratic equation $ax^2 + bx + c = 0$ and found where the graph of $y = ax^2 + bx + c$ cuts the x-axis, for at these points $y = 0$. There are three possibilities:

(1) If the graph of the function $y = ax^2 + bx + c$ cuts the x-axis at $x = m$ and $x = n$, then these are the two solutions (roots) of the corresponding quadratic equation $ax^2 + bx + c = 0$. Therefore $ax^2 + bx + c = 0 \equiv (x - m)(x - n)$.

(2) If the graph of the function $y = ax^2 + bx + c$ is tangent to the x-axis at $x = m$, then this is the single (repeated) root solution of the corresponding quadratic equation $ax^2 + bx + c = 0$. Therefore $ax^2 + bx + c = 0 \equiv (x - m)(x - m)$. The quadratic expression $ax^2 + bx + c$ is then said to be a perfect square.

(3) If the graph of the function $y = ax^2 + bx + c$ neither cuts nor is tangent to the x-axis, then the corresponding quadratic equation $ax^2 + bx + c = 0$ has no (real) roots.

As with the linear functions in chapter 3, the graphical solution to a pair of simultaneous quadratic equations is found where the graphs of the two corresponding functions intersect. There may be 2, 1 or 0 (real) solutions.

4.13 Economic application 1: supply and demand

At the beginning of this chapter we said that we needed to develop some non-linear functions in order to be able to analyse non-linear relationships in economics. We are not yet fully equipped for this analysis, as we first need to look at some other non-linear functions in the next chapter. However, to conclude this chapter on quadratic equations and functions we will briefly indicate some economic relationships that might be quadratic in their form.

We developed the idea of supply and demand functions in the previous chapter, including finding the equilibrium price and quantity and carrying out comparative static exercises, such as finding the effects of a tax on the market equilibrium. There we were restricted to linear supply and demand functions, but this was obviously very limiting. We can now extend the analysis to cases where one or both of the supply and demand functions is quadratic in its form. We will work with the inverse demand and supply functions, as this is common usage among economists (see section 3.15 if you need to revise this point).

EXAMPLE 4.20

A possible inverse demand function which is quadratic in form is

$$p^D = 1.5q^2 - 15q + 35$$

where p^D is the demand price (see section 3.15). Let us also introduce an inverse supply function, which for simplicity we'll assume is linear:

$$p^S = 2q + 7$$

where p^S denotes the supply price. Note that, unlike in section 3.15, we will not distinguish between quantity demanded (q^D) and quantity supplied (q^S). Instead, we will simply take it as given that when we are looking at the demand function, q denotes q^D; and when at the supply function, that q denotes q^S. We assume that in equilibrium $q^D = q^S$. The motive for this slight change in the model, compared to section 3.15, is simply to avoid the superscripts clashing with the powers, such as q^2 in the demand function above.

The inverse demand and supply functions are graphed in figure 4.15. We see that the quadratic demand function has a fairly gentle curvature, becoming flatter as the price falls. This shape is plausible, since it implies a surge in demand as the price becomes very low and the product appears a great bargain. However, we can see straight away that this form of quadratic function is not ideally suited to modelling demand. Quite reasonably, demand seems to reach its maximum at around $q = 3.75$ when p falls to zero, indicating that this is the maximum quantity consumers wish to consume even when the product is free. However, the demand function then mysteriously pops up above the q-axis again at around $q = 6.25$ and from then on is positively sloped, which is very hard to reconcile with any theory of consumer behaviour. To avoid this embarrassment we must restrict q to values less than, say, $q = 4$.

We can find the equilibrium price and quantity in this model in the same way as in section 3.15. Our model consists of the demand and supply functions above, together with the equilibrium condition, $p^D = p^S$. This simply says that the demand price (the price buyers are willing to pay for any given quantity) must equal the supply price (the price sellers are willing to accept for any given quantity). Given $p^D = p^S$, we can set the right-hand sides of the demand and supply functions equal to one another, giving

$$1.5q^2 - 15q + 35 = 2q + 7$$

This rearranges to

$$1.5q^2 - 17q + 28 = 0$$

Solving this quadratic using the formula (rule 4.4) gives $q = 2$ or $9\frac{1}{3}$. From figure 4.15 we can see that $q = 2$ is the appropriate solution. We ignore the solution $q = 9\frac{1}{3}$ which occurs where the upward-sloping branch of the quadratic demand function cuts the supply function for a second time (not shown in figure 4.15). The solution for p is found in the usual way by substituting $q = 2$ into either the demand or supply function, giving $p = 11$. (Check this for yourself.)

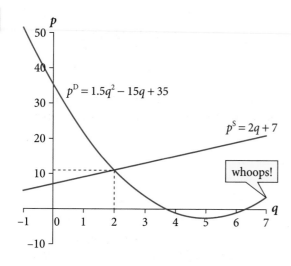

Figure 4.15 A quadratic inverse demand function with a linear inverse supply function.

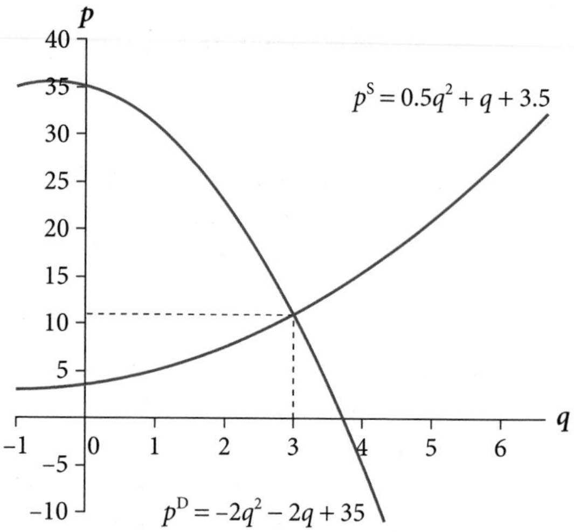

Figure 4.16 Quadratic inverse supply and demand functions.

EXAMPLE 4.21

As a second example we suppose that both the demand and supply functions are quadratic in form. The (inverse) functions are

$$p^D = -2q^2 - 2q + 35 \quad \text{and}$$
$$p^S = 0.5q^2 + q + 3.5$$

and these are graphed in figure 4.16. Note that in this example the coefficient of q^2 in the demand function is negative. This means that the standard U-shape of the quadratic function is inverted, with its turning point a maximum rather than a minimum value as it was in example 4.20. From the graph we can see that this maximum occurs at around $q = -0.5$, which means that when q is positive, which is the part of the curve that interests us, the curve is downward sloping, as a demand function should be.

A feature of the quadratic supply function in this example is that it becomes steeper as q increases. Thus when q is small, the supply curve is quite flat, so a relatively small increase in p induces a relatively large increase in quantity supplied. When q is large, the same size of price increase induces a much smaller increase in quantity supplied. This could be because suppliers are reaching the limits of their productive capacity.

We can find the market equilibrium in the same way as in example 4.20. Given the equilibrium condition $p^D = p^S$, we can set the right-hand sides of the demand and supply functions equal to one another, giving

$$-2q^2 - 2q + 35 = 0.5q^2 + q + 3.5$$

This rearranges to

$$-2.5q^2 - 3q + 31.5 = 0$$

Solving this slightly messy quadratic using the formula (rule 4.4) gives $q = 3$ or -4.2. There are two solutions because the two curves cut twice, the intersection at $q = -4.2$ being off the graph to the left in figure 4.16. We discard this negative solution as having no economic meaning. So we are left with $q = 3$, from which p can be found from either the supply or demand function as $p = 11$.

4.14 Economic application 2: costs and revenue

A firm's total cost function

Another application of the quadratic function is to the relationship between a firm's output and its total costs. At the most general level we can assume that a firm's total costs, TC, are related to its output, q, by some function that we can denote by

$TC = f(q)$ where f( ) denotes some unspecified functional form

What are the most likely shapes for this function? Obviously this is a big subject and here we will attempt only the briefest sketch. An important distinction is between the short run, when the firm is assumed to have entered into certain cost commitments such as leasing factory space and equipment, and the long run when all such commitments expire and therefore all costs are variable. Considering the short run, a possible shape might be linear: that is, of the form $TC = aq + b$. An example with $a = 1.5$ and $b = 100$ is $TC = 1.5q + 100$ (see figure 4.17). We assume b, the intercept term, is positive because of our assumption of cost commitments, or **fixed costs**. The intercept term measures these fixed costs, since they are by definition costs that have to be met even if output is zero and remain constant whatever the level of output. We assume a is positive (or possibly zero, in some special cases) because if a were negative this would mean that a larger output costs less to produce than a smaller output, which is very hard to imagine.

The problem with the linear short-run total cost function is that it implies that the firm can increase output indefinitely. This seems to conflict with our definition of the short run as a period in which the firm has a given plant size, stock of machinery and so on. A quadratic total cost function, of the form $TC = aq^2 + bq + c$, therefore seems more plausible. An example could be

$TC = 0.02q^2 + 1.5q + 100$

(see also figure 4.17). This is the same as the linear TC function in the previous paragraph, but with the addition of a q^2 term with a very small coefficient of 0.02. Comparing this with the linear TC function in figure 4.17, we see that the addition of the quadratic term causes the TC curve to turn upwards, increasing steeply as output increases, reflecting the assumption that costs rise more rapidly as the firm's productive capacity is utilized more and more intensively.

A monopolist's total revenue function

A monopolist is a firm that is the sole source of supply of the product it produces. For reasons that we will examine more fully in chapter 8, a firm with a **monopoly** may find that if it progressively lowers its price and therefore sells more and more of its product, total sales revenue rises at first, reaches a maximum value, and then declines as quantity sold increases further. The decline in total revenue beyond a certain point occurs because the revenue loss from the lower price begins to outweigh the revenue gain from the increase in quantity sold.

A quadratic function of the form $TR = aq^2 + bq$ (where TR = total sales revenue and q = quantity sold) is therefore a suitable functional form for modelling this relationship. We assume that the parameter a is negative, as is required to make the turning point a maximum rather than a minimum value (see section 4.8 above). We also assume that there is no constant term ($c = 0$), because if the firm does not sell any output ($q = 0$), it will not receive any sales revenue ($TR = 0$). A total revenue function with these properties is

$TR = -0.12q^2 + 10q$

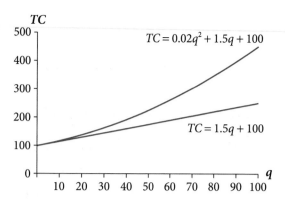

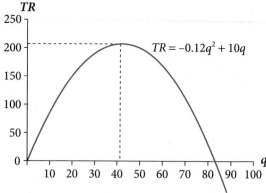

Of the linear and non-linear total cost functions, the non-linear case seems more likely.

Total revenue from sales reaches a maximum when $q = 42$. Larger quantities can be sold, but only by reducing the price by so much that total revenue falls.

Figure 4.17 Linear and quadratic total cost functions.

Figure 4.18 A quadratic total revenue function.

which is shown in figure 4.18. Total revenue from sales reaches a maximum when $q = 42$ (rounded to the nearest whole unit). Larger quantities can be sold, but only by reducing the price by so much that total revenue falls. Hopefully this brief discussion has whetted your appetite for further analysis in chapter 8.

Summary of sections 4.13–4.14

The quadratic function may be applied to any area of economic analysis where we have reason to believe that relationships between economic variables may be non-linear. In section 4.13 we considered the market for a good where either the inverse demand function, the inverse supply function, or both, were quadratic functions. In section 4.14 we considered a firm's short-run total cost function which, for sound economic and technological reasons, may be expected to be non-linear. An inverted quadratic function is also a plausible shape for the total revenue function of a monopolist supplier.

The Online Resource Centre shows how supply and demand models and a firm's total cost function may be graphed using Excel®. www.oxfordtextbooks.co.uk/orc/renshaw3e/

Progress exercise 4.4

1. Solve the following simultaneous equations, and draw sketch graphs of the functions, indicating your solutions.

   (a) $y = 3x + 6; y = x^2 + 2x - 6$ (b) $y = x^2 + 4x - 6; y = -x^2 + 2x + 6$

2. Given the following supply and demand functions for a good, find the equilibrium price and quantity, and draw sketch graphs of the functions, indicating your solution.

   $q^S = 0.5p^2 + p - 8; q^D = -2p^2 - 2p + 100$

3. Given the following inverse supply and demand functions for a good, find the equilibrium price and quantity, and draw sketch graphs of the functions, indicating your solution.

   $p^S = 10q - 30; \ p^D = -0.5q^2 - 8q + 200$

Checklist

Be sure to test your understanding of this chapter by attempting the progress exercises (answers are at the end of the book). The Online Resource Centre contains further exercises and materials relevant to this chapter www.oxfordtextbooks.co.uk/orc/renshaw3e/.

In this chapter we have focused on quadratic equations, the simplest form of non-linear relationship. We solved both single and simultaneous quadratic equations and applied these methods to some economic problems. The main topics covered were:

✔ **Factorizing quadratic expressions.** Understanding what quadratic expressions are and how to factorize them by 'trial and error' methods.

✔ **Solving quadratic equations.** Solving quadratic equations by factorization and also by using the formula. Why some quadratic equations cannot be factorized and why some have a single, repeated root.

✔ **Quadratic functions and their graphs.** Recognizing a quadratic function and sketching its graph from inspection of its parameter values.

✔ **Graphical solution of quadratic equations.** Relationship between the algebraic and graphical solution of a quadratic equation.

✔ **Simultaneous quadratic equations.** Solving simultaneous quadratic equations both algebraically and graphically.

✔ **Economic applications.** Quadratic functions applied to supply and demand models and as total cost and total revenue functions.

If you have absorbed this material on quadratic equations, you are ready to move on to consider some other types of non-linear function in chapter 5.

Chapter 5
Some further equations and techniques

OBJECTIVES

Having completed this chapter you should be able to:

■ Recognize a cubic function and the possible shapes that its graph may have.

■ Draw a rough sketch of a cubic function.

■ Understand why a cubic equation may have one or three distinct real solutions.

■ Recognize the equation of a rectangular hyperbola and draw a rough sketch of its graph.

■ Understand the concepts of a limiting value and a discontinuity in a function.

■ Recognize the equations of a circle and an ellipse, and draw a rough sketch of their graphs.

■ Understand and apply the rules for manipulating inequalities.

■ Understand how these functions and techniques may be used in economic analysis.

5.1 Introduction

In the previous two chapters we examined two types of function and their graphs: the linear function with general form $y = ax + b$ and the quadratic function with general form $y = ax^2 + bx + c$. The first of these gives a straight line when plotted and the second gives a U-shaped curve with one turning point. These two shapes are sufficient for modelling many relationships in economics, but there are many contexts in which curves of other shapes are required.

In this chapter, therefore, we will examine four new types of function. First, we will analyse the cubic function, which produces a characteristically S-shaped graph. This is useful for describing some relationships in economics, such as those in which the dependent variable rises rapidly at first, then more slowly, and finally rapidly once again. Second, we will look at the rectangular hyperbola. This functional form is not only useful to economists in its own right, but also gives us the opportunity to introduce two new and important concepts: the idea of a limiting value to which the dependent variable approaches closer and closer, but never actually reaches; and the idea of a discontinuity or gap in the relationship between x and y. We also look at the equations of the circle and the ellipse, which have some relevance to economics.

Adding these function types to your tool kit will not only be useful in itself but will also hopefully add to your skills and self-confidence in mathematics. We will also look briefly at how these functions can be used in economics. The chapter concludes by looking at the algebra of inequalities, which are used extensively in economics.

5.2 The cubic function

The general form of the cubic function is

$$y = ax^3 + bx^2 + cx + d \quad \text{(where, as usual, } a, b, c, \text{ and } d \text{ are parameters)}$$

We see that a cubic function contains a term involving x^3, a term involving x^2 and a term involving x (that is, x raised to the power 1). The x^3 term is definitional: if the parameter a were zero, the x^3 term would disappear, and we would no longer have a cubic function. However, the x^2 and x terms are optional, as is the constant d. In other words, the parameters b, c, and d may be zero.

Let us now look at some examples of cubic functions and their graphs.

EXAMPLE 5.1

The simplest case is obtained by taking the general form, above, and setting $a = 1$ and $b = c = d = 0$. We are then left simply with $y = x^3$. This is graphed in figure 5.1 for values of x between -3 and $+3$.

The key features of the graph are:

(1) x^3 (which of course equals y) has the same sign as x: that is, x^3 is positive when x is positive, and negative when x is negative.

(2) Because x^3 increases/decreases very rapidly as x increases/decreases, the curve turns sharply up/down, going off very quickly towards $+\infty$ or $-\infty$. For example, when $x = +10$, $x^3 = +1000$; and when $x = -10$, $x^3 = -1000$.

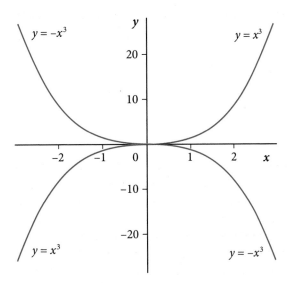

Figure 5.1 Graphs of $y = x^3$ and $y = -x^3$.

Table 5.1 Values for the graph of $y = x^3$.

x	-3	-2	-1	0	1	2	3
y	-27	-8	-1	0	1	8	27

Table 5.2 Values for the graph of $y = x^3$ when x is close to zero.

x	$-\frac{1}{2}$	$-\frac{1}{4}$	$-\frac{1}{8}$	0	$\frac{1}{8}$	$\frac{1}{4}$	$\frac{1}{2}$
y	$-\frac{1}{8}$	$-\frac{1}{64}$	$-\frac{1}{512}$	0	$\frac{1}{512}$	$\frac{1}{64}$	$\frac{1}{8}$

(3) The behaviour of the curve when x is close to zero is interesting. Table 5.2 gives some values of y when x is close to zero. The notable feature is that when x is a fraction, $y = x^3$ is *less* than x in absolute value. So when x is close to zero, y is even closer to zero. For example, when $x = \frac{1}{8}$, $y = (\frac{1}{8})^3 = \frac{1}{512}$. (Note that figure 5.1 is misleading, because it *appears* that the curve touches the axis when x is close to zero. In fact, as long as x is non-zero, y is non-zero too.)

Figure 5.1 also shows, in blue, the graph of $y = -x^3$. This is the mirror image of $y = x^3$: that is, it slopes downwards from left to right, because y is now positive when x is negative and negative when x is positive.

EXAMPLE 5.2

A slightly more complex case arises if, in the standard form, we set $a = 1$ as before, but this time we set $b = 6$, with $c = d = 0$ as before. This gives us $y = x^3 + 6x^2$, which is graphed in figure 5.2.

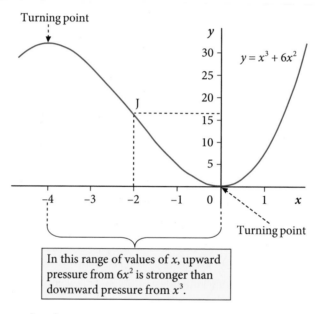

Figure 5.2 Graph of $y = x^3 + 6x^2$.

Table 5.3 Values for graph of $y = x^3 + 6x^2$.

x	−5	−4	−3	−2	−1	0	1	2
y	25	32	27	16	5	0	7	32

The key features of this graph, when compared with the previous case of $y = x^3$, are:

(1) When x is positive, $6x^2$ is positive. So, when x is positive, adding $6x^2$ pushes the curve *up*. Thus when x is positive, the graph of $y = x^3 + 6x^2$ lies above the graph of $y = x^3$.

(2) When x is negative, $6x^2$ is again positive. So, when x is negative, adding $6x^2$ again pushes the curve up. This opposes the effect of x^3, which is negative when x is negative and is therefore pushing the curve *down*. From the graph, we can see that, when x is negative but relatively small in absolute value, the upward pressure from $6x^2$ is stronger than the downward pressure of x^3, hence the curve is rising. For example, when $x = -2$, $x^3 = -8$ and $6x^2 = 24$, so $y = 16$ (see point J in figure 5.2).

Eventually, as we move leftwards and x becomes more and more negative, the downward pressure from x^3 dominates the upward pressure from $6x^2$, and the curve turns down. The turning point is at $x = -4$. To the left of this, the curve falls towards $-\infty$. The shape of this graph, with its two turning points, is a feature of all cubic functions, such as this, when they contain an x^2 term as well as an x^3 term but no x term.

EXAMPLE 5.3

Continuing from example 5.2, $y = x^3 + 6x^2$, we now bring in the term involving x alone, by adding $15x$. So our function is now

$$y = x^3 + 6x^2 + 15x \quad \text{which is graphed in figure 5.3.}$$

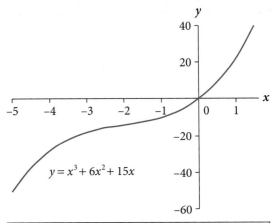

When x is negative, upward pressure from $6x^2$ is weaker than combined downward pressure from x^3 and $15x$, so no turning points.

Figure 5.3 Graph of $y = x^3 + 6x^2 + 15x$.

Table 5.4 Values for graph of $y = x^3 + 6x^2 + 15x$.

x	-5	-4	-3	-2	-1	0	1	2
y	-50	-28	-18	-14	-10	0	22	62

Comparing figures 5.3 and 5.2, we see that adding the $15x$ term has eliminated the two turning points, so that figure 5.3 looks more like $y = x^3$ in figure 5.1 than figure 5.2. This is because the $15x$ term has the same sign as x^3 (that is, both are positive when x is positive, and negative when x is negative). So when x is negative, the $15x$ term is helping the x^3 term to push the curve down, against the force of the $6x^2$ term which is pushing it up. Because we have given the x term a relatively 'heavy' coefficient, 15, its downward pressure is strong enough to eliminate the two turning points. With a smaller coefficient on x (say, 10 instead of 15), the effect would be weaker and the two turning points would still be present.

EXAMPLE 5.4

As a final example, consider $y = -x^3 - 6x^2 - 3x + 10$, which is graphed in figure 5.4.

Looking at figure 5.4, we see that the negative coefficients on all the terms involving x have turned the graph upside down, compared to figures 5.1–5.3. The two turning points are present because the x^2 term has a large coefficient (6), relative to the coefficients on x^3 and x (comparing absolute magnitudes, that is).

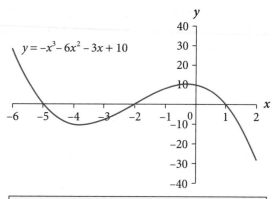

$y = -x^3 - 6x^2 - 3x + 10$

When x is negative and small in absolute value, downward pressure from $-6x^2$ overcomes upward pressure from $-x^3$ and $-3x$.

Figure 5.4 Graph of $y = -x^3 - 6x^2 - 3x + 10$.

An important feature of this example is the constant term, 10. This gives the intercept on the y-axis. You can probably see straight away that the effect of varying the constant term is to shift the whole curve bodily up or down relative to the x-axis. If the constant term in this example were larger, say 25, the whole curve would shift upwards so far that it would cut the x-axis only once, instead of three times.

Table 5.5 Values for graph of $y = -x^3 - 6x^2 - 3x + 10$.

x	-6	-5	-4	-3	-2	-1	0	1	2
y	28	0	-10	-8	0	8	10	0	-28

5.3 Graphical solution of cubic equations

Suppose we want to solve the cubic equation

$$-x^3 - 6x^2 - 3x + 10 = 0$$

Note that this is the function that we graphed in example 5.4, with y set equal to zero. One way of solving this equation would be for us to try to factorize the left-hand side, as we did with quadratic equations in chapter 4. But there is no systematic way of doing this. If we stared at the equation for a long time, we might suddenly see that $x = 1$ satisfies it; for then the left-hand side becomes $-(1^3) - 6(1^2) - 3(1) + 10$, which equals zero. But this would be relying on inspiration rather than method.

It then occurs to us that figure 5.4 gives us the graph of $y = -x^3 - 6x^2 - 3x + 10$. From the graph and also from the table of values, we see that the curve cuts the x-axis at $x = -5$, $x = -2$, and $x = +1$. At these points, $y = 0$. (Recall, $y = 0$ at every point on the x-axis.) Therefore, $-x^3 - 6x^2 - 3x + 10 = 0$ at these points. In other words, $x = -5$, $x = -2$, and $x = +1$ are the solutions (roots) to the cubic equation $-x^3 - 6x^2 - 3x + 10 = 0$. (Check this by substituting each of these values into the equation; in each case it becomes an identity.)

Given that the *solutions* of $-x^3 - 6x^2 - 3x + 10 = 0$ are $x = -5$, $x = -2$, and $x = +1$, it follows that the *factors* are $(x + 5)$, $(x + 2)$, and $(x - 1)$. We can check this by multiplying out

$$(x + 5)(x + 2)(x - 1)$$

We haven't needed to multiply out (expand) three sets of brackets before, but the method follows naturally from the rules for two sets of brackets (see section 4.2). We multiply out the first set of brackets first, and get

$$(x+5)(x+2) = x^2 + 7x + 10$$

Then we multiply the remaining bracket, $(x-1)$, by this. We then see, by following rule 4.1 from section 4.2, that each term in the first pair of brackets multiplies each term in the second pair of brackets:

$$(x^2 + 7x + 10)(x-1) = x^2(x-1) + 7x(x-1) + 10(x-1)$$
$$= x^3 + 6x^2 + 3x - 10$$

Thus we have confirmed that $(x+5)(x+2)(x-1) \equiv x^3 + 6x^2 + 3x - 10$. This is not quite the equation we started with, but if we multiply both sides by -1, it becomes

$$-(x+5)(x+2)(x-1) \equiv -x^3 - 6x^2 - 3x + 10$$

As this is an identity (true for all x), it follows that whenever the left-hand side equals zero, the right-hand side must also equal zero. And the left-hand side equals zero when $x = -5$, -2 or $+1$.

Thus the algebra has confirmed what we learned from the graph. The values of x at which the graph of $y = -x^3 - 6x^2 - 3x + 10$ cuts the x-axis are the solutions to the equation $-x^3 - 6x^2 - 3x + 10 = 0$.

What we have just done is directly analogous to the work we did with quadratic functions and quadratic equations in chapter 4, except we now have three solutions instead of two. Indeed, there is no reason why this should not be true for *any* function. If we consider *any* function, which we can write as $y = f(x)$, where the notation $f(x)$ simply means 'any expression involving x', then the values of x at which its graph cuts the x-axis give us the solutions to the equation $f(x) = 0$.

The above graphical method of solving a cubic equation is not very satisfactory. If in the example above we had picked the wrong values of x (say, $x = 3$ to $x = 100$), our graph would not have cut the x-axis, so we would not have found the roots. We might have to draw a lot of graphs before we stumbled on the right values for x. We could avoid the graphical method if we could find a formula for solving cubic equations, as we did with quadratic equations in chapter 4 (see rule 4.4). Unfortunately, no such formula exists. Apart from the hit-and-miss graphical method, cubic equations can only be reliably solved by 'trial and error', which mathematicians call **iterative methods**. These methods involve lots of laborious calculations when done with pen and paper only, but nowadays they are done very quickly by a computer.

A cubic equation with only one solution

In the previous example, $y = -x^3 - 6x^2 - 3x + 10$, the graph cut the x-axis three times (figure 5.4). Hence the associated cubic equation had three (real) solutions (roots).

However, as we know from examples 5.1–5.3, not all cubic functions cut the x-axis three times. In figure 5.5 we see the graph of $y = x^3 + 6x^2 + 6x + 5$. This cuts the x-axis only once, at $x = -5$. This can also be seen in the table of values (table 5.6).

Thus we can see that $x = -5$ is the only value of x for which $y = x^3 + 6x^2 + 6x + 5 = 0$. Therefore the cubic equation $x^3 + 6x^2 + 6x + 5 = 0$ has only one solution (root); $x = -5$.

Generalizing from this example, we conclude that when the graph of a cubic function cuts the x-axis only once, the corresponding cubic equation has only one solution or root.

Table 5.6 Values for graph of $y = x^3 + 6x^2 + 6x + 5$.

x	-6	-5	-4	-3	-2	-1	-0.5	0	1	2
y	-31	0	13	14	9	4	3.375	5	18	49

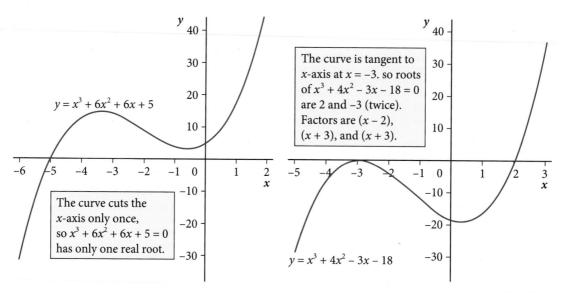

Figure 5.5 Graph of $y = x^3 + 6x^2 + 6x + 5$.

Figure 5.6 Graph of $y = x^3 + 4x^2 - 3x - 18$.

The previous sentence is not quite correct. It can be shown (quite easily, but we won't go into it here) that the factors of $x^3 + 6x^2 + 6x + 5$ are $(x + 5)$ and $(x^2 + x + 1)$. You can check this by multiplying out $(x + 5)(x^2 + x + 1)$.

So $x^3 + 6x^2 + 6x + 5 = 0$ when either $x + 5 = 0$ or $x^2 + x + 1 = 0$. But $x^2 + x + 1 = 0$ is a quadratic equation with no real roots. You can check this by applying the formula for solving a quadratic equation (rule 4.4). Therefore there are two additional roots, but these are complex roots involving imaginary numbers (see section 4.6). So we conclude that $x^3 + 6x^2 + 6x + 5 = 0$ has one real root ($x = -5$) and two complex roots. In this way, using the trick of imaginary numbers, mathematicians are able to conclude that every cubic equation has three roots.)

The boundary case

In the two previous cases we considered first a cubic function that cut the x-axis three times, and then a cubic function that cut the x-axis only once. The boundary or 'knife edge' case that divides these two cases is where the curve just touches the x-axis at one point. This is called a point of tangency. An example is shown in figure 5.6, which is the graph of

$$y = x^3 + 4x^2 - 3x - 18$$

From figure 5.6 we see that the curve is tangent to the axis at $x = -3$, and cuts the axis at $x = +2$. Therefore the equation $x^3 + 4x^2 - 3x - 18 = 0$ apparently has two roots, $x = -3$ and $x = +2$.

However, $x^3 + 4x^2 - 3x - 18$ may be factorized as $(x - 2)(x + 3)(x + 3)$. (Check for yourself by multiplying out.) So we say that there are three roots, $x = -3$ (twice, or repeated) and $x = +2$. This may seem bizarre, but we are forced to repeat the factor $(x + 3)$ because $(x - 2)(x + 3)$ alone does not multiply out to produce $x^3 + 4x^2 - 3x - 18$. The root $x = -3$ is called a repeated root.

This trick, of counting the factor $(x + 3)$ twice, is directly analogous to the case of the perfect square, which we met when solving quadratic equations in section 4.7. The trick helps to make it possible to say that every cubic equation has three roots.

Table 5.7 Values for graph of $y = x^3 + 4x^2 - 3x - 18$.

x	−5	−4	−3	−2	−1	0	1	2	3
y	−28	−6	0	−4	−12	−18	−16	−0	36

Higher-degree polynomials

We have now examined three types of function: first, the linear function $y = ax + b$, in which the highest power to which x is raised is the power 1 (since $x = x^1$); second, the quadratic function $y = ax^2 + bx + c$, in which the highest power to which x is raised is the power 2; and finally the cubic function $y = ax^3 + bx^2 + cx + d$, in which x is raised to the power 3. These are all examples of what are called **polynomial functions**, and in general a polynomial function of degree n follows this pattern, with n being the highest power to which x is raised.

You might be expecting that, since we have considered polynomials of degree 1 (linear), degree 2 (quadratic), and degree 3 (cubic), the logical next step is to consider the degree 4 (quartic) function $y = ax^4 + bx^3 + cx^2 + dx + e$, and then the degree 5 (quintic) function $y = ax^5 + bx^4 + \ldots + \ldots$, and so on. Fortunately this is not necessary, as the linear, quadratic, and cubic functions, together with a few others that we will consider later, give us enough variety of shapes to cover virtually all our needs in economics. Polynomial functions are considered in more detail in chapter W21, available at the Online Resource Centre.

5.4 Application of the cubic function in economics

There are various contexts in economics in which the cubic function is useful. One is the firm's short-run total cost function, discussed in section 4.14. We saw there that a quadratic form for the short-run TC function was likely in many situations to give a better representation of reality than a linear function. Similarly, a cubic version of the TC function might in turn be more realistic than a quadratic function. Consider, for example, the cubic short-run total cost function

$$TC = 2q^3 - 15q^2 + 50q + 50$$

which is graphed in figure 5.7. If we think about the economics of this total cost curve, and the technological and organizational factors lying behind the economics, some interesting conclusions can be drawn. In the range of output from about 1.5 to about 2.5 units, the slope of the curve is very flat, indicating that increases in output result in relatively little increase in total costs. This suggest that the production process is at its most efficient in this range of output. (If 1.5 units seems small, remember this might be 1.5 million tonnes of steel, million of litres of beer, or whatever; and might be daily output, not annual.) At low levels of output, in the range 0–1.5 units, production appears to be relatively inefficient as the TC curve slopes upward relatively steeply, and this is also increasingly true at high levels of output, above about 3.5 units. This might be quite a realistic characterization of some modern production technologies, where the plant is designed to operate most efficiently in a certain range of outputs, and output levels below or above the designed range incur cost penalties. The curve turns upwards increasingly steeply as output increases beyond about 5 units, indicating that the maximum productive capacity of the plant is being approached and extra output can only be produced with an increasing cost penalty.

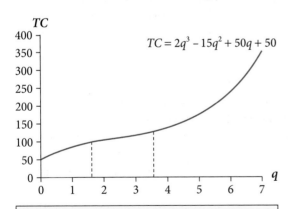

The technology of production is such that at low levels of output it is relatively difficult to increase output, so the TC curve slopes upward relatively steeply. The same is true at high levels of output. In the mid-range, between about 1.5 and 3.5 units of output, it is relatively easy to increase output and the TC curve is relatively flat.

Figure 5.7 A cubic short-run total cost function.

Summary of sections 5.1–5.4

(1) Unlike quadratic equations (see chapter 4), there is no rule or formula for solving the cubic equations of the general form $ax^3 + bx^2 + cx + d = 0$. They can be solved only by trial-and-error methods, which fortunately computers are very good at.

(2) The graph of the cubic function $y = ax^3 + bx^2 + cx + d$ must cut the x-axis at least once, because of its characteristic S-shape. Therefore the cubic equation $ax^3 + bx^2 + cx + d = 0$ always has at least one real solution (root), whatever the values of the parameters a, b, c, and d may be.

(3) Depending on parameter values, the graph of the cubic function $y = ax^3 + bx^2 + cx + d$ may cut the x-axis three times, in which case the cubic equation $ax^3 + bx^2 + cx + d = 0$ has three distinct real solutions (roots). As a boundary case, the parameter values may be such as to cause the cubic function $y = ax^3 + bx^2 + cx + d$ to cut the x-axis once, and be tangent to the x-axis at another point. In this case the cubic equation $ax^3 + bx^2 + cx + d = 0$ has two distinct real roots, one of which is considered to be 'repeated'. (See section 4.3 for repeated roots in the case of the quadratic equation.)

(4) The cubic function might provide a good representation of the short-run total cost curve of a firm where the technology is such that it is relatively costly to increase output when the level of output is low or high.

The Online Resource Centre shows how to graph the cubic function using Excel®, together with further worked examples of the cubic function in economics.

Progress exercise 5.1

1. Plot the graphs of the following cubic functions. Do not aim at a high degree of accuracy, but instead try to draw a sketch which is accurate enough to identify the main features of the graph, especially turning points and intercepts on the x- and y-axes.

(a) $y = x^3 - 2x^2 - 5x + 6$. Take integer values of x between -3 and $+4$.

(b) $y = -x^3 - x^2 + 10x + 20.125$. Take values of x between -3 and $+4$, at intervals of 0.5.

(c) $y = x^3 - 2.5x^2 + 0.5x + 1$. Take values of x between -1 and $+3$, at intervals of 0.5.

(d) $y = x^3 - 3x + 2$. Take integer values of x between -3 and $+3$.

5.5 The rectangular hyperbola

We now consider a type of function called a rectangular hyperbola. This has several important applications in economics; we give an example later. The simplest form of this function is

$$y = \frac{1}{x}$$

Let's construct the graph of this function for values of x between -5 and $+5$. Table 5.8 gives the values and the graph is shown in figure 5.8.

One key feature of this function becomes apparent immediately. When we look at the graph, and also when we look either at the table of values or at the equation $y = \frac{1}{x}$ itself, we see that x and y vary *inversely*: that is, when x is large, y is small, and vice versa. For example, when $x = \frac{1}{10}$, $y = \frac{1}{x} = \frac{1}{1/10} = 10$; and when $x = 5$, $y = \frac{1}{x} = \frac{1}{5}$. The same is true for negative values of x. The reason for this is that x is in the denominator of the equation $y = \frac{1}{x}$, so as x gets larger in absolute value, y becomes smaller in absolute value.

A second key feature may be seen in the table of values: when $x = 0$, $y = \frac{1}{0}$, which we know is undefined (see section 2.8). So there is no value of y corresponding to $x = 0$. This is discussed in the next section.

Table 5.8 Values for graph of $y = \frac{1}{x}$.

x	-5	-4	-3	-2	-1	$-\frac{1}{2}$	$-\frac{1}{5}$	$-\frac{1}{10}$	0	$-\frac{1}{10}$	$\frac{1}{5}$	$\frac{1}{2}$	1	2	3	4	5
y	$-\frac{1}{5}$	$-\frac{1}{4}$	$-\frac{1}{3}$	$-\frac{1}{2}$	-1	-2	-5	-10	$-$	10	5	2	1	$\frac{1}{2}$	$\frac{1}{3}$	$\frac{1}{4}$	$\frac{1}{5}$

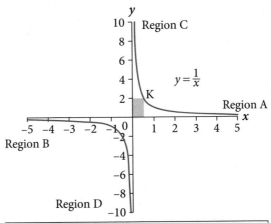

y approaches a limiting value of zero as x approaches + or $-\infty$, and a limiting value of $+\infty$ or $-\infty$ as x approaches zero. There is a discontinuity in the function at $x = 0$. The shaded rectangle has the same area whatever the position of K.

Figure 5.8 Graph of $y = \frac{1}{x}$.

5.6 Limits and continuity

We will now introduce two new ideas: (1) a limit or limiting value of a function; and (2) a discontinuity in a function. We will explain these concepts by looking more closely at the function $y = \frac{1}{x}$. However, these ideas have a much wider range of application than merely to this particular function.

The idea of a limit, or limiting value

We can introduce this concept by looking again at $y = \frac{1}{x}$ and considering what happens to y when x is positive and gets very large; see region A in figure 5.8. As we have just seen, as x becomes larger and larger, y becomes smaller and smaller, and therefore the curve approaches closer and closer to the x-axis. However, in region A the curve never actually touches or cuts the axis. This is because, however large x becomes, $\frac{1}{x}$ (and therefore y) remains positive. For example, when $x = 1{,}000{,}000$, $y = \frac{1}{1{,}000{,}000}$, which is very small but nevertheless positive. Thus we can get y as close to zero as we wish, by making x sufficiently large. But because x can never actually reach infinity, y can never actually *reach zero*, and therefore the curve can never touch the x-axis, however large x becomes.

We describe this situation by saying that, as x approaches plus infinity, y approaches a **limiting value**, or **limit**, of zero. But this limiting value can never be reached because x can never actually *reach* plus infinity. Mathematicians often use the symbol '$\rightarrow$' to mean 'approaches'. We can then write 'y approaches a limiting value of zero as x approaches plus infinity' more compactly as

$y \rightarrow 0$ as $x \rightarrow +\infty$

An alternative notation, with the same meaning, is

$$\lim_{x \to +\infty} y = 0 \qquad (5.1)$$

Note that what is stated here is not that $y = 0$, but that 'lim y' (that is, the limiting value of y) is equal to zero.

We now consider what happens to y when x is negative and becomes larger and larger in absolute value; see region B in figure 5.8. In this case the curve also approaches closer and closer to the x-axis, but this time from below, because y is negative when x is negative. For example, when $x = -1,000,000$, $y = -\frac{1}{1,000,000}$. So we can write

$$y \to 0 \text{ as } x \to -\infty$$

or alternatively

$$\lim_{x \to -\infty} y = 0 \qquad (5.2)$$

In figure 5.8 it is the x-axis that is the limiting value of y, because $y = 0$ everywhere on the x-axis. Thus the curve (y) approaches the x-axis as x approaches plus or minus infinity. The x-axis is therefore called the horizontal **asymptote** of the function. The curve is said to be *asymptotic* to the x-axis, meaning that the curve approaches closer and closer to the axis as x goes off towards plus or minus infinity.

Now let us turn our attention to what happens to $y = \frac{1}{x}$ when x is positive and becomes very small; see region C in figure 5.8. As x becomes smaller and smaller, y becomes larger and larger, and therefore the curve approaches closer and closer to the y-axis. For example, when $x = \frac{1}{50}$, $y = \frac{1}{x} = 50$; and when $x = \frac{1}{1000}$, $y = \frac{1}{x} = 1000$. Thus we can always make y larger by making x smaller. As x gets closer and closer to zero (while remaining positive), so y goes off towards plus infinity.

Using the notation introduced above, we can write 'y approaches a limiting value of plus infinity as x approaches zero' as

$$y \to +\infty \text{ as } x \to 0 \quad \text{(with } x > 0)$$

or, alternatively

$$\lim_{x \to 0} y = +\infty \quad \text{(with } x > 0) \qquad (5.3)$$

In the same way, when x is negative and small in absolute value (region D in figure 5.8), then y heads south towards $-\infty$ as x gets closer and closer to zero. For example when $x = -\frac{1}{1000}$, $y = \frac{1}{x} = -1000$. So we can write

$$y \to -\infty \text{ as } x \to 0 \quad \text{(with } x < 0)$$

or alternatively

$$\lim_{x \to 0} y = -\infty \quad \text{(with } x < 0) \qquad (5.4)$$

Because in figure 5.8 the curve approaches the y-axis as x approaches zero, the y-axis is called the *vertical asymptote* of the function. The curve is said to be asymptotic to the y-axis, meaning that the curve approaches closer and closer to the y-axis as x approaches zero.

You may rightly feel uncomfortable with the idea that either x or y can *approach* infinity, or minus infinity, because infinity is not a number. Given the definition of infinity, it does not really make sense to say, for example, that 1 million is any closer to infinity than is 1 hundred. Both are infinitely far away from infinity. Revealing his fundamental misunderstanding of a closely related point, the comedian Woody Allen once said, 'Eternity lasts a very long time, especially towards the end.'

There are some advanced mathematical devices for dealing with this problem. However, for our purposes it is best to view equations (5.1)–(5.4) above simply as convenient shorthand for describing what happens to y as x gets bigger and bigger, or smaller and smaller, in absolute value.

Discontinuity

While y can never actually reach zero, we can choose the value of x. Therefore you may be wondering what happens when we choose $x = 0$. In explaining this, we introduce the idea of a **discontinuity**.

We have just seen that y approaches plus or minus infinity as x approaches zero, depending on whether x is approaching zero from the left (negative) direction or the right (positive) direction. When x actually reaches zero, we have $y = \frac{1}{x} = \frac{1}{0}$, which is undefined. Thus there is no value of y corresponding to $x = 0$, and we therefore say that there is a discontinuity, or gap, in the function at $x = 0$. (The precise definition of discontinuity is a little more complex, but the detail need not concern us here.)

Examining the function $y = \frac{1}{x}$ has given us the opportunity to introduce these two ideas of limiting values and discontinuities, but they have a much wider applicability and will come up again at many points in later chapters of this book. In general, whatever the function we are looking at, we are often interested in what happens to y as x approaches zero or infinity. We call this the *asymptotic behaviour* of the function.

The area under the curve

Before we leave the function $y = \frac{1}{x}$, there is an intriguing feature that we should look at, and which is relevant when this function is used in economics. Suppose we pick any point on the curve: say, point K in figure 5.8, where $x = 0.5$ and therefore $y = \frac{1}{x} = \frac{1}{0.5} = 2$. Then we draw a rectangle with its corner at K and its opposite corner at the origin (the shaded area). Since this rectangle has a horizontal side of length 0.5 and a vertical side of length 2, its area is $0.5 \times 2 = 1$.

The key point is that the area of this rectangle will equal 1 *irrespective* of the position of K on the curve. We can demonstrate this very easily, as follows. Let us position K at any point where $x = x_0$, where x_0 denotes *any* chosen value of x. Then the corresponding value of y, which we can denote by y_0, will be $y_0 = \frac{1}{x_0}$.

Therefore the rectangle with its corner at K has a horizontal side of length x_0 and a vertical side of length $y_0 = \frac{1}{x_0}$. So its area is $x_0 \times y_0 = x_0 \times \frac{1}{x_0} = 1$. Because x_0 cancels and thus disappears, the area equals 1 whatever value we assign to x_0. Later in this chapter we shall make use of this property in an economic application.

The general form of the rectangular hyperbola

In the previous section we considered the simplest form of a rectangle hyperbola, $y = \frac{1}{x}$. Let us now look at the general form, which is

$$y = \frac{c}{x+b} + a \quad \text{(where } a, b, \text{ and } c \text{ are parameters)}$$

The key points here are that the effect of varying the parameters a and b is to shift the graph around relative to the axes, while parameter c determines the area of any rectangle drawn under the curve. All this is best shown by means of an example.

EXAMPLE 5.5

If we set $a = 3$, $b = 2$, and $c = 5$, we get

$$y = \frac{5}{x-2} + 3$$

which is graphed in figure 5.9.

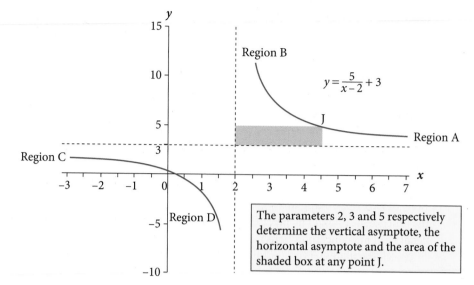

The parameters 2, 3 and 5 respectively determine the vertical asymptote, the horizontal asymptote and the area of the shaded box at any point J.

Figure 5.9 Graph of $y = \frac{5}{x-2} + 3$.

The key features of the graph are:

(1) Looking at region A, we see that y approaches a limiting value of $y = 3$ as x approaches $+\infty$. In other words the horizontal asymptote, shown as a dotted line in figure 5.9, is at $y = 3$. We can check this algebraically. In the function $y = \frac{5}{x-2} + 3$, consider what happens as x approaches $+\infty$. Since x is in the denominator, $\frac{5}{x-2}$ must approach zero. But this means that the right-hand side of the equation must approach 3, which means that y must also approach 3. For example, when $x = 1,000,000$ we have $y = \frac{5}{x-2} + 3 = \frac{5}{999,998} + 3$, which is very slightly greater than 3.

In the same way, in region C we see that y again approaches a limiting value of $y = 3$ as x approaches $-\infty$. The only difference is that the curve now approaches the horizontal asymptote from below. For example, when $x = -1,000,000$ we have $y = \frac{5}{x-2} + 3 = -\frac{5}{1,000,002} + 3$, which is very slightly less than 3.

This finding is easily generalized. Since in this example we have $a = 3$, we conclude that in the general form $y = \frac{c}{x+b} + a$ the horizontal asymptote is given by the parameter a.

(2) Looking now at region B, we see that y goes off towards $+\infty$ as x approaches $+2$ from the right. In other words the vertical asymptote, shown as a dotted line in figure 5.9, is at $x = 2$. We can confirm this algebraically: as x approaches 2, $x - 2$ approaches zero, and therefore $\frac{5}{x-2}$ approaches infinity. Hence the right-hand side of $y = \frac{5}{x-2} + 3$ approaches infinity (because the 3 becomes relatively tiny and thus can be ignored), and therefore y approaches infinity too. The same thing happens in region D, where x is approaching 2 from the left and y is going off towards $-\infty$.

Generalizing this, since in this example we have $b = -2$ and the vertical asymptote is at $x = +2$, in the general form $y = \frac{c}{x+b} + a$, the vertical asymptote is given by $x = -b$.

(3) Finally, what does the 5 in $y = \frac{5}{x-2} + 3$ do? It gives the area of any rectangle bounded by the asymptotes and the curve. For example, in figure 5.9 the shaded rectangle with its corner at J has a vertical side of length 2 ($= 5 - 3$) and a horizontal side of length 2.5 ($= 4.5 - 2$), so its area is $2 \times 2.5 = 5$.

In general, therefore, the parameter c in the general form $y = \frac{c}{x+b} + a$ gives the area of any rectangle bounded by the asymptotes and the curve. As the value of c is increased, the two branches of the curve move further away from the asymptotes in order to make room for larger rectangles to be drawn under the curve. The proof is quite simple, but we will pass over it here.

5.7 Application of the rectangular hyperbola in economics

The rectangular hyperbola pops up quite frequently and in a wide variety of areas of economic analysis, partly because of the fact noted earlier that the area under the curve and bounded by the axes can be constant, with suitable choice of parameters. We will look briefly here at the use of the rectangular hyperbola as a demand function, though it should be stressed that this is only one of many possible applications.

We have already looked at linear and quadratic demand functions in chapters 3 and 4 respectively. Both are somewhat restrictive in that both forms of demand function eventually cut both axes. The rectangular hyperbola is a functional form that offers an escape from this limitation. To take the simplest case, if the inverse demand function is

$$p = \frac{15}{q^D}$$

it will have the shape shown in figure 5.10. (For obvious reasons we consider only positive values of p and q.) The demand function is then asymptotic to both axes. In economic terms this has a threefold significance. First, it means that, as the price gets lower and lower, the demand curve gets closer and closer to the q-axis but never touches or crosses it. Consumers' tastes, which underlie their behaviour, are such that they never become satiated with the good. Instead the quantity demanded expands without limit as the price approaches closer and closer to zero.

Second, at the other extreme, as the price rises higher and higher, demand falls but never falls to zero, however high the price may go. There is something about this good, in the eyes of some consumers at least, which means that they must buy some positive quantity however high the price may be.

Both of these underlying assumptions about consumer tastes become untenable if pushed too far. In reality, demand for any good must surely become satiated if its price is sufficiently low, and

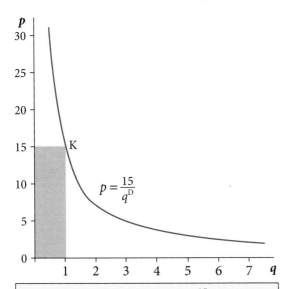

The inverse demand function $p = \frac{15}{q^D}$ never cuts the axes, so quantity demanded is always positive however high the price, and approaches infinity as the price approaches zero. The area of the shaded rectangle is constant whatever the position of K. Consequently, total expenditure is constant whatever the price.

Figure 5.10 Graph of the inverse demand function $p = \frac{15}{q^D}$.

equally it is hard to imagine that a sufficiently high price would not reduce the demand for any good to zero. However, in economic analysis we are often content to adopt models that become unrealistic when pushed to extremes, provided they give us what seem like the right answers over the range of values of the variables in question that we are likely to observe in the real world. There may well be many goods for which demand can appropriately be modelled by a rectangular hyperbola, provided the extremes of price (approaching infinity or zero) are ruled out.

As explained earlier, the area under the curve and bounded by its asymptotes is constant. The third feature of the demand function in this example follows from this. Given the demand function

$$p = \frac{15}{q^D}$$

if we multiply both sides by q^D, we get

$$pq^D = 15$$

Here the left-hand side is price × quantity, which by definition equals consumers' total expenditure on this good. (For example if the price of beer is 3 euros per pint and you buy 5 pints, your total expenditure is $3 \times 5 = 15$ euros.) Therefore this demand function has the intriguing feature that total expenditure on the good is constant, whatever the price may be. Again, this is not an assumption that we would expect to hold without limit.

Variants on the basic version

Above we considered only the simplest case. We know from the previous section that we can move a rectangular hyperbola around, relative to the axes, by varying its parameters. We might, for example, want a demand function in the form of a rectangular hyperbola, but which eventually cuts the q-axis (indicating satiation at a zero price). We can achieve this simply by subtracting a constant, say 2, from the inverse demand curve above. For any given value of x, the value of y will now be 2 units less, shifting the whole curve down by 2 units. The equation of the demand function then becomes

$$p = \frac{15}{q^{\mathrm{D}}} - 2$$

which is graphed in figure 5.11. It now cuts the q-axis, as we wanted, at $q = 7.5$. Further, the introduction of the -2 parameter means that total expenditure is no longer constant. If we multiply both sides of the inverse demand function by q^{D}, we get

$$pq^{\mathrm{D}} = 15 - 2q^{\mathrm{D}}$$

This equation gives total expenditure on the good, which we see decreases as q^{D} increases. As we move down the curve, with p falling and q^{D} rising, total expenditure falls.

As explained in example 5.5, we could also get the curve to cut the p-axis by adding a constant to the denominator of the right-hand side, and move the curve further away from the axes by increasing the constant in the numerator.

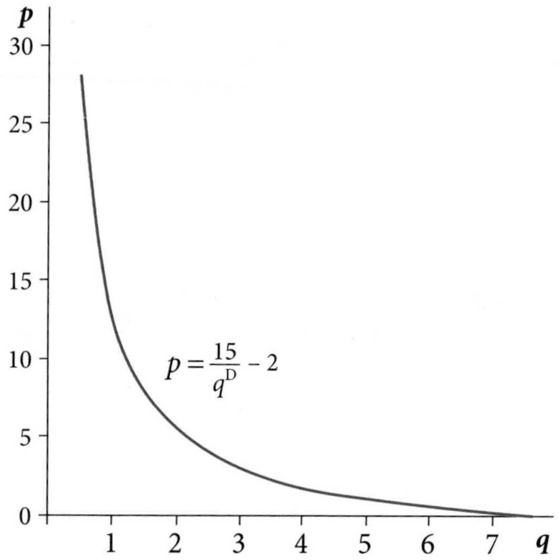

The demand function $p = \frac{15}{q^{\mathrm{D}}} - 2$ cuts the q-axis at $q = 7.5$, so demand becomes satiated. Total expenditure decreases as the price falls.

Figure 5.11 Graph of the inverse demand function $p = \frac{15}{q^{\mathrm{D}}} - 2$.

Summary of sections 5.5–5.7

(1) The general form of the rectangular hyperbola is $y = \frac{c}{x+b} + a$, where a, b, and c are parameters.

(2) The vertical asymptote is at $x = -b$. As x approaches closer and closer to $-b$, y goes off towards $+\infty$ (if x is greater than $-b$) or $-\infty$ (if x is less than $-b$). At $x = -b$, y is undefined and there is a discontinuity in the curve at this point.

(3) The horizontal asymptote is at $y = a$. As x goes off towards $+\infty$, y approaches a limiting value of a but is always greater than a. As x goes off towards $-\infty$, y approaches a limiting value of a but is always less than a.

(4) The parameter c gives the area of any rectangle drawn with one of its corners on the curve and the opposite corner at the intersection of the asymptotes.

(5) The concepts introduced here, of a discontinuity in a function, and a limiting value of y that is approached but never reached, have a wide applicability.

(6) The rectangular hyperbola can be used to represent a non-linear demand function that may or may not cut one or both axes.

The Online Resource Centre shows how to graph the rectangular hyperbola using Excel®.

Progress exercise 5.2

In answering these questions, do not try to produce highly accurate graphs, but try to show the key features of the shape of each function.

1. On the same axes, plot the graphs of $y = \frac{10}{x}$ and $y = \frac{20}{x}$. Take values of x from -20 to -0.1 and from 0.1 to 20. For each function, identify any limiting values and discontinuities.

2. Plot the graph of $y = \frac{90}{x} - 3$. Take values of x from 0 to 45. Identify limiting values and any discontinuity.

3. Plot the graph of $y = \frac{9}{x+1} - 3$. Take values of x from -10 to $+30$. Show that the horizontal asymptote is at $y = -3$, and the vertical asymptote at $x = -1$. Identify the discontinuity. Could this function be a plausible demand function with y as quantity demanded, with y positive? Give reasons for your answer.

5.8 The circle and the ellipse

The last two types of function we will look at in this chapter are the circle and the ellipse. They are worth our attention because their graphs give curves of the correct shape to depict certain economic relationships, of which we give an example below.

The circle

EXAMPLE 5.6

Consider $x^2 + y^2 = 9$

This is an example of an implicit relation. It is implicit because both variables appear on the same side of the equation (see section 3.7). It is defined as a relation, rather than a function, because there are two values of y associated with some values of x (see section 4.9). For example, if $x = 2$, then $y^2 = 9 - x^2 = 5$, and this equation is satisfied when either $y = \sqrt{5} = 2.236$ or $y = -\sqrt{5} = -2.236$ (check for yourself). However, the distinction between a function and a relation is not important for our purposes, so we will continue to refer to $x^2 + y^2 = 9$ as a function.

If we want to draw the graph of $x^2 + y^2 = 9$, then in order to draw up a table of values we must isolate y on one side of the equation. We can achieve this by subtracting x^2 from both sides of the equation, and then taking the square root of each side. We get

$$y = \pm\sqrt{9 - x^2}$$

which is now an explicit function with y treated as the dependent variable. Note that we have inserted the $\pm$ symbol in front of the square root sign in order to pick up the two values of y associated with some values of x. (We say some rather than all values of x because if we choose a value of x such that $x^2 > 9$, then the expression under the square root sign is negative and y is not a real number.) Using this equation, we can now compile table 5.9 and draw the graph (see figure 5.12). We see that the graph is a circle with its centre at the origin.

Although not immediately apparent from the graph, the radius of the circle is 3, the square root of 9. Thus any radius such as the dotted line from the origin to point A has

Table 5.9 Values for the graph of $x^2 + y^2 = 9$.

x	-3	-2	-1	0	1	2	3
$y = \pm\sqrt{9 - x^2}$	0	±2.24	±2.83	±3.00	±2.83	±2.24	0

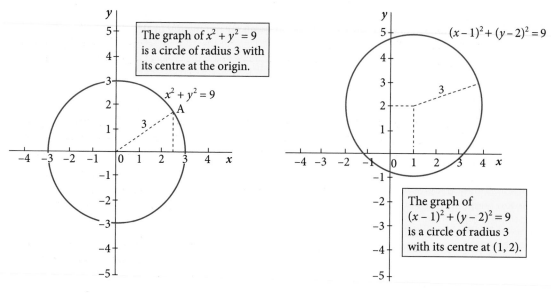

Figure 5.12 Graph of $x^2 + y^2 = 9$.

Figure 5.13 Graph of $(x-1)^2 + (y-2)^2 = 9$.

length 3. The proof of this, using Pythagoras's famous theorem on right-angled triangles, is not difficult but we won't go into it here. However, the fact that the curve is 3 units away from the origin at the intercepts on the x- and y-axes is supporting evidence. Generalizing from this example, we can say that $x^2 + y^2 = c^2$ is the equation of a circle with its centre at the origin and radius given by the parameter c.

EXAMPLE 5.7

If we want a circle, but with its centre at some point other than the origin, we can use a function such as

$$(x-1)^2 + (y-2)^2 = 9$$

In order to construct a table of values, we have to convert this into an explicit function. We can do this by rearranging $(x-1)^2 + (y-2)^2 = 9$ as

$$(y-2)^2 = 9 - (x-1)^2 \quad \text{from which}$$
$$y - 2 = \pm\sqrt{9 - (x-1)^2} \quad \text{and therefore}$$
$$y = 2 \pm \sqrt{9 - (x-1)^2}$$

Note that, as in example 5.6, we insert the $\pm$ sign to capture the two values of y to every feasible value of x. In table 5.10, $y+$ gives the value of y given by

$$y = 2 + \sqrt{9 - (x-1)^2}$$

and $y-$ the value of y given by

$$y = 2 - \sqrt{9 - (x-1)^2}.$$

When graphed in figure 5.13, this is a circle of radius $3 = \sqrt{9}$, as in example 5.6, but now with its centre at the point with coordinates $x = 1, y = 2$.

Generalizing from example 5.7, $(x-a)^2 + (y-b)^2 = c^2$ is the equation of a circle of radius c and with its centre at the point $x = a, y = b$.

Table 5.10 Values for graph of $(x-1)^2 + (y-2)^2 = 9$.

x	−2	−1.5	−1	−0.5	0	0.5	1	1.5	2	2.5	3	3.5	4
$y+$	2	3.66	4.24	4.6	4.83	4.96	5	4.96	4.83	4.6	4.24	3.66	2
$y-$	2	0.34	−0.24	−0.6	−0.83	−0.96	−1	−0.96	−0.83	−0.6	−0.24	0.34	2

The ellipse

We may want an ellipse, rather than a circle. In profile, an ellipse is shaped like a rugby football, or an American football, rather than like a soccer ball, which is circular in profile.

EXAMPLE 5.8

We can turn the circle $x^2 + y^2 = 9$ of example 5.6 into an ellipse by adding the term xy. We then have $x^2 + xy + y^2 = 9$. The graph of this is the ellipse (the solid red line) in figure 5.14.

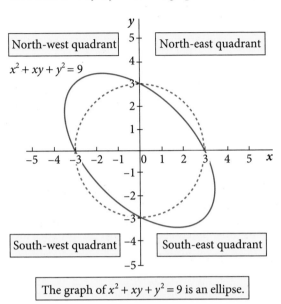

The graph of $x^2 + xy + y^2 = 9$ is an ellipse.

Figure 5.14 Graph of $x^2 + xy + y^2 = 9$ (solid red line).

It is not too difficult to work out why adding the xy term produces the solid line in figure 5.14, compared with the original circle from figure 5.12, which is also shown. In the north-east quadrant, where x and y are both positive, their product xy is positive, so the values of x and y that satisfy $x^2 + xy + y^2 = 9$ are *smaller* than those that satisfy $x^2 + y^2 = 9$. So the ellipse (solid line) *lies inside* the circle. The same thing is true in the south-west quadrant, where both x and y are negative, for then their product is again positive.

In the north-west and south-east quadrants, where x and y have opposite signs, their product xy is negative, so the values of x and y that satisfy $x^2 + xy + y^2 = 9$ are *larger* than those that satisfy $x^2 + y^2 = 9$. So the ellipse lies *outside* the circle.

On the axes, either x or y is zero, so the product xy is zero too. The circle and ellipse therefore coincide.

If we had subtracted xy instead of adding it, thus producing the equation $x^2 - xy + y^2 = 9$, the ellipse would have lain outside of the circle in the north-east and south-west quadrants, and inside it in the north-west and south-east quadrants.

A more general form of the ellipse is $x^2 + axy + y^2 = c^2$, but the parameter a must be restricted to $-2 < a < 2$, otherwise a completely different shape results.

5.9 Application of circle and ellipse in economics

A possible use in economics for the circle or the ellipse is to model an economy's production possibility curve (*PPC*), also known as the transformation curve. We assume that an economy has, in the short run, given (that is, fixed) stocks of labour, capital, and other factors of production. There are also given levels of technological and managerial knowledge, labour force skills and all other factors that influence productive efficiency. We assume that the economy produces only two goods, X and Y, an assumption that is made for simplification only, and which can be relaxed without changing the analysis significantly. The *PPC* is then the function showing the various combinations of quantities of X and Y that the economy can produce. It is likely to have the shape shown in figure 5.15, where the curve is a segment of an ellipse, for positive values of X and Y. By putting all of its resources into producing Y, the economy can produce a maximum of $\bar{Y}$, and similarly by devoting all of its resources into producing X, the

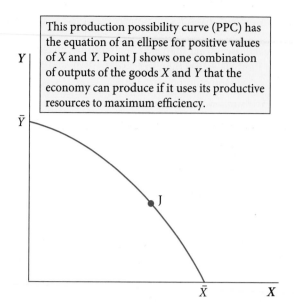

This production possibility curve (PPC) has the equation of an ellipse for positive values of X and Y. Point J shows one combination of outputs of the goods X and Y that the economy can produce if it uses its productive resources to maximum efficiency.

Figure 5.15 Production possibility curve.

economy can produce a maximum of $\bar{X}$. By dividing its resources between X production and Y production, the economy can produce any combination of the two goods traced out by the PPC, such as at J. Of course, there are other functional forms that can produce this shape.

Because the circle and ellipse are implicit functions, they are not easy to graph using Excel®. At the Online Resource Centre we show how implicit functions such as these can be graphed using Maple, a mathematics program used in many higher education institutions.

Summary of sections 5.8–5.9

(1) $(x-a)^2 + (y-b)^2 = c^2$ is the equation of a circle of radius c with its centre at the point $x=a$, $y=b$.

(2) By adding the additional term axy to the equation of a circle, we turn it into an ellipse (a shape in profile like a rugby or American football), but the parameter a must be restricted to $-2 < a < 2$, otherwise a completely different shape results.

(3) A circle or ellipse can be used to represent a production possibility curve (PPC), among several possibilities.

Progress exercise 5.3

On the same axes, sketch the graphs of

(a) $x^2 + y^2 = 16$ (b) $x^2 + xy + y^2 = 16$ (c) $x^2 - xy + y^2 = 16$

Take $x = -4, -3, -2, -1, 0, 1, 2, 3, 4$. Do not try to produce highly accurate graphs, but try to show the general shape of each function.

5.10 Inequalities

We use inequalities frequently in economics, so it's important to be confident about the rules for manipulating them. Since the rules for manipulating inequalities are almost identical to those for manipulating equations, it might seem more logical to have examined inequalities in chapter 3. However, we have delayed looking at inequalities until this point because experience suggests they can easily throw beginners off balance.

Notation for inequalities

$x > y$ means x is greater than y (this is called a 'strong' inequality)

$x \geq y$ means x is greater than, or equal to, y (a 'weak' inequality)

$x < y$ means x is less than y

$x \leq y$ means x is less than, or equal to, y

We can also write a 'double inequality', such as

$$0 < x < 1$$

This says two things: (1) x is greater than zero ($0 < x$); and (2) x is less than 1 ($x < 1$). Note that $0 < x < 1$ and $1 > x > 0$ mean the same thing.

Meaning of an inequality

As a preliminary, let us consider what it means to say that one number is greater or less than another. We can think of numbers as being measured along a line from some arbitrary starting point which we call zero (0); see figure 5.16. Positive numbers are measured off to the right, and negative numbers to the left. The line extends indefinitely in both directions, heading off to the left towards minus infinity (written as $-\infty$), and to the right towards plus infinity (written as $+\infty$). Thus the number ($+5$), or simply 5, is measured off as a distance of 5 units to the right of zero. Similarly, a negative number such as (-3) is measured off as a distance of 3 units to the left of zero.

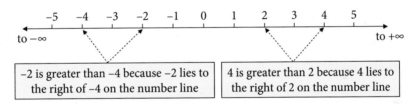

Figure 5.16 The number line.

Any number on the line is considered to be *greater* than any other number lying to the left of it. Thus, in an obvious way, $+4$ is greater than $+2$ because $+4$ lies to the right of $+2$ on the number line. We write this as $4 > 2$.

Slightly less obviously, -2 is greater than -4, again because -2 lies to the right of -4 on the line. So we write $-2 > -4$.

Occasionally we are interested in comparing the *absolute magnitude* of two numbers; that is, without paying any attention to whether they are positive or negative. We write $|x|$ to denote the absolute magnitude of some number x. Since the absolute magnitude ignores sign, it follows that $|-4| > 2$, even though $-4 < 2$.

Rules for manipulating inequalities

With three modifications, the rules for manipulating inequalities are exactly the same as the rules for manipulating equations. (See chapter 2 if you need to refresh your memory on the rules for manipulating equations.) Specifically, the rules are:

Addition and subtraction

We can add or subtract any number provided we add it to or subtract it from both sides of the inequality.

EXAMPLE 5.9

Given 14 > 13, we can add 100 to both sides and get 114 > 113. Alternatively, we can subtract 5 from both sides and get: 9 > 8.

EXAMPLE 5.10

Given $x > y$, we can subtract y from both sides and get $x - y > 0$.

Multiplication and division

Similarly, we can multiply or divide by any number provided we multiply or divide both sides of the inequality by it. However, this is where the first of the three modifications mentioned above comes in:

RULE 5.1 Multiplication and division of inequalities

When we multiply or divide both sides of an inequality by the same variable or constant, then if that variable or constant is negative, the direction of the inequality is reversed.

EXAMPLE 5.11

Given 2 > 1, we can multiply both sides by +5 and get 10 > 5.

EXAMPLE 5.12

Given 2 > 1, we can divide both sides by +5 and get $\frac{2}{5} > \frac{1}{5}$.

In examples 5.11 and 5.12, the direction of inequality is unchanged because we multiplied or divided by a positive number. Below, in examples 5.13 and 5.14 the direction of inequality is reversed because we multiply or divide by a negative number.

EXAMPLE 5.13

Given 2 > 1, we can multiply both sides by −5 and get −10 < −5.

EXAMPLE 5.14

Given 2 > 1, we can divide both sides by −5 and get $-\frac{2}{5} < -\frac{1}{5}$.

We can see that these reversals in the direction of inequality are correct when we look at where these numbers lie on the number line (see figure 5.16 above). For instance, in example 5.14, $-\frac{2}{5}$ lies to the left of $-\frac{1}{5}$.

This rule means that we must be very careful when multiplying or dividing an inequality. The general approach must be that when you are multiplying or dividing an inequality by something, and you don't know whether that something is positive or negative, you must open up two parallel paths of reasoning to cover the two possibilities. This is illustrated in the worked examples below.

Inversion of inequalities

When manipulating inequalities you may find that you want to invert both sides of an inequality. This brings us to the second of the three modifications mentioned above. Suppose we are given the inequality:

$$A > B \tag{5.5}$$

and we want to invert both sides. The correct algebraic procedure is to multiply both sides by $\frac{1}{AB}$, which gives

$$\frac{1}{AB}A <?> \frac{1}{AB}B$$

where the question mark indicates that we are not sure yet how this will affect the direction of the inequality. After cancelling on both sides, this simplifies to

$$\frac{1}{B} <?> \frac{1}{A} \qquad (5.6)$$

Now we can think about the direction of the inequality. Let us assume that $\frac{1}{AB}$ is negative. In that case, from rule 5.1, we know that when we multiply equation (5.5) by $\frac{1}{AB}$ this reverses the inequality. So in this case equation (5.6) becomes

$$\frac{1}{B} < \frac{1}{A} \qquad (5.7a)$$

Comparing equations (5.5) and (5.7a), we see that inverting both sides has reversed the direction of inequality. This is the result when $\frac{1}{AB}$ is negative. In turn, $\frac{1}{AB}$ is negative when A and B have opposite signs (one positive, the other negative).

If we now assume that $\frac{1}{AB}$ is positive, rule 5.1 tells us that when we multiply by $\frac{1}{AB}$ this leaves the direction of inequality unchanged. In that case, equation (5.6) has the same direction of inequality as equation (5.5), so equation (5.6) becomes

$$\frac{1}{B} > \frac{1}{A} \qquad (5.7b)$$

Comparing equations (5.5) and (5.7b), we see that inverting both sides left the direction of inequality unchanged. This is the result when $\frac{1}{AB}$ is positive, which in turn is true when A and B have the same sign (both positive or both negative). Thus we can deduce the following rule:

RULE 5.2 Inverting inequalities

Given $A > B$:

if A and B have opposite signs, then $\frac{1}{B} < \frac{1}{A}$ (equation 5.7a).

if A and B have the same sign, then $\frac{1}{B} > \frac{1}{A}$ (equation 5.7b).

There is no particular need to memorize rule 5.2. When inverting an inequality, you will automatically get the correct result if you understand that inversion of $A > B$ requires multiplying both sides by $\frac{1}{AB}$, and that doing this will (by rule 5.1) reverse the inequality if $\frac{1}{AB} < 0$, which is true if A and B have opposite signs.

Next we give two simple examples of rule 5.2 in action.

EXAMPLE 5.15

Given the inequality: $3 > 2$

To invert this we multiply both sides by $\frac{1}{3 \times 2}$. Because this is positive, rule 5.1 requires us to leave the direction of the inequality unchanged. After cancelling on both sides, we have $\frac{1}{2} > \frac{1}{3}$.

EXAMPLE 5.16

Given the inequality: $3 > -2$

To invert this we multiply both sides by $\frac{1}{3 \times -2} = -\frac{1}{6}$. Because this is negative, rule 5.1 requires us to reverse the direction of the inequality. After cancelling on both sides, we have $-\frac{1}{2} < -\frac{1}{3}$.

Example 5.19 below is a case where inverting both sides of an inequality is a method of solution.

Raising both sides of an inequality to any power

This is the third of the three modifications to the rules of algebra mentioned above. Raising both sides of an inequality to any power is straightforward only if both sides are positive. For example, given $4 > 2$, if we raise both sides to any positive power (say, 3 or $\frac{1}{2}$), the direction of inequality is unchanged, while if we raise both sides to any negative power $\left(\text{say}, -3 \text{ or } -\frac{1}{2}\right)$, the direction of inequality is reversed.

If both sides of the inequality are negative, the effect depends on the power. For example, given $-3 > -4$, raising to the power 2 reverses the inequality ($9 < 16$), but raising to the power 3 leaves the direction unchanged ($-27 > -64$). Raising to a fractional power is not always possible; for example $(-3)^{1/2} \equiv \sqrt{-3}$ does not exist as a real number.

If the two sides of the inequality have opposite signs, the effect in some cases depends on their absolute magnitudes. For example, squaring both sides of $-2 < 3$ gives $4 < 9$ (no reversal); but squaring both sides of $-3 < 2$ gives $9 > 4$ (reversal), while squaring $2 > -2$ gives $4 = 4$ (an equation!).

There are several other cases, but we have said enough to make it clear that raising both sides of an inequality to any power should be avoided if possible and only ever tackled with great care.

5.11 Examples of inequality problems

EXAMPLE 5.17

Solve the inequality $-5 + 8x > 19$.

By 'solve' is meant 'find the values of x that satisfy the inequality': that is, the values for which the inequality is true. This is exactly what we do to solve equations, except that when we solve an equation we find one or more specific values of some variable x, whereas when we solve an inequality we find not one or more specific numerical values for x, but instead a range of values.

To solve the above inequality, the first step is to add 5 to both sides, giving $8x > 24$. Next we divide both sides by 8. Since this is positive, it does not reverse the inequality. We then have

$$x > \frac{24}{8}; \quad \text{that is, } x > 3$$

Hopefully you can see that what we have just done is very similar to solving the equation $-5 + 8 = 19$ (with solution, $x = 3$). The difference is simply that to satisfy the inequality requires x to be greater than 3, while to satisfy the equation requires x to be *equal* to 3.

EXAMPLE 5.18

Solve the inequality $14 - 5x > 34$.

The first step is to subtract 14 from both sides, giving $-5x > 20$.

Next we divide both sides by -5. Since this is negative, this step reverses the inequality. We then have $\frac{-5x}{-5} < \frac{20}{-5}$, which simplifies to $x < -4$.

We can run a spot check on whether this answer is correct by choosing a value for x that is slightly less than -4 and seeing whether the given inequality is then satisfied. For example, if we set $x = -4.1$, the given inequality becomes

$$14 - 5(-4.1) > 34$$

which simplifies to $14 + 20.5 > 34$, which is true.

EXAMPLE 5.19

Find values of x for which $1 + \frac{1}{x} < 0$.

If we subtract 1 from both sides, this becomes $\frac{1}{x} < -1$. We now want to invert both sides of this in order to get x on one side. To do this we multiply both sides by x. This means we must consider two cases according to whether $x > 0$ or $x < 0$.

Case 1. Assume x positive. Then multiplying by x leaves the direction of inequality unchanged. So we get $x(\frac{1}{x}) < -1(x)$, which simplifies to $1 < -x$. Multiplying this by -1 (which reverses the inequality) gives $-1 > x$, which is the same as $x < -1$. But this is a contradiction, since we have assumed $x > 0$. We conclude that there is no positive value of x which satisfies the given inequality. (You may think this is intuitively obvious, but intuition can lead us astray and there's no harm in proving it even if it is obvious!)

Case 2. Assume x negative. Then multiplying by x reverses the direction of inequality. So we get $x(\frac{1}{x}) > -1(x)$, which simplifies to $1 > -x$. Multiplying this by -1 (which reverses the inequality) gives $-1 < x$, which is the same as $x > -1$. But we also know $x < 0$, by assumption. So the full answer is $x > -1$ and $x < 0$. We can write this more compactly as $-1 < x < 0$. In words, x must lie between -1 and 0. For example, $x = -0.5$. Then we have

$$1 + \frac{1}{x} = 1 + \frac{1}{-0.5} = 1 - 2 \quad \text{which is negative, as required.}$$

Note that there are other ways of solving this inequality that do not require inverting both sides. We tackled it this way to illustrate how to invert inequalities reliably.

EXAMPLE 5.20

Find values of x for which $x^2 + x > 2$.

This is more difficult than the previous examples because of the x^2 term. This means that we have a quadratic expression, and therefore we should be thinking that there might be two solutions, just as there are two solutions to a quadratic equation.

First, we subtract 2 from both sides, giving

$$x^2 + x - 2 > 0$$

The left-hand side is now a quadratic expression that factorizes, giving

$$(x + 2)(x - 1) > 0$$

There are at least two ways of proceeding from here.

Method 1. We divide both sides by $(x + 2)$ but then we need to consider two cases, according to whether $(x + 2)$ is positive or negative. This is because if $(x + 2)$ is negative, dividing by it will reverse the inequality.

Case (a). Assume $(x + 2)$ is positive. (This implies $x > -2$.) Divide both sides by $(x + 2)$. This does not reverse the inequality, since $(x + 2)$ is positive by assumption. This gives us

$$\frac{(x + 2)(x - 1)}{(x + 2)} > \frac{0}{(x + 2)}$$

After cancelling on the left-hand side, this leaves us with $x - 1 > 0$ which, after adding 1 to both sides, becomes $x > 1$.

So in case (a), the inequality is satisfied (true) if x is greater than $+1$. (If x is greater than $+1$, this automatically guarantees that $x > -2$, which we saw above was implied by the assumption of case (a) that $(x + 2)$ is positive.)

Case (b). Assume $(x + 2)$ is negative. (This implies $x < -2$.) Divide both sides by $(x + 2)$. This reverses the inequality, since $(x + 2)$ is negative by assumption. So we get

$$\frac{(x + 2)(x - 1)}{(x + 2)} < \frac{0}{(x + 2)}$$

After cancelling on the left-hand side, this leaves us with $x - 1 < 0$ which, after adding 1 to both sides, becomes $x < 1$.

But the condition $x < 1$ is guaranteed by the assumption of case (b) that $(x + 2)$ is negative, since this implies $x < -2$. So in case (b), the inequality is satisfied (true) if $x < -2$.

Combining the results of the two cases, the inequality is satisfied: either if x is positive and greater than $+1$ (case (a)); or if x is negative and less than -2 (case (b)). In symbols, $x > 1$ or $x < -2$.

You may be wondering why we divided by $(x + 2)$ and not $(x - 1)$. The answer is that dividing by $(x - 1)$ gives the same results, as you should check for yourself.

Method 2. This method is quicker but relies more on intuition, which can sometimes lead us astray. Given $(x + 2)(x - 1) > 0$ from above, we see that the left-hand side is positive only if $(x + 2)$ and $(x - 1)$ have the same sign (that is, either both positive or both negative). After some thought we see that $(x + 2)$ and $(x - 1)$ are both positive if $x > +1$ and both negative if $x < -2$.

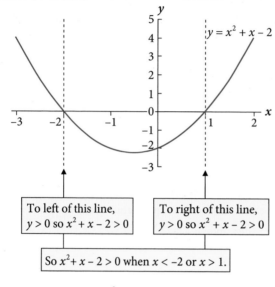

To left of this line, $y > 0$ so $x^2 + x - 2 > 0$

To right of this line, $y > 0$ so $x^2 + x - 2 > 0$

So $x^2 + x - 2 > 0$ when $x < -2$ or $x > 1$.

Figure 5.17 Solution to $x^2 + x - 2 > 0$.

Figure 5.17 gives a graphical illustration of the solution. There we have graphed the function $y = x^2 + x - 2$. The question asks us to find the values of x that make $x^2 + x - 2 > 0$. Given that $y = x^2 + x - 2$, this is equivalent to finding the values of x that make $y > 0$.

We can see from the graph that $y > 0$ when $x < -2$ or $x > 1$: that is, to the left of the left-hand broken line, or to the right of the right-hand broken line. So we conclude that $x^2 + x - 2 > 0$ when $x < -2$ or $x > 1$. This confirms the algebraic results above.

Strong versus weak inequalities

Finally, a few remarks about how to deal with weak inequalities may be useful. The question in example 5.20 was: for what values of x is $x^2 + x > 2$? Let us change this to: for what values of x is $x^2 + x \geq 2$? We now have a weak inequality condition, meaning that $x^2 + x = 2$ also satisfies the given condition on x, in addition to $x^2 + x > 2$.

Given this, we proceed by answering the two questions separately. That is, we first solve the inequality $x^2 + x > 2$, as we have just done above. Then as a separate task we solve the equation $x^2 + x = 2$. By subtracting 2 from both sides we turn this into the quadratic equation $x^2 + x - 2 = 0$, which factorizes immediately as $(x + 2)(x - 1) = 0$, from which $x = -2$ or $+1$. Combining these answers with those we obtained above, we can say that the weak inequality $x^2 + x \geq 2$ is satisfied when $x \leq -2$ or $x \geq 1$. In figure 5.17, the solution now includes the points *on* the broken lines at $x = -2$ and $+1$, as well as points to the left of $x - 2$ and to the right of $x = +1$.

5.12 Applications of inequalities in economics

The economic applications of inequalities are almost as numerous as the economic applications of equations. Here we look at two small examples, the consumption function in macroeconomics and the budget constraint in microeconomics.

The consumption function

In chapter 3 we met the consumption function: the relationship between planned aggregate consumption, $\hat{C}$, and aggregate income, Y. We assumed it was linear, with equation $\hat{C} = aY + b$, where a and b are both positive. We also defined the marginal propensity to consume (MPC) as the slope of the consumption function: in this case, a.

We now introduce a new concept, the average propensity to consume (APC), defined as $\frac{\hat{C}}{Y}$. This measures consumption as a proportion of income (or as a percentage of income, if multiplied by 100). The relationship between the APC and the MPC is of some interest, for reasons that would take too long to explain here. However, suppose we are asked to show that the MPC is less than the APC. Here is how we would proceed:

$$MPC = a \text{ (by definition)} \quad \text{and} \quad APC = \frac{\hat{C}}{Y} \text{ (by definition)}$$

So $MPC < APC$ implies $a < \dfrac{\hat{C}}{Y}$

But $\dfrac{\hat{C}}{Y} = \dfrac{aY + b}{Y}$ (from the definition of $\hat{C}$)

$\qquad = a + \dfrac{b}{Y}$ (by dividing top and bottom by Y)

Thus

$$a < \frac{\hat{C}}{Y} \quad \text{implies} \quad a < a + \frac{b}{Y} \quad \text{which implies} \quad 0 < \frac{b}{Y}$$

This last inequality is true, since b and Y are both positive. However, we can add that, as Y increases, $\frac{b}{Y}$ decreases, so APC falls, while MPC remains constant in the case of a linear consumption function.

The relationship between the MPC and the APC can also be shown graphically. Figure 5.18(a) shows a linear consumption function with a positive slope, a, and positive intercept, b. (See also sections 3.7 and 3.17.) We can read off the MPC immediately as the slope of the consumption

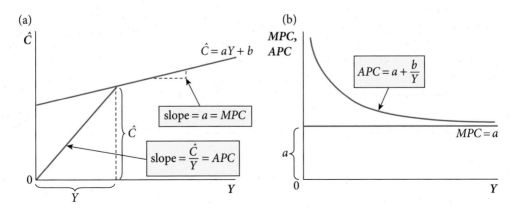

Figure 5.18 Relationship between MPC and APC.

function. We can also show the APC by drawing a straight line from the origin to any given point on the consumption function. That line is the hypotenuse of a triangle with height equal to $\hat{C}$ and base equal to Y. The slope of the hypotenuse therefore equals $\frac{\hat{C}}{Y}$, the APC at that level of income.

When we compare the two slopes, we see that MPC is necessarily less than APC at any income level. The difference between MPC and APC results from the assumption that the consumption function has a positive intercept, b.

Figure 5.18(b) graphs the MPC and APC. The graph of MPC = a is simply a horizontal line, a units above the horizontal axis. The graph of APC = $a + \frac{b}{Y}$ is a rectangular hyperbola (see section 5.5) which approaches the MPC asymptotically as Y approaches infinity.

The consumer's budget constraint

This is a key building block of the theory of consumer behaviour. We suppose that the typical consumer has a given weekly or monthly money income, B (measured in euros, say), and that there are only two goods, X and Y, with current market prices of p_X and p_Y. She spends part or all of her income buying X units of good X and Y units of good Y. As expenditure on any good equals price × quantity, her total expenditure on both goods is $p_X \times X + p_Y \times Y$, which of course in algebra is written as

$$p_X X + p_Y Y$$

We also assume that the consumer has no past savings and cannot borrow. She is therefore subject to the limitation, or constraint, that her total expenditure in each week or month cannot exceed her money income, B, in that time period. We can write this constraint as

$$p_X X + p_Y Y \leq B \tag{5.8}$$

This expression (in maths, a weak inequality) is called the consumer's budget or income constraint. It says that total expenditure in euros must be equal to, or less than, the consumer's income (also known as her budget).

The budget constraint can be shown graphically. The best way of approaching this is to see that the budget constraint is two statements combined:

$$p_X X + p_Y Y = B \quad \text{(if the consumer spends all of her income)} \tag{5.8a}$$

and

$$p_X X + p_Y Y < B \quad \text{(if she spends less than her income)} \tag{5.8b}$$

Equation (5.8a) gives the consumption possibilities when the consumer spends all of her budget, in which case the budget constraint is said to be *binding*. In this case, the budget constraint is an implicit linear function relating the two variables, X and Y, with the prices p_X and p_Y and income B as parameters (see section 3.7 for revision of implicit and explicit functions). We can rearrange equation (5.8a) to express Y, the quantity of good Y purchased, as an explicit function of X. To do this, we subtract $p_X X$ from both sides of equation (5.8a), then divide through by p_Y. This gives

$$Y = \frac{B}{p_Y} - \frac{p_X}{p_Y} X \tag{5.9}$$

This may become clearer with a numerical example. Suppose B = 100, $p_X = 2$ and $p_Y = 5$ (all measured in euros). Then equation (5.9) becomes

$$Y = 20 - \frac{2}{5} X$$

This is a linear function with a Y-intercept of 20 and a slope of $-\frac{2}{5}$, and is graphed in figure 5.19. It is called the **budget line**. Let us think about what this means.

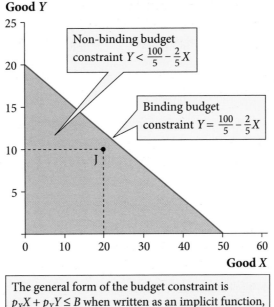

Good Y

Non-binding budget constraint $Y < \frac{100}{5} - \frac{2}{5}X$

Binding budget constraint $Y = \frac{100}{5} - \frac{2}{5}X$

J

Good X

The general form of the budget constraint is $p_X X + p_Y Y \leq B$ when written as an implicit function, and $Y \leq \frac{B}{p_Y} - \frac{p_X}{p_Y}X$ when written in explicit form.

Figure 5.19 Consumer's budget constraint.

First, the Y-intercept of 20 means that, if she buys zero units of X, the consumer can buy 20 units of y. This figure of 20 is her income divided by the price of good Y. We can think of this as measuring her **real income**, in units of Y. Her money income of 100 euros is worth 20 units of Y.

Second, the slope of $-\frac{2}{5}$ means that, as she moves down her budget line, she must give up two-fifths of a unit of Y in order to buy one additional unit of X. This makes sense because, since a unit of Y costs 5, giving up two-fifths of a unit will save two-fifths of 5, which is 2. This is just enough to buy one more unit of X. Thus we can think of the absolute slope of the budget line, which is given by the price ratio $\frac{p_X}{p_Y}$, as measuring the **opportunity cost** of X in terms of Y: that is, the number of units of Y which must be forgone in order to consume one more unit of X. The higher is p_X relative to p_Y, the larger is the number of units of Y which must be forgone.

Finally, let us look at equation (5.8b) above:

$$p_X X + p_Y Y < B$$

Strictly speaking, this is of course an inequality rather than an equation. It simply describes the state of affairs when the consumer does not spend all of her budget. In that case she no longer has to consume less of one good in order to consume more of the other. For example, at point J in figure 5.19 she buys 20 units of X and 10 units of Y. Her total expenditure is therefore $20 \times 2 + 10 \times 5 = 90$, which is less than her income. All the points within the shaded area satisfy this inequality.

Summary of sections 5.10–5.12

In sections 5.10–5.11 we saw that inequalities can be manipulated exactly like equations, except that multiplying or dividing by a negative value reverses the direction of the inequality. We demonstrated how to solve some inequality problems. In section 5.12 we studied the aggregate consumption function and the consumer's budget constraint as two areas of economic analysis involving inequalities.

? **Progress exercise 5.4**

1. If x is restricted by the condition $0 < x < 2$, find the range of values that y can take, given $y = 2x + 1$.

2. Find the values of x for which $\frac{1}{x} + 1 > -5$.

3. Find the values of x for which $\frac{1}{x} + 1 > 2x$.

4. Find the values of x for which $(x - 1)^2 < 2 - 2x$.

5. If x is restricted by the condition $0 < x < 2$, find the range of values that y can take, given $y = (x - 1)^2$.

6. A consumer's income or budget is 120. She buys two goods, x and y, with prices 3 and 4 euros respectively.

   (a) Derive the budget constraint (i) as an implicit function and (ii) with the quantity of y purchased as an explicit function of the quantity of x purchased.

   (b) What is the opportunity cost of x in terms of y?

   (c) What is the consumer's real income, measured in units of (i) good y and (ii) good x?

   (d) If the consumer buys 20 units of x and 12 units of y, is the budget constraint binding?

   (e) If the consumer spends 40% of her income on x and 60% on y, how much of each good does she buy?

Checklist

Be sure to test your understanding of this chapter by attempting the progress exercises (answers are at the end of the book). The Online Resource Centre contains further exercises and materials relevant to this chapter www.oxfordtextbooks.co.uk/orc/renshaw3e/.

In this chapter we have examined some new types of non-linear function, and introduced two new ideas: limits and discontinuities. The specific topics were:

✔ **Cubic functions and equations.** Recognizing a cubic function and the possible shapes its graph may have. Sketching the graph of a cubic function from inspection of its parameters. Understanding why a cubic equation may have one or three real solutions.

✔ **The rectangular hyperbola.** Recognizing the equation of a rectangular hyperbola and drawing a rough sketch of its graph from inspection of its parameters.

✔ **Limits and discontinuities.** Understanding the concepts of limiting values and discontinuity in a function.

✔ **The circle and ellipse.** Recognizing the equations of a circle and an ellipse, and drawing rough sketches of their graphs.

✔ **Inequality expressions.** Understanding the rules for manipulating inequalities. Solving problems involving inequalities.

✔ **Economic applications.** Applying the cubic function, the rectangular hyperbola, the circle and the ellipse in economic modelling. Inequalities in economic analysis.

We have now completed all the necessary preliminary work on algebraic and graphical methods and are ready to move on to part 2 of this book, where we introduce and develop the ideas of differential calculus upon which many of the key results of economic theory depend.

Part Two
Optimization with one independent variable

- Derivatives and differentiation
- Derivatives in action
- Economic applications of functions and derivatives
- Elasticity

Chapter 6
Derivatives and differentiation

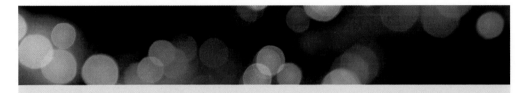

OBJECTIVES

Having completed this chapter you should be able to:

- Understand what is meant by positive, zero, or negative slope of a curve.
- Calculate the difference quotient $\frac{\Delta y}{\Delta x}$ and understand what it means.
- Understand how the derivative $\frac{dy}{dx}$ is obtained and what it measures.
- Find derivatives using the main rules of differentiation.

6.1 Introduction

In several of the economic applications we have considered, it has become apparent that the slope of a function can be of great importance. For example, it is obviously essential for a firm to know whether its total cost curve has a relatively flat slope, or alternatively slopes upwards steeply; or whether the demand curve for its product slopes downwards steeply or gently.

In this chapter, which is probably the most important of the whole book, we develop the necessary theory and technique for measuring the slope of a curve at any point. This technique is called **differentiation**, differential calculus, or simply **calculus**.

To do this we first look in more depth at the slope of a linear function, which we examined in chapter 3. Then we explain how the problem of defining the slope of a curve, which baffled mathematicians for centuries, was finally overcome by Newton and Leibniz in the seventeenth century. They not only defined the slope of a curve at any point as being the slope of a tangent to the curve at that point, but also showed how the slope could be calculated. Then we list and illustrate the rules, based on their theory, for differentiating (finding the slope of) different types of function. Although there are no economic applications in this chapter, the technique of differentiation is used extensively in later chapters: for example, in finding marginal cost and marginal revenue, demand and supply elasticities, and a firm's most profitable output.

6.2 The difference quotient

In section 3.7 we noticed that, whenever we plotted the graph of a function of the general form $y = ax + b$, the slope was always given by the parameter a. But we did not prove this; we simply noted it as what social scientists call 'an empirical regularity'. This is not fully satisfactory, since it remains logically possible that when we next plot a linear function its slope will *not* equal the parameter a. Our first step in this chapter, therefore, is to prove this proposition rigorously, thus ruling out the possibility that a surprise awaits us next time we plot a function of the form $y = ax + b$.

If we take the general form of the linear function, $y = ax + b$, and assume that a and b are both positive, then the curve will have a shape broadly similar to figure 6.1a. (Recall that in maths both straight and curved lines are called curves.) For simplicity we consider only positive values of x.

We can measure the slope of the curve in figure 6.1a in the following way. Let P and Q be any two points on the function, with coordinates (x_0, y_0) and (x_1, y_1) respectively. Then if we imagine that the curve is a hill and we are walking up it from P to Q, we would define the slope as

$$\text{slope(or gradient)} = \frac{\text{distance travelled up}}{\text{distance travelled forward}} = \frac{QK}{PK}$$

where QK is the increase in our height up the hill, and PK is the distance travelled forward. For example, if QK was 50 metres and *PK* was 100 metres, then we have gained 50 metres in height by travelling 100 metres horizontally, and we would say that the slope was 50 divided by 100, which equals 0.5 or 50%.

We will now put this idea of slope on a more rigorous mathematical footing. First, in figure 6.1(a) we see that the distance QK is given by $y_1 - y_0$. (For example, if $y_0 = 3$ and $y_1 = 5$, then y_1 lies 2 units above y_0, and $y_1 - y_0 = 5 - 3 = 2$.)

For compactness of notation, we will write Δy to denote $y_1 - y_0$. We call Δy the **increment** (= increase) in y which occurs when we move from P to Q. (Δ is the Greek upper-case letter 'delta'; see appendix to chapter 2.)

Similarly $PK = x_1 - x_0$. Again, we will write Δx to denote $x_1 - x_0$. We call Δx the increment (= increase) in x which occurs when we move from P to Q. Therefore:

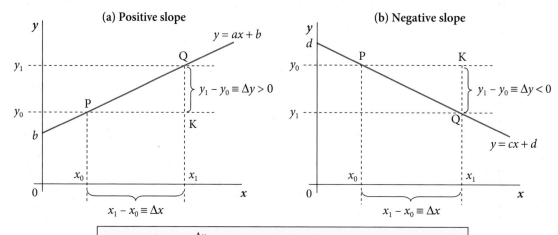

The slope of PQ $\equiv \frac{\Delta y}{\Delta x}$ is called the difference quotient. In figure 6.1(a) it is positive because Δy is positive. It measures the slope of $y = ax + b$.
In figure 6.1(b it is negative because Δy is negative. It measures the slope of $y = cx + d$.

Figure 6.1 The difference quotient for positive and negative slopes.

$$\text{slope} = \frac{QK}{PK} = \frac{y_1 - y_0}{x_1 - x_0} = \frac{\Delta y}{\Delta x} = \frac{\text{increase in } y}{\text{increase in } x} \tag{6.1}$$

We call $\frac{\Delta y}{\Delta x}$ the **difference quotient**. It measures the slope of the function between P and Q. Because the difference quotient is the increase in y divided by the increase in x, it gives us the change in y *per unit of change in x*. We call this the **rate of change** of y.

Thus 'slope', 'gradient', and 'rate of change' are identical in meaning. All are measured by the difference quotient $\frac{\Delta y}{\Delta x}$. To avoid the fuzziness that results from using different words for the same thing, in this book we will mainly use the term 'slope', though occasionally 'rate of change' will be used.

Positive and negative slope

In drawing figure 6.1(a) we assumed the parameter a in $y = ax + b$ to be positive, so the curve slopes upwards from left to right. An increase in x results in an *increase* in y. The slope, or rate of change, is said to be positive. For example, when we move from P to Q we might find that x increases by 6 and y increases by 3. Then we have $\Delta x = 6$ and $\Delta y = 3$, and therefore $\frac{\Delta y}{\Delta x} = \frac{3}{6} = \frac{1}{2}$. Thus the rate of change of y is $\frac{1}{2}$, meaning that if x changes by, say, 10 units, y changes by 5 units. The change in y per unit of change in x is $\frac{1}{2}$.

In figure 6.1(b) we have graphed another linear function, $y = cx + d$. Here we have assumed that the parameter c is negative, so the curve slopes downward from left to right. The effect of this is that Q lies below P, so y_1 is less than y_0. Therefore $\Delta y \equiv y_1 - y_0$ is negative, and consequently so is $\frac{\Delta y}{\Delta x}$. We call this slope negative, because an increase in x leads to a *decrease* in y.

Generalizing the difference quotient

Although we have developed the idea of the difference quotient $\frac{\Delta y}{\Delta x}$ in the context of the linear function $y = ax + b$, there is nothing in the definition to stop us applying it to any function. So we will adopt the following general definition of the difference quotient:

RULE 6.1 The difference quotient

For any function $y = f(x)$, suppose P and Q are any two points on the curve, with coordinates (x_0, y_0) and (x_1, y_1) respectively (see figure 6.1). Then by definition the *difference quotient* (see equation (6.1) above) is

$$\frac{\Delta y}{\Delta x} = \frac{y_1 - y_0}{x_1 - x_0}$$

This measures the ratio of the change in y to the associated change in x. This ratio is called the *rate of change* of y between points P and Q.

6.3 Calculating the difference quotient

In the previous section we defined the difference quotient and saw that it measures the slope (= rate of change) of the function we are looking at. Now we will develop a formula for measuring, or calculating, the difference quotient.

In figure 6.1(a) we know the equation of the line is $y = ax + b$. Consider point P. Since its coordinates are (x_0, y_0) and it lies on the line, it must be true at P that $y_0 = ax_0 + b$. Similarly, at Q, $y_1 = ax_1 + b$. Therefore

$$\Delta y = y_1 - y_0 = (ax_1 + b) - (ax_0 + b) = ax_1 + b - ax_0 - b = ax_1 - ax_0 = a(x_1 - x_0)$$

Thus $\Delta y = a(x_1 - x_0)$. If we substitute this into our expression for the difference quotient, equation (6.1), we get

$$\frac{\Delta y}{\Delta x} = \frac{y_1 - y_0}{x_1 - x_0} = \frac{a(x_1 - x_0)}{x_1 - x_0}$$

After cancelling $x_1 - x_0$ between numerator and denominator, this becomes

$$\frac{\Delta y}{\Delta x} = a \qquad (6.2)$$

Equation (6.2) tells us that:

(1) The slope, $\frac{\Delta y}{\Delta x}$, of the linear function $y = ax + b$ is given by the parameter a. We were already very confident that this was true, from our graphical work in section 3.7. But we have now confirmed it rigorously by algebraic methods.

(2) The slope equals a, wherever we measure it. This is true because equation (6.2) is true whatever the values of x_0 and x_1. Therefore since a is constant, the slope is constant. This proves rigorously that the graph of $y = ax + b$ is a straight line, confirming what we found in our graphical work in section 3.6.

 Progress exercise 6.1

1. For each of the following linear functions, calculate the difference quotient for the specified change in x. In each case, sketch the graph and indicate where you have measured the difference quotient.

(a) $y = 2x + 4$ $\quad x_0 = 2, x_1 = 3$ $\qquad$ (b) $y = \frac{1}{2}x - 2$ $\quad x_0 = -1, x_1 = -\frac{1}{2}$

(c) $y = 50x$ $\quad x_0 = 1, x_1 = 10$ $\qquad$ (d) $y = -3x - 9$ $\quad x_0 = 5, x_1 = 5.1$

(e) $y = 20$ $\quad x_0 = 1, x_1 = 2$ $\qquad$ (f) $y = -\frac{1}{2}x + \frac{1}{4}$ $\quad x_0 = 0, x_1 = 1$

2. (a) For cases (a)–(c) above, show that the value of the difference quotient is independent of the values of x_0 and x_1.

(b) Show that 2(a) is also true for the general linear function $y = ax + b$, where a and b are any constants.

6.4 The slope of a curved line

In the previous section we developed the concept of the difference quotient, $\frac{\Delta y}{\Delta x}$, and showed that we could measure the slope of a straight line by calculating the difference quotient. We found that the slope of the linear function $y = ax + b$ is given by the parameter a.

In this section we will extend these ideas to include curves as well as straight lines. We will use the quadratic function $y = x^2 + 4$ as an example, and again for simplicity we will consider only positive values of x.

Suppose, then, that we want to find the slope of the curve $y = x^2 + 4$, which is sketch graphed in figure 6.2 for positive x.

Of course, the defining feature of a curve is that its slope is constantly changing. In figure 6.2, as x increases and we move up the curve, the slope gets steeper. If this did *not* happen—in other words, if the slope remained constant as we moved up the curve—then it would not be a curve but a straight line. This causes problems in both defining and measuring the slope, as we shall now see.

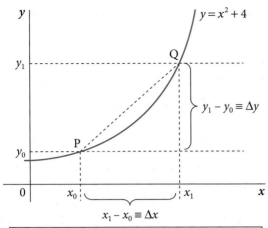

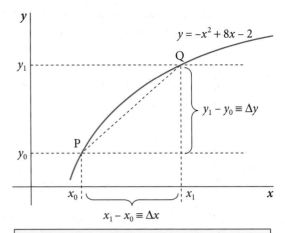

The difference quotient (= slope of dotted chord PQ $\equiv \frac{\Delta y}{\Delta x}$) measures only the average slope of $y = x^2 + 4$ between P and Q.

The average slope between P and Q is the same as in figure 6.2 although the function is very different in its shape.

Figure 6.2 The difference quotient as a measure of the slope of $y = x^2 + 4$.

Figure 6.3 The difference quotient as a measure of the slope of $y = -x^2 + 8x - 2$.

The difference quotient as a measure of slope

Suppose we try to measure the slope of the curve $y = x^2 + 4$ by calculating the difference quotient, as we did with the straight line $y = ax + b$ in the previous section.

Our first step is to pick any two points P and Q on the curve, with coordinates (x_0, y_0) and (x_1, y_1) respectively (see figure 6.2). Then by definition the difference quotient (see rule (6.1) above) is

$$\frac{\Delta y}{\Delta x} = \frac{y_1 - y_0}{x_1 - x_0}$$

But from figure 6.2 we see that the difference quotient, $\frac{\Delta y}{\Delta x}$, is not really measuring the slope of the curve $y = x^2 + 4$, but instead is measuring the slope of the dotted *straight* line joining P to Q. (A straight line connecting two points on a curve is called a **chord**, pronounced 'cord'.) In effect, the difference quotient treats the curve as if it were a straight line between P and Q. We can see this very clearly if we consider another curve, $y = -x^2 + 8x - 2$, which has a very different shape, but which also passes through points P and Q (see figure 6.3). The difference quotient has exactly the same value as in figure 6.2, despite the big difference in shape between the two curves.

Another way of looking at this is to see that the slope of the dotted straight line joining P to Q is equal to the *average* slope of any curve passing through the two points. In figure 6.2, in the neighbourhood of P the curve slopes less steeply than the dotted straight line, while in the neighbourhood of Q it slopes more steeply. In figure 6.3 the reverse if true. But since the straight line and the two curves both start at P and finish at Q, on average they have the same slope between the two points.

Thus the difference quotient is not very satisfactory as a measure of the slope of a curve, as it only gives the average slope between two points on the curve. For some purposes it is sufficiently accurate to measure the average slope of a curve, but in other situations it can lead to serious errors. This troubled mathematicians for centuries.

The tangent as a measure of slope

Eventually, in the seventeenth century, the problem was solved by Newton and Leibniz (who then argued for many years about who had thought of it first). They saw that the slope of a curve

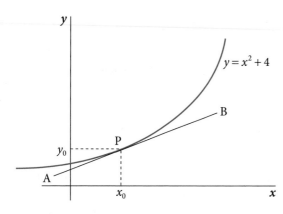

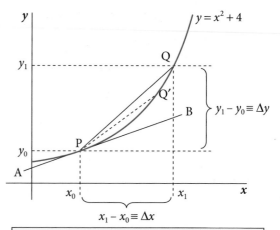

The slope of the tangent AB at P is unique. A tangent with a steeper or less steep slope than AB would touch the curve to the right or left of P.

As Δx approaches zero, Q slides down the curve towards P and $\frac{\Delta y}{\Delta x}$ (= slope of PQ) approaches the slope of tangent APB.

Figure 6.4 The slope of the tangent measures the slope of $y = x^2 + 4$ at a point.

Figure 6.5 Finding the slope of the tangent to the curve $y = x^2 + 4$.

at any point could be measured by drawing the **tangent** to the curve at that point. A tangent to a curve is defined as a straight line that touches the curve at only one point. Thus in figure 6.4 we have drawn the tangent AB to the curve at P. The tangent and the curve therefore have only one point in common: the point P. The slope of the tangent at P is unique; a tangent with a slope slightly steeper or less steep would touch the curve above or below P.

Because of this uniqueness, we can *define* the slope of the curve at P as being equal to the slope of the tangent at P. Strictly speaking, 'the slope at P' is meaningless because P is just a point and a point has no slope. However, what in effect we are saying is that, very close to P, the curve may be treated as linear with a slope equal to the tangent's slope at P. Thus to find the slope of the curve at P we merely have to find the slope of the tangent AB. This we will do next.

6.5 Finding the slope of the tangent

Step 1: Finding the difference quotient

Continuing with the example of $y = x^2 + 4$, our first step towards finding the slope of the tangent at some point P, with coordinates (x_0, y_0) is to find the difference quotient. To do this, we pick a second point on the curve, Q, with coordinates (x_1, y_1) (see figure 6.5). Then by definition (see rule 6.1) the difference quotient is

$$\frac{\Delta y}{\Delta x} = \frac{y_1 - y_0}{x_1 - x_0} \tag{6.3}$$

where (x_0, y_0) and (x_1, y_1) are the coordinates of P and Q respectively (see figure 6.5).

Now, since P lies on the curve $y = x^2 + 4$, we know that $y_0 = (x_0)^2 + 4$. Similarly, since Q lies on the curve $y = x^2 + 4$, we know that $y_1 = (x_1)^2 + 4$. Therefore

$$y_1 - y_0 = [(x_1)^2 + 4] - [(x_0)^2 + 4] \tag{6.4}$$

Now, by definition, $\Delta x = x_1 - x_0$, and we can rearrange this as $x_1 = x_0 + \Delta x$. We use this to substitute for x_1 in equation (6.4) above, which then becomes

$$y_1 - y_0 = [(x_0 + \Delta x)^2 + 4] - [(x_0)^2 + 4] \tag{6.5}$$

If we multiply out $(x_0 + \Delta x)^2$ on the right-hand side, we get

$$(x_0 + \Delta x)^2 = (x_0)^2 + 2x_0(\Delta x) + (\Delta x)^2$$

(See section 4.2 if you are unsure of this step.) Therefore equation (6.5) becomes

$$y_1 - y_0 = [(x_0)^2 + 2x_0(\Delta x) + (\Delta x)^2 + 4] - [(x_0)^2 + 4]$$

which on removing the square brackets simplifies to

$$y_1 - y_0 = 2x_0(\Delta x) + (\Delta x)^2$$

(because the $(x_0)^2$ and the 4 both disappear). If we substitute this into equation (6.3), we get

$$\frac{\Delta y}{\Delta x} = \frac{y_1 - y_0}{x_1 - x_0} = \frac{2x_0(\Delta x) + (\Delta x)^2}{x_1 - x_0} \tag{6.6}$$

The final step is to use the fact that, by definition, $\Delta x = x_1 - x_0$, so equation (6.6) becomes

$$\frac{\Delta y}{\Delta x} = \frac{2x_0(\Delta x) + (\Delta x)^2}{\Delta x} = \frac{\Delta x[2x_0 + \Delta x]}{\Delta x}$$

After cancelling Δx between numerator and denominator, this becomes

$$\frac{\Delta y}{\Delta x} = 2x_0 + \Delta x \tag{6.7}$$

This is the difference quotient for the function $y = x^2 + 4$. It measures the slope of the straight line (the chord) joining P to Q in figure 6.5.

Step 2: Using the difference quotient to find the slope of the tangent

In step 1, the position of Q was chosen arbitrarily (see figure 6.5). Now comes the key step in the reasoning. The position of Q is governed by the size of Δx. It occurs to us that if we reduce the size of Δx, this will cause Q to slide down the curve towards P. Consequently, the slope of the chord PQ (which is what the difference quotient measures) will become more similar to the slope of the tangent AB (which is what we are trying to measure). For example, if we compare Q and Q′ in figure 6.5, the slope of the chord PQ′ is closer than is the slope of the chord PQ to the slope of the tangent AB.

Indeed, it is clear from the diagram that we can continue this process, moving Q closer and closer to P by making Δx smaller and smaller. As we do this, the slope of the chord PQ will approach closer and closer to the slope of the tangent AB.

Following this logic to its limit, we can say that, as Δx approaches zero, so Q will approach closer and closer to P, and therefore the slope of the chord PQ will approach a limiting value. This *limiting value* will be the slope of the tangent AB. (See section 5.6 if you need to refresh your understanding of a limit or limiting value.)

Thus our key conclusion is that, in figure 6.5: *as Δx approaches zero, the slope of the chord PQ approaches the slope of tangent AB.* Using the notation for limits that we introduced in section 5.6, we can write this conclusion more compactly as

$$\lim_{\Delta x \to 0} \text{ of (slope of chord PQ)} = \text{(slope of tangent at P)} \tag{6.8}$$

Since we know from step 1 above that the slope of chord PQ is measured by the difference quotient $\frac{\Delta y}{\Delta x}$, we can write equation (6.8) as

$$\lim_{\Delta x \to 0} \frac{\Delta y}{\Delta x} = \text{slope of tangent at P} \tag{6.9}$$

Now, 'slope of the tangent at P' is so important that a special symbol, $\frac{dy}{dx}$, is used to denote it. Using this symbol, we can write equation (6.9) as

$$\lim_{\Delta x \to 0} \frac{\Delta y}{\Delta x} = \frac{dy}{dx} = \text{slope of tangent at P} \qquad (6.9a)$$

The symbol $\frac{dy}{dx}$ is called the **derivative** of whatever function we started with: in this case, $y = x^2 + 4$. Note that $\frac{dy}{dx}$ must *not* be read as 'd times y, divided by d times x'. Instead, $\frac{dy}{dx}$ is a symbol that must be read as a whole, and means 'the slope of the tangent at P'.

Equation (6.9a) is a fundamental formula. It tells us that to find the slope of the tangent at P, we must first find the difference quotient and then examine it to find what limiting value it approaches as Δx approaches zero. Let us return to our example, $y = 2x^2 + 4$, and see how this works out.

In our example, $y = 2x^2 + 4$, we found (see equation (6.7)) that

$$\frac{\Delta y}{\Delta x} = 2x_0 + \Delta x$$

where x_0 is the value of x at P in figure 6.5.

If we substitute this into our fundamental formula, equation (6.9a), we get

$$\lim_{\Delta x \to 0} (2x_0 + \Delta x) = \frac{dy}{dx} = \text{slope of tangent at P}$$

Looking at $2x_0 + \Delta x$, we see that, as $\Delta x \to 0$, the Δx term becomes negligibly small and can therefore be discarded, leaving only $2x_0$ as the limiting value. So we conclude:

$$\lim_{\Delta x \to 0} \frac{\Delta y}{\Delta x} = \frac{dy}{dx} = \text{slope of tangent at P} = 2x_0 \qquad (6.10)$$

where x_0 is the x-coordinate of P.

Up to this point we have thought of P as a *specific* point on the curve $y = 2x^2 + 4$. If, however, we now think of P as *any* point on the curve, with an x-coordinate simply denoted by x (rather than x_0, which denotes a *specific* point in figure 6.5), then equation (6.10) becomes

$$\lim_{\Delta x \to 0} \frac{\Delta y}{\Delta x} = \frac{dy}{dx} = \text{slope of tangent at P} = 2x \qquad (6.10a)$$

where the value of x is given by the x-coordinate of any point, P, on the curve.

This is a very powerful result. Equation (6.10a) tells us that, for the function $y = x^2 + 4$, the slope of the tangent at *any* point on the curve is given by $2x$, where it is understood that x is the x-coordinate of the point in question. For example, at the point $x = 3$, the slope is $2 \times 3 = 6$; at $x = 5$, the slope is $2 \times 5 = 10$, and so on. (Check this by drawing a fairly accurate graph of $y = x^2 + 4$, drawing tangents at $x = 3$ and $x = 5$ as accurately as you can, then measuring their slopes.)

6.6 Generalization to any function of *x*

In the previous two sections we worked with the example function $y = x^2 + 4$. However, there was no step in our analysis that depended on our choice of function. Therefore our results are valid for *any* function of x. Thus, assume now that we are dealing with some unspecified function of x, which we will write as

$$y = f(x)$$

Here, the notation 'f(x)' should be read as 'a function of x'. We use this notation when we want to talk about a functional relationship between x and y but don't wish to specify, or perhaps don't know, the precise algebraic form of this function. Later, we might decide, or find out, that $f(x) = x^3 + 2x^2 + 3x + 99$, or some other specific functional form.

Thus equation (6.9a), which we derived above for the specific example $y = x^2 + 4$, is valid for any function, so we can generalize the results of the previous section as a rule:

RULE 6.2 **The fundamental formula that defines the derivative**

For any function $y = f(x)$, the derivative $\frac{dy}{dx}$ is:

$$\lim_{\Delta x \to 0} \frac{\Delta y}{\Delta x} = \frac{dy}{dx} = \text{slope of the tangent at any point P on the curve}$$

Note that we often write $f'(x)$ instead of $\frac{dy}{dx}$ as $f'(x)$ looks more tidy and takes up less space on the page. So $f'(x) \equiv \frac{dy}{dx}$ and should also be read as 'the slope of tangent at any point P on the function $y = f(x)$'.

Similarly, if we were looking at some function identified as $y = g(x)$, where g(. . .) denotes some unspecified functional form, we could refer to its derivative as $g'(x)$.

There are two restrictions on the fundamental formula in rule 6.2. First, the function in question must be a smooth curve, without any jumps or spikes in it. Second, if there is a discontinuity in the function at some point, as in the cases we looked at in section 5.6, then obviously the derivative doesn't exist at that point.

Rule 6.2 tells us that, to find the slope of any function $y = f(x)$, we have to follow the two-step procedure of the previous section. Step 1 is to find the difference quotient $\frac{\Delta y}{\Delta x}$, by applying rule 6.1. Step 2 is to find the limiting value that this approaches as Δx approaches zero. By rule 6.2, this limiting value gives us the slope of the tangent to the curve, and therefore the slope of the function in question, at any desired point, which we denote by the symbol $\frac{dy}{dx}$.

However, this method of finding the derivative, $\frac{dy}{dx}$, of a function is lengthy and tedious. Fortunately, there is a set of rules for finding the derivative of various types of function, which speeds up the process of finding a derivative enormously. We will now look at these rules.

6.7 Rules for evaluating the derivative of a function

Below we state the rules for finding the derivative of most types of function. For each rule we include some worked examples to show how the rule is applied. The process of finding a derivative is known as differentiation, differential calculus, or simply calculus. Because these rules are so important, we have numbered them separately from the other rules in this chapter.

RULE D1 **The power rule**

For any function of the form: $y = x^n$

the derivative is: $\frac{dy}{dx} = nx^{n-1}$

Here are some examples to show how the power rule works.

EXAMPLE 6.1

If $y = x^2$, we have $n = 2$, so $\frac{dy}{dx} = 2x^{2-1} = 2x^1 = 2x$

EXAMPLE 6.2

If $y = x^3$, we have $n = 3$, so $\frac{dy}{dx} = 3x^{3-1} = 3x^2$

(and so on, to higher and higher powers of x)

EXAMPLE 6.3

If $y = x^{1/2}$, we have $n = \frac{1}{2}$, so $\frac{dy}{dx} = \frac{1}{2}x^{(1/2)-1} = \frac{1}{2}x^{-1/2}$

(Recall by the way that $x^{1/2}$ is the same thing as $\sqrt{x}$.)

EXAMPLE 6.4

If $y = x^{-2}$, we have $n = -2$, so $\frac{dy}{dx} = -2x^{-2-1} = -2x^{-3}$

(Recall that x^{-2} is the same thing as $\frac{1}{x^2}$, and $-2x^{-3}$ is the same thing as $-\frac{2}{x^3}$.)

EXAMPLE 6.5

If $y = x$, we have $n = 1$, so $\frac{dy}{dx} = 1x^{1-1} = x^0 = 1$

It's easy to get confused over this example. But it makes sense, since the graph of $y = x$ is a line from the origin at an angle of 45 degrees, so its slope is 1.

RULE D2 A multiplicative constant

In words, this rule simply says that a multiplicative constant in a function carries over unchanged into that function's derivative. In formal mathematical language, we write this as

the function ⤑ its derivative

If $y = A\,f(x)$, where A is any constant, then $\frac{dy}{dx} = A\,f'(x)$

multiplicative constant in original function ⤑ reappears in the derivative

Note that we have written $f'(x)$ to denote the derivative of the function $f(x)$. This is a convenient notation which takes up less space on the page than $\frac{dy}{dx}$. We shall use it frequently from now on.

Stated formally like this, rule D2 seems more difficult than it really is. Some examples will show that it is actually very easy to apply.

EXAMPLE 6.6

$y = 10x^2$

Here, $A = 10$ and $f(x) = x^2$. From example 6.1 above, we know that the derivative of x^2 is $2x$. Thus $f'(x) = 2x$. So applying rule D2, we have

Given $y = Af(x) = 10x^2$

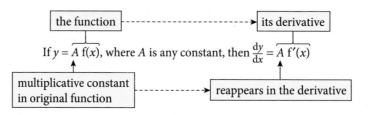

$$\frac{dy}{dx} = 10(2x) = 20x$$

derivative of x^2

multiplicative constant in original function, so reappears in derivative

EXAMPLE 6.7

$y = 15x^3$

Here, $A = 15$ and $f(x) = x^3$. From example 6.2 above, we know that the derivative of x^3 is $3x^2$. Thus $f'(x) = 3x^2$. So applying rule D2 gives

$$\frac{dy}{dx} = 15(3x^2) = 45x^2$$

EXAMPLE 6.8

$$y = \frac{1}{50}x^{1/2}$$

Here, $A = \frac{1}{50}$ and $f(x) = x^{1/2}$. From example 6.3 above, we know that the derivative of $x^{1/2}$ is $\frac{1}{2}x^{-1/2}$. So applying rule D2 gives

$$\frac{dy}{dx} = \frac{1}{50}\left(\frac{1}{2}x^{-1/2}\right) = \frac{1}{100}x^{-1/2}$$

RULE D3 An additive constant

In words, this rule simply says that when a function has an additive constant, that constant does *not* carry over into the derivative. We write this as follows:

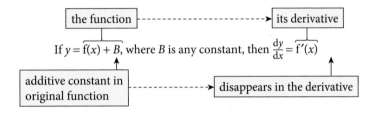

The rule applies equally if the constant is negative: that is, if it is subtracted rather than added. Here is an example of the rule in operation.

EXAMPLE 6.9

$$y = x^2 + 62$$

Here $B = 62$ and $f(x) = x^2$. From example 6.1 above, the derivative of x^2 is $2x$. So we have $f'(x) = 2x$. Therefore, applying rule D3, we get

$$\frac{dy}{dx} = f'(x) = 2x$$

The additive constant, 62, does not reappear in the derivative.

Rule D3 tells us that the additive constant term in any function has no effect on its slope at any point. This fits with our graphical work in chapters 3–5, where we found that varying the additive constant term in any function simply shifted its graph bodily up or down, without affecting its slope.

RULE D4 Sums or differences

In words, this rule is quite simple. It says that where y equals the sum of two functions of x, the derivative $\frac{dy}{dx}$ equals the sum of the derivatives of the two functions. In formal mathematical language, the rule says:

if $y = f(x) + g(x)$ [where $f(x)$ and $g(x)$ are any two functions of x]

then $\frac{dy}{dx} = f'(x) + g'(x)$

[where $f'(x)$ and $g'(x)$ are the derivatives of $f(x)$ and $g(x)$].

This rule becomes clearer in an example.

EXAMPLE 6.10

$$y = 10x^2 + x^{1/2}$$

Comparing this with rule D4 above, we have $f(x) = 10x^2$ and $g(x) = x^{1/2}$.

From example 6.6 above we know that the derivative of $10x^2$ is $20x$. Thus $f'(x) = 20x$.

Similarly, from example 6.3 above we know that the derivative of $x^{1/2}$ is $\frac{1}{2}x^{-1/2}$. Thus $g'(x) = \frac{1}{2}x^{-1/2}$.

Therefore rule D4 tells us that

$$\frac{dy}{dx} = f'(x) + g'(x) = 20x + \frac{1}{2}x^{-1/2}$$

Note that rule D4 applies in the same way when there is a minus sign instead of a plus sign on the right-hand side. To illustrate this:

EXAMPLE 6.11

If $y = 10x^2 - x^{1/2}$

then $\dfrac{dy}{dx} = 20x - \frac{1}{2}x^{-1/2}$

Note also that rule D4 holds for any number of terms on the right-hand side provided they are separated only by plus or minus signs (and not by multiplication or division signs). Thus:

EXAMPLE 6.12

If $y = x^3 + 5x^2 - 3x$

then rule D4 tells us to apply rules D1 and D2 to each term separately on the right-hand side, so we get

$$\frac{dy}{dx} = 3x^2 + 10x - 3$$

? **Progress exercise 6.2**

Using the rules of differentiation D1–D4 above, find the derivatives of:

(a) $y = 5x^2$ (b) $y = x^4$ (c) $y = 12x^3$

(d) $y = x^{0.5}$ (e) $y = x^3 + x^2$ (f) $y = 3x^4 + 10x^2$

(g) $y = 4x^3 - 16x + 9$ (h) $y = 10x^3 + 5x^2 - 9x + \frac{5}{x}$ (Hint: remember $\frac{1}{x} \equiv x^{-1}$)

(i) $y = \frac{1}{5}x^3 + \frac{1}{2}x^2 + 5x + 3$ (j) $y = x^{1/2} + x^{-2}$ (k) $y = x^{1/2} - x^{-2}$

(l) $y = -2x^{-2} + x^{-0.5} + x$ (m) $y = 1 + \frac{1}{x} - x^{-0.25}$ (n) $y = x + \frac{1}{x^2}$

(o) $y = -2x^5 - 0.3x^{-0.3} + \frac{1}{2}x^{1/2}$

(p) $y = ax + b$ (where a and b are parameters)

(q) $y = ax^2 + bx + c$ (where a, b, and c are parameters)

(r) $q = 2p^3 + 5p^2$ (Don't be thrown by the switch to p and q instead of x and y; this changes nothing)

(s) $q = Ap^{-\alpha}$ (where q and p are variables, A and α are parameters)

(t) $z = 0.5t^3 + 4t + 15$

If you have completed exercise 6.2 successfully, you are now ready to move on to some slightly more difficult rules of differentiation.

RULE D5 **The function of a function rule (also known as the chain rule)**

In formal mathematical language, this rule says:

If $y = f(u)$ where $u = g(x)$, then $\dfrac{dy}{dx} = \dfrac{dy}{du}\dfrac{du}{dx}$

where $\dfrac{dy}{du}$ is the derivative of the function $y = f(u)$, and $\dfrac{du}{dx}$ is the derivative of the function $u = g(x)$. (Of course, $\dfrac{dy}{du}\dfrac{du}{dx}$ means $\dfrac{dy}{du} \times \dfrac{du}{dx}$.)

We will explain this using examples.

EXAMPLE 6.13

$$y = (x^2 + 5x)^3$$

To find the derivative, we first create a new variable, which we will call u, and which we define as $u = x^2 + 5x$. We then substitute this into the given equation to get

$$y = u^3 \quad \text{where} \quad u = x^2 + 5x$$

(You may find this new variable, u, a little confusing at first sight. But keep in mind that u is something we have created ourselves to help us analyse the problem at hand. It is just a convenient shorthand for $x^2 + 5x$. It has no independent life of its own, and definitely cannot bite! Creating new variables in this way is something that mathematicians do frequently.)

We now have two functions, $y = u^3$ and $u = x^2 + 5x$. Thus we see that y depends on u via the function $y = u^3$. In turn, u depends on x via the function $u = x^2 + 5x$. So y is a function of something (u) which is itself a function of x. Thus y is a *function of a function* of x.

Another way of expressing the same idea is to say that there is a *chain* of causation running from x to y via u. As shown in figure 6.6, any change in x changes u via the relationship $u = x^2 + 5x$. This in turn causes y to change via the relationship $y = u^3$.

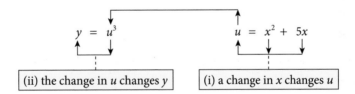

Figure 6.6 The chain of causation running from x to y via u.

Now, how are we to find the derivative $\dfrac{dy}{dx}$? Rule D5 tells us that

$$\frac{dy}{dx} = \frac{dy}{du}\frac{du}{dx}$$

where $\dfrac{dy}{du}$ is the derivative of $y = u^3$ and $\dfrac{du}{dx}$ is the derivative of $u = x^2 + 5x$.

In words, this says that we take the derivatives of the two functions above, that is $\dfrac{dy}{du}$ and $\dfrac{du}{dx}$, and multiply one by the other to get $\dfrac{dy}{dx}$.

At first sight, you may be a little surprised to see derivatives like $\dfrac{dy}{du}$ and $\dfrac{du}{dx}$ rather than the more familiar $\dfrac{dy}{dx}$. However, there is really nothing new here. When we look at a function such as $y = u^3$, what is different is that the independent variable is denoted by the symbol u rather than the x that we are accustomed to. Given this difference, it follows naturally that the derivative of this function is $\dfrac{dy}{du}$ rather than $\dfrac{dy}{dx}$. And similarly with $u = x^2 + 5x$, where the dependent variable is u rather than the more familiar y and therefore the derivative is $\dfrac{du}{dx}$ rather than $\dfrac{dy}{dx}$.

We find these two derivatives by applying the rules that we already know. Given $y = u^3$, the derivative by rule D1 is $\frac{dy}{du} = 3u^2$. Similarly, given $u = x^2 + 5x$, the derivative by rules D1, D2, and D4 is $\frac{dy}{du} = 2x + 5$. Therefore on substitution into rule D5 above we get

$$\frac{dy}{dx} = \frac{dy}{du}\frac{du}{dx} = 3u^2(2x + 5)$$

This result is rather messy because it contains both u and x. To tidy up, we replace u with what it equals: $x^2 + 5x$. This gives us

$$\frac{dy}{dx} = \frac{dy}{du}\frac{du}{dx} = 3(x^2 + 5x)^2(2x + 5)$$

which is our final result.

Let us work through another example of rule D5:

EXAMPLE 6.14

$$y = (x^2 + 1)^5$$

First we define a new variable, u, as $u = x^2 + 1$. We then have

$$y = u^5 \quad \text{where} \quad u = x^2 + 1$$

This conforms to the pattern of rule D5, with $y = f(u) = u^5$ and $u = g(x) = x^2 + 1$. Therefore

$$\frac{dy}{du} = f'(x) = 5u^4 \quad \text{(using rule D1), and}$$

$$\frac{du}{dx} = g'(x) = 2x \quad \text{(rules D1 and D3)}$$

Therefore in rule D5 we have

$$\frac{dy}{dx} = \frac{dy}{du}\frac{du}{dx} = 5u^4(2x)$$

Our final step is to replace u using our definition, $u = x^2 + 1$. So our final answer is

$$\frac{dy}{dx} = \frac{dy}{du}\frac{du}{dx} = 5(x^2 + 1)^4(2x)$$

which simplifies slightly to

$$\frac{dy}{dx} = 10x(x^2 + 1)^4$$

RULE D6 The product rule

This rule says that, given a function of the general form:

$$y = uv$$

where u and v are functions of x, the derivative is

$$\frac{dy}{dx} = u\frac{dv}{dx} + v\frac{du}{dx}$$

where $\frac{du}{dx}$ and $\frac{dv}{dx}$ are the derivatives of the functions u and v.

As usual, we will explain this rule by means of examples.

EXAMPLE 6.15

Given:

$$y = (x^2 + 1)(x^3 + x^2) \qquad (6.11)$$

we see that on the right-hand side we have a *product*: that is, $(x^2 + 1)$ multiplied by $(x^3 + x^2)$. Therefore rule D6 can be applied.

To implement the rule, we first create two new variables, which we'll call u and v, and which we define as

$$u = x^2 + 1 \quad \text{and} \quad v = x^3 + x^2$$

As we explained in discussing rule D5, you should not be afraid of these new variables. We bring them into existence purely to help our thinking; they have no independent life of their own. When we substitute these into equation (6.11), it becomes

$$y = uv \quad \text{where} \quad u = x^2 + 1 \quad \text{and} \quad v = x^3 + x^2$$

The function $u = x^2 + 1$ is a function like any other, and therefore has a derivative, $\frac{du}{dx}$. Applying rules D1, D3, and D4 to $u = x^2 + 1$, we find this derivative as $\frac{du}{dx} = 2x$. Similarly, with $v = x^3 + x^2$, applying rules D1 and D4 we find the derivative as $\frac{dv}{dx} = 3x^2 + 2x$.

So, collecting these bits and pieces together, we have

$$u = x^2 + 1 \quad \text{with derivative} \quad \frac{du}{dx} = 2x \text{ and}$$

$$v = x^3 + x^2 \quad \text{with derivative} \quad \frac{dv}{dx} = 3x^2 + 2x$$

Substituting these four into rule D6, we get

$$\frac{dy}{dx} = u\frac{dv}{dx} + v\frac{du}{dx} = \overbrace{(x^2 + 1)}^{u}\underbrace{(3x^2 + 2x)}_{\frac{dv}{dx}} + \overbrace{(x^3 + x^2)}^{v}\underbrace{(2x)}_{\frac{du}{dx}}$$

After multiplying out the brackets this simplifies to

$$\frac{dy}{dx} = 5x^4 + 4x^3 + 3x^2 + 2x$$

EXAMPLE 6.16

As a second example of rule D6, consider

$$y = (5x^2 + 3x)(x^4 + x)$$

We can write this as $y = uv$, where we define

$$u = 5x^2 + 3x \quad \text{and} \quad v = x^4 + x$$

We now have y as equal to the product of two variables, u and v, both of which are functions of x. So we can apply the product rule. To do this, we have to find the derivatives of $u = 5x^2 + 3x$ and $v = x^4 + x$. Applying rules D1–D4, the derivatives are

$$\frac{du}{dx} = 10x + 3 \quad \text{and} \quad \frac{dv}{dx} = 4x^3 + 1$$

Now we can substitute this information into the formula given by rule D6 above, and get

$$\frac{dy}{dx} = u\frac{dv}{dx} + v\frac{du}{dx} = \overbrace{(5x^2 + 3x)}^{u}\underbrace{(4x^3 + 1)}_{\frac{dv}{dx}} + \overbrace{(x^4 + x)}^{v}\underbrace{(10x + 3)}_{\frac{du}{dx}}$$

After multiplying out the brackets, this simplifies to

$$\frac{dy}{dx} = 30x^5 + 15x^4 + 15x^2 + 6x$$

Note that the product rule extends to cases where three or more functions are multiplied together, but such cases are rare in our environment so we won't trouble with this here.

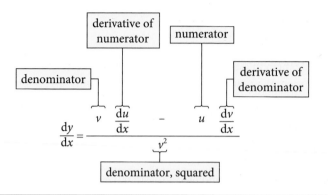

RULE D7 Quotient rule

This rule says that, if $y = \frac{u}{v}$, where u and v are functions of x:

This seems a very complicated rule at first sight, but with some practice you can hack it. Let's do an example.

EXAMPLE 6.17

$$y = \frac{x^2 + 1}{x^3 + x^2} \tag{6.12}$$

The key characteristic is that on the right-hand side we have a *quotient*: that is, $x^2 + 1$ *divided* by $x^3 + x^2$. So we can apply rule D7.

Our first step is to create two new variables, u and v, defined as

$$u = x^2 + 1 \quad \text{and} \quad v = x^3 + x^2$$

When we substitute these back into equation (6.12) we get

$$y = \frac{u}{v} \quad \text{where} \quad u = x^2 + 1 \quad \text{and} \quad v = x^3 + x^2$$

Next we find the derivatives of $u = x^2 + 1$ and $v = x^3 + x^2$. Applying rules D1–D4, the derivatives are

$$\frac{du}{dx} = 2x \quad \text{and} \quad \frac{dv}{dx} = 3x^2 + 2x$$

Substituting all of these into rule D7 above gives

$$\frac{dy}{dx} = \frac{v \frac{du}{dx} - u \frac{dv}{dx}}{v^2} = \frac{(x^3 + x^2)(2x) - (x^2 + 1)(3x^2 + 2x)}{(x^3 + x^2)^2}$$

After multiplying out the brackets in the numerator and cancelling where possible, this simplifies to

$$\frac{dy}{dx} = \frac{-x^3 - 3x - 2}{x^3(x + 1)^2}$$

EXAMPLE 6.18

As a second example of rule D7, consider

$$y = \frac{2x+1}{3x^2+3x+1} \tag{6.13}$$

On the right-hand side we again have a *quotient*: $2x+1$ *divided* by $3x^2+3x+1$. So we can apply rule D7.

Our first step is to create two new variables, u and v, defined as

$$u = 2x+1 \quad \text{and} \quad v = 3x^2+3x+1$$

When we substitute these back into equation (6.13) we get

$$y = \frac{u}{v} \quad \text{where} \quad u = 2x+1 \quad \text{and} \quad v = 3x^2+3x+1$$

Next we find the derivatives of $u = 2x+1$ and $v = 3x^2+3x+1$. Applying rules D1–D4, the derivatives are

$$\frac{du}{dx} = 2 \quad \text{and} \quad \frac{dv}{dx} = 6x+3$$

Substituting all of these into rule D7 above gives

$$\frac{dy}{dx} = \frac{v\frac{du}{dx} - u\frac{dv}{dx}}{v^2} = \frac{(3x^2+3x+1)(2)-(2x+1)(6x+3)}{(3x^2+3x+1)^2}$$

After multiplying out the brackets in the numerator and cancelling where possible, this simplifies to

$$\frac{dy}{dx} = \frac{-6x^2-6x-1}{(3x^2+3x+1)^2}$$

RULE D8 The derivative of the inverse function

This rule says that, for any function $y = f(x)$, $\dfrac{dy}{dx} = \dfrac{1}{\frac{dx}{dy}}$

The meaning of this rule is best seen in an example.

EXAMPLE 6.19

Given $y = 3x+2$, find the derivative, $\frac{dx}{dy}$, of the inverse function.

Using rules D1–D4, we can quickly find the derivative of $y = 3x+2$ as $\frac{dy}{dx} = 3$. Then, applying rule D8, we have

$$\frac{dx}{dy} = \frac{1}{\frac{dy}{dx}} = \frac{1}{3}$$

In this example we can check the answer by finding the inverse function. Given $y = 3x+2$, by subtracting 2 from both sides and dividing by 3, we obtain the inverse function as

$$x = \frac{1}{3}y - \frac{2}{3}$$

Using rules D1–D4, the derivative of the inverse function is quickly found as $\frac{dx}{dy} = \frac{1}{3}$. This is the same answer as we found above by using the inverse function rule.

EXAMPLE 6.20

Given $x = y^2$, find $\frac{dy}{dx}$.

Using the power rule (rule D1 above), the derivative of $x = y^2$ is $\frac{dx}{dy} = 2y$. Therefore, applying rule D8 gives

$$\frac{dy}{dx} = \frac{1}{\frac{dx}{dy}} = \frac{1}{2y}$$

In this case, rule D8 gives the slightly odd result that the derivative $\frac{dy}{dx}$ is expressed as a function of y rather than x. This can be fixed, but as this rule is seldom needed, we won't pursue the point.

6.8 Summary of rules of differentiation

We conclude this chapter by listing the rules of differentiation explained in this chapter. These rules cover all of the types of function used in this book, except for one function type that we will look at in chapter 13.

✱ **RULE D1 Power rule**

If $y = x^n$, then $\frac{dy}{dx} = nx^{n-1}$

✱ **RULE D2 Multiplicative constant**

If $y = Af(x)$, where A is any constant, then $\frac{dy}{dx} = Af'(x)$

✱ **RULE D3 Additive constant**

If $y = f(x) + B$, where B is any constant, then $\frac{dy}{dx} = f'(x)$

This rule applies also if the constant is negative: that is, if B is subtracted rather than added.

✱ **RULE D4 Sums or differences**

If $y = f(x) + g(x)$, where $f(x)$ and $g(x)$ are any two functions of x,

then $\frac{dy}{dx} = f'(x) + g'(x)$

where $f'(x)$ and $g'(x)$ are the derivatives of $f(x)$ and $g(x)$

This rule applies also if we have $y = f(x) - g(x)$; see example 6.11.

✱ **RULE D5 Function of a function (or chain) rule**

If $y = f(u)$ where $u = g(x)$,

then $\frac{dy}{dx} = \frac{dy}{du}\frac{du}{dx}$

where $\frac{dy}{du}$ is the derivative of the function $f(u)$, and $\frac{du}{dx}$ is the derivative of the function $g(x)$

✱ **RULE D6 Product rule**

If $y = uv$, where u and v are functions of x

then $\frac{dy}{dx} = u\frac{dv}{dx} + v\frac{du}{dx}$

where $\frac{du}{dx}$ and $\frac{dv}{dx}$ are the derivatives of the functions u and v

RULE D7 **Quotient rule**

If $y = \frac{u}{v}$ (where u and v are functions of x),

then $\dfrac{dy}{dx} = \dfrac{v\frac{du}{dx} - u\frac{dv}{dx}}{v^2}$

where $\frac{du}{dx}$ and $\frac{dv}{dx}$ are the derivatives of the functions u and v

RULE D8 **Derivative of the inverse function**

If $y = f(x)$,

then $\dfrac{dy}{dx} = \dfrac{1}{\frac{dx}{dy}}$

where $\frac{dx}{dy}$ is the derivative of the function inverse to the function $y = f(x)$

Progress exercise 6.3

Using rules of differentiation D5–D8 above, find the derivative $\frac{dy}{dx}$ of each of the following functions:

(a) $y = (1 + 2x)^2$ (Use the function of a function rule; do not multiply out!)

(b) $y = (1 - x^2)^{1/2}$

(c) $y = (x^2 - 2x^3)^5$ (Use the function of a function rule; do not multiply out!)

(d) $y = (x^2 - x)^{0.5}$

(e) $y = (3x - 2)(2x + 1)$ (Use the product rule; do not multiply out!)

(f) $y = (4x^2 - 3)(2x^5 + x)$

(g) $y = \frac{x^2 + 1}{1 - x^2}$ (h) $y = \frac{(x^2 + 1)^{1/2}}{1 - x^2}$ (i) $x = (y + 1)^{0.5}$ (j) $y = \frac{1}{1 - x}$

Checklist

Be sure to test your understanding of this chapter by attempting the progress exercises (answers are at the end of the book). The Online Resource Centre contains further exercises and materials relevant to this chapter www.oxfordtextbooks.co.uk/orc/renshaw3e/.

This has been an important and demanding chapter, in which we have introduced and developed the idea of the derivative as a measure of the slope of a curve at any point, and made the idea operational in the form of a set of rules for finding derivatives. More specifically, the topics we covered were:

✔ **The slope of a curve.** What is meant by positive, zero, or negative slope of a curve.

✔ **The difference quotient.** Calculating the difference quotient $\frac{\Delta y}{\Delta x}$ and understanding what it measures.

✔ **The derivative.** How the derivative $\frac{dy}{dx}$ measures the slope of the tangent to the curve, and is

obtained as the limiting value of $\frac{\Delta y}{\Delta x}$ as Δx approaches zero.

✔ **Rules of differentiation.** Finding derivatives using the eight main rules of differentiation: the power rule, the rules for multiplicative and additive constants, the rule for sums and differences of function, the function of a function (or chain) rule, the rules for products and quotients, and the inverse function rule.

We have covered a lot of completely new and difficult material in this chapter. Hopefully you are still on board and ready for the next two chapters. In chapter 7 we will show how derivatives may be used in a purely mathematical context. Then in chapter 8 we will begin to apply this knowledge of differentiation to some questions in economic analysis.

Chapter 7
Derivatives in action

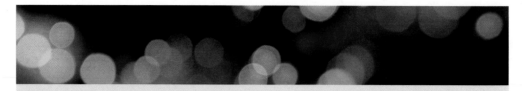

OBJECTIVES

Having completed this chapter you should be able to:

- Determine whether a function is monotonic.
- Use the first derivative to find the stationary points of a function.
- Understand what the second derivative measures and use it to distinguish between maximum and minimum values of a function.
- Understand what points of inflection are and use the second and third derivatives to locate them.
- Use the information obtained by differentiation to sketch the graph of a function.
- Understand the definitions of convexity and concavity.
- Use derivatives to determine whether a function is convex or concave.
- Understand what the differential of a function is and how it can be used as a linear approximation.

7.1 Introduction

In the previous chapter we developed the concept of a derivative, $\frac{dy}{dx}$, as a way of measuring the slope of any function at any point. We also learned the main rules for finding the derivatives of the most common types of function. After this heavy but necessary investment in mathematical technique, in this chapter we will show how the derivative can be used to obtain important information about the shape of a function, without the tedium and inevitable imprecision of having to plot its graph.

In economics we often want to know whether a function has a maximum or minimum value. For example, a firm wants to know the output level at which its profits are at a maximum, or its average costs are at a minimum. We will show how the derivative developed in the previous chapter can be used to find such points.

First we will introduce the concept of a monotonic function: that is, a function that is always either positively or negatively sloped. Then we will look at what are called stationary points, where a function reaches a maximum or minimum value, or a point of inflection. Identifying stationary points is known as optimization. To distinguish between these cases we introduce the second derivative of a function, which tells us whether the slope is increasing or decreasing, and the third derivative, which also conveys important information.

Finally, we introduce a new tool, the differential of a function, which is based on the derivative and which is used to approximate small changes in the function itself.

7.2 Increasing and decreasing functions

In economics, when we are looking at some function it is often important to know whether the function is positively or negatively sloped. For example, it is important to know whether a demand function slopes downwards and a supply curve upwards. One way of answering this question of course is to plot the graph of the function, but this is a tedious and error-prone activity. Moreover, since we can never plot the graph of a function for *all* values of the variables, graphical methods can never reveal the slope at all possible points on the curve.

A quicker and more powerful method is to examine the derivative of the function. From chapter 6 we know that the derivative, $\frac{dy}{dx}$, of any function $y = f(x)$ measures the slope at any point (provided that the function is smooth and there is no discontinuity at the point in question). It therefore follows that if we find that $\frac{dy}{dx}$ is positive at a point where x takes some specific value, say $x = x_0$, then we know that the function is positively sloped at that point. And conversely if $\frac{dy}{dx}$ is negative at some value of x, say $x = x_1$, then we know that the function is negatively sloped at that value of x. (Recall from chapters 3 and 6 that positive or negative slope, also known as gradient, means that the graph of the function slopes respectively upwards or downwards from left to right.)

For example, in figure 7.1(a), the function $y = f(x)$ is positively sloped between $x = 1$ and $x = 10$. We then say that the function is **strictly increasing** between $x = 1$ and $x = 10$. Consequently, any tangent drawn to the curve at any point is positively sloped too. And since the derivative measures the slope of the tangent, the derivative $\frac{dy}{dx}$ must be positive over this range of values of x.

Figure 7.1(b) shows the opposite case, where some function $y = g(x)$ is negatively sloped between $x = 1$ and $x = 10$. In this case we say that the function is **strictly decreasing** between $x = 1$ and $x = 10$. Any tangent drawn to the curve at any point is negatively sloped, and therefore the derivative $\frac{dy}{dx}$ must be negative over this range of values of x.

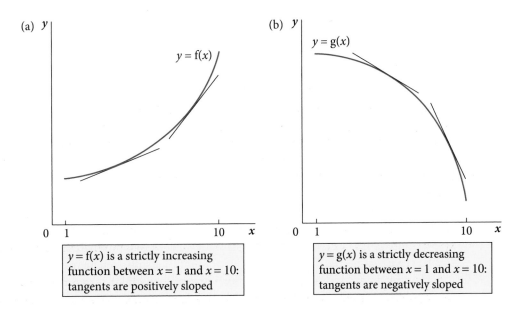

(a) $y = f(x)$ is a strictly increasing function between $x = 1$ and $x = 10$: tangents are positively sloped

(b) $y = g(x)$ is a strictly decreasing function between $x = 1$ and $x = 10$: tangents are negatively sloped

Figure 7.1 Increasing and decreasing functions.

In figure 7.1, if the function $y = f(x)$ had been mostly upward sloping but with a horizontal section in it too, we would say that it was increasing (rather than strictly increasing) between $x = 1$ and $x = 10$. Similarly in figure 7.2 if the function $y = g(x)$ had been mostly downward sloping but also with a horizontal section, we would say that it was decreasing (rather than strictly decreasing). Here we will ignore these cases and consider only *strictly* increasing or decreasing functions. There is more discussion of this topic in chapter W21, available at the Online Resource Centre (www.oxfordtextbooks.co.uk/orc/renshaw3e).

Figure 7.1 examines the slopes of two functions only between $x = 1$ and $x = 10$. How these functions behave outside of this range of values is not specified. In general, a function may be strictly increasing or strictly decreasing for *all* values of x. Alternatively, a function may be strictly increasing for some values of x and strictly decreasing for other values. We examine these cases later in this chapter.

Let us now look at two examples to see how the algebra of strictly increasing and decreasing functions works out.

EXAMPLE 7.1

Suppose we are asked to show that the function $y = 3x + 4$ is strictly increasing for all values of x. We can answer this by looking at the derivative.

Given $y = 3x + 4$, the derivative is $\frac{dy}{dx} = 3$. Since $\frac{dy}{dx}$ is thus positive for all values of x, we conclude that the graph of $y = 3x + 4$ is always positively sloped, so this function is strictly increasing. Any increase in x always results in an increase in y.

EXAMPLE 7.2

We are asked to show that the function $y = -x^3 - 5x$ is strictly decreasing. Again we can answer this by examining the derivative.

Given $y = -x^3 - 5x$, the derivative is $\frac{dy}{dx} = -3x^2 - 5$. So we have to show that

$$-3x^2 - 5 < 0 \quad \text{for all } x$$

If we add 5 to both sides of this inequality and then divide by -3, the inequality becomes

$$x^2 > -\frac{5}{3} \quad \text{(Note that the inequality is reversed by dividing by -3.)}$$

Since x^2 cannot be negative, this inequality is always satisfied, so $\frac{dy}{dx}$ is always negative. Therefore the function is strictly decreasing. Figure 7.2 confirms this for values of x between -3 and $+3$, though as we remarked earlier a graph can never be completely conclusive by itself because it can never cover all values of x.

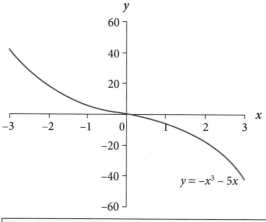

$\frac{dy}{dx}$ is negative for all x, so graph is strictly decreasing.

Figure 7.2 The function $y = -x^3 - 5x$ is monotonic decreasing.

Before leaving strictly increasing and decreasing functions, we should note that it is possible for a function to be strictly increasing (decreasing) even if $\frac{dy}{dx}$ is not positive (negative) at every point. For example, $y = x^3$ is strictly increasing, because y increases whenever x increases; but $\frac{dy}{dx}$ is not always positive, since at $x = 0$ we have $\frac{dy}{dx} = 3x^2 = 0$. Thus if $\frac{dy}{dx}$ is positive (negative), this is a sufficient, but not necessary, condition for a function to be strictly increasing (decreasing). (On necessary and sufficient conditions, see section 2.13.)

Summary of sections 7.1–7.2

If the derivative $\frac{dy}{dx}$ of a function is positive, the function is strictly *increasing* (sloping upward from left to right). If the derivative $\frac{dy}{dx}$ is negative, the function is strictly *decreasing*. These are sufficient but not necessary conditions.

Progress exercise 7.1

By examining the sign of $\frac{dy}{dx}$, show that:

(a) $y = -2x + 10$ is strictly decreasing

(b) $y = x^3 + x$ is strictly increasing

(c) $y = -3x^2 + 4x + 2$ is strictly decreasing when $x > \frac{2}{3}$

(d) $y = x^3 + 6x^2 + 12x + 15$ is strictly increasing for all $x > -2$

(e) $y = 2x^3 + 21x^2 + 60x + 10$ is strictly increasing when $x < -5$ or $x > -2$

7.3 Optimization: finding maximum and minimum values

If a function $y = f(x)$ is not strictly increasing or decreasing, then it may have a maximum or minimum value, as in figure 7.3. The process of finding maximum and minimum values in order to solve a problem has become known as **optimization**. Let us look at specific examples of a maximum and minimum value.

(a)　　A maximum of $y = f(x)$ at J

(b)　　A minimum of $y = g(x)$ at P

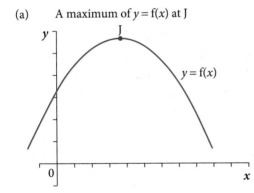
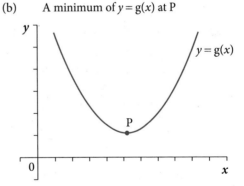

Figure 7.3 Maximum and minimum values of a function.

7.4 A maximum value of a function

EXAMPLE 7.3

Consider $y = -x^2 + 10x$

This function is graphed in figure 7.4(a). We see that it reaches its maximum value at J, where $x = 5$. At this point the tangent is horizontal, and its slope is therefore zero. At this point, y is neither increasing nor decreasing, but is stationary. This maximum is called a **stationary point**.

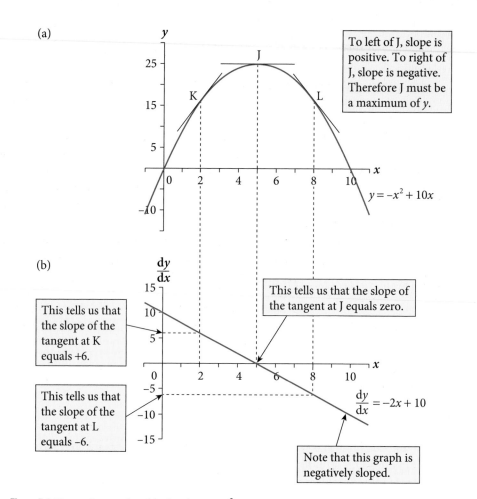

(a)

To left of J, slope is positive. To right of J, slope is negative. Therefore J must be a maximum of y.

$y = -x^2 + 10x$

(b)

This tells us that the slope of the tangent at K equals +6.

This tells us that the slope of the tangent at J equals zero.

This tells us that the slope of the tangent at L equals –6.

$\frac{dy}{dx} = -2x + 10$

Note that this graph is negatively sloped.

Figure 7.4 The maximum value of the function $y = -x^2 + 10x$.

To the left of J, where x is less than 5, the curve is positively sloped and any tangent drawn to the curve, such as the tangent at K, is therefore positively sloped. To the right of J, where x is greater than 5, the curve is negatively sloped and any tangent drawn to the curve, such as the tangent at L, is negatively sloped.

Therefore our expectation is that $\frac{dy}{dx}$, which measures the slope of the tangent, will be positive when $x < 5$, zero when $x = 5$, and negative when $x > 5$. Let us check this out algebraically.

Given the function $y = -x^2 + 10x$, the derivative is

$$\frac{dy}{dx} = -2x + 10$$

To find point J (the maximum) we set $\frac{dy}{dx}$ equal to zero, giving $-2x + 10 = 0$. The solution to this equation is $x = 5$, as we expected from looking at the graph. When $x = 5$, we have $y = -x^2 + 10x = -(5)^2 + 10(5) = 25$. This is therefore the maximum value of y.

To find where $\frac{dy}{dx}$ is positive, we solve the inequality $-2x + 10 > 0$. Subtracting 10 from both sides and dividing by -2 (which reverses the inequality), we arrive at $x < 5$. Thus we have proved that $\frac{dy}{dx}$ is positive when x is less than 5, as we expected from looking at the graph. Similarly, to find where $\frac{dy}{dx}$ is negative, we solve the inequality $-2x + 10 < 0$. The solution is $x > 5$, as expected. (See section 5.10 if you need to revise solving inequalities.) So we have proved algebraically that the function $y = x^2 + 10x$ has a positive slope when $x < 5$, a zero slope when $x = 5$, and a negative slope when $x > 5$. Figure 7.4(a) is therefore correctly drawn.

7.5 The derivative as a function of *x*

Now we introduce a new idea: *the derivative $\frac{dy}{dx}$ is itself a function of x*. This is true because, in general, $\frac{dy}{dx}$ varies as *x* varies; and any expression that varies when *x* varies is, by definition, a function of *x*.

Therefore the derivative in the previous example, $\frac{dy}{dx} = -2x + 10$, is a function of *x* and we can plot its graph just like any other graph. The only difference is that the dependent variable (as usual, on the vertical axis) is now $\frac{dy}{dx}$ rather than *y* as we have been accustomed to. This graph is plotted in figure 7.4(b). We see that it is a linear function with an intercept of 10 on the vertical axis and a slope of −2. The key point is that the *slope* of the curve in figure 7.4(a) is now measured as a *distance* up the vertical axis in figure 7.4(b).

The graph of the derivative confirms the information we have already extracted by looking at the function itself in figure 7.4(a) and also by algebraic means. From figure 7.4(b) we see that when *x* is less than 5, $\frac{dy}{dx}$ is positive. For example, when $x = 2$, $\frac{dy}{dx} = -2x + 10 = 6$. Similarly, when *x* is greater than 5, $\frac{dy}{dx}$ is negative. For example, when $x = 8$, $\frac{dy}{dx} = -2x + 10 = -6$. Finally, when $x = 5$, we see in figure 7.4(b) that $\frac{dy}{dx}$ is zero (that is, its graph cuts the *x*-axis at $x = 5$).

Note that the graph of $\frac{dy}{dx}$ is negatively sloped. We will explain the significance of this fact shortly.

7.6 A minimum value of a function

EXAMPLE 7.4

To study the case of a minimum, we repeat the analysis of example 7.3 for the function

$$y = x^2 - 10x + 30 \quad \text{with derivative} \quad \frac{dy}{dx} = 2x - 10$$

This function is graphed in figure 7.5(a), and its derivative in figure 7.5(b). From figure 7.5(a) we see that this quadratic function has a minimum value at P, where $x = 5$. The tangent at P is horizontal, and therefore has a slope of zero. As in the case of the maximum in example 7.3, this minimum point is called a stationary point because at this point *y* is neither increasing nor decreasing.

To the left of P the curve is negatively sloped and any tangent drawn to the curve, such as at Q, is therefore negatively sloped. Similarly, to the right of P the curve is positively sloped and any tangent drawn to the curve, such as at R, is positively sloped.

Therefore our expectation is that $\frac{dy}{dx}$, which measures the slope of the tangent, will be negative when $x < 5$, zero when $x = 5$, and positive when $x > 5$. Let us check the algebra to see if this is indeed the case.

From above, the derivative of $y = x^2 - 10x + 30$ is

$$\frac{dy}{dx} = 2x - 10$$

To find point P (the minimum) we set this equal to zero, giving $2x - 10 = 0$. The solution to this equation is $x = 5$, as we expected. When $x = 5$, we have $y = x^2 - 10x + 30 = 5$. This is therefore the minimum value of *y*.

To find where $\frac{dy}{dx}$ is negative, we solve the inequality $2x - 10 < 0$. The solution is $x < 5$, as we expected. Similarly, to find where $\frac{dy}{dx}$ is positive, we solve the inequality $2x - 10 > 0$. The solution is $x > 5$, as we expected. This algebra thus confirms that the graph of $y = x^2 - 10x + 30$ is negatively sloped to the left of $x = 5$ and positively sloped to the right.

As in the previous example, the graph of the derivative in figure 7.5(b) confirms what we have just deduced algebraically. Thus from figure 7.5(b) we can read off the

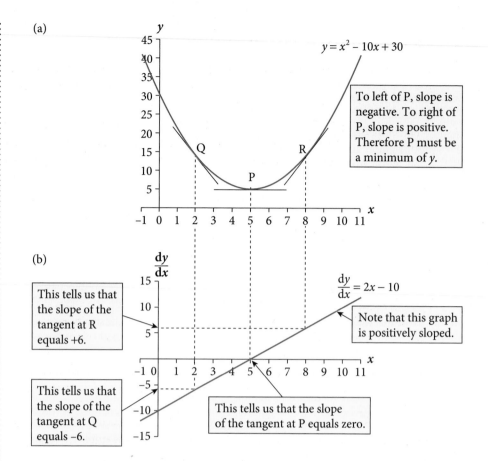

Figure 7.5 The minimum value of the function $y = x^2 - 10x + 30$.

information that at Q, where $x = 2$, the slope of the function is negative (specifically, the slope equals -6). At P, where $x = 5$, we can see that the slope is zero (the graph of $\frac{dy}{dx}$ cuts the x-axis at this point). Finally at R, where $x = 8$, we can read off the slope from figure 7.5(b) as positive (specifically, $+6$).

Note that the graph of $\frac{dy}{dx}$ is positively sloped. We will explain the significance of this shortly.

Maximum or minimum?

Comparing examples 7.3 and 7.4, we see that a maximum and a minimum value of a function have the common feature that $\frac{dy}{dx} = 0$ at the point in question. This is because in both cases the tangent is horizontal at that point. This is confirmed in figures 7.4(b) and 7.5(b), where in both cases we see the graph of $\frac{dy}{dx}$ crossing the x-axis at the value of x where the function reaches its maximum or minimum value.

Thus knowing that $\frac{dy}{dx} = 0$ at a point tells us that the function could be at either a maximum or a minimum at that point. What we need now is some means of distinguishing the maximum from the minimum, without having to go to the trouble of plotting the graph each time or fiddling around looking at the sign of $\frac{dy}{dx}$ on each side of the point as we did in the previous two examples.

The key difference between the two is that in the case of the maximum the graph of $\frac{dy}{dx}$ is negatively sloped (see figure 7.4(b)), while in the case of the minimum the graph of $\frac{dy}{dx}$ is positively sloped (see figure 7.5(b)). So what we need is a way of checking, by some algebraic technique, whether the graph of $\frac{dy}{dx}$ is positively or negatively sloped.

Summary of sections 7.3–7.6

In sections 7.3–7.5 we examined how the derivative $\frac{dy}{dx}$ could help us find a maximum value of a function. We saw that $\frac{dy}{dx}$ itself was a function of x, which can be graphed. Specifically, we examined point J in figure 7.4, a maximum value of y. At J:

(1) $\frac{dy}{dx} = 0$; and (2) $\frac{dy}{dx}$ is decreasing, since it is positive to the left of the maximum and negative to the right of it. This means also that the graph of $\frac{dy}{dx}$ is *negatively* sloped at maximum.

Similarly, in section 7.6 we considered how the derivative could help us locate a minimum value of a function. Specifically, we saw that point P in figure 7.5 is a minimum value of y. At P:

(2) $\frac{dy}{dx} = 0$; and (2) $\frac{dy}{dx}$ is increasing, since it is negative to the left of the minimum and positive to the right. This means also that the graph of $\frac{dy}{dx}$ is *positively* sloped at the minimum.

All this will be true for any function $y = f(x)$ at a maximum or minimum value.

7.7 The second derivative

The algebraic technique we are looking for is as follows. We can find the slope of the graph of $\frac{dy}{dx}$ by carrying out the process of differentiation again, but this time on $\frac{dy}{dx}$ rather than on the original function. Since we have already seen that $\frac{dy}{dx}$ is itself a function of x, this function must have a derivative, which we can find simply by applying the rules of differentiation that we learned in chapter 6.

By repeating the process of differentiation in this way, we will obtain 'the derivative of the derivative'. Let us write $\frac{d}{dx}$ as a shorthand for 'the derivative of'. Then we can write 'the derivative of the derivative' as $\frac{d}{dx}(\frac{dy}{dx})$. This will measure the slope of the graph of $\frac{dy}{dx}$.

Applying this to example 7.3, we know that $\frac{dy}{dx} = -2x + 10$, so when we differentiate this (using the rules of differentiation in the usual way) we get

$$\frac{d}{dx}\left(\frac{dy}{dx}\right) = \frac{d}{dx}(-2x + 10) = -2$$

This tells us that the slope of the graph of $\frac{dy}{dx} = -2x + 10$ in figure 7.4(b) is −2. The fact that the slope of the graph of $\frac{dy}{dx}$ is negative tells us that $\frac{dy}{dx}$ decreases as x increases. As a logical necessity, if $\frac{dy}{dx} = 0$ at J, and $\frac{dy}{dx}$ decreases as x increases, then it must be true that $\frac{dy}{dx}$ is greater than zero to the left of J, and less than zero to the right of J. In other words $\frac{dy}{dx}$, which measures the slope of $y = x^2 + 10x$, is positive to the left of J and negative to the right of J. Therefore J must be a maximum.

Repeating this for example 7.4, we know that $\frac{dy}{dx} = 2x - 10$, so

$$\frac{d}{dx}\left(\frac{dy}{dx}\right) = \frac{d}{dx}(2x - 10) = 2$$

This tells us that the slope of the graph of $\frac{dy}{dx} = 2x - 10$ in figure 7.5(b) is +2. The fact that this is positive tells us that $\frac{dy}{dx}$ increases as x increases. If $\frac{dy}{dx} = 0$ at P, and $\frac{dy}{dx}$ increases as x increases, then it must be true that $\frac{dy}{dx}$ is less than zero to the left of P, and greater than zero to the right of P. In other words, the slope of $y = x^2 - 10x$ is negative to the left of P and positive to the right of P. Therefore P must be a minimum.

Notation for the second derivative

Before going any further let us take a moment to tidy up our notation a little. Above, we wrote $\frac{d}{dx}(\frac{dy}{dx})$ to denote 'the derivative of the derivative'. As there are two ds in the numerator and two xs in the denominator, we can write $\frac{d}{dx}(\frac{dy}{dx})$ more compactly as

$$\frac{d^2y}{dx^2}$$

and we shall refer to it as 'the second derivative'. Of course, the d^2 and the x^2 do not mean 'd squared' and 'x squared'. Instead we must read the expression as a whole, as meaning 'the derivative of the derivative', or 'what you get when you perform the operation of differentiation for a second time'. The key point to hang on to is that $\frac{d^2y}{dx^2}$ measures the slope of the graph of $\frac{dy}{dx}$.

7.8 A rule for maximum and minimum values

We can now express our findings up to this point in the form of a general rule. This rule holds for any function that is smooth and continuous at the point in question. The rule provides us with an algebraic technique for locating the maximum or minimum value(s) of any function.

RULE 7.1 Maximum and minimum values

For any function $y = f(x)$, at any point P on that function:

if at that point $\frac{dy}{dx} = 0$ and $\frac{d^2y}{dx^2} < 0$, the point is a maximum of y

if at that point $\frac{dy}{dx} = 0$ and $\frac{d^2y}{dx^2} > 0$, the point is a minimum of y

The condition $\frac{dy}{dx} = 0$ is called the **first order condition** for a maximum or minimum value and must be satisfied in both cases. The condition $\frac{d^2y}{dx^2} < 0$ (for a maximum) or $\frac{d^2y}{dx^2} > 0$ (for a minimum) is called the **second order condition** and it is by reference to this that we distinguish the two cases.

Note that rule 7.1 does not cover the case where $\frac{d^2y}{dx^2} = 0$. We will examine this case shortly.

7.9 Worked examples of maximum and minimum values

To show how to apply rule 7.1 above we will now work through two examples.

EXAMPLE 7.5

Suppose we are given the function

$$y = x^2 + 2x + 9$$

and asked (a) to determine whether this function has a maximum or minimum value, and (b) to draw a sketch graph of the function.

To answer (a), we first find $\frac{dy}{dx}$, set it equal to zero, and solve the resulting equation. This gives

$$\frac{dy}{dx} = 2x + 2 = 0 \quad \text{with solution } x = -1$$

We can find the value of y at this point by substituting $x = -1$ into the function $y = x^2 + 2x + 9$. This gives $y = (-1)^2 + 2(-1) + 9 = 8$.

So now we know that the tangent to the curve is horizontal at the point $x = 1$, $y = 8$, but this could be either a maximum or a minimum of y (or neither, as we shall see later).

To determine whether the point is a maximum or minimum, we find $\frac{d^2y}{dx^2}$ and examine its sign. This gives

$$\frac{d^2y}{dx^2} = \frac{d}{dx}\left(\frac{dy}{dx}\right) = \frac{d}{dx}[2x + 2] = 2$$

Since this is a positive constant, it is positive when $x = -1$. So, applying rule 7.1, we conclude that there is a *minimum* of $y = x^2 + 2x + 9$ at $x = -1$. (In this example, it happens that $\frac{d^2y}{dx^2}$ is positive for *all* values of x, but its value when $x = -1$ is what actually matters.) Furthermore, this is the *only* maximum or minimum, since there are no other values of x which make $\frac{dy}{dx} = 0$. So on each side of the minimum, the function must go on rising forever.

To answer (b), we can easily draw a rough sketch graph of the function now that we know that it has a minimum at $x = 1$, $y = 8$. We can also improve the accuracy of our sketch a little by noting that the function $y = x^2 + 2x + 9$ cuts the y-axis at $y = 9$ (because $y = 9$ when $x = 0$). To sketch the graph we mark these two points [minimum at $(-1, 8)$ and y-intercept at $y = 9$] on the graph paper and then join them up with a freehand sketch, such as is shown in figure 7.6. It is not very accurate, but nevertheless it tells us everything important about the function. Of course, accuracy can be improved if desired by compiling a table of values.

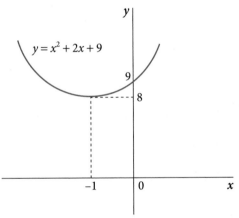

Figure 7.6 Sketch graph of $y = x^2 + 2x + 9$ (not to scale).

EXAMPLE 7.6

In this example we will look at a more complex case in which the function has both a minimum and a maximum value. Given

$$y = x^3 + 6x^2 + 9x + 6$$

suppose we are asked to find any maximum and/or minimum values, and sketch the graph. Following rule 7.1, we start by finding $\frac{dy}{dx}$, setting it equal to zero, and solving the resulting equation. We get

$$\frac{dy}{dx} = 3x^2 + 12x + 9 = 0$$

Hopefully you will recognize this as a quadratic equation (see section 4.4 if you need to revise this). To solve it we first divide both sides by 3, giving

$$x^2 + 4x + 3 = 0$$

This then factorizes quite easily as $(x + 3)(x + 1) = 0$, so the roots are $x = -3$ and $x = -1$. (If you find factorizing difficult, it is quite OK to use the formula for solving a quadratic equation (see section 4.5), which of course gives the same answer.)

So now we know that $\frac{dy}{dx} = 0$ when $x = -3$ or $x = -1$. The corresponding values of y, obtained by substituting $x = -3$ and $x = -1$ into $y = x^3 + 6x^2 + 9x + 6$, are $y = 6$ and $y = 2$ respectively.

To find out whether these points are maximum or minimum values of the function, we have to take the second derivative and examine its sign at the two values of x. The second derivative is

$$\frac{d^2y}{dx^2} = \frac{d}{dx}[3x^2 + 12x + 9] \quad \text{(recall that } \frac{d}{dx} \text{ means 'the derivative of')}$$
$$= 6x + 12$$

We now have to examine the sign of this second derivative when $x = -3$ and $x = -1$. By substitution, when $x = -3$, we have:

$$\frac{d^2y}{dx^2} = 6(-3) + 12 = -6; \text{ that is, negative.}$$

So, applying rule 7.1, we conclude that the function reaches a maximum at $x = -3$. Similarly, when $x = -1$:

$$\frac{d^2y}{dx^2} = 6(-1) + 12 = +6; \text{ that is, positive.}$$

So, again using rule 7.1, we conclude that the function reaches a minimum at $x = -3$.

To sketch the graph we first mark the positions of the maximum ($x = -3$, $y = 6$) and minimum ($x = -1$, $y = 2$) on the graph paper. It is also helpful to mark the y-intercept; in this example, $y = 6$ when $x = 0$. We can then draw a rough freehand sketch such as is shown in figure 7.7(a). You may be wondering whether the curve goes on rising forever to the right of $x = -1$, and falling forever to the left of $x = -3$. The answer is yes to both. We know this because, when we solved the equation $\frac{dy}{dx} = 3x^2 + 12x + 9 = 0$ above, we automatically found *all* the stationary points of this function. So there are no stationary points other than at $x = -3$ and -1.

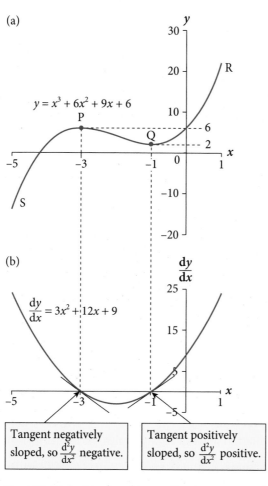

Although it was not necessary to answer the question, we have also plotted the graph of $\frac{dy}{dx}$ in figure 7.7(b). The shape of this confirms that the algebra we did above is correct. We see that $\frac{dy}{dx} = 0$ at both $x = -3$ and $x = -1$. At $x = -3$ the graph of $\frac{dy}{dx}$ is negatively sloped, so $\frac{d^2y}{dx^2}$ is negative (and equal to -6, as we found in the algebra above). At $x = -1$ the graph of $\frac{dy}{dx}$ is positively sloped, so $\frac{d^2y}{dx^2}$ is positive (and equal to $+6$, as we found in the algebra above).

| Tangent negatively sloped, so $\frac{d^2y}{dx^2}$ negative. | Tangent positively sloped, so $\frac{d^2y}{dx^2}$ positive. |

Figure 7.7 Graph of $y = x^3 + 6x^2 + 9x + 6$ showing its maximum and minimum values.

Local versus global maxima or minima

Before moving on there is one methodologically important point that we should note. Rule 7.1 for locating maximum and minimum values give us information about the function's shape only in the immediate vicinity of the point in question. In other words, the rule permits us to locate only *local* maximum or minimum values.

It is perfectly possible that the function may have a *local* maximum or minimum at some point, but nevertheless reaches higher or lower values elsewhere. Thus in example 7.6 we found a maximum at P (see figure 7.7) but this is only a *local* maximum since *y* reaches higher values in region R. Similarly we found a *local* minimum at Q, but there are lower values of *y* in region S.

Summary of sections 7.7–7.9

In section 7.7 we introduced the idea of the second derivative, $\frac{d^2y}{dx^2}$, which is found by applying the rules of differentiation to the first derivative. The second derivative measures the slope of the graph of the first derivative, and therefore its sign tells us whether the graph of the first derivative is positively or negatively sloped. From this we derived a rule for finding a maximum or minimum value of a function. To do this, we find $\frac{dy}{dx}$, set it equal to zero, and solve the resulting equation for *x*. We then examine the sign of $\frac{d^2y}{dx^2}$ at the solution values of *x*. If the sign is negative (positive), the point is a maximum (minimum). These points are only *local*, not *global*, maxima or minima. The case where $\frac{d^2y}{dx^2} = 0$ is analysed in the next section.

Progress exercise 7.2

Find any local maximum and/or minimum value(s) of the following functions. In each case, draw a sketch graph of the function.

(a) $y = x^2 - 6x + 12$ (b) $y = -x^2 + 10x + 11$ (c) $y = 2x^2 - 6x + 15$

(d) $y = x^3 - 9x^2 + 15x + 5$ (e) $y = -x^3 + 15x^2 - 27x + 50$

7.10 Points of inflection

In the previous section we showed how to find a maximum or minimum value of any function $y = f(x)$. Both are called **stationary points** because in both cases the tangent is horizontal. Thus the curve is stationary in the sense that it is neither rising nor falling at that point.

However, there are also stationary points that are neither a maximum nor a minimum value of the function. These stationary points are called **points of inflection**, and are of two types:

(1) Point K in figure 7.8(a) is a stationary point because the curve is neither rising nor falling at this point. Therefore the tangent is horizontal at K and $\frac{dy}{dx} = 0$. It is called a point of inflection because the curve is *rising* on both sides of point K, and therefore a tangent drawn to the curve on either side of K is positively sloped. Thus $\frac{dy}{dx}$ is positive on both sides of point K, and zero at K.

(2) Point Q in figure 7.9(a) is a stationary point because the tangent is horizontal at Q, and therefore $\frac{dy}{dx} = 0$. It is a point of inflection because the curve is *falling* on both sides of point K, and therefore a tangent drawn to the curve on either side of K is negatively sloped. Thus $\frac{dy}{dx}$ is negative on both sides of point Q, and zero at Q.

Our next task is to examine how we can identify points of inflection such as K and Q. We shall see that a key insight is that a point of inflection is a point at which the *slope* of the function (not the function itself) reaches a maximum or minimum value. We will explain this with two examples.

EXAMPLE 7.7

Consider the function

$$y = \tfrac{1}{3}x^3 - 3x^2 + 9x + 5$$

which is graphed in figure 7.8(a). There is a point of inflection in the curve at K, where $x = 3$. The slope is positive on both sides of K. We therefore expect to find that $\frac{dy}{dx}$ is positive to the left of K, zero at K, then positive again to the right of K.

To understand better what is happening here, it may help to imagine that the curve in figure 7.8(a) is a hill that you are walking up from left to right, starting from J. As you walk from J to K the slope becomes less and less steep (the slope is decreasing), and is at its flattest at K, where $x = 3$. Once you get beyond K and head towards L, you find that the slope is becoming steeper and steeper (the slope is increasing). Thus point K is where the slope of the curve is at its *minimum*: that is, the least positively sloped point on the curve.

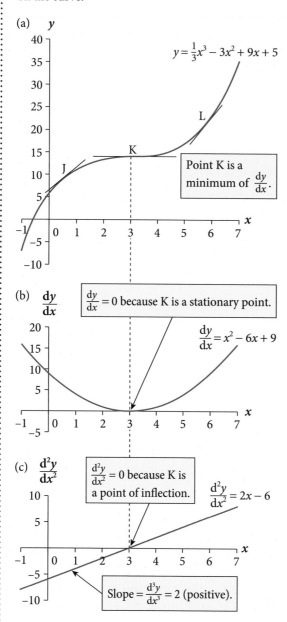

That K is a minimum of the slope of the curve is confirmed by examining the slopes of the tangents drawn to the curve at J, K, and L. The tangents at J and L have a relatively steep slope, whereas the tangent at K is flatter, and indeed is horizontal. It is the least steep (flattest) tangent to the curve that we can draw between J and L.

Note that any tangent drawn to the left of K lies above the curve, while any tangent drawn to the right of K lies below the curve. At K itself, the tangent *cuts* the curve. This is one way of defining a point of inflection: that is, as a point where the tangent cuts the curve as the tangent crosses over from above to below the curve or vice versa.

That K is a minimum of the slope of the curve is also confirmed when we look at the graph of $\frac{dy}{dx}$. Given $y = \tfrac{1}{3}x^3 - 3x^2 + 9x + 5$, the derivative is

$$\frac{dy}{dx} = x^2 - 6x + 9$$

This quadratic function is graphed in figure 7.8(b). We see that, as x increases and we move from left to right, $\frac{dy}{dx}$ is initially positive but decreasing, reaches its minimum value of zero at $x = 3$, and then to the right of this point becomes positive once more.

Figure 7.8 Graph of $y = \tfrac{1}{3}x^3 - 3x^2 + 9x + 5$ showing a stationary point that is also a point of inflection.

The fact that K is a minimum of $\frac{dy}{dx}$ suggests an algebraic technique for locating K. We can locate K by finding the minimum value of $\frac{dy}{dx}$, using rule 7.1 that we developed in section 7.8 for finding maximum and minimum values. Let us show how the algebra works out for this example.

We want to find the minimum of the function $\frac{dy}{dx} = x^2 - 6x + 9$. Since this is a function of x like any other, rule 7.1 applies, so its minimum is found where its first derivative is zero and its second derivative is positive. What is new about this situation is that the function we are looking at is already $\frac{dy}{dx}$, so its *first* derivative is $\frac{d^2y}{dx^2}$. We can evaluate this derivative as

$$\frac{d^2y}{dx^2} = \frac{d}{dx}[x^2 - 6x + 9] = 2x - 6$$

Setting this equal to zero gives $2x - 6 = 0$, with solution $x = 3$.

Like the first derivative, this second derivative is itself a function of x, and is graphed in figure 7.8(c). The graph is a positively sloped straight line in this example. As we have just found algebraically, it cuts the x-axis at $x = 3$.

Continuing with the algebra, rule 7.1 tells us that to show that we have a minimum of $\frac{dy}{dx}$ we must show that its second derivative is positive at the point in question. As we have just seen, the *first* derivative of $\frac{dy}{dx}$ is already $\frac{d^2y}{dx^2}$. Therefore the *second* derivative is the derivative of $\frac{d^2y}{dx^2}$: that is, $\frac{d}{dx}[\frac{d^2y}{dx^2}]$.

The third derivative

This leads us to the idea of a *third* derivative, $\frac{d}{dx}[\frac{d^2y}{dx^2}]$, which we can write more compactly as $\frac{d^3y}{dx^3}$. There is really nothing fundamentally new in this concept. It is found by applying the rules of differentiation (see chapter 6) to $\frac{d^2y}{dx^2}$, and measures the slope of the graph of $\frac{d^2y}{dx^2}$. Putting this into effect, we have, from above:

$$\frac{d^2y}{dx^2} = 2x - 6 \quad \text{therefore:}$$

$$\frac{d^3y}{dx^3} = \frac{d}{dx}[2x - 6] = 2$$

This third derivative measures the slope of the graph of $\frac{d^2y}{dx^2}$ in figure 7.8(c). Since it is positive, we now know that at $x = 3$ the conditions for a minimum of $\frac{dy}{dx}$ are fulfilled: that is, $\frac{d^2y}{dx^2}$ (the first derivative of $\frac{dy}{dx}$) equals zero, and $\frac{d^3y}{dx^3}$ (the second derivative of $\frac{dy}{dx}$) is positive.

EXAMPLE 7.8

Consider $y = -x^3 + 6x^2 - 12x + 50$

which is graphed in figure 7.9(a). There is a point of inflection at Q, where $x = 2$. To the left and right of Q the curve is negatively sloped. As always at a point of inflection, the tangent cuts the curve at Q. To the left of Q any tangent lies below the curve, and to the right of Q any tangent lies above it.

Tangents drawn on each side of Q, at points P and R, are negatively sloped. The tangent at Q has a slope of zero. Since zero is greater than any negative number, the slope at Q is greater than the slope on either side of Q, and therefore Q is a *maximum* of $\frac{dy}{dx}$. If we imagine the curve as a hill, which we are walking down from left to right, then Q is the least negatively sloped (= most positively sloped) point on the hill.

We can confirm graphically that Q is a maximum of $\frac{dy}{dx}$ by plotting the graph of $\frac{dy}{dx}$. From $y = -x^3 + 6x^2 - 12x + 50$, we have

$$\frac{dy}{dx} = -3x^2 + 12x - 12$$

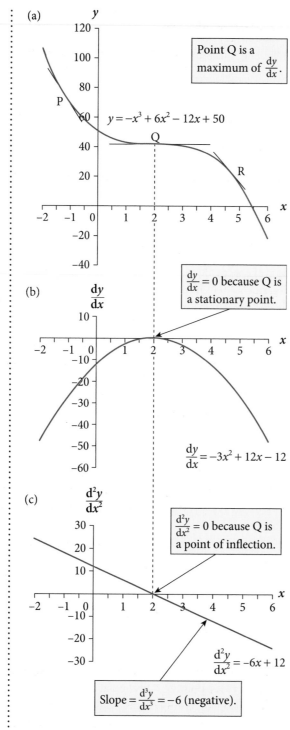

(a)

Point Q is a maximum of $\frac{dy}{dx}$.

$y = -x^3 + 6x^2 - 12x + 50$

(b)

$\frac{dy}{dx} = 0$ because Q is a stationary point.

$\frac{dy}{dx} = -3x^2 + 12x - 12$

(c)

$\frac{d^2y}{dx^2} = 0$ because Q is a point of inflection.

$\frac{d^2y}{dx^2} = -6x + 12$

Slope $= \frac{d^3y}{dx^3} = -6$ (negative).

Figure 7.9 Graph of $y = -x^3 + 6x^2 - 12x + 50$ showing a stationary point that is also a point of inflection.

The graph of this quadratic function is shown in figure 7.9b. We see that, as x increases and we move from left to right, $\frac{dy}{dx}$ is initially negative but increasing (that is, becoming less negative). It reaches its maximum value of zero at $x = 2$, and then to the right of this point becomes negative once more (and indeed increasingly negative as x increases).

In addition to these graphical methods, we can also show that Q is a maximum of $\frac{dy}{dx}$ by the algebraic method we developed in the previous example. To do this we must show that at Q the first derivative of $\frac{dy}{dx}$ is zero and the second derivative of $\frac{dy}{dx}$ is negative.

First, we take the first derivative of $\frac{dy}{dx}$, which of course is $\frac{d^2y}{dx^2}$, set it equal to zero and solve. From above we have

$$\frac{dy}{dx} = -3x^2 + 12x - 12 \quad \text{so:}$$

$$\frac{d^2y}{dx^2} = -6x + 12$$

Setting this equal to zero gives $-6x + 12 = 0$, with solution $x = 2$. This second derivative is graphed in figure 7.9c.

Second, to confirm that $x = 2$ is a maximum of $\frac{dy}{dx}$ we take $\frac{d^3y}{dx^3}$ (the second derivative of $\frac{dy}{dx}$), which is

$$\frac{d^3y}{dx^3} = \frac{d}{dx}[-6x + 12] = -6$$

Since this is negative, we know from rule 7.1 that we have a maximum of $\frac{dy}{dx}$ at Q, confirming that figure 7.9 is correctly drawn.

7.11 A rule for points of inflection

We can now derive from examples 7.7 and 7.8 a general rule for locating stationary points that are also points of inflection by algebraic means. The rule holds for any function that is smooth and continuous at the point in question.

RULE 7.2 Stationary points that are points of inflection

For any function $y = f(x)$, a stationary point that is a point of inflection is a point on the curve at which the slope is at a minimum value (such as example 7.7) or a maximum value (example 7.8). At such a point:

(1) $\frac{dy}{dx} = 0$ (since the point is a stationary point)

(2) $\frac{d^2y}{dx^2} = 0$ (since the point is a point of inflection)

(3) if $\frac{d^3y}{dx^3} > 0$, the point is a point of inflection that is a minimum of the slope of the function (example 7.7)

(4) if $\frac{d^3y}{dx^3} < 0$, the point is a point of inflection that is a maximum of the slope of the function (example 7.8)

As with rule 7.1 for maximum and minimum values, condition (2), that $\frac{d^2y}{dx^2} = 0$, is the *first order condition* for a point of inflection. The condition $\frac{d^3y}{dx^3} > 0$ or $\frac{d^3y}{dx^3} < 0$ is the *second order condition*, which enables us to distinguish the two types of point of inflection.

Note that rule 7.2 does not tell us what to infer if $\frac{d^3y}{dx^3} = 0$ at the point in question. This is addressed in chapter W21, section 21.2, which is available at the Online Resource Centre (www.oxfordtextbooks.co.uk/orc/renshaw3e/).

We haven't quite finished with points of inflection, as there exist also points of inflection that are not stationary points. We will deal with these in the next section.

Progress exercise 7.3

Find the stationary points of each of the following functions, and determine whether each stationary point is a local maximum, local minimum or point of inflection. In each case, draw a sketch graph of the function.

(a) $y = x^3 + 6x^2 + 12x + 15$ (b) $y = 2x^3 - 21x^2 + 60x + 10$ (c) $y = 4x^3 - 6x^2 + 3x + 12$

(d) $y = x^3 + 9x^2 + 15x + 10$ (e) $y = -x^3 + 6x^2 - 12x + 50$

7.12 More about points of inflection

In the previous section we examined stationary points (where $\frac{dy}{dx} = 0$) that were also points of inflection. When a stationary point is also a point of inflection, we found that in addition to $\frac{dy}{dx} = 0$ we have $\frac{d^2y}{dx^2} = 0$, with $\frac{d^3y}{dx^3}$ either positive or negative (rule 7.2). We looked at two examples where a stationary point was also a point of inflection (examples 7.7 and 7.8).

However, a point of inflection can also occur at a point that is not a stationary point. This can happen in four distinct ways, shown in figure 7.10.

The first case is shown in figure 7.10(a), where there is a point of inflection at P. This is clearly not a stationary point, since the curve is positively sloped at P and therefore $\frac{dy}{dx} > 0$. But P is clearly a point of inflection, since the slope reaches a maximum there; the slope is flatter on both sides of P. Thus P is a maximum of $\frac{dy}{dx}$ and therefore $\frac{d^2y}{dx^2} = 0$ and $\frac{d^3y}{dx^3} < 0$.

The second case is shown in figure 7.10(b), where there is a point of inflection at Q. Again this is not a stationary point, since the curve is positively sloped at Q and therefore $\frac{dy}{dx} > 0$. But Q is clearly a point of inflection, since the slope reaches a minimum there; the slope is steeper on both sides of Q. Thus Q is a minimum of $\frac{dy}{dx}$ and therefore $\frac{d^2y}{dx^2} = 0$ and $\frac{d^3y}{dx^3} > 0$.

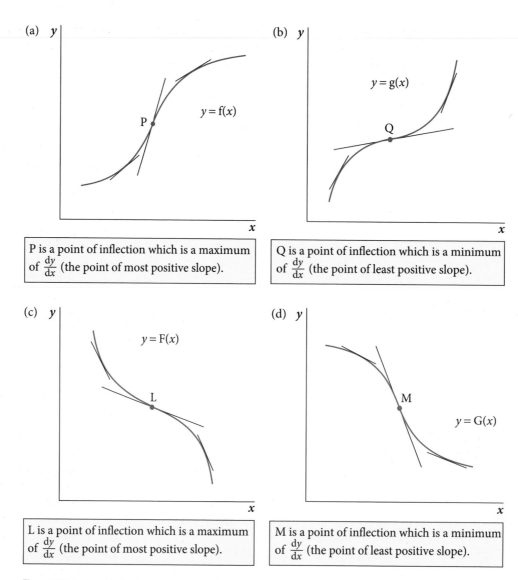

(a) P is a point of inflection which is a maximum of $\frac{dy}{dx}$ (the point of most positive slope).

(b) Q is a point of inflection which is a minimum of $\frac{dy}{dx}$ (the point of least positive slope).

(c) L is a point of inflection which is a maximum of $\frac{dy}{dx}$ (the point of most positive slope).

(d) M is a point of inflection which is a minimum of $\frac{dy}{dx}$ (the point of least positive slope).

Figure 7.10 Points of inflection that are not stationary points.

The third case is shown in figure 7.10(c), where there is a point of inflection at L. This is not a stationary point, since the curve is negatively sloped at L and therefore $\frac{dy}{dx} < 0$. But L is a point of inflection, since the slope reaches a maximum there; the slope is more negative ($=$ less positive) on both sides of L. Thus L is a maximum of $\frac{dy}{dx}$ and therefore $\frac{d^2y}{dx^2} = 0$ and $\frac{d^3y}{dx^3} < 0$.

The final case is shown in figure 7.10(d), where there is a point of inflection at M. This is not a stationary point, since the curve is negatively sloped at M and therefore $\frac{dy}{dx} < 0$. But M is a point of inflection, since the slope reaches a minimum there; the slope is flatter (less negative) on both sides of M. Thus M is a minimum of $\frac{dy}{dx}$ and therefore $\frac{d^2y}{dx^2} = 0$ and $\frac{d^3y}{dx^3} > 0$.

We will now offer two examples to show how identifying these points of inflection works out in practice.

EXAMPLE 7.9

Suppose we are asked to find the points of inflection in the function

$$y = -x^3 + 9x^2 + 50$$

We know that points of inflection are points where $\frac{dy}{dx}$ is at a maximum or minimum value, and therefore $\frac{d^2y}{dx^2} = 0$ and $\frac{d^3y}{dx^3}$ is either negative (if the point is a maximum of $\frac{dy}{dx}$) or positive (if the point is a minimum of $\frac{dy}{dx}$).

So to find any point(s) of inflection on this function we first find the second derivative, set it equal to zero and solve the resulting equation. Given $y = x^3 + 9x^2 + 50$, the first derivative is $\frac{dy}{dx} = -3x^2 + 18x$, so the second derivative is

$$\frac{d^2y}{dx^2} = -6x + 18$$

Setting this equal to zero gives $-6x + 18 = 0$, with solution $x = 3$.

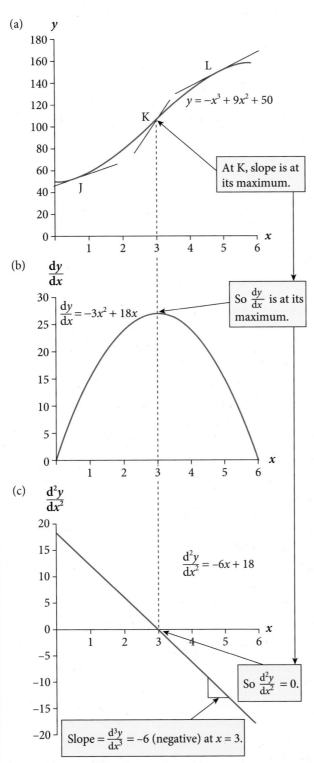

Next we examine the sign of $\frac{d^3y}{dx^3}$ at this point. We have

$$\frac{d^3y}{dx^3} = \frac{d}{dx}[-6x + 18] = -6$$

This is negative when $x = 3$ (and indeed at all values of x, but it is only $x = 3$ that matters here).

So we now know that, at $x = 3$, $\frac{d^2y}{dx^2} = 0$ and $\frac{d^3y}{dx^3} < 0$. This tells us that there is a point of inflection at $x = 3$ which is a maximum of $\frac{dy}{dx}$.

However, we don't yet have a completely clear picture of what the curve is doing at $x = 3$. We could have a curve shaped like figure 7.10(a), or like figure 7.10(c), since both have points of inflection where $\frac{dy}{dx}$ is at a maximum. (Point P is where the slope is at its most positive, while L is where the slope is at its least negative; each is a local maximum of the slope.)

To distinguish between these two possibilities we need to know whether, at $x = 3$, the curve is positively or negatively sloped. From above, we have $\frac{dy}{dx} = -3x^2 + 18x$, which equals $+27$ when $x = 3$. Thus the function is positively sloped and we therefore have a curve like figure 7.10(a) and not like figure 7.10(c).

The correctness of all this algebra is confirmed by the graphs drawn in figure 7.11. Part (a) shows the function $y = -x^3 + 9x^2 + 50$ with its

Figure 7.11 Graph of $y = -x^3 + 9x^2 + 50$ showing points of inflection.

point of inflection at K. Notice that it is not easy to see the exact position of K merely by studying the graph, because the curve is almost (but not quite) linear around K. Therefore we would need to plot a very large-scale graph to find K by purely graphical methods. This is a concrete illustration of the unreliability of purely graphical methods. It is the algebra that tells us exactly where K is.

That K is a maximum of the slope of the curve is also confirmed when we look at the graph of $\frac{dy}{dx} = -3x^2 + 18x$. This quadratic function is graphed in figure 7.11(b). We see that, as x increases and we move from left to right, $\frac{dy}{dx}$ is initially increasing, reaches its maximum value at $x = 3$, and then to the right of this point begins to decrease. The maximum value of $\frac{dy}{dx}$, which we see from figure 7.11(b) to be 27, is of course the slope of the tangent at K. (The slope looks a lot less than 27 in figure 7.11(a), but this is deceptive because the scale on the vertical axis is about 30 times smaller than on the horizontal.)

The second derivative, $\frac{d^2y}{dx^2} = -6x + 18$, is graphed in figure 7.11(c). The graph is a negatively sloped straight line. As we have just found algebraically, it cuts the x-axis at $x = 3$.

Finally the third derivative, $\frac{d^3y}{dx^3}$, is of course the slope of the graph of the second derivative, which is -6.

EXAMPLE 7.10

Consider the function $y = x^3 - 12x^2 + 300$

As in the previous example, to find any point(s) of inflection of this function we first find the second derivative, set it equal to zero and solve the resulting equation. Given $y = x^3 - 12x^2 + 300$, the first derivative is

$$\frac{dy}{dx} = 3x^2 - 24x \quad \text{so the second derivative is}$$

$$\frac{d^2y}{dx^2} = 6x - 24$$

Setting this equal to zero gives $6x - 24 = 0$, with solution $x = 4$.

Next, we examine the sign of $\frac{d^3y}{dx^3}$ at this point. We have

$$\frac{d^3y}{dx^3} = \frac{d}{dx}[6x - 24] = 6$$

This is positive when $x = 4$ (and indeed at all values of x). So we now know that, at $x = 4$, $\frac{d^2y}{dx^2} = 0$ and $\frac{d^3y}{dx^3} > 0$. This tells us that there is a point of inflection at $x = 4$ which is a minimum of the slope. This could be a curve shaped like figure 7.10(b) or figure 7.10(d), since both are points of inflection where $\frac{dy}{dx}$ is at a minimum.

To determine which of these two we have, we look at the sign of $\frac{dy}{dx}$ at $x = 4$. From above, we have $\frac{dy}{dx} = 3x^2 - 24x$, which equals -48 when $x = 4$. Thus the function is negatively sloped at $x = 4$ and we therefore have a curve like figure 7.10d and not like figure 7.10b.

The above results are confirmed by the graphs in figure 7.12. Part (a) shows the function $y = x^3 - 12x^2 + 300$ with its point of inflection at M, where $x = 4$. The tangents at L, M, and N are all negatively sloped, but the tangent at M is the most steeply negative that can be drawn in this section of the curve. The slope at M is thus at its most negative. But to say that something is at its most negative is equivalent to saying it is at its least positive. Therefore at M the slope is at its *minimum* value.

That M is a minimum of the slope is confirmed by the graph of $\frac{dy}{dx} = 3x^2 - 24x$ in figure 7.12(b). The graph of this quadratic function shows that the slope of the function is always negative and reaches its minimum (its most negative, or least positive, value) at $x = 4$. The graph shows that this minimum value is -48, which you can check by substituting $x = 4$ into $3x^2 - 24x$. This is the slope of the tangent at M. (The slope looks

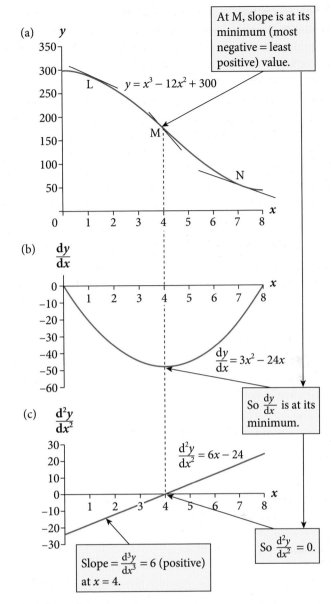

At M, slope is at its minimum (most negative = least positive) value.

(a) y

$y = x^3 - 12x^2 + 300$

L

M

N

(b) $\dfrac{\mathrm{d}y}{\mathrm{d}x}$

$\dfrac{\mathrm{d}y}{\mathrm{d}x} = 3x^2 - 24x$

So $\dfrac{\mathrm{d}y}{\mathrm{d}x}$ is at its minimum.

(c) $\dfrac{\mathrm{d}^2y}{\mathrm{d}x^2}$

$\dfrac{\mathrm{d}^2y}{\mathrm{d}x^2} = 6x - 24$

So $\dfrac{\mathrm{d}^2y}{\mathrm{d}x^2} = 0$.

Slope $= \dfrac{\mathrm{d}^3y}{\mathrm{d}x^3} = 6$ (positive) at $x = 4$.

Figure 7.12 Graph of $y = x^3 - 12x^2 + 300$ showing points of inflection.

a lot flatter than -48 in figure 7.12(a), but this is deceptive because the scale on the vertical axis is about 50 times smaller than on the horizontal.)

Finally, in figure 7.12c we see the graph of $\dfrac{\mathrm{d}^2y}{\mathrm{d}x^2} = 6x - 24$, which equals zero (cuts the x-axis) at $x = 4$. The slope of this graph is given by $\dfrac{\mathrm{d}^3y}{\mathrm{d}x^3} = 6$.

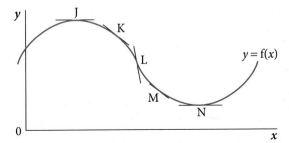

Figure 7.13 A maximum and minimum separated by a point of inflection at L.

A final point on points of inflection

Before leaving points of inflection, it is important to note that any function that has both a minimum and a maximum must have a point of inflection somewhere between the two. We can demonstrate this by looking at the sketch graph in figure 7.13.

In figure 7.13 we see that there is a maximum of the function at J and a minimum at N, with a point of inflection lying between them at L. To see why the point of inflection must be there, suppose the curve is a path that we are walking along, starting from the maximum at J. As we begin to walk down the curve from J towards N, the slope is, of course, negative and initially becomes *increasingly* negative until we reach L, the point of inflection. Point L is where the slope is at its minimum (that is, its greatest negative value). Beyond L, as we continue towards N, the slope remains negative but now becomes *decreasingly* negative until we reach N.

Given, then, that the slope is at first increasingly negative but later becomes decreasingly negative, there must be a point at which it is at its most negative. This point is, by definition, a point of inflection.

The same conclusion can also be reached by looking at the tangents to the curve. At J, the maximum, the tangent lies above the curve. As we move down the curve towards N, the tangent initially lies above the curve at points such as K. But at L, the point of inflection, the tangent cuts the curve, and between L and N the tangent lies below the curve, such as at M. Given, then, that the tangent lies above the curve at J but below it at N, the tangent must cross over at some point, and that point is the point of inflection.

If we flip the curve over so that J becomes the minimum and N the maximum, nothing changes except that the point of inflection at L is now the point where the slope is at its most positive, rather than most negative.

We can state our conclusions on points of inflection as a rule, which holds for any function that is smooth and continuous at the point in question.

RULE 7.3 Points of inflection

For any function $y = f(x)$, a point of inflection is a point on the curve at which the *slope* is at a maximum value or a minimum value. At such a point:

(1) $\dfrac{d^2y}{dx^2} = 0$

(2) if $\dfrac{d^3y}{dx^3} < 0$, the point is a point of inflection that is a maximum of the slope of the function (such as figures 7.10(a) and (c)).

(3) if $\dfrac{d^3y}{dx^3} > 0$, the point is a point of inflection that is a minimum of the slope of the function (such as figures 7.10(b) and (d)).

Condition (1), that $\dfrac{d^2y}{dx^2} = 0$, is called the *first order condition* for a point of inflection, while the condition $\dfrac{d^3y}{dx^3} > 0$ or $\dfrac{d^3y}{dx^3} < 0$ is the *second order condition*, which enables us to distinguish the two types of point of inflection.

Note that $\dfrac{dy}{dx}$ may or may not equal zero at a point of inflection. However, if $\dfrac{dy}{dx} = 0$ at a point of inflection, that point is also a stationary point (see rule 7.2).

Note also that rule 7.3 does not tell us what to infer if $\dfrac{d^3y}{dx^3} = 0$ at the point in question. For example, suppose $y = x^4$. Then we have $\dfrac{dy}{dx} = 4x^3$, which equals 0 at $x = 0$. We also have $\dfrac{d^2y}{dx^2} = 12x^2$, which also equals 0 at $x = 0$. So far, $x = 0$ partly satisfies the conditions for a stationary value that is a point of inflection (see rule 7.2). However, $\dfrac{d^3y}{dx^3} = 24x$, which also equals 0 at $x = 0$. So this point does not satisfy the conditions of rule 7.1 (which require $\dfrac{d^2y}{dx^2} \neq 0$); nor the conditions of either rule 7.2 or 7.3 (both of which require $\dfrac{d^3y}{dx^3} \neq 0$). So our rules do not permit us to say whether $x = 0$ is a maximum, minimum, or point of inflection of $y = x^4$; or none of these. In fact, as you can easily check by sketching the graph, say, for $x = -3$ to $x = +3$, $y = x^4$ has a minimum at $x = 0$.

This example shows that applying rules 7.1–7.3 will not identify every case of a maximum, minimum or point of inflection. Thus the rules give us *sufficient*, but not *necessary*, conditions for identifying these points (see section 2.13). This gap in our analysis is addressed in chapter W21, section 21.2, which can be downloaded from the Online Resource Centre (www.oxfordtextbooks.co.uk/orc/renshaw3e/).

Summary of sections 7.10–7.12

In these sections we introduced the point of inflection, which we saw is a point at which the *slope* of the function reaches a maximum or minimum value. Given this, it follows (see rule 7.3) that we find points of inflection by setting the second derivative equal to zero and solving the resulting equation. We then examined the sign of the third derivative at these points; the sign must be positive or negative. The case where the third derivative equals zero at the point in question is left open.

In section 7.11 we examined points of inflection that are also stationary points of the function itself (so that the first derivative equals zero); and in section 7.12 points of inflection that are *not* also stationary points.

To round off the chapter up to this point, figure 7.14 gives a summary, in the form of a flow chart, of the criteria for both stationary points and points of inflection set out in rules 7.1–7.3. Note the logical gap when the third derivative equals zero, as discussed immediately above.

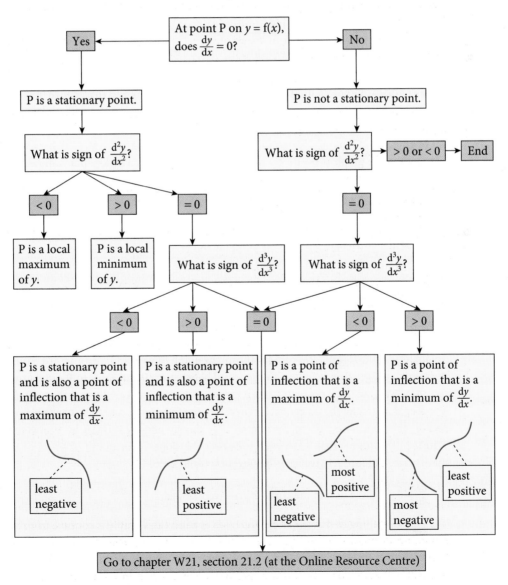

Figure 7.14 Flow chart for identifying stationary points and points of inflection.

This means that the rules do not identify the maximum, minimum, or points of inflection of *all* functions; some functions can slip through the net, as in the case of $y = x^4$ considered immediately above.

Progress exercise 7.4

Find all local maximum and minimum points, and points of inflection, in the following functions. Use your findings to draw a sketch graph of each function.

(a) $y = x^3 - 9x^2 + 15x + 5$ (b) $y = -x^3 + 15x^2 - 27x + 50$

(c) $y = x^3 - 27x^2 + 216x + 10$ (d) $y = 8x^3 - 6x + 25$

(e) $y = 2x^3 - 33x^2 + 60x + 15$

7.13 Convex and concave functions

We began this chapter by showing how, by looking at the sign of $\frac{dy}{dx}$, we could learn whether the function in question was increasing (positively sloped) or decreasing (negatively sloped), either at a specific point or over a range of values of x.

We will now show how, by looking at the sign of $\frac{d^2y}{dx^2}$, we can learn something useful about the curvature of the function: that is, whether the function in question is convex or concave. Here we will explore what mathematicians call *strict* convexity and *strict* concavity, though for brevity we will refer to them simply as convexity and concavity. We will explain convexity/concavity by four examples, shown in figures 7.15–7.18.

Looking first at figure 7.15, part (a) shows the graph of $y = 10x^2 + 5x + 20$, for $x = 0$ to $x = 5$. We see that the graph is positively sloped, so $\frac{dy}{dx}$ is positive. The key point, however, is that the slope increases as x increases: for example, the slope of the tangent at K is steeper than the tangent at J. When the slope increases as x increases, the function is said to be **convex from below**. (Imagine you are looking up at the curve from the x-axis; that is, from below.) Because the slope is increasing, the graph of $\frac{dy}{dx}$, shown in figure 7.15(b), is positively sloped. In turn this means that $\frac{d^2y}{dx^2}$, which measures the slope of the graph of $\frac{dy}{dx}$, is positive. To summarize, in figure 7.15 we have, between $x = 0$ and $x = 5$:

(1) $\frac{dy}{dx} > 0$ telling us that the curve is positively sloped;

(2) $\frac{d^2y}{dx^2} > 0$ telling us that the slope increases as x increases (curve is convex from below).

Moving next to figure 7.16, part (a) shows the graph of $y = -x^2 + 12x + 10$, for $x = 0$ to $x = 5$. We see that the graph is again positively sloped between $x = 0$ and $x = 5$, so $\frac{dy}{dx}$ is positive. In this case, however, the slope decreases as x increases: for example, the slope of the tangent at M is less steep than that of the tangent at L. Therefore the function is said to be **concave from below**. Because the slope is decreasing, the graph of $\frac{dy}{dx}$, shown in figure 7.16(b), is negatively sloped.

In turn this means that $\frac{d^2y}{dx^2}$, which measures the slope of the graph of $\frac{dy}{dx}$, is negative. To summarize, in figure 7.16 we have, between $x = 0$ and $x = 5$:

(1) $\frac{dy}{dx} > 0$,   telling us that the curve is positively sloped;

(2) $\frac{d^2y}{dx^2} < 0$,   telling us that the slope decreases as x increases (curve is concave from below).

Our third example, $y = x^2 - 12x + 50$, is graphed in figure 7.17(a). This curve is negatively sloped, so $\frac{dy}{dx}$ is negative. The key point in this case is that the slope increases (that is, becomes decreasingly negative or increasingly positive) as x increases: for example, the slope of the

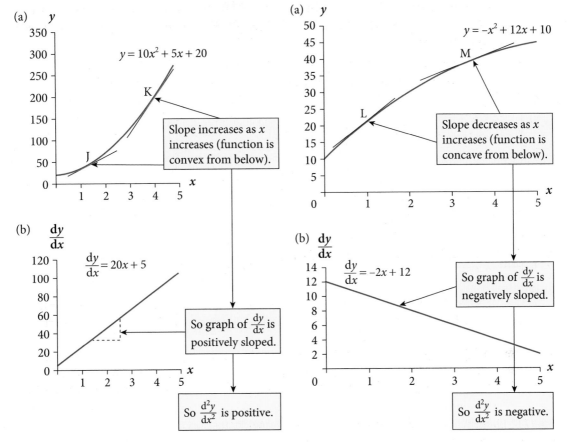

Figure 7.15 The function $y = 10x^2 + 5x + 20$ (convex between $x = 0$ and $x = 5$).

Figure 7.16 The function $y = -x^2 + 12x + 10$ (concave between $x = 0$ and $x = 5$).

tangent at Q is less negative (= more positive) than that of the tangent at P. This curve is said to be *convex from below*.

Because the slope is increasing, the graph of $\frac{dy}{dx} = 2x - 12$, shown in figure 7.17(b), is positively sloped. In turn this means that $\frac{d^2y}{dx^2}$ is positive. To summarize, in figure 7.17 we have, between $x = 0$ and $x = 5$:

(1) $\frac{dy}{dx} < 0$, telling us that the curve is negatively sloped;

(2) $\frac{d^2y}{dx^2} > 0$, telling us that the slope increases as x increases (curve is convex from below).

Our fourth and final example, $y = -x^2 - 2x + 50$, is shown in figure 7.18(a). This curve is again negatively sloped, so $\frac{dy}{dx}$ is negative in figure 7.18(b) (that is, its graph lies below the x-axis). The important thing in this case is that, as x increases, the slope of the function becomes steeper (that is, increasingly negative or decreasingly positive): for example, the tangent at S is more negatively sloped (= less positively sloped) than the tangent at R. We say that the curve is *concave from below*.

Because the slope is decreasing, the graph of $\frac{dy}{dx} = -2x - 2$, shown in figure 7.18(b), is negatively sloped. In turn this means that $\frac{d^2y}{dx^2}$ is negative. To summarize, in figure 7.18 we have, between $x = 0$ and $x = 5$:

(1) $\frac{dy}{dx} < 0$, telling us that the curve is negatively sloped;

(2) $\frac{d^2y}{dx^2} < 0$, telling us that the slope decreases as x increases (curve is concave from below).

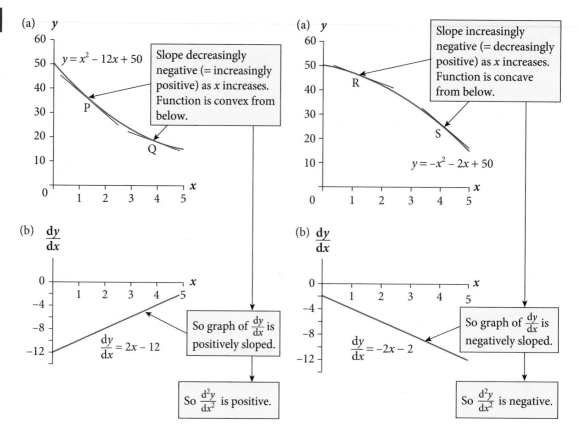

Figure 7.17 The function $y = x^2 - 12x + 50$ (convex between $x = 0$ and $x = 5$).

Figure 7.18 The function $y = -x^2 - 2x + 50$ (concave between $x = 0$ and $x = 5$).

To summarize, if the second derivative is positive, the curve is convex from below (figures 7.15 and 7.17); while if the second derivative is negative, the curve is concave from below (figures 7.16 and 7.18).

Graphically, a curve is convex from below if any tangent drawn to the curve lies below the curve; and concave from below if any tangent drawn to the curve lies above the curve. Linking with the previous section, we can now see that a point of inflection (where the tangent cuts the curve) is a point where the curve switches from concave to convex or vice versa.

Note however that although a positive/negative second derivative tells us that the curve is convex/concave, it is not an infallible guide. It is possible for a curve to be convex or concave even where the second derivative is zero. As in the previous section, $y = x^4$ illustrates this. We can easily check by sketching the graph between, say, $x = -3$ and $+3$, that the curve is convex for all values of x; a tangent drawn to the curve at any point lies below the curve. Therefore we expect to find that the second derivative is positive for all values of x. However, we have $\frac{d^2y}{dx^2} = 12x^2$, which equals zero when $x = 0$ (but is otherwise positive, as expected). Thus $y = x^4$ is convex at $x = 0$, even though the second derivative is zero at that point. We conclude that a positive/negative second derivative is a *sufficient* condition for convexity/concavity at any point, but not a *necessary* condition.

If you have trouble distinguishing convexity from concavity, it might help to remember that if you were inside a church such as St Paul's in London or St Peter's in Rome, and looking up at the dome, you would see a concave surface.

The question of whether a function is convex or concave is extremely important in more advanced economic analysis, and is taken up again in chapter W21, which is to be found at the Online Resource Centre (www.oxfordtextbooks.co.uk/orc/renshaw3e).

Summary of section 7.13

In this section we explored the concept of convexity and concavity of a function $y = f(x)$. To summarize everything in one sentence, $\frac{dy}{dx}$ positive/negative tells us that the slope of the function is positive/negative, while $\frac{d^2y}{dx^2}$ positive/negative tells us that that the function is convex/concave from below. We can examine the signs of these derivatives either at a point or over some range of values of x that interests us.

Table 7.1 collects and generalizes our findings. Note that the definition of convexity and concavity we have used here is as viewed from below, which is the definition most commonly used. Viewing from above would, of course, reverse the interpretation of the sign of the second derivative.

Table 7.1 Convex and concave functions for any function $y = f(x)$. Signs of first and second derivatives as indicators of slope and curvature at any point or over a range of points.

Example	$\frac{dy}{dx}$	Slope of function	$\frac{d^2y}{dx^2}$	Curvature of function
Figure 7.15	>0	upward sloping	>0	convex
Figure 7.16	>0	upward sloping	<0	concave
Figure 7.17	<0	downward sloping	>0	convex
Figure 7.18	<0	downward sloping	<0	concave

7.14 An alternative notation for derivatives

This is a good moment to develop further an alternative and widely used notation for functions and their derivatives which we introduced in section 6.6. Suppose we are examining two functions at the same time, say

$$y = 2x^3 + 5x^2 + 50 \quad \text{and} \quad y = x^3 + 6x^2 + 9x + 6$$

There is scope for muddle here because we have two ys which refer to different relationships with x and in general have different values.

The derivatives of these functions are

$$\frac{dy}{dx} = 6x^2 + 10x \quad \text{and} \quad \frac{dy}{dx} = 3x^2 + 12x + 9$$

This is similarly unsatisfactory, as we have two expressions for $\frac{dy}{dx}$ and we could easily get into a muddle about which derivative relates to which function.

An alternative notation which avoids these possible muddles is to replace y in the first function with $f(x)$. Here $f(x)$ means simply 'a function of x'. Similarly, we can replace y in the second function with $g(x)$, meaning 'another function of x, different from $f(x)$'. Then we can write the two functions as

$$f(x) = 2x^3 + 5x^2 + 50 \quad \text{and} \quad g(x) = x^3 + 6x^2 + 9x + 6$$

You may find it a bit alarming to see the familiar y disappear from the left-hand side of the equation, but nothing has really changed. It is simply that the dependent variables have been relabelled and this now avoids the possible muddle from having two ys floating around.

Having done this, a step that suggests itself naturally is to write the first derivative of each function as $f'(x)$ and $g'(x)$. Thus we have

$$f'(x) = 6x^2 + 10x \quad \text{and} \quad g'(x) = 3x^2 + 12x + 9$$

Again, although the familiar $\frac{dy}{dx}$ has disappeared from the expression for the derivative, nothing has really changed. It is simply a relabelling. We are now writing f'(x) and g'(x) in place of $\frac{dy}{dx}$. This avoids the possible muddling of the $\frac{dy}{dx}$ that is the derivative of $2x^3 + 5x^2 + 50$ and the $\frac{dy}{dx}$ that is the derivative of $x^3 + 6x^2 + 9x + 6$.

A natural extension of this notation is to write the second and third derivatives of the two functions as f″(x), f‴(x), g″(x), g‴(x), and so on.

Often, Greek letters are used to denote functions, such as φ(x) or θ(x). We have avoided Greek letters as far as possible, but you need to try to overcome your fear of them as you will find them frequently in writings of economists.

This alternative notation is also attractive because of its compactness, and is widely used. We will make increasing use of it as we progress through the book.

7.15 The differential and linear approximation

Here we will show that, for any function, $y = $ f(x), the first derivative $\frac{dy}{dx}$ can be used as an approximation to the function itself. We will demonstrate this with the help of figure 7.19.

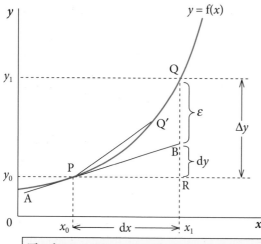

The distance QB = ε gives the error when we measure the change in y as dy using the derivative as a linear approximation to the function itself. As dx becomes smaller, this error becomes smaller, as Q then approaches B.

Figure 7.19 The first derivative as an approximation to the function $y = $ f(x).

In what follows we will use the familiar notation Δx and Δy to denote changes (of any size) in x and y. We also introduce a new notation: we will write dx and dy to denote *small* changes in x and y.

In figure 7.19 the initial position is at P with co-ordinates (x_0, y_0). Then we suppose x increases by a small amount dx and we move to a new position at Q with coordinates (x_1, y_1). The resulting increase in y is the distance QR, which we label Δy.

Now consider the distance BR, which we have labelled dy. We can use the distance BR = dy as an approximation to QR = Δy, provided we are prepared to tolerate the error in this approximation. The error is the distance QB, which we have labelled ε (the Greek letter epsilon). Thus we have

$$QR = BR + QB, \quad \text{or} \quad \Delta y = dy + \varepsilon \qquad (7.1)$$

Next, the slope of the tangent PB, which equals $\frac{BR}{PR}$, gives us the derivative $\frac{dy}{dx}$ of the function, evaluated at P. Also, the distance PR equals dx. Therefore

$$BR = \frac{BR}{PR}PR, \quad \text{or} \quad dy = \frac{dy}{dx}dx \qquad (7.2)$$

Equation (7.2) is potentially confusing because it seems to invite cancelling the two dxs on the right-hand side, causing the equation to collapse into the identity, d$y = $ dy. This cancellation would not be correct because, as explained in chapter 6, $\frac{dy}{dx}$ does not mean dy divided by dx. It is a special notation that means 'the derivative of the function' (the function being $y = $ f(x) in this case). In contrast, the free-standing dx and dy in equation (7.2) denote small changes in x and y, to which ordinary rules of algebra apply.

Equation (7.2) is called the **differential** of the function. ('Differential' simply means 'difference', and both dx and dy in equation (7.2) are differences, or changes.)

Combining equations (7.1) and (7.2), we have

$$\Delta y = \frac{dy}{dx}dx + \varepsilon \qquad (7.3)$$

Equation (7.3) says that if we are willing to ignore the error, ε, we can use the differential to measure the change in y, Δy, that occurs when x changes by a small amount dx. Can we justify ignoring the error? The answer is that if we feel the error is too large to be ignored, we can always make it smaller by making dx smaller. We can see this in figure 7.19. By making dx smaller, we cause point Q to slide down the curve to a point such as Q′. It is clear that when we recalculate equation (7.3) at Q′ instead of at Q, we will find that the error ε is now smaller. Therefore, by making dx sufficiently small, we can always reduce the error to a size so small that we are willing to ignore it. In other words, however fussy we may be about accuracy, there must come a point where the error is so small that we are prepared to ignore it.

The differential (equation (7.2)) is called a **linear approximation** to the true change, Δy. Figure 7.19 also helps us to see why it is so called. If the function $y = f(x)$ were linear, Q would coincide with B, and the error QB would truly be zero. By ignoring the error QB, we are effectively treating the function as if it were linear between x_0 and x_1. This is a valuable simplification because linear functions are much easier to manipulate. Linear approximation is examined more fully in chapter W21, section 21.2, which is available at the Online Resource Centre.

To summarize, we can state a rule for the differential of a function:

RULE 7.4 The differential of a function

Given any function $y = f(x)$, if x changes by a small amount dx, the resulting change in y, Δy, is approximately given by the differential

$$dy = \frac{dy}{dx} dx \qquad\qquad (7.2) \text{ repeated}$$

The differential gives a linear approximation to the true change in y following a small change in x. The smaller is the change in x, the closer is the approximation.

Note that, using the notation discussed in section 7.14, we can also write equation (7.2) as

$$dy = f'(x) dx \qquad\qquad (7.2a)$$

EXAMPLE 7.11

This simple example may help you to see more clearly the use of the differential as a linear approximation. Suppose the function in question is $y = x^2 + 4$, and x increases from 3 to 3.01. So we have $dx = 0.01$, and $\frac{dy}{dx} = 2x = 6$ when $x = 3$.

So, using rule 7.4, the resulting change in y is given by

$$dy = \frac{dy}{dx} dx = 6 \times 0.01 = 0.06$$

We know this answer involves a small error, ε. Let's find ε by calculating the true change in y, Δy, using the function itself rather than the derivative. When $x = 3$, $y = x^2 + 4 = 3^2 + 4 = 13$. When $x = 3.01$, $y = x^2 + 4 = (3.01)^2 + 4 = 13.0601$. Therefore $\Delta y = 13.0601 - 13 = 0.0601$. So, using rule 7.4, in this example the error that results from using the differential as a linear approximation to the true change is $\varepsilon = \Delta y - dy = 0.0601 - 0.06 = 0.0001$.

Summary of sections 7.14–7.15

In section 7.14 we noted that a more compact notation for writing derivatives of the function $y = f(x)$ was to write $f'(x)$ instead of $\frac{dy}{dx}$, $f''(x)$ instead of $\frac{d^2y}{dx^2}$, and so on. We will use this notation frequently in this book.

In section 7.15 we developed the concept of the differential of a function $y = f(x)$. The differential formula $dy = \frac{dy}{dx} dx$ gives a linear approximation to the true change in y following a small change in x (see rule 7.4).

Progress exercise 7.5

1. For each of the following functions, find the slope (positive or negative) and the curvature (convex or concave) for the range of values of x specified. Draw a rough sketch of each function for the specified range of values.

   (a) $y = x^2 - 6x + 12$, for $x \geq 4$
   (b) $y = -x^2 + 10x + 11$, for $x \leq 4$

   (c) $y = 2x^2 - 6x + 15$, for $x < 1.5$
   (d) $y = x^2 + 4x + 4$, for $x > -2$

   (e) $y = -x^3 + 15x^2 - 27x + 50$, for $x > 9$ and $x < 1$

2. Show that $y = x^3$ has a stationary point and a point of inflection at $x = 0$, and is concave when $x < 0$ and convex when $x > 0$. Sketch the graph of this function.

3. Given, $y = x^3 + x^2 + x + 10$, suppose x increases from 1 to 1.1. Find the absolute and percentage errors that result if we approximate the resulting change in y using the differential formula $dy = \frac{dy}{dx} dx$.

Checklist

Be sure to test your understanding of this chapter by attempting the progress exercises (answers are at the end of the book). The Online Resource Centre contains further exercises and materials relevant to this chapter www.oxfordtextbooks.co.uk/orc/renshaw3e/.

In this chapter we have shown how the derivative can be used to obtain important information about the shape of a function without the necessity of plotting its graph. We have also extended the idea of a derivative to the second and third derivatives. The key topics were:

✔ **Strictly increasing/decreasing functions.** A function is strictly increasing (decreasing) if its slope is always positive (negative).

✔ **Stationary points.** Using the first derivative to find the stationary points of a function: that is, where the slope is zero. A stationary point may be a maximum, minimum or point of inflection.

✔ **The second derivative.** What the second derivative measures and how to use it to distinguish between maximum and minimum values of a function.

✔ **Curve sketching.** Using information obtained by differentiation to sketch the graph of a function.

✔ **Points of inflection and the third derivative.** What points of inflection are and how to use the second and third derivatives to locate them.

✔ **Convex and concave functions.** The meaning of convexity and concavity of a function. Using the second derivative to determine whether a function is convex or concave.

✔ **The differential and linear approximation.** Understanding what the differential of a function $y = f(x)$ is and how it can be used as a linear approximation of changes in y.

After two chapters of rather heavy investment in mathematical techniques, in the next two chapters we will focus on application of these techniques to economic analysis.

Chapter 8
Economic applications of functions and derivatives

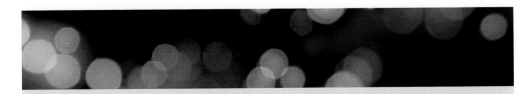

OBJECTIVES

Having completed this chapter you should be able to:

- Understand the firm's short-run total cost function—key characteristics and probable shapes.
- Derive the average cost and marginal cost functions; understand their definitions and how their shapes are related to that of the total cost function from which they are derived.
- Prove that the marginal cost curve cuts the average cost curve at the point of minimum average cost, and understand why this is true.
- Understand the market demand function for a good, and derive from it the inverse demand function, the total revenue, and marginal revenue functions.
- Prove that marginal revenue is zero at the point of maximum total revenue, and interpret this in words.
- Understand the relationship between the firm's profit function and its total cost and total revenue functions, for both monopoly and perfect competition.
- Find the most profitable level of output, and understand how this relates to the marginal cost and marginal revenue functions, for both monopoly and perfect competition.
- Understand the meaning of the second order condition for maximum profit, and when it may not be satisfied.

8.1 Introduction

This chapter is entirely devoted to economic applications, using everything we have learned so far in this book but especially the technique of differentiation and its applications to finding stationary points, which we studied in chapters 6 and 7. Hopefully, this chapter is where all the hard work you have put in on the earlier chapters begins to pay off.

There are three areas of economic application in this chapter. First we look at how a firm's costs vary with output in the short run. By the short run, economists mean a period of time in which the firm is committed to certain expenditures that are, by assumption, unavoidable. These expenditures are called fixed costs or overhead costs. We find that a quadratic function provides a suitable functional form (though by no means the only such form) for representing the main features of a firm's short-run total cost function. From this we can derive the firm's average cost function and its marginal cost function, the latter defined as the first derivative of the total cost function. We explore the relationship between marginal and average cost, showing that when they are equal to one another, average cost is at its minimum.

The second area of economic application is the market demand function for a good. Here we explore the relationship between the demand function and the total revenue and marginal revenue functions, all considered as functions of output. We consider both the monopoly and perfect competition cases. Finally, we use calculus to examine the profit-maximizing behaviour of the firm again under both monopoly and perfect competition. We will use calculus to prove rigorously that the most profitable output is where marginal cost equals marginal revenue, provided certain other conditions (known as second order conditions) are also satisfied.

8.2 The firm's total cost function

We briefly discussed the firm's total cost function in chapter 4. We saw that economists believe that it is reasonable to suppose that, in the short run, the functional relationship between a firm's total costs (TC) and its output (q) is likely to have roughly the shape shown in figure 8.1. We write $TC = f(q)$ to denote some unspecified or general functional relationship between TC and q. Here TC is measured in, say, euros per week, while output is measured in, say, tonnes per week.

The key features of the TC function in figure 8.1 are as follows:

(1) There is a positive intercept on the TC-axis, assumed here for the sake of concreteness to be €500 per week. This means that the firm incurs costs of €500 per week even if $q = 0$: that is, if it produces nothing. Costs that are incurred even if the firm produces nothing are called *fixed costs*. A good example of fixed costs would be the rent the firm has to pay for its land and buildings. (Fixed costs are also known as overhead costs.)

(2) The TC curve is positively sloped. This is easy to justify. It means that total costs increase as output increases, reflecting the reasonable assumption that in order to increase output the firm has to hire more workers, buy more raw materials and so on.

(3) The TC curve becomes increasingly steeply sloped as output increases. This results from our assumption that we are looking at the firm's situation *in the short run*. We assume that in the short run some inputs (for example, number and type of machines and other capital equipment, size of factory, land) are fixed in supply. The reason for this is that it takes time to buy and install additional machines, build an extension to the factory and so on. Thus in the short run, the firm expands its output by hiring more and more

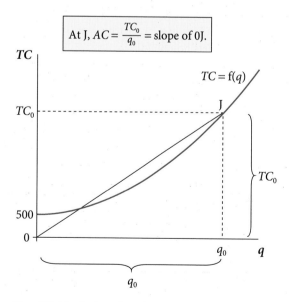

Figure 8.1 The firm's total cost function.

workers to operate a fixed number of machines in a factory of a fixed size. This causes efficiency to decline and therefore costs rise disproportionately as output increases. We will examine this phenomenon more closely in chapter 14.

(4) The *TC* curve is smooth and continuous, implying that output is continuously variable. How realistic this assumption is depends on the nature of the firm's product. If the firm produces paint, it is reasonable to suppose it can increase or reduce its annual output by 1 millilitre or even by 0.1 or 0.001 millilitres, at least in principle. If the firm produces cars, it can increase its output by 1 car, but not by 0.1 of a car since this would be useless. However, if the firm produces 500,000 cars per year, an increase of 1 car is so small relatively that it is a reasonable approximation to view its output as being continuously variable.

We can encapsulate these remarks about the general shape of the short-run total cost function in a very compact and precise algebraic form. We can write the *TC* function in a completely general way as

$$TC = f(q)$$

where $f(q)$ denotes some function of output, q. Reflecting our assumption that output is continuously variable, we must assume that $f(q)$ is a smooth and continuous function. Then, to get a curve with the shape shown in figure 8.1 we need to add the restrictions that $f(q)$ has a positive intercept on the *TC*-axis, is positively sloped and slopes upwards increasingly steeply as q increases. In mathematical notation we can write these restrictions as $f(0) > 0$, $f'(q) > 0$ and $f''(q) > 0$, where f' and f'' are the first and second derivatives of $f(q)$, and $f(0)$ is the value of the function when $q = 0$.

EXAMPLE 8.1

As a specific example of a *TC* curve that obeys these restrictions on its shape, we could use

$$TC = 5q^2 + 5q + 2000$$

This gives *TC* as a quadratic function of q and is graphed in figure 8.2(a). We can see that fixed costs are the intercept term, 2000, while the component of total costs which varies with output (known as **variable costs**) is given by $5q^2 + 5q$.

We should emphasize that the shape of the short-run *TC* curve shown in figures 8.1 and 8.2 is only one of several possibilities, some of which are examined in section 8.6 below. For a fuller explanation, see any microeconomics textbook.

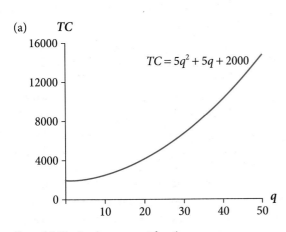

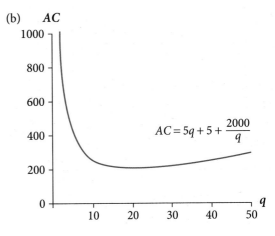

Figure 8.2 Total and average cost functions.

8.3 The firm's average cost function

We define *average* cost (*AC*) in a perfectly straightforward way, as total cost divided by output. In algebraic form, we write this definition as

$$AC \equiv \frac{TC}{q}$$

To make this concept more concrete, suppose a firm produces output of, say, $q = 5$ (tonnes per week). Suppose also that it finds that its total costs are then $TC = 635$ (euros per week). Then its average cost, or cost per unit of output, is

$$AC = \frac{635 \text{ euros per week}}{5 \text{ tonnes per week}} = 127 \text{ euros per tonne.}$$

Note that the time unit, 'per week', effectively cancels between numerator and denominator. This is because average cost depends only on the quantity produced and the total cost, not on the time taken. Notice that average cost is the same thing as cost per tonne, or more generally as cost per unit of output. The term 'cost per unit of output' is often abbreviated to 'unit cost'.

It is interesting to note that AC can be inferred from the graph of TC. For example, in figure 8.1, at point J output is at some level we will call q_0 and total cost is, say, TC_0. If we draw the straight line 0J from the origin to point J, then we can see that the slope of this line is TC_0 divided by q_0, so its slope measures AC at that level of output. (A straight line from the origin, such as this, is called a **ray**.) In general, and for any TC curve, we can read off the level of AC at any point on the TC curve by examining the slope of a line from the origin, or ray, to the point in question.

EXAMPLE 8.2

Using the TC function from the previous example:

$$TC = 5q^2 + 5q + 2000$$

we can apply the definition of AC and get

$$AC \equiv \frac{TC}{q} = \frac{5q^2 + 5q + 2000}{q}$$

which can also be written as

$$AC = 5q + 5 + \frac{2000}{q} \tag{8.2}$$

Figure 8.2(b) graphs this AC function. Note that it is, very roughly, U-shaped, and thus there is a minimum level of average cost, which seems to occur at about $q = 20$. This AC function is a hybrid functional form, for it is the sum of the linear function $5q + 5$ and the rectangular hyperbola $\frac{2000}{q}$ (see section 5.5 if you need to revise the rectangular hyperbola).

Minimum average cost

It seems likely that the firm in example 8.2 will be interested in finding the level of output at which its average or unit costs are at the minimum level, which we have seen from figure 8.2b is at around $q = 20$. We do this by applying rule 7.1 from chapter 7 for finding the minimum of a function. We find the derivative of the AC curve, set it equal to zero and solve the resulting equation. Our starting point is equation (8.2) above:

$$AC = 5q + 5 + \frac{2000}{q} \quad \text{which we can write as} \quad AC = 5q + 5 + 2000q^{-1}$$

Applying the power rule of differentiation (see section 6.7), we get

$$\frac{\mathrm{d}AC}{\mathrm{d}q} = 5 - 2000q^{-2} = 5 - \frac{2000}{q^2}. \quad \text{Setting this equal to zero gives}$$

$$5 - \frac{2000}{q^2} = 0$$

At first sight this doesn't look easy to solve, but if we multiply both sides by q^2 and divide both sides by 5 we get

$$q^2 = 400$$

from which $q = +20$ or -20. We ignore the solution $q = -20$, as this would mean a negative output, which doesn't really make sense.

We need to check that $q = 20$ is indeed a minimum of the AC function (rather than a maximum or a point of inflection). We do this by looking at the sign of the second derivative (see rule 7.1). Given, from above:

$$\frac{\mathrm{d}AC}{\mathrm{d}q} = 5 - 2000q^{-2}$$

we differentiate this, using the power rule, and get

$$\frac{\mathrm{d}^2 AC}{\mathrm{d}q^2} = 4000q^{-3} = \frac{4000}{q^3}$$

which is positive when $q = 20$. So $q = 20$ is the output at which AC is at its minimum, confirming figure 8.2(b).

It is also of interest to find the level of AC at this point, by substituting $q = 20$ into the AC function. This gives

$$AC = 5q + 5 + \frac{2000}{q} = 205 \quad \text{when } q = 20$$

(If you are sceptical about whether 205 is truly the minimum average cost, try calculating AC when q is a little above or below 20: for example, $q = 19.9$ and $q = 20.1$. You will find in both cases that $AC > 205$. Note however that this is not a rigorous method, because it's logically possible that we would have got a different answer if we had chosen different values for q; say, $q = 19.95$ and 20.05.)

Asymptotic behaviour of *AC*

The point of minimum average cost is not the only interesting feature of the AC curve shown in figure 8.2b. We also need to think about what happens to average cost when output is very large or very small. Consider first what happens as output becomes larger and larger, and ultimately approaches infinity (this is a theoretical, not a real-world possibility). Given that

$$AC = 5q + 5 + \frac{2000}{q}$$

we can see that as q becomes larger and larger, so $\frac{2000}{q}$ becomes smaller and smaller, and indeed approaches zero as q approaches infinity. This means that AC will approach a limiting value of $5q + 5$.

There is an economic logic behind this. Since in this example 2000 is the level of fixed costs, $\frac{2000}{q}$ can be interpreted as average fixed costs, or fixed costs per unit of output. Thus when output is very large, fixed costs are spread over such a large output that fixed costs per unit of output (which measures the burden of fixed costs that has to be 'carried' by each unit of output) becomes very small. This is why firms with large fixed costs like to produce a large output.

At the other extreme, as output, q, becomes very small and ultimately approaches zero, we can see that $\frac{2000}{q}$ gets larger and larger, and ultimately approaches infinity. This causes AC to become larger and larger, and ultimately to approach infinity. Thus the AC curve goes off towards $+\infty$ as q approaches zero. Graphically, this means that the AC curve gets closer and closer to the vertical axis but never touches it. The economic logic is that when fixed costs are spread over a very small output, the burden of fixed costs that has to be carried by each unit of output is very large.

Finally, recall our remark earlier that the shape of the TC curve of figures 8.1 and 8.2(a) is only one of several possible shapes. And of course, if the TC curve has a different shape, the derived AC curve has a different shape too.

8.4 Marginal cost

In introductory economics textbooks, marginal cost is defined as the increase in total cost that results from an increase in output of 1 unit. We will call this definition MC_1 in order to distinguish it from another definition that we will introduce shortly.

We can illustrate and develop this definition of marginal cost using the total cost function that we used above:

$$TC = 5q^2 + 5q + 2000$$

The graph of this total cost function is repeated in figure 8.3.

We will start by finding the difference quotient, $\frac{\Delta TC}{\Delta q}$, for this total cost function (see section 6.3 if you need to revise this). Suppose initially that we are at point A in figure 8.3, producing output q_0 with total cost TC_0, and then allow output to increase by some amount Δq. The new level of total cost is then TC_1. Thus the increase in total cost is $\Delta TC = TC_1 - TC_0$. Note that ΔTC gives us the increase in total cost that occurs when output increases by the amount Δq.

The difference quotient, $\frac{\Delta TC}{\Delta q}$, then equals the distance BC divided by the distance AC, which in turn equals the slope of the chord AB.

A special case of this arises if we now impose the restriction that $\Delta q = 1$ (that is, we consider the effect of an increase in output by 1 unit). In this case, we have

$$\frac{\Delta TC}{\Delta q} = \Delta TC \quad (\text{since } \Delta q = 1)$$

In words, this says that the difference quotient equals ΔTC, which is the increase in total cost that occurs when output increases by 1 unit. But we noted above that marginal cost (MC_1) was defined in introductory textbooks as exactly this: the increase in total cost that occurs when output increases by 1 unit. Thus MC_1 equals the difference quotient, $\frac{\Delta TC}{\Delta q}$, with Δq set equal to 1.

However, this definition of marginal cost, MC_1, has several defects in the eyes of theoretical economists. First, to find MC_1 we must evaluate the difference quotient $\frac{\Delta TC}{\Delta q}$, which as we saw in chapter 6 is a tedious business. Second, the calculation is affected by the units in which output is measured. For example, if we were looking at a sugar factory, the effect on total cost of a 1-unit

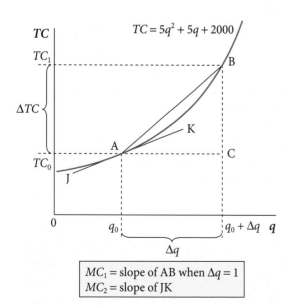

$$MC_1 = \text{slope of AB when } \Delta q = 1$$
$$MC_2 = \text{slope of JK}$$

Figure 8.3 Two definitions of marginal cost.

increase in output would obviously depend heavily on whether the unit of output was tonnes, kilos or perhaps grains! Perhaps most importantly, as we shall see later, certain theoretical results are difficult and untidy to demonstrate using this definition.

Because of these defects, a second definition, which we will call MC_2, is used in theoretical economics and is defined as $\frac{dTC}{dq}$, the derivative of the total cost function, evaluated at the point we are interested in. In figure 8.3, MC_2 is measured by the slope of the tangent JK. So, collecting results, our two definitions of marginal costs are

$$MC_1 \equiv \frac{\Delta TC}{\Delta q} \quad \text{with } \Delta q = 1; \text{ and}$$

$$MC_2 \equiv \frac{dTC}{dq}$$

How are the two definitions, MC_1 and MC_2, related? From section 6.5 we know that the difference quotient $\frac{\Delta TC}{\Delta q}$ approaches the derivative $\frac{dTC}{dq}$ as Δq becomes smaller. So we can conclude that provided output, q, is measured in sufficiently small units, so that Δq can be both small and equal to 1, the difference between the two definitions of marginal cost is small enough, in most situations, to be ignored.

Why do two alternative definitions continue to exist? Why don't economists settle on one definition and scrap the other? The answer is that each definition has merits. Definition MC_1 is easier to explain to beginners in economics and to those who have not learned how to differentiate a function. It is also the definition we would use in any real-world situation where we were actually trying to calculate marginal cost. This is because we can only ever observe a finite change in output, and so can only ever calculate the difference quotient, not the derivative. Definition MC_2 in contrast is better suited to theoretical work and is the definition we shall use throughout this book. With MC_1 no longer in the picture, we shall refer to MC_2 simply as MC.

We therefore define marginal cost as follows:

RULE 8.1 Definition of marginal cost

For any total cost function $TC = f(q)$, we define marginal cost (MC) at any level of output q as

$$MC \equiv \frac{dTC}{dq}$$

In graphical terms, MC is measured by the slope of the tangent to the TC function at the point in question, such as the tangent JK in figure 8.3.

Independence of marginal cost and fixed costs

An important point that we can demonstrate immediately is that MC is independent of the level of fixed costs. Suppose we write the total cost function in general form as

$$TC = f(q) + c$$

where c denotes fixed costs (by definition, cost that are independent of output) and f(q) denotes total variable costs (costs that vary with output). Then marginal cost is

$$MC \equiv \frac{dTC}{dq} = f'(q)$$

where f$'(q)$ is the derivative of the function f(q). The key point is that the additive constant, c, disappears on differentiation (see section 6.7, rule D3 if you are uncertain on this point). The economic logic of this is that, since by definition fixed costs do not change when output increases, they have no effect on marginal cost, which measures the *additional* costs arising from an increase in output.

EXAMPLE 8.3

Continuing with the total cost function from the previous example, we have

$$TC = 5q^2 + 5q + 2000 \quad \text{so the marginal cost function is}$$

$$MC \equiv \frac{dTC}{dq} = 10q + 5$$

Note that this expression for MC is unaffected by the level of fixed costs, which are 2000 in this example. If fixed costs rose to 5000, this would leave MC unchanged. The graph of MC against output, for this particular total cost function, is shown in figure 8.4. (Ignore the AC curve in figure 8.4 for the moment.)

If you are puzzled about the relationship between figures 8.2(a) and 8.4, remember that marginal cost is measured by the slope of a tangent drawn to the TC curve. So, when we are looking at the MC curve, what is measured up the vertical axis in figure 8.4 is the *slope* of the TC curve in figure 8.2(a), for any given value of q. The upward-sloping MC curve in figure 8.4 reflects the fact that the slope of the TC curve in figure 8.2(a) increases as q increases.

We can use this example to demonstrate our assertion above that the difference between MC_1 and MC_2 is likely to be small. Suppose output increases from 100 to 101 units. When $q = 100$, $TC = 5(100)^2 + 5(100) + 2000 = 52,500$. When $q = 101$, $TC = 5(101)^2 + 5(101) + 2000 = 53,510$. So $MC_1 = \Delta TC = 1010$. When $q = 100$, $MC_2 \equiv \frac{dTC}{dq} = 10q + 5 = 10(100) + 5 = 1005$. For most purposes, the difference between 1010 and 1005 is likely to be small enough to be ignored.

Finally, note yet again that the TC function we have considered in figures 8.2–8.4 is only one of several possible shapes for the TC function (and therefore for the AC and MC functions derived from it).

8.5 The relationship between marginal and average cost

Figure 8.4 shows AC (from figure 8.2(b)) as well as MC (from example 8.3). Recall that both curves are derived from the total cost function $TC = 5q^2 + 5q + 500$. The key point is that the marginal cost (MC) curve cuts the average cost (AC) curve at the point of minimum average cost. We know from section 8.3 above that minimum AC is at $q = 20$, $AC = 205$. To the right of this point (that is, when $q > 20$), we see that the MC curve lies above the AC curve (that is, $MC > AC$). To the left of this point (that is, when $q < 20$), we see that the MC curve lies below the AC curve (that is, $MC < AC$).

Let us check algebraically that figure 8.4 is indeed drawn correctly. First, we'll check that $MC = AC$ when AC is at its minimum. We'll do this by setting MC equal to AC and solving the resulting equation for output, q.

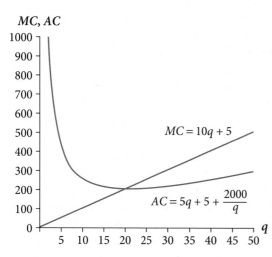

Figure 8.4 Marginal and average cost curves and their relationship.

Since $MC = 10q + 5$ and $AC = 5q + 5 + \dfrac{2000}{q}$,

we will have $MC = AC$ when

$$10q + 5 = 5q + 5 + \frac{2000}{q}$$

If we subtract $(5q + 5)$ from both sides, multiply both sides by q and divide both sides by 5, this becomes $q^2 = 400$ and hence $q = 20$ (as usual, we discard the negative solution, $q = -20$).

Since we have already found in section 8.3 that $q = 20$ is where the AC curve reaches its minimum, we have confirmed algebraically what figure 8.4 is telling us, namely that $MC = AC$ when AC is at its minimum.

We still have a little more work to do, however, because we need to confirm algebraically that the MC curve lies above the AC curve when q is greater than 20, and below the AC curve when q is less than 20. We can do this using inequalities. (Refer to section 5.10 if you need to revise the techniques for manipulating inequalities.)

To show this we first set $MC > AC$, which is true when

$$10q + 5 > 5q + 5 + \frac{2000}{q}$$

We then subtract $5q + 5$ from both sides, multiply both sides by q, and divide both sides by 5. (Since both q and 5 are positive, this does not reverse the direction of the inequality.) We are then left with $q^2 - 400 > 0$. Factorizing the left-hand side (see section 4.3) we obtain $(q + 20)(q - 20) > 0$. This inequality is satisfied when $q > 20$ or $q < -20$. We discard the negative solution, so $q > 20$.

Similarly, we set $MC < AC$ and follow the same steps as above but with the inequality sign now reversed. We then arrive at $q < 20$ (ignoring as before the negative output solution).

To summarize, when $q < 20$, $MC < AC$; when $q = 20$, $MC = AC$; and when $q > 20$, $MC > AC$. We also know that when $q < 20$, AC is falling; and when $q > 20$, AC is rising. Thus the key features of figure 8.4 are confirmed algebraically. When output is less than 20 units, marginal cost is below average cost, and average cost is falling. When output is exactly 20 units, marginal and average cost are equal, and average cost is momentarily stationary at its minimum value: that is, it is neither rising nor falling. When output is greater than 20 units, marginal cost is above average cost, and average cost is rising.

Although we have derived these results with a specific total cost function, $TC = 5q^2 + 5q + 2000$, in fact they hold completely generally for any total cost function (and indeed for any function, whether it relates to a firm's total costs or anything else). We prove this in appendix 8.1 at the end of this chapter, but meanwhile a purely verbal argument might help to persuade you of the generality of the above results.

Consider a football team's goal-scoring average. Suppose the team has played 20 games this season and scored a total of 60 goals. Clearly its average score, or score per game, is $60/20 = 3$. Now suppose it plays its twenty-first game of the season. We can call the number of goals scored in this game its 'marginal score'. Suppose it scores four goals in its twenty-first game: that is, its marginal score is 4. The team's total score is now 64 goals in 21 matches, so its new average is $64/21 = 3.05$. Thus because its marginal score (4) is greater than its previous average score (3), its average score rises. If the team's marginal score had been less than three goals, its average would have fallen. And if its marginal score had been equal to its average score (three goals), its average would have neither risen nor fallen. (Check this for yourself with some numerical examples. It doesn't have to be football.)

As a second example, suppose there are 10 people in a room and their total (combined) weight is 750 kilos. Their average weight is therefore $750/10 = 75$ kilos. Now suppose an additional, or 'marginal', person weighing 130 kilos enters the room. This raises the total weight to 880 kilos and the average weight to $880/11 = 80$ kilos. Thus because the marginal person's weight was greater than the average weight, the average weight rises. If the marginal person had weighed less than the average of 75 kilos, the average would have fallen (check this for yourself).

We will now work through three examples of possible relationships between total, average and marginal cost.

8.6 Worked examples of cost functions

Examples 8.1–8.3, which we have just finished working through, assumed that the firm's total cost function was a quadratic function. We will now consider another quadratic function, for additional practice, and two other possible functional forms.

EXAMPLE 8.4

In this example we assume that the total cost function is again a quadratic function:

$$TC = 2q^2 + 5q + 1800$$

which is graphed in figure 8.5(a). The average cost function is therefore

$$AC \equiv \frac{TC}{q} = \frac{2q^2 + 5q + 1800}{q} \quad \text{which can also be written as}$$

$$AC = 2q + 5 + \frac{1800}{q} \quad \text{or as} \quad AC = 2q + 5 + 1800q^{-1}$$

and which is graphed in figure 8.5(b).

To find the output at which average cost is at its minimum, we take the derivative of AC, set it equal to zero and solve the resulting equation for q. From above:

$$AC = 2q + 5 + 1800q^{-1}. \quad \text{Therefore}$$

$$\frac{dAC}{dq} = 2 - 1800q^{-2} = 2 - \frac{1800}{q^2}$$

Setting this equal to zero gives

$$2 - \frac{1800}{q^2} = 0$$

If we multiply both sides by q^2 and divide both sides by 2, we get $q^2 = 900$, from which $q = 30$ (ignoring the $q = -30$ solution).

This is a minimum rather than a maximum because

$$\frac{d^2AC}{dq^2} = 3600q^{-3} = \frac{3600}{q^3} \quad \text{which is positive when } q = 30$$

The level of AC at this point is found by substituting $q = 30$ into the AC function. This gives

$$\text{minimum } AC = 2(30) + 5 + \frac{1800}{30} = 125$$

Turning to the marginal cost function, this is found as

$$MC \equiv \frac{dTC}{dq} = 4q + 5$$

which is graphed, with AC, in figure 8.5(b).

To show that $MC = AC$ when AC is at its minimum, we set MC equal to AC and solve the resulting equation for q, as follows:

$$4q + 5 = 2q + 5 + \frac{1800}{q}$$

If we subtract $2q + 5$ from both sides, multiply both sides by q, and divide both sides by 2, this becomes $q^2 = 900$ and hence $q = 30$ (discarding the other solution, $q = -30$).

We have already found that $q = 30$ is where the AC curve reaches its minimum, so we have confirmed that $MC = AC$ when AC is at its minimum.

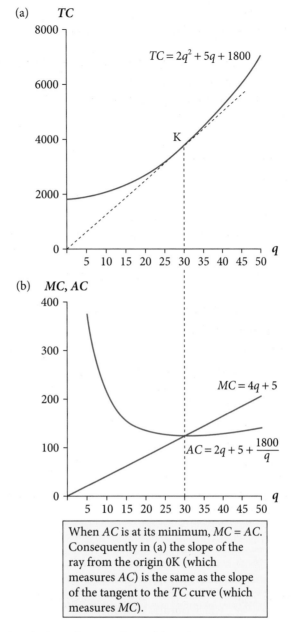

(a) *TC*

$TC = 2q^2 + 5q + 1800$

K

(b) *MC, AC*

$MC = 4q + 5$

$AC = 2q + 5 + \dfrac{1800}{q}$

When *AC* is at its minimum, *MC = AC*. Consequently in (a) the slope of the ray from the origin 0K (which measures *AC*) is the same as the slope of the tangent to the *TC* curve (which measures *MC*).

Figure 8.5 Average and marginal costs from a quadratic total cost function.

To show that the *MC* curve lies above the *AC* curve when *q* is greater than 30, and below the *AC* curve when *q* is less than 30, we first set *MC > AC*, which is true when

$$4q + 5 > 2q + 5 + \frac{1800}{q}$$

We then subtract $2q + 5$ from both sides, multiply both sides by *q*, and divide both sides by 2. (Since both *q* and 2 are positive, this does not reverse the direction of the inequality.) We are then left with $q^2 - 900 > 0$. This factorizes as $(q + 30)(q - 30) > 0$, hence $q > 30$ or $q < -30$. Rejecting the negative solution we arrive at $q > 30$.

Similarly, we set *MC < AC* and follow the same steps as above but with the inequality sign now reversed. We then arrive at $q < 30$.

Thus the key features of figure 8.5 are confirmed algebraically. When output is less than 30 units, marginal cost is below average cost, and average cost is falling. When output is 30 units, marginal and average cost are equal, and average cost is momentarily stationary at its minimum value: that is, neither rising nor falling. When output is greater than 30 units, marginal cost is above average cost, and average cost is rising.

We can take the opportunity here to point out an interesting diagrammatic feature. In figure 8.5(a), at point K where $q = 30$, the straight broken line from the origin is tangent to the TC curve. This is the only point on the TC curve where this is true. You can probably see immediately the explanation for this. We know from section 8.2 (figure 8.1) above that the slope of 0K measures AC when $q = 30$. We also know that marginal cost is defined as $\frac{dTC}{dq}$, which means that the tangent to the TC curve at K measures MC when $q = 30$. Since $MC = AC$ at K, the tangent at K must have the same slope as the ray from the origin 0K, as indeed is the case in figure 8.5(a).

Thus at the point where $MC = AC$, the slope of the tangent to the TC curve = slope of the straight line (ray) from the origin to the TC curve.

Further, we can see that at any point to the left of K in figure 8.5(a), the slope of a tangent drawn to the TC curve is flatter (less steep) than the slope of a ray from the origin to the same point. Since the former slope measures MC and the latter AC, we can infer that $MC < AC$ at any point to the left of K. By the same reasoning, at any point to the right of K we have $MC > AC$.

EXAMPLE 8.5

In this example we assume that total cost is given by the linear function

$$TC = 2q + 20$$

which is graphed in figure 8.6(a). In section 4.14 we said that a linear short-run total cost function was not likely because it implied that output could be expanded indefinitely without cost penalties, despite the assumption of a fixed physical productive capacity. This remains a valid argument, but it could nevertheless be true that the total cost function was linear up to a certain critical level of output. So in figure 8.6a we assume that this critical level is greater than $q = 25$.

Given $TC = 2q + 20$, marginal cost is

$$MC \equiv \frac{dTC}{dq} = 2$$

Thus marginal cost is constant. Average cost is

$$AC \equiv \frac{2q + 20}{q} \quad \text{which can also be written as}$$

$$AC = 2 + \frac{20}{q} \quad \text{or as} \quad AC = 2 + 2q^{-1}$$

The MC and the AC functions are graphed in figure 8.6(b). As MC is constant, its graph is simply a horizontal line, while AC is a rectangular hyperbola (see section 5.5 for revision of the rectangular hyperbola). The AC function has no minimum value, but instead approaches closer and closer to the horizontal MC function as q increases. We can see this in the algebra: since $MC = 2$, this can be substituted into our expression for AC, giving

$$AC = 2 + \frac{20}{q} = MC + \frac{20}{q}$$

As q gets larger and larger, $\frac{20}{q}$ gets smaller and smaller, so AC approaches MC from above. However, this approach is asymptotic: that is, because $\frac{20}{q}$ is always positive however large q may be, AC is always greater than MC.

The economic logic of this is very simple. Because the TC function is linear, it is only the presence of fixed costs that causes AC (which includes fixed costs) to exceed MC (which

(a) *TC*

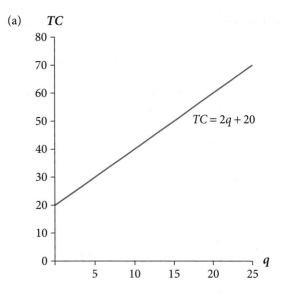

(b) *MC, AC*

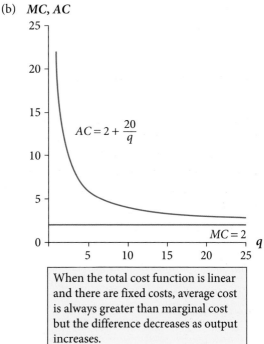

When the total cost function is linear
and there are fixed costs, average cost
is always greater than marginal cost
but the difference decreases as output
increases.

Figure 8.6 Average and marginal cost from a linear total cost function.

does not). As output increases, the fixed costs are spread over a larger and larger output, so
fixed costs per unit of output, $\frac{20}{q}$, decrease but always remain positive.

EXAMPLE 8.6

As a final example we assume that total cost is given by the cubic function

$$TC = 3q^3 - 54q^2 + 500q + 2592$$

which is graphed in figure 8.7(a). The average cost function is therefore

$$AC = \frac{TC}{q} = \frac{3q^3 - 54q^2 + 500q + 2592}{q}$$

which can also be written as

$$AC = 3q^2 - 54q + 500 + \frac{2592}{q} \quad \text{or as} \quad AC = 3q^2 - 54q + 500 + 2592q^{-1}$$

Marginal cost is

$$MC \equiv \frac{dTC}{dq} = 9q^2 - 108q + 500$$

AC and MC are graphed in figure 8.7(b). Looking at the TC function first, we see that this has the property, discussed in section 5.4, that it has a relatively steep slope at low levels of output, becomes less steep in the intermediate range of outputs, then becomes steeper once more as output increases further. The underlying rationale for this could be that, given the plant size, this productive process is relatively technologically inefficient when producing low and high levels of output. This gives the TC curve a slight S-shape.

Looking next at the MC graph, we see that the S-shape in the TC curve means that as output increases, marginal cost, which measures the slope of the TC curve, initially

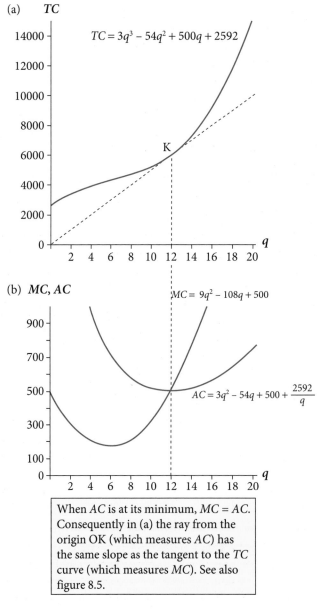

When AC is at its minimum, $MC = AC$. Consequently in (a) the ray from the origin OK (which measures AC) has the same slope as the tangent to the TC curve (which measures MC). See also figure 8.5.

Figure 8.7 Average and marginal costs from a cubic total cost function.

falls, reaches a minimum at $q = 6$, and then rises again. In mathematical terms, you have probably already spotted that this is a point of inflection in the TC curve. We know from sections 7.10–7.12 that a point of inflection in TC occurs where $\frac{dTC}{dq} = MC$ is at a maximum or minimum value, which in turn requires $\frac{dMC}{dq} = 0$ and $\frac{d^2MC}{dq^2}$ either negative (for a maximum) or positive (for a minimum). To find the point of inflection we therefore take the derivative of MC, set it equal to zero and solve the resulting equation. We have

$MC = 9q^2 - 108q + 500$ therefore

$\frac{dMC}{dq} = 18q - 108 = 0.$ The solution to this linear equation is $q = 6$.

The second derivative is $\frac{d^2MC}{dq^2} = 18$, which is positive, so $q = 6$ is a minimum of MC.

The AC curve is not much affected by the point of inflection in the TC curve, and has the usual roughly U-shape. The AC function in this case is a hybrid of $3q^2 + 54q + 500$, which is a quadratic function, and $\frac{2592}{q}$, which is a rectangular hyperbola.

To find the output at which average cost is at its minimum, we take the derivative of AC, set it equal to zero, and solve the resulting equation for q. From above:

$$AC = 3q^2 - 54q + 500 + \frac{2592}{q}, \quad \text{so}$$

$$\frac{dAC}{dq} = 6q - 54 - 2592q^{-2} = 6q - 54 - \frac{2592}{q^2}$$

Setting this equal to zero gives

$$6q - 54 - \frac{2592}{q^2} = 0$$

If we multiply both sides by q^2 and divide both sides by 6, we get

$$q^3 - 9q^2 - 432 = 0$$

This cubic equation has one positive solution, $q = 12$, a solution that can be found with the help of the Maple computer algebra program (see section 5.3 and also the Online Resource Centre). This is a minimum rather than a maximum of AC because

$$\frac{d^2AC}{dq^2} = 6 + 5184q^{-3} = 6 + \frac{5184}{q^3} \quad \text{which is positive when } q = 12.$$

The level of AC at this point is found by substituting $q = 12$ into the AC function. This gives

$$\text{minimum } AC = 3(12)^2 - 54(12) + 500 + \frac{2592}{12} = 500$$

When $q = 12$ we also have

$MC = 9q^2 - 108q + 500 = 500$

Thus marginal and average cost are equal when average cost is at its minimum.

Note that the rule we derived above, which says that when marginal cost is below (above) average cost, average cost is falling (rising) remains valid in this case, and indeed in every case, since it follows from the definitions of 'total', 'average', and 'marginal' in this context. The fact that between $q = 6$ and $q = 12$ marginal cost is rising, while average cost is falling, is logically distinct from the fact that marginal cost is less than average cost when q is less than 12.

Note also that figure 8.7(a) has the same property as figure 8.5(a): at point K, where marginal and average cost are equal, the ray 0K from the origin is tangent to the TC curve. This is because the slope of the ray 0K measures AC, while the slope of the tangent measures MC. Since $AC = MC$ at K, these slopes are equal; that is, the ray and the tangent have the same slope.

Summary of sections 8.1–8.6

We will now briefly summarize what we have learned from the discussion and examples above concerning the relationship between total cost, average cost, and marginal cost. A general algebraic analysis is provided in appendix 8.1 at the end of this chapter.

(1) The short-run total cost function, $TC = f(q)$, has two key characteristics: first, a positive intercept due to the existence of fixed costs, which, almost by definition, are present in the short run; second, a slope that becomes increasingly steep, due to capacity constraints, as output becomes very large.

(2) Because marginal cost, MC, by definition measures the slope, $\frac{dTC}{dq}$, of the TC function, the second characteristic above causes MC to increase as q increases, at least when q is very large. So the graph of MC will slope upwards, at least at large outputs. Over some range of outputs MC may be decreasing, but is unlikely ever to fall below zero as this would imply a negatively sloped TC curve, which seems extremely unlikely.

(3) The two features of the TC curve noted above cause the average cost function, $\frac{TC}{q}$, to be roughly U-shaped. Initially, as q increases, AC falls as the fixed costs are being spread over a larger output. Beyond a certain point, however, this is more than offset by capacity limitations, which cause variable costs to rise faster than output. Therefore the AC function must have a minimum value.

(4) When MC is less than AC, AC is falling. When $MC = AC$, AC is at its minimum value. When MC is greater than AC, AC is rising. We demonstrated this graphically and verbally above (the football team's goal scoring average). It is proved algebraically in appendix 8.1.

The Online Resource Centre gives more explanation and worked examples of total, average, and marginal cost, including using Excel® to graph these functions.

Progress exercise 8.1

1. (a) Sketch the graph of the total cost function

    $TC = 0.5q + 2$, for $q = 0$ to $q = 10$.

   (b) Find the marginal and average cost functions. Sketch their graphs.

   (c) Explain in words why average cost is greater than marginal cost at all levels of output.

   (d) If the TC function is linear, on what assumptions can $MC = AC$?

2. (a) Sketch the graph of the total cost function

    $TC = 3q^2 + 5q + 48$, for $q = 0$ to $q = 5$.

   (b) Find the marginal and average cost functions.

   (c) Show that AC is at its minimum when $q = 4$, and that $MC = AC$ at this output.

   (d) Sketch the graphs of the MC and AC functions, on the same axes.

3. Given the total cost function $TC = 2q^3 - 2q^2 + 5q + 24$:

   (a) Show that average cost is at its minimum when $q = 2$.

   (b) Show that $MC = AC$ when AC is at its minimum.

   (c) Find the output at which marginal cost is at its minimum.

   (d) Sketch the graphs of the MC and AC functions, on the same axes.

   (e) Use the information obtained above to sketch the graph of the TC function.

4. Given the total cost function $TC = 6q^3 - 20q^2 + 25q + 144$:

   (a) Show that average cost is at its minimum when $q = 3$.

   (b) Show that $MC = AC$ when AC is at its minimum.

   (c) Find the output at which marginal cost is at its minimum.

   (d) Sketch the graphs of the MC and AC functions, on the same axes.

   (e) Use the information obtained above to sketch the graph of the TC function.

8.7 Demand, total revenue, and marginal revenue

Having examined a firm's cost functions, we will now look at the demand curve for a firm's product and the firm's total and marginal revenue functions. We shall consider both the monopoly firm and the perfectly competitive firm.

8.8 The market demand function

We discussed the demand function briefly in sections 3.14 and 4.13. We noted that economic theory tells us that the quantity of any good that a consumer will wish to buy (per day, say) is likely to vary with its price, if all other factors remain constant. The quantity the consumer wishes to buy is called his or her demand for the good, and the relationship between the consumer's demand and price is known as his or her demand curve or demand function. If we aggregate the demands of all consumers at any given price, we arrive at what is called the *market* demand function. This function is likely to be negatively sloped, meaning that a reduction in price is associated with an increase in quantity demanded. For a detailed discussion, see any introductory microeconomics textbook.

EXAMPLE 8.7

Suppose we are told that the market demand function for a good is

$q = -2p + 200$

where q = quantity demanded and p = price. We see immediately that this is a linear function with a negative slope (see figure 8.8(a)). Note that this graph follows the mathematician's convention of putting the *dependent* variable (in this case, q) on the vertical axis. However, as explained in section 3.15, in economics we conventionally put price, p, on the vertical axis when we graph a demand (or supply) function. To do this, we derive the *inverse* demand function. By elementary algebraic operations we rearrange

$q = -2p + 200$

to get

$p = -0.5q + 100$

The inverse demand function is graphed in figure 8.8(b). It should be stressed that both functions and their graphs convey exactly the same information, but arranged in two different ways. Each demand function is the inverse of the other. (Refer to section 3.5 if you need to refresh your knowledge of what an inverse function is.) Economists soon become proficient at switching between a demand (or supply) function and its inverse. (For simplicity, here we denote quantity demanded by q rather than q^D as in section 3.14; and similarly we denote the demand price by p rather than p^D as in sections 3.15 and 4.13.)

(a)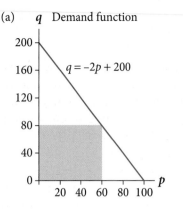

(b)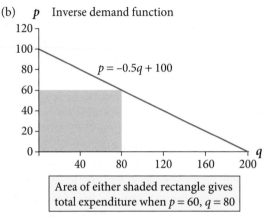

Area of either shaded rectangle gives total expenditure when $p = 60$, $q = 80$

Figure 8.8 Demand and inverse demand functions.

An important feature of figure 8.8 is that total expenditure by all buyers is measured by the area of any rectangle drawn under the market demand function (or its inverse) with its corner at any point on the function. This is because, with price and quantity on the axes, the price gives the length of one side of the rectangle, and quantity the length of the other side. The area of the rectangle is therefore price × quantity, which by definition equals total expenditure. In figure 8.8, when the price is 60 and the quantity demanded is 80, total expenditure is $60 \times 80 = 4800$ euros. (To be precise, this is not *actual* expenditure, but what buyers would spend if the price were 60.)

This has an important corollary. As a matter of logical necessity, total expenditure by all buyers of a product must equal total sales revenue received by all sellers of the product. Thus the area of any rectangle drawn under the demand function (or its inverse) also measures the total revenue of sellers. We can express this algebraically as

$$TR \equiv pq \tag{8.3}$$

where TR = total revenue received by seller = total expenditure by buyers; p = price and q = quantity bought and sold.

This equation in itself does not take us very far, since we don't yet know the values of either p or q. However, we do know that p and q are not independent of one another. On the contrary, they are related to one another through the market demand function, $q = -2p + 200$. This means that we cannot have any values of p and q we please in equation (8.3), but only values that satisfy the demand function. In order to restrict ourselves to these values, we can use the demand function to substitute for q in equation (8.3). This gives

$$TR \equiv pq = p(-2p + 200) = -2p^2 + 200p$$

This gives us the sellers' total revenue as a function of price. However, we could with equal validity take the inverse demand function, which in this example is $p = -0.5q + 100$, and use it to substitute for p in equation (8.3). This gives

$$TR \equiv pq = (-0.5q + 100)q = -0.5q^2 + 100q \tag{8.4}$$

This gives us the sellers' total revenue as a function of quantity sold, and is the form of total revenue function most used by economists. If there are many sellers, then the total revenue, TR, has to be divided between them by some process which we will not attempt to analyse here. Instead, in the next section we will consider the simplest case, when there is only one seller; that is, monopoly.

8.9 Total revenue with monopoly

Although the word 'monopoly' may carry overtones of ruthless exploitation, it may mean only that products are differentiated and therefore each producer is the sole supplier of its own brand. For example, Volkswagen has a monopoly in producing Volkswagen cars and has the ability to set their price. But the ability of this producer to exploit consumers is limited by the availability of many close substitutes produced by other car manufacturers. In general this means that, while a monopoly firm has the power to set its selling price, nevertheless only price/quantity combinations consistent with the demand function for the firm's product are feasible. This key point is sometimes overlooked by beginning students and non-economists, who make the misleading statement that 'a monopoly firm can charge what it likes'.

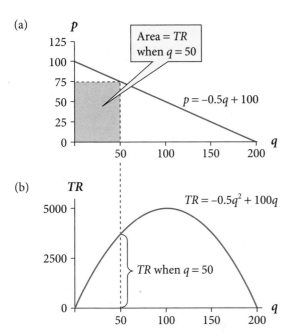

Figure 8.9 Demand and total revenue curves under monopoly.

The key fact for a monopoly supplier is that the demand function for its product is given by the *market* demand function for the product. Therefore, continuing with the previous example, if we now assume that the product is supplied by a monopoly firm, the firm's total revenue function is given by equation (8.4) above. This equation is graphed in figure 8.9(b), while the inverse demand function $p = 0.5q + 100$ from which it is derived is repeated in figure 8.9(a). Note that equation (8.4) has no constant term; consequently TR is zero when either price or quantity is zero. The explanation is simple: the monopolist's revenue is zero if it either gives the product away ($p = 0$) or sells nothing ($q = 0$).

Incidentally, recall that in figure 8.8 we saw that total expenditure by buyers was measured by the area of a rectangle drawn under the demand curve. For example, when $q = 50$, TR is measured by the area of the shaded rectangle in figure 8.9(a). It therefore follows that what we are plotting in figure 8.9b is how the area of a rectangle drawn under the demand curve in figure 8.9(a) changes as q changes.

Maximum total revenue

The point where TR reaches its maximum is obviously of some interest. From figure 8.9b we can see that this is at $q = 100$, with maximum TR being 5000. We can use differentiation to check algebraically that this is correct. To find a maximum of TR we take the first derivative, $\frac{dTR}{dq}$, and set it equal to zero. Thus, given $TR = -0.5q^2 + 100q$ from equation (8.4), we get

$$\frac{dTR}{dq} = -q + 100 = 0 \quad \text{with solution } q = 100. \text{ The second derivative is}$$

$$\frac{d^2TR}{dq^2} = -1$$

which is negative, so we have a maximum and figure 8.9 is correctly drawn.

The level of TR at its maximum can then be found by substituting $q = 100$ into the TR function. This gives

$$TR = -0.5q^2 + 100q = -0.5(100)^2 + 100(100) = 5000$$

We can also find the price at which TR is at its maximum, by substituting $q = 100$ into the inverse demand function $p = -0.5q + 100$. This gives us $p = -0.5(100) + 100 = 50$.

To summarize, we have found that for this particular demand function, a price of $p = 50$ results in quantity demanded being $q = 100$, and this combination of price and quantity yields maximum total revenue, $TR = 5000$, to the seller.

Why does total revenue reach a maximum in this way? The intuition is that when the price is very high, TR is low because the quantity demanded is very low. Conversely, when the quantity demanded is very high, TR is low because the price is very low. Somewhere in between these extremes, there is a unique combination of price and quantity that maximizes TR. Using differentiation, we have found this combination.

If you remain sceptical about these results, a non-rigorous test of their correctness is to consider prices and quantities that differ slightly from the revenue-maximizing values, $q = 100$ and $p = 50$. For example, if $p = 49$, then from the demand function we see that $q = 200 - 2p = 200 - 2(49) = 102$. Therefore, using the definition of TR, we have

$$TR \equiv pq = 49(102) = 4998$$

Thus a slightly lower price gives lower TR, even though quantity demanded is higher. Similarly, a slightly higher price, say $p = 51$, with associated quantity of $q = 98$, also results in lower TR (check this for yourself). However, this method is not rigorous because of the logical possibility that slightly different values, for example $p = 49.5$ or $p = 50.5$, would give different results.

8.10 Marginal revenue with monopoly

As with marginal cost (section 8.4 above), there are two definitions of marginal revenue to be considered, which we will call MR_1 and MR_2 to avoid confusion. In introductory economics textbooks, marginal revenue is defined as the increase in the seller's total revenue that results from selling 1 more unit of quantity. We will call this MR_1.

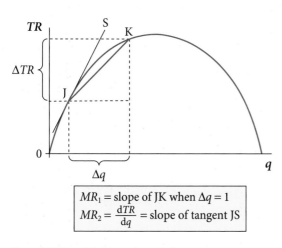

$$MR_1 = \text{slope of JK when } \Delta q = 1$$
$$MR_2 = \frac{dTR}{dq} = \text{slope of tangent JS}$$

Figure 8.10 Two definitions of marginal revenue.

The definition MR_1 closely parallels that of MC_1 in section 8.4 above. To explain MR_1 we have drawn, in figure 8.10, a typical monopolist's TR function. We start by finding the difference quotient, $\frac{\Delta TR}{\Delta q}$. Suppose we are initially at point J in figure 8.10, where quantity sold is, say, q_0. Then we assume that quantity increases by some amount Δq, so that we move to K and total revenue increases by ΔTR. Note that ΔTR gives us the increase in total revenue which occurs when quantity sold increases by the amount Δq.

In a way that is now quite familiar, we can say that the difference quotient, $\frac{\Delta TR}{\Delta x}$, measures the slope of the chord JK. If we now set $\Delta q = 1$ (that is, we consider a 1-unit increase in quantity), then we have

$$\frac{\Delta TR}{\Delta q} = \Delta TR \quad (\text{since } \Delta q = 1)$$

Thus $\Delta TR = $ additional revenue from selling one more unit. But as noted above, this is also the definition MR_1 of marginal revenue used in introductory textbooks. Combining these two facts, we can write, when $\Delta q = 1$:

$$\frac{\Delta TR}{\Delta q} = MR_1$$

In words, this says that MR_1, the increase in total revenue which occurs when quantity sold increases by 1 unit, is given by the difference quotient, $\frac{\Delta TR}{\Delta q}$, evaluated with Δq set equal to 1.

However, this definition of marginal revenue, MR_1, has the same two defects that we found in the definition MC_1 of marginal cost that we considered in section 8.4 above. It is tedious to calculate, and its numerical value in any particular situation depends on the units in which quantity sold is measured.

Therefore in theoretical economics, definition MR_2 is preferred, and is defined as $MR_2 \equiv \frac{dTR}{dq}$, the derivative of the total revenue function. In figure 8.10, this measures the slope of the tangent JS at J.

Since the difference quotient $\frac{\Delta TR}{\Delta q}$ and the derivative $\frac{dTR}{dq}$ are approximately equal when Δq is small (see section 6.5), we can treat the two definitions of MR as equivalent provided quantity sold, q, is measured in sufficiently small units that Δq can be both small and equal to 1.

The two definitions of marginal revenue co-exist because both have their uses. Definition MR_1 is easier to explain to beginners in economics and can be used in real-world situations where we were actually trying to calculate marginal revenue. Definition MR_2 on the other hand is better suited to theoretical work and is the definition we shall use throughout this book.

Thus from now on we will define marginal revenue (MR) as in rule 8.2:

RULE 8.2 Definition of marginal revenue (MR)

$MR \equiv \dfrac{dTR}{dq}$ (the derivative of the total revenue function)

What do we use the concept of marginal revenue for?

In example 8.7 above we had a TR function:

$$TR = -0.5q^2 + 100q \qquad \text{(8.4) repeated}$$

so, from the definition, we obtain MR by differentiating the TR function. In this case

$$MR = \frac{dTR}{dq} = -q + 100$$

In fact we have already made use of MR, without realizing it, when we were seeking the maximum value of the TR function in section 8.9 above. To find the maximum, we took the first derivative of the TR function and set it equal to zero. Now we see that this first derivative is, by definition, marginal revenue. Thus maximum TR occurs where

$$MR = \frac{dTR}{dq} = -q + 100 = 0$$

It follows immediately from this that $MR = 0$ is the first order condition for maximum TR. We also need the second order condition, $\frac{d^2TR}{dq^2} < 0$, to be satisfied too, since otherwise the point could be a minimum or a point of inflection of the TR function. (See section 7.8 if you need refreshment on first and second order conditions.) Since $MR = \frac{dTR}{dq}$, it follows automatically that $\frac{d^2TR}{dq^2} = \frac{dMR}{dq}$, which measures the slope of the MR curve. Thus the second order condition $\frac{d^2TR}{dq^2} < 0$, which we require for a maximum of TR, means that the marginal revenue curve must be negatively sloped at the point where $MR = 0$ (that is, where the MR curve cuts the q-axis). This condition is satisfied in example 8.7, for we have

$$\frac{dMR}{dq} = \frac{d^2TR}{dq^2} = \frac{d}{dq} \text{ of } -q + 100 = -1$$

8.11 Demand, total, and marginal revenue functions with monopoly

We are now ready to bring together the demand function, total revenue function, and marginal revenue function for the monopoly seller in sections 8.9 and 8.10, and thereby show their relationship more clearly. Figure 8.11(a) shows the inverse demand function

$$p = -0.5q + 100$$

and the marginal revenue function

$$MR = -q + 100$$

which we obtained above by differentiating the total revenue function (equation (8.4)). Note that both functions have the same intercept (100) on the vertical axis, but the *MR* function slopes down more steeply than the demand function. In this example, the *MR* function has a slope of −1, while the demand function has a slope of −0.5, so the *MR* function slopes downward exactly *twice* as steeply as the demand function; hence its intercept on the *q*-axis is *half* that of the demand function (100 versus 200). This property of the *MR* function having a slope that is twice as steep as the demand function is in fact a general property of all *linear* demand functions, as is proved in example 8.9 in section 8.13 below.

Figure 8.11(b) shows the total revenue function (from figure 8.9) which, as we found earlier, reaches its maximum at *q* = 100. Comparing figures 8.12(a) and 8.12(b), we can examine how the graphs of *MR* and *TR* are related.

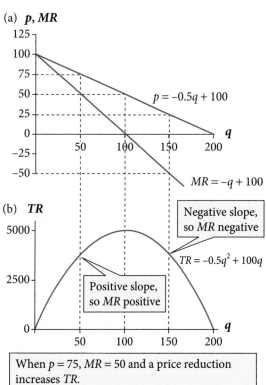

When $p = 75$, $MR = 50$ and a price reduction increases *TR*.
When $p = 50$, $MR = 0$ and *TR* is at its maximum.
When $p = 25$, $MR = -50$ and a price reduction reduces *TR*.

Figure 8.11 Marginal and total revenue functions under monopoly.

The relationship between MR and TR

Since by definition $MR = \frac{\mathrm{d}TR}{\mathrm{d}q}$, the graph of *MR* is simply the graph of the slope of the *TR* function.

Once this essential point is grasped, it should come as no surprise to see that when *q* is less than 100 (such as *q* = 50, with *p* = 75, in figure 8.11), *MR* is positive (and equals 50). This is because *TR* is increasing, hence the slope of the *TR* curve is positive. The economic significance of this is that, at any point where *q* is less than 100, such as *q* = 50, a price reduction (and the associated increase in quantity sold) results in an *increase* in total revenue.

Conversely, when *q* is greater than 100 (such as *q* = 150, with *p* = 25), *MR* is negative (−50). This is because *TR* decreases as *q* increases. A price reduction (and the associated increase in quantity sold) results in a *decrease* in total revenue. The boundary between these two cases is where *q* = 100. Here, *TR* is at its maximum and *MR* is zero.

The neat and tidy curves of figure 8.11 result from our choice of a linear demand function. Other forms of demand function can result in more irregular curves, some of which we will examine in section 8.13 below. Nevertheless the key result remains valid. If *MR* is positive at some value of q, then a price reduction results in an increase in total revenue. Conversely, if MR is negative at some value of q, then a price reduction results in a decrease in total revenue. Finally, if $MR = 0$ at the point in question, total revenue is at its maximum. (The second order condition for a maximum, $\dfrac{d^2TR}{dq^2} < 0$, must also be satisfied, as illustrated in the previous section. This is usually the case, but as an exercise you could try to sketch a demand function such that TR had a *minimum* value.)

These results on the relationship between price, marginal revenue, and total revenue, which we have obtained for the specific inverse demand function $p = -0.5q + 100$, are generalized in appendix 8.2 at the end of this chapter. This appendix may be skipped without loss of continuity.

8.12 Demand, total, and marginal revenue with perfect competition

A firm is said to be operating under conditions of **perfect competition** if it is one of a large number of relatively small firms producing and selling a homogeneous product. The quantity supplied by the firm is then so small in relation to market supply and demand that it cannot drive up the market price by reducing its own supply to the market, or drive down the market price by increasing its supply. It therefore must take the ruling market price as given (that is, as an exogenous variable in the terminology we introduced in section 3.17). However, because its supply is negligible relative to market supply and demand, we assume the firm can sell as much or as little as it wishes at this price.

In graphical terms, this last point means that the inverse demand function for the firm's product is simply a horizontal straight line at the ruling market price, which we can label $\bar{p}$ (see figure 8.12(a)). We can interpret this as meaning that if the firm attempted to set a price greater than $\bar{p}$, it would lose literally all its customers; while the attempt to set a price less than $\bar{p}$ would result in virtually infinite demand, since the firm would then attract all buyers in the market. The ruling market price, $\bar{p}$, is determined by the point of intersection of the *market* supply and demand functions, as analysed briefly in section 3.14. We will assume that this price is constant.

We can now derive the perfectly competitive firm's total revenue function. As always, because it is a matter of definition, we have

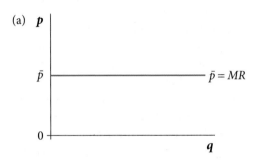

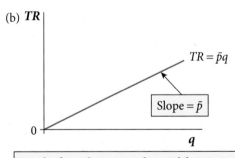

For the firm, the inverse demand function and the marginal revenue function coincide under perfect competition. Consequently, the total revenue function is linear.

Figure 8.12 Marginal and total revenue functions under perfect competition.

$$TR = pq$$

We also have the condition, or constraint, that $p = \bar{p}$. The obvious next step is to combine these two equations by substituting $p = \bar{p}$ into the definition of TR, giving us

$$TR = \bar{p}q$$

The graph of this total revenue function is simply a straight line from the origin with a slope given by the parameter (or exogenous variable) $\bar{p}$ (see figure 8.12(b)). Unlike the monopoly case, there is no maximum to the perfectly competitive firm's TR.

Finally, we can find marginal revenue. By definition, this is $\frac{dTR}{dq}$. In this case, $TR = \bar{p}q$, so we have

$$MR = \frac{dTR}{dq} = \bar{p}$$

If this puzzles you momentarily, recall that this firm can sell as much as it wishes at the ruling market price. If it sells 1 more unit of output, the additional revenue is simply $\bar{p}$. This differs from the case of the monopolist in section 8.10 above. The monopolist can only sell more units by reducing the price, causing revenue from pre-existing sales to fall. The marginal revenue function for the monopolist is a more complex algebraic expression because it takes this loss of revenue from pre-existing sales into account.

Since MR and p are both equal to $\bar{p}$ for the firm under perfect competition, in figure 8.12a the firm's inverse demand function and its marginal revenue function are both given by the horizontal line $\bar{p}$ units above the horizontal axis. In effect, one horizontal function sits on top of the other. If there is some shift in the market supply and demand functions, causing, say, a rise in $\bar{p}$, this horizontal line simply shifts up. A rise in $\bar{p}$ also causes the TR function in figure 8.12(b) to become more steeply sloped.

8.13 Worked examples on demand, marginal, and total revenue

The following three examples all assume that the firm is a monopolist. This is because for the perfectly competitive firm, there is very little to analyse; figure 8.12 says it all.

EXAMPLE 8.8

Suppose the market demand function for a good (which is also the firm's demand function, with monopoly) is given by

$$q = -5p + 1000$$

Then we can find the monopolist's total revenue function as follows. First, we derive the inverse demand function by rearranging the demand function above as

$$p = -0.2q + 200$$

Then we use this to substitute for p in the definition of total revenue, and thereby get

$$TR = pq = (-0.2q + 200)q = -0.2q^2 + 200q$$

Note that TR is conventionally defined as a function of q (see example 8.7 above). It would be equally possible to define it as a function of p, but the conventional definition helps with some manipulations and proofs that we will see later.

To find the point of maximum TR, we differentiate the TR function to get marginal revenue, MR, set it equal to zero, and solve the resulting equation:

$$MR = \frac{dTR}{dq} = -0.4q + 200 = 0 \quad \text{with solution } q = 500$$

This is a maximum because

$$\frac{d^2TR}{dq^2} = \frac{dMR}{dq} = -0.4$$

which is negative when $q = 500$, as required for a maximum. And when $q = 500$, $p = -0.2q + 200 = 100$ from the inverse demand function.

The graphs of the inverse demand function, the MR function and the TR function are identical to those in figure 8.11, apart from the calibration of the axes.

EXAMPLE 8.9

In this example we generalize the case of the linear demand function for a monopolist's product. Consider the linear inverse demand function

$$p = -aq + b$$

(where the parameters a and b are both positive, and thus the function is negatively sloped with a positive intercept). We use this to substitute for p in the definition of total revenue, and thereby get

$$TR = pq = (-aq + b)q = -aq^2 + bq \quad \text{so marginal revenue is}$$

$$MR = \frac{dTR}{dq} = -2aq + b$$

Comparing the inverse demand function and the MR function, we see that both have an intercept of b on the vertical (price) axis, but the slope of the MR function ($-2a$) is twice as steep as that of the inverse demand function ($-a$). This generalizes our findings in the particular case of $p = -0.5q + 100$ that we examined in section 8.11 above. By setting first $MR = 0$ and then $p = 0$ we can easily show that the intercept of the MR function on the q-axis is exactly half way between zero and the intercept of the inverse demand function on the q-axis, as in figure 8.11. (Check this for yourself.)

Finally, maximum TR is where

$$MR = \frac{dTR}{dq} = -2aq + b = 0 \quad \text{The solution to this equation is}$$

$$q = \frac{b}{2a}$$

This point is where the MR function cuts the q-axis, as at $q = 100$ in figure 8.11. This is a maximum of TR because

$$\frac{d^2TR}{dq^2} = \frac{dMR}{dq} = -2a \quad \text{which is negative because we assumed } a > 0.$$

EXAMPLE 8.10

As a final example we'll consider the non-linear inverse demand function

$$p = -q^2 - q + 208$$

Proceeding exactly as in the two previous examples, we get

$$TR = pq = (-q^2 - q + 208)q = -q^3 - q^2 + 208q$$

This cubic function is slightly intimidating, but we simply proceed as before. To find maximum TR we find the marginal revenue function, set it equal to zero and solve:

$$MR = \frac{dTR}{dq} = -3q^2 - 2q + 208 = 0$$

Solving this quadratic using the formula, we get $q = 8$ or $-8\frac{2}{3}$. We discard the negative solution. At $q = 8$ the second order condition for a maximum, $\frac{dMR}{dq} < 0$, is satisfied (check for yourself). When $q = 8$, $p = 136$ (from the inverse demand function) and $TR = 1088$.

The inverse demand, marginal, and total revenue functions are graphed in figure 8.13.

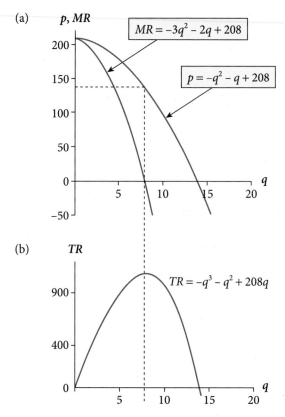

Figure 8.13 Inverse demand, marginal revenue, and total revenue functions for example 8.10.

> **!** **Summary of sections 8.7–8.13**
>
> Before moving on, let us briefly summarize the relationship between the demand function, total revenue, and marginal revenue.
>
> (1) A key concept is the demand function $q = f(p)$ or its inverse, which we can call $p = g(q)$. The slope of both of these functions is negative for a monopoly (a firm with market power). For a perfectly competitive firm, the inverse demand function is $p = \bar{p}$, where $\bar{p}$ is the ruling market price. Price does not vary with quantity.
>
> (2) Total revenue, TR, by definition is $p \times q$. Because a monopolist faces a negatively sloped demand function, q can increase only if p falls. It is therefore likely to be true (though not necessarily so) that the TR function has a maximum at some value of q (and therefore p, since p and q are connected via the demand function). For a perfectly competitive firm, we have $TR = \bar{p}q$, with $\bar{p}$ a parameter. Thus the firm can sell any quantity at the ruling market price, so the TR function has no maximum.
>
> (3) Marginal revenue, MR, by definition is $\frac{dTR}{dq}$, the derivative of the TR function. Therefore maximum TR for the monopolist is found where $MR = 0$. If $MR > 0$, then by definition an increase in q (which requires a reduction in p because the demand curve slopes downward) will *increase* TR. If $MR < 0$, then by the same reasoning an increase in q (and associated reduction in p) will *decrease* TR.
>
> A rigorous algebraic analysis is provided in appendix 8.2 at the end of this chapter.

 The Online Resource Centre gives more examples of demand, total revenue, and marginal revenue functions, including using Excel® to graph these functions.
www.oxfordtextbooks.co.uk/orc/renshaw3e/

Progress exercise 8.2

1. The inverse demand function for a good is $p = -\frac{1}{8}q + 250$.

(a) Find the total revenue function and the marginal revenue function.

(b) Find the price and quantity at which total revenue is maximized.

(c) Sketch the graphs of the inverse demand function and the marginal revenue functions on the same axes, indicating your solution on the graph.

(d) Sketch the graph of the total revenue function.

2. The demand function for a good is $q = 120 - 2.5p$.

(a) Find the inverse demand function, the total revenue function, and the marginal revenue function.

(b) Find the price and quantity at which total revenue is maximized.

(c) Sketch the graphs of the inverse demand function and the marginal revenue functions on the same axes, indicating your solution on the graph.

(d) Sketch the graph of the total revenue function.

3. Given the demand function $q = 300 - 4p^2$

(a) Find the total and marginal revenue functions.

(b) Find the price and quantity at which TR is at its maximum. (Hint: it's easier to start by finding the price rather than the quantity at which TR is maximized.)

(c) Sketch the graph of the demand function, the MR and the TR functions (the last is not easy!).

4. Given the inverse demand function $p = \frac{16}{q+2} - 2$:

(a) Find the intercepts on the p and q axes, and sketch the function. (Hint: you should recognize this as a rectangular hyperbola.)

(b) Find the TR and MR functions.

(c) Find the price and quantity at which total revenue is at its maximum.

(d) Sketch the functions, indicating your solution.

5. With p on the vertical axis and q on the horizontal, the graph of the demand for a firm's product is a horizontal line 5 units above the q-axis. Find the total and marginal revenue functions and sketch their graphs.

6. (a) Explain in words (with the help of any algebra and/or graphs you consider necessary) why marginal revenue is necessarily less than price for any demand function which is negatively sloped.

(b) Prove this result algebraically. (Hint: start by assuming an inverse demand function $p = f(q)$ with $\frac{dp}{dq} = f'(q) < 0$.)

8.14 Profit maximization

In sections 8.1–8.6 we examined the firm's total, average, and marginal cost functions and their interrelationships. In sections 8.7–8.13 we looked at the market demand curve for a good and introduced the concepts of total and marginal revenue. Now we can bring cost and demand conditions together and thus examine how a firm should proceed if it wishes to achieve maximum profit. Because we have already done most of the hard work earlier in this chapter, this turns out to be less difficult than you might expect.

As usual we shall develop the analysis by means of a worked example, and then generalize our findings. We will consider first the case of monopoly, then perfect competition.

8.15 Profit maximization with monopoly

EXAMPLE 8.11

Suppose there is a monopolist firm with total cost function $TC = q^2 + 2q + 500$, and the market demand function for the product is $q = -0.5p + 100$.

Given that the firm is a monopolist—that is, it is the only supplier of the product in question—the *market* demand function is also the demand function for the monopoly firm's product, as noted earlier. If the monopolist chooses to fix the price at which it offers the product for sale, then the demand function determines the quantity that consumers will buy at that price. For example, if the price is fixed at $p = 20$, then q is determined by the demand function as $q = -0.5p + 100 = 90$. Alternatively, if the monopolist chooses to produce a certain quantity and offer it for sale, then the demand function determines the price that consumers are willing to pay for that quantity. For example, if the monopolist produces and sells $q = 80$, then in the demand function we have $80 = -0.5p + 100$, and the solution to this equation is $p = 40$. Thus the monopolist can fix either the price or the quantity sold, but not both.

To obtain the firm's total revenue function we first rearrange the demand function above to derive the inverse demand function, which is $p = -2q + 200$.

Given the definition of total revenue as $TR = pq$, we then use the inverse demand function to eliminate p from the TR expression. This gives

$$TR = pq = (-2q + 200)q = -2q^2 + 200q$$

In figure 8.14(a) we have graphed the total cost and total revenue functions. In this example they are both quadratic functions.

Next we define the firm's profits, which economists usually denote by the Greek letter Π (pronounced 'pie' as in apple pie; see appendix 2.1 for the full Greek alphabet). In a common-sense way, we define profits, Π, as simply the firm's total revenue from sales minus its total production costs. Thus we have, as a definition:

$$\Pi \equiv TR - TC$$

What can we infer about profits from an examination of figure 8.14a? First, at any given level of production and sales, such as $q = 20$, we can follow the dotted lines to read off the firm's total revenue, which appears from figure 8.14(a) to be about 3100, and its total costs, which appear to be about 900. The difference, $TR - TC$, between these two values gives us profit at that output, which appears to be about 2200. (All costs, revenues, and profits are, of course, measured in the relevant currency, such as euros.) Thus, in general, at any output the vertical distance between the TR and TC functions in figure 8.14(a) measures Π. (You may have noticed that we have implicitly assumed that the quantity produced and the quantity sold are identical. In other words, the firm does not hold any stock of output as a cushion between production and sales.)

Second, from figure 8.14(a) we can also see that there are two levels of output, q_0 and q_1, at which the total revenue and total cost functions intersect, and therefore $TR = TC$. At these two points, profits are zero. These are known as **break-even points.** While there are two break-even points in this example, other examples can produce only one such point (or, more rarely, three or more). In economic terms, the left-hand break-even point arises from the existence of fixed costs, which give the TC curve a positive intercept. The right-hand break-even point is there because when quantity sold is large, the firm's costs are

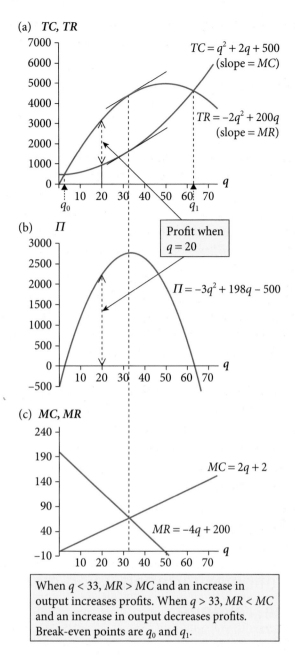

Figure 8.14 Profit maximization under monopoly.

increasing rapidly due to pressure on plant capacity, while total revenue is declining because the firm can sell large quantities only at relatively low prices due to the constraint of the downward-sloping demand curve.

At any level of output (= quantity sold) between q_0 and q_1, we see from figure 8.14(a) that the TR curve lies above the TC curve, and therefore profits are positive. Conversely, when output is less than q_0 or greater than q_1, TR lies below TC and profits are negative (that is, the firm loses money).

We are now ready to address the key question for the firm: what level of output will yield maximum profits? We can answer this algebraically as follows. We have a total cost function $TC = q^2 + 2q + 500$, and a total revenue function $TR = -2q^2 + 200q$. If we substitute these into the profit function defined above, we get

$$\Pi = TR - TC = (-2q^2 + 200q) - (q^2 + 2q + 500)$$

On removing the brackets and simplifying, this expression becomes

$$\Pi = -3q^2 + 198q - 500$$

(Note how, on removing the brackets, the fixed costs of 500 euros are subtracted from the profits, as they should be.)

Hopefully you will recognize this as a quadratic function, which is graphed in figure 8.14(b). Figure 8.14(b) shows profits peaking at around $\Pi = 2800$ when q is about 32 or 33. We can find the exact value of q algebraically by applying rule 7.1 from chapter 7. This rule tells us that Π is at a (local) maximum at any point where the first derivative $\frac{d\Pi}{dq}$ equals zero (the first order condition) and the second derivative $\frac{d^2\Pi}{dq^2}$ is negative (the second order condition).

To find the maximum we therefore take the first derivative of $\Pi = -3q^2 + 198q - 500$, set it equal to zero and solve the resulting equation, giving

$$\frac{d\Pi}{dq} = -6q + 198 = 0 \quad \text{with solution } q = 33.$$

Then we take the second derivative, which is $\frac{d^2\Pi}{dq^2} = -6$. This is negative for all q, including $q = 30$. So at $q = 33$ we have $\frac{d\Pi}{dq} = 0$ and $\frac{d^2\Pi}{dq^2} < 0$. Thus both the first order and the second order conditions for a maximum of Π are satisfied at $q = 33$, confirming that figure 8.14 is correctly drawn. The maximized level of profits is

$$\Pi = -3q^2 + 198q - 500 = -3(33)^2 + 198(33) - 500 = 2767$$

In terms of figure 8.14(a), the profit-maximizing output, $q = 33$, is the point at which the vertical distance between the TR and TC curves reaches its maximum. The significance of the tangents drawn to the TR and TC curves at this point will become clear shortly.

We can also find by algebra the two break-even points. In figure 8.14(a) we can see they occur where $TR = TC$. Since we have $TR = -2q^2 + 200q$ and $TC = q^2 + 2q + 500$, we can find the break-even points by setting these two equal to one another and solving the resulting equation. Alternatively, we see in figure 8.14(b) that the break-even points occur where $\Pi = 0$. Since we have $\Pi = -3q^2 + 198q - 500$ we can find these points by setting this expression equal to zero and solving the resulting equation. This is left to you as an exercise.

8.16 Profit maximization using marginal cost and marginal revenue

There is an alternative approach to finding the firm's most profitable output which uses the concepts of marginal cost (MC) and marginal revenue (MR) that we developed in sections 8.4 and 8.10. If you have studied economics before, you may have already met this approach, but don't worry if you haven't as all will be explained.

We can demonstrate this approach by going back to the definition of profits (Π) as simply total revenue (TR) minus total cost (TC):

$$\Pi = TR - TC$$

To find the output at which profits are at a maximum, we look for a point or points where the first derivative $\frac{d\Pi}{dq}$ equals zero and the second derivative $\frac{d^2\Pi}{dq^2}$ is negative. We begin by finding $\frac{d\Pi}{dq}$ and setting it equal to zero. Given $\Pi = TR - TC$, we get

$$\frac{d\Pi}{dq} = \frac{dTR}{dq} - \frac{dTC}{dq} = 0 \tag{8.5}$$

(If you are not completely happy about the validity of this step, refer back to rule D4 of differentiation in chapter 6.)

Looking at equation (8.5), on the right-hand side we have $\frac{dTR}{dq}$, which is, by definition, marginal revenue (MR), as defined in section 8.10 above. Similarly we have $\frac{dTC}{dq}$, which is marginal cost (MC), as defined in section 8.4 above. Substituting these into equation (8.5), the condition for profit maximization becomes

$$\frac{d\Pi}{dq} = MR - MC = 0 \qquad (8.5a)$$

This equation is satisfied when $MR = MC$. Thus from equation (8.5) we have learned that the condition $\frac{d\Pi}{dq} = 0$ and the condition $MR = MC$ are equivalent to one another. In words, the first order condition for maximum profit is that marginal revenue must equal marginal cost. This is true not only for the monopoly firm but also for the perfectly competitive firm, as we will show in the next section.

Returning to the previous example, in figure 8.14(a) the slope of any tangent to the TR curve measures MR, and the slope of any tangent to the TC curve measures MC. When $q = 33$, we see that these tangents are parallel, so they have the same slope. Therefore $MR = MC$ when $q = 33$.

To the left of $q = 33$, at any level of q the tangent to the TR curve is steeper than the tangent to the TC curve, so we must have $MR > MC$. To the right of $q = 33$, the opposite is true: $MC > MR$.

Second order conditions

It is tempting to assume that the condition, marginal revenue equals marginal cost, guarantees that profits are maximized. However, the condition $MR = MC$, or its equivalent $\frac{d\Pi}{dq} = 0$ is only the first order condition for maximum profit. To be sure that the point in question is indeed a point of maximum profit (rather than a minimum or a point of inflection in the profit function) we require the second order condition, $\frac{d^2\Pi}{dq^2} < 0$, to be satisfied. By differentiating equation (8.5a) we obtain $\frac{d^2\Pi}{dq^2}$ as

$$\frac{d^2\Pi}{dq^2} = \frac{dMR}{dq} - \frac{dMC}{dq}$$

and we require this to be negative for a maximum of the profit function. We will postpone consideration of what this implies for MC and MR until section 8.20 below.

Diagrammatic treatment of the *MC* = *MR* condition

We have just shown how equating marginal cost and marginal revenue gives us an algebraic technique for finding the most profitable output. We now show how the same method can be used diagrammatically.

Returning to the previous example, the TC function is $TC = q^2 + 2q + 500$. By differentiating this, we get the MC function as

$$MC = \frac{dTC}{dq} = 2q + 2$$

In the same way, the TR function is $TR = -2q^2 + 200q$. Differentiating this, we get

$$MR = \frac{dTR}{dq} = -4q + 200$$

The graphs of the MC and MR functions are shown in figure 8.14c. The $MC = 2q + 2$ function is positively sloped, reflecting rising marginal cost in this example. The $MR = -4q + 200$ function is negatively sloped, with a slope of -4 and an intercept of 200. It should be no surprise to see that the MC and MR functions intersect at $q = 33$, the point of maximum profit. (Check the algebra for yourself.)

The relationship between price and marginal cost

At the monopolist's profit-maximizing output of $q = 33$, we can find the level of MC (and therefore of MR, since the two are equal) by substituting for q in the MC function. We get

$MC = 2q + 2$ which equals 68 when $q = 33$

We can also find the price by substituting $q = 33$ into the inverse demand function. We get

$p = -2q + 200$ which equals 134 when $q = 33$

Thus this monopolistic firm is able to sell its output at a price (134) that is almost double its marginal production cost (68). This contrasts with perfect competition, where price and marginal cost are equal when profits are maximized, as we shall see in the next section. Economic theory shows this difference between monopoly and competition to be of great significance, and we examine it further in section 8.18 below.

8.17 Profit maximization with perfect competition

We will now look at the profit-maximizing behaviour of a perfectly competitive firm and compare the equilibrium with that of the monopolist in the previous example.

EXAMPLE 8.12

We assume that the perfectly competitive firm has the same short-run total cost function as the monopolist in the previous example:

$TC = q^2 + 2q + 500$

This assumption may well be an over-simplification, but it will serve to introduce this area of economics. As discussed earlier, the perfectly competitive firm can sell all it wishes at the ruling market price, which we will arbitrarily assume to be $\bar{p} = 72$. The firm's total revenue function is therefore

$TR = \bar{p}q = 72q$

This TR function and the TC function above are graphed in figure 8.15(a). Comparing this with figure 8.14(a), we see that the difference is that the perfectly competitive firm's TR function is linear because it doesn't need to reduce its price in order to sell a large quantity. As in the monopoly case there are two break-even points, now at about $q = 8$ and $q = 60$. Since profit is given by the vertical distance between the TR and TC functions, the profit-maximizing output is where this vertical distance is at its greatest, which is at $q = 35$, as we will now show.

From the TC and TR functions we can get the profit function as

$\Pi = TR - TC = 72q - (q^2 + 2q + 500)$ which simplifies to

$\Pi = -q^2 + 70q - 500$

This is graphed in figure 8.15(b). This profit function has the same general shape as in the monopoly case (example 8.11), but the parameters are different.

To find maximum profit we set the derivative of the profit function equal to zero. Thus

$\dfrac{d\Pi}{dq} = -2q + 70 = 0$ with solution $q = 35$

This is a maximum because $\dfrac{d^2\Pi}{dq^2} = -2$, which is negative when $q = 35$. Maximized profits are found from the profit function with $q = 35$. This gives $\Pi = 725$.

We can alternatively find the profit-maximizing output by setting marginal cost and marginal revenue equal to one another. Looking first for marginal cost, as we have assumed that the perfectly competitive firm and the monopoly firm have the same TC

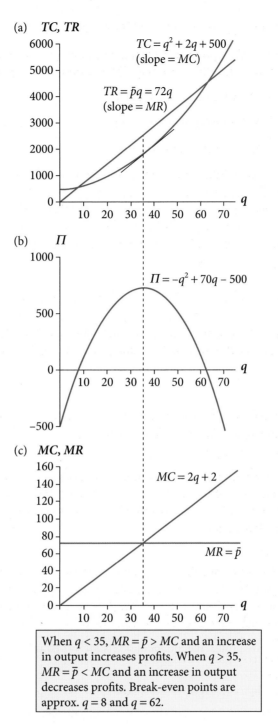

(a) *TC, TR*

(b) *Π*

(c) *MC, MR*

When $q < 35$, $MR = \bar{p} > MC$ and an increase in output increases profits. When $q > 35$, $MR = \bar{p} < MC$ and an increase in output decreases profits. Break-even points are approx. $q = 8$ and $q = 62$.

Figure 8.15 Profit maximization under perfect competition.

function, they obviously have the same *MC* function, found by differentiating the total cost function. Thus, given

$TC = q^2 + 2q + 500$ we have

$$MC = \frac{\mathrm{d}TC}{\mathrm{d}q} = 2q + 2$$

Looking now for marginal revenue, the perfectly competitive firm's *TR* function is

$TR = \bar{p}q = 72q$ so marginal revenue is

$$MR = \frac{\mathrm{d}TR}{\mathrm{d}q} = \bar{p} = 72$$

The key point to see in this last equation is that, for the perfectly competitive firm, marginal revenue and price are necessarily equal, as we saw in section 8.12.

To find the profit-maximizing output we set $MR = MC$, which gives

$MR = \bar{p} = 72 = MC = 2q + 2$ with solution $q = 35$

Thus we see that $\bar{p} = MC$. This is because $\bar{p} = MR$ under perfect competition, and $MR = MC$ when profits are maximized.

The MR and MC functions are graphed in figure 8.15(c). Their intersection gives the profit-maximizing output, $q = 35$. Note that $MR = \bar{p}$ is a horizontal straight line.

8.18 Comparing the equilibria under monopoly and perfect competition

It is tempting to compare the levels of output, profits, and so on for the monopoly firm and the perfectly competitive firm in examples 8.11 and 8.12 above. The assumption that they have the same cost function would seem to make such comparisons appropriate. However, the demand conditions differ in a completely arbitrary way, for we simply plucked the perfectly competitive firm's ruling market price, $\bar{p} = 72$, out of thin air. Therefore we cannot directly compare the two cases, though one generalization is possible and important.

The monopolist and the perfectly competitive firm both seek to maximize their profits and do so by producing at the output where $MC = MR$. The key difference between the two producers is that, for the monopolist, price is always greater than marginal revenue, because the demand function is negatively sloped. So we have $p > MR = MC$ in the case of monopoly. (To see why $p > MC$, see figure 8.11a and appendix 8.2.)

In the case of the perfectly competitive firm we have $p = MR = MC$ (see section 8.12). This is an inherent difference, independent of any assumptions about demand or cost conditions, other than the strictly definitional assumptions that the monopolist faces a downward-sloping demand function while the perfectly competitive firm faces a horizontal (inverse) demand function: that is, $p = \bar{p}$.

We can make a direct comparison of the monopolist (example 8.11) and the perfectly competitive firm (example 8.12) in the following way. Let us start by looking again at the monopolist. The inverse market demand function (which is also, under monopoly, the demand for the firm's output) is $p = -2q + 200$. Therefore total revenue is $TR = -2q^2 + 200q$ and marginal revenue is $MR = -4q + 200$. The monopolist's total costs are $TC = q^2 + 2q + 500$, so marginal cost is $MC = 2q + 2$.

To maximize profits, the monopolist sets $MC = MR$ and solves the resulting equation for q. This gives

$2q + 2 = -4q + 200$ from which $q = 33$

We can then find p by substituting $q = 33$ into the inverse demand function, giving

$p = -2(33) + 200 = 134$

We will denote these monopolist's equilibrium values as $q_M = 33$ and $p_M = 134$.

Now let us suppose that the monopolist producer is overnight replaced by a large number of firms, all too small to be able to influence price, and with the same total cost and therefore marginal cost functions as the monopolist. The demand function $p = -2q + 200$ remains the market demand function, but the demand function for each firm's product is now $p = \bar{p}$, where $\bar{p}$ is the ruling market price. So $TR = \bar{p}q$ and $MR = \frac{dTR}{dq} = \bar{p}$.

Each perfectly competitive firm seeks to maximize profits by setting $MC = MR$ and solving the resulting equation for q. From above, we have $MC = 2q + 2$, so $MC = MR$ implies

$2q + 2 = \bar{p}$ (8.6)

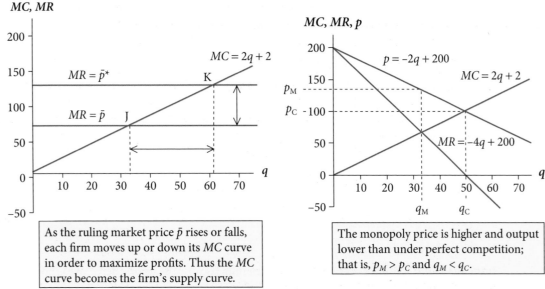

As the ruling market price $\bar{p}$ rises or falls, each firm moves up or down its MC curve in order to maximize profits. Thus the MC curve becomes the firm's supply curve.

Figure 8.16 Supply under perfect competition.

The monopoly price is higher and output lower than under perfect competition; that is, $p_M > p_C$ and $q_M < q_C$.

Figure 8.17 Equilibrium under monopoly and perfect competition.

but this equation has two unknowns, q and $\bar{p}$, and therefore has no unique solution. In section 8.17 above we steered around this problem by arbitrarily assuming $\bar{p} = 72$. Now we need to consider how $\bar{p}$ is determined.

The key to solving this is that equation (8.6) is effectively the supply function of each firm. To see this, look at figure 8.16. We suppose that a value for $\bar{p}$ has somehow been determined, and a typical firm is in equilibrium at J with $MC = MR$ and therefore equation (8.6) satisfied. Now suppose $\bar{p}$ rises to a new level, $\bar{p}^*$. To restore equilibrium (that is, $MC = MR$) the firm must move up its marginal cost curve from J to K. Thus the slope of the MC curve tells us the firm's response, in terms of increase in quantity supplied, to a rise in the market price. But this is exactly what a supply curve tells us; so the firm's MC curve *is* its supply curve, defined as a relationship between market price and quantity supplied. Moreover, since this must hold for each and every firm, it must hold for all firms in aggregate, so the aggregate of all firms' MC functions is the *industry* supply function under perfect competition.

If we assume for simplicity that all firms have the same MC function, we can find the industry supply at any price by simply looking at the quantity supplied by a single firm, at that price, and multiplying this quantity by the number of firms in the industry. This means that in equation (8.6) we can, with appropriate change of scale on the q-axis, treat q as the output of all firms, or industry output.

In addition, we also know that the equilibrium market price, $\bar{p}$, and aggregate quantity sold, q, must satisfy the inverse market demand function $p = -2q + 200$, since otherwise buyers are not in equilibrium. Therefore we must have

$$\bar{p} = -2q + 200 \tag{8.7}$$

Equations (8.6) and (8.7) can now be solved as simultaneous equations, with solution $q = 49.5$ and $\bar{p} = 101$. (Be sure to check this solution for yourself.) We'll label these equilibrium values under perfect competition as $q_C = 49.5$ and $p_C = 101$.

The equilibria under monopoly and perfect competition are shown in figure 8.17. Thus we have a robust conclusion that, under monopoly, output is lower than under perfect competition ($q_M = 33$; $q_C = 49.5$), while price is higher ($p_M = 134$; $p_C = 101$). However, to derive this result we assumed that the total cost function and the market or industry demand function are the same under perfect competition and monopoly, which may well be incorrect. Clearly, we have only scratched the surface of a large subject area of economics.

8.19 Two common fallacies concerning profit maximization

First fallacy: profit maximization requires that average costs be minimized

In examples 8.11 and 8.12 above, we have $TC = q^2 + 2q + 500$, so the AC function is

$$AC = \frac{TC}{q} = \frac{q^2 + 2q + 500}{q} = q + 2 + 500q^{-1}$$

To find minimum AC we find the derivative, set it equal to zero and solve the resulting equation. We obtain

$$\frac{dAC}{dq} = 1 - 500q^{-2} = 0$$

from which $q^2 = 500$ and $q = 22.36$. This is a minimum because the second derivative is positive (check for yourself).

Thus the profit-maximizing output under monopoly, $q = 33$, and the profit-maximizing price with perfect competition, $q = 49.5$, are both much greater than the output that minimizes AC. Therefore when profits are maximized the level of AC is above the minimum—although, as it happens in this example, not very much above the minimum because the TC curve is not strongly convex and therefore AC rises quite slowly. (Check this for yourself.) We conclude that, for both a monopoly firm and a perfectly competitive firm, maximizing profit does not imply or entail any desire to minimize average cost. Indeed, we can make the stronger statement that any movement by the firm away from the output that maximizes profits and towards the output that minimizes AC would, by definition, result in a reduction in profits.

However, it may be that under perfect competition entry of new firms attracted by profits in the industry will drive down the ruling market price, causing the output that maximizes the individual firm's profits to approach the output that minimizes its AC. But this raises issues of economic analysis that are beyond the scope of this book.

Second fallacy: profit maximization by a monopolist requires that total revenue be maximized

That this is mistaken is apparent when we recall that when TR is at its maximum, $\frac{dTR}{dq} = MR = 0$. Since profit maximization requires $MR = MC$, the output where $MR = 0$ could be the profit-maximizing output only if it happened that MC was equal to zero at this output also. Although not impossible, it is unlikely that MC is ever zero. (Try to sketch the shapes of the TC and TR curves that would produce this result.) In figure 8.14 the profit-maximizing output is $q = 33$, where $MC = MR$, while the revenue maximizing output is $q = 50$, where $MR = 0$.

8.20 The second order condition for profit maximization

It is easy to slip into the assumption that when we find a point where $MR = MC$, this point is a maximum of the profits function. This may not be true, however, since $MR = MC$ is merely the first order condition for a maximum. As noted in section 8.15 above, we must also check that the second order condition for a maximum is satisfied, namely that the appropriate second derivative is negative at the point in question, in order to be sure that we have located a maximum for profits, rather than a minimum or a point of inflection.

The second order condition has certain implications for the marginal cost and marginal revenue functions, which we will now explore. From section 8.15 we know that the profit function is $\Pi \equiv TR - TC$ with derivative

$$\frac{d\Pi}{dq} = \frac{dTR}{dq} - \frac{dTC}{dq} = MR - MC$$

When we differentiate again to find the second derivative of the profit function, we get

$$\frac{d^2\Pi}{dq^2} = \frac{d^2TR}{dq^2} - \frac{d^2TC}{dq^2} = \frac{dMR}{dq} - \frac{dMC}{dq}$$

At a maximum of profits, this will be negative. We will therefore have

$$\frac{dMR}{dq} - \frac{dMC}{dq} < 0 \quad \text{which rearranges as}$$

$$\frac{dMC}{dq} > \frac{dMR}{dq} \tag{8.8}$$

In diagrammatic terms, the left-hand side of this inequality is the slope of the marginal cost curve, while the right-hand side is the slope of the marginal revenue curve. Thus the second order condition requires that the slope of the MC curve must be greater than the slope of the MR curve. This will automatically be true if the MC curve is positively sloped and the MR curve negatively sloped, since any positive number is greater than any negative number. This condition is fulfilled in figures 8.14 and 8.15, but not in figure 8.18.

In figure 8.18(a) the firm is assumed for simplicity to be in a perfectly competitive market, so its total revenue function is linear. The constant slope of the TR function measures both the ruling market price, $\bar{p}$, and MR. The key feature is that the total cost function has a point of inflection at K (the point of minimum MC), with the consequence that there are two points, L and J, at which marginal cost and marginal revenue are equal. We know that $MC = MR$ at these points because the tangents at these points are parallel with the TR function. Thus the first order condition for maximum profit, $MC = MR$, is fulfilled at both L and J.

Let us now consider the second order condition, equation (8.8) above. We can examine this with the help of figure 18.19(b). There we have graphed the MR and MC functions. We know that MR is constant (and equal to $\bar{p}$), so its graph is horizontal. Therefore $\frac{dMR}{dq} = 0$, and consequently the second order condition, equation (8.8), is satisfied if $\frac{dMC}{dq} > 0$.

The shape of the MC function in figure 8.18b requires a little more thought. As output increases from zero, we can see that the TC function is initially concave from below and therefore its slope ($= MC$) is *decreasing*. The graph of the MC function is therefore downward sloping and its derivative, $\frac{dMC}{dq}$, is negative. At K, the point of inflection,

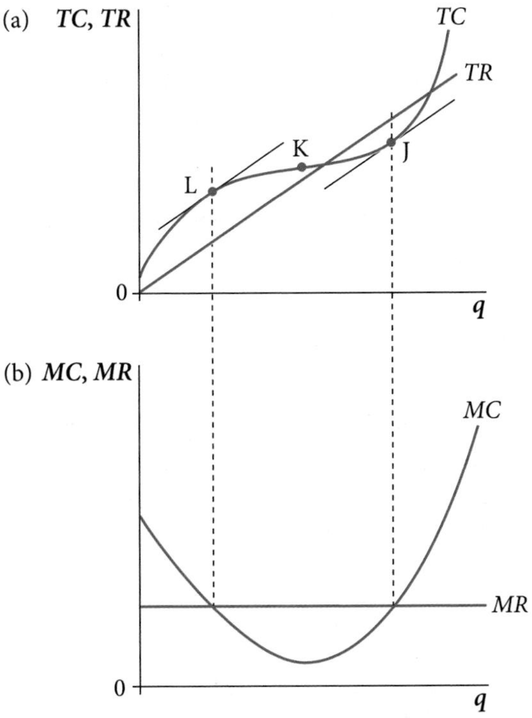

At L, $MC = MR$ but $\frac{dMC}{dq} < \frac{dMR}{dq}$ so second order condition for profit maximization is not satisfied. At J, $MC = MR$ and $\frac{dMC}{dq} > \frac{dMR}{dq}$, so second order condition for profit maximization is satisfied.

Figure 8.18 Satisfying the second order condition for profit maximization.

the slope of the *TC* function (= *MC*) reaches its minimum value and its derivative, $\frac{dMC}{dq}$, equals zero. To the right of K, the *TC* curve is convex from below and its slope is increasing. So *MC* is rising and its derivative, $\frac{dMC}{dq}$, is positive. This explains why the *MC* function in figure 18.19(b) is broadly U-shaped, with a minimum corresponding to the point of inflection in the *TC* function.

The key conclusion from the previous paragraph is that $\frac{dMC}{dq}$ is negative at L but positive at J. So the second order condition for maximum profit is satisfied only at J. Point L is actually a point of profit *minimization* (or loss maximization). To test your understanding, sketch the graph of the profit function.

The importance of the second order condition is therefore clear. If you were the economic adviser to this firm, failure to check that it was satisfied might lead you to advise the firm to produce at L, merely because *MC* = *MR* at that point. This could cost you your job!

In this example, we adopted the simplifying assumption that the firm was in a perfectly competitive market, and the *TR* function was therefore linear. If the firm were a monopoly, the *TR* function would be concave in figure 8.18(a) and the *MR* function would be downward sloping in figure 8.18(b). This would not significantly change the analysis or conclusions, as you should check for yourself by redrawing figure 8.18(b).

Summary of sections 8.14–8.20

These sections were concerned with profit maximization. The key points were:

(1) The point of departure is the definition of a firm's profits, Π, as

$\Pi = TR - TC$ (that is, total revenue minus total cost).

It follows directly that we find maximum profit by taking the derivative of this function and setting it equal to zero, which gives

$$\frac{d\Pi}{dq} = \frac{dTR}{dq} - \frac{dTC}{dq} = 0$$

Since by definition $\frac{dTR}{dq} = MR$ and $\frac{dTC}{dq} = MC$, this implies that the *first order* condition for profit maximization is *MR* = *MC*: that is, marginal cost equals marginal revenue.

(2) The above condition holds straightforwardly for a monopolist. For a perfectly competitive firm, however, there is a slight twist to the story. The perfectly competitive firm's *TR* function is $TR = \bar{p}q$ (where $\bar{p}$ is the ruling market price), so marginal revenue is

$$MR = \frac{dTR}{dq} = \bar{p}$$

that is, marginal revenue and price are identical. Therefore profit maximization requires a perfectly competitive firm to seek the level of output at which $MC = MR = \bar{p}$.

(3) Comparing a product supplied by a single (monopoly) firm with the same product supplied by many perfectly competitive firms, with the same cost and market demand functions, the monopoly firm will always sell a smaller output at a higher price.

(4) The condition *MR* = *MC* guarantees only that a stationary point of the profit function has been found. The stationary point could be a maximum, minimum or a point of inflection in the profit function. To ensure that the firm has found a maximum, the *second order* condition must be fulfilled: that is, $\frac{d^2\Pi}{dq^2} < 0$. This translates into the condition that the slope of the *MC* curve must be greater (more positive) than the slope of the *MR* curve. This condition is not always satisfied.

The Online Resource Centre gives more discussion and examples of profit maximization by the firm, including using Excel® to graph the relevant functions.
www.oxfordtextbooks.co.uk/orc/renshaw3e/

Progress exercise 8.3

1. A firm's short-run total cost function is $TC = q^2 - 3q + 500$. The firm sells in a perfectly competitive market and the ruling price is $p = 67$.

(a) Find the most profitable level of output, and the profits at that output.

(b) Does the firm produce at minimum average cost? Explain your answer.

(c) Sketch the graphs of total cost and total revenue with the same axes, and do the same with marginal cost and marginal revenue.

(d) Sketch the graph of the profit function. Indicate in all diagrams the equilibrium values of the variables.

2. Continuing with the firm in question 1, suppose the ruling market price rises, first to 68, then 69, then 70.

(a) What is the firm's response to these price increases?

(b) Can you deduce from this the firm's supply function (that is, a relationship between market price and the quantity the firm chooses to supply)? Illustrate with a sketch graph of the supply function.

(c) What price would induce this firm to produce at minimum average cost?

3. (a) Continuing with the firm in question (1), with $p = 67$ suppose the firm's fixed costs rise from 500 to 1000. What effect will this have on the firm's chosen level of output? What effect will it have on the firm's profits?

(b) Repeat question 3(a) for an increase in fixed costs to 1250.

4. A firm's short-run total cost function is $TC = q^2 - 3q + 500$. The firm is a monopolist and the inverse demand function for its product is $p = -q + 105$.

(a) Find the most profitable level of output, and the profits at that output.

(b) When maximizing profit, does the firm produce at minimum average cost?

(c) Sketch the graphs of the inverse demand function, marginal revenue, and marginal cost function with the same axes, showing the equilibrium price, MR and MC, and output.

(d) Sketch the graphs of total cost and total revenue with the same axes, showing the equilibrium output and profits. Sketch the graph of the profit function, showing the equilibrium output and profits.

5. Suppose the inverse demand function for the product of the firm in question 4 shifts from $p = -q + 105$ to $p = -2q + 159$ (as the result of advertising, perhaps).

(a) Show that this has no effect on the firm's output, but increases its price and its profits.

(b) Explain why this is so.

(c) Illustrate graphically.

6. A firm's short-run total cost function is $TC = 20q + 500$. The firm is a monopolist and the inverse demand function for its product is $p = -3q + 146$.

(a) Find the most profitable level of output, and the profits at that output.

(b) Could the firm under any circumstances produce at minimum average cost while maximizing profit?

(c) Sketch the graphs of the inverse demand function, marginal revenue, and marginal cost function with the same axes, showing the equilibrium price, MR and MC, and output.

(d) Sketch the graphs of total cost and total revenue with the same axes, showing the equilibrium output and profits.

(e) Sketch the graph of the profit function, showing the equilibrium output and profits.

7. Suppose a firm had the same *TC* function as in question 6, but sold its product in a perfectly competitive market where the ruling market price was $p = 25$.

(a) What would be the equilibrium in this case?

(b) What is the break-even output level?

(c) What would be the equilibrium if the ruling market price were $p = 15$?

(d) Are your answers to (a), (b), and (c) affected by the level of fixed costs?

(e) Explain your answers with the aid of algebra and appropriate graphs.

8. A monopoly firm's total cost function is $TC = 2q^3 - 48q^2 + 700q + 100$. It sells in a perfectly competitive market and the ruling market price is $p = 532$.

(a) Find the most profitable and least profitable outputs (i) using the marginal cost and marginal revenue functions, and (ii) using the profit function. What are the profit levels at these outputs?

(b) In both (i) and (ii) illustrate your solutions with sketch graphs. Draw sketch graphs also of the *TC* and *TR* functions.

(c) Explain the role of the second order condition for a maximum in this case.

(d) Would this firm ever choose an equilibrium output at which marginal cost was decreasing?

9. Repeat the previous question for a monopoly firm with a total cost function $TC = 4q^3 - 185q^2 + 4500q + 500$. The demand function for its product is $q = -\frac{1}{5}p + 600$.

10. Are there any conditions or assumptions that would result in firm finding that its most profitable output was also the output at which its average costs were minimized? Consider both (a) a monopoly firm and (b) a perfectly competitive firm. Explain your answer with the aid of graphs, algebra or preferably both.

Checklist

Be sure to test your understanding of this chapter by attempting the progress exercises (answers are at the end of the book). The Online Resource Centre contains further exercises and materials relevant to this chapter www.oxfordtextbooks.co.uk/orc/renshaw3e/.

This has been a long chapter but one in which you have at last seen some reward for the heavy investment in mathematical technique in earlier chapters. The chapter has focused entirely on economic applications in the area of microeconomics. We have looked at the firm's short-run cost functions, its revenue, and its search for maximum profit. Specifically we have analysed:

✔ **Total cost.** The firm's short-run total cost function; its key characteristics and the possible shapes of its graph.

✔ **Average and marginal cost.** Understanding the definitions of average cost and marginal cost, how to derive their equations from the total cost

function, and how the shapes of their graphs are related to that of the total cost function.

✔ **When average and marginal cost are equal, and why.** The proof that the marginal cost curve cuts the average cost curve at the point of minimum average cost, and why this is true.

✔ **The market demand function.** Understanding the market demand function for a good, and how to derive from it the inverse demand function.

✔ **The total revenue and marginal revenue functions.** Proof that marginal revenue is zero at the point of maximum total revenue, and explaining in words why this is true.

✔ **The profit function.** The firm's profit function and its shape as the difference between its total cost and total revenue functions, for both monopoly and perfect competition.

✔ **Profit maximization.** Finding the most profitable level of output by equating marginal cost and marginal revenue, for both monopoly and perfect competition. Understanding the second order condition for maximum profit, and when it may not be satisfied.

We now move on, in chapter 9, to a study in depth of the concept of elasticity of a function, which plays a major role across a wide range of economic analysis.

Appendix 8.1 The relationship between total cost, average cost, and marginal cost

(This appendix can be omitted without loss of continuity.)

To generalize the results of section 8.5, we assume now that total cost (TC) is some unspecified function of q, which we will call f(q). Thus we have

$$TC = f(q)$$

If we use the notation f$'(q)$ to denote the derivative of this function, then from the definition of marginal cost we have

$$MC \equiv \frac{dTC}{dq} \equiv f'(q) \qquad (A1)$$

Also, from the definition of average cost we have

$$AC \equiv \frac{TC}{q} = \frac{f(q)}{q} \qquad (A2)$$

To find the point of minimum AC, we find the derivative of equation (A2) and set it equal to zero. Using the quotient rule of differentiation (see section 6.7, rule D7), this gives

$$\frac{dAC}{dq} = \frac{qf'(q) - f(q)}{q^2} = 0 \qquad (A3)$$

We assume that the denominator of this expression, q^2, is *not* zero. Then, equation (A3) will equal zero only if the numerator is zero: that is, if

$$qf'(q) - f(q) = 0 \quad \text{which can be rearranged as}$$
$$f'(q) = \frac{f(q)}{q} \qquad (A4)$$

But the left-hand side of this expression is, from equation (A1), marginal cost, while the right-hand side, from (A2), is average cost. When these definitions are substituted into equation (A4) it becomes

$$MC = AC$$

Thus we have proved that, for any total cost function $TC = f(q)$, where f(q) is any function of q, average cost is at its minimum when marginal and average cost are equal.

Actually the previous sentence is not quite true. To be sure that we have found the point of *minimum* average cost (rather than a maximum or point of inflection), the second order condition for a minimum must be satisfied. For a minimum of AC, the second derivative of the AC function must be positive (see section 7.8, rule 7.1). We will now examine what must be true if this condition is to be satisfied.

From above, the first derivative of the AC function is

$$\frac{dAC}{dq} = \frac{qf'(q) - f(q)}{q^2} \quad \text{so the second derivative is}$$

$$\frac{d^2AC}{dq^2} = \frac{d}{dq} \text{ of } \frac{qf'(q) - f(q)}{q^2}$$

To differentiate this we have to use the quotient rule (see section 6.7, rule D7). Let $u = qf'(q) - f(q)$ and $v = q^2$. Then we have

$$\frac{du}{dq} = [qf''(q) + f'(q)] - f'(q)$$

(using the product rule to differentiate $qf'(q)$; see rule D6); and

$$\frac{dv}{dq} = 2q$$

Therefore, using the quotient rule,

$$\frac{d^2AC}{dq^2} = \frac{v\frac{du}{dq} - u\frac{dv}{dq}}{v^2}$$

$$= \frac{q^2[qf''(q)] - 2q[qf'(q) - f(q)]}{(q^2)^2} \qquad (A5)$$

From equation (A3), $qf'(q) - f(q) = 0$; and from equation (A1), $f''(q) = \frac{dMC}{dq}$, the slope of the marginal cost curve. Substituting these into (A5) gives

$$\frac{d^2AC}{dq^2} = \frac{1}{q}\frac{dMC}{dq}$$

Since q is positive, this expression is positive if $\frac{dMC}{dq}$ > 0; that is, if the marginal cost curve is positively sloped at the point where it cuts the average cost curve. Another way of describing this condition is to say that, at the point of minimum AC, the MC curve must cut the AC curve from below.

We now consider the relationship between MC and AC when average cost is increasing. When AC is increasing, the AC curve is upward sloping, and therefore $\frac{dAC}{dq}$ is positive. From equation (A3), we see that this is true when $\frac{qf'(q) - f(q)}{q^2} > 0$. As before, since q^2 is positive, this requires

$qf'(q) - f(q) > 0$ which easily rearranges as

$f'(q) > \frac{f(q)}{q}$

But, as noted above, the left-hand side of this last expression is marginal cost, while the right-hand side is average cost. So we conclude that

$$\frac{dAC}{dq} \quad \text{is positive if} \quad MC > AC$$

In words, average cost is increasing (that is, the AC curve is upward sloping) when marginal cost is greater than average cost. In exactly the same way, but with the inequality sign reversed, we can show that average cost is falling when marginal cost is less than average cost. The details of this are left to you.

We can summarize our results as a rule:

RULE 8.3 **The relationship between marginal cost (MC) and average cost (AC)**

(1) If $MC = AC$ and $\frac{dMC}{dq} > 0$, this point is a (local) minimum of AC.

(2) If $MC > AC$, AC is rising; if $MC < AC$, AC is falling.

We should emphasize how general these results are. To derive them, we did not need to place any prior restrictions on the shape of the total cost curve. We assumed only that output, q, was positive.

Appendix 8.2 The relationship between price, total revenue, and marginal revenue

(This appendix can be omitted without loss of continuity.)

Our analysis in section 8.11 of the relationship between price, total revenue, and marginal revenue for the monopolist was based on the example demand function $q = -2p + 200$. Here we show that our findings are valid for any demand function, and extend them to the case of perfect competition that we considered in section 8.12.

Suppose that there is some unspecified demand function, which we'll write as

$q = g(p)$, (where $g(p)$ denotes an unspecified functional form)

We shall need to use the inverse demand function in a moment, so we'll assume this is given by

$p = h(q)$ (B1)

where $h(q)$ is whatever we get when we invert the demand function $q = g(p)$ above. Note that the derivative of this function is

$$\frac{dp}{dq} = h'(q)$$

which we assume to be negative because the essence of monopoly is that the demand function is negatively sloped.

By definition, $TR = pq$. If, as we did in section 8.9, we substitute for p using equation (B1) above, we get

$TR = h(q)q$ (B2)

(If this puzzles you, just suppose $h(q)$ is, say, $-0.5q + 100$ as in section 8.11 above.)

We can then differentiate equation (B2), using the product rule, and get marginal revenue as

$MR = h(q) + qh'(q)$ (B3)

From above, we have $p = h(q)$ and $\frac{dp}{dq} = h'(q)$. Substituting these into equation (B3), we get

$$MR = p + q\frac{dp}{dq}$$

This is true of any demand function, $q = g(p)$, so we have a general rule:

RULE 8.4 Price, marginal revenue, and the slope of the inverse demand function

For any demand function, price, marginal revenue, and the slope of the inverse demand function are related by the equation

$$MR = p + q\frac{dp}{dq}$$

There are then two cases:

(1) Monopoly, where the inverse demand function is downward sloping, so $\frac{dp}{dq} < 0$. In that case, since q is positive, in rule 8.4 we must have $MR < p$.

(2) Perfect competition, where the inverse demand function is horizontal, so $\frac{dp}{dq} = 0$. In that case, in rule 8.4 we must have $MR = p$.

We can derive a second, related rule. This rule says that, for a monopolist, a price reduction will increase (decrease) total revenue if marginal revenue is positive (negative).

The proof is as follows. We take the demand function from above, $q = g(p)$, assumed downward sloping, so $\frac{dq}{dp}$ (which we can also write as $g'(p)$) is negative. We assume also some unspecified total revenue function:

$$TR = f(q)$$

So we have

$$TR = f(q), \quad \text{where} \quad q = g(p) \qquad (B4)$$

This is a 'function of a function' situation, in which TR depends on q, which in turn depends on p. We can therefore apply the function of a function or chain rule of differentiation (see chapter 6, rule D5 if you need to revise this) to obtain $\frac{dTR}{dp}$ as

$$\frac{dTR}{dp} = \frac{dTR}{dq}\frac{dq}{dp} \quad \text{or, equivalently,}$$

$$\frac{dTR}{dp} = f'(q)g'(p)$$

Since, by definition, $\frac{dTR}{dq} \equiv MR$, we can substitute this into the above expression and get

$$\frac{dTR}{dp} = MR\frac{dq}{dp} \qquad (B5)$$

Next, using (B4), we can write

$$TR = f(g(p))$$

which gives us TR as a function of p alone. Its differential (see section 7.15) is:

$$dTR = \frac{dTR}{dp}dp$$

Using (B5), this becomes $dTR = MR\frac{dq}{dp}dp$. This gives us the rule:

RULE 8.5 The effect of a change in price on total revenue

Following a small change in price, dp, the change in TR is given by

$$dTR = MR\frac{dq}{dp}dp$$

Since we have assumed $\frac{dq}{dp} < 0$, a price *reduction* ($dp < 0$) means that $\frac{dq}{dp}dp > 0$. (That is, a price reduction increases quantity sold.) It follows immediately that dTR has the same sign as MR. So if MR is positive, a price reduction increases total revenue, and if MR is negative, a price reduction decreases total revenue. This is exactly what we found in the case of the monopolist in section 8.11.

Note that rule 8.5 is relevant only to a monopolist, and not to a perfectly competitive firm. A perfectly competitive firm has, effectively, no influence on the selling price of its product. By definition, if a perfectly competitive firm raises its price (above the ruling market price), then it loses all its customers so TR drops to zero; while if it lowers its price (below the ruling market price), then it attracts the whole market so TR becomes indefinitely large. Selling, or even attempting to sell, at any price other than the ruling market price is thus incompatible with equilibrium for the firm and is therefore ruled out. (This last sentence raises the question of how, in that case, the ruling market price can ever change, but we will leave this to the microeconomics textbooks.)

Chapter 9
Elasticity

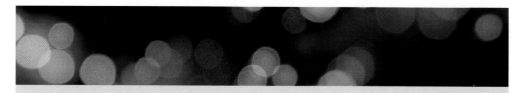

OBJECTIVES

Having completed this chapter you should be able to:

- Understand the distinction between absolute, proportionate, and percentage changes in a variable.
- Understand what the arc elasticities of supply and demand measure and how to estimate them from a graph.
- Understand what the point elasticities of supply and demand measure and why they are preferred to the arc elasticities.
- Understand and derive the relationship between the elasticity of demand and marginal revenue, under monopoly and perfect competition.
- Understand how elasticity can be applied to other functions in economics, such as the total cost function and the consumption function.

9.1 Introduction

This chapter is entirely devoted to the concept of elasticity, a concept that figures prominently in almost every area of economic analysis. The elasticity of any function $y = f(x)$ is a way of measuring the relationship between a given proportionate change in x and the resulting proportionate change in the dependent variable, y. Economists' preoccupation with the elasticity of functions reflects the fact that we are often more concerned with the relationship between proportionate changes in x and y than with absolute changes. For example, if we are told that the minimum wage in Cambodia has increased by 5 dollars per month (an absolute change), this does not in itself tell us very much. But if we are told that this represents an increase of one-tenth (a proportionate change), we are in a better position to judge whether the increase is large or small.

In this chapter we examine two key elasticities in economics: the elasticity of supply and the elasticity of demand. In both cases we consider two alternative measures of elasticity: the *arc elasticity*, which relates to discrete changes in the variables, and the *point elasticity*, which incorporates the derivative and relates to indefinitely small changes in the variables. In the case of the demand function, we identify and explore an important relationship that exists between price, marginal revenue, and the elasticity of demand. Finally we consider how the elasticity concept may be applied to other functions in economics and indeed to functions in general.

9.2 Absolute, proportionate, and percentage changes

In economics we are often more concerned with proportionate or percentage changes in an economic variable than with absolute changes in that variable. Let us therefore consider first exactly what is the difference between an absolute change and a proportionate change in any variable.

Suppose my income, Y, is initially $Y_0 = 200$ (measured in euros per week, say). Then suppose it goes up by 40 to a new level $Y_1 = 240$. The absolute change is

$Y_1 - Y_0 = 240 - 200 = 40$ (euros per week)

Following our usual notation we denote this absolute change by ΔY. Thus we have

absolute change $\equiv \Delta Y \equiv Y_1 - Y_0 = 40$ (euros per week)

The proportionate change is defined as the absolute change in the variable, divided by its initial value. Thus we have

proportionate change $\equiv \dfrac{\Delta Y}{Y_0} = \dfrac{40}{200} = \dfrac{1}{5} = 0.2$ (in decimal format)

In words, when we calculate the proportionate change we find that my income has gone up by one-fifth. The increase, 40, is one-fifth of my initial income level, 200. The proportionate change thus expresses the absolute change as a proportion of the initial level. The key point to note is that we convert an absolute change in a variable into a proportionate change by dividing by the initial value of the variable.

The proportionate change often conveys information that is more interesting and more useful than that conveyed by the absolute change. This is because we often don't know whether to view the absolute change as large or small until we have something to compare it with. For example, to someone earning only €80 per week, an increase of €40 per week is a proportionate change of one-half, which we would view as large. But to someone earning €800 per week the same absolute increase, of €40 per week, is a proportionate change of only one-twentieth, which we would view as quite small.

Another reason for looking at proportionate rather than absolute changes is that a proportionate change is independent of the units in which the variable is measured. Thus if we rework the first example in the previous paragraph with income measured in US dollars at an exchange rate of, say, €1 = $1.5, an income of €80 per week becomes an income of $120 per week, and an increase of €40 per week becomes an increase of $60. So the proportionate change is $60/$120 = one-half, as before.

We may also be interested in the percentage change in my income (or any other variable). This is easy to calculate, since we know from section 1.12 that to convert any fraction into a percentage we simply multiply by 100.

Thus to find the percentage change we simply multiply the proportionate change (which is a fraction) by 100. In the example of my income above this gives

percentage change = proportionate change $\times 100 = \dfrac{\Delta Y}{Y_0}(100) = \dfrac{1}{5}(100) = 20(\%)$

So, to summarize, if the *absolute* change in my income is €40 per week and my initial income level is €200 per week, then the *proportionate* change is one-fifth (or 0.2 in decimal form), and the *percentage* change is 20(%).

Note that, since

$$\text{percentage change} = \text{proportionate change} \times 100$$

the proportionate change and the percentage change differ only in *scale*: that is, we can always convert one into the other by multiplying or dividing by 100. For this reason, in the remainder of this book we will treat the terms 'proportionate change' and 'percentage change' as equivalent.

In this example we have considered changes in my income, Y, but it should be obvious that the concept of a proportionate change can be applied to any variable that is changing in value for any reason. We can thus generalize as follows:

RULE 9.1 Proportionate and percentage changes in a variable

If any variable y changes by a finite amount, Δy, for any reason, the proportionate change in y is measured by $\frac{\Delta y}{y_0}$ (where y_0 is the initial value of y). The percentage change is measured by $\frac{\Delta y}{y_0} \times 100$.

When a proportionate change is meaningless

There is an important qualification to the generalization in rule 9.1. When the variable in question is measured from an arbitrary zero point, or origin, then the proportionate change conveys no useful information, and indeed can be considered meaningless.

There is one such variable which we use frequently in economics: time. For example, if you had an appointment at 4 p.m. but in fact arrived 2 hours late at 6 p.m., it makes no sense to say that your proportionate lateness is $\frac{2 \text{ hours}}{4 \text{ hours}} = \frac{1}{2}$ or 50%. To see why, suppose we were using the 24-hour clock. Then 4 p.m. would be 16.00 hours, and if you arrived 2 hours late at 18.00 hours your proportionate lateness is $\frac{2 \text{ hours}}{16 \text{ hours}} = \frac{1}{8}$ or $12\frac{1}{2}$%. In the first case, time is measured from a zero that is set at 12 o'clock midday, and in the second case it is measured from a zero set at 12 midnight. Thus you can change your proportionate lateness by choosing a different origin (zero point) from which to measure time. The origin (zero point) from which we measure time is arbitrary.

However, a different situation arises when we are talking about *elapsed* time. For example, if you caught a train that was supposed to take 2 hours to reach its destination but in fact took 3, it would make sense to say that the journey took 50% longer than intended. This is meaningful because we are looking at the time that elapsed between the train's departure and its arrival, and this is a purely relative measure, which is the same whether the train left at 10 a.m. or 3 p.m.

Graphical representation

We can show absolute and proportionate changes graphically. In figure 9.1 we show some function $y = f(x)$. Suppose the initial position is at point A, where $x = x_0$ and $y = y_0$. Then x increases to x_1 and consequently y increases to y_1. The resulting *absolute change* in y, $\Delta y = y_1 - y_0$, is the distance BC. The initial level of y, y_0, is given

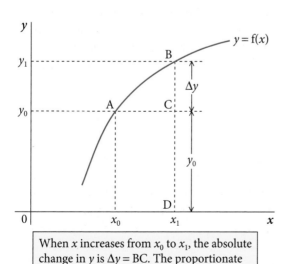

When x increases from x_0 to x_1, the absolute change in y is $\Delta y = \text{BC}$. The proportionate change is $\frac{\Delta y}{y_0} = \frac{\text{BC}}{\text{CD}}$.

Figure 9.1 Absolute and proportionate changes.

by the distance CD. Therefore the *proportionate change* in y is given by the ratio of these two:

$$\frac{\Delta y}{y_0} = \frac{BC}{CD}$$

In the diagram, the distance BC is roughly one-third of the distance CD, so we can infer that the proportionate change in y when x increases from x_0 to x_1 is about one-third or $33\frac{1}{3}\%$.

9.3 The arc elasticity of supply

In this section we introduce the idea of elasticity, using supply elasticity as an example. There are two similar but nevertheless distinct definitions of elasticity: arc elasticity and point elasticity. We will examine each of them and show how they are related. The concept of elasticity has many applications in economics, and we will discuss some of these later in this chapter.

As explained briefly in sections 3.14 and 4.13, the market supply function for a good gives the quantity of the good which sellers in aggregate choose to offer for sale as a function of the price they receive for each unit of the good. (For more detail, see any introductory microeconomics textbook.) We can write this function as

$$q = f(p)$$

where q denotes the quantity supplied (measured in some physical unit) and p denotes the selling price of each unit. Thus if the good in question is apples, the physical unit might be kilos and the price might be measured in euros per kilo. We assume a competitive market, meaning that individual sellers have no influence on the price, but simply decide what quantity to offer for sale at any ruling market price. Thus p is the independent variable and q the dependent variable. We assume this function to be smooth, continuous, and positively sloped.

Given this supply function, we can now define the **arc elasticity** of supply, which we will denote by E_A^S. In introductory economics textbooks the definition used is

$$E_A^S \equiv \frac{\text{percentage change in quantity supplied}}{\text{percentage change in price}} \tag{9.1}$$

The idea behind this is to measure the *responsiveness* of quantity supplied by sellers to a change in the price they receive. Suppose for example that there is a 10% increase in price. Given the assumption that the supply function is positively sloped, this induces sellers in aggregate to increase the quantity supplied. Let us suppose that quantity supplied increases by 20%. Then

$$E_A^S \equiv \frac{20\%}{10\%} = \frac{20}{10} = 2$$

Notice that the elasticity is a pure number because the units of measurement (in this case, percentage points) cancel between the numerator and denominator. Thus $E_A^S = 2$ tells us that the percentage change in quantity supplied is twice as large as the percentage price increase that induced it. In general, if E_A^S is large, this means that the response of quantity (the numerator) is large relative to the change in price which triggered it (the denominator). Note that E_A^S is also positive if the price change is negative, for then the quantity change is negative too, due to our assumption that the supply function is positively sloped. We then have both numerator and denominator negative, hence E_A^S is positive.

We now want to add precision to this idea of responsiveness which underlies the elasticity concept. Before going further we shall make a small modification to equation (9.1). From section 9.2 above we know that a percentage change is simply a proportionate change multiplied by 100. Thus equation (9.1) can be written as

$$E_A^S \equiv \frac{\text{proportionate change in quantity supplied} \times 100}{\text{proportionate change in price} \times 100}$$

and since the 100 cancels between numerator and denominator, we have

$$E_A^S \equiv \frac{\text{proportionate change in quantity supplied}}{\text{proportionate change in price}} \qquad (9.2)$$

Using the notation of section 9.2 above, we can write the numerator of equation (9.2) as $\frac{\Delta q}{q_0}$, and the denominator as $\frac{\Delta p}{p_0}$, where it is understood that q_0 and p_0 are the initial quantity and price (that is, before any price change occurs). Substituting these into equation (9.2) gives us a new rule:

RULE 9.2 Arc elasticity of supply

Given any supply function $q = f(p)$, the arc elasticity of supply, E_A^S, is defined as

$$E_A^S \equiv \frac{\dfrac{\Delta q}{q_0}}{\dfrac{\Delta p}{p_0}} = \frac{\text{proportionate change in quantity supplied}}{\text{proportionate change in price}}$$

Because writing the supply elasticity in this way takes up a lot of space, from now on we will write it as

$$E_A^S \equiv \frac{\Delta q / q_0}{\Delta p / p_0}$$

(This makes use of the convention that any ratio, $\frac{a}{b}$, may be written as a/b.)

9.4 Elastic and inelastic supply

In rule 9.2, if $\frac{\Delta q}{q_0}$ is greater than $\frac{\Delta p}{p_0}$, this means that the numerator is greater than the denominator, and hence that E_A^S is greater than 1. In this case we say that supply is **elastic**.

Conversely, if $\frac{\Delta q}{q_0}$ is less than $\frac{\Delta p}{p_0}$, this means that the numerator is less than the denominator, and hence that E_A^S is less than 1. In this case we say that supply is **inelastic**.

The boundary case between these two occurs when $\frac{\Delta q}{q_0}$ and $\frac{\Delta p}{p_0}$ are equal, so that $E_A^S = 1$. In this case we say that the supply function has unitary or **unit elasticity**, or the supply elasticity is unity.

9.5 Elasticity as a rate of proportionate change

We saw in section 6.2 that a ratio of two absolute changes is called a rate of change. Thus, given any function $y = f(x)$, if x changes by the absolute amount Δx and y consequently changes by the absolute amount Δy, then the ratio $\frac{\Delta y}{\Delta x}$ measures the change in y per unit of change in x. We call this ratio of two absolute changes the *rate of change* of y. We also give it a special name: the difference quotient.

By the same reasoning, the ratio of two proportionate changes is a rate of *proportionate* change. Thus, given the supply function $q = f(p)$, if p changes by the proportionate amount $\frac{\Delta p}{p_0}$ and q consequently changes by the proportionate amount $\frac{\Delta q}{q_0}$, then the ratio

$$\frac{\Delta q / q_0}{\Delta p / p_0} \quad \text{(that is, the elasticity)}$$

measures the proportionate change in q per unit of proportionate change in p. We call this ratio of two proportionate changes the **rate of proportionate change** of q. Since this relates to the supply function $q = f(p)$, then this rate of proportionate change has a special name: the elasticity of supply. But we can calculate the rate of proportionate change, or elasticity, of any function, not merely a supply function, in the same way. We will look at other elasticities later in this chapter.

Why are economists often more interested in the rate of proportionate change (the elasticity) than in the rate of absolute change (the difference quotient)? As we explained in section 9.2 above, absolute changes, and therefore rates of absolute change, are often hard to interpret unless we have something to compare them with and thus provide a meaningful scale.

For example, suppose $q = f(p)$ is the supply function for beer, where q is measured in millions of pints (per day) and p is the price in euros per pint. Suppose the price rises by 0.5 euros, and the resulting increase in quantity supplied is 50 million pints. Thus we have $\Delta p = 0.5$ euros and $\Delta q = 50$ million pints, so the rate of change of quantity is

$$\frac{\Delta q}{\Delta p} = \frac{50 \text{ million pints}}{0.5 \text{ euros}} = 100 \text{ million pints per euro}$$

This tells us that the rate of change of quantity is such that a 1 euro increase in price increases supply by 100 million pints. Is this a large increase or a small increase? We can't say. Thus the information content of $\frac{\Delta q}{\Delta p}$, the rate of change, is low.

However, if we also know that the initial price was $p_0 = 2$ euros and the initial quantity was $q_0 = 500$ million pints, then we can calculate the elasticity (the rate of proportionate change of quantity) as

$$E_A^S \equiv \frac{\Delta q / q_0}{\Delta p / p_0} = \frac{50 \text{ million pints}/500 \text{ million pints}}{0.5 \text{ euros}/2 \text{ euros}} = \frac{0.1}{0.25} = 0.4$$

This tells us that the rate of proportionate change is only 0.4 or 40%. In other words, the proportionate increase in quantity supplied (0.1 or 10%) is less than half as large as the proportionate increase in price (0.25 or 25%); supply is inelastic. This tells us more about the supply function of beer than we learned from $\frac{\Delta q}{\Delta p}$, the difference quotient.

Another advantage of the elasticity measure is that, as noted above, its value is independent of the units in which price and quantity are measured. This is because, in E_A^S above, the units of measurement cancel in both the numerator and the denominator. This would be an advantage if we were comparing the supply functions for beer in two different countries that used different currencies and/or different units of quantity (for example, litres instead of pints).

9.6 Diagrammatic treatment

The arc elasticity of supply can be shown diagrammatically. Figure 9.2 shows the graph of some unspecified supply function $q = f(p)$. Note that we have drawn the graph in accordance with the economist's convention of putting p on the vertical axis, rather than the mathematician's convention of putting the dependent variable (in this case, q) on the vertical axis. (See section 8.8 if you need to revise this point.)

In figure 9.2, suppose suppliers are initially at J with price p_0 and quantity supplied q_0. Then suppose that for some reason the price increases from p_0 to p_1. Suppliers respond by increasing the quantity supplied from q_0 to q_1.

On the horizontal axis, we can see that the distance HR = JF gives us the change in quantity, Δq, while the distance 0H = NJ gives us the initial quantity, q_0. Thus

$$\frac{JF}{NJ} = \frac{\Delta q}{q_0} = \text{proportionate change in quantity supplied}$$

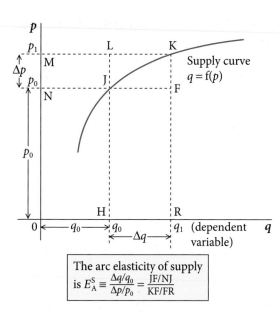

Figure 9.2 Arc elasticity of supply.

In the same way, on the vertical axis the distance $NM = FK$ gives us the change in price, Δp, while the distance $0N = RF$ gives us the initial price, p_0. Thus

$$\frac{FK}{RF} = \frac{\Delta p}{p_0} = \text{proportionate change in price}$$

If we now take the ratio of these two ratios, we get

$$\frac{JF/NJ}{FK/RF} = \frac{\Delta q/q_0}{\Delta p/p_0} = E_A^S$$

As figure 9.2 is drawn, $\frac{JF}{NJ}$ is about 0.875 and $\frac{FK}{RF}$ is about 0.25. (Check for yourself with a ruler.) If we substitute these values into the previous equation we get

$$\frac{JF/NJ}{FK/RF} = \frac{\Delta q/q_0}{\Delta p/p_0} = \frac{0.875}{0.25} = 3.5$$

Thus the elasticity of supply equals roughly 3.5, telling us that the proportionate change in q (0.875, or 87.5%) is 3.5 times as large as the proportionate change in p (0.25, or 25%). Supply is highly elastic.

An alternative formulation

There is an alternative algebraic formulation of the arc elasticity of supply that in some ways is preferable. The definition given in rule 9.2 was

$$E_A^S \equiv \frac{\Delta q/q_0}{\Delta p/p_0} \quad \text{which is the same thing as } \frac{\Delta q}{q_0} \div \frac{\Delta p}{p_0}$$

We can rearrange this expression a little by making use of the fact that dividing by a fraction is the same as multiplying by the reciprocal of that fraction (see section 1.8). Thus instead of dividing by $\frac{\Delta p}{p_0}$ we can multiply by $\frac{p_0}{\Delta p}$ and so get

$$E_A^S \equiv \frac{\Delta q}{q_0} \frac{p_0}{\Delta p} \tag{9.3}$$

Second, since in the denominator $q_0 \times \Delta p$ is the same thing as $\Delta p \times q_0$ (because 3×2 is the same thing as 2×3), we can write equation (9.3) in terms of rule 9.2a:

RULE 9.2a Alternative forms of the arc elasticity of supply

The arc elasticity of supply, E_A^S, as defined in rule 9.2, may be written as

$$E_A^S \equiv \frac{\Delta q}{\Delta p} \frac{p_0}{q_0} \tag{9.4}$$

or, equivalently (since multiplying by $\frac{p_0}{q_0}$ is the same thing as dividing by $\frac{q_0}{p_0}$):

$$E_A^S \equiv \frac{\Delta q/\Delta p}{q_0/p_0} \tag{9.4a}$$

From equation (9.4) we see that E_A^S consists of our old friend the difference quotient, $\frac{\Delta q}{\Delta p}$, multiplied by $\frac{p_0}{q_0}$, the ratio of price to quantity at the initial position. (See section 6.2 if you need to revise the difference quotient.)

A common mistake made by students new to economics is to suppose that the elasticity depends only on the slope of the supply function. From rule 9.2a we can see clearly that the arc elasticity depends partly on slope, as measured by the difference quotient $\frac{\Delta q}{\Delta p}$. But it also depends on the position of the supply function and the chosen position on it, since these determine the values of p_0 and q_0 and therefore $\frac{p_0}{q_0}$. Logically, $\frac{\Delta q}{\Delta p}$ and $\frac{p_0}{q_0}$ have equal importance in determining the elasticity.

9.7 Shortcomings of arc elasticity

Arc elasticity has three shortcomings that have led economists to develop the concept of *point* elasticity as an alternative definition, which, as we shall see, is particularly useful in theoretical work.

The first shortcoming of arc elasticity is that its value varies according to the size of the assumed price change, Δp. In figure 9.2, if Δp were smaller than as shown, so that K was closer to J, this would change the value of $\frac{\Delta q}{\Delta p}$. From rule 9.2a we can see that this would change the value of the elasticity. The only exception to this would arise if the supply function were linear between J and K, for in that case the difference quotient, $\frac{\Delta q}{\Delta p}$, is unchanged by variation in the size Δp. Apart from this exception, therefore, two independent researchers who chose the same values for p_0 (and therefore q_0) but different values for Δp would arrive at different, but equally valid, estimates of the supply elasticity of whatever good they were studying. This is clearly unsatisfactory.

The second shortcoming of the arc elasticity is its asymmetry. If in figure 9.2 the initial position is at J, a price increase of Δp takes us to K and the arc elasticity, as we saw above, is

$$E_A^S \equiv \frac{\Delta q}{\Delta p}\frac{p_0}{q_0} \quad \text{(a movement from J to K)}$$

If there is now a price reduction of Δp which takes us back to J again, the initial values of p and q are now p_1 and q_1 so that the arc elasticity in this case is

$$E_A^S \equiv \frac{\Delta q}{\Delta p}\frac{p_1}{q_1} \quad \text{(a movement from K to J)}$$

As $\frac{p_0}{q_0}$ and $\frac{p_1}{q_1}$ are generally not equal, the elasticity for a movement from J to K differs from the elasticity for a movement from K to J. This asymmetry, or irreversibility as it is sometimes called, is clearly untidy. The only exception arises if $\frac{p_0}{q_0} = \frac{p_1}{q_1}$, which in figure 9.2 would be true if J and K lay on a straight line from the origin. (The proof of this is left to you.)

A third shortcoming of the arc elasticity is that it is laborious to calculate; see the worked examples in section 9.10 below.

9.8 The point elasticity of supply

To escape from these shortcomings, in theoretical economics we use a second definition of elasticity, the point elasticity. We derive this by starting from the arc elasticity, $\frac{\Delta q}{\Delta p}\frac{p_0}{q_0}$, and then allowing Δp to approach zero. As Δp approaches zero, the difference quotient $\frac{\Delta q}{\Delta p}$ approaches, by definition, the derivative $\frac{dq}{dp}$. (See section 6.5 if you need to revise why this is true.) Thus the **point elasticity** of supply, which we will denote by E^S, is defined in terms of rule 9.3:

OPTIMIZATION WITH ONE INDEPENDENT VARIABLE

RULE 9.3 The point elasticity of supply

The point elasticity of supply, E^S, is defined as

$$E^S \equiv \lim_{\Delta p \to 0} \frac{\Delta q}{\Delta p} \frac{p_0}{q_0} \equiv \frac{dq}{dp} \frac{p_0}{q_0} \tag{9.5}$$

or, equivalently (since multiplying by $\frac{p_0}{q_0}$ is the same thing as dividing by $\frac{q_0}{p_0}$):

$$E^S \equiv \frac{dq/dp}{q_0/p_0} \tag{9.5a}$$

Looking at equation (9.5) we see that, at any given point on the supply curve, the value of E^S is determined solely by p_0 and q_0 (the coordinates of that point), and by $\frac{dq}{dp}$, the slope of the tangent at that point. The point elasticity at any point is thus unique and unambiguous. It is also symmetric, in that it has the same value whether we consider an (indefinitely small) price increase or an (indefinitely small) price reduction.

A diagrammatic interpretation of the point elasticity is shown in figure 9.3. Here we have graphed the supply function from figure 9.2 but this time obeying the maths convention by putting the dependent variable, q, on the vertical axis. At point J the derivative $\frac{dq}{dp}$ is given by the slope of the tangent BJC, while $\frac{q_0}{p_0}$ is given by the slope of 0J. The point elasticity of supply at J, as defined in equation (9.5a), is therefore given by the ratio of these two slopes. The important thing to see is that both slopes are uniquely determined at any point such as J on the supply curve.

As figure 9.3 is drawn, the slope of BJC is larger (steeper) than the slope of the ray 0J, so $\frac{dq}{dp}$ is greater than $\frac{q_0}{p_0}$ at point J. Therefore the numerator of equation (9.5a) is greater than the denominator, and the elasticity of supply is greater than 1. Supply is elastic.

When the supply function is drawn with p on the vertical axis, as in figure 9.2, the point elasticity can be inferred in the same way as in figure 9.3, but the task is complicated by the fact that interchanging the axes causes all the slopes to become inverted. This is left to you as an exercise.

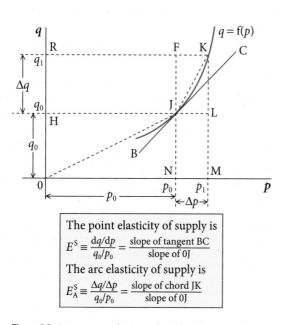

The point elasticity of supply is
$$E^S \equiv \frac{dq/dp}{q_0/p_0} = \frac{\text{slope of tangent BC}}{\text{slope of 0J}}$$
The arc elasticity of supply is
$$E_A^S \equiv \frac{\Delta q/\Delta p}{q_0/p_0} = \frac{\text{slope of chord JK}}{\text{slope of 0J}}$$

Figure 9.3 Comparison of point and arc elasticity of supply.

Incidentally, the arc elasticity at J can also be read off figure 9.3 in the same way. Assuming that a price increase of Δp moves suppliers from J to K, the difference quotient $\frac{\Delta q}{\Delta p}$ is given by the slope of the broken line JK, while $\frac{q_0}{p_0}$ is given by the slope of 0J, as for the point elasticity. From equation (9.4a) the arc elasticity is

$$E_A^S \equiv \frac{\Delta q/\Delta p}{q_0/p_0} = \frac{\text{slope of chord JK}}{\text{slope of 0J}}$$

This should equal the arc elasticity that we measured on figure 9.2: that is, about 3.5. Check for yourself with a ruler.

9.9 Reconciling the arc and point supply elasticities

So far in this chapter we have developed two definitions of supply elasticity:

(1) arc elasticity: $E_A^S \equiv \dfrac{\Delta q}{\Delta p} \dfrac{p_0}{q_0}$ $\hspace{4cm}$ (9.4) repeated

(2) point elasticity: $E^S \equiv \dfrac{dq}{dp} \dfrac{p_0}{q_0}$ $\hspace{4cm}$ (9.5) repeated

The difference between the two definitions is that the arc elasticity uses the difference quotient, while the point elasticity uses the derivative of the supply function. Note that the initial values of p and q, which we have written as p_0 and q_0, are normally written simply as p and q, and this is the notation we will use in later chapters.

You may find it a little worrying that we now have two apparently competing definitions of the elasticity of supply. How can we reconcile the two? One special case arises when the supply function that we are looking at is linear. In that case, $\frac{\Delta q}{\Delta p}$ and $\frac{dq}{dp}$ are identically equal (see sections 6.1–6.4 if you can't see why) and the distinction between arc elasticity and point elasticity disappears. But this is not a very interesting case, as it seems likely that many, if not most, supply functions are non-linear. We must consider the general case where the supply function is non-linear.

The point elasticity overcomes the three shortcomings of the arc elasticity that we identified in section 9.7. On the other hand, the point elasticity suffers from the apparent disadvantage that it is only strictly valid in the limiting case as Δp approaches zero, and price changes this small are not observed in the real world. However, as we have already done at several points in this book (in particular when defining marginal cost and marginal revenue in chapter 8), we can make use of the fact that the difference quotient, $\frac{\Delta q}{\Delta p}$, approaches closer and closer to the derivative, $\frac{dq}{dp}$, as Δp approaches zero. (See section 6.5 if you need to revise this.) Therefore, if we are willing to restrict ourselves to considering only very small price changes, the difference between $\frac{\Delta q}{\Delta p}$ and $\frac{dq}{dp}$ is small enough to be ignored, in which case we can treat the two definitions as identical for all practical purposes.

Given this identity, we can use whichever definition is the more convenient. In theoretical economics the point elasticity is invariably the more convenient because it's easier to calculate and algebraically more manipulable than the arc elasticity (see the worked examples below). These attractions outweigh the disadvantage of the point elasticity, namely that it is only strictly valid for indefinitely small price changes.

In the real world, too, the point elasticity is more convenient than the arc elasticity and can safely be used for analysing small price changes. However, if greater precision is required, or to analyse the relatively large price changes that sometimes occur in the real world, the arc elasticity is the appropriate definition to use. In using it, though, we must keep in mind that the value of the arc elasticity at any point on the supply curve will vary with the size of the price change, unless the supply function is linear.

9.10 Worked examples on supply elasticity

EXAMPLE 9.1

Given the supply function: $q = 10p^3 - 1000$

(a) Find the arc elasticity of supply when p increases from 10 to 11.

(b) Find the point elasticity when $p = 10$.

(c) Draw a sketch graph of the supply function, illustrating your answers to (a) and (b).

Answer:

(a) Given $p_0 = 10$, $q_0 = 10(10)^3 - 1000 = 9000$

Given $p_1 = 11$, $q_1 = 10(11)^3 - 1000 = 12{,}310$

Therefore $\Delta p \equiv p_1 - p_0 = 11 - 10 = 1$, and $\Delta q \equiv q_1 - q_0 = 12{,}310 - 9000 = 3310$

Therefore the arc elasticity of supply between $p_0 = 10$ and $p_1 = 11$ is

$$E_A^S \equiv \frac{\Delta q / q_0}{\Delta p / p_0} = \frac{3310/9000}{1/10} = 3.678 \quad \text{(see rule 9.2)}$$

Alternatively, the arc elasticity can be calculated as

$$E_A^S \equiv \frac{\Delta q / \Delta p}{q_0 / p_0} = \frac{3310/1}{9000/10} = 3.678 \text{ (as before)} \quad \text{(see equation (9.4))}$$

(b) Given $q = 10p^3 - 1000$, $\frac{dq}{dp} = 30p^2$, so the point elasticity when $p = 10$ is

$$E^S = \frac{dq/dp}{q_0/p_0} = \frac{30(10)^2}{9000/10} = 3.333 \quad \text{(see equation (9.5))}$$

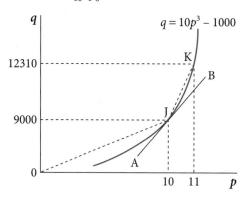

(c) See figure 9.4. Using equation (9.4), the arc elasticity is given by the slope of the chord JK (3310/1) divided by the slope of 0J (9000/10). Using equation (9.5), the point elasticity is given by the slope of the tangent AJB (which we know from (b) above equals $30(10)^2$) divided by the slope of 0J = (9000/10). Because JK is steeper than AJB, the arc elasticity (3.678) is greater than the point elasticity (3.333).

Figure 9.4 Answer to example 9.1(c) (not to scale).

EXAMPLE 9.2

Given the supply function: $q = -0.06p^2 + 2p - 5$

(a) Find the arc elasticity of supply when p increases from 10 to 11.

(b) Find the arc elasticity of supply when p decreases from 11 to 10.

(c) Find the point elasticity when $p = 10$.

(d) Draw a sketch graph of the supply function, illustrating your answers to (a)–(c).

Answer:

(a) Given $p_0 = 10$, $q_0 = -0.06(10)^2 + 2(10) - 5 = 9$

Given $p_1 = 11$, $q_1 = -0.06(11)^2 + 2(11) - 5 = 9.74$

Therefore $\Delta p \equiv p_1 - p_0 = 11 - 10 = 1$, and $\Delta q \equiv q_1 - q_0 = 9.74 - 9 = 0.74$

Therefore the arc elasticity of supply when price increases from $p_0 = 10$ to $p_1 = 11$ is

$$E_A^S = \frac{\Delta q / q_0}{\Delta p / p_0} = \frac{0.74/9}{1/10} = 0.822 \quad \text{(see rule 9.2)}$$

Alternatively, the arc elasticity may be calculated as

$$E_A^S \equiv \frac{\Delta q / \Delta p}{q_0 / p_0} = \frac{0.74/1}{9/10} = 0.822 \text{ (as before)} \quad \text{(see equation (9.4))}$$

(b) We now have $p_0 = 11$, so $q_0 = -0.06(11)^2 + 2(11) - 5 = 9.74$

and $p_1 = 10$, so $q_1 = -0.06(10)^2 + 2(10) - 5 = 9$

Therefore $\Delta p \equiv p_1 - p_0 = 10 - 11 = -1$ and $\Delta q \equiv q_1 - q_0 = 9 - 9.74 = -0.74$

Therefore the arc elasticity of supply when price decreases from $p_0 = 11$ to $p_1 = 10$ is

$$E_A^S \equiv \frac{\Delta q/q_0}{\Delta p/p_0} = \frac{-0.74/9.74}{-1/11} = 0.836$$

Comparing answers to (a) and (b), we see that this illustrates the point that the arc elasticity is asymmetric; the elasticity for a price increase from 10 to 11 differs from the elasticity for a price decrease from 11 to 10.

(c) Given $q = -0.06p^2 + 2p - 5$, $\frac{dq}{dp} = -0.12p + 2$. So the point elasticity when $p = 10$ is

$$E^S \equiv \frac{dq/dp}{q_0/p_0} = \frac{-0.12(10) + 2}{9/10} = 0.889 \quad \text{(see equation (9.5a))}$$

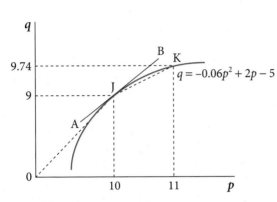

(d) See figure 9.5. Using equation (9.4a), the arc elasticity in (a) is given by the slope of the chord JK (0.74/1) divided by the slope of 0J (9/10). Using equation (9.5a), the point elasticity is given by the slope of the tangent AJB (which we know from (c) above equals 0.12(10) + 2) divided by the slope of 0J (9/10). Because JK is flatter than AJB, the arc elasticity (0.822) is less than the point elasticity (0.889).

Figure 9.5 Answer to example 9.2(d) (not to scale).

Note that in example 9.1, both the arc elasticity and the point elasticity are greater than 1, because in figure 9.4 the chord JK and the tangent AJB are both steeper than 0J. In example 9.2, however, both the arc elasticity and the point elasticity are less than 1, because in figure 9.5 the chord JK and the tangent AJB are both flatter than 0J.

Summary of sections 9.1–9.10

So far in this chapter we have defined the arc elasticity and the point elasticity of supply (see below). The elasticity measures the rate of proportionate change of q with respect to p. A large numerical value of the supply elasticity tells us that the proportionate change in quantity supplied is large relative to the proportionate change in price that induced it. This conveys more useful information than knowledge only of the slope of the supply curve can do. When the proportionate change in quantity supplied is greater than (less than) the proportionate change in price that induced it, the elasticity is greater than (less than) 1. We then say that supply is elastic (inelastic). The boundary case of unit or unitary elasticity occurs when the two proportionate changes are equal.

Definitions of elasticity of supply:

(1) arc elasticity: $E_A^S \equiv \dfrac{\Delta q}{\Delta p}\dfrac{p_0}{q_0}$ \hfill (9.4) repeated

(2) point elasticity: $E^S \equiv \dfrac{dq}{dp}\dfrac{p_0}{q_0}$ \hfill (9.5) repeated

We saw that these are approximately equal to one another for small price changes, and that the point elasticity is preferred for theoretical work.

The Online Resource Centre gives more worked examples of the elasticity of supply, including some graphics exercises. www.oxfordtextbooks.co.uk/orc/renshaw3e/

Progress exercise 9.1

1. Given the supply function: $q = 2p - 5$

(a) Find the arc elasticity of supply when p increases (i) from 5 to 6; (ii) from 50 to 51.

(b) Find the point elasticity (as a function of price) (i) when $p = 5$; (ii) when $p = 6$, and compare your answers with (a) above.

(c) Show that the point elasticity is always greater than 1, but approaches a limiting value of 1 as p approaches infinity. (Assume $2p > 5$.)

(d) Sketch the graph of the supply function, showing how the elasticities may be measured geometrically.

2. Given the supply function: $q = 0.1p^2 + 5p - 10$.

(a) Find the arc elasticity when p increases from 10 to 11, and compare your answer with the point elasticity when $p = 10$. Why are the arc and point elasticities so similar in this case?

(b) Sketch the graph of the supply function, showing how the elasticities may be measured geometrically.

3. Repeat question 2 for the supply function $q = -0.03p^2 + 5p - 10$. (In this case it is necessary to restrict p to being less than 80, since the curve is negatively sloped when $p > 80$, which of course is inappropriate for a supply function.)

9.11 The arc elasticity of demand

We now move on to consider the elasticity of demand. The analysis of the elasticity of demand is closely analogous to that of the elasticity of supply. Therefore we can move a little faster now that the basic principles have been established.

The basic features of the market demand function were examined in sections 4.13 and 8.8. The market demand function is the functional relationship between the quantity of a good which buyers, in aggregate, demand (that is, wish to buy) and the market price. We can write the demand function as

$$q = g(p)$$

where q, the dependent variable, denotes the quantity demanded (measured in some physical unit) and p denotes the purchase price of each unit. We have written $g(p)$ to denote the demand function so as to avoid any danger of confusing it with the supply function $q = f(p)$ examined earlier in this chapter. Economic theory suggests that the function $g(p)$ is normally negatively sloped, because buyers will wish to buy smaller quantities at higher prices. If there is only one (monopoly) supplier of the good, the market demand function is also the demand function for that firm's product. However, under perfect competition a typical firm can sell as much output as it wishes at the ruling market price, and thus faces a horizontal inverse demand function. As with the supply function, we assume that the demand function is smooth and continuous. For more detail on the demand function, see any introductory microeconomics textbook.

Given the market demand function, introductory economics textbooks define the arc elasticity of demand, which we will denote by E_A^D, as:

$$E_A^D \equiv \frac{\text{percentage change in quantity demanded}}{\text{percentage change in price}}$$

By following the same steps as in section 9.3 above, we can re-express this as in rule 9.4:

 RULE 9.4 The arc elasticity of demand

The arc elasticity of demand is defined as

$$E_A^D \equiv \frac{\frac{\Delta q}{q_0}}{\frac{\Delta p}{p_0}} = \frac{\text{proportionate change in quantity demanded}}{\text{proportionate change in price}} \qquad (9.6)$$

or equivalently, since $\frac{\Delta q}{q_0}$ can be written as $\Delta q/q_0$, and $\frac{\Delta p}{p_0}$ can be written as $\Delta p/p_0$:

$$E_A^D \equiv \frac{\Delta q/q_0}{\Delta p/p_0} = \frac{\text{proportionate change in quantity demanded}}{\text{proportionate change in price}} \qquad (9.6a)$$

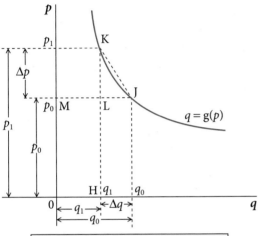

The arc elasticity of demand is:
$$E_A^D \equiv \frac{\Delta q/q_0}{\Delta p/p_0} = \frac{LJ/MJ}{KL/LH}, \text{ or } \frac{\Delta q/\Delta p}{q_0/p_0} = \frac{LJ/KL}{MJ/LH}$$

Figure 9.6 Arc elasticity of demand.

We will often use the notation of equation (9.6a), to save space on the page.

The definition in rule 9.4 is exactly the same as our definition of the arc elasticity of supply (see rule 9.2), except that in E_A^D the superscript D indicates that it refers to a demand function. Analogously to the supply function, the basic idea is to measure the responsiveness of quantity of the good demanded by buyers to a change in the price they have to pay for the good.

Figure 9.6 gives a diagrammatic treatment of the demand elasticity. The graph of the demand function $q = g(p)$ is downward sloping, reflecting the assumption that an increase in price leads to a reduction in quantity demanded. Note that figure 9.6 observes the economist's convention of putting the dependent variable, q, on the *horizontal* axis.

In figure 9.6, suppose buyers are initially at J with price p_0 and quantity demanded q_0. Then we suppose that for some reason the price increases from p_0 to p_1. Buyers respond by reducing the quantity demanded from q_0 to q_1 and thus move to point K on the demand curve.

On the vertical axis we see that the distance KL gives us the change in price, Δp, while the distance LH gives us the initial price, p_0. Thus

$$\frac{KL}{LH} = \frac{\Delta p}{p_0} = \text{proportionate change in price}$$

Similarly, on the horizontal axis, we see that the distance LJ gives us the change in quantity, Δq, while the distance MJ gives us the initial quantity, q_0. Thus

$$\frac{LJ}{MJ} = \frac{\Delta q}{q_0} = \text{proportionate change in quantity supplied}$$

 Hint At this point it is very important to note that, because the demand curve is assumed to be negatively sloped, the quantity demanded falls when the price increases: that is, q_1 lies to the left of q_0 in figure 9.6. Thus $q_1 < q_0$ and therefore $q_1 - q_0$ is negative. This means that Δq, which by definition equals $q_1 - q_0$, is negative.

If we now take the ratio of the two proportionate changes above, we get

$$\frac{\frac{LJ}{MJ}}{\frac{KL}{LH}} = \frac{\frac{\Delta q}{q_0}}{\frac{\Delta p}{p_0}} = E_A^D \quad \text{(see equation (9.6) above)}$$

Note that, because Δq is negative while all the other terms in the equation above are positive, *the elasticity of demand is negative*. This is necessarily the case when we assume, as we normally do, that the market demand function is negatively sloped. It remains equally true if we consider a price reduction instead of an increase, for then it is Δp rather than Δq which is negative. Thus either Δq or Δp must be negative, and equation (9.6) above is therefore always negative.

We can roughly estimate the demand elasticity in figure 9.6. If we use a ruler to measure LJ and MJ, we find that $\frac{LJ}{MJ}$ is about -0.4 (the minus sign reflecting the fact that $\Delta q = $ LJ is negative), while $\frac{KL}{LH}$ is about 0.5 (check this for yourself). Therefore at point J, and for the given price change Δp, the arc elasticity of demand is approximately $\frac{-0.4}{0.5} = -0.8$. This means that the proportionate *reduction* in quantity demanded is about eight-tenths, or 80%, as large as the proportionate *increase* in price.

An alternative formulation

As we did with the supply elasticity in section 9.6 above (see rule 9.2a), we can rearrange equation (9.6) as

$$E_A^D \equiv \frac{\Delta q}{\Delta p} \frac{p_0}{q_0} \tag{9.7}$$

or, equivalently (since multiplying by $\frac{p_0}{q_0}$ is the same thing as dividing by $\frac{q_0}{p_0}$):

$$E_A^D \equiv \frac{\Delta q / \Delta p}{q_0 / p_0} \tag{9.7a}$$

where $\frac{\Delta q}{\Delta p}$ is the difference quotient of the demand function, which is negative because we assume the demand function is downward sloping; and $\frac{q_0}{p_0}$ is the initial ratio of quantity to price.

9.12 Elastic and inelastic demand

Let us now consider more broadly how we should interpret information about the elasticity of demand. The object of the elasticity formula is to provide us with a measure of the responsiveness of quantity to a change in price. The analysis is basically the same as for the elasticity of supply, but the fact that the demand elasticity is negative can be a source of confusion. For this reason many economists prefer to focus on the *absolute value* of the elasticity, written as $|E_A^D|$, where the two vertical bars instruct us to treat whatever lies between them as positive, whether it is actually positive or not (for example, $|-2| = 2$).

At any point on any demand curve, following a price change there are three possible responses of quantity demanded:

(1) The proportionate change in quantity is *larger* in absolute magnitude than the proportionate change in price. For example, suppose that the proportionate price change is +10%, and in response to this, a proportionate change in quantity demanded of −20% occurs. Then, although −20 is less than +10 (since any negative number is less than any positive number), $|-20|$, the absolute magnitude of −20, is larger than 10. Then we have

$$\frac{\Delta p}{p_0} = +10\% = 0.1 \quad \text{and} \quad \frac{\Delta q}{q_0} = -20\% = -0.2$$

So the arc elasticity of demand is

$$E_A^D \equiv \frac{\Delta q/q_0}{\Delta p/p_0} = \frac{-0.2}{0.1} = -2 \quad \text{and therefore } |E_A^D| = 2$$

We interpret $E_A^D = -2$ as meaning that the proportionate change in quantity demanded (−20%) is twice as large in absolute magnitude as the proportionate change in price (+10%), and of opposite sign. In all cases such as this, where the proportionate change in quantity is absolutely larger than the proportionate change in price, we say that demand is *elastic*. In all such cases, $E_A^D < -1$, and therefore $|E_A^D| > 1$. In words, *elastic* demand is when the elasticity is less than −1, and the absolute elasticity is greater than +1.

(2) The proportionate change in quantity is *smaller* in absolute magnitude than the proportionate change in price. For example, suppose again that the price increases by +10%, but now quantity demanded changes by only −5%. Then we have

$$\frac{\Delta p}{p_0} = +10\% = 0.1 \quad \text{and} \quad \frac{\Delta q}{q_0} = -5\% = -0.05$$

So the arc elasticity of demand is

$$E_A^D \equiv \frac{\Delta q/q_0}{\Delta p/p_0} = \frac{-0.05}{0.1} = -\frac{1}{2} \quad \text{and therefore } |E_A^D| = \frac{1}{2}$$

We interpret $E_A^D = -\frac{1}{2}$ as meaning that the proportionate change in quantity demanded (−5%) is half as large in absolute magnitude as the proportionate change in price (+10%), and of opposite sign. In all cases such as this, where the proportionate change in quantity is smaller in absolute magnitude than the proportionate change in price, we say that demand is *inelastic*. In all such cases, $E_A^D > -1$. But note also that since E_A^D is always negative, we also have $E_A^D < 0$. Combining these two inequalities, with inelastic demand we have $-1 < E_A^D < 0$. In words, E_A^D lies between −1 and zero; that is, it is a negative fraction (such as $-\frac{1}{2}$ in this example). In turn, this implies that $|E_A^D|$ is a positive fraction; that is $0 < |E_A^D| < 1$.

(3) The dividing line between the two previous cases is when the proportionate change in quantity demanded is *equal* in absolute magnitude to the change in price. Then, with a price increase of 10% we have

$$\frac{\Delta p}{p_0} = +10\% = 0.1 \quad \text{and} \quad \frac{\Delta q}{q_0} = -10\% = -0.1$$

So the arc elasticity of demand is

$$E_A^D \equiv \frac{\Delta q/q_0}{\Delta p/p_0} = \frac{-0.1}{0.1} = -1 \quad \text{and therefore } |E_A^D| = 1$$

We interpret $E_A^D = -1$, or equivalently $|E_A^D| = 1$, as meaning that the proportionate *reduction* in quantity demanded is equal in absolute size to the proportionate *increase* in price. We say that demand is of *unit elasticity*, or that the demand elasticity is unity.

If this discussion has left you a bit confused, figure 9.7 may help. Figure 9.7(a) shows, on a horizontal number scale, the possible values of the elasticity of demand as we have defined it, E_A^D. Because E_A^D is always negative, these range from 0 to −∞. When $-1 < E_A^D < 0$ (that is, the elasticity lies between zero and −1) we say that demand is inelastic. When $E_A^D < -1$ (that is, the elasticity is less than −1) we say that demand is elastic. The boundary case of unit or unitary elasticity is where $E_A^D = -1$.

Figure 9.7(b) shows, on a horizontal number scale, the possible values of the absolute elasticity of demand, $|E_A^D|$. Because $|E_A^D|$ is always positive, its possible values range from 0 to +∞. When $0 < |E_A^D| < 1$ (that is, the absolute elasticity lies between zero and +1) we say that demand is inelastic. When $|E_A^D| > 1$ (that is, the absolute elasticity is greater than 1), we say that demand is elastic. The boundary case of unit or unitary elasticity is where $|E_A^D| = 1$.

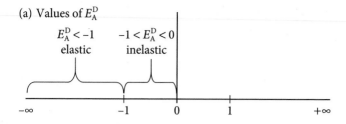

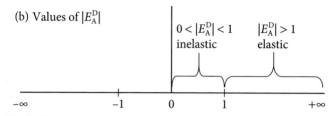

Figure 9.7 Range of values for the demand elasticity and the absolute demand elasticity.

Clearly Figures 9.7(a) and 9.7(b) convey the same information, but in slightly different ways. The existence of these two ways of discussing the demand elasticity, E_A^D and $|E_A^D|$, is inevitably confusing, especially since in any particular discussion it is not always completely clear which definition of demand elasticity the author or speaker is using. This is something that we economists have to live with.

9.13 An alternative definition of demand elasticity

As we have seen above, the fact that demand elasticities are normally negative can be a source of confusion. Discussing the elasticity in terms of its absolute value is one way of trying to avoid this confusion. Another method used by some economists is to redefine the demand elasticity with an extra minus sign, thus ensuring that it is positive. Thus their definition, which we will denote by E_A^{D*}, is

$$E_A^{D*} \equiv -\left(\frac{\Delta q/q_0}{\Delta p/p_0}\right) \equiv -E_A^D$$

where E_A^D is our definition (equation 9.6a). The additional minus sign means that the demand elasticity is now necessarily positive (provided, as always, that the demand function is negatively sloped). This means that E_A^{D*} will always equal $|E_A^D|$, so that the possible values of E_A^{D*} are the same as those of $|E_A^D|$ in figure 9.7 and table 9.1 below.

> **Summary of sections 9.11–9.13**
>
> In section 9.11 we introduced the concept of the arc elasticity of demand, and in section 9.12 we examined the three possible cases of demand elasticity (elastic, inelastic, and unit elasticity) with examples showing the associated numerical values of the elasticity, and the absolute elasticity, in each case. Our conclusions are summarized in table 9.1. Finally, in section 9.13 we noted an alternative definition of demand elasticity used by some economists, which is equivalent to the absolute value of our definition.

Table 9.1 Interpreting demand elasticity.

Value of arc elasticity, E_A^D	Implied absolute elasticity, $|E_A^D|$	Interpretation	Terminology
$E_A^D < -1$ (such as -2 as in case (1) above)	$|E_A^D| > 1$	Proportionate change in quantity is *larger* in absolute magnitude than the proportionate change in price, and of opposite sign	Demand is elastic
$E_A^D > -1$ (such as $-1/2$ as in case (2) above) but also $E_A^D < 0$ (because D-curve has negative slope); so $-1 < E_A^D < 0$	$|E_A^D| < 1$ but also $|E_A^D| > 0$ (because D-curve has negative slope); so $0 < |E_A^D| < 1$	Proportionate change in quantity is *smaller* in absolute magnitude than the proportionate change in price, and of opposite sign	Demand is inelastic
$E_A^D = -1$ (as in case (3) above)	$|E_A^D| = 1$	Proportionate change in quantity is *equal* in absolute magnitude to the proportionate change in price, and of opposite sign	Demand is of unit elasticity or demand elasticity is unity

9.14 The point elasticity of demand

Let us now move on to examine the *point* elasticity of demand. By analogy with rule 9.3 for supply elasticity, this is defined as follows:

RULE 9.5 **The point elasticity of demand**

The point elasticity of demand, E^D, is defined as

$$E^D \equiv \lim_{\Delta p \to 0} \frac{\Delta q}{\Delta p} \frac{p_0}{q_0} \equiv \frac{\mathrm{d}q}{\mathrm{d}p} \frac{p_0}{q_0} \qquad (9.8)$$

or, equivalently (since multiplying by $\frac{p_0}{q_0}$ is the same thing as dividing by $\frac{q_0}{p_0}$):

$$E^D \equiv \frac{\mathrm{d}q/\mathrm{d}p}{q_0/p_0} \qquad (9.8a)$$

where $\frac{\mathrm{d}q}{\mathrm{d}p}$ is the derivative of the demand function and q_0 and p_0 are the coordinates of the point on the demand curve at which the elasticity is being measured.

Looking at equations (9.8) and (9.8a), we see that at any given point on the demand curve, the value of E^D is determined by p_0 and q_0 (the coordinates of that point), and by $\frac{\mathrm{d}q}{\mathrm{d}p}$, the slope of the tangent at that point. The point elasticity of demand at any point is thus unique and unambiguous. This is perfectly analogous to the point elasticity of supply (see sections 9.8 and 9.9 above).

In figure 9.8 we have redrawn the demand function from figure 9.6, now with q on the vertical axis. From this graph we can infer the point elasticity at J. The derivative $\frac{\mathrm{d}q}{\mathrm{d}p}$ is given by the slope of the tangent AJB, while $\frac{q_0}{p_0}$ is given by the slope of the broken line (the ray) from the origin, 0J, which has a slope given by $\frac{\mathrm{JM}}{0\mathrm{M}}$. From equation (9.8a), the point elasticity is the ratio of these two slopes.

It is interesting to compare this with the arc elasticity of demand at J, which we can also measure. If we assume a price increase of Δp, buyers move from J to K and quantity falls from q_0 to q_1. The difference quotient is then

$$\frac{\Delta q}{\Delta p} = \frac{\mathrm{JL}}{\mathrm{LK}} \qquad \text{which equals the slope of the chord JK. In addition:}$$

$$\frac{q_0}{p_0} = \text{slope of } 0J = \frac{JM}{0M} \quad \text{(the same as for the point elasticity above)}$$

So, using equation (9.7b), the arc elasticity is

$$E_A^D \equiv \frac{\Delta q / \Delta p}{q_0 / p_0} = \frac{JL/LK}{JM/0M}$$

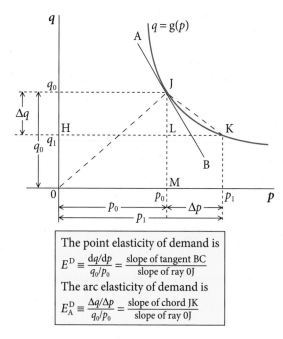

The point elasticity of demand is
$$E^D \equiv \frac{dq/dp}{q_0/p_0} = \frac{\text{slope of tangent } BC}{\text{slope of ray } 0J}$$
The arc elasticity of demand is
$$E_A^D \equiv \frac{\Delta q / \Delta p}{q_0 / p_0} = \frac{\text{slope of chord } JK}{\text{slope of ray } 0J}$$

Figure 9.8 Comparison of point and arc elasticity of demand.

Comparing the point and arc elasticities, we see that both have the same denominator, $\frac{q_0}{p_0} = \frac{JM}{0M}$. They differ in that the point elasticity has $\frac{dq}{dp}$ (the slope of the tangent AJB) as its numerator; while the arc elasticity has the difference quotient, $\frac{\Delta q}{\Delta p}$ (the slope of the chord JK) as its numerator.

Thus we can see that, in general, these two slopes are not equal and thus the arc elasticity will always differ from the point elasticity unless the demand function is linear between J and K. From figure 9.8 we can see that this difference between the point and arc elasticities is greater, the larger is Δp and the greater is the curvature of the demand function in the neighbourhood of J. (These conclusions apply equally to the supply elasticity; see figure 9.3.)

In figure 9.8 the arc elasticity is smaller in absolute magnitude than the point elasticity because the slope of the chord JK is flatter than that of the tangent AJB. However, if the demand function were concave rather than convex in the vicinity of J, the arc elasticity would be greater in absolute magnitude than the point elasticity. A price increase rather than a reduction, starting from J, would also give a different result.

A key point that we can also see in figure 9.8 is that, as Δp gets smaller, K moves up the demand curve towards J and so $\frac{\Delta q}{\Delta p}$ approaches $\frac{dq}{dp}$. The arc elasticity approaches the point elasticity as Δp approaches zero.

9.15 Reconciling the arc and point demand elasticities

The relative merits of the arc elasticity and the point elasticity of demand are perfectly analogous to those of the arc and point elasticities of supply (see sections 9.8 and 9.9 above), and therefore need little discussion. The arc elasticity of demand is imprecise because its value at any point on the demand curve depends on the size of the assumed price change, Δp; and because its value is different for a price increase than for a price reduction. The point elasticity is precise in the sense that, for any given demand function, its value at any point is uniquely determined by the coordinates of that point. For this reason the point elasticity is invariably preferred for use in theoretical work, although it is only strictly valid for indefinitely small price changes. However, the point elasticity is so much easier to calculate and manipulate that it is often used to analyse real-world situations where the price change, Δp, is sufficiently small that the inevitable error is small enough to be ignored. In a real-world situation in which price changes are too large for this error to be overlooked, the arc elasticity is the appropriate definition to use. (See example 9.4 below for an illustration of this point.)

9.16 Worked examples on demand elasticity

EXAMPLE 9.3

Given the demand function: $q = -5p + 100$

(a) Find the arc elasticity of demand when p increases (i) from 5 to 6; (ii) from 13 to 14.

(b) Find the point elasticity (i) when $p = 5$; (ii) when $p = 13$.

(c) Sketch the demand function, indicating the solutions to (a) and (b) above. Why is demand elastic when p is high and inelastic when p is low, in this example?

Answer: (a)

(i) Given $p_0 = 5$, $q_0 = -5(5) + 100 = 75$

Given $p_1 = 6$, $q_1 = -5(6) + 100 = 70$

Therefore $\Delta p \equiv p_1 - p_0 = 6 - 5 = 1$ and $\Delta q \equiv q_1 - q_0 = 70 - 75 = -5$

Therefore the arc elasticity of demand between $p_0 = 5$ and $p_1 = 6$ is

$$E_A^D \equiv \frac{\Delta q / q_0}{\Delta p / p_0} = \frac{-5/75}{1/5} = -0.333 \quad \text{(see equation (9.6a))}$$

Alternatively, the arc elasticity may be calculated as

$$E_A^D \equiv \frac{\Delta q / \Delta p}{q_0 / p_0} = \frac{-5/1}{75/5} = -0.333 \text{ (as above)} \quad \text{(see equation (9.7a))}$$

(ii) Given $p_0 = 13$, $q_0 = -5(13) + 100 = 35$

Given $p_1 = 14$, $q_1 = -5(14) + 100 = 30$

Using the same method as in (i) above gives

$$E_A^D \equiv \frac{\Delta q / q_0}{\Delta p / p_0} = \frac{-5/35}{1/13} = -1.857$$

(b) (i) Given $q = -5p + 100$, $\frac{dq}{dp} = -5$. So the point elasticity when $p = 5$, $q = 75$ is

$$E^D \equiv \frac{dq/dp}{q_0/p_0} = \frac{-5}{75/5} = -0.333 \quad \text{(see equation (9.8a))}$$

(ii) Repeating (i) with $p = 13$, $q = 35$ gives

$$E^D \equiv \frac{dq/dp}{q_0/p_0} = \frac{-5}{35/13} = -1.857$$

Note from (a) and (b) that in this example the arc elasticity and the point elasticity are equal at any point on the demand function. This is because the demand function is linear, for then $\frac{dq}{dp} = \frac{\Delta q}{\Delta p}$ and hence $E^D = E_A^D$.

(c) See figure 9.9. As noted above, a key feature of this example is that the demand function is linear, and therefore $\frac{dq}{dp} = \frac{\Delta q}{\Delta p}$ and hence $E^D = E_A^D$.

Using equation (9.7a), at point J both the arc elasticity and the point elasticity are given by the slope of JK divided by the slope of 0J. Similarly, at L both the arc elasticity and the point elasticity are given by the slope of LM divided by the slope of 0L. The elasticity at L is larger in absolute value than at J because the slope of 0L is less than the slope of 0J.

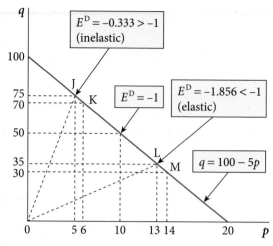

Figure 9.9 Answer to example 9.3(c) (not to scale).

Another way of grasping this is to see that, at J, the price increase from 5 to 6 is large in proportionate terms: that is, $\frac{\Delta p}{p_0}$ is large. Conversely, at J the quantity reduction from 75 to 70 is small in proportionate terms: that is, $\frac{\Delta q}{q_0}$ is small (ignoring its negative sign). Thus the arc elasticity, which is $\frac{\Delta q/q_0}{\Delta p/p_0}$, has a small numerator and a large denominator, and is therefore small in absolute value. At L the reverse is true: the price increase from 13 to 14 is small in proportionate terms, while the quantity reduction from 35 to 30 is large in proportionate terms (ignoring its negative sign). Thus the elasticity is large in absolute value. This example also illustrates the point that the elasticity of a function does not depend merely on its slope, as the slope is constant in this case.

Note in figure 9.9 that when $p = 10$, $q = 50$ and therefore $E^D = \frac{-5}{50/10} = -1$.

EXAMPLE 9.4

Given the demand function: $q = 130p^{-1} - 0.2p + 5$

(a) Find the arc elasticity of demand when p increases from 9 to 9.1.

(b) Find the point elasticity of demand when $p = 9$.

(c) When p increases from 9 to 9.1, what size of error arises if we use the point elasticity as an approximation to the arc elasticity?

(d) Draw a sketch graph of the demand function.

Answer:

(a) Given $p_0 = 9$, $q_0 = 130(9)^{-1} - 0.2(9) + 5 = 17.644$

Given $p_1 = 9.1$, $q_1 = 130(9.1)^{-1} - 0.2(9.1) + 5 = 17.465$

Therefore $\Delta p \equiv p_1 - p_0 = 9.1 - 9 = 0.1$, and $\Delta q \equiv q_1 - q_0 = 17.465 - 17.644 = -0.179$.

Therefore the arc elasticity of demand between $p_0 = 9$ and $p_1 = 9.1$ is

$$E_A^D \equiv \frac{\Delta q/q_0}{\Delta p/p_0} = \frac{-0.179/17.644}{0.1/9} = -0.913 \quad \text{(see equation (9.6a))}$$

Alternatively, the arc elasticity may be calculated as

$$E_A^D \equiv \frac{\Delta q/\Delta p}{q_0/p_0} = \frac{-0.179/0.1}{17.644/9} = -0.913 \quad \text{(as above)} \quad \text{(see equation (9.7a))}$$

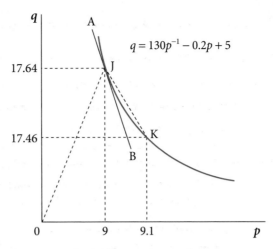

Figure 9.10 Answer to example 9.4(d) (not to scale).

(b) Given $q = 130p^{-1} - 0.2p + 5$, $\frac{dq}{dp} = -130p^{-2} - 0.2$

So the point elasticity when $p = 9$ is

$$E^D \equiv \frac{dq/dp}{q_0/p_0} = \frac{-130(9)^{-2} - 0.2}{17.644/9} = -0.921 \quad (\text{see equation (9.8a)})$$

(c) From (b) and (a) above, the error is $E^D - E^D_A = -0.921 - (-0.913) = -0.008$. In most cases, we would feel that this error was negligibly small, but it's important to keep in mind that the error increases with the size of the price change and with the curvature of the demand function. This is apparent from inspection of figure 9.10.

(d) See figure 9.10. Because the slope of the tangent AJB is steeper than the chord JK, the point elasticity at J exceeds in absolute value the arc elasticity between J and K. But the difference is very small because the assumed price change is only $\Delta p = 0.1$, which means that K is very close to J.

9.17 Two simplifications

Before moving on to some further applications of the elasticity concept, we take this opportunity to simplify our discussion in two ways:

(1) For reasons given in sections 9.9 and 9.15, economists prefer to use the point elasticity rather than the arc elasticity, except very occasionally when analysing large price changes. Therefore in the remainder of this book we will focus entirely on the point elasticity, and we shall refer to it simply as 'the elasticity' (of supply, demand or whatever the case may be). However, there are some questions on arc elasticity in progress exercise 9.2. These have been included to familiarize you with the concept even though we will make no further use of it.

(2) As noted in section 9.9, it is customary to simplify the notation for the point elasticity definition by dropping the zero subscripts on p_0 and q_0. Thus from now on we will write p and q in the point elasticity expressions to denote the coordinates of the point at which we are examining the elasticity: for example, point J in figure 9.8 or 9.10. We will retain the subscripts for the arc elasticities in progress exercise 9.2.

With the simplifications above, the definitions of supply and demand elasticity that we shall use in the remainder of this book are as follows:

For any supply function, $q = f(p)$, with derivative $\frac{dq}{dp} = f'(p)$, the point elasticity of supply is

$$E^S \equiv \frac{dq}{dp}\frac{p}{q} \qquad\qquad\qquad \text{(9.5) repeated}$$

which can be rearranged as

$$E^S \equiv \frac{dq/dp}{q/p} \qquad\qquad\qquad \text{(9.5a) repeated}$$

For any demand function, $q = g(p)$, with derivative $\frac{dq}{dp} = g'(p)$, the point elasticity of demand is

$$E^D \equiv \frac{dq}{dp}\frac{p}{q} \qquad\qquad\qquad \text{(9.8) repeated}$$

which can be rearranged as

$$E^D \equiv \frac{dq/dp}{q/p} \qquad\qquad\qquad \text{(9.8a) repeated}$$

We will also make occasional reference to $|E^D|$, the absolute value of the demand elasticity.

Progress exercise 9.2

1. Given the demand function: $q = -2p + 160$

 (a) Find the arc elasticity of demand when p increases (i) from 16 to 17; (ii) from 48 to 49.

 (b) Find the point elasticity (i) when $p = 16$; (ii) when $p = 48$.

 (c) Sketch the demand function, indicating the solutions to (a) and (b) above.

 (d) Why is demand elastic when p is high and inelastic when p is low, in this example?

2. Given the demand function: $q = -4p^2 - 1.5p + 792$

 (a) Find the arc elasticity of demand when p increases (i) from 5 to 6; (ii) from 12 to 13.

 (b) Find the point elasticity (i) when $p = 5$; (ii) when $p = 12$.

 (c) Sketch the demand function, indicating the solutions to (a) and (b) above.

 (d) For what *mathematical* reasons does demand become more elastic (that is, E^D decrease or $|E^D|$ increase) as p increases, in this example?

3. Given the inverse demand function: $p = \frac{50}{q} - 2$

 (a) Find the demand function (that is, the inverse of the inverse demand function).

 (b) Find the point elasticity of demand (either as a function of p or as a function of q) and show that demand is inelastic at any price.

 (c) Sketch the demand function. Does the graph help us to see why demand is always inelastic? (Hint: section 5.5 might help you with the graph.)

4. Repeat question 3 for the inverse demand function: $p = \frac{50}{q} + 2$. In this case, show that demand is elastic at any price.

5. Repeat question 3 for the inverse demand function: $p = \frac{50}{q}$. In this case, show that demand has unit elasticity at any price.

9.18 Marginal revenue and the elasticity of demand

In section 8.11 we examined the relationship between the demand function for a good (assumed to be negatively sloped) and the associated total revenue and marginal revenue functions. We saw there that, at any point on the demand function, if marginal revenue is positive this means that a small increase in quantity sold (resulting from a small price reduction) leads to an *increase* in total revenue. Conversely, at any point on the demand function, if marginal revenue is negative this means that a small increase in quantity sold (again, resulting from a small price reduction) leads to a *decrease* in total revenue. The boundary between these two cases occurs at a point on the demand function where marginal revenue is zero. At this point, total revenue is at its maximum.

We now want to derive an important relationship between marginal revenue and the (point) elasticity of demand. We will first develop a general rule, then illustrate the rule by applying it to the inverse demand function $p = -0.5q + 100$ that we used in section 8.11.

We know from section 8.8 that, for any inverse demand function $p = f(q)$, total revenue (TR) is given by

$$TR = pq \tag{9.9}$$

Here, p and q are not just plucked out of the air. Any combination of p and q that appears in the TR function must satisfy the inverse demand function $p = f(q)$ (or, equivalently, must satisfy the demand function itself). Therefore we can substitute $f(q)$ in place of p in equation 9.9 and thereby eliminate p, resulting in

$$TR = f(q)q \tag{9.10}$$

The right-hand side is now a product of $f(q)$ and q. As both are functions of q, we can apply the product rule of differentiation (chapter 6, rule D6) to equation 9.10. If we let $u = f(q)$ and $v = q$, then $\frac{du}{dq} = f'(q)$ and $\frac{dv}{dq} = 1$. Using rule D6 we then get:

$$\frac{dTR}{dq} = u\frac{dv}{dq} + v\frac{du}{dq} = f(q) + qf'(q) \tag{9.11}$$

From the inverse demand function we have $p = f(q)$ and therefore $\frac{dp}{dq} = f'(q)$. From the definition of marginal revenue (rule 8.2) we have $\frac{dTR}{dq} = MR$. After substituting these three pieces of information into equation 9.11 it becomes

$$MR = p + q\frac{dp}{dq}$$

On the right-hand side we can take p outside a pair of brackets, and the equation above then becomes

$$MR = p\left(1 + \frac{q}{p}\frac{dp}{dq}\right) \tag{9.12}$$

(Be sure to check the algebra for yourself.)

From the definition of the elasticity of demand (rule 9.5), we have $E^D \equiv \frac{p}{q}\frac{dq}{dp}$, so $\frac{q}{p}\frac{dp}{dq} = \frac{1}{E^D}$. Substituting this into equation 9.12, we get rule 9.6:

RULE 9.6 Marginal revenue, price, and the elasticity of demand

The relationship between marginal revenue, price, and the elasticity of demand is

$$MR = p\left(1 + \frac{1}{E^D}\right)$$

EXAMPLE 9.5

In this example we will check that rule 9.6 applies to the demand function $q = -2p + 200$ that we used in section 8.8. We will first find $p(1 + \frac{1}{E^D})$, then show that it equals MR, thus verifying rule 9.6.

From the demand function, we have $\frac{dq}{dp} = -2$, and therefore $E^D = \frac{p}{q} \frac{dq}{dp} = -\frac{2p}{q}$.

Therefore: $p\left(1 + \dfrac{1}{E^D}\right) = p\left(1 - \dfrac{q}{2p}\right) = p - 0.5q$ (9.13)

Given the demand function above, we can quickly obtain the inverse demand function as $p = -0.5q + 100$. Using this to eliminate p from equation (9.13), we have

$$p\left(1 + \dfrac{1}{E^D}\right) = p - 0.5q = -0.5q + 100 - 0.5q = -q + 100$$ (9.14)

But from section 8.10, we have $MR = -q + 100$. Combining this with equation (9.14), we arrive at $MR = -q + 100 = p(1 + \frac{1}{E^D})$, thus verifying rule 9.6.

Let us now explore what rule 9.6 is telling us. Suppose demand is elastic, meaning that E^D is less than -1. We'll assume a specific value; say, $E^D = -2$. Substituting this into rule 9.6, we get

$$MR = p\left(1 - \dfrac{1}{2}\right) = \dfrac{1}{2}p$$

Thus MR is positive provided p is positive, which we assume it to be. Since MR measures the slope of the total revenue curve, we must therefore be somewhere on the upward-sloping section of the total revenue curve. This means that a small price reduction, which leads to an increase in quantity, will *increase* total revenue. Although we have shown this only for $E^D = -2$, it is easy to see that whenever demand is elastic, so that $E^D < -1$, the right-hand side of rule 9.6 is positive, and so therefore is MR.

Suppose now that demand is inelastic, meaning that E^D lies between -1 and zero; say, $E^D = -\frac{1}{2}$. Substituting this into rule 9.6, we get

$$MR = p(1 - 2) = -p$$

Thus MR is now negative, and by the same reasoning as above we infer that we are on the downward-sloping section of the total revenue curve; hence a price reduction, which leads to an increase in quantity, will *decrease* total revenue.

The boundary between these two cases is when $E^D = -1$. It is then easy to see that the right-hand side of rule 9.6 equals zero, and so therefore does MR. We are at the peak of the total revenue curve (provided the second order condition for a maximum is satisfied, which we can normally assume is the case). Any price change, whether positive or negative, decreases TR.

EXAMPLE 9.6

Figure 9.11 may help you to see more clearly the relationship we have just derived between the elasticity of demand, marginal revenue, and total revenue, in the case of a linear demand function. Figure 9.11 repeats figure 8.11 from chapter 8. The demand function is $q = -2p + 200$, which we also used in example 9.5. Panel (a) of the figure indicates that demand is elastic in the upper section of the inverse demand function, where MR is positive and the TR curve (in panel (b) of the figure) is upward sloping. Similarly, demand is inelastic in the lower section of the inverse demand function, where MR is negative and the TR curve is downward sloping. The boundary point, at $p = 50$, $q = 100$, is where $E^D = -1$ (the demand elasticity is unity) and $MR = 0$ because TR is at its maximum.

Let us check out the algebra to see whether the information in figure 9.11 concerning the demand elasticity is correct.

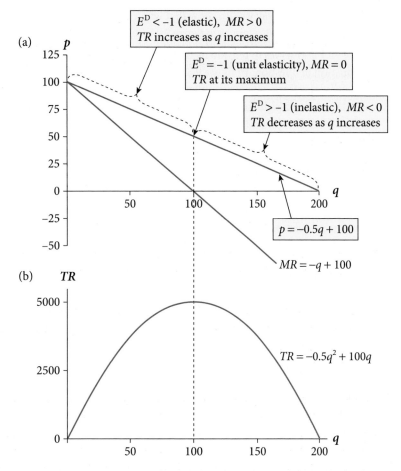

Figure 9.11 Elasticity of demand under monopoly, *TR* and *MR*, assuming a linear demand function.

Our first step is to find the elasticity of demand. Given the demand function

$q = 200 - 2p$ we have $\dfrac{dq}{dp} = -2$, therefore

$$E^D = \frac{p}{q}\frac{dq}{dp} = -\frac{2p}{q} \tag{9.15}$$

From the demand function, we can get the inverse demand function as $p = -0.5q + 100$. We use this to substitute for p in equation (9.15), and get

$$E^D = \frac{p}{q}\frac{dq}{dp} = \frac{-2(-0.5q + 100)}{q} = \frac{q - 200}{q} \tag{9.16}$$

Equation (9.16) gives the demand elasticity as a function of q (p having been substituted out). By definition, when demand is elastic $E^D < -1$. From equation (9.16), $E^D < -1$ implies

$$\frac{q - 200}{q} < -1$$

We can multiply both sides of this by q (which does not reverse the inequality since $q > 0$) and get

$$q - 200 < -q$$

If we add 200 and also add q on both sides, this becomes $2q < 200$, from which $q < 100$.

So we have shown that demand is elastic when $q < 100$, consistent with figure 9.11.

By repeating the steps above with the inequality sign reversed, we can similarly find that when $q > 100$, demand is inelastic ($E^D > -1$), also as shown in figure 9.11.

Finally, in the same way we can show that when $q = 100$, demand has unit elasticity ($E^D = -1$) and $MR = 0$, again as in figure 9.11.

Figure 9.11 also serves to remind us that elasticity does not depend only on the slope of the function in question. In figure 9.11 the demand function has a constant slope, but its elasticity varies from $-\infty$ at the top left-hand end of the demand function (where $q = 0$) to zero at the bottom right-hand end (where $p = 0$). Check that you understand why this is so. Any linear function has an elasticity that varies as we move along it, as is discussed further below.

Summary

Let us collect together our findings from this important section of the book. We have shown that the general relationship between marginal revenue (MR), price (p), and the elasticity of demand (E^D) is given by rule 9.6:

$$MR = p\left(1 + \frac{1}{E^D}\right)$$

The relationship between MR and E^D that is implied by this key equation is summarized in Table 9.2. It is assumed that p is positive and that the demand function is negatively sloped. Be sure to check for yourself, by means of numerical examples, that you fully understand the implications of rule 9.6 and table 9.2.

Table 9.2 Relation between marginal revenue and demand elasticity implied by rule 9.6.

Marginal revenue (MR)	Elasticity of demand (E^D)	Absolute elasticity $|E^D|$	Economic significance
$MR > 0$	$E^D < 1$ (that is, *elastic* demand)	$|E^D| > 1$	Small price reduction increases total revenue
$MR = 0$	$E^D = -1$ (that is, *unit* elasticity of demand)	$|E^D| = 1$	Total revenue is at its maximum*
$MR < 0$	$E^D > -1$ (that is, *inelastic* demand) (Also, $E^D < 0$ because the demand function is negatively sloped. So when $MR < 0$, we have $-1 < E^D < 0$)	$|E^D| < 1$	Small price reduction reduces total revenue

* Subject to second order condition for a maximum being satisfied.

9.19 The elasticity of demand under perfect competition

Up to this point we have assumed that the demand function has a negative slope. The important exception to this is the demand function for the product of a single firm operating under conditions of perfect competition (see section 8.12). In this case, although the *industry* demand curve (that is, the total demand for the product) may be downward sloping, each firm is assumed to be so small that it is able to sell as much output as it wishes at the ruling market price, which we denote by $\bar{p}$.

Figure 9.12 shows the inverse demand function implied by this assumption. Because this is the *inverse* demand function, p is the dependent variable and therefore appears on the vertical

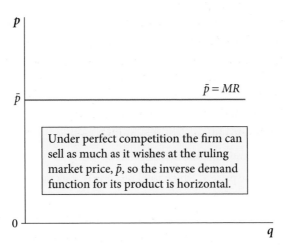

Figure 9.12 Inverse demand function under perfect competition.

axis. The inverse demand function is horizontal and its equation is $p = \bar{p}$, indicating that the firm can sell any quantity it chooses at the price $\bar{p}$. The slope of any horizontal line is zero, so the slope of this inverse demand function is $\frac{dp}{dq} = 0$.

The problem with this case is that it implies a discontinuous relationship between q and p. When $p = \bar{p}$ the quantity demanded, q, is whatever the firm chooses to sell, so q is indeterminate without further information. When $p > \bar{p}$ the firm sells nothing, so q jumps to zero. When $p < \bar{p}$, q is indefinitely large as the firm attracts the whole market demand. Thus q, quantity demanded, is not a smooth and continuous function of p, and therefore the derivative $\frac{dq}{dp}$ does not exist. This means that the elasticity of demand, $E^D = \frac{dq}{dp}\frac{p}{q}$, does not exist either. This is inconvenient because we want to be able to say something about the elasticity of demand for the firm's product under perfect competition.

A way of working round this problem is to view the horizontal inverse demand function of figure 9.12 as a limiting case. We can imagine an initial situation in which each firm's product is slightly different, in the eyes of buyers, from other firms' products. Therefore the firm does not lose *all* of its customers if its price exceeds $\bar{p}$; nor does it attract the *whole* market demand if its price is below $\bar{p}$. The firm's inverse demand function therefore has a small negative slope. Then we imagine the product becoming increasingly homogeneous across firms, as a result of which the inverse demand curve becomes flatter and flatter. As the inverse demand function becomes flatter, its slope, $\frac{dp}{dq}$, approaches zero. As $\frac{dp}{dq}$ approaches zero, $\frac{dq}{dp}$, the slope of the demand function itself (which is negative) approaches $-\infty$. (This is an example of the inverse function rule of differentiation, which says that $\frac{dq}{dp} = \frac{1}{(dp/dq)}$; see chapter 6, rule D8.) And as $\frac{dq}{dp}$ approaches $-\infty$, so also does $E^D = \frac{dq}{dp}\frac{p}{q}$, since p and q both remain finite.

To summarize, we can say that the horizontal inverse demand function is a limiting case which is approached as $\frac{dp}{dq}$ approaches zero. And as $\frac{dp}{dq}$ approaches zero, so the demand elasticity approaches a limiting value of $-\infty$.

Therefore, looking at rule 9.6, with $p = \bar{p}$:

$$MR = \bar{p}\left(1 + \frac{1}{E^D}\right)$$

we see that as E^D approaches $-\infty$, so $\frac{1}{E^D}$ approaches zero, and therefore MR approaches $\bar{p}$. Given this, we can say that $MR = \bar{p}$ is a limiting value. Figure 9.12 shows this limiting value. The MR function and the inverse demand function coincide: one sits on top of the other. As we saw in section 8.12, the intuition here is that the firm does not need to lower its price in order to sell one more unit of output, so the marginal revenue (the *net* addition to total revenue from selling that additional unit) is simply its price. This limiting case is the one typically analysed in economic theory, although, because it is a limiting case, it can never be observed in the real world.

9.20 Worked examples on demand elasticity and marginal revenue

EXAMPLE 9.7

This example is a continuation of example 9.3 above.

Given the demand function: $q = -5p + 100$

(a) (i) Find the price at which total revenue is at its maximum.

(ii) Show that demand has unitary elasticity at this price. Show also that demand is elastic at higher prices and inelastic at lower prices.

(b) Draw a sketch graph of the inverse demand function, marginal revenue, and total revenue functions.

Answer:

(a) (i) We find maximum total revenue as follows. Given the demand function $q = -5p + 100$, we rearrange this as $p = 20 - 0.2q$. This is the inverse demand function (see section 8.8 if you need to revise this point). We then take the definition of total revenue, $TR \equiv pq$, and substitute for p from the inverse demand function. This gives us

$$TR \equiv pq = (20 - 0.2q)q = 20q - 0.2q^2$$

This gives us total revenue as a function of q alone. Next, we differentiate this to obtain marginal revenue, as

$$MR \equiv \frac{dTR}{dq} = 20 - 0.4q$$

Finally, we set MR equal to zero and solve the resulting equation for q:

$$MR \equiv \frac{dTR}{dq} = 20 - 0.4q = 0 \quad \text{with solution } q = 50.$$

The second order condition for a maximum at $q = 50$ is that the second derivative be negative at this point. Here $\frac{d^2TR}{dq^2} = -0.4$, which is negative for any value of q and therefore for $q = 50$ as required.

When $q = 50$, $p = 20 - 0.2(50) = 10$, so maximum TR is $TR \equiv pq = 10(50) = 500$.

(ii) When $q = 50$, $p = 10$, and given $\frac{dq}{dp} = -5$, the elasticity of demand is

$$E^D \equiv \frac{dq/dp}{q/p} = \frac{-5}{50/10} = -1 \quad \text{(as the question asks us to show)}$$

To show that demand is elastic (that is, $E^D < 1$) when $p > 10$, we have from above

$$E^D \equiv \frac{dq/dp}{q/p} = \frac{-5}{q/p} = \frac{-5p}{q} = \frac{-5p}{100 - 5p} \quad \text{(using demand function } q = -5p + 100)$$

By dividing numerator and denominator by -5, this simplifies to

$$E^D = \frac{p}{p - 20}$$

When demand is elastic we have $E^D < -1$, which implies

$$\frac{p}{p - 20} < -1$$

We can assume $p < 20$, since, if $p > 20$, in the demand function $q = 100 - 5p$ we then have $q < 0$, which is impossible. Given $p < 20$, we have $p - 20 < 0$. This means that when we multiply both sides of the inequality above by $p - 20$ the direction of the inequality is reversed. This multiplication therefore gives

$$p > -(p - 20) = -p + 20$$

Adding p to both sides and dividing by 2 gives

$$p > 10$$

So we have shown that demand is elastic ($E^D < -1$ or $|E^D| > 1$) when $p > 10$. To show that demand is inelastic ($E^D > 1$ or $|E^D| < 1$) when $p < 10$, we simply repeat the above but with the direction of the inequality reversed. That is, our first line is

$$\frac{p}{p - 20} > -1$$

From here, following the same steps as above leads us to $p < 10$. That is, demand is inelastic ($E^D > -1$ or $|E^D| < 1$) when $p < 10$.

(b) It is unnecessary to draw these graphs as they would be identical in general shape to figure 9.11, the only difference being in parameter values and hence solution values. In this example the inverse demand function is $p = 20 - 0.2q$ and, as we found above, at the value of q at which the total revenue curve reaches its maximum, marginal revenue is zero and the elasticity of demand is unity: that is $q = 50$ rather than $q = 100$ as in figure 9.11. The associated price is $p = 10$ and maximized total revenue is $TR = 500$.

EXAMPLE 9.8

This example is a continuation of example 9.4 above.

Given the demand function: $q = 130p^{-1} - 0.2p + 5$

(a) (i) Find the price, p^*, at which total revenue is at its maximum.

(ii) Show that demand is elastic when $p > p^*$ and inelastic when $p < p^*$.

(b) Draw a sketch graph of the demand function and the total revenue function (as a function of p).

Answer: (a)

(i) There is a problem in answering this. In example 9.7, we found maximum total revenue by the conventional route. Starting from the definition of total revenue, $TR \equiv pq$, we used the inverse demand function to substitute for p and thus found an expression for total revenue, TR, as a function of q. Then we differentiated this to obtain marginal revenue, MR, defined as $MR \equiv \frac{dTR}{dq}$. Finally we set MR equal to zero and solved the resulting equation for q.

In the present example this method won't work because the inverse demand function cannot easily be found. So we have to use a different approach. We start from the definition of total revenue, $TR \equiv pq$. Then we substitute for q (rather than p) in this expression, using the given demand function $q = 130p^{-1} - 0.2p + 5$. This gives

$$TR \equiv pq = p(130p^{-1} - 0.2p + 5) = 130 - 0.2p^2 + 5p$$

To find the maximum we now find the derivative of this function, set it equal to zero, and solve

$$\frac{dTR}{dp} = -0.4p + 5 = 0 \qquad \text{with solution } p = \frac{-5}{-0.4} = 12.5$$

This is a maximum because $\frac{d^2TR}{dp^2} = -0.4$ (that is, negative).

Note, though, that $\frac{dTR}{dp}$ is not MR, since MR is defined as $\frac{dTR}{dq}$.

When $p = 12.5$, $q = 130p^{-1} - 0.2p + 5 = 12.9$, so maximum TR is

$$TR \equiv pq = 12.5(12.9) = 161.25.$$

(ii) Demand is elastic when

$$E^D \equiv \frac{p}{q}\frac{dq}{dp} = \frac{p(-130p^{-2} - 0.2)}{130p^{-1} - 0.2p + 5} < -1$$

This rearranges as

$$-130p^{-1} - 0.2p < -(130p^{-1} - 0.2p + 5)$$

from which $-0.4p < -5$, from which $p > \frac{-5}{-0.4} = 12.5$ (where of course $p^* = 12.5$).
Note that dividing both sides by -0.4 reverses the inequality sign. So demand is
elastic when $p > 12.5$. By repeating the above steps with the reverse inequality, we
can show that demand is inelastic when $p < 12.5$ and finally that demand elasticity
is unity ($E^D = -1$ or $|E^D| = 1$) when $p = 12.5$, the price that maximizes total
revenue. The corresponding values of q can easily be found from the demand
function.

(b) See figure 9.13. Comparing this figure with figure 9.11, we see that in this case we
have the graph of the demand function (rather than the inverse demand function)
and therefore have p (rather than q) on the horizontal axis. Similarly, we have both
TR and $\frac{dTR}{dp}$ as functions of p rather than q. However, this makes no fundamental
difference as we find maximum TR where $\frac{dTR}{dp} = 0$, giving $p = 12.5$. At this point the
demand elasticity is unity. When $p > 12.5$, demand is elastic, meaning that a price
reduction increases total revenue (and a price increase reduces it). When $p < 12.5$,
demand is inelastic, meaning that a price reduction reduces total revenue (and a
price increase increases it).

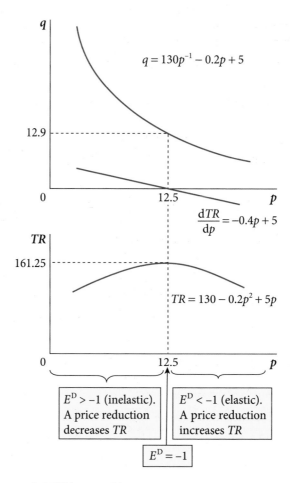

Figure 9.13 Answer to example 9.8(b) (not to scale).

Summary of sections 9.14–9.20

In section 9.14, equations (9.8) and (9.8a), we defined the point elasticity of demand as

$$E^D \equiv \frac{dq}{dp}\frac{p_0}{q_0} \quad \text{or, equivalently:} \quad E^D \equiv \frac{dq/dp}{q_0/p_0}$$

In section 9.15 we saw that the arc elasticity approaches the point elasticity as the size of the price change approaches zero, and that the point elasticity is the definition invariably used in all theoretical and most applied economics. As noted earlier in discussing the supply elasticity, the elasticity measure (whether arc or point elasticity) depends on the values of p and q at the point where it is measured, as well as on the slope of the demand function. In section 9.16 there are some worked examples on all this.

In section 9.18, rule 9.6 established a key relationship between marginal revenue, price, and the elasticity of demand:

$$MR = p\left(1 + \frac{1}{E^D}\right)$$

The implications of this rule are set out in table 9.2 at the end of section 9.18. In section 9.20, example 9.7 and figure 9.11 illustrate all of the above for the particular demand function $q = 200 - 2p$. However, the above relationships are of course valid for any demand function that is negatively sloped.

In section 9.19 we considered how rule 9.6 applies to a perfectly competitive firm. For such a firm, as a limiting case the inverse demand function is horizontal at the ruling market price, $\bar{p}$. Total revenue is $TR = \bar{p}q$, and marginal revenue is $MR = \frac{dTR}{dq} = \bar{p}$. Thus MR is always positive and therefore the TR function has no maximum. Because p is constant, $\frac{dp}{dq} = 0$ as a limiting value. Therefore, by the inverse function rule of differentiation, $\frac{dp}{dq}$ has a limiting value of $-\infty$. Thus the limiting value of the demand elasticity, $E^D \equiv \frac{dq}{dp}\frac{p}{q}$, is also infinite.

The Online Resource Centre gives more worked examples of the elasticity of demand and its relationship with marginal revenue, including some graphics exercises.

Progress exercise 9.3

These questions are a continuation of exercise 9.2.

1. Using the demand function from exercise 9.2, question 1; that is, $q = -2p + 160$:

(a) Find the total revenue and marginal revenue functions (as functions of q).

(b) Find the price and quantity, p^* and q^*, at which total revenue is at its maximum.

(c) Show that, when $p = p^*$ and $q = q^*$, the demand elasticity is unity.

(d) Show that, when $p > p^*$ and $q < q^*$, demand is elastic (that is, $E^D < -1$ or $|E^D| > 1$).

(e) Similarly, show that, when $p < p^*$ and $q > q^*$, demand is inelastic (that is, $E^D > -1$ or $|E^D| < 1$).

(f) Sketch the graphs of the demand function, total revenue, and marginal revenue functions. Can you explain, using words and diagrams only, the relationship between marginal revenue and the elasticity of demand?

2. Using the demand function from exercise 9.2, question 2; that is, $q = -4p^2 - 1.5p + 792$:

(a) Find total revenue as a function of p. (Hint: don't attempt to find total revenue as a function of q, as to do so requires use of the inverse demand function, which is difficult to find in this case.)

(b) Show that when $-12p^2 - 3p + 792 = 0$, total revenue is at its maximum. Find the price and quantity that maximize total revenue. (Ignore the negative solution to this quadratic equation.)

(c) Show that the demand elasticity, as a function of p, is given by

$$E^D = \frac{-8p^2 - 1.5p}{-4p^2 - 1.5p + 792}$$

(d) Show that when demand has unit elasticity (that is, $E^D = -1$ or $|E^D| = 1$), total revenue is at its maximum.

(e) For what values of p is demand (i) elastic; (ii) inelastic?

3. Using the inverse demand function $p = \frac{50}{q} - 2$ from exercise 9.2, question 3:

(a) Find total revenue and marginal revenue as functions of q, and sketch their graphs.

(b) Show that marginal revenue is always negative, and therefore that a price increase always increases total revenue. How does this conclusion relate to your findings in exercise 9.2, question 3, regarding the elasticity of demand?

4. Repeat question 3 above for the inverse demand function $p = \frac{50}{q} + 2$ from exercise 9.2, question 4.

(a) In this case you should find that marginal revenue is always positive, and therefore that a price reduction always increases total revenue. How does this conclusion relate to your findings in exercise 9.2, question 4, regarding the elasticity of demand?

(b) Is a demand function with the characteristics of this question or the previous question likely to be found in the real world?

9.21 Other elasticities in economics

So far in this chapter we have examined the application of the elasticity concept to the supply function and the demand function. These are the most common applications in economics. In the next two sections we will consider two other applications; and, finally, we will generalize the concept of elasticity.

9.22 The firm's total cost function

We examined the firm's short-run total cost function in sections 4.14 and 8.2. If we take any total cost function $TC = f(q)$, where $q = $ output, we can follow the pattern of the supply and demand elasticities above and define the elasticity of total cost with respect to output as

$$E^{TC} \equiv \frac{dTC}{dq} \frac{q}{TC} \quad \text{or equivalently:} \quad E^{TC} \equiv \frac{dTC/dq}{TC/q} \tag{9.17}$$

The interpretation of this elasticity is closely analogous to the supply and demand elasticities. E^{TC} measures the rate of proportionate change of total cost with respect to output. If E^{TC} is greater than 1, this means that following a small increase in output the proportionate change in total cost is greater than the proportionate change in output. Conversely, if E^{TC} is less than 1, this means that following a small increase in output the proportionate change in total cost is less than the proportionate change in output. The boundary between these two possibilities is when E^{TC} is equal to 1, meaning that following a small increase in output the proportionate change in total cost is equal to the proportionate change in output.

There is an interesting feature of equation (9.17). From section 8.4 we know that the numerator of equation (9.17), $\frac{\mathrm{d}TC}{\mathrm{d}q}$, the derivative of the total cost function, is by definition marginal cost, *MC*. We also know from section 8.3 that the denominator, $\frac{TC}{q}$, again by definition is average cost, *AC*. Therefore if we substitute these two definitions into equation (9.17), we get

$$E^{TC} \equiv \frac{\mathrm{d}TC/\mathrm{d}q}{TC/q} \equiv \frac{MC}{AC} \qquad (9.18)$$

Thus from equation (9.18) we see that elasticity of the total cost function is given by the ratio of marginal cost to average cost. Thus when $MC < AC$, the numerator of equation (9.18) is less than the denominator and therefore the elasticity is less than 1. Conversely, when $MC > AC$, the numerator of equation (9.18) is greater than the denominator and therefore the elasticity is greater than 1. Finally, when $MC = AC$, the numerator of equation (9.18) is equal to the denominator and therefore the elasticity is equal to 1. Thus any statement made about the elasticity can be translated into an equivalent statement about the relationship between marginal and average cost. For example, to say that $E^{TC} = 2$ is equivalent to saying that marginal cost is twice as large as average cost.

This analysis may be taken a stage further. We know from section 8.5 that when $MC < AC$, average cost is falling; and when $MC > AC$, average cost is rising; while when $MC = AC$, average cost is at its minimum (provided of course that the second order condition for a minimum is also fulfilled). Combining these results with those of the previous paragraph, we can now say that when E^{TC} is less than 1, marginal cost is less than average cost, and average cost is falling. When E^{TC} is greater than 1, marginal cost is greater than average cost, and average cost is rising. Finally, when E^{TC} is equal to 1, marginal cost is equal to average cost, and average cost is at its minimum.

These relationships may also be shown diagrammatically. In figure 9.14 we show a firm's short-run total cost curve, with equation $TC = f(q)$. At output level q_0, average cost is given by the slope of 0B, while marginal cost is given by the slope of the tangent ABC. Since ABC is flatter than 0B, marginal cost is less than average cost. From this it follows that average cost is falling and the elasticity of total cost, E^{TC}, is less than 1.

At output q_1, average cost is given by the slope of 0D, while marginal cost is also given by the slope of 0D, since 0D is tangent to the *TC* curve at D. Therefore marginal cost equals average cost. From this it follows that average cost is at its minimum and the elasticity of total cost, E^{TC}, is equal to 1. Finally, at output level q_2, average cost is given by the slope of 0F (which equals the slope of 0B, but this is accidental and irrelevant), while marginal cost is given by the slope of the tangent EFG. Since EFG is steeper than 0F, marginal cost is greater than average cost. From this it follows that average cost is rising and the elasticity of total cost, E^{TC}, is greater than 1.

The concept of elasticity of total cost is not used very frequently in economics, although some researchers have tried to calculate cost elasticities for certain industries and firms.

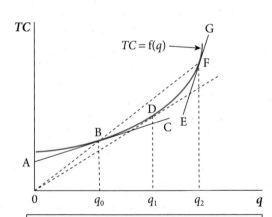

At B slope of tangent AC is flatter than slope of ray 0B, so $MC < AC$ and $E_{TC} \equiv \frac{MC}{AC} < 1$

At D slope of tangent 0D equals slope of ray 0D, so $MC = AC$ and $E_{TC} \equiv \frac{MC}{AC} = 1$

At F slope of tangent EG is steeper than slope of ray 0F, so $MC > AC$ and $E_{TC} \equiv \frac{MC}{AC} > 1$

Figure 9.14 Elasticity of total cost as a relationship between *MC* and *AC*.

9.23 The aggregate consumption function

As we saw in section 3.17, one important macroeconomic relationship is the consumption function, in which aggregate consumption expenditure in an economy, C, is said to be a function of aggregate income, Y. Thus we can write

$C = f(Y)$ where $f(Y)$ is some functional relationship.

Strictly speaking, C is planned consumption, but for simplicity we will take it as actual consumption here. For reasons we will not examine here, it is normally assumed that the consumption function $f(Y)$ is positively sloped and has a positive intercept, such as is shown in figure 9.15. Note that we are now allowing for a possibly non-linear consumption function, unlike in chapter 3. (For more detail, see any introductory macroeconomics textbook.)

Two related concepts are the average propensity to consume (APC) and the marginal propensity to consume (MPC). The APC is easily defined; it is simply consumption expressed as a proportion of income, or $\frac{C}{Y}$.

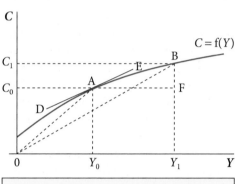

Slope of ray 0A measures APC at A. Slope of tangent DE measures MPC at A. Both APC and MPC fall as Y increases.
Elasticity of consumption with respect to income $\equiv \frac{MPC}{APC}$.
For this consumption function $\frac{MPC}{APC} < 1$ at all income levels.

Figure 9.15 A consumption function and its elasticity.

Thus, for example, in figure 9.15 if income is Y_0, then consumption is C_0 and the proportion of income consumed is $\frac{C_0}{Y_0}$, which equals the slope of the broken line from the origin, 0A. If income rises to Y_1, then consumption rises to C_1 and the proportion of income consumed is then $\frac{C_1}{Y_1}$, which equals the slope of the broken line 0B. Since 0B is flatter than 0A, we see that the APC falls as income rises. More specifically, as figure 9.15 is drawn, the slope of 0A is only slightly less than 1, say about 0.9, meaning that at income level Y_0 consumers spend about 90% of their income on consumption. The slope of 0B, however, is only about 0.6, meaning that at income level Y_1 consumers spend only about 60% of their income on consumption.

There are two definitions of the marginal propensity to consume (MPC). In introductory macroeconomics textbooks it is usually defined as the increase in consumption resulting from a 1-unit increase in income. We can show this in figure 9.15 as follows. First, we can define ΔY as $Y_1 - Y_0$ and similarly $\Delta C = C_1 - C_0$. Then we can form the difference quotient $\frac{\Delta C}{\Delta Y}$ in the usual way (refer to section 6.2 if you have momentarily forgotten what a difference quotient is). We then set $\Delta Y = 1$: that is, a 1-unit increase in income. With $\Delta Y = 1$, $\frac{\Delta C}{\Delta Y} = \Delta C =$ the increase in consumption following a 1-unit increase in income —that is, the MPC. (This way of developing the MPC is of course closely analogous to our development of marginal cost and marginal revenue in chapter 8.)

In figure 9.15, when Y increases from Y_0 to Y_1, the difference quotient is $\frac{BF}{AF}$, which equals the slope of the chord AB (not shown). If we assume that ΔY, which is the distance AF, is equal to 1, then $\frac{BF}{AF} = BF =$ the increase in income following a 1-unit increase in income—that is, the MPC.

As we saw earlier when defining marginal cost and marginal revenue, this definition of the MPC is somewhat unsatisfactory. It is not easy to compute or manipulate, and its value will vary according to how a 1-unit increase in income is defined and the unit in which income is measured (pounds or euros, say). For these reasons, in theoretical economics the MPC is invariably defined as the derivative of the consumption function, $\frac{dC}{dY}$. On this definition, the MPC at income Y_0 is given by the slope of DAE, the tangent at A. For sufficiently small changes in income, the slope of the tangent at A may be treated as insignificantly different from the slope of the chord AB, so that the two definitions of the *MPC* are equivalent.

We now define the point elasticity of consumption with respect to income, which we will write as E^C, in exactly the same way as we have defined earlier point elasticities:

$$E^C \equiv \frac{dC}{dY}\frac{Y}{C} \equiv \frac{dC/dY}{C/Y}$$

Since dC/dY is the *MPC* and C/Y is the *APC*, we see that the elasticity of consumption is given by the ratio of the *MPC* to the *APC*:

$$E^C \equiv \frac{MPC}{APC} \tag{9.19}$$

At a point such as A in figure 9.15, the *MPC* is given by the slope of the tangent DAE and the *APC* is given by the slope of 0A. Clearly the *MPC* is less than the *APC* and the elasticity of consumption is therefore less than 1. This is also true at point B, and indeed at any point on the consumption function as it is drawn in figure 9.15. Writing in the 1930s, the great economist J.M. Keynes argued that a 'psychological law' ensured that the *MPC* was always less than the *APC*.

How useful is the concept of the elasticity of the consumption function? We could take the view that it is a somewhat redundant concept, since, as is clear from equation (9.19), to say that E^C is greater (or less) than 1 is simply another way of saying that the *MPC* is greater (or less) than the *APC*. But information about E^C can be more immediately usable than information about the *MPC* and the *APC*. For example, if we know that the $E^C = 2$, say, then we know that a 1% rise in income will lead to an increase in consumption of 2%. This is very useful for macro-economic management and forecasting.

There is another reason for being interested in whether E^C is greater or less than 1. If E^C is greater (less) than 1, then the *MPC* is greater (less) than the *APC*. As we have seen earlier, if the *MPC* is greater (less) than the *APC*, then the *APC* rises (falls) as income increases. This is also an important issue in macroeconomics. If you have trouble seeing this, think of the football team's marginal and average score, as discussed in section 8.5.

We return to the consumption function in section 15.15.

9.24 Generalizing the concept of elasticity

The concept of elasticity is a perfectly general one and can be applied to any function, whether in economics or any other field of study. Given any function $y = f(x)$, we can define the elasticity, E^y, of the function as

$$E^y \equiv \frac{dy/dx}{y/x} \equiv \frac{dy}{dx}\frac{x}{y}$$

For any function $y = f(x)$, the elasticity measures the rate of *proportionate* change of y. This is to be distinguished from the derivative, $\frac{dy}{dx}$, which measures the rate of *absolute* change.

If we define $M \equiv \frac{dy}{dx}$ as the marginal value of the function, and $A \equiv \frac{y}{x}$ as the average value, then the definition above becomes

$$E^y \equiv \frac{dy/dx}{y/x} \equiv \frac{\text{marginal value of } y}{\text{average value of } y} \equiv \frac{M}{A}$$

From this equation we can see that $E^y > 1$ if $M > A$, and $E^y < 1$ if $M < A$. (We assume M and A are both positive.)

It is not difficult to prove that when the marginal value is greater (less) than the average value, the average value is rising (falling). To prove this, we first find the maximum or minimum of A, by taking the derivative with respect to x and setting it equal to zero. Since $A \equiv \frac{y}{x}$ is a quotient, we have to use the quotient rule of differentiation. This gives

$$\frac{\mathrm{d}A}{\mathrm{d}x} = \frac{x\frac{\mathrm{d}y}{\mathrm{d}x} - y\frac{\mathrm{d}x}{\mathrm{d}x}}{x^2} = 0 \tag{9.20}$$

Assuming $x \neq 0$, and since $\frac{\mathrm{d}x}{\mathrm{d}x} = 1$, this expression will equal zero when the numerator equals zero: that is, when

$$x\frac{\mathrm{d}y}{\mathrm{d}x} - y = 0 \quad \text{which is true when}$$

$$\frac{\mathrm{d}y}{\mathrm{d}x} = \frac{y}{x}.$$

But the left-hand side of this equation is the marginal value, M, and the right-hand side is the average value of the function, A. So the average value, A, reaches a maximum (or minimum) when $M = A$: that is, when the marginal and average values are equal.

Using equation (9.20) with the equals sign replaced by > or <, we can also show quite easily that when A is rising (that is, when $\frac{\mathrm{d}A}{\mathrm{d}x} > 0$) we must have $M > A$; and that when A is falling we must have $M < A$. This is left to you as an exercise.

Summary of sections 9.21–9.24

In section 9.21 we broadened the concept of elasticity and in 9.22 applied it to a firm's total cost function. We can summarize as follows. For any total cost function $TC = f(q)$, where $q =$ output, at any given output level:

$MC < AC$	implies	AC falling	and	$E^{TC} < 1$
$MC = AC$	implies	AC at its minimum*	and	$E^{TC} = 1$
$MC > AC$	implies	AC rising	and	$E^{TC} > 1$

* provided, as always, that the relevant second order condition is satisfied.

In section 9.23 we looked briefly at the elasticity of the aggregate consumption function and saw that it equals the ratio of the marginal propensity to consume (MPC) to the average propensity to consume (APC). Finally, in section 9.24 we saw that elasticity was a completely general concept that could be applied to any function (whether in economics or elsewhere), and measures the ratio of the marginal to the average value of the function in question.

Progress exercise 9.4

1. Given the total cost function: $TC = 9q^2 + 2q + 8100$:

(a) Find marginal cost (MC) and average cost (AC) as functions of q.

(b) Find the output at which average cost is minimized, and show that $MC = AC$ at this output.

(c) Show that when $MC < AC$, AC is falling; and when $MC > AC$, AC is rising.

(d) Find the elasticity of total cost, E_{TC}, with respect to output, and show that it equals $\frac{MC}{AC}$.

(e) Show that when $E_{TC} < 1$, AC is falling; and when $E_{TC} > 1$, AC is rising.

(f) Illustrate graphically.

2. Repeat question 1 for the general quadratic total cost function, $TC = aq^2 + bq + c$. (Assume $a > 0, b > 0, c > 0$.)

3. Given the aggregate consumption function: $C = 0.9Y + 100$ (where C = aggregate consumption and Y = aggregate income):

(a) Find the marginal propensity to consume (MPC) and the average propensity to consume (APC).

(b) Show that the MPC is always less than the APC, but that the difference decreases as Y increases.

(c) Find the elasticity of consumption with respect to income, and show that it equals $\frac{MPC}{APC}$. Hence show that the elasticity is always less than 1.

(d) Illustrate graphically, showing the consumption function and the MPC and APC.

4. Given a general linear aggregate consumption function: $C = aY + b$:

(a) What values of the parameters a and b would reverse the results of question 3 above: that is, would result in $MPC > APC$ and an elasticity greater than 1?

(b) What parameter values would result in $MPC = APC$ and an elasticity equal to 1?

Checklist

Be sure to test your understanding of this chapter by attempting the progress exercises (answers are at the end of the book). The Online Resource Centre contains further exercises and materials relevant to this chapter www.oxfordtextbooks.co.uk/orc/renshaw3e/.

In this chapter we have focused almost entirely on the concept of elasticity as the relationship between proportionate rather than absolute changes. We have examined both supply and demand elasticities, and the relationship between demand elasticity and marginal revenue. The main topics were:

✔ **Absolute and proportionate changes.** The distinction between absolute, proportionate, and percentage changes in a variable.

✔ **Arc elasticities.** Definition of the arc elasticities of supply and demand, what they measure, and how to estimate them from a graph.

✔ **Point elasticities.** Definition of the point elasticities of supply and demand, what they measure, and why they are generally preferred to the arc elasticities.

✔ **Demand elasticity and marginal revenue.** Understanding of, and how to derive, the relationship between the elasticity of demand and marginal revenue, under monopoly and perfect competition. Proof that elastic (inelastic) demand implies that marginal revenue is positive (negative).

✔ **Generalized elasticity.** How elasticity can be applied to other functions in economics such as the total cost function and the consumption function.

We are now ready to move on to part 3 of the book, which is concerned with the mathematics of finance and growth. We return to the economics of the firm and of the consumer in part 4.

Part Three
Mathematics of finance and growth

- Compound growth and present discounted value
- The exponential function and logarithms
- Continuous growth and the natural exponential function
- Derivatives of exponential and logarithmic functions and their applications

Chapter 10
Compound growth and present discounted value

OBJECTIVES

Having completed this chapter you should be able to:

■ Understand and use the formulae for an arithmetic or geometric series.

■ Understand the formula for compound growth and use it to solve problems involving growth of economic variables.

■ Understand and apply the formula for calculating the present discounted value of a series of future receipts.

■ Understand how the market value of a perpetual bond is calculated.

■ Understand and calculate nominal and effective interest rates.

■ Understand and calculate how a loan may be repaid in equal instalments.

10.1 Introduction

In economics we often want to analyse variables that are increasing or decreasing as time passes. A common example of a variable increasing with time is an amount of money you might deposit with (= lend to) a bank or building society and which earns interest at a given annual percentage rate. An example of a variable decreasing with time is the value of a machine or some other capital asset that depreciates (= loses value) with time, either due to wear and tear or because it becomes outdated due to technological advance. When an economic variable is increasing with time, it is natural to refer to this process as growth. When a variable is decreasing with time, we refer to this as decline, or negative growth.

This chapter is the first of four that are concerned with the mathematics of borrowing and lending, and of economic variables that are growing through time. We begin by examining arithmetic and geometric series, and this leads naturally into the distinction between simple and compound interest, from which we derive the formula giving the value of any variable that is growing at a constant compound rate. We demonstrate the application of this formula to problems involving growth in economic variables. This formula also introduces the distinction between the nominal and effective rates of interest.

In this chapter we will also develop the formula for calculating the present discounted value of a series of future payments or receipts. This formula provides a criterion for choosing between alternative investment projects by reference to the expected future series of profits. A special case of this is the valuation of a government bond that pays its owner a fixed amount of money each year for ever, which is known as a perpetual bond. Finally, we derive a formula for calculating the repayments of a loan in equal instalments.

10.2 Arithmetic and geometric series

Arithmetic series

An arithmetic series (also known as an arithmetic progression or AP) is a series of numbers in which you get from one number to the next by *adding* a constant.

EXAMPLE 10.1

Examples of arithmetic progression are:

$1, 2, 3, 4$ (added constant = 1)

$0, -25, -50, -75$ (added constant = −25)

$^1/_2, ^3/_4, 1, 1^1/_4$ (added constant = $^1/_4$)

Generalizing from these examples, we can see that the general form is

$A, A + d, A + 2d, A + 3d, \ldots$

where A = first term of the series and d = added constant, known as the 'common difference'.

So, for example, if we are told that an AP has $A = 3$, $d = 2$, we know that the series is $3, 5, 7, 9, \ldots$

Two formulae relating to the arithmetic progression are given as rule 10.1 below. You need to be aware of these two formulae but as background knowledge only.

RULE 10.1 **Two formulae for the arithmetic progression (AP)**

Given an AP of the general form: $A, A + d, A + 2d, A + 3d, \ldots$

Rule 10.1a

The nth term of the series is given by $A + (n - 1)d$

Rule 10.1b

The sum of the first n terms of the series, denoted by $\sum_n$, is given by

$$\sum_n = \frac{n}{2}[2A + (n - 1)d]$$

Here 'Σ' is the letter sigma, in upper case, from the Greek alphabet (see appendix to chapter 2). As a mathematical symbol, it means 'sum of'.

EXAMPLE 10.2

Given the series $1, 3, 5, 7, \ldots$

(a) Find the 50th term; (b) Find the sum of the first 50 terms.

Answer

(a) Here $A = 1$, $d = 2$. So, using rule 10.1a, the 50th term is

$\qquad A + (n - 1)d = 1 + (50 - 1)2 = 99$

(b) Using rule 10.1b, the sum of first 50 terms is

$$\sum_n = \frac{n}{2}[2A + (n - 1)d] = \frac{50}{2}[2 + (50 - 1)2] = 2500$$

You might want to check these answers using Excel®. You can use the 'Insert Function' command to generate the series. The Online Resource Centre explains how to do this. www.oxfordtextbooks.co.uk/orc/renshaw3e/

Geometric series

A geometric series (or geometric progression, GP) is a series of numbers in which you get from one number to the next by *multiplying* by a constant.

EXAMPLE 10.3

Examples of geometric progressions are:

1, 2, 4, 8, 16, . . .   (multiplicative constant = 2)

$\frac{1}{2}, \frac{1}{4}, \frac{1}{8}, \frac{1}{16}, \ldots$ (multiplicative constant = $\frac{1}{2}$)

Generalizing from these examples, we can see that the general form is

$$A, AR, AR^2, AR^3, \ldots$$

where A = first term, and R = multiplicative constant, known as the 'common ratio'.

So if, for example, we are told that a GP has $A = 5$, $R = -2$, we know that the series is

$$5, -10, 20, -40, \ldots$$

Rule 10.2 below comprises two formulae relating to the geometric progression. You need to be aware of rule 10.2a, but as background knowledge only. Rule 10.2b, however, is used quite frequently by economists, so you should look at it carefully and preferably memorize it. One of its uses is analysing the Keynesian multiplier, which we will consider shortly.

 RULE 10.2 Two formulae for the geometric progression (GP)

Given a GP of the general form: $A, AR, AR^2, AR^3, \ldots$

Rule 10.2a
The nth term of the series is given by AR^{n-1}

Rule 10.2b
The sum of the first n terms of the series, denoted by $\sum_n$, is given by

$$\sum_n = \frac{A(1 - R^n)}{1 - R}$$

By multiplying numerator and denominator by -1, this can also be written as

$$\sum_n = \frac{A(R^n - 1)}{R - 1}$$

Here 'Σ' is the letter sigma, in upper case, from the Greek alphabet. As a mathematical symbol, it means 'sum of'.

EXAMPLE 10.4

Given the series 1, 2, 4, 8, 16, . . .

(a) Find the 10th term.

(b) Find the sum of the first 10 terms.

Answer

(a) Here $A = 1$, $R = 2$, so the 10th term is $AR^{n-1} = 1(2^9) = 512$

(b) Sum of first 10 terms is

$$\sum_{10} = \frac{A(R^n - 1)}{R - 1} = \frac{1(2^{10} - 1)}{2 - 1} = 2^{10} - 1 = 1024 - 1 = 1023$$

(Use the $\wedge$ or y^x key on your calculator to find 2^9 and 2^{10}; see section 1.14.)

A special case of the geometric progression

There is a special case of the formula for the sum of a GP (rule 10.2b above) which is important in economics. This arises when the common ratio is a positive fraction (that is, $0 < R < 1$), and the number of terms is indefinitely large: that is, n approaches infinity ($n \to \infty$). Let us see what effect this has on rule 10.2b, which was

$$\sum_{n} = \frac{A(1 - R^n)}{1 - R}$$

In this formula, when R is a positive fraction, R^n gets smaller and smaller as n increases (for example, if $R = \frac{1}{2}$ and $n = 3$, then $R^n = \frac{1}{8}$). So we can see that R^n will approach zero as n approaches infinity. Thus, using the notation for limits that we developed in section 5.6, we have

as $n \to \infty$, $R^n \to 0$, therefore

as $n \to \infty$, $\sum_{n} = \frac{A(1 - R^n)}{1 - R} \to \frac{A(1 - 0)}{1 - R} = \frac{A}{1 - R}$

Thus we can deduce a special case of rule 10.2b, as follows:

RULE 10.3 A special case of rule 10.2b

Given a *GP* in which the common ratio, R, is a positive fraction and the number of terms, n, approaches infinity, then the sum of the series is

$$\sum_{\infty} = \frac{A}{1 - R}$$

EXAMPLE 10.5

Given the series: $\frac{1}{2}, \frac{1}{4}, \frac{1}{8}, \frac{1}{16}, \ldots$

we see that $A = \frac{1}{2}$ and $R = \frac{1}{2}$. So the sum, with an infinite number of terms, is

$$\sum_{\infty} = \frac{A}{1 - R} = \frac{1/2}{1 - 1/2} = 1$$

10.3 An economic application

An important economic application of rule 10.3 is the Keynesian investment multiplier, which we examined in section 3.17. There we showed how the equilibrium level of aggregate income is determined in the basic Keynesian model, and we also conducted a comparative statics exercise to show how the equilibrium level changed in response to an exogenous disturbance such as a fall in business confidence. (All this is, of course, discussed in more depth in any introductory macroeconomics textbook.)

Here we will set up a simple yet instructive dynamic model to show not just how the equilibrium level of income changes following a disturbance, but also the process whereby income gets from its old equilibrium level to its new equilibrium.

The first two equations of our model are the same as those in section 3.17. We assume that, in any month, aggregate income (Y) received by households is necessarily equal to aggregate demand or expenditure (E). Aggregate expenditure comprises consumption expenditure (C) by households plus investment expenditure by firms (I). This gives us the two equations:

$$Y \equiv E \tag{10.1}$$

$$E \equiv C + I \tag{10.2}$$

The next equation is the consumption function. We assume that households spend 90% of their income on consumption (and save the remaining 10%). Thus their marginal propensity to consume is 0.9. However, here we depart from the model of section 3.22 by assuming that, because households receive their pay cheques at the end of each month, in any month they are spending the income they earned in the *previous* month. So the consumption function is

$$\hat{C} = 0.9Y_{-1} \tag{10.3}$$

where $\hat{C}$ is planned consumption and Y_{-1} denotes the income of the previous month. (Note that there is no constant term in this consumption function, but this has no significant effects on the workings of the model.)

The next two equations are the same as in section 3.17: first, the equilibrium condition that planned and actual consumption must be equal; second, that investment is exogenous; that is, determined in some way that is outside our model.

$$\hat{C} = C \tag{10.4}$$

$$I = \bar{I} \tag{10.5}$$

This gives us five equations, but compared to the model of section 3.17 we have gained an additional, sixth unknown, Y_{-1}. So we need another equation to make the model solvable. We get this equation by imposing a further equilibrium condition, that Y must not be changing from one month to the next. Thus we have

$$Y = Y_{-1} \tag{10.6}$$

The first five equations above can easily be reduced to two. Combining (10.1), (10.2), and (10.5) gives

$$Y = C + \bar{I} \tag{10.7}$$

and combining equations (10.3) and (10.4) gives

$$C = 0.9Y_{-1} \tag{10.8}$$

Equations (10.7) and (10.8) constitute our 'reduced form': that is, we have reduced the model from six equations to only two.

In chapter 20 we will show how to solve this reduced-form model algebraically. Here we will content ourselves with a *simulation* or numerical example to show how this economy works (see table 10.1).

When setting up this table, we arbitrarily assumed that income in month 0 was $Y = 100$. We need to know this in order to calculate consumption in month 1. The level of $\bar{I}$ was also assumed to be 10 for the first 3 months. (Incidentally, the choice of months as the time unit is somewhat arbitrary.)

Looking at the table, we see that for the first three months this economy is in equilibrium, with consumption, investment, and income all constant, and consumption equal to 90% of income. Then in month 4 we assume that this equilibrium is disturbed by an increase in

Table 10.1 A dynamic investment multiplier process.

Month (1)	$C = 0.9Y_{-1}$ [equation (10.8)] (2)		$\bar{I}$ (3)	$Y = C + \bar{I}$ [equation (10.7)] (4)	Increase in Y since previous period (5)
1	90	$(= 0.9 \times 100)$	10	100	0
2	90	$(= 0.9 \times 100)$	10	100	0
3	90	$(= 0.9 \times 100)$	10	100	0
4	90	$(= 0.9 \times 100)$	20	110	10
5	99	$(= 0.9 \times 110)$	20	119	9
6	107.1	$(= 0.9 \times 119)$	20	127.1	8.1
7	114.39	$(= 0.9 \times 127.1)$	20	134.39	7.29
8	120.95	$(= 0.9 \times 134.39)$	20	147.51	6.56
9			20		
10			20		
11			20		
12			20		

Note: to calculate C in month 1, we assume $Y = 100$ in month 0.

investment from 10 to 20, for reasons we do not attempt to explain. This causes income to rise by 10 (to 110) in month 4, and this in turn causes consumption to increase in month 5, to 90% of 110 instead of 90% of 100. But this increase in consumption causes income to rise to 119 in month 5, and this in turn causes consumption to increase in month 6, to 90% of 119. (Note that investment is assumed to remain at its new level of 20.) To test your understanding, try to fill in the blanks in table 10.1. You will need only your calculator to do this.

Thus the result of the assumed jump in the level of investment is that a series of increases in consumption and income is triggered off. Two interesting questions suggest themselves:

(1) Is there a way of calculating the total increase in income after, say, 5 years, without the tedium of having to extend table 10.1 until we have the necessary number of rows?

(2) Does income go on rising forever or does it eventually stabilize at some new level ('converge', in economists' terminology)?

We can answer these questions using the GP formulae (rule 10.2 above). Look at the increases in Y in column (5) of table 10.1. The pattern here is that the initial increase is 10, and then each subsequent increase is 90% of the previous month's increase. So the series of increases in Y is

$$10, \quad 10 \times 0.9, \quad (10 \times 0.9) \times 0.9, \quad (10 \times 0.9 \times 0.9) \times 0.9, \quad \text{and so on.}$$

This series is a GP with first term $(A) = 10$ and common ratio $(R) = 0.9$. The cumulative total increase in Y is the sum of all these monthly increases. So to answer question (1), after 60 months (5 years) the cumulative increase in income is given by rule 10.2b:

$$\sum_n = \frac{A(1 - R^n)}{1 - R} \quad \text{with } n = 60, \quad A = 10, \text{ and } R = 0.9$$

Substituting these values into the formula, we get

$$\sum_{60} = \frac{10(1 - 0.9^{60})}{1 - 0.9} = \frac{10(1 - 0.9^{60})}{0.1} = 100(1 - 0.9^{60}) = 100(1 - 0.00179701)$$

$$= 99.820299 \tag{10.9}$$

(Use the $\wedge$ or y^x key on calculator to find 0.9^{60})

To answer question (2), since R is a positive fraction, the sum to infinity of this series is given by rule 10.3 above:

$$\sum_{\infty} = \frac{A}{1-R} = \frac{10}{1-0.9} = \frac{10}{0.1} = 100$$

This tells us that as an unlimited amount of time passes, the increase in income approaches a limiting value of 100. Thus 'convergence' occurs, but only as a limiting value. From the economist's point of view, it is interesting and important to note that the increase in income, $\frac{A}{1-R}$, is the increase in investment, A, divided by 1 minus R, the marginal propensity to consume.

The fact that the full increase in income, 100, is achieved only as a limiting value is also interesting. If this model is valid, it means that the successive increases in world income resulting from investment in the form of the building of the Great Pyramid of Giza in 2500 BCE are continuing to this day, and will continue forever. However, the increases now must be very small. In our answer to question (1) in the example above the increase in income after 5 years is already 99.820299. Thus after as few as 5 years, the sum is already very close to the limiting value of 100, even though the full 100 is reached only after an infinite amount of time has passed—that is, never. The reason for this rapid convergence is that 0.9^{60} in equation (10.9) is very close to zero, hence there is very little difference between the values of Y given by the two equations above $\left(\sum_{60} \text{ and } \sum_{\infty}\right)$. When convergence is rapid, as in this case, economists invariably use the limiting value as an approximation to the actual value reached after, say, 5 years or whatever fairly short period is being considered. The main reason for doing so is that the limiting value is much easier and quicker to calculate, as we have just seen in the example above. Note, however, that the accuracy of such an approximation depends on the value of R (the marginal propensity to consume, in this model) as well as on the number of years. The smaller is R, the smaller is R^n for any given n, and the closer therefore is the approximation.

Progress exercise 10.1

1. Find the sum of: $1 + 2 + 3 + \ldots + 50$

2. Find the 11th term of the series: $25, 50, 75, \ldots$

3. Find the sum of the first 10 terms of the series: $1, 2, 4, 8, \ldots$

4. Find the sum of the first 10 terms of the series: $^1/_2, {}^1/_4, {}^1/_8, {}^1/_{16}, \ldots$ What limiting value does this sum approach as the number of terms goes to infinity?

5. Given the following macroeconomic model:

   $$Y = C + I \quad C = 0.75Y_{-1}$$

   where Y = national output (= national income); I = aggregate investment (an exogenous variable); C = aggregate planned and actual consumption (all in the current month); Y_{-1} = national income in the previous month.

   (a) Suppose initially (month 1) $C = 75$, $I = 25$, and $Y = 100$. Then in month 2, I rises to 35 and thereafter remains at this level. Trace the path of C, I, and Y over the next 5 months.

   (b) By how much has Y risen after one year?

   (c) What limiting value does Y approach as the number of months goes to infinity?

   (d) What is the practical usefulness of knowing what Y would become if an infinite amount of time could pass?

10.4 Simple and compound interest

In this section we will take our first steps in financial mathematics, which is basically concerned with borrowing and lending money, and with valuing assets. The starting point is that people are not usually willing to lend money unless they are rewarded by receiving interest. Interest is usually calculated as an annual percentage rate, so if you borrow, say, €1000 for 1 year at an interest rate of 5% per year, then after 1 year you will have to repay the original €1000, which is known as the capital or **principal**—an abbreviation of 'principal sum', which is why it has the spelling of the adjective (principal) rather than the noun (principle). You will also have to pay interest of 5% of 1000. From chapter 1 we know that $5\% = \frac{5}{100} = 0.05$, so 5% of 1000 is most easily written and calculated as 0.05(1000). So your total payment (principal + interest) will be $1000 + 0.05(1000)$, which factorizes into $1000(1 + 0.05)$.

Generalizing this, if the annual interest rate is $100r\%$ and you borrow €Z for 1 year, then after 1 year you will have to repay the capital, Z, plus interest of $(100r/100)Z = rZ$. So your total payment (principal + interest) will be $Z + rZ = Z(1 + r)$.

This way of formulating the interest rate in algebraic form, as $100r\%$, is a little confusing at first but its attraction is that we only have to multiply the capital by r to find the interest. If instead we defined the interest rate as $r\%$, then we would have to multiply the capital by $\frac{r}{100}$ to calculate the interest, which is less neat. Another way of seeing this is to see that r is the interest rate expressed as a proportion rather than as a percentage (see sections 1.11 and 1.12 if you need to revisit percentages and proportions). So if the interest rate is 10%, $r = 0.1$.

Simple interest

Suppose you open an account at a bank and deposit money in it. The bank rewards you for lending it your money by paying interest once a year, calculated at some annual percentage rate. With 'simple' interest, the key assumption is that you withdraw the interest from your account as soon as it is paid (and presumably spend it). Let's look at a simple numerical example of how this might work out.

EXAMPLE 10.6

Assume you lend the bank €100, with an interest rate of 10% per annum. This initial loan, or investment, of €100 is the principal. Since $10\% = \frac{10}{100} = 0.1$, we can calculate the interest by multiplying the principal by 0.1.

The evolution over time of the account is shown in table 10.2. This table is very boring because of the key assumption of simple interest, which is that the interest is *withdrawn*

Table 10.2 Simple interest with €100 lent at 10% per year (example 10.6).

(1)	(2)	(3)	(4)
Year	Balance (= cumulative value of investment) at beginning of year (€)	Interest added at end year (€)	Cumulative value of investment at end year (= principal + total interest received) (€)
1	100	$(100)0.1 = 10$	$100 + 10 = 110$
2	100	$(100)0.1 = 10$	$100 + 10 + 10 = 120$
3	100	$(100)0.1 = 10$	$100 + 10 + 10 + 10 = 130$
⋮	⋮	⋮	⋮
x	100	$(100)0.1 = 10$	$100 + 10x$

as soon as it is credited to the account. Consequently the amount invested (the principal) remains constant at €100. So interest earned in every year is simply 10% of €100 = €10.

The cumulative value of the investment (column 4) forms the series 100 (in year 0, when the account is opened), 110, 120, 130, . . . This is an arithmetic progression (AP) with first term 100 and common difference 10. The total return after x years—that is, principal plus cumulative interest—would be $100 + 10x$ euros.

Compound interest

With compound interest, the key assumption is that the interest credited to the account at the end of each year is not withdrawn but left in the account (that is, it is reinvested) and thus added to the opening credit balance of the following year. Let's look at an example of how this works out.

EXAMPLE 10.7

We assume compound interest with an initial sum deposited (principal) of €100 and an interest rate of 10% per year, as in the previous example. The growth of capital and interest is shown in table 10.3.

The key point to see in table 10.3 is that the value of the investment (column 2) increases each year by the amount of the previous year's interest. The effect of this is that the *amount* of interest added each year increases, even though the interest *rate* is constant. This gives the 'compounding' effect. To test your understanding of this, complete the blank cells in the table.

Table 10.3 Compound interest with €100 lent at 10% per year (example 10.7).

(1)	(2)	(3)	(4)
Year	Balance (= cumulative value of investment) at beginning of year (€)	Interest added at end year (€)	Cumulative value of investment at end year (= principal + total interest received) (€)
1	100	(100)0.1 = 10	100 + 10 = 110
2	110	(110)0.1 = 11	110 + 11 = 121
3	121	(121)0.1 = 12.1	121 + 12.1 = 133.1
4	133.1	(133.1)0.1 = 13.31	133.1 + 13.31 = 146.41
5			
6			177.16

The compounding effect makes a huge difference to the cumulative value of the investment (principal + interest). For example, after 6 years you have received €77.16 interest, compared with only €60 with simple interest (= 6 years at €10 per year). The compounding effect is especially large when either the number of years or the annual interest rate is large.

Note that the cumulative values (column 4 of table 10.3) form a geometric progression with first term 100 (the initial investment) and common ratio 1.1.

The compound interest formula

Now we want to generalize the previous example, by considering €a invested at $100r\%$. Thus in the previous example we had $a = 100$ and $r = 0.1$ (so that $100r = 10$, the percentage rate of interest).

EXAMPLE 10.8

We assume compound interest (= compound growth) with an initial deposit of €a and an interest rate of $100r$% per year. The growth of capital and interest is shown in table 10.4.

In table 10.4, we can see a pattern emerging. After 2 years, the cumulative value of the investment is $a(1 + r)^2$. After 3 years it is $a(1 + r)^3$, and so on. Be sure to check carefully that you understand how all the entries in table 10.4 are arrived at, and check your understanding by completing the row for year 5. You may find it helpful to make a careful comparison of tables 10.3 and 10.4.

Table 10.4 Compound interest with €a invested at $100r$% per year.

(1)	(2)	(3)	(4)
Year	Balance (= cumulative value of investment) at beginning of year (€)	Interest added at end year (€)	Cumulative value of investment at end year (= principal + total interest received) (€)
1	a	ar	$a + ar$ $= a(1 + r)$
2	$a(1 + r)$	$\{a(1 + r)\}r$	$a(1 + r) + \{a(1 + r)\}r = a(1 + r)(1 + r)$ $= a(1 + r)^2$
3	$a(1 + r)^2$	$\{a(1 + r)^2\}r$	$a(1 + r)^2 + \{a(1 + r)^2\}r = a(1 + r)^2(1 + r) = a(1 + r)^3$
4	$a(1 + r)^3$	$\{a(1 + r)^3\}r$	$a(1 + r)^3 + \{a(1 + r)^3\}r = a(1 + r)^3(1 + r) = a(1 + r)^4$
5			
$\vdots$	$\vdots$	$\vdots$	$\vdots$
x			$= a(1 + r)^x$

Looking at the pattern revealed in table 10.4, we can infer that after x years the cumulative value of the bank deposit equals $a(1 + r)^x$. We state this as a rule below. Moreover, this rule will apply equally to any variable, not just a bank deposit, that is growing at a constant compound rate. We will refer to it throughout this book as the compound growth formula.

 RULE 10.4 The compound growth formula

The formula is: $y = a(1 + r)^x$

where

a = 'principal' (initial sum invested, or initial value of the variable in question)
r = annual proportionate interest rate (for example, if the interest rate is 10%, the proportionate interest rate is $r = 10 \div 100 = 0.1$)
x = number of years
y = future compounded value (= value of initial sum + cumulative interest, after x years).

Note that the word 'compound' is often dropped when referring to the above formula, it being understood that growth follows compound rather than simple interest. It is also understood that 'growth' can be negative as well as positive, so the formula can deal equally well with a variable that is *decreasing* at a constant compound rate. We give an example of this below.

It is important for you to be comfortable with the terminology here. In particular, note that y, the value of the initial principal plus cumulative interest, is called the **future compounded value**. It is what the principal, a, which is an amount of money now, will become in the future as the result of adding compound interest. Of course, it's equally possible that a might be an amount of money invested in the past, which has grown to a value of y now. But as a matter of convention and to avoid muddle, we always refer to y as the future value. The importance of sticking to this convention will be seen shortly.

10.5 Applications of the compound growth formula

In the next two sections we will illustrate how the compound growth formula (rule 10.4) can be applied to problems involving the growth of money lent at interest and, more generally, to any economic variable growing through time at a constant compound growth rate.

Growth of bank deposits

EXAMPLE 10.9

If I deposit €100 with a bank at an annual interest rate of 4%, with interest reinvested, how much will I get back at the end of 5 years?

Answer: We use the compound growth formula $y = a(1 + r)^x$ with $a = 100$, $r = 0.04$, and $x = 5$, giving

$$y = 100(1 + 0.04)^5 = 100(1.04)^5 = 100(1.2167)$$
$$= 121.67 \text{ (using the } \wedge \text{ or } y^x \text{ key on calculator to find } (1.04)^5)$$

Table 10.5 Growth of a bank deposit.

x (completed years)	0	1	2	3	4	5
$y = 100(1.04)^x$	100	104	108.16	112.49	116.99	121.67

The solution to example 10.9 can also be shown graphically. Using a calculator, we can quickly calculate the values of y for each year by setting x equal to 0, 1, 2, 3, 4, and 5 and calculating the corresponding values of y (see table 10.5). You may find it confusing that the first value for x in table 10.5 is $x = 0$. This is because x measures the number of *completed* years since the money was invested. Thus during the first year, the number of completed years is zero, so no interest has yet been added, hence we have $x = 0$ and $y = 100(1.04)^0 = 100$ euros. Similarly, $x = 1$ in the second year; 2 in the third year, and so on until we have $x = 5$ in the sixth year. The value of y in the sixth year is its value *after* 5 completed years: 121.67 euros.

Table 10.5 gives us the values for y from which figure 10.1 below, called a bar chart, is drawn. The height of each bar gives the (constant) value of my bank deposit during the year, which remains unchanged throughout the year. Then, at the end of each year, interest is added and my deposit jumps to a new level. So the value of the investment *after* 5 years is given by the level that it jumps up to *at the end of* year 5; that is, its level in the sixth year.

Note that table 10.5 tells us that the size of the jumps increases as time passes, though this is not very obvious in figure 10.1 because the growth rate is quite small and the time period short; hence the compounding effect is not very strong.

The tops of the bars form a step function (see section 3.6 if you have forgotten what this is) which gives the value of my deposit at any moment of time. We have a step function because y does not vary continuously but in jumps, as x takes the integer values 1, 2, . . . , 5.

Thus over 5 years my bank deposit grows from $y_0 = 100$ at the beginning of year 1 to $y_5 = 121.67$ at the beginning of year 6. The cumulative proportionate growth (see section 9.2 if you need to revise this) is

$$\frac{\Delta y}{y_0} = \frac{y_5 - y_0}{y_0} = \frac{121.67 - 100}{100} = 0.2167$$

The cumulative percentage growth is $0.2167 \times 100 = 21.67\%$.

EXAMPLE 10.10

This is an example of negative growth. A certain country once experienced an inflow of 'hot money' from abroad which threatened to destabilize its economy. To discourage the inflow, the central bank ordered banks to offer a *negative* interest rate of 3% per year on deposits received from abroad. If a foreigner deposited 1000 for 3 years at this interest rate, how much would she get back?

Answer: Here $r = -0.03$, $a = 1000$, $x = 3$. So in the compound growth formula (rule 10.4) we have

$$y = 100(1 - 0.03)^3 = 1000(0.97)^3 = 912.7$$

(using the $\wedge$ or y^x key on calculator to find $(0.97)^3$). Using the same method as in the previous example, we can calculate the cumulative proportionate growth as -0.0873 and the percentage growth as -8.73%. (Notice that with simple interest the growth would be $-$ -3% per year for 3 years, totalling $-3\% \times 3 = -9\%$. Thus when the growth rate is negative, the compounding effect works to *reduce* the decline.)

Extension to other economic variables

The compound growth formula $y = a(1 + r)^x$ can also be applied to any economic variable that is growing at a constant compound rate and in discrete (periodic) jumps, such as once per year, once per month, and so on.

In economics we almost invariably assume that variables that are growing through time obey the laws of compound interest, not simple interest. To see why we assume this, consider the growth of a price index (inflation). Suppose the price index is initially 100 in year 0 and inflation is 10% per year. In year 1 the price index will therefore rise by 10% of 100 (= 10) to a new level of 110. In year 2 the price index will again rise by 10% but since it was already 110 it must rise by 10% of 110 (= 11) to a new level of 121. Clearly the price index is following the path of table 10.3 (compound interest) rather than table 10.2 (simple interest). Growth is compounded because each year's 10% rise is applied to a level that is itself increasing year by year. The same reasoning applies to other economic variables such as national production (GDP) because again the growth rate is applied to a level that itself is rising.

In the next example, the economic variable we consider is a price index called the GDP deflator, which is an average of all prices in the economy in a given year. As explained in section 1.13, it is customary to choose units of measurement for a price index so as to make its initial value 100. It is then easy to calculate the percentage change in subsequent years.

EXAMPLE 10.11

Suppose we are told that the price index (the GDP deflator) equalled 100 in year 0 and then rose by 4% per year for the next 5 years. Then we can use the formula $y = a(1 + r)^x$, with $a = 100$, $r = 0.04$, and $x = 0, 1, 2, 3, 4, 5$ to calculate the price index in each of the 6 years (that is, the base year together with the next five). Since the numbers in this example are the same as in the example 10.9, the values for the price index will be the same as in table 10.5 above, and as shown in figure 10.1. The only

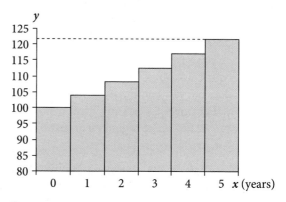

Figure 10.1 Growth of a bank deposit (example 10.9) or a price index (example 10.11).

difference is that the vertical axis now measures the price index, rather than the value of my investment in euros, as was the case in example 10.9.

Another small difference is that we should , in the case of the price index, view the step function in figure 10.1 a little differently. My bank deposit in example 10.9 jumps at the end of each year when the bank's computer calculates and adds on the interest due. But the price index does not jump at the end of each year. Instead, we interpret the height of each bar as measuring the *average* level of prices over the whole year, relative to the base year. There is a step function in average prices but not in actual prices; these latter change every day.

EXAMPLE 10.12

As another example of an economic variable, consider GDP (gross domestic product, a measure of the total value of goods and services produced in the economy in a year). If GDP was initially 100 (billion euros, say) and grew at 4% per year for 5 years, we would have the same numbers as in the previous example, and we could show the path of GDP graphically exactly as in table 10.5 and figure 10.1. The heights of the bars would then give us the total value of goods and services produced in each year. As in the previous example, if we view the tops of the bars as forming a step function, the step between, say, year 1 and year 2 tells us, not that output suddenly jumped at the end of each year 1, but that the total amount produced in year 2 as a whole was higher than the corresponding total for year 1.

 Hint Beware of a mistake that is often made in handling growth problems like those above. The mistake is to assume that if a variable grows at 4% per year for 5 years, the total growth is $4\% \times 5 = 20\%$. This would be correct if growth followed the simple interest formula (see section 10.4); but, as explained above, growth almost invariably follows the compound growth formula (rule 10.4).

10.6 Discrete versus continuous growth

Up to this point we have considered variables that increase in a jump from one year to the next. In the case of the bank deposit, the jump occurs because interest is added at the end of the year. In the examples of the price index and GDP, the jump occurs because the variable in question is recalculated every year using new data. Growth that occurs in jumps like this is called **discrete growth** ('discrete' means discontinuous or separate—not to be confused with 'discreet'!). For this reason we will from now on refer to the formula $y = a(1 + r)^x$ in rule 10.4 as the discrete, or discontinuous, growth formula. We examine continuous growth in chapter 12.

10.7 When interest is added more than once per year

The compound growth formula $y = a(1 + r)^x$ assumes that interest is added, or more generally that growth occurs, in a single jump at the end of each year. Now we examine how the formula must be modified when growth occurs in more than one jump each year. We can make this idea more concrete by thinking of lending money to a bank where the interest is added to the account every 6 months.

EXAMPLE 10.13

We assume €100 is deposited with a bank with an interest rate of 10% per year, as in example 10.7, but now with interest added every 6 months. The effect of this is that, every 6 months, the bank must credit half of a full year's interest to the account. It calculates this by using an interest rate of half the annual rate: that is, 5%. This process is shown in table 10.6.

Looking at the first row of table 10.6, we see that the first payment of interest, €5, is received after 6 months. This is then added to the principal, so the amount invested for the second half of the first year is €105. Therefore the interest added at the end of the second half year is 5% of 105, which is €5.25. Therefore the value of the investment grows from €100 to €110.25 by the end of the first year, an increase of 10.25%, compared with only 10% if interest was added once a year (see table 10.3).

This process is then repeated in the second year. The investment grows from 110.25 to 121.55 during the second year, again an increase of 10.25%. In every year, in fact, growth is 10.25% compared with 10% when interest is added only once per year.

We can summarize this example as follows. When interest is added twice per year instead of once per year, the interest rate is divided by 2 but the frequency with which interest is added is multiplied by 2. These two differences do not cancel one another out, for the investment grows at a faster annual percentage rate.

Table 10.6 Compound interest added every 6 months.

(1)	(2)	(3)	(4)
No. of years since account opened	Balance (€) (= cumulative value of investment) at beginning of each half-year	Interest added at end of half-year (€)	Balance at end of half-year (= balance at beginning of half-year + interest) (€)
0.5	100	$(100)0.05 = 5$	$100 + 5 = 105$
1	105	$(105)0.05 = 5.25$	$105 + 5.25 = 110.25$
1.5	110.25	$(110.25)0.05 = 5.5125$	$110.25 + 5.5125 = 115.76$
2	115.76	$(115.76)0.05 = 5.7881$	$115.76 + 5.7881 = 121.55$
2.5			
3			$= 134.01$

(Test your understanding by filling in the blanks in this table.)

In example 10.13 we saw that when interest is added twice per year instead of once per year, the interest rate is divided by 2 but the frequency with which interest is added is multiplied by two. Rule 10.4 therefore becomes $y = a(1 + \frac{r}{2})^{2x}$, because the interest rate, r, is divided by 2; but after x years, interest will have been added $2x$ times.

This example can easily be generalized to any case where interest is added n times per year, for then the interest rate is divided by n; but after x years, interest will have been added nx times. The compound growth formula is therefore modified as follows:

RULE 10.5 Compound growth when interest is added n times per year

When interest is added n times per year, our earlier compound growth formula (rule 10.4), $y = a(1 + r)^x$, becomes

$$y = a(1 + \frac{r}{n})^{nx}$$

Using rule 10.5 we can handle cases where interest is added monthly, weekly, daily, or at any other frequency, as we shall see below.

Nominal and effective interest rates

In example 10.13 we saw that when interest was added twice per year the investment grew to 110.25 by the end of the first year, compared with 110 when interest was added once per year. This means that the *effective* annual interest rate is 10.25% in this example, while the *nominal* interest rate is 10%. The effective annual interest rate gives the actual growth in the investment each year, while the nominal interest rate gives the growth that would occur if interest were added only at the end of each year. The effective annual interest rate is greater than the nominal annual interest rate whenever interest is added more than once per year.

This is because adding interest, say, twice yearly has two effects. First, the amount of interest added is halved (from 10% to 5% in the example above) because only half a year's interest is earned in each half-year. Second, interest is added twice as often (for example, four times in 2 years). The net effect is to increase the value of y after x years, compared with when interest is added only once per year (compare tables 10.6 and 10.3).

We can derive a formula for the effective annual interest rate (EAR). Let us set $a = €1$ and $x = 1$, so we are considering €1 lent for 1 year. If interest is added n times in the year, from rule 10.5 we will get back $y = 1(1 + \frac{r}{n})^n = (1 + \frac{r}{n})^n$ euros at the end of the year. But this includes our original investment of €1, so we must deduct this to find the interest alone. The *amount* of interest is therefore $(1 + \frac{r}{n})^n - 1$ euros, and to get the *rate* of interest we must divide this by our initial investment, which was €1. Dividing by €1 enables us to cancel the currency unit between numerator and denominator, leaves us with the pure number, $(1 + \frac{r}{n})^n - 1$ which is the effective interest rate. We have therefore deduced rule 10.6:

> **RULE 10.6 Nominal and effective interest rates**
>
> When interest is added n times per year, the effective annual interest rate (EAR) is given by the formula
>
> $$EAR = (1 + \tfrac{r}{n})^n - 1$$
>
> where r is the nominal annual interest rate.
>
> Note that if interest is added only once per year, we have $n = 1$ and the formula collapses to
>
> $$EAR = (1 + \tfrac{r}{1})^1 - 1 = r$$
>
> The nominal and effective interest rates are thus equal.

The distinction between nominal and effective interest rates or growth rates is important and we will refer to it frequently in this and later chapters.

EXAMPLE 10.14

As a test run of rule 10.6, we will calculate the EAR in example 10.13. We had $r = 0.1$ and $n = 2$, so rule 10.6 gives the effective rate as

$$(1 + \tfrac{0.1}{2})^2 - 1 = (1.05)^2 - 1 = 0.1025 = 10.25\% \text{ (as we found in example 10.13)}.$$

EXAMPLE 10.15

Some banks and other financial institutions add interest on a daily basis, so in this case $n = 365$. Then if €1 is invested for 1 year at a nominal rate of $r = 0.1$, rule 10.6 gives the effective interest rate as

$$(1 + \tfrac{r}{n})^n - 1 = (1 + \tfrac{0.1}{365})^{365} - 1 = 10.5156\% \quad \text{(using the } \wedge \text{ or } y^x \text{ key).}$$

The distinction between the effective and the nominal interest rate is important in the real world. In the UK, the government requires banks and other financial institutions to publish the effective interest rates that they charge for lending, known as the 'annual percentage rate' (APR) and calculated using a formula specified by the Office of Fair Trading. The APR is formally identical to the EAR in rule 10.6 above, though the real-world calculation is often more complex than our simple example here. Another real-world interest rate is the AER, the annual effective rate, which measures the effective interest rate paid on deposits with these financial institutions and is calculated in a way very similar to our EAR.

Although our examples above considered a bank paying interest, rule 10.6 is equally valid for any variable growing in periodic jumps. For example, if data on GDP are published quarterly (that is, for the periods January–March, April–June, and so on) we could use the rule with $n = 4$ to analyse its effective annual growth.

In chapter 12 we consider what happens to the formula in rule 10.6 when n gets very large.

Worked examples

To conclude our study of compound growth we will look at some worked examples using rules 10.4–10.6.

EXAMPLE 10.16

What is the future compounded value of €100 invested (= lent) for 20 years at 5% p.a. compound, interest added annually?

Answer: Here we use $y = a(1 + r)^x$ and solve for y, with $a = 100$, $r = 0.05$ (because $100r = 5$), $x = 20$.

$$y = a(1 + r)^x = 100(1 + 0.05)^{20} = 100(1.05)^{20}$$

Using the $\wedge$ or y^x key on calculator, $(1.05)^{20} = 2.6533$. So

$$y = 100(2.6533) = 265.33$$

Note that with simple interest y would be only 200 (= interest of €5 per year for 20 years, plus the initial sum of €100), so we can see the powerful effect of compound growth over a long period.

EXAMPLE 10.17

UK real GDP in 1980 was £293,325 million, and grew at an average rate of 2.6% p.a. in 1980–90. What was the level of GDP in 1990?

Answer: $y = a(1 + r)^x$ with $a = 293,325$, $r = 0.026$, $x = 10$. So

$$y = 293,325(1.026)^{10} = 379,160$$

EXAMPLE 10.18

If I deposit €100 in a bank that pays interest at a nominal rate of 10% per year, how much will I get back after 5 years if interest is added (a) annually; (b) monthly?

Answer:

(a) Using rule 10.4, $y = a(1+r)^x$ with $a = 100$, $r = 0.1$ (because $100r = 10$), $x = 5$, we get

$$y = a(1+r)^x = 100(1+0.1)^5 = 161.05$$

(b) Using rule 10.5, $y = a(1+\frac{r}{n})^{nx}$ with $a = 100$, $r = 0.1$, $x = 5$, $n = 12$, we get

$$y = a(1+\frac{r}{n})^{nx} = 100(1+\frac{0.1}{12})^{60} = 164.53$$

EXAMPLE 10.19

I paid €35,000 for my house 20 years ago. It is now worth €350,000. What was the average annual growth rate of its value?

Answer

$y = a(1+r)^x$ with $a = 35,000$, $y = 350,000$, $x = 20$ and r unknown. This is slightly more difficult than the previous examples because we are trying to find r rather than y. We can do this by taking the formula $y = a(1+r)^x$ and performing two operations: (1) divide both sides of the equation by a; (2) raise both sides to the power $\frac{1}{x}$. The first operation gives

$$\frac{y}{a} = (1+r)^x \quad \text{and the second gives}$$

$$\left[\frac{y}{a}\right]^{1/x} = [(1+r)^x]^{1/x}$$

The right-hand side of this simplifies to $1 + r$ (using rule 2.4), so

$$r = \left[\frac{y}{a}\right]^{1/x} - 1$$

Substituting for y, a, and x on the right-hand side, we get

$$r = \left[\frac{350,000}{35,000}\right]^{1/20} - 1 = [10]^{0.05} - 1 = 0.122 = 12.2\% \text{ (per year)}$$

Summary of sections 10.1–10.7

In section 10.2 we derived rule 10.2b, which gives us the formula for the sum of the terms of a geometric series, with rule 10.3 as an important special case. In section 10.3 we applied the latter rule to derive the Keynesian investment multiplier (see also section 3.17). In sections 10.4 and 10.5 we developed rule 10.4, the compound interest or compound growth formula, which is used where growth is in annual (or other periodic) jumps, known as discrete growth. We showed some of its economic and financial applications. In sections 10.6 and 10.7 we extended this to rule 10.5, which deals with cases where interest is added, or growth occurs, more than once a year. Rule 10.6 tells us how to calculate the effective interest or growth rate in such cases.

The Online Resource Centre gives further discussion and worked examples of compound growth problems, and shows how Excel® can be used for both the calculations and graphical presentation of results. www.oxfordtextbooks.co.uk/orc/renshaw3e/

Progress exercise 10.2

1. If national income (GDP) is 100 now and grows at 2.5% per year for the next 25 years, what will be its level at the end of this period? (Assume growth in annual jumps.)

2. Aggregate real consumption by all households in the UK rose from £411 billion in 1992 to £538 billion in 2000. What was the average annual growth rate? (Assume growth in annual jumps.)

3. A child's height has increased from 100 cm to 115 cm in the past year. If this growth rate continues, how tall will she be in 5 years' time? (Assume growth in annual jumps.)

4. If €100 is invested at 6% per annum for 5 years, find its future value when interest is added (a) annually, (b) twice a year, (c) monthly.

5. Bank A pays a nominal interest rate of 5% per year on deposits, credited annually in arrears. Bank B pays a nominal rate of 4.9% per year, but interest is credited at the end of each month. Calculate the return over a 1-year horizon from lending to A and B. What are the effective interest rates offered by A and B?

6. A firm owns a machine which when new produces annual output worth €100,000. However, as the machine gets older its productivity falls due to wear and tear, and after 5 years its annual output will be worth only €20,000. What is the average annual rate of depreciation? (Assume depreciation causes the machine to lose a fixed proportion of its productive capability, in a jump at the end of each year.)

10.8 Present discounted value

Up to this point, except in example 10.19, we have assumed that a (the principal, or initial value of the variable) was known and we wanted to find y, the future compounded value of a. Let's now instead consider a case where we know y and want to find a.

EXAMPLE 10.20

Suppose I wish to invest €a now, so that I will have €15,000 in 10 years' time, with interest added annually at $100r = 5\%$ per year compound (that is, with interest reinvested). What is the value of a?

Answer: This is a compound growth problem like those considered above, with $r = 0.05$, $x = 10$. What is new is that now we know y ($y = 15,000$) but a is unknown. However, using rule 10.4, $y = a(1 + r)^x$, we can solve for a by dividing both sides by $(1 + r)^x$, giving

$$a = \frac{y}{(1+r)^x} = \frac{15,000}{(1.05)^{10}} = 9208.7$$

Here €9208.7 is called the **present discounted value** (present value, or PV, for short) of €15,000 in 10 years' time, discounted at 5% p.a. It is the sum I need to invest now, in order to have €15,000 in 10 years' time, at the given interest rate. This is true because $9208.7(1.05)^{10} = 15,000$. Note that discounting is the reverse of compounding. Discounting is when we *divide* by $(1 + r)^x$, resulting in a smaller number (assuming r is positive). Compounding is when we *multiply* by $(1 + r)^x$, resulting in a larger number. When we are discounting, r is often called the discount rate (rather than the interest rate).

Generalizing from example 10.20, when we are given y, r, and x and wish to find a, we divide both sides of rule 10.4 by $(1 + r)^x$ and thus obtain a as

$$a = \frac{y}{(1+r)^x} \tag{10.10}$$

where a is the present discounted value (PV). Thus equation (10.10) is simply the compound growth formula $y = a(1 + r)^x$ but rearranged to isolate a, the unknown, on the left-hand side. In other words, it's the inverse function to the compound growth formula $y = a(1 + r)^x$. (See section 3.5 if you need to revise the inverse function.)

A confusing change of notation

To comply with the maths convention that y denotes the *dependent* variable, a and y must swap places in equation (10.10) above. This is potentially very confusing, but keep in mind that this is merely a relabelling that doesn't really change anything. It is simply that in equation (10.10) above, the unknown PV, which was previously labelled a, must now be relabelled y. And the known future value, previously y, is relabelled a.

After this interchange of a and y, equation (10.10) above is written as

$$y = \frac{a}{(1+r)^x} \tag{10.11}$$

where now y is the unknown PV and a the known future value. This formula is used to solve for y when a (and r and x) are known.

If discounting occurs n times per year, equation (10.11) becomes

$$y = \frac{a}{(1+\frac{r}{n})^{nx}} \tag{10.12}$$

This exactly parallels the way in which the compound growth formula $y = a(1 + r)^x$ becomes $y = a(1+\frac{r}{n})^{nx}$ when interest is added (or, more generally, when growth occurs) n times per year. We thus have two rules for finding a present discounted value.

The present discounted value (PV) of a receipt of a due in x years' time, discounted at $100r\%$ per year, is given by y in the following:

RULE 10.7

$$y = \frac{a}{(1+r)^x} \quad \text{also written as} \quad y = a(1+r)^{-x}$$

(when discounting occurs once per year)

RULE 10.8

$$y = \frac{a}{(1+\frac{r}{n})^{nx}} \quad \text{also written as} \quad y = a(1+\frac{r}{n})^{-nx}$$

(when discounting occurs n times per year)

Here a is the known future value and y is the unknown present value.

In rule 10.7 and 10.8, y is the *present discounted value*. In rules 10.4 and 10.5, which for ease of comparison we repeat below, y is the *future compounded value*.

RULE 10.4 Compound growth formulae

$$y = a(1 + r)^x$$

(when growth occurs once per year)

RULE 10.5

$$y = a(1+\frac{r}{n})^{nx} \quad \text{(when growth occurs } n \text{ times per year)}$$

Here a is the known present value and y is the unknown future value.

It is mainly rules 10.4 and 10.7, where growth or discounting occur once per year, that are used by economists. So the two key formulae to remember are the growth formula $y = a(1 + r)^x$ (where y = future compounded value) and the *PV* formula $y = \frac{a}{(1+r)^x}$ (where y = present discounted value).

To avoid confusing these two, remember that when you invest, you get growth. So in the compound growth formula, y is bigger than a, so you *multiply a* by $(1 + r)^x$ to get y. In the *PV* formula, the opposite is true; a large sum in the future is equivalent to only a smaller sum today, so you *divide a* by $(1 + r)^x$ to get y.

10.9 Present value and economic behaviour

This is a good moment to bring a little economic analysis into the question of present and future values. In the example above we found that €9208.7 was the present value (*PV*) of €15,000 in 10 years' time, discounted at 5% p.a. The *PV* is the sum I need to invest now in order to have €15,000 in 10 years' time, given the interest rate.

Economic theory takes the view that these two amounts of money, €9208.7 now and €15,000 in 10 years' time, are *equivalent* to one another provided certain assumptions hold.

By equivalent, we mean that the offer of *either* €9208.7 now *or* €15,000 in 10 years' time has the same value to me, if I can borrow or lend as much as I wish at 5% p.a. (and there is no risk or uncertainty, such as the fact that, tomorrow, interest rates might change or I might die). Why do they have the same value? Because (1) if someone gives me €9208.7 now, I can lend it for 10 years and get back €15,000 (including accumulated interest); (2) if someone promises to give me €15,000 in 10 years' time with no risk of default, I can borrow €9208.7 now and repay it in 10 years' time (including accumulated interest) with the €15,000 that I will then have. So by lending or borrowing I can *convert* €9208.7 now into €15,000 ten years from now, and vice versa, with no cost or risk. Hence they are equivalent because have the same **market value** to me. (In the language of microeconomics we say that €9208.7 now, and €15,000 ten years from now, lie on the same inter-temporal budget constraint.)

This idea of a future sum and its *PV* having the same value to a person or firm that can borrow or lend freely at the given interest rate (and for whom there is no risk or uncertainty) is used very frequently in economics.

Note, though, that while €9208.7 now and €15,000 ten years from now may have the same market value to me, which of the two I will choose depends on my *preferences*. In example 10.20, I preferred to give up €9208.7 now in order to receive €15,000 ten years from now. Another individual might prefer to do the opposite. If so, she and I can meet in the market place and hopefully strike a mutually beneficial deal.

10.10 Present value of a series of future receipts

EXAMPLE 10.21

In the previous section we calculated the present discounted value (*PV*) of a single future receipt or payment, €15,000 in 10 years' time (discounted at 5% per year).

Suppose now I have to make *two* future payments, say €15,000 in 10 years' time and €20,000 in 20 years' time. (This could arise for example if I have two children, one expected to start university in 10 years' time, the other in 20 years' time.)

Suppose I wish to know what lump sum (single amount) I must invest now in order to have *both* of these amounts in future. This lump sum is the combined present value of the two future amounts.

The solution is simple: we just treat each future payment separately, find its *PV*, and add the results, as follows:

(1) *PV* of 15,000 in 10 years' time $= \dfrac{15,000}{(1.05)^{10}} = 9208.7$

(as found in section 10.8 above)

(2) *PV* of 20,000 in 20 years' time $= \dfrac{20,000}{(1.05)^{20}} = 7537.8$

So combined *PV* = 9208.7 + 7537.8 = 16,746.5.

Thus I need to invest now a lump sum of 16,746.5. Although the total of 16,746.5 may be in a single bank account, I can view it conceptually as being composed of two distinct 'tranches' (a word used in finance, meaning 'slices' in French). The first tranche, 9208.7, will grow to 15,000 over 10 years and will then be withdrawn (to pay the first child's fees, say). The second tranche, 7537.8, will grow to 20,000 over 20 years and will then be withdrawn (to pay the second child's fees). See figure 10.2 for a diagrammatic treatment.

The idea of present values and future amounts having equal value, as discussed above, is equally valid in this example. If I can borrow or lend freely at 5% p.a. (and all risk and uncertainty are ignored), I can make a deal in the market place in which I exchange 16,746.5 now for a package consisting of promises of 15,000 in 10 years *and* of 20,000 in 20 years. And I could also make the reverse deal. Thus the two alternatives have equal market value to me. Which of them I choose depends on my preferences.

And of course it doesn't really matter whether we are talking about future payments or future receipts. Receiving 16,746.5 now and receiving 15,000 in 10 years plus 20,000 in

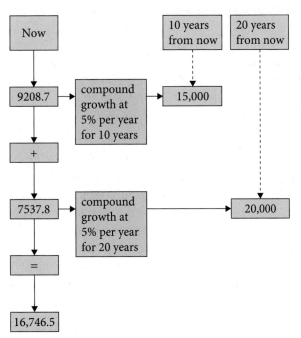

I invest 9208.7 now, which grows to 15,000 in 10 years. I also invest 7537.8 now, which grows to 20,000 in 20 years. Thus 16746.5 is the total amount I need to invest now in order to produce a future income series comprising 1500 in 10 years' time and 2000 in 20 years' time (discounted at 5% per year). Therefore by definition 1674.64 is the *PV* of this future income series.

Figure 10.2 Combined present value of €15,000 in 10 years' time and €20,000 in 20 years' time.

20 years have equal value to me. And equally, paying 16,746.5 now and paying 15,000 in 10 years plus 20,000 in 20 years have equal value to me. Of course, I would always rather receive money than pay money, but that's not the choice I face here!

The key concept to grasp is the notion of some future amount(s) and their *PV* having equal value, in the sense we have defined. But equal value in this sense does assume complete certainty about the future and the ability to borrow or lend unlimited amounts at a constant interest rate. These are strong assumptions! They can be relaxed to some extent, but this would take us beyond the scope of this book.

Present value of a future income series: generalization

Using the method of the previous example, we can find the *PV* of any series of future payments or receipts. (Note that we refer to a sequence of future receipts or payments as an income or expenditure 'series', but the terms 'income stream' or 'expenditure stream' are also commonly used. Here we try to avoid the term 'stream', as it may give the impression that the receipts or payments are a continuous flow when in fact they are discrete amounts.)

EXAMPLE 10.22

Find the *PV* of the payments: €500 one year from now, €400 two years from now, €300 three years from now, €200 four years from now and €100 five years from now. Assume a discount rate of 5% per year.

Answer:

(1) the *PV* of 500 one year from now is $\frac{500}{(1.05)} = 476.19$

(2) the *PV* of 400 two years from now is $\frac{400}{(1.05)^2} = 362.81$

(3) and so on . . .

So the combined *PV* is simply the sum of these individual *PVs*; that is:

$$PV = \frac{500}{(1.05)} + \frac{400}{(1.05)^2} + \frac{300}{(1.05)^3} + \frac{200}{(1.05)^4} + \frac{100}{(1.05)^5}$$
$$= 476.19 + 362.81 + 259.15 + 164.54 + 78.35 = 1341.04$$

Analogously to the previous example, the €1341.04 can be thought of as being composed of five slices or tranches. The first tranche, €476.19, grows to €500 after 1 year and is then withdrawn. The second, €362.81, grows to €400 after 2 years and is then withdrawn. The third, €259.15, grows to €300 after 3 years and is then withdrawn. The fourth, €164.54, grows to €200 after 4 years and is then withdrawn. Finally the fifth, €78.35, grows to €100 after 5 years and is then withdrawn. We show this process in full detail in table 10.7. Be sure to check all of the calculations in the table before moving on.

Table 10.7 Tranches of *PV* and their growth.

	Invested now (€)	Value (€) at end of:				
		Year 1	Year 2	Year 3	Year 4	Year 5
Tranche 1	476.19	500.00				
Tranche 2	362.81	380.95	400.00			
Tranche 3	259.15	272.11	285.71	300.00		
Tranche 4	164.54	172.77	181.41	190.48	200.00	
Tranche 5	78.35	82.27	86.38	90.70	95.24	100.00
Total invested now	1341.04					

Generalization

Extending the example above, the general formula is presented in rule 10.9.

RULE 10.9 **Present value of a future series of payments or receipts**

A future series of payments or receipts comprising: a_1 one year from now, a_2 two years from now, $\ldots$, a_n n years from now, discounted at $100r\%$ p.a., has a present discounted value (PV) given by

$$PV = \frac{a_1}{(1+r)} + \frac{a_2}{(1+r)^2} + \cdots + \cdots + \frac{a_n}{(1+r)^n}$$

As discussed in section 10.9, we view the series of payments or receipts and their PV as having equal value, in the sense that, if you could borrow or lend unlimited amounts at $100r\%$ p.a., and there was no uncertainty, you could exchange either for the other at no cost.

Note that the process of finding a PV is sometimes called **capitalization**, because the PV is the capital sum now which is equivalent in value to the given future series of payments or receipts.

10.11 Present value of an infinite series

A special case of the PV formula above, which is important in economics, arises when (1) the payments or receipts are equal and (2) the series continues forever, this being called a perpetual income series (or income stream).

Since the payments or receipts are equal, we have $a_1 = a_2 = \cdots = \cdots = a_n$. With equal payments or receipts the subscripts are redundant and we can drop them. The PV expression then becomes

$$PV = \frac{a}{(1+r)} + \frac{a}{(1+r)^2} + \cdots + \cdots + \frac{a}{(1+r)^n} \quad \text{(a sum with an infinite number of terms)}$$

Here the right-hand side is the sum of the terms of a GP with first term $A = \frac{a}{1+r}$, common ratio $R = \frac{1}{1+r}$ (which is a positive fraction), and an infinite number of terms. Using rule 10.3 above, the sum is therefore

$$\sum_{\infty} = \frac{A}{1-R} = \frac{\frac{a}{1+r}}{1 - \frac{1}{1+r}} = \frac{\frac{a}{1+r}}{\frac{1+r-1}{1+r}} = \frac{a}{1+r} \cdot \frac{1+r}{r} = \frac{a}{r}$$

Thus we conclude:

RULE 10.10 **Present value of an infinite series**

A series of equal payments or receipts of £a per year forever has a present value, when discounted at $100r\%$ per year, given by: $PV = \frac{a}{r}$

Here is an example of rule 10.10 in operation:

EXAMPLE 10.23

Find the PV of €100 a year forever, discounted at 5%.

Answer: We have $a = 100$, $r = 0.05$, so $PV = \frac{a}{r} = \frac{100}{0.05} = 2000$

This answer makes sense, for if you had €2000 you could invest it at 5% and draw an income of €100 a year *forever* (because the capital would remain intact). As discussed earlier, you would therefore be indifferent between receiving (1) a lump sum of €2000 now or (ii) €100 per year forever, provided you could borrow or lend freely at 5% per year. (Of course, you personally can't actually receive €100 a year forever, due to your life being finite; but we can work around this slight problem by assuming that you plan to bequeath the right to receive €100 a year to your heirs, and they to their heirs, and so on.)

10.12 Market value of a perpetual bond

Rule 10.10 has an important role in macroeconomics, where we often assume that the only asset that can be bought and sold in the economy is a bond issued by the government, which promises to pay its owner a fixed interest payment every year, *forever*. This is known as a **perpetual bond**. We use rule 10.10 to find the *PV* of such a perpetual bond. If for example the fixed interest payment (sometimes known as the 'coupon') is €10 per year and the market rate of interest is 5%, then we have $a = 10$ and $r = 0.05$, so from rule 10.10 the *PV* of a bond is

$$PV = \frac{a}{r} = \frac{10}{0.05} = 200$$

Assuming that many such bonds are held by, and bought and sold between, members of the public, the question is, how is their market price determined? If we assume that everyone can borrow or lend freely at the market rate of interest, the answer is that the market price of a bond, P_B, must equal its *PV*. The explanation is as follows. Suppose the market rate of interest is 5% and the 'coupon' is €10 per year. Then, if P_B were, say, €210, every bond owner would wish to sell, since she could lend the €210 cash received at the market rate of interest of 5% and enjoy a return of 5% of €210, which equals €10.5, better than the €10 return from the bond. This pressure of selling would drive bond prices down whenever P_B was greater than the *PV*.

Similarly, if P_B were, say, €190, then everyone would borrow in the market at 5% and use the money to buy bonds, as their percentage return would be €10 ÷ €190 = 5.26%, thereby yielding a profit of 0.26%. This pressure from buyers would drive bond prices up whenever the price was less than the *PV*. We conclude that buying or selling pressure will drive the market price of bonds up or down until the price equals the *PV*. Therefore we have, as a condition for equilibrium in the bond market, that

$$P_B = PV = \frac{a}{r} \tag{10.13}$$

where a is the fixed interest payment on the bond (the 'coupon'), and $100r$ is the market rate of interest. Since a is a parameter, fixed when the bond was first issued (perhaps many years ago), this equation gives us the relationship between two variables: the market rate of interest, r, and the market price of bonds, P_B. Since r is the denominator on the right-hand side of equation (10.13), we recognize this as the equation of a rectangular hyperbola, graphed in figure 10.3. (See section 5.5 if you need to revise the rectangular hyperbola.) Thus P_B and r vary in *inverse proportion*: a doubling of the market rate of interest results in a halving of the bond price.

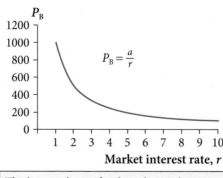

The 'coupon', a, is fixed, so the market price of a perpetual bond, P_B, and the market rate of interest, r, vary in inverse proportion.

Figure 10.3 Relation between the price of a perpetual bond and the market rate of interest, given the 'coupon', a.

Progress exercise 10.3

1. Assuming that I can borrow and lend any amount at 5% per year, with interest paid or received annually in arrears:

   (a) What is the present discounted value (PV) of €20,000 due 10 years from now?

   (b) What lump sum (= capital sum) must I invest (lend) now in order to have €20,000 in 10 years' time?

   (c) Suppose I know that I will inherit €20,000 in exactly 10 years' time. I want to borrow €x now and repay it (together with accumulated interest charges) in 10 years' time, from my inheritance. Determine the value of x such that my inheritance will be just sufficient.

   Hint: the answers to the three questions above are closely related.

2. Assuming that I can borrow and lend any amount at 5% per year, with interest paid or received annually in arrears:

   (a) What is the present discounted value to me of a flow of receipts of €10,000 each year for 10 years, received in annual instalments and beginning 1 year from now?

   (b) Suppose I am due to retire from work tomorrow. My employer offers me the choice between a pension of €10,000 per year for 10 years (paid annually at the beginning of each year) or a lump sum of €x now. Determine the minimum value of x that would induce me to accept the lump sum. (Assume that I am certain to live for 10 more years and that I do not care what happens after that.)

3. From the following table, calculate which of series A and B has the higher PV, when discounted at 10% per year. What can we infer from this answer regarding the 'front-loading' or otherwise of a series of payments or receipts?

Years from now	1	5	10
Series A (€m)	50	100	150
Series B (€m)	40	100	165

4. (This question illustrates a very important theoretical point in economics.) Consider the following two income series:

Years from now	1	2	3	4	5
Series A (€m)	20	40	50	50	120
Series B (€m)	80	70	50	40	8

   (a) Which series has the larger present discounted value, if the discount rate is (i) 5% per year and (ii) 10% per year? Try to explain your answers, in terms of 'front-loading'.

   (b) Suppose series A gives the costs of building and operating a nuclear power plant, with the high costs in year 5 being costs of nuclear decontamination. Suppose that series B gives the costs of a coal-fired power plant, where the costs are front-loaded. Both plants produce the same amount of electricity. How does your answer to (a) help to decide whether to build the nuclear or the coal-fired plant? (Assume no risks such as nuclear melt-down.)

10.13 Calculating loan repayments

EXAMPLE 10.24

Suppose that, after you finish your education and at last start to earn a salary, you borrow €5000 to buy a car and agree to repay the principal (the €5000) in five annual instalments of €1000, the first instalment to be paid 1 year from now. You also agree to pay, at the end of each year, interest at 10% per year on the amount owed during that year. Your repayments of capital, payments of interest and total payments would then be as shown in table 10.8.

Table 10.8 Repaying a loan with equal capital repayments (€).

		Year				
		1	2	3	4	5
Row 1	Owed at beginning of year	5000	4000	3000	2000	1000
Row 2	Interest paid at end-year (10% of row 1)	500	400	300	200	100
Row 3	Repayment of capital at end-year	1000	1000	1000	1000	1000
Row 4	End-year total payment (= row 2 + row 3)	1500	1400	1300	1200	1100

Total capital repaid (sum of row 3): 5000

Total payments (sum of row 4): 6500

Notice that in this repayment schedule the total amount paid declines each year, from 1500 at the end of year 1 to only 1100 in year 5. This is because, as you repay the principal, the annual interest charges decrease (see row 2). This repayment schedule is said to be 'front-loaded', meaning that early payments are larger than later payments. You might well find this inconvenient, and propose instead to the lender that you repay in equal annual payments. The problem is to calculate what these equal annual payment should be. You might think this is easily done, by simply averaging the annual payments in table 10.8. Let's see what happens when we do this.

The total payments in table 10.8 are 1500 + 1400 + 1300 + 1200 + 1100 = 6500. Dividing this by 5 gives us the average annual payment, 1300. If this were agreed as the annual payment, your repayment schedule would then be as shown in table 10.9.

We see from table 10.9 that repaying €1300 per year creates a new problem: the capital is not fully repaid by the end of the repayment period: there is still 5000 − 4884.08 = €115.92 owing. The reason for this is that simply averaging the total payments from table 10.8

Table 10.9 Repayment in equal instalments (€).

		Year				
		1	2	3	4	5
Row 1	Owed at beginning of year	5000	4200	3320	2352	1287.2
Row 2	Interest paid at end-year (10% of row 1)	500	420	332	235.2	128.72
Row 3	Repayment of capital at end-year (= row 4 minus row 2)	800	880	968	1064.8	1171.28
Row 4	End-year total payment	1300	1300	1300	1300	1300

Total capital repaid (sum of row 3): 4884.08

Total payments (sum of row 4): 6500

ignores the fact that repaying €1 of capital in year 1 is not the same as repaying €1 of capital in year 5. It is different because early repayments of capital mean that less interest has to be paid in subsequent years and therefore, given the total annual payment, more capital is paid off in subsequent years. In table 10.9, too little capital is repaid in the first year, which means that more interest has to be paid in later years, which means that too little capital is repaid in later years too; hence the shortfall of €115.92 at the end. This type of repayment scheme, where there is an extra or residual payment at the end, is called a **sinking fund**. (The residual payment may in principle be negative, of course.)

Of course, a pragmatic solution to this problem is simply to pay the extra €115.92 at the end of the final year and thereby clear the debt. But this is mathematically inelegant and anyway undermines the whole point of the exercise, which was to devise a repayment scheme with equal annual payments.

A formula for repayment in equal instalments

We will now develop a mathematical technique for finding how to pay off the loan with equal annual payments. To do this we must rework the example above in algebraic form. Suppose now that you borrow K (= 5000 in the example above) with an interest rate of $100r\%$ (= 10% in the example above), and repayment in five equal annual instalments as before. We also write P_1 as your repayment of capital at the end of year 1, P_2 as your repayment of capital at the end of year 2, and so on. Our task is to determine the P values. Your repayment schedule would then be as shown in table 10.10.

Looking first at year 1, we see that at the end of the year you have not yet made any repayments, so you owe what you initially borrowed, K (row 1). You therefore pay interest of rK (row 2). You also pay P_1 as repayment of capital (row 3), so your total payment at the end of year 1 is $P_1 + rK$ (row 4).

In year 2, your debt is now reduced to $K - P_1$ thanks to your repayment of capital of P_1 in year 1. So your interest is reduced to $r(K - P_1)$. You pay off P_2 of the capital, so your total payment at the end of year 2 is $P_2 + r(K - P_1)$. Years 3, 4, and 5 are constructed in the same way. (As an exercise, check that you can fill in the gaps indicated by question marks in the table for years 4 and 5.)

Now we come to the key step. By assumption, the total repayment in each year is the same. So the contents of each of the cells of row 4 in table 10.10 are equal to one another. Therefore, for years 1 and 2, we have from row 4:

$$P_1 + rK = P_2 + r(K - P_1)$$

By simple algebraic manipulation (check that you can do this) this rearranges as

$$P_2 = P_1(1 + r) \tag{10.14}$$

Table 10.10 Developing a formula for repayment in equal instalments.

		Year				
		1	2	3	4	5
Row 1	Owed at beginning of year	K	$K - P_1$	$K - P_1 - P_2$	?	?
Row 2	Interest paid at end-year (100r% of row 1)	rK	$r(K - P_1)$	$r(K - P_1 - P_2)$	?	?
Row 3	Repayment of capital at end-year	P_1	P_2	P_3	P_4	P_5
Row 4	End-year total payment (= row 2 + row 3)	$P_1 + rK$	$P_2 + r(K - P_1)$	$P_3 + r(K - P_1 - P_2)$	?	?

Table 10.11 Modification to Table 10.10 to give equal total annual repayments.

	Year				
	1	2	3	4	5
Table 10.10, row 3: Capital repayment	P_1	$P_1(1+r)$	$P_1(1+r)^2$	$P_1(1+r)^3$	$P_1(1+r)^4$

By following the same steps with years 2 and 3, we get from row 4

$$P_2 + r(K - P_1) = P_3 + r(K - P_1 - P_2) \text{ which rearranges as}$$

$$P_3 = P_2(1+r) \tag{10.15}$$

Comparing equations (10.14) and (10.15), we see a pattern developing, from which we can deduce, without bothering to work it out, that for years 4 and 5 we have

$$P_4 = P_3(1+r) \tag{10.16}$$

$$P_5 = P_4(1+r) \tag{10.17}$$

Next, by substituting equation (10.14) into equation (10.15) we can eliminate P_2 and get

$$P_3 = P_1(1+r)^2 \tag{10.15a}$$

and by substituting in the same way into equations (10.16) and (10.17) we get

$$P_4 = P_1(1+r)^3 \quad \text{and} \tag{10.16a}$$

$$P_5 = P_1(1+r)^4 \tag{10.17a}$$

Substituting equations (10.14), (10.15a), (10.16a), and (10.17a) into row 3 of table 10.10 above, the repayments of capital in years 1 to 5 become as shown in table 10.11.

Hopefully, you will recognize the series of repayments in table 10.11 as a GP with first term P_1 and common ratio $1 + r$. (See section 10.2 if you need to revise this.)

Now, we know that the total of these capital repayments must exactly equal the initial loan, K. Therefore we must have

$$P_1 + P_1(1+r) + P_1(1+r)^2 + P_1(1+r)^3 + P_1(1+r)^4 = K$$

Factoring out P_1, this becomes

$$P_1[1 + (1+r) + (1+r)^2 + (1+r)^3 + (1+r)^4] = K \tag{10.18}$$

The expression in brackets on the left-hand side of (10.18) is the sum of a GP with first term $A = 1$, common ratio $R = 1 + r$, and number of terms $n = 5$. Using rule 10.2, this sum is

$$\sum_n = \frac{A(R^n - 1)}{R - 1} = \frac{1[(1+r)^5 - 1]}{1 + r - 1} = \frac{(1+r)^5 - 1}{r}$$

Substituting this back into equation (10.18), we get

$$P_1\left[\frac{(1+r)^5 - 1}{r}\right] = K \quad \text{and therefore}$$

$$P_1 = K\left[\frac{r}{(1+r)^5 - 1}\right] \tag{10.19}$$

Equation (10.19) at last gives us what we want, because it tells us how to calculate the *first* repayment of capital, P_1, given the amount borrowed, K, and the interest rate, r. Once we have found P_1 we can calculate all the later payments of capital from table 10.11. We can also calculate the total instalment (capital plus interest) in the first year as $P_1 + rK$. Since by assumption the total instalment is the same in every year, this also gives us the instalments for the other years. This information can then be entered into a revised version of table 10.10 and the interest payments in years after year 1 can be calculated.

Admittedly, equation (10.19) is not a completely general formula because it assumes a repayment period of 5 years, but there was nothing in our derivation that depended on the repayment period being of this specific length. Thus the completely general formulation, found by simply replacing 5 in equation (10.19) with, say, x to denote the length of repayment period in any particular case, is as follows:

RULE 10.11 **Formula for calculating repayment of a loan in equal instalments**

$$P_1 = K\left[\frac{r}{(1+r)^x - 1}\right]$$

where P_1 = capital repaid in first instalment, K = amount borrowed, $100r$ = percentage interest rate per period, x = number of periods (usually years or months). The amount of each instalment (capital plus interest) is then $P_1 + rK$.

EXAMPLE 10.25

To test-drive this formula, let's try it out on the previous example (example 10.24), where $K = 5000$ and $r = 0.1$ (that is, $100r = 10\%$). The number of years was 5. Substituting these values into rule 10.11 gives

$$P_1 = 5000\left[\frac{0.1}{(1.1)^5 - 1}\right] = 5000\left[\frac{0.1}{0.61051}\right] = 5000[0.16379748] = 818.99$$

So the first repayment of capital needs to be 818.99 to exactly pay off the loan in five equal instalments of capital plus interest. In table 10.12 we have calculated the numbers to check that this actually gives the right answer. We have rounded up the first capital repayment to 819.

Looking at year 1, the first entry is simply the amount borrowed, 5000. Given the interest rate of 10%, this then determines the interest charged on the debt, 500. Then comes our

Table 10.12 Repayment in equal instalments (€).

		Year				
		1	2	3	4	5
Row 1	Owed at beginning of year	5000	4181	3280.1	2289.11	?
Row 2	Interest paid at end-year (10% of row 1)	500	418.1	328.01	228.911	?
Row 3	Repayment of capital at end-year (in year 1, calculated using rule 10.11; in remaining years, equals row 4 − row 2)	819	900.9	?	1090.089	1199.098
Row 4	End-year total payment (in year 1, equals row 2 + row 3; in remaining years, equals same as year 1)	1319	1319	1319	1319	1319
	Total capital repaid (from row 3): 5000.077					

calculation of the first year's capital repayment, 819. The total payment in year 1 is then determined as $500 + 819 = 1319$.

Moving to year 2, the first entry is the amount owed, now $5000 - 819 = 4181$. This determines the interest paid, 418.1. The total repayment is, by construction, the same as the first year, 1319. So the capital repayment in the second year is found as $1319 - 418.1 = 900.9$. The remaining years are filled in column by column in the same way; a few question marks have been left for you to fill in for yourself as a test of your understanding. As required, the total repaid equals the total borrowed, except for a negligible rounding error (less than 10 cents in a loan of €5000).

Instalment

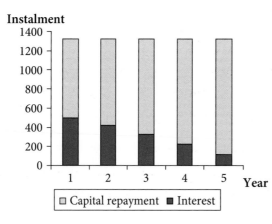

Figure 10.4 Instalment structure.

The structure of the instalments is shown in figure 10.4. We see that, within the equal instalments of €1319, the interest element declines progressively from its peak of €500 in year 1 to €119.90 in year 5, while the capital repayment element correspondingly increases from its lowest value of €819 in year 1 to €1199.10 in year 5.

The type of repayment schedule that we have examined in this section is very common in all types of borrowing, including consumer credit, mortgages, and business finance. The key characteristic is that instalments are equal but successive instalments include an increasing capital element and a diminishing interest element. This method of calculating the repayments (that is, using rule 10.11) is called the **annuity method**. It is so called because the method was originally developed to calculate annuity payments, an interesting type of financial transaction which unfortunately we don't have space to discuss here.

 Summary of sections 10.8–10.13

In section 10.8 we introduced the concept of a present discounted value, or present value, which in essence involves 'reversing' the compound growth formula. In section 10.9 we examined the underlying assumptions of the *PV* concept and formula. The present value of a future series of payments or receipts, comprising: a_1 one year from now, a_2 two years from now, ... a_n n years from now is calculated according to rule 10.9 (section 10.10).

In section 10.11 we examined an important special case of rule 10.9, arising when the payments or receipts are equal and continue forever. This gives us rule 10.10. In section 10.12 we used this rule to show that the market value of a perpetual bond varies in inverse proportion to the market rate of interest.

Finally, in section 10.13 we developed a formula to calculate the repayment of a loan in equal instalments (the annuity method), expressed as rule 10.11.

 The Online Resource Centre gives further explanation and examples of present value and loan repayment problems, and shows how Excel® can be used for both the calculations and graphical presentation of results. www.oxfordtextbooks.co.uk/orc/renshaw3e/

Progress exercise 10.4

1. Suppose in example 10.24 the interest rate was 5% instead of 10%. Calculate the equal annual instalments (interest plus capital repayment) in this case.

2. Suppose you are thinking of buying a new mobile phone priced at €200. The shop offers you the option of paying by 24 equal monthly instalments with the first payment 1 month after purchase. Interest will be charged monthly at 1.5% of the outstanding debt. Calculate the amount of each monthly payment and the total amount of interest paid.

3. Suppose I borrow €500,000 to buy a house, and agree to repay by 25 equal annual instalments with the first payment 1 year after purchase. Interest will be charged annually at the end of each year at 6% of the debt outstanding at the beginning of that year.

    (a) Calculate the amount of each total annual payment (capital repayment plus interest).

    (b) How much of the amount borrowed (known as a mortgage) will I have paid off after 5 years?

    (c) How much interest will I pay in total over the 25 years?

    (d) How are the calculations changed by the fact that mortgages are normally paid off in monthly rather than annual instalments? Give a general explanation only; do not calculate exact values (unless you are really keen).

Checklist

Be sure to test your understanding of this chapter by attempting the progress exercises (answers are at the end of the book). The Online Resource Centre contains further exercises and materials relevant to this chapter www.oxfordtextbooks.co.uk/orc/renshaw3e/.

In this chapter we have looked at the mathematics of compound growth, applied initially to the growth of interest-earning assets and then extended to growth of any economic variable. We also introduced a key economic concept, the present discounted value of a future receipt or payment, and developed a formula for calculating repayment of a loan. Specific topics were:

✔ **Arithmetic and geometric series.** Understanding and use of the formulae for an arithmetic and geometric series.

✔ **Compound growth.** The formula for compound growth and its use to solve problems involving growth of economic variables.

✔ **Present value.** Understanding the concept of present discounted value and using the *PV* formula to calculate the present discounted value of a series of future receipts.

✔ **Perpetual bonds.** What a perpetual bond is and how its the market value is calculated.

✔ **Nominal and effective interest rates.** Understanding the distinction between the nominal and effective interest rates, and how to calculate the effective rate.

✔ **Loan repayments.** Understanding and using the formula for calculating how a loan may be repaid in equal instalments, known as the annuity method.

These concepts and formulae relate to discontinuous growth and to discrete series of future payments or receipts. In the next chapter we develop the maths necessary to extend these ideas to continuous growth and continuous flows of future payments and receipts.

Chapter 11
The exponential function and logarithms

OBJECTIVES

Having completed this chapter you should be able to:

- Understand the exponential function $y = 10^x$ and sketch its graph.
- Understand how the logarithmic function $y = \log x$ is derived from the exponential function and sketch its graph.
- Use a calculator to find exponents and logs to base 10.
- Understand and apply the rules for manipulating logs to base 10.
- Use logs in solving growth and other problems.
- Understand how the exponential and logarithmic functions may be generalized and sketch the relevant graphs.

11.1 Introduction

In this chapter we will introduce a type of function that we haven't met before, called the exponential function. We will also introduce the function that is inverse to the exponential function, called the logarithmic function. Despite their rather scary names, these functions are based on fairly simple ideas and are quite user-friendly. They are really no more than a natural extension of the two basic formulae that we developed in chapter 10: the formula for compound interest or compound growth; and its inverse, the formula for the present discounted value of a series of future payments or receipts. Although the main focus of this chapter is on new mathematical developments, we are able to put the logarithmic function to work straight away as a means of solving compound growth problems in which the unknown is the number of years over which a given amount of growth occurs. The logarithmic function is also used quite widely in economics in other contexts than that of variables growing through time: for example, in a logarithmic specification for the demand function.

It is important to understand these new functions because they are an essential tool in any economic analysis involving variables that are growing through time, as we shall illustrate with some examples later in this chapter.

Preliminary: revision of the inverse function

Your understanding of this chapter will be greatly helped by a clear picture of exactly what an inverse function is. Therefore we begin with a brief revision of the inverse function (see section 3.5).

EXAMPLE 11.1

Suppose we take any function, say

$$y = 3x + 4$$

which is graphed in figure 11.1(a) for values of x between zero and 3 (see also table 11.1a). By convention, y is considered to be the dependent variable because it is isolated on the left-hand side of the equation. Figure 11.1(a), obeying this convention, puts the dependent variable y on the vertical axis.

Now consider the inverse function. By definition, this expresses x as a function of y. So we find the inverse function by finding an expression in which x is isolated on the left-hand side.

Given $y = 3x + 4$, by subtracting 4 from both sides and dividing both sides by 3, we obtain the inverse function as:

$$x = \tfrac{1}{3}y - \tfrac{4}{3}$$

It is now x that is the dependent variable, because it is isolated on the left-hand side. This function is graphed in figure 11.1(b), with x on the vertical axis so as again to obey the convention of putting the dependent variable on the vertical axis. Since in plotting a graph we may choose any values we wish for the independent variable, y, we have taken the values 4, 7, 10, and 13 (see table 11.1b).

Comparing figures 11.1(a) and 11.1(b), we see that the graphs of $y = 3x + 4$ and its inverse $x = \tfrac{1}{3}y - \tfrac{4}{3}$ are identical, except that the axes swap places in moving from one graph to the other. This swap occurs because in $y = 3x + 4$ the dependent variable is y, while in $x = \tfrac{1}{3}y - \tfrac{4}{3}$ the dependent variable is x, and by convention we put the dependent variable on the vertical axis.

However, it's important to see that this swapping of the axes does not change the relationship between x and y. This is because the function $y = 3x + 4$ and its inverse,

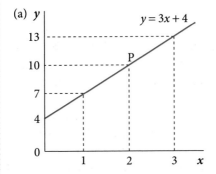

Figure 11.1 Inverse functions (not to scale).

Table 11.1a Values for $y = 3x + 4$.

x	0	1	2	3
y	4	7	10	13

Table 11.1b Values for $x = \tfrac{1}{3}y - \tfrac{4}{3}$.

y	4	7	10	13
x	0	1	2	3

$x = \frac{1}{3}y - \frac{4}{3}$, express, in two alternative ways, *exactly the same relationship* between x and y. Thus any pair of values of x and y which satisfies one function automatically satisfies the other. For example, the coordinates of point P in figure 11.1(a) are also the coordinates of point Q in figure 11.1(b).

In a nutshell, $y = 3x + 4$ and $x = \frac{1}{3}y - \frac{4}{3}$ are two ways of expressing the same relationship between x and y, just as 'John is Ann's brother' and 'Ann is John's sister' are two ways of expressing the same relationship between John and Ann. This will obviously be true of any pair of mutually inverse functions, not just the pair we have used as an example here.

Given, then, that a function and its inverse convey the same information, you may be wondering what is the purpose of the inverse function. The answer is as follows. The function $y = 3x + 4$ gives us a formula, or rule, for finding y for any *given* value of x. For example, given $x = 2$, we plug this number into the function and find the corresponding value of y as $y = 3(2) + 4 = 10$. In contrast, the inverse function $x = \frac{1}{3}y - \frac{4}{3}$ gives us a rule for finding x for any *given* value of y. For example, given $y = 10$, we can find the corresponding value of x as $x = \frac{1}{3}(10) - \frac{4}{3} = 2$. Thus whether we wish to work with a function or its inverse depends on whether it is the value of x or of y that is either given to us or that we are free to choose.

11.2 The exponential function $y = 10^x$

An example of an exponential function is

$$y = 10^x$$

The key feature is that the independent variable, x, appears as the power (= exponent) to which a constant base, 10 in this example, is raised. Don't confuse 10^x with x^{10}, which is a completely different animal.

There is no reason why the base should be restricted to 10. For example we can have $y = 3^x$, $y = 5^x$, and in general $y = a^x$ where a is any constant. (For reasons that we won't explain here, a is restricted to being greater than 1.) These cases are examined briefly in section 11.10 below, but here we will focus on $y = 10^x$.

Let's see what the graph of $y = 10^x$ looks like. Table 11.2 gives values for x between -3 and $+3$, and the graph is shown in figure 11.2(a).

In the table, let us first look at $x = 0$. Then we see $y = 10^0 = 1$. Thus the y-intercept of the graph is at $y = 1$. Then, as x takes increasing positive values, so y increases very rapidly and the curve turns up very sharply. In fact for $x = 1, 2,$ and 3, figure 11.2a is deliberately drawn with the y-axis out of scale, because otherwise the curve turns up so steeply that it quickly goes off the page (try to draw it to scale for yourself). Note that this is a continuous function because x can take any value. For example, if $x = 1.5 = {}^3/_2$, we have $y = 10^{3/2} = \sqrt[2]{10^3} = 31.623$.

Now let us consider negative values of x. From table 11.2 we see that as x becomes increasingly negative, y rapidly gets smaller but always remains positive. For example when $x = -2$,

Table 11.2 Values for $y = 10^x$.

x	-3	-2	-1	0	1	2	3
y	$10^{-3} = \frac{1}{10^3}$	$10^{-2} = \frac{1}{10^2}$	$10^{-1} = \frac{1}{10^1}$	$10^0 =$	$10^1 =$	$10^2 =$	$10^3 =$
	$= \frac{1}{1000}$	$= \frac{1}{100}$	$= \frac{1}{10}$	1	10	100	1000
	$= 0.001$	$= 0.01$	$= 0.1$				

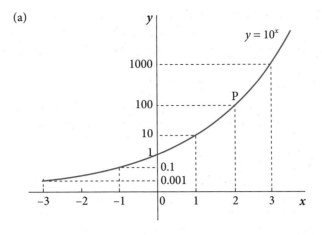

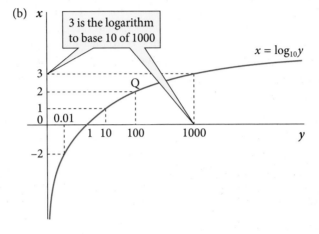

Figure 11.2 Graphs of $y = 10^x$ and $x = \log_{10} y$ (not to scale).

$y = 10^{-2} = \frac{1}{10^2} = 0.01$. And when $x = -3$, $y = 10^{-3} = \frac{1}{10^3} = 0.001$. Thus the curve becomes asymptotic (= approaches closer and closer) to the x-axis as x goes off towards $-\infty$. Again for $x = -1, -2$, and -3, the y-axis of figure 11.2a is deliberately drawn out of scale because otherwise the curve quickly gets so close to the y-axis that it seems to merge with it.

To summarize, the key features of the graph of the exponential function $y = 10^x$ are: (1) the curve never cuts the y-axis because 10^x is always positive whatever the value of x; (2) the y-intercept is at $y = 1$, because $10^0 = 1$; (3) between $x = -3$ and $x = +3$, a comparatively small range, y goes from 0.001 to 1000, a very large range.

11.3 The function inverse to $y = 10^x$

Now we come to a key step in the analysis. Suppose we want to find the function that is inverse to $y = 10^x$. Recall that the inverse function will express the same relationship between y and x, but rearranged so that x is isolated on the left-hand side and thus becomes the dependent variable. The function $y = 10^x$ gives a rule for getting from a given value of x to the corresponding value of y: for example, given $x = 2$, then $y = 10^2 = 100$. The inverse function, if we can find it, will similarly give us a rule for getting from a given value of y to the corresponding value of x.

Now, there's no problem in drawing the graph of the inverse function to $y = 10^x$. We know from section 11.1 that the graph of any inverse function is simply the same graph as the original function but with the axes interchanged. Following this rule, the inverse function to $y = 10^x$ is graphed as figure 11.2(b). We see that x is now on the vertical axis (because x is now the dependent variable) and y on the horizontal axis, but the relationship between x and y is unchanged. For example, point P in figure 11.2(a) has the same coordinates as point Q in figure 11.2(b).

A problem arises however when we try to find the equation of the function inverse to $y = 10^x$ by algebraic manipulation. In the case of $y = 3x + 4$ considered in section 11.1, this was a simple task: we subtracted 4 from both sides and divided through by 3, and thus obtained the inverse function as $x = \frac{1}{3}y - \frac{4}{3}$. But when we try to manipulate $y = 10^x$ in a similar way in order to arrive at something of the form

$x = f(y)$ where f(y) denotes some functional form

we find that no amount of algebraic manipulation of $y = 10^x$, using the rules of algebra developed so far in this book, can succeed in isolating x on one side of an equation (try it for yourself!).

The only way forward is to define a new concept, a logarithm. Specifically, when $y = 10^x$, we define x as being the logarithm of y, with a base of 10. Thus the logarithm x of any number y is the power to which 10 must be raised so that it equals y. This definition means that we can now write the function inverse to $y = 10^x$ as

$x =$ the logarithm of y, with a base of 10

We can write this definition in more concise mathematical language as in rule 11.1:

> **RULE 11.1 Definition of the logarithmic function**
>
> Given the exponential function: $y = 10^x$
>
> the inverse function is defined as: $x = \log_{10} y$
>
> where '$\log_{10}$' means 'the logarithm to base 10'.

In words, rule 11.1 says 'Given any number y, then its logarithm to base 10 is x, where x is the power to which 10 must be raised in order to equal y'.

The definition given by rule 11.1 means that the function inverse to $y = 10^x$ is

$x = \log_{10} y$

so this is the name of the function graphed in figure 11.2(b), and we have labelled that graph accordingly.

This means that in figure 11.2(b), we now interpret the x values as being the logarithms, to base 10, of the y values. Logarithms to base 10 are called **common logarithms**. Later we will consider logarithms with other bases.

You may think that there's something of a trick in overcoming a difficulty (finding the function inverse to $y = 10^x$) by defining a new concept (a logarithm). But there are many good precedents for this in maths: for example, negative numbers were defined to solve the problem that arose when people tried to subtract 8 from 5.

Hint We saw earlier that figures 11.1(a) and 11.1(b) express the same underlying relationship between x and y. In the same way, figures 11.2(a) and 11.2(b) also express the same relationship between x and y. This is the key to understanding what a logarithm is. Study the four figures carefully until you are clear on this fundamental point.

11.4 Properties of logarithms

Many students find logarithms deeply mysterious, at least initially. To dispel some of this mystery, look at figure 11.2(b) where we have the graph of $x = \log_{10} y$. By definition the xs are the logs (to base 10) of the ys. So, at point Q in figure 11.2(b) we see that 2 is the log of 100. This is true *because*, at point P in figure 11.2(a) we have $10^2 = 100$. If you can grasp this point, that 2 is the log of 100 *because* $10^2 = 100$; then hopefully you can grasp, with some practice, the generalization of this point, which is that x is the log of y *because* $10^x = y$.

An alternative way of trying to understand the relationship between a number and its log is to look solely at the exponential function $y = 10^x$, graphed in figure 11.2a. Since this is the same graph as figure 11.2(b) but with axes interchanged, we can also use this graph to find the logs of numbers. For example in figure 11.2(a) we see that when $x = 2$, $y = 10^2 = 100$. And since in this graph the xs are, by definition, the logs of the ys, it follows that $2 = \log_{10}100$.

The key thing to see here is that, as we said earlier, the statements '$y = 10^x$' and '$x = \log_{10} y$' are equivalent, in exactly the same way that 'John is Ann's brother' and 'Ann is John's sister' are equivalent. They are simply alternative ways of describing the same relationship.

If you have problems understanding what a log is, or perhaps have flashes of understanding interspersed with bouts of incomprehension, it is probably best to think graphically. Try to keep a picture in your mind of the graph of either $x = \log_{10} y$ (figure 11.2b); or of $y = 10^x$ (figure 11.2(a)), and remember that in either case the xs are the logs of the ys.

Important features of logs

From figure 11.2(b), we can see that:

(1) The curve never cuts the x-axis. So it is only when y is positive that there is a corresponding value of x. Since the xs are the logs of the ys, this means that only *positive* numbers have logs.

(2) When $y = 1$, $x = 0$. Since the xs are the logs of the ys, this means that $0 = \log_{10}$ of 1. In words, the log of 1 is zero. (This is because $10^0 = 1$.)

(3) When $0 < y < 1$, x is negative. For example, when $y = 0.01$, $x = -2$. Thus the log of any positive fraction is negative.

It is easy to get in a muddle over these three points, but again the best way of avoiding this is to keep figure 11.2(b), or 11.2(a) if you prefer, clearly in your head.

11.5 Using your calculator to find common logarithms

We noted earlier that logarithms to base 10 are called common logarithms, usually abbreviated to 'logs'. We can write '$\log_{10} x$' to denote the common logarithm of any number x, but we usually omit the subscript and write simply '$\log x$'. As we have just seen, the logs of some numbers can be read off the graph of $x = \log_{10} y$ in figure 11.2(b). For example, from figure 11.2(b) we see that $\log_{10}$ of 100 = 2, $\log_{10}$ of 0.01 = −2, and so on. Thus in principle if we plotted figure 11.2(b) sufficiently accurately and for a sufficiently large range of values of y, we could use the graph to read off the log to base 10 of any positive number.

However, in order to plot the graph we would first have to compile a table of values, which would necessitate calculating the logs. So the table of values would gives us the logs directly, making the graph superfluous. Each entry in the table of values is compiled by assigning a value to y, say $y = y_0$, and then finding the value of x that satisfies the equation $y_0 = 10^x$. These laborious calculations were first done nearly 400 years ago and the results published. Until a few decades ago, we found the logs of numbers by looking them up in these log tables. Now we just reach for our calculators, which are programmed to calculate common logs.

Thus on most calculators, to find $\log_{10}$ of 100, you just hit the key marked 'log', then key in 100 and press 'equals', and the answer, 2, is displayed. (Ignore the key marked 'ln' on your calculator for now.) Of course, 2 is the log of 100 because $10^2 = 100$.

Because it's very easy to make mistakes when using a calculator, it's a good idea to try to roughly check that the answer displayed is correct. Suppose for example we wanted to find the log of 29. We know that the log of 10 is 1 (because $10^1 = 10$). We also know that the log of 100 is 2 (because $10^2 = 100$). Since 29 lies between 10 and 100, its log must lie between 1 (the log of 10) and 2 (the log of 100). Using a calculator we get $\log_{10} 29 = 1.462$, which lies between 1 and 2, so is probably the right answer.

You can also reverse the process of finding a log. This process is sometimes called 'finding the anti-log'. Depending on the make and model of your calculator, you may have to press 'SHIFT', 'INV', or '2nd F', followed by the 'log' key, then key in, say, 2 and press the equals key, and the answer 100 is displayed. Here you have asked the calculator to find the number of which 2 is the log (answer: 100, because $10^2 = 100$). Thus asking your calculator to find the anti-log of 2 is equivalent to asking your calculator to find 10^2. (You can of course also find 10^2 directly by using the ^ or y^x key.)

Don't forget that the logs of fractions are negative: for example, $\log 0.1 = -1$ (because $10^{-1} = 0.1$). When reversing this example to find the anti-log of -1, hit 'SHIFT', 'INV', '2nd F' (or equivalent key), then 'log', then the key marked '–' or '+/–', then 1. (This procedure may vary a little according to the make and model of your calculator.) You should then see 0.1 on the display, this being the number of which -1 is the log.

Generalizing the example above, given $y = \log_{10} x$, we have two possibilities. First, where x is given, we use the 'log' key to find y, the log of x. Second, and reversing this, when y is given we use 'SHIFT', 'INV', or '2nd F' and 'log' keys to find x, the anti-log of y. This operation finds 10^y. At the end of this chapter there are some exercises aimed at making you confident at using your calculator to find logs and their reverse.

11.6 The graph of $y = \log_{10} x$

In figure 11.2b we drew the graph of the logarithmic function $x = \log_{10} y$. Here x is the dependent variable, but this violates the mathematical convention of denoting the dependent variable by y. To comply with the convention we must interchange the names or labels of the two variables, so that the variable called 'x' in figure 11.2b is now called 'y', and vice versa. This gives us the function $y = \log_{10} x$, which is graphed in figure 11.3, and which is identical to figure 11.2b apart from the relabelling of the axes.

The graphs in figures 11.2(b) and 11.3 both depict the same mathematical relationship: that is, between numbers and their logs. The only difference is that in figure 11.2(b) it is the xs that are the logs of the ys, while in figure 11.3 it is the ys that are the logs of the xs.

This relabelling can sometimes cause confusion. The function graphed in figure 11.2(b) ($x = \log_{10} y$) is the inverse function to $y = 10^x$ graphed in figure 11.2(a). Both express the same relationship between y and x. The function graphed in figure 11.3 ($y = \log_{10} x$) is not related to $y = 10^x$. The function inverse to $y = \log_{10} x$ is $x = 10^y$.

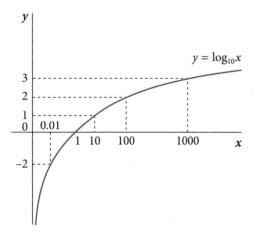

Figure 11.3 Graph of $y = \log_{10} x$ (x-axis not to scale).

Hint As explained above, we conventionally write 'log' to denote a 'common' logarithm: that is, a logarithm to base 10. Your calculator obeys this convention, so you should press the key marked 'log' to find a common logarithm. To find an anti-log (that is, to get from a log to the number of which it is the log) the procedure varies a little from one calculator to another, but typically you have to press 'SHIFT', 'INV', or '2nd F' followed by 'log'. You can also find the common anti-log of, say, 2.5 by finding $10^{2.5}$, using the ∧ or y^x key (answer: 316.228). It's a good idea to practise these operations until you know what to key in without too much prior thought and you cease to be surprised by what appears on the display of your calculator!

Progress exercise 11.1

1. *Without* using your calculator, try to find each of the following. In some cases your answer will necessarily be a rough estimate. Looking at figure 11.2(b) or 11.3 might help. Then use your calculator and check your answers. If you're not brave enough to do that, go straight to your calculator but at least look carefully at each answer and ask yourself whether you understand it. (Hint: by definition, if $x = \log y$, then $10^x = y$.)

   (a) $\log 10$ (b) $\log 100$ (c) $\log 2$ (d) $\log 0.5$

   (e) How are the answers to (c) and (d) related, and why?

2. (a) Draw a sketch graph of $y = 10^x$ for values of x between -3 and $+3$. Don't worry too much about getting the scale on the y-axis right.

   (b) Use the graph to estimate the approximate values of $\log(500)$ and $\log(\frac{1}{500})$. Compare these values with those given by your calculator (using the log key).

   (c) Use the graph to explain how $\log \frac{1}{y_0}$ and $\log y_0$ are related (where y_0 is any value of any variable, y).

   (d) Use the graph to explain why (i) the logs of positive fractions are negative, and (ii) the logs of negative numbers do not exist.

   (e) Use the graph to explain why $\log 10 = 1$ and $\log 1 = 0$.

3. (a) If $y = 10^{0.379}$, what is $\log y$? (Don't use a calculator!)

   (b) If $y_0 = 10^{x_0}$ and $y_1 = 10^{-x_0}$, how are y_0 and y_1 related?

4. (a) Sketch the graph of $y = \log x$ for $x > 0$. Don't worry too much about getting the scale on the x-axis right.

   (b) Use your graph to estimate the approximate values of $\log(50)$ and $\log(\frac{1}{50})$. Compare these values with those given by your calculator (using the 'log' key).

   (c) Explain how this graph is related to the graphs of $x = 10^y$ and $y = 10^x$.

11.7 Rules for manipulating logs

The manipulation of logs is governed by some simple rules, which we now list. We won't attempt to prove any of these rules, but we will try to explain them and give some examples to show them in action. From this point onwards we will simplify our notation by writing 'log x' instead of 'log$_{10} x$'. This simplification is standard practice in mathematics.

✱

RULE 11.2 **Rules for manipulating common logs**

For any two positive numbers, A and B:

Rule 11.2a $\log(AB) = \log A + \log B$

Example: let $A = 100$, $B = 1000$, so $AB = 100 \times 1000 = 100{,}000$. Then we have

   $\log(AB) = \log 100{,}000 = 5$ (because $10^5 = 100{,}000$)
   $\log A = \log 100 = 2$ (because $10^2 = 100$)   and
   $\log B = \log 1000 = 3$ (because $10^3 = 1000$)

Thus $\log(AB) = 5 = 2 + 3 = \log A + \log B$, confirming the rule in this example.

   Note that rule 11.2a extends: $\log(ABC) = \log A + \log B + \log C$, and so on.

Rule 11.2b $\log(A^n) = n \log A$

This is a special case of rule 11.2a. In rule 11.2a, if $A = B$, then $AB = A^2$. Then rule 11.2a becomes

   $\log(A^2) = \log A + \log A = 2(\log A)$

Example: let $A = 10$. Then $\log A = \log 10 = 1$, so $2 \log A = 2 \times 1 = 2$.

Also, $A^2 = 100$ and $\log A^2 = \log 100 = 2$.

Thus we have $\log A^2 = 2 \log A = 2$

Generalizing,

$\log(A^n) = n(\log_{10} A)$ for any value of n

Note that the brackets on the right-hand side are normally omitted.

Rule 11.2b is an important rule. It means, for example, that $\log(x^2) = 2(\log x)$, written as $2 \log x$.

Rule 11.2c $\log(\frac{A}{B}) = \log A - \log B$

Example: let $A = 100{,}000$, $B = 1000$, so $\frac{A}{B} = \frac{100{,}000}{1000} = 100$. Then we have

   $\log A = \log 100{,}000 = 5$ (because $10^5 = 100{,}000$)
   $\log B = \log 1000 = 3$ (because $10^3 = 1000$)
   $\log(\frac{A}{B}) = \log 100 = 2$ (because $10^2 = 100$)

Since $2 = 5 - 3$, we have $\log(\frac{A}{B}) = \log A - \log B$, confirming the rule.

   Note that rule 11.2c simply reverses rule 11.2a, in the sense that division reverses multiplication.

Rule 11.2d $\log 1 = 0$

This is a special case of rule 11.2c. If $A = B$, rule 11.2c becomes

$$\log\left(\frac{A}{A}\right) = \log A - \log A$$

but, $\frac{A}{A} \equiv 1$, so on the left-hand side we have $\log(1)$.

Also, $\log A - \log A \equiv 0$, so on the right-hand side we have 0. Combining these, we have $\log(1) = 0$

(This rule also follows directly from the fact that $10^0 = 1$, so 0 is the log of 1.)

Rule 11.2e $\log 10 = 1$

This is true because $10^1 = 10$, so 1 is the log of 10.

It's easy to confuse rules 11.2d and 11.2e. Rule 11.2d is true because $10^0 = 1$, and rule 11.2e is true because $10^1 = 10$.

11.8 Using logs to solve problems

In the next chapter we will develop the theory and application of logs in economics. However, there is one important practical application that we can explain now. This is to solve an exponential equation: that is, an equation in which the unknown is a power (= exponent).

EXAMPLE 11.2

Given $3^x = 91$, suppose we want to find x.

At first sight, it is tempting to imagine that we can solve this equation by reaching for our trusty calculator and using the $\wedge$ or y^x key. However, this key can only be used if we already know the value of x, which we don't in a problem of this kind.

The method of solution is to take logs on both sides of the equation. This is permissible because if two things are equal, so are their logs. This gives

$$\log(3^x) = \log 91$$

From rule 11.2b above, $\log(3^x) = x(\log 3)$. Substituting this in, we get

$$x(\log 3) = \log 91$$

Then, dividing both sides by log 3 gives

$$x = \frac{\log 91}{\log 3} = \frac{1.9590}{0.4771} = 4.1061 \quad \text{(using log key on calculator to find log 91 and log 3)}$$

As a rough check that we haven't got the answer completely wrong (usually, from mistakes in using the calculator), this answer means that $3^{4.1061} = 91$. Since $3^4 = 81$, the answer seems correct. You can check the answer exactly by using the $\wedge$ or y^x key on your calculator to find $3^{4.1061}$, which should result in 91 (or something very close to 91, due to rounding) appearing on the display.

EXAMPLE 11.3

The same type of problem as example 11.2 arises in economics when we want to solve for the number of years, x, in the compound growth formula $y = a(1 + r)^x$. Note that, for any given value of r, $y = a(1 + r)^x$ is an exponential function with variables x and y, a multiplicative constant a, and base $1 + r$.

Suppose we are told that prices are rising by 3% per year, and asked how many years it will take for the price level to double.

This is a compound growth problem, so the relevant equation is $y = a(1 + r)^x$ (see section 10.4 if you need to refresh your memory on this).

We are not given the current (= initial) price level, but since the price level is an index number (see section 1.13) we can assume its initial level is $a = 100$. Therefore the price level x years from now will be $y = 200$ (by assumption that prices double), and we have to solve for x, given $r = 0.03$. Therefore in the formula

$$y = a(1 + r)^x \qquad \text{we have}$$

$$200 = 100(1 + 0.03)^x \quad \text{Dividing both sides by 100 gives}$$

$$2 = (1.03)^x$$

Taking logs on both sides (a legitimate step because if two things are equal, then so are their logs) gives

$\log 2 = \log\{(1.03)^x\}$ Therefore, using rule 11.2b

$\log 2 = x \log(1.03)$ Dividing both sides by $\log(1.03)$, we get

$$x = \frac{\log 2}{\log 1.03} = \frac{0.30103}{0.0128372} = 23.4498 \text{ (using log key on calculator to find log 2 and log 1.03)}$$

It seems sensible to round this answer down to 23 years. Hopefully you can see that we would get the same answer whatever value we assigned to the initial price level, a, provided we set $y = 2a$. Notice the power of compound growth in this answer: without compounding, it would take $33\frac{1}{3}$ years of inflation at 3% per year to double the price level.

11.9 Some more exponential functions

The function $y = 10^x$ is the simplest form of exponential function. In economics we may sometimes meet more complex forms, such as

$y = 10^{ax}$ where a is a parameter

How does the parameter a affect the shape of the graph? This depends on whether a is positive or negative, and whether a is greater or less than 1 in absolute magnitude. We will consider three cases.

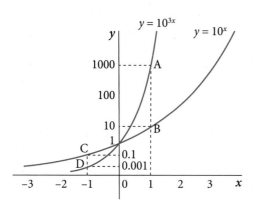

Figure 11.4 Graphs of $y = 10^x$ and $y = 10^{3x}$ compared (y-axis not to scale).

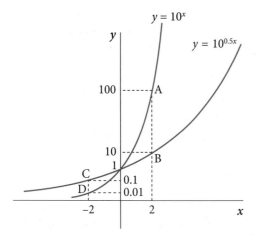

Figure 11.5 Graphs of $y = 10^x$ and $y = 10^{0.5x}$ compared (y-axis not to scale).

Case 1 is when a is greater than 1: for example, $a = 3$. Then we have $y = 10^{3x}$. This is graphed in figure 11.4, together with the graph of $y = 10^x$ as a benchmark. We see that when x is positive, $3x > x$, so $10^{3x} > 10^x$, and the graph of $y = 10^{3x}$ therefore lies *above* the graph of $y = 10^x$ (compare points A and B in figure 11.4).

When x is negative, the reverse is true: $3x < x$, so $10^{3x} < 10^x$. Consequently the graph of $y = 10^x$ lies *below* the graph of $y = 10^x$ (compare points C and D in figure 11.4). Note however that the two curves have the same y-intercept, at $y = 1$, because when $x = 0$, $10^{3x} = 10^x = 10^0 = 1$.

Case 2 is when a is positive but less than 1: for example, $a = 0.5$. Then we have $y = 10^{0.5x}$. This is graphed in figure 11.5, again with the graph of $y = 10^x$ as a benchmark. In this case, when x is positive, $0.5x < x$, so the graph of $y = 10^{0.5x}$ lies below the graph of $y = 10^x$ (compare points A and B in figure 11.5).

When x is negative, the reverse is true: $0.5x > x$, so the graph of $y = 10^{0.5x}$ lies above the graph of $y = 10^x$ (compare points C and D in figure 11.5). As in case 1, the two curves have the same y-intercept, at $y = 1$, because when $x = 0$, $10^{0.5x} = 10^x = 10^0 = 1$.

Case 2, where $y = 10^{ax}$ with a being a fraction, arises frequently in economics, as we shall see in chapter 12.

Case 3 is when a is negative. Let us take the simplest possible case, where $a = -1$. Then we have $y = 10^{-x}$. This is graphed in figure 11.6, together with our benchmark case, $y = 10^x$. Compared with the benchmark case we see that the effect of the minus sign is that every point is transposed from the positive x to the negative x

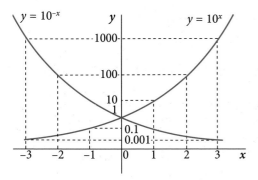

Figure 11.6 Graphs of $y = 10^x$ and $y = 10^{-x}$ compared (y-axis not to scale).

quadrant, and vice versa. Thus $y = 10^{-x}$ is a mirror image of $y = 10^x$. (If you hold figure 11.6 up to a mirror you will see that the two curves swap places on the page.) We can easily generalize this result and say that any function of the form $y = 10^{-ax}$ has a graph that is the mirror image of the graph of $y = 10^{ax}$.

Another functional form that we may meet is

$$y = A(10^x) + B \quad \text{where } A \text{ and } B \text{ are constants}$$

Let us focus first on the parameter A, assuming $B = 0$. If A is positive and greater than 1, then y will be *larger* for any given value of x, so the graph of $y = A(10^x)$ will lie entirely above the graph of the benchmark form, $y = 10^x$. For example, if $A = 5$, then for any given value of x the value of y from $y = A(10^x)$ will be 5 times larger than the value of y from $y = 10^x$. The y-intercept is now at $y = A(10^0) = A > 1$.

Conversely, if A is positive but less than 1 (that is, $0 < A < 1$), y will be *smaller* for any given value of x, so the graph of $y = A(10^x)$ will lie entirely below the graph of the basic form. The y-intercept is now at $y = A(10^0) = A < 1$.

If A is negative, then y will now be *negative* for any given value of x (recall that 10^x is positive for any x). The graph will thus lie entirely below the x-axis, and y will head off towards $-\infty$ (instead of $+\infty$) as x increases. The y-intercept of the function will be at $y = A$, which of course is negative by assumption.

We now turn to the parameter B, with A assumed equal to 1. Compared with $y = 10^x$, the effect of adding B is to shift the whole graph bodily upwards (if B is positive) or bodily downwards (if B is negative). The y-intercept will be at $y = 1 + B$.

Finally, there is another direction in which we could extend the basic idea of $y = 10^x$. For example, consider

$$y = 10^{3x^2 + 2x + 1}$$

Here the innovation is that the base, 10, is raised to a power which is a non-linear function of x, rather than the linear functions considered up to this point. Fortunately, cases like this do not come up very often in economics, so we can simply note these possibilities and move on with a sigh of relief. There is one case of this kind that is important in statistics—the normal distribution—but this lies outside the scope of this book.

Generalizing the exponential and logarithmic functions

The main focus of this chapter has been on the exponential function $y = 10^x$ and its inverse $x = \log_{10} y$. However, $y = 10^x$ is just one member of a family of exponential functions with general form $y = a^x$ (where a is any constant greater than 1). For example, if we choose $a = 2$, we get

$$y = 2^x$$

As you can easily check for yourself, the graph of $y = 2^x$ has the same shape as that of $y = 10^x$. The only difference is in the calibration of the y-axis. For example, when $x = 0, 1, 2, 3$, we have $10^x = 1, 10, 100, 1000$ and $2^x = 1, 2, 4, 8$.

We can also define the function inverse to $y = 2^x$ as

$$x = \log_2 y$$

and thereby define logs to base 2. These logs to base 2 obey all the rules relating to logs that we have already developed for base 10. To obey the convention that the dependent variable is denoted by y, we write the logarithmic function as $y = \log_2 x$. This is merely a relabelling.

We can obtain the graph of $y = 2^x$ from the graph of $y = 10^x$ by applying the appropriate conversion rule. Similarly we can obtain the graph of $y = \log_2 x$ from the graph of $y = \log_{10} x$ by applying the appropriate conversion rule. These conversion rules may be found at the Online Resource Centre. Given the ease with which we can convert both the exponential and logarithmic functions from one base into another, the choice of which base to work with is somewhat arbitrary. In practice, only two bases are normally used. One is the base 10: that is, $y = 10^x$ and its inverse, $x = \log_{10} y$, the focus of this chapter. A second base, which has some special characteristics, will be introduced in the next chapter.

! **Summary of chapter 11**

In section 11.2 we introduced a new functional form, the exponential function $y = 10^x$. Its graph has the characteristic shape shown in figure 11.2(a). From this we defined the function inverse to $y = 10^x$ as $x = \log_{10} y$ (see figure 11.2(b)). As with every inverse function, this expresses the same relationship between the two variables, but with the x- and y-axes interchanged. In section 11.4 we examined the properties of logs, which follow from the definition of the logarithmic function. Thus, for example, $\log_{10} 100 = 2$, *because* $10^2 = 100$. In section 11.5 we explained how to find logs and exponentials on a calculator.

The rules for manipulating logarithms also follow from the definition of the logarithmic function, and are listed in section 11.7. The basic rule, from which all others follow, is that $\log_{10}(AB) = \log_{10} A + \log_{10} B$. In section 11.8 we saw that logs are essential to solve problems involving the compound growth formula $y = a(1 + r)^x$ when x is the unknown.

In section 11.9 we considered some cases of the more general exponential function, such as $y = 10^{5x+4}$ and $y = 10^{-x}$.

? **Progress exercise 11.2**

1. Use your calculator to find log 2. Then, without using a calculator, find:

(a) $\log 4$ (b) $\log(0.5)$ (c) $\log(\sqrt{0.5})$ (d) $\log(\frac{1}{8})$ (e) $\log(800)$

(Hint: use the rules in section 11.7.)

2. At present my accumulated stock of savings is €50,000 and is growing at 15% per year. If I plan to retire when my stock of savings is €250,000, in how many years from now will I be able to retire?

3. Every year, 5% of the world's remaining stock of trees is cut down. Assuming no new trees are planted, how long will it be before the stock is half of its current level?

4. For values of x between −2 and +2, sketch the graphs of:

(a) $y = 10^x$ (b) $y = 10^{0.25x}$ (c) $y = 10^{2x}$

(Use an elastic scale on the vertical axis.) For any given value of x, what is the exact relationship between the values of y given by each of these functions? (Hint: $a^{nm} = (a^n)^m$; see section 2.10.) Why do all of these functions have the same y-intercept?

5. For values of x between −2 and +2, sketch the graphs of:

(a) $y = 10^{-x}$ (b) $y = 50(10^{0.5x})$

(c) $y = 25(10^x) + 5$ (d) $y = 3(10^{-0.25x})$

(Use an elastic scale on the vertical axis.) In each case, explain how the curve is related to the graph of $y = 10^x$.

Checklist

Be sure to test your understanding of this chapter by attempting the progress exercises (answers are at the end of the book). The Online Resource Centre contains further exercises and materials relevant to this chapter www.oxfordtextbooks.co.uk/orc/renshaw3e/.

This has been a short but quite difficult chapter. We have introduced the idea of the exponential function and its inverse, the logarithmic function. We have explained the rules for manipulating logs to base 10 and how to find these logs using a calculator. The specific topics were:

✔ **The exponential function** $y = 10^x$. Understanding the exponential function $y = 10^x$ and its graph.

✔ **The logarithmic function** $y = \log x$. Understanding how the logarithmic function $y = \log x$ is derived from the exponential function and the shape of its graph.

✔ **Calculator.** Using a calculator to find exponents and logs to base 10.

✔ **Rules of logs.** The rules for manipulating logs to base 10 and how to apply them.

✔ **Solving growth problems.** Using logs to solve growth and other problems.

✔ **Generalization.** How the exponential and logarithmic functions may be generalized, and sketching the relevant graphs.

In the next chapter we develop these ideas further by deriving a formula for continuous growth and looking at some of its applications.

Chapter 12

Continuous growth and the natural exponential function

OBJECTIVES

Having completed this chapter you should be able to:

■ Understand the concepts of discrete and continuous growth.

■ Understand the derivation of the natural exponential and logarithmic functions.

■ Sketch the graphs of the natural exponential and logarithmic functions.

■ Use the natural exponential and logarithmic functions to solve growth problems.

■ Understand and calculate continuously discounted future payments or receipts.

■ Understand why graphs of economic and other variables are often plotted with a log scale on the vertical axis, and draw such graphs.

12.1 Introduction

In this chapter we draw on and combine the material of chapters 10 and 11. We take the compound interest or compound growth formula from chapter 10 and develop it into a new concept, the natural exponential function. This is a member of the family of exponential functions described in chapter 11 and describes any variable growing continuously through time at a constant compound rate. In the same way, we derive the natural logarithmic function (and hence natural logs) as the function inverse to the natural exponential function. From this we derive a formula for calculating the present value of a future receipt or payment, when discounting is continuous.

In the latter part of the chapter we demonstrate some applications of these two new formulae to economic problems and compare continuous growth and discounting with their discontinuous counterparts from chapter 10. Finally, we show by means of an example why time series graphs of economic variables often use a logarithmic scale for the dependent variable. This has the effect that the growth rate can be inferred from the slope of the graph.

12.2 Limitations of discrete compound growth

In section 10.4 we developed the compound growth formula

$$y = a(1 + r)^x$$

where

a = initial value of the variable

r = annual interest rate or growth rate (in proportionate terms, so for example an interest rate
   or growth rate of 10% per year means $r = 0.1$)

x = number of years

y = 'future compounded value' (= value of the variable after x years).

We used this formula to solve various problems involving the growth of money deposited with a bank or the growth of economic variables such as national production (GDP) and the retail price index (RPI).

When interest is added (or more generally, when growth occurs) n times per year, the compound growth formula becomes

$$y = a(1 + \tfrac{r}{n})^{nx}$$

In this formula we found that at any given *nominal* annual interest or growth rate, r, the *effective* annual interest rate or growth rate increases as n increases. For example, in section 10.7 we looked at an example with a *nominal* interest rate of 10% per year, and found that if interest was added every 6 months, the *effective* interest rate was 10.25% per year. If interest was added daily (which is true of some bank accounts) the effective annual interest rate rose to 10.52%.

However, this formula is cumbersome and unsatisfactory because, whatever the value of n, it can deal only with variables that grow in discrete jumps. Yet there are many variables in economics that may be viewed as varying continuously. Consider, for example, GDP (national production). Data are commonly published in annual or quarterly form—that is, the total amount of output produced in a year or a quarter is reported—and thus vary discretely. But the underlying process consists of a continuous flow of output of goods and services emerging from factories, shops, and offices on a '24/7' basis. Similarly, in the case of prices, an index of the average level of prices is published monthly, and thus varies in monthly jumps. (There are two such indexes for the UK: the retail price index (RPI) and, since December 2003, the consumer price index (CPI). See www.statistics.gov.uk.) But again the underlying data—the prices of thousands of different goods and services sold every day—are varying in a way that is virtually continuous. Therefore we can take the view that these variables would be better described by a continuous growth formula. We will now explain how such a formula is derived.

12.3 Continuous growth: the simplest case

First, let us be clear about what we mean by continuous growth. In nature, we view plants and animals as growing continuously. However, when we say that a blade of grass is growing continuously, we mean that its growth takes the form of a very large number of very small jumps, each jump being no more than an additional new cell. These jumps are so small that we view this growth as a smooth and continuous process. If we are considering the growth of an economic variable such as a price index, the analogue of a new cell in a blade of grass is a small increase in one of the many prices from which the index is constructed. Therefore it occurs to us that if, in the formula $y = a(1 + \tfrac{r}{n})^{nx}$, we allow n to become very large, this will describe the situation that exists when the variable in question grows *continuously*.

For example, if we are looking at the growth of national production (GDP), as remarked above, it seems appropriate to view this as a variable that grows continuously, even though data on GDP may be collected and published only discretely. We could try to capture this idea of continuous growth by setting $n = 365$ in the formula $y = a(1 + \frac{r}{n})^{nx}$, thus assuming that GDP increases by a small jump every day (as the result of increases in some of the individual outputs that make up the total). The total of the 365 daily jumps then gives the annual increase in GDP. This is clearly a lot more realistic than using the formula $y = a(1 + r)^x$, which assumes that GDP increases in a single jump at the end of each year.

But why stop at 365 increases in GDP per year? Why not allow n to become even larger so that GDP is assumed to increase perhaps once per minute or even once per second? By doing this we will approach closer to the reality that GDP is a continuous flow of output pouring off production lines throughout the country.

To explore this question, let us examine the effects of allowing n to become larger and larger, in the compound growth formula. To focus on n, we will push the other variables into the background, by setting $a = 1, x = 1, r = 1$ (that is, 100%). So our formula

$$y = a(1 + \tfrac{r}{n})^{nx} \quad \text{simplifies to} \quad y = (1 + \tfrac{1}{n})^n$$

If we want to think of this concretely in terms of money lent, we can suppose that with $a = x = r = 1$ we are looking at €1 lent for 1 year at 100% a year nominal interest. With these assumptions, we are considering how y, the end-year value, is affected by increasing the number of times per year that interest is added. But of course our conclusions will be valid for the growth of *any* variable, not just money lent.

Table 12.1 Value of $y = (1 + \frac{1}{n})^n$ as n varies.

n	$y = (1 + \frac{1}{n})^n$
1	$(1 + 1)^1 = 2$
2	$(1 + \frac{1}{2})^2 = 2.25$
10	$(1 + \frac{1}{10})^{10} = 2.5937$
100	$(1 + \frac{1}{100})^{100} = 2.7048$
1000	$(1 + \frac{1}{1000})^{1000} = 2.7169$
10000	$(1 + \frac{1}{10000})^{10000} = 2.7181459$
100000	$(1 + \frac{1}{100000})^{100000} = 2.7182546$

Table 12.1 shows how y varies with n. The table tells us that, for example, if interest is added (or growth occurs) ten times per year, then €1 grows to €2.5937 after 1 year. To test your understanding, check the numbers in the table for yourself, using your calculator. From the table, we can see that increasing the number of times that growth occurs in the year increases the end-year value, y, *but with diminishing effect* as n gets larger and larger. For example, increasing n from 10,000 to 100,000 has very little effect on the value of y. So it *looks* from the table as if $y = (1 + \frac{1}{n})^n$ is approaching a limiting value of 2.71828 ... as n approaches infinity, and indeed this is correct, though we won't prove it here. Thus using the notation introduced in section 5.6, we can write

$$\text{as } n \to \infty, y = a(1 + \tfrac{1}{n})^n \to 2.71828 \ldots$$

Note that the dots at the end of the number 2.71828 ... are there to indicate missing digits. The digits are missing because, by choosing a larger value for n, we can always add some more digits (as in table 12.1). But we can never find the limiting value exactly because n can never actually reach infinity.

For compactness in writing, we denote this limiting value by the symbol e. Then, using the notation for a limiting value introduced in section 5.6, we can write:

RULE 12.1 Definition of the constant, e

By definition:

$$e \equiv \lim_{n \to \infty} (1 + \tfrac{1}{n})^n$$

You may find it a little disturbing and unsatisfactory that, however large a value we choose for n, we can only ever calculate e approximately. However, this is also true of another constant known as π, used in calculating the area of a circle. The true value of π can never be calculated, as the decimal places go on forever. Nevertheless we can calculate the area of a circle reasonably accurately using an approximation to π, such as 3.142. If we feel the need for a more accurate answer, we can simply recalculate using a closer approximation to π such as 3.14159. So it is with e; we can always calculate e more accurately by making n larger, as in table 12.1.

Concretely in terms of money lent, e euros is what you would get back if you lent €1 for 1 year at a nominal interest rate of 100% per year, with interest added continuously. Because no bank computer, however powerful, could ever add interest continuously, what you would get back could only ever be an approximation to e. However, if the bank's computer added interest even 10,000 times per year (roughly every hour, well within the computer's capability), then we can see in table 12.1 that you would get back €2.7181459, which is very close to e.

It's worth noting in passing (though we return to this in the next section) that in this example we can calculate the effective interest rate, as we did in section 10.7. You get back €2.7181459, but €1 of this is your original investment, so the interest alone is 2.7181459 − 1. As a proportion of your original investment, which was €1, the interest is (2.7181459 − 1) ÷ 1 = 1.7181459. To convert this into a percentage return we multiply by 100, giving 171.81459%. (See section 10.7 to refresh your memory on nominal and effective interest rates, and section 9.2 on proportionate and percentage growth.) So we conclude that when the nominal interest rate, r, is 100% per year, and interest is added continuously, the effective interest rate is approximately 172%.

Graphical treatment

You may find it helpful to look at the meaning of e graphically. In figure 12.1 we have plotted the graph of $y = (1 + \frac{1}{n})^{nx}$, with $x = 1$, for various values of n. We can think of these graphs as showing the growth of €1 invested for 1 year at a nominal interest rate of 100% per year, for various values of n (though of course, as we have already said, the analysis is equally valid for any variable growing for any reason).

In figure 12.1(a), we have taken $n = 2$. So y grows in two steps or jumps. The first step occurs after 6 months (when $n = 1$), when 6 months' interest is added. Since the annual interest rate is 100%, the interest after 6 months is 50%, so y increases from 1 to 1.5. The second step occurs at the end of the year (when $n = 2$), when a further 50% interest is added, so y increases from 1.5 to 2.25. (See table 12.1 for $n = 2$.)

In figure 12.1(b), we have taken $n = 10$. So y grows in 10 steps. Each step occurs after one-tenth of a year, and at each step one-tenth of a year's interest is added. Since the annual interest rate is 100%, this means that growth of 10% occurs at each step. So for example at the first step ($n = 1$), y increases from 1 to 1.1. At the second step ($n = 2$), y increases from 1.1 to 1.21 (= 1.1 plus one-tenth of 1.1). And so on. The value of y at the end of the year is given by $y = (1 + \frac{1}{10})^{10} = 2.5937$. (See table 12.1 for $n = 10$.)

Finally, in figure 12.1(c) we have taken $n = 100$. So y grows in 100 steps, and the growth at each step is one-hundredth of 100%, or 1%. So, for example, at the first step ($n = 1$), y increases from 1 to 1.01. At the second step ($n = 2$), y increases from 1.01 to 1.0201 (= 1.01 plus one-hundredth of 1.01). And so on. The value of y at the end of the year is given by $y = (1 + \frac{1}{100})^{100} = 2.7048$. (See table 12.1 for $n = 100$.)

Figure 12.1 demonstrates two key points:

(1) The graph of $y = (1 + \frac{1}{n})^{n}$ as a function of n is a step function when n is finite (see section 3.6 if you have forgotten what a step function is). The steps are obvious in figures 12.1(a) and 12.1(b), and if you look closely at figure 12.1(c) you will see the steps there too (100 of them), even though at first glance it looks like a smooth curve.

(2) As n approaches infinity, the graph will approximate more and more closely to a smooth curve. And as n approaches infinity, the end-year value of y approaches a limiting value. This limiting value is

$$y = \lim_{n \to \infty} (1 + \frac{1}{n})^{n} = e$$

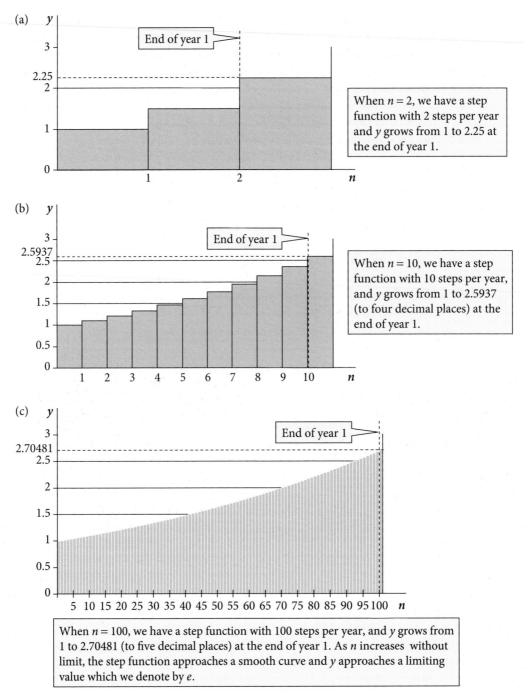

(a) When $n = 2$, we have a step function with 2 steps per year and y grows from 1 to 2.25 at the end of year 1.

(b) When $n = 10$, we have a step function with 10 steps per year, and y grows from 1 to 2.5937 (to four decimal places) at the end of year 1.

(c) When $n = 100$, we have a step function with 100 steps per year, and y grows from 1 to 2.70481 (to five decimal places) at the end of year 1. As n increases without limit, the step function approaches a smooth curve and y approaches a limiting value which we denote by e.

Figure 12.1 Values of $y = (1 + \frac{1}{n})^{nx}$ after 1 year for different values of n.

12.4 Continuous growth: the general case

In the previous section we considered €1 invested (lent) for 1 year at 100% per year (nominal), with interest added (or growth occurring) n times per year. Thus our growth equation was $y = (1 + \frac{1}{n})^{n}$. We found that if n increases without limit (so that interest is added, or growth occurs, continuously), the value of the investment, y, approaches a limiting value of $y = e$ (where e is as defined in rule 12.1 above).

Obviously we don't want to be restricted to considering €1 invested for 1 year at a nominal interest rate of 100% per year. So we now want to extend our analysis to consider the general case, $y = a(1 + \frac{r}{n})^{nx}$, in which some given amount €a is invested for a variable period of x years at a given nominal interest rate of r. This seems at first sight a complex task, as we have introduced two additional parameters (a and r) and an additional variable, x.

Fortunately, it is easy to show that in the general case where $y = a(1 + \frac{r}{n})^{nx}$, if n increases without limit the value of the investment approaches a limiting value of

$$y = ae^{rx}$$

This is a new and very important result. We now have a formula that can handle any variable that is growing *continuously* at a constant nominal rate. And although in the example above we considered money lent to a bank, the formula is of course valid for *any* variable growing continuously at a constant rate.

To summarize, we now have the following:

RULE 12.2 Two compound growth formulae

Rule 12.2a

$$y = a(1 + \frac{r}{n})^{nx} \quad \text{(from section 10.7, rule 10.5)}$$

(appropriate when y grows in discrete jumps, n times per year)

Rule 12.2b

$$y = ae^{rx}$$

(appropriate when y grows *continuously*)

In both cases, a is the initial value of the variable, x is the number of years, and $100r\%$ is the nominal annual growth rate (that is, the growth that would occur if growth were in annual jumps).

At this point you may be a little uncertain as to which of the two formulae above should be used in any particular situation. We can be quite relaxed on this question, because it turns out that in many cases they give very similar answers, so we can use whichever formula is more convenient or seems more appropriate in the particular case. Moreover, when using the formula in rule 12.2a, the value chosen for n rarely changes the answer by very much, so in most cases we can set $n = 1$. Some examples will be given in section 12.9 below, but a little more theoretical development is necessary first.

A big attraction of rule 12.2b above is that it allows time (x) to vary *continuously* and this means that the graph of $y = ae^{rx}$ is a smooth and continuous curve, as we shall now see. This contrasts with the graph of $y = a(1 + \frac{r}{n})^{nx}$, which as we saw in section 12.3 is a step function.

12.5 The graph of $y = ae^{rx}$

In the continuous growth equation $y = ae^{rx}$ the independent variable is x (the number of years); the dependent variable is y (the future compounded value); and a and r are parameters. In this section we will examine the shape of this function for various values of the parameters a and r. In section 12.8 we explain how you can draw such graphs for yourself using your calculator.

The simplest case

Let's begin with the simplest case by setting $a = 1$, $r = 1$, so we have

$$y = e^x$$

Since e is a constant, you will hopefully recognize this as an exponential function (like $y = 10^x$ considered in chapter 11). It is called the *natural* exponential function because it was first developed to describe the growth found in nature, where plants and animals grow continuously.

The function $y = e^x$ is graphed in figure 12.2. As expected, the graph has the same general shape as $y = 10^x$ considered in chapter 11, with only the scale on the y-axis different. For example, when $x = 1$, $y = 10^x$ has the value $10^1 = 10$, while $y = e^x$ has the value $e^1 = e \approx 2.718$. Similarly, when $x = 2$, $y = 10^x$ has the value $10^2 = 100$, while $y = e^x$ has the value $e^2 \approx (2.718)^2 = 7.39$. (The symbol $\approx$ means 'approximately equals'.)

Two points should be noted:

(1) Because e can never be *precisely* calculated (even by a calculator!), the graph of $y = e^x$ can never be drawn completely accurately. But this doesn't really matter because we can always increase the accuracy of our graph, if we feel the need, by taking a closer approximation to e (see table 12.1).

(2) Because we are considering here only cases where x denotes time, the position of $x = 0$ (time zero) is arbitrary. Normally, we denote the earliest date to be considered as time zero, so that we don't need to consider negative values of x. This will become clearer when we look at some examples.

The general case

We now consider the general case, $y = ae^{rx}$. So we must bring in the parameters a and r. The following discussion closely parallels section 11.9 where we considered variants of $y = 10^x$, so we can be brief here.

(1) Bringing in first the parameter r, we now have

$$y = e^{rx}$$

What does the graph of this function look like? Considering first the case where $r > 0$ (a positive growth rate), there are two sub-cases.

Sub-case 1 is where r is greater than 1. For example, $r = 2$, so we have $y = e^{2x}$. This is graphed in figure 12.3, together with $y = e^x$ as a benchmark. We see that the two curves have the same y-intercept because when $x = 0$, $e^x = e^{2x} = e^0 = 1$. However, $y = e^{2x}$ turns up more steeply as x increases and therefore lies above $y = e^x$ when x is positive, but below it when x is negative.

Sub-case 2 is where r is less than 1. For example, $r = 0.5$, so we have $y = e^{0.5x}$. This is graphed in figure 12.4, together with $y = e^x$ as a benchmark. Again the two curves have the same y-intercept, but in this case $y = e^{0.5x}$ turns up less steeply as x increases and therefore

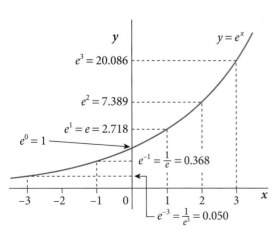
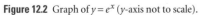

Figure 12.2 Graph of $y = e^x$ (y-axis not to scale).

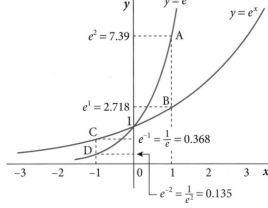

Figure 12.3 Graphs of $y = e^{2x}$ and $y = e^x$ compared (y-axis not to scale).

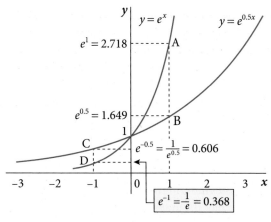

Figure 12.4 Graphs of $y = e^{0.5x}$ and $y = e^x$ compared (y-axis not to scale).

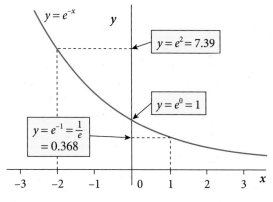

Figure 12.5 Graph of $y = e^{-x}$ (y-axis not to scale).

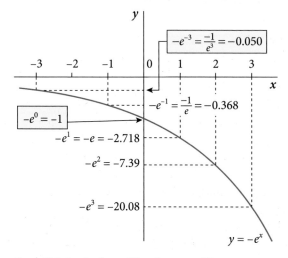

Figure 12.6 Graph of $y = -e^x$ (y-axis not to scale).

lies below $y = e^x$ when x is positive, but above it when x is negative.

We now consider the case where $r < 0$: that is, a negative growth rate. We will consider the simplest case where $r = -1$, so we have $y = e^{-x}$.

The effect of the minus sign is that y now has the same value at, say, $x = 3$ as $y = e^x$ has at $x = -3$. So the graph of $y = e^x$ is simply the mirror image of $y = e^x$, as in figure 12.5.

(2) Bringing in the parameter a, we now have

$$y = ae^x$$

How does the parameter a change the graph, compared with $y = e^x$ in figure 12.2? If a is positive, then it merely changes the scale on the y-axis. For example, if $a = 5$, then we have $y = 5e^x$. Compared with $y = e^x$, for any given x the value of y is now 5 times larger, which is equivalent to scaling up the y-axis by a factor of 5 (so that, for example, when $x = 3$, $y = 100.4$ instead of 20.08). This applies equally of course to the y-intercept, which becomes $y = 5e^0 = 5$, instead of $y = e^0 = 1$.

Similarly, if $a = 0.5$, then we have $y = 0.5e^x$. Compared with $y = e^x$, the value of y is halved for any x, which is equivalent to scaling down the y-axis by a factor of 2 (so that, for example, when $x = 3$, $y = 10.04$ instead of 20.08). The y-intercept becomes $y = 0.5e^0 = 0.5$, instead of $y = e^0 = 1$.

If a is negative, then $y = ae^x$ becomes negative for any given x, and the curve is flipped downwards into the negative quadrants (see figure 12.6 for $a = -1$). The scale on the y-axis is also stretched or compressed according to the absolute value of a.

12.6 Natural logarithms

In section 11.3 we saw that the function inverse to $y = 10^x$ is, by definition

$$x = \log_{10} y$$

In the same way, we can define the function inverse to $y = e^x$ as follows:

RULE 12.3 **The natural logarithmic function**

By definition, the function inverse to $y = e^x$ is

$$x = \log_e y$$

and this defines x as the **natural logarithm** of y.

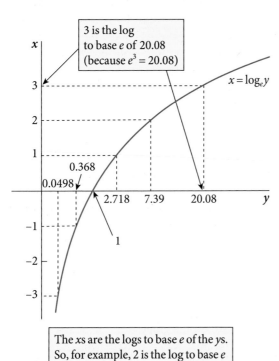

3 is the log to base e of 20.08 (because $e^3 = 20.08$)

$x = \log_e y$

0.368

0.0498

2.718 7.39 20.08

1

The xs are the logs to base e of the ys. So, for example, 2 is the log to base e of 7.39, because $e^2 = 7.39$ (approx.).

Figure 12.7 Graph of $x = \log_e y$ (y-axis not to scale).

As a natural extension of the discussion in chapter 11 in relation to $y = 10^x$, we see immediately that because $x = \log_e y$ is the function inverse to $y = e^x$, it expresses the same relationship between x and y. So if we plot the graph of $x = \log_e y$, with x on the vertical axis (because x is the dependent variable), we get figure 12.7. This is the same graph as $y = e^x$ (figure 12.2) but rotated and flipped over so that x is now on the vertical axis.

This means that in both figure 12.2 and figure 12.7, the x values are the logs to base e of the y values. Thus in both figures we see that $\log_e 20.086 = 3$, $\log_e 0.368 = -1$, and so on.

Many students find logs to base e even more mysterious than logs to base 10. To achieve de-mystification, try to fix figure 12.7 (or figure 12.2 if you prefer) in your mind, and remember that in both figures the xs are the logs to base e of the ys.

Alternatively, if you're happier with algebra than with graphs, try to fix in your mind one of the following (but probably not more than one, for fear of muddle!):

(1) The basic definition of a log to base e is: if $y = e^x$, then by definition $x = \log_e y$.

(2) From (1), by substitution for x in $y = e^x$, we get $y = e^{\log_e y}$. In words, this says that the natural log of y is the power that e must be raised to, to equal y.

(3) Another variant obtainable from (1) is that $\log_e(e^x) = x$. This is fairly easy to remember, at least.

So for example if you want to find $\log_e 15$, you have to find a number x such that $e^x = 15$. Shortly we will explain how to find x using your calculator, but for the moment let's try to make an educated guess. We know that e is a little less than 3. We also know that $3^2 = 9$ and $3^3 = 27$, so $e^2 \approx 9$ and $e^3 \approx 27$. So perhaps $e^{2.5} \approx 15$, and therefore $\log_e 15 \approx 2.5$. (The symbol $\approx$ means 'approximately equals'.)

We can make another rough estimate of $\log_e 15$ from figure 12.7 (or figure 12.2). From the graph, we see that $\log_e 7.39 = 2$ and $\log_e 20.08 = 3$. We roughly estimate where 15 lies on the y-axis (though remember the axis is not to scale), and from there read off the log to base e of 15 as a number that is probably nearer to 3 than to 2: say, about 2.7 or 2.8. Shortly, we will see how close these guesstimates are.

The graph of $y = \log_e x$

The function $x = \log_e y$, which we have just graphed in figure 12.7, violates the convention in mathematics of denoting the dependent variable with the symbol y and the independent

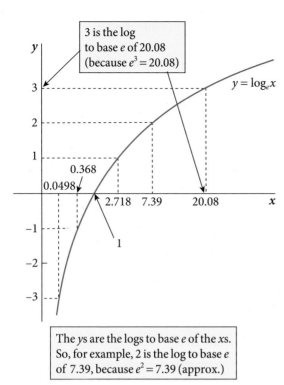

3 is the log to base e of 20.08 (because $e^3 = 20.08$)

$y = \log_e x$

The ys are the logs to base e of the xs. So, for example, 2 is the log to base e of 7.39, because $e^2 = 7.39$ (approx.)

Figure 12.8 Graph of $y = \log_e x$ (x-axis not to scale).

variable with x. In order to comply with this convention we must interchange the labels of the two variables. The function now becomes

$$y = \log_e x$$

which is the standard formulation of the natural logarithmic function and is graphed in figure 12.8. It should be fairly obvious that figure 12.8 is just figure 12.7 with the labels on the axes interchanged. (What we have done here corresponds exactly with what we did to get from $x = \log_{10} y$ to $y = \log_{10} x$ in section 11.6.)

A small notational change

In chapter 11, we said that we usually write $\log_{10}$ as just 'log'. In the same way, we usually write $\log_e$ as just 'ln'. This is how it is written on the appropriate key of your calculator. So from now on we will write $\ln x$ to mean $\log_e x$.

12.7 Rules for manipulating natural logs

As you might expect, natural logs obey all of the manipulative rules for common logs listed in section 11.7. We restate them briefly here:

> **❋ RULE 12.4 Rules for manipulating natural logs**
>
> For any two positive numbers, A and B:
>
> **Rule 12.4a**
> $\ln(AB) = \ln A + \ln B$
>
> **Rule 12.4b**
> $\ln(A^n) = n(\ln A)$
>
> **Rule 12.4c**
> $\ln(A/B) = \ln A - \ln B$
>
> **Rule 12.4d**
> $\ln 1 = 0$ (because $e^0 = 1$)
>
> **Rule12.4e**
> $\ln e = 1$ (because $e^1 = e$)

12.8 Natural exponentials and logs on your calculator

We noted in section 12.3 above that any calculation of e can only ever be an approximation, such as 2.71828, of the 'true' or limiting value. This is equally true of the value of e stored in your calculator; it is an approximation, but one that is accurate enough for almost any purpose except possibly a landing on Mars.

In section 11.5 we explained how to find common logs on a calculator. The procedure for finding natural logs is almost identical. Conventionally 'ln' denotes a natural logarithm. Your calculator obeys this convention, so to find the natural log of, say, 100 you press the key marked 'ln', then key in 100, then press 'equals' and the answer, 4.605170, is displayed.

To reverse this process, press 'SHIFT' (or '2nd F', or 'INV' or whatever is the equivalent key on your calculator), then 'ln', then key in 4.605170. The answer, 100, is then displayed. This reverse process actually calculates $e^{4.605170}$, with answer 100. (The answer displayed by your calculator will not be exactly 100 because of rounding error and because the value of e stored in your calculator is an approximation.)

The values used in the graphs of e^x, $\ln x$, and others in this chapter were found in the way just described. To test your understanding, use your calculator to check these values. For example, check that $e^{0.5} = 1.649$ in figure 12.4.

We can also check the crude estimates of $\log_e 15$ that we made at the end of section 12.6 above. If we hit the 'ln' key, then 15, we get 2.708, not far from our estimates.

Don't forget to use the '(−)', or '+/−' key (or equivalent) when keying in a negative power, otherwise you will get the wrong answer. For example, to find e^{-2}, key in 'SHIFT' (or equivalent), 'ln', '(−)', 2. The answer is 0.1354. (If you forget to key in the minus sign, you will get $e^2 = 7.39$.) As it's very easy to make mistakes with calculators, we can run a rough check that 0.1354 is right. Recall that e^{-2} is the same as $1/e^2$, and since e is close to 3, $1/e^2$ must be reasonably close to $\frac{1}{9}$, which equals 0.1111 in decimal. This is quite close to 0.1354, so our answer seems correct.

If you are curious to know what value of e is stored in your calculator, you can find out by pressing 'SHIFT' (or equivalent), then 'ln', then 1. This will find e^1, which is, of course, e. (If you do the same but press 'log' instead of 'ln' you will find 10^1; that is, 10.)

The Online Resource Centre shows how natural exponentials and logs can be found in the 'formula' commands of Excel®, and how the graphs of these functions can be plotted using Excel®.

Progress exercise 12.1

1. (a) First, *without* using your calculator, try to make a rough estimate of the values of each of the following. Then key each into your calculator and see whether your estimates were broadly correct. If you're not brave enough to do that, at least look carefully at each answer and ask yourself whether you understand it and whether it seems about right.

   (1) $\ln 30$

   (2) $\ln 900$. How does this relate to the answer to (1)?

   (3) $\ln 10$

   (4) $\ln 0.1$. How does this relate to the answer to (3)?

   (5) $\ln 2.71828$

   (6) e^3

   (7) e^1. How does this relate to the answer to (5)?

   (8) $\ln 1$

   (9) $\ln e$. (Use 2.71828 as an approximation to e.)

   (10) e^0

   (b) Find the inverses of each of the above. For example, the inverse to (1) is found as 'SHIFT' or 'INV' or '2nd F', then 'ln', then 3.4012.

2. (a) Draw a sketch graph of $y = e^x$ for values of x between −3 and +3. Don't worry about getting the scale on the y-axis right.

   (b) Use the graph to estimate the approximate values of $\ln(20)$ and $\ln(\frac{1}{20})$. Compare these values with those given by your calculator.

3. (a) Sketch the graph of $y = \ln(x)$ for x between −3 and 3. Don't worry about getting the scale on the x-axis right.

   (b) Use your graph to estimate the approximate values of $\ln(2.5)$ and $\ln(0.4)$. Compare these values with those given by your calculator.

   (c) Explain how this graph is related to the graphs of $x = e^y$ and $y = e^x$.

4. Sketch the graphs of (a) $y = e^{0.05x}$; (b) $y = e^{-x}$; (c) $y = 100e^{2x}$; (d) $y = \ln(x)$.

12.9 Continuous growth applications

In this section we will show how the natural exponential function can be used to solve problems involving variables that are growing continuously through time at a constant rate. First, though, we will show how to calculate the effective interest rate (or growth rate) of a variable that is growing continuously at a constant rate.

The effective interest rate

In section 10.7 we introduced the important distinction between the nominal and effective interest rate, a distinction that arises when interest is added more than once a year. We saw that in the discontinuous growth formula $y = a(1 + \frac{r}{n})^{nx}$, the effective interest rate is given by $(1 + \frac{r}{n})^n - 1$ (see rule 10.6). When $n = 1$, this collapses down to r, the nominal interest rate (which by definition is the return when interest is added once a year).

Now let us consider the effective interest rate for the continuous growth case, $y = ae^{rx}$. After 1 year (that is, with $x = 1$) you get back $y = ae^r$. But a of this is your original investment, so the interest alone is $ae^r - a = a(e^r - 1)$. To calculate the proportionate return, we divide this by your original investment of a, which gives $[a(e^r - 1)] \div a = e^r - 1$. This is sufficiently important to be worth stating as a rule:

> **RULE 12.5** The effective growth rate when growth is continuous
>
> In the continuous growth formula $y = ae^{rx}$, the effective growth rate is
>
> $e^r - 1$
>
> where r is the nominal growth rate; that is, the annual growth that would occur if growth were in annual jumps.
>
> (From rule 10.6, the effective growth rate for a variable that is growing *discontinuously* is $(1 + \frac{r}{n})^n - 1$. See section 10.7 to revise the distinction between nominal and effective interest or growth rates.)

EXAMPLE 12.1

Suppose a variable is growing continuously at a nominal rate of 10% per year. What is the effective growth rate?

Answer: Here $r = 0.1$ (that is, 10%), so from rule 12.5 we have

$e^r - 1 = e^{0.1} - 1 = 1.105170918 - 1 = 0.105170918$

To turn this proportionate return into a percentage return, we multiply by 100 and get 10.5171% (to four decimal places). (Note that we calculate $e^{0.1}$ by keying in 'SHIFT' (or equivalent), then 'ln', then 0.1, then '='.)

Thus the effective return is 0.52% (to 2 d.p.) higher than the nominal return of 10% in this example. This compares with example 10.14 in section 10.7, where we found, using rule 10.6, that with a nominal growth rate of 10%, when growth occurred in two jumps per year the effective growth rate was 10.25%, or 0.25% higher than the nominal rate.

Examples of continuous growth

Recall from rule 12.2 in section 12.4 above that we now have two formulae for solving compound growth problems:

(1) $y = a(1 + \frac{r}{n})^{nx}$ used when y grows in n jumps per year

(2) $y = ae^{rx}$ used when y grows continuously

We will now use these formulae to solve some problems. Before we start, there is a useful simplification that we can make. In practice, in formula (1) we almost invariably put $n = 1$: that is, we consider growth only in annual jumps. This is because the continuous growth formula, (2), is usually a close enough 'fit' to cover any case where growth occurs more frequently than once a year. Indeed, the continuous growth formula is sometimes used even where actual growth is in annual jumps. We will illustrate this point in example 12.2.

EXAMPLE 12.2

Suppose GDP grows at 4% per year. What is its level after 5 years? What is the cumulative percentage growth over 5 years? There are two methods for answering this.

Method 1: As discussed at the beginning of this chapter, there is a case for treating GDP as a variable that grows in annual jumps because data on GDP are published annually. In that case, we use formula (1), with $n = 1$, so we have

$$y = (1 + r)^x$$

We are not given the initial level of GDP, a, so we will arbitrarily assume $a = 100$ (billions of euros, say). We are given $r = 0.04$, $x = 5$. Substituting these values gives

$$y = 100(1.04)^5$$

Using the $\wedge$ or y^x key on our calculator, we find $(1.04)^5 = 1.2166$, so

$$y = 100(1.2166) = 121.66 \quad (= \text{level of GDP in €bn after 5 years})$$

We find the cumulative percentage growth as follows. The absolute growth is

$$y - a = 121.66 - 100 = 21.66 \text{ (bn euros)}$$

The proportionate growth (see section 9.2 if you need to refresh on proportionate growth) is

$$\frac{y - a}{a} = \frac{121.66 - 100}{100} = 0.2166$$

So the percentage growth is proportionate growth times 100, which equals $0.2166 \times 100 = 21.66$. That is, over 5 years GDP grows by 21.66%.

Method 2: It may also be argued that GDP itself grows continuously, even though the data are only collected annually. If we now assume that GDP grows continuously, formula (2) is used, so we have

$$y = ae^{rx}$$

with $a = 100$, $r = 0.04$, $x = 5$ (as before). Substituting these values gives

$$y = e^{0.04(5)} = 100e^{0.2}$$

To find $e^{0.2}$ we key into our calculator 'SHIFT' (or equivalent), then 'ln', then 0.2, giving $e^{0.2} = 1.2214$. So $y = 100(1.2214) = 122.14$ (= level of GDP, in euros, after 5 years). Using the same method as before, the percentage growth is 22.14%.

The two answers are illustrated graphically in figure 12.9. The graph of $y = 100(1.04)^5$ is a step function because y increases in jumps (or steps) at the end of each year. Notice how small is the difference between the two answers: 22.14% for continuous growth versus 21.66% for growth in annual jumps. This small difference also explains why there is not much point in using formula (1) with $n = 4$ (for growth in 4 jumps per year), or $n = 12$ (monthly growth). We would simply get an answer somewhere between 22.14% and 21.66%, a difference too small to be worth the trouble of calculating, for most purposes.

This small difference in our two answers (21.66% and 22.14%) is not just a quirk of this example but is found whenever r is fairly small. The reason for this is as follows. From rule 12.5 in section 12.9 above we know that when the continuous growth formula $y = ae^{rx}$ is used, the effective annual interest rate is $e^r - 1$. When the annual growth formula

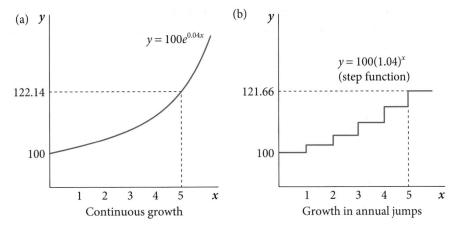

Figure 12.9 Cumulative percentage growth in GDP after 5 years (y-axis not to scale).

$y = a(1 + r)^x$ is used, the effective annual interest rate is, of course, r. As you should check on your calculator, $e^r - 1$ and r are approximately equal when r is small. That is why our two answers above are so similar.

Which of the two answers should we prefer? We argued earlier that *actual* GDP is a continuous flow of output, therefore growth occurs when the flow increases. For example, the number of cars rolling off the end of a production line might increase from 2 to 3 per minute. This would suggest that formula (2) (continuous growth) is the correct one to use. However, GDP is measured by the national accounts statisticians as an annual quantity. So *measured* GDP may be viewed as a discrete variable: that is, one that grows in annual jumps. This suggest that the discrete growth formula should be used. Thus there is no overwhelming case for or against either formula, and in practice researchers seem to choose whichever they feel is most convenient and appropriate to the data in question. These two ways of looking at GDP growth can be applied to other variables, such as the growth of a child. The child's actual growth is continuous, but its measured growth is necessarily discrete.

EXAMPLE 12.3

GDP (or some other variable) grows at 5% per year. How long does it take to double?

Answer 1: Assuming growth in annual jumps, we use

$$y = a(1 + r)^x$$

We are not given the initial value of the variable, so we will set it arbitrarily as $a = 1$. Since the variable doubles, this means $y = 2$. We are given $r = 0.05$, and x is to be found. Substituting these values into the growth formula, we get

$$2 = 1(1.05)^x$$

To solve this equation, we take logs on both sides of the equation. We did this earlier, in section 11.8, example 11.2. Recall that it's a valid step because if any two things are equal, their logs are equal too. These logs may be either common logs (base 10) or natural logs (base e). We will take logs to base 10, giving

$$\log 2 = \log(1.05)^x$$

But $\log(1.05)^x = x[\log(1.05)]$ (by rule 11.2b; see section 11.7 above). Hence

$$\log 2 = x\log(1.05) \quad \text{(Note we have dropped the redundant square brackets)}$$

Dividing both sides by $\log(1.05)$ gives

$$x = \frac{\log 2}{\log 1.05} = \frac{0.3010}{0.0211893} = 14.2067 \quad \text{(years) (using 'log' key on calculator)}$$

Answer 2: Assuming continuous growth, we use

$$y = ae^{rx}$$

with, as before, $a = 1$, $y = 2$, $r = 0.05$ and x to be found. On substitution:

$$2 = 1e^{0.05x}$$

and again we must take logs on both sides in order to solve. Again these logs may be either common logs (base 10) or natural logs (base e). Natural logs are slightly more convenient in this case. Thus taking natural logs on both sides gives

$$\ln 2 = \ln(e^{0.05x})$$

But $\ln(e^{0.05x}) = 0.05x(\ln e)$ (by rule 12.4b, section 12.7 above). Also, we know that $\ln e = 1$ (by rule 12.4e, section 12.7 above), so it disappears from the right-hand side. Thus the equation above becomes

$$\ln 2 = 0.05x \quad \text{Hence, dividing both sides by 0.05:}$$

$$x = \frac{\ln 2}{0.05} = \frac{0.6931}{0.05} = 13.8629 \quad \text{(years)}$$

where we found $\ln 2$ by pressing the 'ln' key, followed by 2 and '=' on our calculator.

As in example 12.2, the difference between the two answers is quite small (14.2 vs. 13.9 years), with the doubling period obviously being shorter when growth is assumed continuous because the effective annual growth rate is higher.

EXAMPLE 12.4

UK gross domestic product (GDP), measured in year 2000 prices, rose from £256,501 million in 1948 to £988,338 million in 2002. What was the average annual growth rate?

Method 1: Assuming continuous growth, we use the formula $y = ae^{rx}$ which we can also write as

$$\frac{y}{a} = e^{rx}$$

We have $a = 256,501$, $y = 988,338$, $x = 54$, with r to be found. Substituting these values gives

$$\frac{988,338}{256,501} = e^{54r} \quad \text{from which}$$

$$3.853 = e^{54r}$$

Taking natural logs on both sides of this equation, we get

$\ln 3.853 = \ln(e^{54r}) = 54r(\ln e) = 54r$ (using rules 12.4b and 12.4e from section 12.7 above). So

$$r = \frac{\ln 3.853}{54} = \frac{1.34885}{54} = 0.024973$$

This is the proportionate growth rate, so we multiply by 100 to get the percentage growth rate as 2.4973% per year.

Method 2: Assuming growth in annual jumps, we use the formula $y = a(1 + r)^x$ which we can also write as

$$\frac{y}{a} = (1 + r)^x$$

We have $a = 256,501$, $y = 988,338$, $x = 54$, with r to be found. Substituting in these values gives

$$\frac{988,338}{256,501} = (1 + r)^{54} \quad \text{from which}$$

$$3.853 = (1 + r)^{54}$$

There are two methods of solving this equation.

Method (a) (easier) Raise both sides of the equation to the power 1/54. (We previously did this trick in example 10.19 in section 10.7.) This gives

$$(3.853)^{1/54} = [(1 + r)^{54}]^{1/54}$$

The right-hand side then simplifies to $1 + r$, so we have

$$(3.8532)^{1/54} = 1 + r$$

Using our calculator, $1/54 = 0.0185$, so $(3.8532)^{1/54} = (3.8532)^{0.0185} = 1.025294$

Thus we have $1.025294 = 1 + r$, so $r = 0.025294$

This is the proportionate growth rate, so we multiply by 100 to get the percentage growth rate as 2.5294% per year.

Method (b) Take common logs (though natural logs would work equally well) on both sides. This gives

$$\log 3.853 = \log[(1 + r)^{54}] = 54[\log(1 + r)] \quad \text{by rule 11.2b, so}$$

$$\log(1 + r) = \frac{\log 3.853}{54} = \frac{0.5858}{54} = 0.010548129$$

This tells us that $1 + r$ is a number whose log is 0.010848129. So we want to find the anti-log of 0.010848129. To do this on our calculator we press the 'SHIFT' (or equivalent) key, followed by 'log' and 0.010848129. The answer displayed should be 1.025293 or something close to it. This equals $1 + r$, so $r = 0.025293$ and the percentage growth rate is 2.5293%.

(As an exercise, try method (b) but taking natural logs. You should get the same answer.)

We conclude that the average annual growth rate of UK GDP 1948–2002, to 4 decimal places, was 2.4973% if GDP is assumed to grow continuously (method 1); and 2.5293% if growth is assumed to be in annual jumps. For most purposes this difference is small enough to be ignored. This again illustrates the point that it matters little whether we use the discrete or the continuous growth formula, except when we have some combination of a long time period and a high growth rate.

EXAMPLE 12.5

How long does a variable take to double, at various continuous growth rates?

Answer: The answer to this question illustrates an intriguing relationship known as the 'rule of 69'. We assume the variable has an initial value of $a = 1$ and doubles, so $y = 2$. Growth is continuous, so the formula is $y = ae^{rx}$. On substitution this becomes

$2 = 1\, e^{rx}$

Taking natural logs on both sides gives

$\ln 2 = \ln(e^{rx}) = rx(\ln e) = rx$ (using rules 12.4b and 12.4e from section 12.7 above)

From a calculator, $\ln 2 = 0.6931$.

Thus we have $rx = 0.6931$. This result is stated as rule 12.6:

RULE 12.6 The 'rule of 69'

When a variable that grows continuously is known to double in value, we have $a = 1$ and $y = 2$, so $y = e^{rx} = 2$, and therefore

$$\ln y = rx = \ln 2 = 0.6931$$

This says that if a variable is known to double in value, the product of its growth rate, r, and the number of years, x, must equal 0.6931, or approximately 0.69. This is known as the 'rule of 69'.

As an example of the rule of 69, if the growth rate is 3% ($r = 0.03$), we have

$$0.6931 = 0.03x \quad \text{so}$$

$$x = \frac{0.6931}{0.03} = 23.1033$$

Thus the time required for the variable to double, when it grows continuously at 3% per year, is approximately 23 years (because $3 \times 23 = 69$).

In the same way, if the growth rate were 5%, the number of years required for the variable to double would be $\frac{0.6931}{0.05} = 14$ (approx.). Thus you need only remember the number 69, and you can quickly calculate in your head *either* the doubling period for a given growth rate, *or* the growth rate required for doubling within a given number of years. This will greatly impress people who don't know the trick.

As a final example, a leading British politician once promised that, provided his party remained in power, the standard of living (= real income per person) could double every 25 years. This may seem an ambitious promise, but from the rule of 69 we can immediately calculate that it requires only that real income per person rise at $69/25 = 2.8\%$ (approx.) per year, a fairly modest growth rate (although actual growth has been even more modest since this promise was made).

Further examples of using the natural exponential function to solve continuous growth problems are given at the Online Resource Centre.

Progress exercise 12.2

1. Suppose GDP is initially 100 and is growing at 3% per year (nominal rate).

(a) Assuming it grows in annual jumps, find its level after 1 year and after 25 years.

(b) Assuming it grows continuously, find its level after 1 year and after 25 years. Compare your answers to (a) and (b).

2. Show that if GDP grows *continuously* at 3% per year, it will double after approximately 23 years. Show that if it grows *continuously* at 23% per year, it will double in approximately 3 years. Compare and explain these results. (See example 12.5.)

12.10 Continuous discounting and present value

In this section we will show how the natural exponential function can be used to find the present value of a single future payment or receipt when growth, and therefore discounting (which is the same process running backwards), is continuous.

In section 10.8 we developed the present discounted value (PV) formula, rule 10.7, which showed that the PV of a single payment or receipt of a euros due in x years' time, discounted once per year at $100r\%$ per year, was given by

$$y = \frac{a}{(1+r)^x} \quad [\text{often written as } y = a(1+r)^{-x}] \text{ where } y \text{ is the } PV$$

Table 12.2 uses this equation to show how the PV of a single payment or receipt, a, varies according to when it is received or paid (that is, as x varies) for a given r. Note that we are not here considering a series of payments or receipts of a euros *per year*. We are considering how the PV of a *single* payment of a euros varies according to how far into the future it is to be received or paid. We have arbitrarily set the number of years at 4.

Table 12.2 Variation in present value of €a according to when received (annual intervals).

x (= number of years from now until payment of a received)	0	1	2	3	4
y (= PV)	$\frac{a}{(1+r)^0} = a$	$\frac{a}{(1+r)^1}$	$\frac{a}{(1+r)^2}$	$\frac{a}{(1+r)^3}$	$\frac{a}{(1+r)^4}$

Let us briefly recall from section 10.9 the underlying rationale for the *PV* formula. It is assumed that I can borrow or lend freely at the given interest rate, r. This means that if I have $a(1 + r)^{-2}$ euros now, I can lend this at interest rate r, and its value in 2 years' time will be $[a(1+r)^{-2}](1+r)^2 = a$ euros. Alternatively, if I have the right to receive a in 2 years' time, I can borrow $a(1 + r)^{-2}$ now, and after 2 years I will owe $[a(1+r)^{-2}](1+r)^2 = a$, which I can repay when I receive my a. Therefore having $a(1 + r)^{-2}$ now and having a two years from now are of equal monetary value to me, since I can convert one into the other by borrowing or lending. That is why in table 12.2 the right to receive, or the obligation to pay, a in two years' time has a present value of $a(1 + r)^{-2}$, or $\frac{a}{(1+r)^2}$.

(It's important to note, though, that although $a(1+r)^{-2}$ now and a in two years' time may be of equal monetary value to me in the sense just described, that doesn't necessarily mean I am indifferent between them. I may prefer one to the other, and therefore wish to borrow or lend.)

To make things more concrete we will now repeat table 12.2 with $a = 100$ and $r = 0.2$ (that is, 20%) per year. This gives us table 12.3. (Check the values in table 12.3 for yourself.) To avoid constantly writing 'payment or receipt' we will simply refer to a as a receipt.

Table 12.3 Variation in present value of €100 according to when received ($r = 0.2$, annual intervals).

x (= number of years from now that payment of €100 received)	0	1	2	3	4
y (= PV)	100.00	83.33	69.44	57.87	48.23

The values from table 12.3 are graphed in figure 12.10(a). The graph is not a smooth curve, and not even a step function, but simply a series of points or vertical bars at 1-year intervals. This is because we have considered only alternative dates for the receipt of a that are 1 year apart, and assumed that discounting occurs only once a year. (Recall from section 10.8 that 'discounting' means dividing by $(1 + r)^x$, and is the reverse of 'compounding', which means multiplying by $(1 + r)^x$.)

Table 12.4 repeats the calculations of table 12.3, but now assuming that the alternative dates on which a might be received are 6 months apart, or half-yearly. We must therefore use rule 10.8:

$$y = \frac{a}{(1 + \frac{r}{n})^{nx}}$$

to calculate the present values, with $n = 2$ and, as before, $r = 0.2$.

Table 12.4 Variation in present value of €100 according to when received ($r = 0.2$, half-yearly intervals).

x (= number of years from now that payment of €100 received)	0	0.5	1	1.5	2	2.5	3	3.5	4
y (= PV)	100.00	90.91	82.64	75.13	68.30	62.09	56.45	51.32	46.65

Table 12.5 Present value of €100 with continuous discounting ($r = 0.2$).

x (= years until received)	0	1	2	3	4
y (PV)	100.00	81.87	67.03	54.88	44.93

(a)

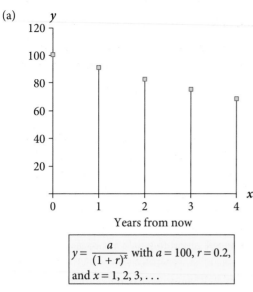

$$y = \frac{a}{(1 + r)^x} \text{ with } a = 100, r = 0.2,$$
and $x = 1, 2, 3, \ldots$

(b)

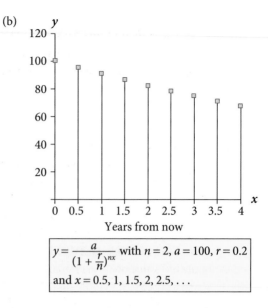

$$y = \frac{a}{(1 + \frac{r}{n})^{nx}} \text{ with } n = 2, a = 100, r = 0.2$$
and $x = 0.5, 1, 1.5, 2, 2.5, \ldots$

(c)

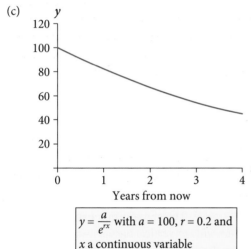

$$y = \frac{a}{e^{rx}} \text{ with } a = 100, r = 0.2 \text{ and}$$
x a continuous variable

Figure 12.10 Present value of a single receipt or payment of €100 according to when received, when $r = 0.2$. (a) Received at annual intervals with annual discounting; (b) received at half-yearly intervals with twice-yearly discounting; (c) received at any time, continuous discounting.

Thus we are now considering the *PV* of a single payment of 100 received at alternative half-yearly intervals, with discounting at 20% per year twice yearly. Figure 12.10(b) shows the resulting graph of the values in table 12.4. (Check the values in table 12.4 for yourself.)

Comparing figure 12.10(b) with figure 12.10(a), in 12.10(b) we now have twice as many values for the *PV*, and the distance between them is halved. Clearly we can go on increasing the value of n, and as n increases so does the number of alternative points in time at which the single payment is assumed to be received or paid, and correspondingly the frequency with which discounting occurs. As we do so, each successive graph will have more and more data points, which will become closer and closer together until they start to merge with one another to form a continuous curve.

There is no good reason to stop this process at any finite value of n, since the single payment could in principle be received at any moment in time, and discounting can occur virtually continuously thanks to computers replacing accounts clerks. Therefore it seems reasonable to allow n to approach infinity. Now, we know from section 12.4 above that, as n approaches infinity, $(1 + \frac{r}{n})^{nx}$ approaches e^{rx}. Therefore, in our PV formula (rule 10.8):

$$y = \frac{a}{(1 + \frac{r}{n})^{nx}} \quad \text{approaches} \quad y = \frac{a}{e^{rx}} \quad \text{(where } y \text{ is the } PV\text{)}$$

Thus we have a new rule:

RULE 12.7 Present value with continuous discounting

The *PV* of a single receipt or payment of *a*, to be received *x* years from now, when *x* varies continuously and discounting is continuous at the nominal rate of 100*r*% per year, is given by

$$y = \frac{a}{e^{rx}} \quad \text{(which, to save space on the page, is usually written as } y = ae^{-rx}\text{)}$$

We have used this new formula, $y = ae^{-rx}$, with the same parameters as before ($a = 100$, $r = 0.2$, $x = 4$) to create figure 12.10c. Since *x* is now a continuous variable, the graph is a smooth and continuous curve. It gives the *PV* of a single payment of 100 receivable at any moment in the next 4 years, when discounting is continuous. The *PV* values for integer values of *x* are shown in table 12.5. (Check these for yourself, using the 'SHIFT' (or equivalent) key on your calculator. To get you started, the figure of 81.87 in table 12.5 equals $100 \div e^{0.2}$. Feel free to calculate some non-integer values too, as we needed these in order to draw the graph.)

As a final point, note that the *PV*s change when the frequency of discounting changes. For example, from tables 12.3, 12.4, and 12.5 we see that the *PV* of 100 due in 4 years' time is 48.23, 46.65, and 44.93 respectively as the discounting occurs once yearly, twice yearly or continuously. This exactly parallels our finding in section 12.3, that the cumulative value of a growing variable is greater, the more frequently growth occurs. In most circumstances these differences are small enough to be ignored. It happens that the differences noted above are relatively large, but this is because we have chosen a relatively high discount rate, 20% per year.

12.11 Graphs with semi-log scale

You may have noticed that graphs in serious newspapers such as *The Economist* and the *Financial Times*, and in economics books and journals, often use a logarithmic scale on the vertical axis. We will state the reason for this as a rule:

RULE 12.8 Graphs with logarithmic scale

If a variable *y* is growing at a constant proportionate rate through time, then the graph of the log of *y* against time is a straight line.

This is true whatever base for the logs is used.

Here we will show the truth of this rule, and its significance, by means of examples. We will prove it later, in section 13.8.

EXAMPLE 12.6

Consider an economic variable *y* that is growing through time, *x*, as shown in table 12.6. We see that from an initial value of 1, *y* doubles every year, to 2, 4, and 8 in years 1, 2, and 3. The year-to-year (proportionate) growth rate is found as follows:

From year 0 to year 1: $\dfrac{\Delta y}{y_0} = \dfrac{y_1 - y_0}{y_0} = \dfrac{2 - 1}{1} = 1$

From year 1 to year 2: $\dfrac{\Delta y}{y_1} = \dfrac{y_2 - y_1}{y_1} = \dfrac{4 - 2}{2} = 1$

and so on. (See section 9.2 if you need to revise proportionate changes.) So the year-to-year proportionate growth is constant and equal to 1. The percentage growth is found as the proportionate growth × 100 = 1 × 100 = 100 (% per year). (The choice of growth rate was arbitrary.) Be sure you understand these concepts and calculations before you move on.

Table 12.6 Data for figure 12.11.

x (years)	0	1	2	3
y	1	2	4	8
$\log y$	0	0.30103	0.60206	0.90309
$\ln y$	0	0.69315	1.38629	2.07944

(a)

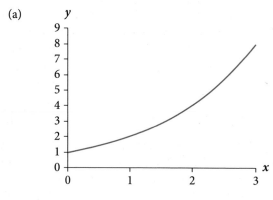

(b)

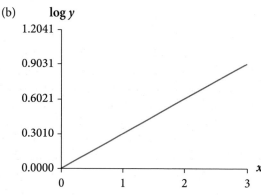

(c)

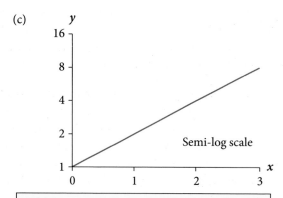

In (a), y appears to be growing quite rapidly through time but we can't tell whether the growth rate is constant, increasing or decreasing; in (b), with $\log y$ on the vertical axis, the linear graph tells us that the growth rate of y is constant; (c) is identical to (b) except that the values of the logs of y have been replaced by the values of y itself, conveying more information

Figure 12.11 An economic variable, y, growing at a constant rate.

In figure 12.11(a) we have used the data in table 12.6 to plot the graph of y against x, assuming the relationship to be smooth and continuous. The result is a curve that turns upwards quite steeply, but other than that the graph does not convey very much useful information. In particular, the graph does not tell us whether the growth rate of y is constant, increasing or decreasing—though we know from table 12.6 that in fact the growth rate is constant (at 100% per year).

Now look at row 3 of table 12.6, where we show the common logs of the y values. If you check these numbers you will find that, every year, $\log y$ increases by a constant *absolute* amount, 0.30103. This is the key point: the constant *proportionate* changes in y are associated with constant *absolute* changes in $\log y$. Consequently, when we plot the graph of $\log y$ against x, as in figure 12.11(b), we get a straight line with a slope of 0.30103. The fact that this graph is a straight line tells us immediately that y is increasing at a constant proportionate or percentage rate. (The graph passes through the origin because the log of 1 is zero.)

What is true of the common logs is also true of the natural logs of the y values, shown in the bottom row of table 12.6. These also increase from year to year by a constant absolute amount, 0.69315 (check this for yourself). So the graph of $\ln y$ against x (not drawn) would also be a straight line, though with a steeper slope than figure 12.11(b). Check this for yourself by sketching this graph, using rows 1 and 4 of table 12.6 as data.

Next, we note that although figure 12.11(b) is superior to figure 12.11(a) because its linearity tells us that the growth rate of y is constant, on the other hand having the logs of the ys on the vertical axis is an obvious disadvantage of figure 12.11b because the graph then tells us nothing about the values of y itself. We can overcome this by replacing the logs of the ys with the corresponding values of y itself, taken from table 12.6. The result is figure 12.11(c), where the log y labels on the vertical axis in figure 12.11b (0, 0.3010, 0.6021, 0.9031) have been replaced by the corresponding y values (1, 2, 4, 8), but with the scale unchanged.

To avoid potential confusion we have followed what is considered good practice by making a note on figure 12.11(c), that we have used a logarithmic scale on the vertical axis (called a semi-log scale, because it affects only one axis). Seeing this, an informed person looking at the graph knows immediately that the linear relationship implies a constant growth rate. However, if we should forget to flag up the log scale, it is anyway revealed by the fact that, on the vertical axis, the distance from 2 to 4 is the same as the distance from 4 to 8 and the same as the distance from 8 to 16. In other words, equal *proportionate* changes in y are measured as equal *absolute* distances on the vertical axis. This is the key characteristic of a log scale.

EXAMPLE 12.7

In this example we consider a case where over a 10-year period the growth rate of the variable y is *not* constant. In row 2 of table 12.7 we give the values of some variable, y, over 10 years. The proportionate growth of y in each year ($\Delta y/y$) is calculated in row 3. We see that the growth of y is 0.4 ($= 40\%$) in year 1, 0.38 ($= 38\%$) in year 2, and so on down to 22% in year 10 (see example 12.6 above if you need to check how these values are calculated). Thus the growth rate is *decreasing* from one year to the next.

However, when we draw the graph of y (figure 12.12(a)), again assuming that y varies continuously with time, we see that the graph is not merely upward sloping, but the slope is *increasing*. Anyone looking at this graph might well be tempted to infer from this that the growth rate is increasing, the opposite of the truth. The truth is revealed when we raw the graph of the logarithm of y as a function of x. In the previous example we used common logs, but in this example we have used natural logs in order to demonstrate that the outcome is the same. Thus figure 12.12(b) plots ln y against time, using the data from row 4 of table 12.7. In figure 12.12(b) the decreasing slope now tells us that the growth rate is decreasing. Finally, figure 12.12(c) repeats figure 12.12(b) but with the values of y pasted over the values of ln y. (That is, $0.6931 = \ln 2$, and so on.) As in the previous example, in figure 12.12(c) equal *absolute* distances on the vertical axis correspond to equal *proportionate* changes in y. For example, the distance from 2 to 4 is the same as the distance from 4 to 8, because both correspond to a doubling of y.

From these two examples we now see why serious journalists and researchers often use a logarithmic scale on the vertical axis of their graphs of time series data. The logarithmic scale means that simply by looking at the slope of the curve we can say whether the growth rate of the variable is constant, increasing or decreasing. Keep in mind that although in figures such as 12.12(c) the *scale* on the vertical axis is logarithmic, the *labelling* of the axis is in values of the variable itself, since these values obviously convey a lot more information than the logs would.

Table 12.7 Values for figure 12.12.

x (time)	0	1	2	3	4	5	6	7	8	9	10
y	1	1.40	1.93	2.63	3.52	4.65	6.04	7.73	9.74	12.08	14.74
$\Delta y/y$		0.40	0.38	0.36	0.34	0.32	0.30	0.28	0.26	0.24	0.22
ln y	0	0.3365	0.6586	0.9660	1.2587	1.5363	1.7987	2.0456	2.2767	2.4918	2.6906
Δln y		0.3365	0.3221	0.3075	0.2927	0.2776	0.2624	0.2469	0.2311	0.2151	0.1989

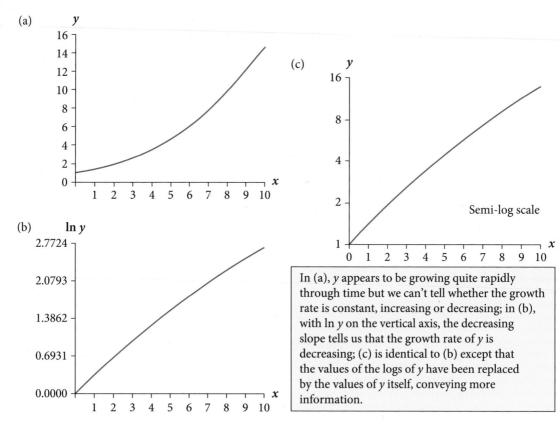

Figure 12.12 An economic variable, y, growing at a decreasing rate.

In (a), y appears to be growing quite rapidly through time but we can't tell whether the growth rate is constant, increasing or decreasing; in (b), with $\ln y$ on the vertical axis, the decreasing slope tells us that the growth rate of y is decreasing; (c) is identical to (b) except that the values of the logs of y have been replaced by the values of y itself, conveying more information.

EXAMPLE 12.8

As a final example we use some 'live' data (see figure 12.13). Graph (a) shows a time series of UK real GDP from 1948 to 2002. (By real GDP we mean GDP adjusted for inflation, the adjustment being achieved in this case by revaluing GDP in every year using the prices of the year 2000.) We have data only for each year's GDP but have joined up these annual observations to make a continuous curve, as in the two previous examples. Although graph (a) conveys useful information, such as that real GDP increased more than four-fold over the period, from just over £200 billion to almost £1000 billion, it tells us almost nothing about the growth rate of GDP. The curve appears to become steeper after about 1984, but we know from example 12.7 above that we must resist the temptation to infer that the growth rate increased in more recent years.

Figure 12.13(b) shows the same data but with a logarithmic scale on the vertical axis. We have used natural logs in this example. Note, though, that the labels on the vertical axis (100, 200, 400, . . .) are in units of GDP, not the log of GDP. As already explained, the log scale means that the distance between 100 and 200 is the same as the distance between 400 and 800, because both correspond to a doubling of GDP.

In graph (b) variations in the slope of the graph reflect variations in the growth rate of GDP. There is some short-term variation, especially between about 1972 and 1992, but the growth rate appears remarkably constant over the period as a whole. You can confirm this impression by using a ruler and pencil and drawing a straight line connecting the two end points. This straight line is often referred to as a trend line, though there are other, more sophisticated ways of measuring trend. You will find that the actual graph stays remarkably close to the trend line, apart from a brief acceleration in growth beginning in the early 1960s, which was reversed around 1980.

12 CONTINUOUS GROWTH AND THE NATURAL EXPONENTIAL FUNCTION

(a)   Natural scale   (b)   Log scale

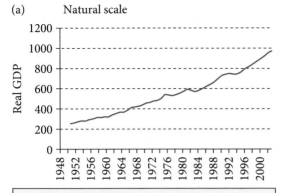

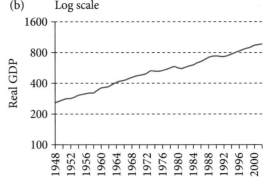

In (a) we can see that the level of GDP has occasionally fallen, but it is impossible to say whether the average growth rate has been constant, increasing or declining over any sub-period or over the period as a whole.

In (b), with a log scale on the vertical axis, the slope is now a measure of the growth rate. There have been short-run variations in the growth rate, but the average growth rate over the period appears remarkably constant, as the graph appears almost linear.

Figure 12.13 Real GDP, 1948–2002 (£bn in year 2000 prices).

To summarize the three previous examples, we have found that if the growth rate of a variable y through time is constant, then the graph of y with a logarithmic scale on the vertical axis is a straight line, thus confirming rule 12.8. The slope of this type of graph increases (decreases) when the growth rate increases (decreases).

There is one final point to be made about the graph of a time series with a logarithmic scale on the vertical axis. We know from above that the slope of this graph is constant, increasing or decreasing as the growth rate of y is constant, increasing or decreasing. In addition, it is also true that if a natural log scale is used, and y varies continuously with time, the slope of the graph actually *measures* the growth rate of y, to an acceptable degree of approximation. We will prove this in the next chapter.

As noted earlier, the graphs we have considered here—with a log scale on the vertical axis— are said to have a 'semi-log' scale. When the horizontal axis is also in logarithmic scale, this is called 'double-log' or 'log-log' scale. In the case of a time-series graph, where time is measured on the horizontal axis, we don't use a 'double-log' scale because it doesn't make sense to take the log of time. You should think about why (and see also section 9.2). In the next chapter we will look at cases where using logarithmic scales on both axes is useful.

 Chapter summary

In sections 12.2–12.4 of this chapter we took the compound growth formula $y = a(1 + \frac{r}{n})^{nx}$ and showed that, as n approaches infinity, $(1 + \frac{1}{n})^n$ approaches a limiting value, which we label e. This gave us a new compound growth formula $y = ae^{rx}$. The two growth formulae were summarized in rule 12.2:

(1) $y = a(1 + \frac{r}{n})^{nx}$ (from section 10.7, rule 10.5)
      (appropriate when y grows in discrete jumps, n times per year)

(2) $y = ae^{rx}$ (appropriate when y grows *continuously*)

In both cases, a is the initial value of the variable, and $100r\%$ is the nominal annual growth rate (that is, the growth that would occur if growth were in annual jumps).

Next, in sections 12.5–12.6 we examined the graph of $y = e^x$, which is called the natural exponential function. We also derived the inverse function $x = \log_e y$, called the natural logarithmic function, which defines natural logs (see rule 12.3). We noted the rules for manipulating these logs, and also looked at some applications of the continuous growth formula $y = ae^{rx}$ and of natural logs (sections 12.7–12.9).

Then in section 12.10 we derived the *PV* formula for a single future payment or receipt, a, when discounting is continuous. This formula is given by

$$y = \frac{a}{e^{rx}} \quad \text{usually written as } y = ae^{-rx} \quad \text{(see rule 12.7)}$$

Finally, in section 12.11 we examined the use of a logarithmic scale for the vertical axis (the dependent variable) when graphing a variable that is growing through time (a time series). We showed by worked examples that the slope of the graph then tells us whether the growth rate is constant, increasing or decreasing. We return to this in chapter 13.

The Online Resource Centre contains more examples of the application of the natural exponential and logarithmic functions discussed in this section.

Progress exercise 12.3

1. (a) A variable y is growing continuously at a nominal rate of 5% per year from an initial value of 100. Write down the equation giving y as a function of time (x), and sketch the function.

   (b) Repeat for a variable declining continuously at 5% per year.

   (c) For (a) and (b) above, write down the equation giving $\ln y$ as a function of x and sketch the function (that is, with $\ln y$ on the vertical axis and x on the horizontal). (Hint: you need to use the rules of logs; see section 12.7.)

2. A variable (y) has an initial value of 100 and grows continuously for 30 years at a nominal rate of 5% per year. Sketch a graph showing the path of y through time (x). What value does y reach after 30 years? What is the percentage growth of y over that time?

3. Repeat the previous question for a variable (y) that grows continuously for 30 years at a nominal rate of *minus* 5% per year (that is, y declines at 5% per year).

4. A variable (y) has an initial value of 125 and is growing through time (x) as shown in the table below:

At end year:	1	2	3	4	5
Level of y:	187.5	281.25	421.88	632.81	949.22

   (a) Draw a reasonably accurate graph of y as a function of x, for the above values.

   (b) From inspection of the graph, what can you infer about the growth rate of y? Does it appear to be constant, increasing, or declining?

   (c) Draw another reasonably accurate graph, this time taking $\ln y$ as the dependent variable.

   (d) From inspection of this graph, what can you infer about the growth rate of y? Does it appear to be constant, increasing, or declining?

   (e) Use your calculator to check the actual year-to-year growth rate(s) of y, from the table (if you haven't done so already!).

   (f) Hence explain briefly why we often use an ln scale to graph a variable growing through time.

   (Hint: it doesn't really matter whether you treat time as a continuous or discrete variable— that is, whether you assume that y grows continuously or in annual jumps. If you choose the latter assumption, then in order to assess whether the growth rate is constant, increasing, or declining you will need to join up your points to give a quasi-continuous curve.)

5. Repeat the previous question with a variable with initial value of 125, which then grows as follows:

At end year:	1	2	3	4	5
Level of y:	225	382.5	612	918	1285.2

6. In a certain economy the ratio of government expenditure to GDP is currently 40%. In the future, government expenditure is expected to grow at 3% per year and GDP is expected to grow at 2% per year. Calculate in how many years from now the ratio of government expenditure to GDP will reach 50%, assuming (a) continuous growth; (b) growth in annual jumps. (Hint for (a): $\frac{e^{ax}}{e^{bx}} = e^{(a-b)x}$.)

7. The level of GDP of the republic of Stagnatia was twice as high as that of the republic of Dynamica at the end of 1995 but only 50% higher 10 years later.

(a) Find the difference in their average annual growth rates.

(b) If these growth rates continue, when will Dynamica's GDP overtake that of Stagnatia? (Hint: assume GDP grows continuously. Write down a separate equation for each country's GDP as a function of time. You need only to find their *relative* growth rates. The hint from the previous question is also relevant.)

Checklist

Be sure to test your understanding of this chapter by attempting the progress exercises (answers are at the end of the book). The Online Resource Centre contains further exercises and materials relevant to this chapter www.oxfordtextbooks.co.uk/orc/renshaw3e/.

In this chapter we have moved on from the base 10 to the base e for our exponential and logarithmic functions, and shown that base e provides us with a formula for describing continuous growth and discounting. The topics were:

✔ **Continuous growth.** Understanding the concept of continuous growth and how it relates to discrete growth.

✔ **The natural exponential and logarithmic functions.** Understanding how these are derived, and the shapes of their graphs for various parameter values.

✔ **Problem solving.** Using the natural exponential and logarithmic functions to solve growth problems.

✔ **Continuous present value.** Understanding the concept and calculating the present value of a future payment or receipt when continuously discounted.

✔ **Logarithmic scales.** How plotting graphs of economic and other variables with a log scale on the vertical axis reveals the growth rate, and how to plot these graphs.

We now have an understanding of the continuous exponential and logarithmic functions and their relevance to analysing various problems of growth and present value. In the next chapter we find the derivatives of these two functions and develop their applications further.

Chapter 13

Derivatives of exponential and logarithmic functions and their applications

OBJECTIVES

Having completed this chapter you should be able to:

- Find the derivatives of the natural exponential and logarithmic functions.
- Understand the concept of the rate of proportionate change and measure it for discrete and continuous variables.
- Understand why and how we plot graphs with a logarithmic scale on the vertical axis.
- Understand and demonstrate the relationship between the slope of a function plotted with log scales on both axes and the elasticity of the function.

13.1 Introduction

In chapter 6 we stated the rules for finding the derivatives of many, but not all, functions. We now need to add to our toolkit the derivatives of the natural exponential and logarithmic functions. As we saw in chapter 12, both of these function types are smooth and continuous, so they must have derivatives. We therefore begin the chapter by stating the rules for finding the derivatives of these two functions. Using these derivatives, we are then able to prove rigorously something that in chapter 12 we showed only by numerical example: that in the natural continuous growth formula $y = ae^{rx}$ the parameter r gives the *instantaneous* growth rate of y; that is, the growth rate at a moment in time, of a variable that is growing continuously. We also explain how it is true that this same r also measures the nominal growth rate in the discontinuous growth formula $y = a(1 + r)^x$. We can also now prove rigorously that when a time-series graph is plotted with a logarithmic scale on the vertical axis, the slope of the graph measures the instantaneous growth rate. We showed this also in chapter 12 but only by means of numerical examples. Finally, we show by an example that when log scales are taken on both axes, the slope of the resulting graph measures the elasticity of the function.

13.2 The derivative of the natural exponential function

We simply state without proof (though the proof is not difficult) our first rule:

RULE 13.1 The derivative of the natural exponential function

Given $y = e^x$, then $\frac{dy}{dx} = e^x$

Note that, since $y = e^x$, we can also write the derivative as

$$\frac{dy}{dx} = e^x = y$$

This is a remarkable result; this is the only function which has the property that the derivative at any point on the curve is given by the value of the function itself at that point. A graphical interpretation can be seen in figure 13.1. At point J, the derivative is given by the slope of the tangent RS. This is true for any function, but rule 13.1 tells us that in the case of $y = e^x$, the derivative is also given by the distance JM, since this gives the value of y, y_0, which in turn equals e^{x_0}, since in general $y = e^x$. Thus the slope of RS and the distance JM are equal, wherever J is positioned on the curve.

Rule 13.1 deals only with the simple case where the power to which e is raised is simply x. But we may occasionally encounter more complex cases such as $y = e^{3x^2+9x}$. Such cases require rule 13.2, which is a generalization of rule 13.1:

RULE 13.2 Generalization of rule 13.1

Given $y = e^{f(x)}$, then $\frac{dy}{dx} = e^{f(x)}f'(x)$

where as usual $f'(x)$ denotes the derivative of $f(x)$

EXAMPLE 13.1

$$y = 10e^{x^2+3x} + 50$$

Here we have $f(x) = x^2 + 3x$, so $f'(x) = 2x + 3$. Applying rule 13.2, we get

$$\frac{dy}{dx} = 10e^{f(x)}f'(x) = 10e^{x^2+3x}(2x + 3)$$

Notice that, as usual, the multiplicative constant (10) reappears in the derivative, while the additive constant (50) disappears. (See section 6.7 if you need to revise this.)

EXAMPLE 13.2

Consider the continuous growth formula

$$y = ae^{rx} \quad \text{(where } a \text{ and } r \text{ are constants).}$$

Here we have $f(x) = rx$, so $f'(x) = r$. Applying rule 13.2, we get

$$\frac{dy}{dx} = ae^{f(x)}f'(x) = ae^{rx}(r), \text{ which may be written more tidily as } \frac{dy}{dx} = rae^{rx}.$$

Note that the multiplicative constant, a, reappears. Recall that $y = ae^{rx}$ describes any variable y growing continuously at a constant nominal rate of r for x years (with an initial value $= a$). (See sections 12.3–12.4 if you need to revise this further.)

Hint It's worth looking at example 13.2 carefully, because economists use the exponential function $y = ae^{rx}$ and its derivative very frequently. We return to this function in section 13.6 below.

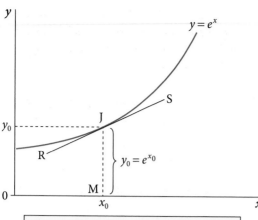

If $y = e^x$, $\frac{dy}{dx} = e^x$. This means that at any point J the slope of the tangent RS and the distance JM are both equal to $\frac{dy}{dx}$.

Figure 13.1 The derivative of $y = e^x$.

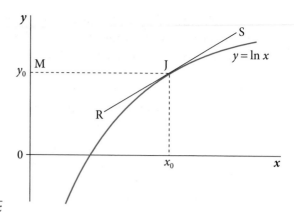

If $y = \ln x$, $\frac{dy}{dx} = \frac{1}{x}$. At any point J the slope of RS $= \frac{dy}{dx}$. Also $\frac{1}{x_0} = \frac{1}{\text{distance JM}}$.
So $\frac{dy}{dx} =$ slope of RS $= \frac{1}{\text{distance JM}}$

Figure 13.2 The derivative of $y = \ln x$.

13.3 The derivative of the natural logarithmic function

We simply state the following rule without proof:

> ✱ **RULE 13.3 The derivative of the natural logarithmic function**
> Given $y = \ln x$, then $\frac{dy}{dx} = \frac{1}{x}$

This is another surprising result. It says that the slope of the function $y = \ln x$ at any point is inversely proportional to the value of x at that point. So for example doubling x halves the slope. Figure 13.2 shows this graphically. Rule 13.2 tells us that the slope of the tangent RS equals the reciprocal of the distance JM (because distance JM gives us the value of x at J).

Rule 13.3 deals only with the simple case of a logarithmic function. We also need to be able to deal with more complex cases such as $y = \ln(2x^2 + 9x)$ or, more generally, $y = \ln f(x)$ where, as usual, $f(x)$ denotes any function of x. Such cases require rule 13.4, which is a generalization of rule 13.3:

> ✱ **RULE 13.4 Generalization of rule 13.3**
> Given $y = \ln f(x)$, then $\frac{dy}{dx} = \frac{f'(x)}{f(x)}$ where $f'(x)$ denotes the derivative of $f(x)$.

EXAMPLE 13.3

$y = \ln(x^3 + 2x)$

Here we have $f(x) = x^3 + 2x$, so $f'(x) = 3x^2 + 2$. Applying rule 13.4, we get

$$\frac{dy}{dx} = \frac{f'(x)}{f(x)} = \frac{3x^2 + 2}{x^3 + 2x}$$

The next rule, rule 13.5, is not really a new rule but merely an extension of rule 13.4. However, this extension is so important in economics that we have given it the status of a rule.

First, note that rule 13.4, like any rule of differentiation, is a rule for finding the derivative of the expression on the right-hand side of the given equation. The rule therefore holds, whatever symbol is used to label the variable on the left-hand side. So, for example, suppose we are given

$$\zeta = \ln f(x)$$

where ζ is some variable. We have deliberately chosen an unusual Greek letter (zeta) to emphasize that it doesn't matter what is on the left-hand side. Then applying rule 13.4 we get

$$\frac{d\zeta}{dx} = \frac{f'(x)}{f(x)}$$

Thus the right-hand side is the derivative of whatever happens to be on the left-hand side. This preliminary may help you to understand rule 13.5:

RULE 13.5 **Extension of rule 13.4**

For any function $y = f(x)$, by taking natural logs on both sides we get

$$\ln y = \ln f(x)$$

As just explained with the example of ζ, the rules of differentiation may be applied to the right-hand side, whatever is on the left-hand side. So we can treat $\ln y$ on the left-hand side in exactly the same way as we treated ζ in the example above.

So, applying rule 13.4 to $\ln y = \ln f(x)$, we get

$$\frac{d \ln y}{dx} = \frac{f'(x)}{f(x)}$$

Further, since $y = f(x)$ and therefore $\frac{dy}{dx} = f'(x)$, on substitution we get

$$\frac{d(\ln y)}{dx} = \frac{f'(x)}{f(x)} = \frac{\frac{dy}{dx}}{y} = \frac{1}{y}\frac{dy}{dx}$$

This is true of any function, $y = f(x)$.

The significance of this rule will be explained in section 13.8 below.

13.4 The rate of proportionate change, or rate of growth

In chapters 10–12 we considered variables that were growing through time, such as bank deposits, GDP, the price level, and so on. We mainly considered variables that were growing at a constant compound rate, which we denoted by r. If the variable grew in discrete jumps, we used the compound growth formula $y = a(1 + \frac{r}{n})^{nx}$, while if growth was continuous, we used the formula $y = ae^{rx}$.

We now want to consider how to define and measure growth when the variable in question is not necessarily growing at a constant rate. First we will consider cases where growth is discrete, for example in annual jumps.

13.5 Discrete growth

Suppose we are looking at data for some economic variable, y, that varies with time, x, in annual jumps. If we plot the graph of y against x, we will get a step function, such as in figure 13.3. The growth rate of y may or may not be constant.

Suppose we want to measure the **rate of growth** of y in year 5. As discussed in section 9.2, we can measure the absolute change in y as the distance $\Delta y = y_1 - y_0$ in figure 13.3, where y_1 denotes

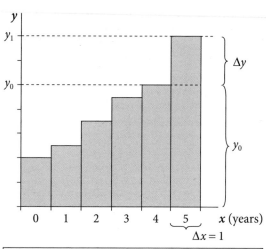

For a data series that varies in annual jumps, the year-to-year growth rate is measured by $\frac{1}{\Delta x}\frac{\Delta y}{y_0}$, with $\Delta x = 1$ (see rule 13.6), but this measure is inappropriate for measuring growth over a longer period, when $y = a(1 + r)^x$ or $y = ae^{rx}$ should be used.

Figure 13.3 Measuring the growth rate of a discrete variable.

the value of y in year 5 and y_0 is the value in year 4. This change is usually written more compactly as Δy (Δ as usual denoting 'the change in'). The proportionate change is then defined as

$$\frac{y_1 - y_0}{y_0} \quad \text{or equivalently} \quad \frac{\Delta y}{y_0}$$

The interpretation is that, by taking the *absolute* change, $y_1 - y_0$, in y and dividing it by the initial value, y_0, we express the change as a proportion of its initial value: that is, as a *proportionate* change.

For example, if in year 5 y increases from $y_0 = 200$ to $y_1 = 240$, the absolute change is $240 - 200 = 40$, and the proportionate change is

$$\frac{\Delta y}{y_0} = \frac{y_1 - y_0}{y_0} = \frac{240 - 200}{200} = \frac{40}{200} = 0.2$$

The percentage change is then $0.2 \times 100 = 20$ (%).

Next, we divide $\frac{\Delta y}{y_0}$ by Δx. This is the same thing as multiplying by $\frac{1}{\Delta x}$, so the result is

$$\frac{\frac{\Delta y}{y_0}}{\Delta x} = \frac{1}{\Delta x}\frac{\Delta y}{y_0} \quad \text{which is usually written as} \quad \frac{1}{y_0}\frac{\Delta y}{\Delta x} \tag{13.1}$$

Equation (13.1) gives three ways of writing the same thing: the proportionate change in y divided by the absolute change in x. Equation (13.1) therefore measures the proportionate change in y per unit of change in x. This is called the *rate of proportionate change* of y, often simply referred to as the rate of growth (which may be either positive or negative).

In our example, $\Delta x = 1$ because we considered a change in y from year 4 to year 5. It's important to see that equation (13.1) is an unacceptably crude measure of the growth rate if the time interval chosen, Δx, is longer than 1 year in the case of data published annually, 1 month in the case of data published monthly, and so on. The reason for this is that equation (13.1) distributes the total growth equally over the time interval without allowing for any compounding effect (see the next example for an illustration of this mistake). This is not satisfactory because, as discussed in section 10.5, it is reasonable to suppose that the growth of economic variables follows the laws of compound interest rather than simple interest. To measure the average growth rate of a discrete variable over a longer period than 1 year, month or whatever the time unit may be, we should therefore use the compound growth formula $y = a(1 + r)^x$. We can illustrate this with a simple example.

EXAMPLE 13.4

Table 13.1 gives the actual values of Britain's GDP in the period 1995–2009, in millions of pounds sterling (repeated from section 1.13). Column (3) gives the year-to-year growth, calculated using equation 13.1 with $\Delta x = 1$. For example, the growth rate between 1995 and 1996 is calculated as

$$\frac{1}{y_0}\frac{\Delta y}{\Delta x} = \frac{\frac{\Delta y}{y_0}}{\Delta x} = \frac{\frac{992,617 - 964,780}{964,780}}{1} \times 100 = 2.89 \text{ per cent}$$

Similarly, the growth between 1999 and 2000 (a year of rapid growth) is calculated as

$$\frac{1}{y_0}\frac{\Delta y}{\Delta x} = \frac{\frac{\Delta y}{y_0}}{\Delta x} = \frac{\frac{1142,372 - 1099,327}{1099,327}}{1} \times 100 = 3.92 \text{ per cent}$$

Table 13.1 UK GDP, 1995–2009.

Year (1)	GDP (£m) (2)	Year-to-year growth (%) (3)
1995	964,780	–
1996	992,617	2.89
1997	1025,447	3.31
1998	1062,433	3.61
1999	1099,327	3.47
2000	1142,372	3.92
2001	1170,489	2.46
2002	1195,035	2.10
2003	1228,595	2.81
2004	1264,852	2.95
2005	1292,335	2.17
2006	1328,363	2.79
2007	1364,029	2.68
2008	1363,139	−0.07
2009	1296,390	−4.90
Cumulative growth 1995–2009: 34.37%		

Source: Office of National Statistics; author's calculations.

However, if we want the average annual growth rate over the 14 years as a whole, it would be inappropriate to use this method. If we do so, we get

$$\frac{1}{y_0}\frac{\Delta y}{\Delta x} = \frac{\frac{\Delta y}{y_0}}{\Delta x} = \frac{\frac{1296,390 - 964,780}{964,780}}{14} \times 100 = 2.46 \text{ per cent}$$

This is inappropriate because it simply takes the total growth over the 14 years, 34.37% of the base year, and distributes it equally over the 12 years. The result is $34.37 \div 14 = 2.46\%$. It assumes thereby that GDP grew each year by an equal *absolute* amount, 2.46% of its base year value, that is, $964,780 \times 0.0246 = 23,733.59$ (million pounds). Such growth is not impossible, but growth at a constant *absolute* rate implies that the *proportionate* rate was declining year by year, which seems unlikely.

An averaging procedure that is likely to be closer to the actual path of GDP is to assume that it grew by an equal *proportionate* amount in each year. With this assumption, the average annual growth rate is calculated using our old friend from chapter 10, $y = a(1 + r)^x$, with $y = 1296,390$, $a = 964,780$, and $x = 14$. Solving for r gives $r = 2.13\%$. (We could also have used $y = ae^{rx}$, which would have given a slightly different answer.) Note that this is a significantly lower average than the 2.46% obtained in the previous paragraph, as we are now assuming constant *proportionate* growth (compound interest) rather than constant absolute growth (simple interest) (see section 10.4). (This method of calculating the average growth rate is itself subject to criticism because it considers only the levels of GDP in 1995 and 2009, but we will not pursue this here.)

It may have occurred to you that another way of calculating the average annual growth rate would be to take the arithmetic average of the year-to-year growth rates in column 3 of table 13.1. The error in this again arises from a confusion between absolute and proportionate growth, or equivalently between simple and compound growth. Suppose for example that a variable has an initial value of 100, then grows by 1% in the first year, to 101; and by 100% in the second year, to 202. Then the average of these two year-to-year growth rates is 50.5%. However, if we apply this growth rate to the variable, it grows by 50.5% in the first year, to 150.5; and by 50.5% in the second year, to 150.5 + 50.5% of 150.5, which equals 226.5. This of course over-shoots the true final value, 202. To reach the final value of 202, growth must be 50.5% of the *base* value in each year, not 50.5% of the *current* value. But this is exactly the distinction between simple and compound growth.

Conclusions on discrete growth

For discrete data any one of the three expressions in equation (13.1) is appropriate for measuring the growth rate from one period to the next. We will state this as a rule:

RULE 13.6 Discrete one-period growth

If a variable y varies discretely with time, x, the rate of proportionate growth of y from one period to the next is measured by

$$\frac{\frac{\Delta y}{y_0}}{\Delta x} = \frac{1}{\Delta x}\frac{\Delta y}{y_0} \quad \text{which is usually written as} \quad \frac{1}{y_0}\frac{\Delta y}{\Delta x}$$

where y_0 is the initial value of y, and Δx is the time interval for which data are published (that is, 1 year for annual data, 1 month for monthly data, and so on).

To calculate the average compound growth rate over more than one time period, one of the compound growth formulae $y = a(1 + r)^x$ or $y = ae^{rx}$ (see rules 10.4 and 12.2) should be used.

Progress exercise 13.1

1. Find $\frac{dy}{dx}$ for each of the following functions, and in cases (a) to (d) draw a sketch graph of the function:

   (a) $y = 15e^{0.4x}$ (b) $y = 100e^{-0.05x}$ (c) $y = 2\ln(x)$ (d) $y = -0.5[\ln(x)]$

   (e) $y = \ln(x^2)$ (f) $y = \ln(\frac{1}{x^{0.5}})$ (g) $y = e^{-0.5x^2}$ (h) $y = \frac{1}{\sqrt{e^x}}$

2. The table below gives the annual average consumer price index (CPI) for the UK 1990–2010, with 2005 = 100. (This index measures the average *level* of prices of all goods and services used in consumption by households.)

   (a) Calculate the cumulative growth in the price index between 1990 and 2010.

   (b) Calculate the year-to-year growth in this price index (that is, the inflation rate) for the years 1990–2000.

   (c) Calculate the average annual growth rate of prices for the periods (i) 1990–97; (ii) 1997–2010.

Year	CPI	Year	CPI
1990	71.5	2000	93.1
1991	76.8	2001	94.2
1992	80.1	2002	95.4
1993	82.1	2003	96.7
1994	83.8	2004	98.0
1995	86.0	2005	100.0
1996	88.1	2006	102.3
1997	89.7	2007	104.7
1998	91.1	2008	108.5
1999	92.3	2009	110.8
		2010	114.5

Source: UK Office of National Statistics.

13.6 Continuous growth

We now want to develop a formula for measuring the growth rate of a continuous variable. Suppose some economic variable, y, varies with time, x, according to some smooth and continuous function $y = f(x)$, as shown in figure 13.4.

We take as our starting point the one-period discrete growth formula, rule 13.6 from above:

$$\frac{1}{y_0}\frac{\Delta y}{\Delta x}$$

Looking at this expression, we recognize $\frac{\Delta y}{\Delta x}$ as our old friend, first defined in chapter 6, the difference quotient. We know that in figure 13.4 this measures the slope of the chord (the straight broken line) joining J to K. So the rate of proportionate growth can be measured as the slope of JK divided by y_0, the vertical distance JM. Thus

$$\frac{1}{y_0}\frac{\Delta y}{\Delta x} = \frac{\text{slope of chord JK}}{\text{distance JM}}$$

We now see diagrammatically that the formula in rule 13.6 is a rather crude measure of the growth rate when y is a continuous variable. It is crude because it depends for its value on the difference quotient. In figure 13.4 the difference quotient measures the slope of the dotted line JK, which gives the *average* slope of the function between J and K (see section 6.4 for refreshment on this). That is why, as we saw in the previous section, the formula in rule 13.6 should be used only for calculating growth of a discrete variable from one period to the next, as in example 13.4.

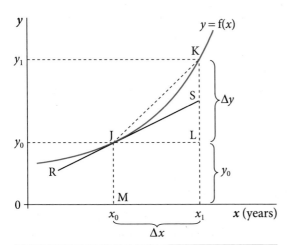

$\frac{1}{y_0}\frac{\Delta y}{\Delta x}$ measures the average rate of proportionate change (= growth rate) between x_0 and x_1. It equals the slope of the chord JK divided by the distance JM.

$\frac{1}{y_0}\frac{dy}{dx}$ measures the instantaneous rate of proportionate change (= growth rate) at x_0. It equals the slope of the tangent RS divided by the distance JM.

Figure 13.4 Measuring the instantaneous growth rate of a continuous variable.

The slope of the dotted line JK and therefore the difference quotient have the same value whatever the shape of the curve $y = f(x)$ may be, provided only that the curve passes through J and K. The formula in rule 13.6 would therefore be a precise measure of the growth rate of y between x_0 and x_1 only if the function itself were linear between J and K. (Check all of this for yourself with some back-of-an-envelope sketch graphs.)

However, consider what happens if we allow the time interval, Δx, over which growth is measured to become shorter and shorter and eventually approach zero. As Δx approaches zero, so the difference quotient $\frac{\Delta y}{\Delta x}$ approaches the derivative of the function, $\frac{dy}{dx}$. (See chapter 6 if you need to revise this.) Therefore in rule 13.6 we can replace with $\frac{\Delta y}{\Delta x}$ with $\frac{dy}{dx}$, and the formula then becomes

$$\frac{1}{y_0}\frac{dy}{dx}$$

What does this measure? Looking at figure 13.4, we see that $\frac{dy}{dx}$ is the slope of the tangent RS at J, which measures the rate of absolute change of y at J. Thus $\frac{1}{y_0}\frac{dy}{dx}$ consists of the rate of absolute change of y, divided by y_0, the distance JM. By dividing $\frac{dy}{dx}$ by y_0 we convert the rate of absolute change into the rate of proportionate change (= rate of growth) of y at point J. Thus

$$\frac{1}{y_0}\frac{dy}{dx} = \frac{\text{slope of tangent RS}}{\text{distance JM}}$$

This is a precise measure because $\frac{dy}{dx}$ is uniquely defined by the slope of the tangent at J. Unlike the formula in rule 13.6, which measures the average growth rate over some discrete time interval Δx, this new formula measures the rate of growth at a moment in time, and for this reason is often referred to as the **instantaneous rate of growth**. Thus we can formally state rule 13.7:

RULE 13.7 The instantaneous growth rate

If a variable y is related to time, x, by a smooth and continuous function, $y = f(x)$, then the instantaneous rate of growth of y at any moment in time is measured by

$$\frac{1}{y}\frac{dy}{dx}$$

Note that we have dropped the zero subscript from y, it being now understood that this is the value of y at the point in time when the growth rate is being measured.

EXAMPLE 13.5

Let's give the formula in rule 13.7 a little test run. Suppose we believe that some economic variable y is related to time, x, by some continuous linear function $y = ax + b$, where a and b are positive constants. (Such a relationship is not very likely, but this is purely for example.) Then we have $\frac{dy}{dx} = a$, so the instantaneous rate of growth is

$$\frac{1}{y}\frac{dy}{dx} = \frac{1}{y}a = \frac{a}{ax+b} \quad \text{(since } y = ax + b\text{)}$$

What do we learn from this? We learn that, since x appears in the denominator, $\frac{a}{ax+b}$ declines as x increases (that is, as time passes). So the instantaneous rate of growth of y declines through time. Depending on the context, this could be important.

EXAMPLE 13.6

Now let's consider a more important case. In chapter 12, rule 12.2, we derived the formula for continuous compound growth at a constant rate, r, as

$$y = ae^{rx}$$

From example 13.2 above we know that the derivative of this function is $\frac{dy}{dx} = rae^{rx}$. But $ae^{rx} = y$, from the function itself. Using this to substitute for ae^{rx}, the derivative becomes

$$\frac{dy}{dx} = ry$$

If we multiply both sides by $\frac{1}{y}$ the left-hand side becomes the instantaneous growth rate, and the right-hand side becomes $\frac{1}{y}ry = r$. We state this result as a rule:

RULE 13.8 The instantaneous growth rate when $y = ae^{rx}$

For the continuous growth formula $y = ae^{rx}$, the instantaneous growth rate is given by

$$\frac{1}{y}\frac{dy}{dx} = r$$

Figure 13.4 illustrates this result. At J, the slope of the tangent RS equals $\frac{dy}{dx}$, while the distance JM equals y_0, the value of y at the point in question (in rule 13.8 we have dropped the subscript on y). So the instantaneous growth rate is given by the slope of RS divided by the distance JM. This is true of any function, but in the case of $y = ae^{rx}$ the slope of RS divided by the distance JM is constant (irrespective of the position of J on the curve) and equal to r.

13.7 Instantaneous, nominal, and effective growth rates

(This section may be skipped without loss of continuity.)

In section 12.4 we said that in the continuous growth formula $y = ae^{rx}$, r is the *nominal* growth rate. The nominal growth rate is defined as the growth rate that occurs when growth is in annual jumps: that is, when $y = a(1 + r)^x$.

But rule 13.8 above says that r is also the *instantaneous* growth rate. This is puzzling at first sight, because it's hard to see how the nominal and instantaneous growth rates can be equal. On the contrary, they must surely be different, for we know that continuous growth at any given positive rate always results in a higher value of y than discrete growth at the same rate.

This apparent contradiction is resolved by closer consideration of what the instantaneous growth rate measures. It measures not the *actual* growth rate over any period, but the growth rate *at a moment in time*. The fact that the instantaneous growth rate and the nominal growth rate are both equal to r is illustrated in figure 13.5, where the curve $y = ae^{rx}$ shows y to be growing continuously at the nominal rate r. At J (the moment in time x_0) the instantaneous growth rate is:

$$\frac{1}{y_0}\frac{dy}{dx} = r \quad \text{(from rule 13.8)}$$

In figure 13.5, $y_0 = LM$ and $\frac{dy}{dx} = \frac{SL}{JL} = SL$ (since $JL = 1$). So the instantaneous growth rate is:

$$r = \frac{1}{y_0}\frac{dy}{dx} = \frac{1}{LM}SL = \frac{SL}{LM}$$

Now suppose that, instead of growing continuously, y grows discretely in annual jumps at the same rate r from the initial position J. Nothing would happen for 1 year, so y would simply move along the dotted line from J to L as time passed (a horizontal section of a step function). On reaching L at the end of 1 year, y would jump to S. We know that y jumps to S because we know that the instantaneous and nominal growth rates are equal. The growth rate of y when growth is in annual jumps is thus given by $\frac{SL}{LM} = r$, which is also the instantaneous growth rate when growth is continuous.

We can also use figure 13.5 to show the actual growth rate when growth is continuous. In one year, from the moment in time x_0 to the moment in time $x_0 + 1$, y grows from y_0 to y_1, so the actual growth rate of y over this period is $\frac{y_1 - y_0}{y_0} = \frac{TL}{LM}$. But we can also write:

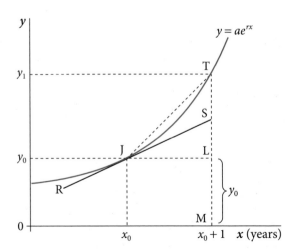

$\frac{SL}{LM} = r$ measures both the nominal and the instantaneous rates of growth of y, while $\frac{TL}{LM} = e^r - 1$ measures the effective (annual) rate of growth.

Figure 13.5 Instantaneous, nominal, and effective growth rates.

$$\frac{y_1 - y_0}{y_0} = \frac{y_1}{y_0} - 1 = \frac{ae^{r(x_0+1)}}{ae^{rx_0}} - 1 = e^{r(x_0+1)-rx_0} - 1 = e^r - 1$$

Hopefully you will recognize $e^r - 1$ as the effective growth rate when growth is continuous (see rule 12.5). Thus what we have here called the actual growth rate is simply the effective growth rate that we have already met.

To summarize, in figure 13.5 $\frac{SL}{LM} = r$ measures both the instantaneous and the nominal rates of growth of y, while $\frac{TL}{LM} = e^r - 1$ measures the effective (annual) rate of growth. These growth rates are, of course, independent of the position of J.

13.8 Semi-log graphs and the growth rate again

In section 12.11 we considered two numerical examples in which a variable y was growing smoothly and continuously through time, x. We found that when we plotted the graph of $\ln y$ against x, the slope of the graph told us whether the growth rate was constant, increasing, or decreasing. We are now ready to show why this is true.

Suppose $y = f(x)$ is the function relating the variable, y, and time, x. Suppose we plot this function using a natural logarithmic scale on the vertical axis (recall from section 12.11 that this is called a semi-log graph). First, since in this graph we have $\ln y$ on the vertical axis and x on the horizontal, then by definition the slope of the graph at any point is given by

$$\frac{d(\ln y)}{dx}$$

Note that this derivative, like any other, measures a rate of (absolute) change. In this case, it's the rate of change of $\ln y$ with respect to x.

Second, from rule 13.5 above we know that, if $y = f(x)$, then

$$\frac{d(\ln y)}{dx} = \frac{1}{y}\frac{dy}{dx}$$

Third, we also know from rule 13.7 that $\frac{1}{y}\frac{dy}{dx}$ measures the rate of *proportionate* change of y at any moment in time (also known as the instantaneous growth rate of y). Combining these three facts, we can conclude as follows:

RULE 13.9 The slope of the semi-log graph and the growth rate

For any function $y = f(x)$

$$\frac{d(\ln y)}{dx} = \frac{1}{y}\frac{dy}{dx} \quad \text{where}$$

$\frac{d(\ln y)}{dx}$ is the slope of the semi-log graph, and

$\frac{1}{y}\frac{dy}{dx}$ measures the instantaneous growth rate of y.

The key fact in rule 13.9 is that $\frac{d(\ln y)}{dx}$, the rate of *absolute* change of the natural log of y, equals $\frac{1}{y}\frac{dy}{dx}$, the rate of *proportionate* change of y itself. (Note that this is only true of the natural log of y.)

This explains why, in section 12.11, example 12.6, where y was growing at a constant rate, the semi-log graph was a straight line; and why in example 12.7, where y was growing at a decreasing rate, the slope of the semi-log graph decreased as x increased. And finally we could have considered an example (but didn't) where y was growing at an increasing rate, when we would have found that the slope of the semi-log graph increased as x increased.

Rule 13.9 is, of course, much more powerful than anything we can learn from numerical examples. The result proves two things:

(1) A constant, increasing, or decreasing slope of the semi-log graph tells us whether the growth rate of y is constant, increasing, or decreasing—something that *appeared* to be true from the two numerical examples in section 12.11, but something that we could not until now be certain of as a universal truth.

(2) The slope of the semi-log graph at any point actually *measures* the instantaneous growth rate of y at that point (provided the logs are natural logs, as assumed in deriving the results above—if the logs are to some other base, the slope is proportional to the growth rate but not equal to it).

Point (2) was not apparent from our numerical examples. This is because in any numerical example we can consider only discrete changes in x, which therefore result in discrete changes in $\ln y$. Consequently, we cannot measure the slope at a point, but only the average slope between two points. This average slope is measured by the difference quotient, $\frac{\Delta \ln(y)}{\Delta x}$. This gives only an approximation to the instantaneous growth rate $\frac{d(\ln y)}{dx}$. Nevertheless the approximation is quite close if Δx (and therefore Δy) is small. Check the following two examples with your calculator. Both assume that y is a function of time, x.

EXAMPLE 13.7

Suppose Δx is sufficiently small that it causes y to increase from, say, 2 to 2.2 (a proportionate increase of 0.1). Since $\ln 2 = 0.693147$ and $\ln 2.2 = 0.788457$, the absolute increase in $\ln y$ is 0.095310.

So the difference between $\frac{\Delta y}{y}$ (the proportionate change in y) and $\Delta \ln y$ (the absolute change in $\ln y$) is $0.1 - 0.09531 = 0.00469$. This is quite small.

EXAMPLE 13.8

Suppose Δx is even smaller so that it causes y to increase from 2 to 2.02 (a proportionate increase of 0.01). Then $\ln y$ increases from 0.693147 to 0.703098 (an absolute increase of 0.009950). So the difference between the proportionate change in y and the absolute change in $\ln y$ is $0.01 - 0.009950 = 0.000050$. This is very small.

We can see from these two examples that as the proportionate increase in y gets smaller, so the difference between the proportionate change in y and the absolute change in $\ln y$ gets smaller. Rule 13.9 tells us that the difference disappears as Δx approaches zero. Try this with a numerical example of your own.

13.9 An important special case

Rule 13.9 above is valid for any function $y = f(x)$ where y is growing continuously through time (x) at any rate, whether that rate is constant or not. However, there is a special case of the above result that is of particular interest because it is also the most common case. This is the function

$$y = ae^{rx}$$

From rule 13.8 we know that the rate of proportionate growth of y is given by

$$\frac{1}{y}\frac{dy}{dx} = r \qquad (13.2)$$

From rule 13.9 we also know that $\frac{d(\ln y)}{dx} = \frac{1}{y}\frac{dy}{dx}$ for any smooth and continuous function. Let's confirm that this is indeed true for the function $y = ae^{rx}$.

Given $y = ae^{rx}$, if we take natural logs on both sides, we get

$$\ln y = \ln(ae^{rx})$$
$$= \ln a + rx(\ln e) = \ln a + rx \text{ (using rules 12.4a, 12.4b and 12.4e from section 12.7)}$$

Thus we have

$$\ln y = \ln a + rx$$

This defines a functional relationship between $\ln y$ and x. Though it may not be immediately obvious, this relationship is *linear*, with a constant (intercept) term of $\ln a$ and a slope of r. (See Hint below.)

Since $\ln a$ is an additive constant and r is a multiplicative constant, the derivative of this function is

$$\frac{d(\ln y)}{dx} = r \qquad\qquad (13.3)$$

Combining equations (13.2) and (13.3), we have

$$\frac{d(\ln y)}{dx} = \frac{1}{y}\frac{dy}{dx} = r$$

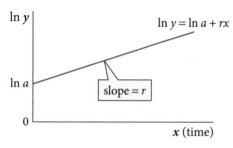

Figure 13.6 The function $\ln y = \ln a + rx$ is a linear relationship between $\ln y$ and x.

Figure 13.6 shows this graphically. We see that the graph of $\ln y = \ln a + rx$ is linear, with slope equal to r.

The function $y = ae^{rx}$ is often described as being **log linear**, though it should really be described as 'semi-log linear', since it becomes linear when a log scale is taken on the y-axis only. The x-axis remains calibrated in units of time, x.

In section 13.10 we will look at a function that is 'double log linear': that is, a linear function when log scales are taken on *both* axes.

 Hint If you have trouble seeing that $\ln y = \ln a + rx$ is a linear function, it may help if we define two new variables: $Y = \ln y$; and $A = \ln a$. When these are substituted into $\ln y = \ln a + rx$, it becomes

$$Y = A + rx$$

which hopefully you can see is a linear relationship between the two variables Y and x, with an intercept of A and a slope of r.

Conclusions on continuous growth

Collecting results from sections 13.8 and 13.9, we have shown that:

(1) For any function $y = f(x)$, $\frac{1}{y}\frac{dy}{dx} = \frac{d(\ln y)}{dx} =$ the instantaneous growth rate of y.

(2) In the special case of the function $y = ae^{rx}$, $\frac{1}{y}\frac{dy}{dx} = \frac{d(\ln y)}{dx} = r$ (where r is, of course, the constant growth rate).

(3) For any function $y = f(x)$, $\frac{1}{y}\frac{dy}{dx}$ measures the rate of *proportionate* change of y (= the growth rate), while $\frac{d(\ln y)}{dx}$ measures the rate of *absolute* change of the natural log of y. From (1) above, we know that these are equal.

13.10 Logarithmic scales and elasticity

So far we have considered, in the context of logarithms, only cases where the dependent variable, y, was related to time, x. Now we want to consider cases where we have economic variables on both axes.

Although a general proof of the following result is quite easy, we shall merely illustrate with a specific example. Consider the demand function

$$q = Ap^{-\alpha}$$

where q = quantity demanded, p = price and A and α are positive constants.

We first find the elasticity of demand in the usual way. The derivative of $q = Ap^{-\alpha}$, by the power rule, is

$$\frac{dq}{dp} = -\alpha Ap^{-\alpha-1}$$

So the elasticity of demand is

$$\frac{p}{q}\frac{dq}{dp} = \frac{p}{q}(-\alpha Ap^{-\alpha-1}) = \frac{p(-\alpha Ap^{-\alpha-1})}{q} = \frac{-\alpha Ap^{-\alpha}}{q} = -\alpha\frac{Ap^{-\alpha}}{q} = -\alpha$$

(since $q = Ap^{-\alpha}$ from the demand function). Thus we have

$$\frac{p}{q}\frac{dq}{dp} = -\alpha$$

Thus the elasticity of demand is constant, and equal to $-\alpha$.

Now let us go back to the demand function

$$q = Ap^{-\alpha}$$

and see what happens when we take natural logs on both sides. Using rules 12.4a and 12.4b (see section 12.7), we get

$$\ln q = \ln A + \ln(p^{-\alpha}) = \ln A - \alpha(\ln p)$$

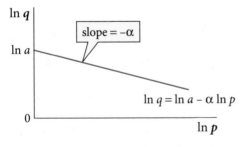

Figure 13.7 graph: ln q axis, ln a, slope = −α, ln q = ln a − α ln p, ln p axis, 0

Figure 13.7 The function $\ln q = \ln A - \alpha \ln p$ is a linear relationship between $\ln q$ and $\ln p$.

Thus we have

$$\ln q = \ln A - \alpha(\ln p) \tag{13.4}$$

Since $\ln A$ and α are both constants, this defines a *linear* relationship between the two variables $\ln q$ and $\ln p$ (see figure 13.7). The intercept is $\ln A$ and the slope is $-\alpha$ (which we assume to be negative, as demand curves usually slope downwards).

When we apply the standard rules of differentiation to this linear function, the additive constant $\ln A$ drops out and the multiplicative constant $-\alpha$ is retained. The derivative is therefore

$$\frac{d(\ln q)}{d(\ln p)} = -\alpha = \text{the elasticity of demand, from above.} \tag{13.5}$$

Equation (13.4) is graphed in figure 13.7. With logarithmic scales on both axes, the slope is, by definition, given by $\frac{d(\ln q)}{d(\ln p)}$. The graph confirms that in this example the slope is constant and equal to the demand elasticity, $-\alpha$.

Hint If you have trouble following the above steps, it may help to define three new variables: $Q = \ln q$; $B = \ln A$; and $P = \ln p$. When these are substituted into equation (13.4) above, it becomes

$$Q = B - \alpha P$$

which hopefully you can see is a linear relationship between the two variables Q and P, with an intercept of B and a slope of $-\alpha$. The derivative of this function is

$$\frac{dQ}{dP} = -\alpha$$

and since $Q = \ln q$ and $P = \ln p$, this is the same as

$$\frac{d(\ln q)}{d(\ln p)} = -\alpha$$

Conclusions and generalization

We have shown both algebraically and graphically that when we plot a demand function with natural log scales on *both* axes, the slope of the resulting graph measures the elasticity of demand. This is called a 'double-log' or 'log-log' scale.

Admittedly, we have demonstrated this only for a specific demand function, and moreover a demand function with constant elasticity of demand. But in fact it holds generally. For any function $y = f(x)$, whatever x and y denote, when we take natural log scales on both axes the slope of the resulting graph, which of course is measured by the derivative $\frac{d(\ln y)}{d(\ln x)}$, equals the elasticity of the function. This is true whether or not the elasticity is constant. Moreover, it is true if we use logs to base 10 or any other base. Thus:

RULE 13.10 The slope of the log-log graph and the elasticity

For any function $y = f(x)$:

$$\frac{d(\ln y)}{d(\ln x)} = \frac{x\,dy}{y\,dx} = \text{the elasticity of the function}$$

$$= \text{the rate of proportionate change in } y \text{ per unit of proportionate change in } x$$
$$= \text{the slope of the graph with natural log scales on both axes}$$

This is also true if we take logs to base 10 or any other base.

And finally . . .

Two things are easily confused:

(1) $\dfrac{d(\ln y)}{d(\ln x)} = \dfrac{x\,dy}{y\,dx} = $ elasticity of the function $y = f(x)$

This measures the rate of proportionate change of y as x changes *proportionately*. It is the proportionate change in y divided by the (very small) proportionate change in x that caused it.

(2) $\dfrac{d(\ln y)}{dx} = \dfrac{1}{y}\dfrac{dy}{dx}$

This measures the rate of proportionate change of y as x changes *absolutely*. It is the proportionate change in y divided by the (very small) absolute change in x that caused it. This is used when x denotes time, and measures the instantaneous growth rate of y. We use the absolute change when x denotes time because a proportionate change in time is meaningless, as explained in section 9.2. (Think about it—if you were due at a lecture at 10 a.m. and you arrived 2 hours late, would you be 20% late because 2 is 20% of 10?)

Summary

In sections 13.1–13.3 we stated the rules of differentiation for the natural exponential and logarithmic functions (see rules 13.1–13.5).

Then in sections 13.4 and 13.5 we considered how to measure the growth rate (= rate of proportionate change) of a variable growing in discrete jumps from one period to the next at a varying rate. We arrived at the formula for discrete one-period growth (rule 13.6). This says that if a variable y varies discretely with time, x, the rate of proportionate growth of y from one period to the next is measured by

$$\frac{\frac{\Delta y}{y_0}}{\Delta x}, \text{ or, equivalently, } \frac{1}{\Delta x}\frac{\Delta y}{y_0}, \text{ or } \frac{1}{y_0}\frac{\Delta y}{\Delta x}$$

where y_0 is the initial value of y, and Δx is the time interval for which data are published (that is, 1 year for annual data, 1 month for monthly data, and so on).

We then showed in section 13.6 how, when growth is continuous, the difference quotient $\frac{\Delta y}{\Delta x}$ can be replaced by the derivative $\frac{dy}{dx}$, giving us the formula for the instantaneous growth rate (rule 13.7). This formula says that if a variable y is related to time, x, by a smooth and continuous function, $y = f(x)$, then the instantaneous rate of growth of y at any moment in time is measured by

$$\frac{1}{y}\frac{dy}{dx}$$

A special case of this formula is the function $y = ae^{rx}$, which we know from chapter 12, rule 12.2b describes a variable growing continuously *at a constant rate*. We showed that for this function, $\frac{1}{y}\frac{dy}{dx} = r$, thus confirming that r measures the (constant) instantaneous growth rate for this function (see rule 13.8).

In section 13.8 we returned to the use of a logarithmic scale on the vertical axis of a time-series graph. This is done in order to reveal the growth rate. In section 12.11 we demonstrated this by numerical example; in section 13.8 we proved it, showing that, for any function $y = f(x)$, $\frac{d(\ln y)}{dx}$, the slope of the semi-log graph equals $\frac{1}{y}\frac{dy}{dx}$, which measures the instantaneous growth rate of y (rule 13.9).

Finally, in section 13.10 we showed that if any function $y = f(x)$ is graphed with log scales on both axes, the slope of the graph measures the elasticity of the function (rule 13.10).

Progress exercise 13.2

1. (a) Show that, in the continuous growth formula $y = ae^{rx}$, the instantaneous rate of proportionate growth (usually referred to as the growth rate) is given by the parameter r.

   (b) Illustrate by means of a sketch graph how the growth rate can be inferred from the graph, in the case of a variable with an initial value of 100 and growing continuously at (i) 5% per year; (ii) −3% per year.

2. (a) Given the continuous growth formula $y = ae^{rx}$ (where y is any variable and x is time), show that

   $$\frac{d(\ln y)}{dx} = \frac{1}{y}\frac{dy}{dx}$$

   and explain the meaning of this result. (Hint: given $y = ae^{rx}$, you need to find $\frac{d}{dx}$ of ln f(x).)

   (b) Illustrate (a) with sketch graphs of ln y against time and y against time.

3. The table below gives the consumer price index (CPI) for the economy of Ruritania, 1985–2009, with 1985 = 100.

   (a) Treating the CPI as a continuous variable, plot a reasonably accurate graph of the CPI time series (that is, by joining up the data points to form a continuous curve). From inspection of the graph, try to assess whether the annual growth rate of prices (= inflation rate) is constant, increasing, or decreasing through time.

   (b) Repeat (a) above, but this time plotting the log of the CPI. (Hint: it may help your assessment to fit a rough trend line to this graph, using a ruler and pencil.)

   (c) Explain in words (preferably supported by the relevant maths) why assessing the growth rate is easier in (b) than in (a).

Year	CPI	Year	CPI
1985	100	1998	492
1986	115	1999	590
1987	132	2000	708
1988	152	2001	850
1989	175	2002	1020
1990	201	2003	1173
1991	231	2004	1348
1992	492	2005	1551
1993	590	2006	1783
1994	708	2007	2051
1995	850	2008	2359
1996	1020	2009	2712
1997	1173		

4. Given the demand function $q = 16p^{-0.5}$:

   (a) Find an expression for the price elasticity of demand.

   (b) Take natural logs on both sides and hence show that the function is 'log-log linear'.

   (c) Differentiate your expression in (b) above to find $\frac{d(\ln q)}{d(\ln p)}$. (Hint: just apply the normal rules of differentiation.)

   Hence verify that $\frac{d(\ln q)}{d(\ln p)} = \frac{p}{q}\frac{dq}{dp}$ and explain this result.

   (d) Sketch the graph of the demand function, (i) with q and p on the axes, and (ii) with ln q and ln p on the axes. What presentational advantage does (ii) have over (i)?

Checklist

Be sure to test your understanding of this chapter by attempting the progress exercises (answers are at the end of the book). The Online Resource Centre contains further exercises and materials relevant to this chapter www.oxfordtextbooks.co.uk/orc/renshaw3e/.

This has been a short but somewhat intense chapter. We have seen how to differentiate the natural exponential and logarithmic functions and the significance of their derivatives. We have also developed the idea of logarithmic scales and how these relate to elasticity. The topics were:

✔ **Derivatives of the natural exponential and logarithmic functions.** Rules for differentiating the natural exponential and logarithmic functions, and understanding what they measure.

✔ **Measuring growth rates.** Understanding the concept of a rate of proportionate change and how to measure it for discrete and continuous variables. Relation between instantaneous, nominal and effective growth rates.

✔ **Semi-logarithmic scales.** Proof that when we plot time series graphs with a natural log scale on the vertical axis, the slope gives the growth rate.

✔ **'Log-log' scales.** Understanding and proving that the slope of a function plotted with log scales on both axes gives the elasticity of the function.

We have now completed part 3 of the book, which has been concerned with the mathematics of finance and growth. Before moving on to part 4, it might be advisable to review part 3 and check your understanding of the key concepts, which are an essential part of every economist's toolkit.

Part Four
Optimization with two or more independent variables

- Functions of two or more independent variables

- Maximum and minimum values, the total differential, and applications

- Constrained maximum and minimum values

- Returns to scale and homogeneous functions; partial elasticities; growth accounting; logarithmic scales

Chapter 14
Functions of two or more independent variables

OBJECTIVES

Having completed this chapter you should be able to:

■ Understand how a function with two independent variables corresponds to a surface in three-dimensional space, and with *n* independent variables to a hyper-surface in $(n + 1)$-dimensional space.

■ Take sections through a three-dimensional surface in order to analyse how any two variables are related.

■ Understand and evaluate first and second order partial derivatives.

■ Understand the assumptions of the neoclassical production function and its resulting shape.

■ Use partial derivatives from the production function to find the marginal products of the inputs and the slope of an isoquant, including testing for a decreasing marginal rate of substitution in production.

■ Use partial derivatives from the utility function to find the marginal utilities of the goods and the slope of an indifference curve, including testing for a decreasing marginal rate of substitution in consumption.

■ Apply these techniques to the Cobb–Douglas production and utility functions.

14.1 Introduction

Up to now we have considered only functions with one independent variable, usually labelled *x* and usually on the right-hand side of the equation. This is obviously very limiting because in economics it is common for a variable to depend on two or more other variables. An example would be a demand function where the quantity demanded depends not just on the price of that good but also on the incomes of buyers. In that case we have a dependent variable (quantity demanded) and two independent variables (the price of the good and buyers' incomes).

Therefore in this chapter and the following three we will examine functions with two or more independent variables. We begin by considering functions of the form $z = f(x, y)$: that is, where the dependent variable, z, is determined by the values of two independent variables, x and y. We can visualize this as a surface in three-dimensional space with every point uniquely identified by the values of x, y, and z at that point. We consider several examples of such functional forms and the shapes of the surfaces that they describe. With more than two independent variables on the right-hand side of the equation, we can no longer envisage any corresponding physical surface but the basic ideas remain valid.

Next we consider how the concept of a derivative can be extended and applied to the function with two or more independent variables. The crucial insight is to see that if we hold y constant in the equation $z = f(x, y)$, it becomes a relationship between z and x alone, and a derivative can be found in the usual way. Such derivatives are called partial derivatives. We develop this idea further to include second order partial derivatives. In the last part of the chapter we consider the application of these ideas to two important relationships in economics: the production function and the utility function.

14.2 Functions with two independent variables

First, let's consider functions of the general form

$$z = f(x, y)$$

where the dependent variable is labelled z and there are two independent variables, x and y. As usual f( ) denotes some unspecified functional form. Examples could be

$$z = 100 - 2x + 5y; \quad z = 3x^2 - 9y; \quad z = e^{2x+3y}$$

Let's think about how we might draw graphs of functions such as these. Since we have three variables in all, we need three axes to measure them along and therefore three dimensions, one for each axis. It occurs to us that although this book is written on two-dimensional sheets of paper, we can nevertheless represent three dimensions by means of a perspective drawing, as in figure 14.1(a).

In this figure we have drawn three axes, which meet at right angles at the origin, 0, where x, y, and z are all zero. As we move away from the origin along the x-axis, x becomes increasingly positive (as we move in the direction labelled x^+ in the figure) or negative (as we move in the direction labelled x^-), while y and z remain zero. Similarly, as we move away from the origin along the y-axis, y becomes increasingly positive (labelled y^+) or negative (labelled y^-), while x and z remain zero. And in the same way, along the z-axis, z is varying while x and y remain zero. The axes, of course, extend indefinitely far in their respective directions.

The three axes give us a three-dimensional space in which every point is uniquely identified by its coordinates. For example, we see that point J has coordinates (x_0, y_0, z_0), meaning that J is located x_0 units away from the origin in the x^+ direction, y_0 units away from the origin in the y^+ direction, and z_0 units away from the origin in the z^+ direction. We have positioned J in the zone where x, y, and z are all positive, but we could equally have placed it where one or more of the three variables was negative. (See section 3.6 for refreshment on coordinates.)

Later, we will be particularly interested in points in space where one of the three variables is constant. Three cases deserve special attention. First, consider all the points in figure 14.1(a) at which $z = 0$. Point K is one such point, with coordinates $(x_0, y_0, 0)$, and so is point L, which has coordinates $(x_1, y_1, 0)$. Some other points with a z-coordinate of zero are indicated by the blue shaded area. (In principle the blue shaded area extends indefinitely far in the x and y directions, but of necessity we have given it arbitrary boundaries.) These points collectively form a horizontal flat surface called the **0xy plane**. The 0xy plane cuts the z-axis at right angles, and includes the x- and y-axes themselves as well as the origin, 0.

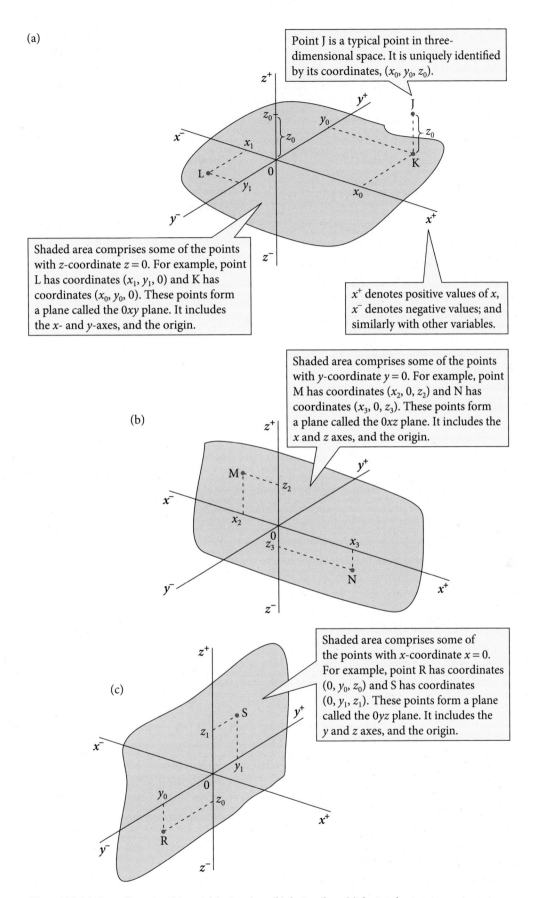

Figure 14.1 (a) Three-dimensional axes and the $0xy$ plane; (b) the $0xz$ plane; (c) the $0yz$ plane.

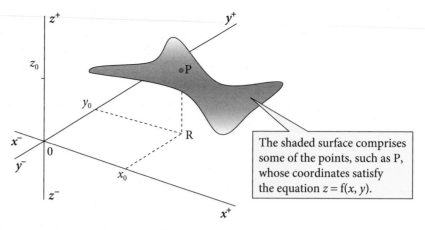

The shaded surface comprises some of the points, such as P, whose coordinates satisfy the equation $z = f(x, y)$.

Figure 14.2 A function $z = f(x, y)$ as a surface in three-dimensional space.

Other horizontal planes, parallel to the $0xy$ plane, may be defined by choosing other fixed values of z (other than $z = 0$, that is). For example point J in figure 14.1(a) lies on a plane that is parallel to the $0xy$ plane and at a distance z_0 units above it.

In the same way, all the points in the three-dimensional space that have a y-coordinate of $y = 0$, such as M and N in figure 14.1(b), form collectively a flat surface called the **$0xz$ plane**. Some of this plane is shown by the vertical shaded area in figure 14.1(b), cutting the y-axis at right-angles. It includes the x and z axes themselves as well as the origin. Other planes may be defined for other fixed values of y and these are parallel to the $0xz$ plane.

Finally, the **$0yz$ plane** consists of all points with an x-coordinate of zero. Some of this plane is shown by the vertical shaded area in figure 14.1(c) and includes points such as R and S, as well as the y and z axes and the origin. It cuts the x-axis at right angles. Note that the $0xz$ and $0yz$ planes are both vertical but at right angles to each other, and to the horizontal $0xy$ plane. We can define any other plane parallel to the $0yz$ plane by choosing a fixed value of x other than zero.

Now that we have defined our three-dimensional space, we can consider what a function located in this space might look like. As an example, consider the shaded surface in figure 14.2, which looks a little like a magic carpet floating in space above the 'floor' formed by the $0xy$ plane. This represents a possible shape for the function $z = f(x, y)$. Note that the function $z = f(x, y)$ is a *surface* in three-dimensional space, in the same way that the function $y = f(x)$ is a *line* in two-dimensional space. Every point lying on the surface defined by the function is uniquely identified by its coordinates. For example, point P is located x_0 units away from the origin in the positive x direction, y_0 units away from the origin in the positive y direction, and z_0 units away from the origin in the positive z direction. It therefore has coordinates (x_0, y_0, z_0).

The algebraic requirement is that the coordinates of every point on the surface, such as (x_0, y_0, z_0), must satisfy the equation $z = f(x, y)$. Conversely, if the coordinates of a point do not satisfy the equation $z = f(x, y)$, then that point does not lie on the surface. We will explore this further below.

Extension to many variables

We can just about cope with two independent variables by means of perspective drawing, as in figure 14.2. However, to model the complexities of the real world we need to be able to handle functions with more than two independent variables. If there is an unspecified number, n, of independent variables that influence a dependent variable, z, we can write the functional relationship between them as

$$z = g(x_1, x_2, x_3, \ldots, x_n)$$

where $x_1, x_2, \ldots$ are the n independent variables.

Visual representation, such as a perspective drawing, is now impossible as there are simply not enough dimensions in physical space to cope with all the variables. However, just as we see the function $z = f(x, y)$ as representing a surface in three-dimensional space, so we can view the function $z = g(x_1, x_2, x_3, \ldots, x_n)$ as representing a *hyper-surface* in $(n + 1)$-dimensional space. This sounds like something out of *Star Trek*, but works surprisingly well and without mind-blowing complexity. From time to time in the remainder of this book we will touch upon these cases where more than two independent variables are involved. But for the most part we will keep things as simple as possible by considering functions with only two independent variables. We will now look at some examples of these.

14.3 Examples of functions with two independent variables

Let us now consider what some specific functions look like in three-dimensional space.

EXAMPLE 14.1

$z = 150 - 2x - 3y$

The graph of this function is a plane (a flat surface) in three-dimensional space: see the shaded surface in figure 14.3. For simplicity, in this graph we have not considered any negative values of x, y, and z, so the shaded surface is shown only in the zone where all three variables are zero or positive. In figure 14.3 and subsequently, on the x-axis we will label the direction for positive x as simply x rather than x^+, leaving the direction for negative x unlabelled; and similarly for other variables. Figure 14.3 is only a sketch; to draw the graph accurately we would need to draw up a table of values, assigning values to x and y and then calculating the associated value of z. Or, we could resort to a graph-drawing computer program such as Maple.

 The Online Resource Centre explains how to use Maple.
www.oxfordtextbooks.co.uk/orc/renshaw3e/

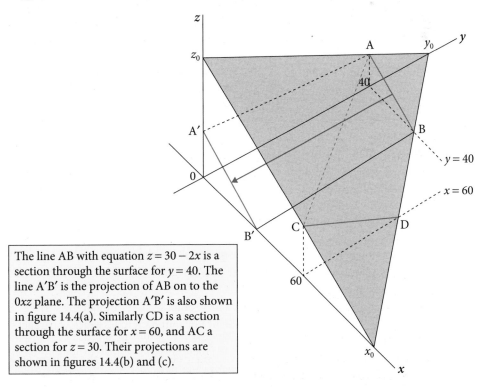

The line AB with equation $z = 30 - 2x$ is a section through the surface for $y = 40$. The line A′B′ is the projection of AB on to the 0xz plane. The projection A′B′ is also shown in figure 14.4(a). Similarly CD is a section through the surface for $x = 60$, and AC a section for $z = 30$. Their projections are shown in figures 14.4(b) and (c).

Figure 14.3 Graph of $z = 150 - 2x - 3y$ is a plane in three-dimensional space (not to scale).

The graph of $z = 150 - 2x - 3y$ is a plane because neither x, y, nor z is raised to any other power than 1. This is fully analogous to what we found in chapter 3 for a linear function such as $y = 5x + 9$. The graph of $y = 5x + 9$ is a straight line because neither x nor y is raised to any other power than 1. A plane is thus the counterpart in three-dimensional space of a line in two-dimensional space.

Let us explore some of the properties of the function $z = 150 - 2x - 3y$. First, we will look at the intercepts of the function on the three axes. We have labelled the intercepts x_0, y_0, and z_0 in figure 14.3.

Consider the x-intercept ($x = x_0$) in figure 14.3. We know that anywhere on the x-axis we have $y = 0$ and $z = 0$ (examine figure 14.3 carefully if you are unsure of this point). So the coordinates of the x-intercept are $x = x_0$, $y = 0$, and $z = 0$.

We also know that point x_0 lies on the surface and therefore satisfies the equation

$$z = 150 - 2x - 3y$$

Therefore we can find x_0 by substituting $x = x_0$ and $y = z = 0$ into the equation $z = 150 - 2x - 3y$. This gives

$$0 = 150 - 2x_0 - 3(0) \quad \text{from which } x_0 = 75.$$

In the same way, we can find the y-intercept ($y = y_0$) by substituting $y = y_0$ and $x = z = 0$ into the equation $z = 150 - 2x - 3y$, giving $y_0 = 50$.

The z intercept ($z = z_0$) is similarly found by setting $x = y = 0$, giving $z_0 = 150$.

Sections through the surface

In the case of $z = 150 - 2x - 3y$, it is fairly easy to visualize and construct a three-dimensional perspective drawing because the shape of the surface, a plane, is a very simple one.

But generally, three-dimensional perspective drawing is difficult and imprecise, so we often simplify by taking **sections** (slices) through the surface for fixed values of x, y, or z. Normally (but not necessarily) we consider sections through the surface that are parallel to either the $0xy$ plane, the $0yz$ plane, or the $0xz$ plane.

In figure 14.3 the section AB is obtained by taking a slice through the surface with y fixed at $y = 40$. Because y is fixed, the section AB is parallel with the $0xz$ plane and is the locus or collection of points on the surface with a y-coordinate of 40. The section AB thus shows all the combinations of positive values of x and z that satisfy the equation $z = 150 - 2x - 3y$ when $y = 40$. (If we allowed x to be negative, the line would continue beyond A, and if we allowed z to be negative, the line would continue beyond B.)

To find the equation of the section AB we substitute $y = 40$ into the function $z = 150 - 2x - 3y$. This gives $z = 150 - 2x - 3(40)$, which simplifies to $z = 30 - 2x$. Note that $z = 30 - 2x$ is a linear function, confirming that AB is correctly drawn as a straight line in figure 14.3. The constant, 30, is the value of z when $x = 0$. This is at point A. The coordinates of A are thus $x = 0$, $y = 40$, $z = 30$.

Similarly, the section CD is obtained by taking a slice through the surface with x fixed at $x = 60$. Since x is fixed, the section CD is parallel with the $0yz$ plane and is the locus of points on the surface with an x-coordinate of 60. The section CD shows all the combinations of positive values of y and z that satisfy the equation $z = 150 - 2x - 3y$ when $x = 60$.

The equation of the section CD is found by substituting $x = 60$ into $z = 150 - 2x - 3y$, giving $z = 30 - 3y$. This is a linear function. The constant, 30, is the value of z when $y = 0$ (and $x = 60$), which is true at point C.

Finally, in figure 14.3 we have taken a horizontal section through the surface, by fixing the value of z at $z = 30$. This defines the straight line joining A to C, which is parallel with the $0xy$ plane. The equation of the line AC is $30 = 150 - 2x - 3y$, which rearranges as $y = 40 - \frac{2}{3}x$. At point A, $x = 0$, so $y = 40$. At point C, $y = 0$, so $x = 60$. (To simplify the diagram we have chosen values such that A lies both on section AB and on section AC; and similarly C lies both on section CD and on section AC.)

It is worth spending some time to be sure that you understand these calculations.

Projected sections

Because in each case above there are only two variables (the third being held constant), the three sections AB, CD, and AC can be *projected* as lines in two-dimensional space. For example, in figure 14.3 the line A′B′ is the projection of the section AB on to the 0xz plane.

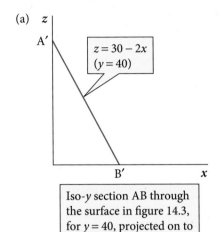

(a)

Iso-*y* section AB through the surface in figure 14.3, for y = 40, projected on to the 0xz plane

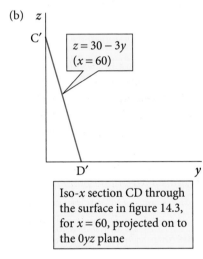

(b)

Iso-*x* section CD through the surface in figure 14.3, for x = 60, projected on to the 0yz plane

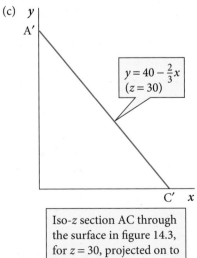

(c)

Iso-*z* section AC through the surface in figure 14.3, for z = 30, projected on to the 0xy plane

Figure 14.4 Projected sections from figure 14.3 (not to scale).

This projection process is analogous to the way in which an overhead projector in a lecture hall throws the image (AB) from a slide on to the screen as A′B′ (though in our case without enlarging the image). We could imagine taking very sharp scissors, cutting out the (infinitely narrow) line AB from the surface defined by the function, and placing it on the glass of the overhead projector. Then we could do the same with the x and z axes. Finally we rotate the image around the vertical axes so that positive values of x are measured to the right instead of to the left as in figure 14.3 when viewed from the north-east direction. When we switch on the projector, what appears on the screen is figure 14.4(a).

In figure 14.4(a) we see the projected section A′B′ with equation $z = 30 - 2x$. This gives us z as a function of x alone, when y is held constant at 40. Note that in the figure we have flagged the fact that y = 40, to remind ourselves that the relationship $z = 30 - 2x$ holds only when y is held constant at 40. If we took another fixed value for y, greater or less than 40, this would identify a new section across the surface, parallel with and to the right or left of AB in figure 14.3.

Similarly, figure 14.4(b) shows the projected section C′D′, the projection of the line CD on to the 0yz plane. The equation of CD is $z = 30 - 3y$. It gives us z as a function of y alone, when x is held constant at 60. Note that it is now y that appears on the horizontal axis.

Finally, figure 14.4(c) shows the section A′C′, the projection of the horizontal line AC on to the (horizontal) 0xy plane. Its equation is $y = 40 - \frac{2}{3}x$. It gives us y as a function of x alone, when z is held constant at 30.

The sections A′B′, C′D′, and A′C′ in figure 14.4 can be described as showing us three 'partial' relationships between each of the three pairs of variables, with the other variable held constant. There is an infinite number of such projections for all the possible fixed values of x, y, or z.

Iso sections

For ease of reference we'll call a section such as AB in figure 14.3 an **iso-*y* section** ('iso' meaning 'equal'), since it shows points on the surface where y is constant and equal to some fixed value (in the case of AB, y = 40). Similarly we'll call a section such as CD, where x is constant, an **iso-*x* section** and a section such as AC, where z is constant, an **iso-*z* section**.

The three types of section through the surface that we have just discussed are all obtained by assigning a fixed value to one of the three variables: that is, the sections were all parallel to one axis. A range of further possibilities is opened up if we fix instead the *relative* values of two variables. For example, we could fix x as, say, $x = 2y$. This would identify a locus of points on the surface which, unlike the cases considered above, would not be parallel with any of the $0xy$, $0yz$ or $0xz$ planes. We will consider this possibility further in chapter 16.

EXAMPLE 14.2

As a second example we consider the function

$$z = x^2 + y^2$$

This gives a more complicated surface than the previous example because both x and y are raised to the power 2. Perhaps the best way to work out the shape of the surface that this function describes is to take sections through the surface for fixed values of z. When we assign a fixed value to z, say $z = z_0$, the equation becomes

$$x^2 + y^2 = z_0 \quad \text{(a constant)}$$

We know from section 5.8 that this is the equation of a circle in the $0xy$ plane with its centre at the origin of x and y measurement and radius $\sqrt{z_0}$. If we assign to z the values 0, 9, 16, and 25, we get a series of circles of different radius with their centres at the origin, as shown in table 14.1.

Thus we can think of the three-dimensional surface, $z = x^2 + y^2$, as being made up of a large number of circles stacked on top of one another, with their centres on the z-axis, and with increasing radius as z increases. This shape is a cone with its vertex (point) at the origin (see figure 14.5). Think of an ice-cream cone.

Table 14.1 Iso-z sections of $z = x^2 + y^2$.

z_0	$\sqrt{z_0}$	Resulting section
0	0	Circle of zero radius: that is, a point located at the origin
9	3	Circle of radius 3, centred on z-axis
16	4	Circle of radius 4, centred on z-axis
25	5	Circle of radius 5, centred on z-axis

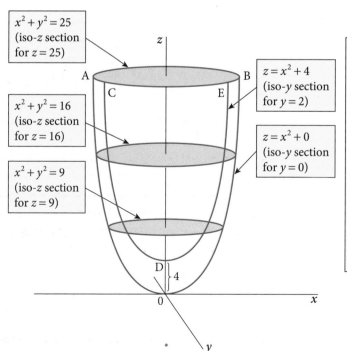

$x^2 + y^2 = 25$ (iso-z section for $z = 25$)

$x^2 + y^2 = 16$ (iso-z section for $z = 16$)

$x^2 + y^2 = 9$ (iso-z section for $z = 9$)

$z = x^2 + 4$ (iso-y section for $y = 2$)

$z = x^2 + 0$ (iso-y section for $y = 0$)

$z = x^2 + y^2$ is a cone with its vertex (point) at the origin. Iso-z sections through it are circles with their centres on the z-axis. Two of these are projected on to the $0xy$ plane in figure 14.6. Iso-y sections are quadratic functions, with the shape of a parabola. Two of these are projected on to the $0xz$ plane in figure 14.7.

Figure 14.5 Three-dimensional perspective sketch of the surface $z = x^2 + y^2$, showing some iso-y and iso-z sections.

Taking sections through the cone

We can take sections through the surface for various fixed values of z (that is, iso-z sections) and project them on to the $0xy$ plane. In effect we are cutting out the shaded circles in figure 14.5 and pasting them on to the $0xy$ plane. The result is shown in figure 14.6. If we think of z as measuring height, we can imagine we are flying over the cone in a helicopter and looking down into it from above. Viewed from above in this way, these 'horizontal' sections through the cone then form a series of circles with a common centre, the origin. They are conceptually identical to the contour lines on a relief map.

Similarly we can take sections through the surface by assigning various fixed values to y. Then we get iso-y sections through the surface that are parallel to the $0xz$ plane (since any line parallel with the $0xz$ plane has the property that y is constant). Figure 14.5 shows the shape of two of these sections. When we fix y at the value $y = 0$, we get the section A0B, with equation $z = x^2 + 0$. When we fix y at the value $y = 2$, we get the section CDE, with equation $z = x^2 + 4$. These sections are projected on to the $0xz$ plane in figure 14.7.

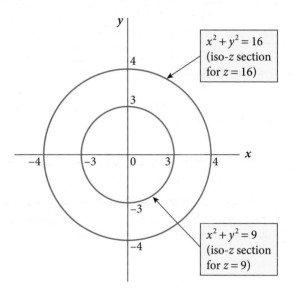

Figure 14.6 Iso-z sections of $z = x^2 + y^2$ for $z = 4$ and $z = 9$, projected on to the $0xy$ plane (not to scale).

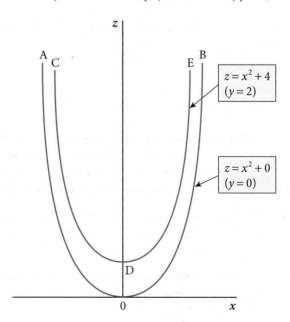

Figure 14.7 Iso-y sections of $z = x^2 + y^2$, projected on to the $0xz$ plane (not to scale).

Hopefully you will recognize $z = x^2 + 0$ and $z = x^2 + 4$ as quadratic functions and therefore members of the parabola family (see chapter 4). The sections in figure 14.8 below have the familiar U-shape of the parabola. Each time we vary the fixed value of y we get a new parabola in the $0xz$ plane.

If we had chosen to fix x instead of y, we would also have obtained sections with the shape of a parabola because, in the function $z = x^2 + y^2$, x and y affect z in exactly the same way. The function is said to be a **symmetrical function** in x and y.

The Online Resource Centre shows how to plot three-dimensional graphs using Maple.
www.oxfordtextbooks.co.uk/orc/renshaw3e/

Progress exercise 14.1

1. Given the surface $z = 120 - 3x - 0.5y$, with x and y assumed not negative:

(a) Show that the iso-x and iso-y sections are all linear and have slopes of -0.5 and -3, respectively. Find the slope of the iso-z sections. What does this information about slopes tell us about the shape of the surface?

(b) Find the intercepts on the three axes and, using this information together with your answers to (a), make a three-dimensional sketch graph of the surface.

2. Given the surface $z = x^2 + 3y$, with x and y assumed not negative.

(a) Show that the iso-y sections are quadratic in form, and that the iso-x sections are linear. Find the shape of the iso-z sections. What does this tell us about the shape of the surface?

(b) Find the intercepts on the three axes and, using this information together with your answers to (a), make a three-dimensional sketch graph of the surface.

3. Given the surface $z = xy$, with x and y assumed positive:

(a) Show that the iso-x and iso-y sections are linear.

(b) Sketch some iso-z sections and hence make a three-dimensional sketch of the surface.

14.4 Partial derivatives

We will now develop a technique for measuring the slope of a surface in three-dimensional space. At first sight this seems a very difficult task because at any point on a three-dimensional surface, such as point P in figure 14.2, the slope varies according to whether we move from P in the positive x direction, the positive y direction, or one of the infinite number of combinations of the two. Partial derivatives, which are similar to the derivatives that we first met in chapter 6, are a way of tackling this problem. (From here on we will refer to the positive x direction as simply the x direction, and similarly for other variables.)

Step 1: defining the slope of a surface

Our first task is to define clearly what we mean by the slope of a surface. Suppose in figure 14.8 that P is some point on a surface defined by some function $z = f(x, y)$. The coordinates of P are (x_0, y_0, z_0). The distance $0z_0 =$ distance JP.

We have taken two sections through the surface at P. One section, FPG, is an iso-y section parallel to the $0xz$ plane with y fixed at $y = y_0$. The other, NPM, is an iso-x section, parallel to the $0yz$ plane with x fixed at $x = x_0$. (The positions of M and G, where the sections end, are arbitrary.)

To each of these sections we can draw a tangent at P. The slope of the tangent AB measures the slope of the surface at P as we move away from P in the x direction (with y constant). If this

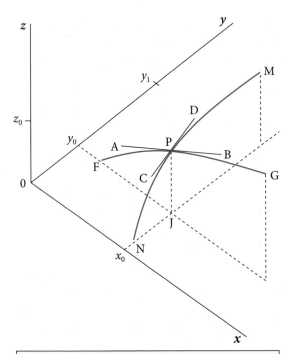

The tangent APB measures the slope of the surface at P in the x direction (y constant). The tangent CPD measures the slope of the surface at P in the y direction (x constant).

Figure 14.8 The slope of a surface in the x and y directions.

tangent has a large, positive slope it means that, at P, z increases rapidly as x increases with y constant. We will provisionally use the symbol $\frac{dz}{dx}$ to label the slope of this tangent, a choice of symbol obviously inspired by our earlier work on derivatives.

Similarly, the slope of the tangent CD measures the slope of the surface at P as we move away from P in the y direction (with x constant). If this tangent has a large, positive slope it means that, at P, z increases rapidly as y increases with x constant. We will provisionally use the symbol $\frac{dz}{dy}$ to label the slope of this tangent. As figure 14.8 is drawn, both slopes are positive but this obviously need not be the case.

Step 2: measuring the slopes of the tangents

Next, we need an algebraic technique for measuring the slopes of these two tangents at P. We'll focus first on the tangent AB, which measures the slope at P in the x direction with y held constant at $y = y_0$.

To sharpen our thinking, we take the section FPG, which is parallel to the $0xz$ plane (since y is constant), together with its tangent APB, and project them on to the $0xz$ plane (see figure 14.9(a)). Our task is to measure the slope of this tangent.

The way we do this closely parallels the method used in chapter 6 to find the derivative of $y = f(x)$. From an initial position at P, we first increase x by an arbitrary amount Δx so that we move from P to Q on the section FPG. Consequently, z increases by an amount Δz. We then form the difference quotient $\frac{\Delta z}{\Delta x}$, which measures the slope of the chord PQ. Finally, we allow

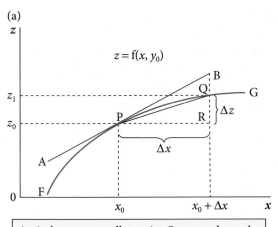

As Δx becomes smaller, point Q moves down the curve towards P and $\frac{\Delta z}{\Delta x}$ approaches slope of tangent APB. The other variable, y, remains constant throughout.

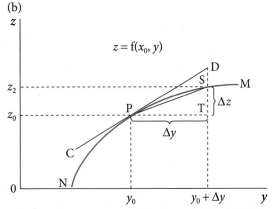

As Δy becomes smaller, point S moves down the curve towards P and $\frac{\Delta z}{\Delta y}$ approaches slope of tangent CPD. The other variable, x, remains constant throughout.

Figure 14.9 Measuring the slopes of the tangents in the x and y directions.

Δx to shrink, so that Q slides down the curve towards P. As it does so, the slope of the chord PQ approaches closer and closer to the slope of the tangent APB which we are trying to measure. We conclude that, in the limit, as Δx approaches zero, the two slopes will be equal. So

$$\lim_{\Delta x \to 0} \text{(slope of chord PQ)} = \text{(slope of tangent at P)}$$

But the slope of the chord PQ is given by the difference quotient $\frac{\Delta z}{\Delta x}$, while above we labelled the slope of the tangent at P as $\frac{dz}{dx}$. Therefore we can write

$$\lim_{\Delta x \to 0} \frac{\Delta z}{\Delta x} = \frac{dz}{dx} = \text{slope of tangent APB in figure 14.8}$$

At this point you may be wondering what has happened to the other variable, y, which seems to have disappeared from the discussion. The reason for this apparent disappearance is that, by definition, y has a constant value of y_0 everywhere on the section FPG, hence this section is a relationship between only two variables, z and x. As explained in section 14.3 above, the equation of the section FPG is $z = f(x, y_0)$. Because y is constant in this equation, there is no variation in y that needs to be taken into account in calculating either $\frac{\Delta z}{\Delta x}$ or its limiting value, $\frac{dz}{dx}$. (The algebraic demonstration of this is quite straightforward, but we will not trouble with it here.) Note, however, that although we need not worry about *variation* in y along FPG, the *level* of y nevertheless affects the slope of the surface in the x direction, for any given value of x. The reason is easy to see. If we had chosen a different fixed value for y, say $y = y_1$ in figure 14.8, then we would not be at P on the iso-y section FPG, but at some other point on another iso-y section where the slope would, in general, be different for any given value of x.

We can deal with the slope of the tangent CPD, which measures the slope at P in the y direction, in exactly the same way. We take the section NPM together with its tangent CPD and project it on to the $0yz$ plane (see figure 14.9(b)).

Starting from the same initial position at P, we now increase y by an arbitrary amount Δy so that we move from P to S on the section NPM (with x held constant). The difference quotient is now $\frac{\Delta z}{\Delta y}$, which measures the slope of the chord PS. When we allow Δy to shrink, S slides down the curve towards P. As it does so, the slope of the chord PS approaches closer and closer to the slope of the tangent CPD which we are trying to measure. We conclude that, in the limit, as Δy approaches zero, the two slopes will be equal. So

$$\lim_{\Delta y \to 0} \frac{\Delta z}{\Delta y} = \frac{dz}{dy} = \text{slope of tangent CPD in figure 14.8}$$

Thus we have obtained two new derivatives. One, $\frac{dz}{dx}$, measures the slope at P of the section FPG: that is, the slope in the x direction. The other, $\frac{dz}{dy}$, measures the slope at P of the section NPM: that is, the slope in the y direction.

These two derivatives are called **first order partial derivatives**. They are 'partial' because they give us partial information about the slope of the surface, namely the slope in the x and y directions. The question of the slope when x and y both vary simultaneously will be examined in chapter 15.

Notation

Before going further we need to tidy up our notation a little. Above we wrote $\frac{dz}{dx}$ and $\frac{dz}{dy}$ to denote the partial derivatives. However, the convention is that we write the partial derivatives as $\frac{\partial z}{\partial x}$ and $\frac{\partial z}{\partial y}$. The 'curly d' is a variant of the Greek letter δ (delta), which we say out loud as 'partial dee' or 'partial differential'. It is a reminder that these are partial derivatives: that is, there is another variable in each case that is being held constant.

To summarize, we have obtained a rule for partial derivatives:

RULE 14.1 **Partial derivatives**

For any function $z = f(x, y)$ there are two first order partial derivatives:

(1) The partial derivative $\frac{\partial z}{\partial x}$, defined as $\frac{\partial z}{\partial x} = \lim\limits_{\Delta x \to 0} \frac{\Delta z}{\Delta x}$, measures the slope of the surface in the x direction. As explained below, it is evaluated by applying the rules of differentiation to $z = f(x, y)$ with y treated as a constant.

(2) The partial derivative $\frac{\partial z}{\partial y}$, defined as $\frac{\partial z}{\partial y} = \lim\limits_{\Delta y \to 0} \frac{\Delta z}{\Delta y}$, measures the slope of the surface in the y direction. As explained below, it is evaluated by applying the rules of differentiation to $z = f(x, y)$ with x treated as a constant.

These ideas can be extended to functions with three or more independent variables. The function $z = f(x, y, v)$ has three partial derivatives $\frac{\partial z}{\partial x}$, $\frac{\partial z}{\partial y}$, and $\frac{\partial z}{\partial v}$, measuring the slope of the four-dimensional hyper-surface $z = f(x, y, v)$ in the x, y, and v directions. Each partial derivative is evaluated by treating the other two independent variables as constants.

14.5 Evaluation of first order partial derivatives

The mechanics of finding partial derivatives are best shown by means of specific examples.

EXAMPLE 14.3

Suppose the function $z = f(x, y)$ that we are looking at is actually

$$z = x^3 + 3x^2 y^2 + y^3 \tag{14.1}$$

It's hard to visualize what the shape of this surface is, but it doesn't matter, as we evaluate partial derivatives in a completely mechanical way, as we will now show.

(1) Finding $\frac{\partial z}{\partial x}$. To find $\frac{\partial z}{\partial x}$, the partial derivative with respect to x, we must treat y as a constant, then differentiate equation (14.1) with respect to x, using the rules from section 6.7. (That these rules apply equally to finding partial derivatives can easily be proved, but we won't bother with the proof here.)

With y viewed as a constant in equation (14.1), we see that in the second term y^2 appears as a *multiplicative* constant (multiplying x^2), and in the third term y^3 appears as an *additive* constant. Since we know from chapter 6 that multiplicative constants reappear in the derivative, but additive constants do not, this means that the y^2 term will reappear in the derivative, but the y^3 term will not. The 3 in the second term is also a multiplicative constant, which therefore reappears. So we have

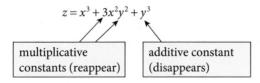

Therefore the partial derivative is

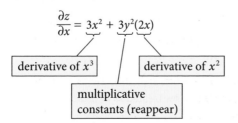

(2) Finding $\frac{\partial z}{\partial y}$. To find $\frac{\partial z}{\partial y}$, the partial derivative with respect to y, we must proceed exactly as above, but this time treating x as the constant. We see that x^3 in the first term of equation (14.1) is an additive constant (and will therefore disappear), while it is x^2 that is now the multiplicative constant in the second term and will therefore reappear. As before, 3 is also a multiplicative constant and reappears.

Thus we have

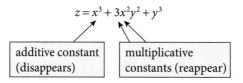

$$z = x^3 + 3x^2y^2 + y^3$$

additive constant (disappears) multiplicative constants (reappear)

So therefore the partial derivative is

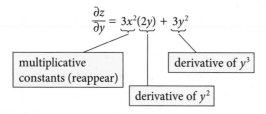

$$\frac{\partial z}{\partial y} = 3x^2(2y) + 3y^2$$

multiplicative constants (reappear) derivative of y^3

derivative of y^2

EXAMPLE 14.4

Suppose the function is

$z = x^{0.5}y^{0.5} - 10$ (with only positive values of x, y, and z considered)

This function would, in fact, give a surface with roughly the shape implied by the sections through the surface shown in figures 14.8 and 14.9. So we can refer once again to these figures, assuming them to be generated by the function $z = x^{0.5}y^{0.5} - 10$. This is not strictly necessary because, as we saw in the previous example, we evaluate partial derivatives in a completely mechanical way. Looking at figures 14.8 and 14.9 will, however, give us some feel for what the partial derivatives that we obtain actually measure.

(1) Finding $\frac{\partial z}{\partial x}$. To find $\frac{\partial z}{\partial x}$ we apply the rules of differentiation, treating y as a constant. Looking at $z = x^{0.5}y^{0.5} - 10$, we see that $y^{0.5}$ is a multiplicative constant, which will therefore reappear in the derivative. The additive constant, -10, disappears. Therefore the derivative is

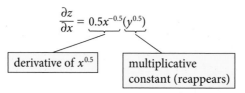

$$\frac{\partial z}{\partial x} = 0.5x^{-0.5}(y^{0.5})$$

derivative of $x^{0.5}$ multiplicative constant (reappears)

Because the surface $z = x^{0.5}y^{0.5} - 10$ has a shape similar to figure 14.8, we can view the partial derivative that we have just found as measuring the slope of the tangent drawn to any iso-y section, such as the section FPG, at any point. To find the numerical value of the slope of the surface in the x direction at any point, we simply substitute the x- and y-coordinates of the point into our expression for $\frac{\partial z}{\partial x}$ above. For example, if at the point P in figure 14.8 we have $x_0 = 25$ and $y_0 = 9$, the slope of the tangent APB is

$$\frac{\partial z}{\partial x} = (0.5)(25)^{-0.5}(9)^{0.5} = \frac{3}{10}$$

(2) Finding $\frac{\partial z}{\partial y}$. To find $\frac{\partial z}{\partial y}$ we proceed in exactly the same way but with the roles of x and y now interchanged. With x now treated as a (multiplicative) constant, the derivative we obtain is

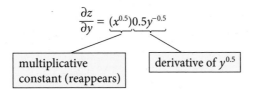

$$\frac{\partial z}{\partial y} = \underbrace{(x^{0.5})}_{}\underbrace{0.5y^{-0.5}}_{}$$

multiplicative constant (reappears) — derivative of $y^{0.5}$

In figure 14.9, this partial derivative measures the slope of any iso-x section, such as NPM, at any point. Thus if at the point P we have $x_0 = 25$ and $y_0 = 9$, the slope of the tangent CPD is

$$\frac{\partial z}{\partial y} = (25^{0.5})0.5(9)^{-0.5} = \frac{5}{6}$$

Summary of sections 14.1–14.5

The function $z = f(x, y)$, assuming that it is smooth and continuous, corresponds to a surface in three-dimensional space. We can examine the relationship between z and x by taking an iso-y section through the surface for any fixed value of y; and similarly examine the relationship between z and y by taking an iso-x section through the surface for any fixed value of x.

Any smooth and continuous function $z = f(x, y)$ has two first order partial derivatives, $\frac{\partial z}{\partial x}$ and $\frac{\partial z}{\partial y}$. The partial derivative $\frac{\partial z}{\partial x}$ measures the slope of the surface in the x direction and the partial derivative $\frac{\partial z}{\partial y}$ measures the slope of the surface in the y direction. Each is evaluated by applying the standard rules of differentiation to the function, treating the other variable (or variables if there are more than two independent variables) as a constant.

Progress exercise 14.2

For each of the following functions, find $\frac{\partial z}{\partial x}$ and $\frac{\partial z}{\partial y}$. (Hint: all the rules of differentiation from chapters 6 and 13 for a function of one independent variable apply straightforwardly, with the other independent variable(s) treated as constants.)

(a) $z = x^2 + 2xy + y^2$ (b) $z = 3x^3 + 2x^2y + y^2 + y$ (c) $z = \frac{1}{3}x^3 + y^{0.5} + \frac{1}{xy}$

(d) $z = (x^3 + y^2)^{0.5}$ (e) $z = \frac{x^3 + y^2}{x - y}$

(f) $z = (x^2 + y)(x - y^2)$ (Use the product rule; don't multiply out!)

(g) $z = 100e^{2x+3y}$ (h) $z = \ln(x^3 + y^2) - \ln(xy)$

14.6 Second order partial derivatives

For a function of the form $z = f(x, y)$ there are two types of **second order partial derivative**: direct partial and 'cross partial' second derivatives.

Direct second derivatives

In chapter 7 we examined the function $y = f(x)$ and saw that if the second derivative $\frac{d^2y}{dx^2}$ was positive or negative at some point P, this told us that $\frac{dy}{dx}$ was increasing or decreasing (respectively) as x increased. The direct second order partial derivatives of the function $z = f(x, y)$ have an exactly analogous role.

Refer to figure 14.10(a), which repeats part of figure 14.9(a), showing the section FPG through the surface in the x direction. In section 14.4 above, we showed that the partial derivative $\frac{\partial z}{\partial x}$ measures the slope of the tangent APB, the slope in the x direction at P.

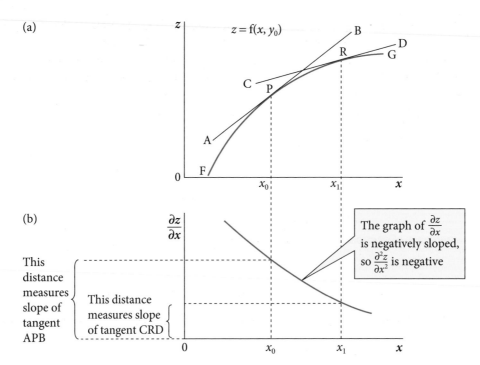

Figure 14.10 Interpreting the direct second order partial derivatives.

Looking at the section FPG, we see that it is concave from below, which in other words means that its slope decreases as x increases (see section 7.13 to revise convexity and concavity). Thus the tangent at R is flatter than the tangent at P. Therefore the graph of $\frac{\partial z}{\partial x}$, which measures this slope, is decreasing (downward sloping from left to right), as shown in figure 14.10b.

If we now partially differentiate for a second time, continuing to treat y as a constant; that is we evaluate $\frac{\partial}{\partial x}$ of $\frac{\partial z}{\partial x}$, we obtain what is called the direct second order partial derivative with respect to x. Just as we found with functions of one independent variable in section 7.7, this second derivative measures the slope of the graph of $\frac{\partial z}{\partial x}$. Because this graph is negatively sloped in figure 14.10b, this direct second order derivative is negative. If conversely the section FPG had been convex from below, the graph of $\frac{\partial z}{\partial x}$ would have been positively sloped and the direct second order derivative would have been positive. All this is perfectly analogous with the case of a function with only one independent variable discussed in chapter 7. The analogy is perfect because we are looking only at the section FPG, along which y is constant, and therefore there is effectively only one independent variable, x.

In the same way, referring now to figure 14.9(b), we can see that the slope in the y direction also decreases as y increases, and therefore the graph of $\frac{\partial z}{\partial y}$ (not shown) is also negatively sloped.

If we partially differentiate for a second time, continuing to treat x as a constant; that is we evaluate $\frac{\partial}{\partial y}$ of $\frac{\partial z}{\partial y}$, we obtain the direct second order partial derivative with respect to y. This measures the slope of the graph of $\frac{\partial z}{\partial y}$ and therefore in this example is negative.

Thus we have the following rule:

RULE 14.2 Direct second order partial derivatives

Any function $z = f(x, y)$ has two direct second order partial derivatives:

(1) $\frac{\partial}{\partial x}$ of $\frac{\partial z}{\partial x}$, which we write more compactly as $\frac{\partial^2 z}{\partial x^2}$. The sign ($+/-$) of this tells us whether the slope in the x direction increases or decreases as x increases.

(2) $\frac{\partial}{\partial y}$ of $\frac{\partial z}{\partial y}$, which we write more compactly as $\frac{\partial^2 z}{\partial y^2}$. The sign ($+/-$) of this tells us whether the slope in the y direction increases or decreases as y increases.

Notice once more that, both in concept and notation, rule 14.2 follows almost exactly the second derivative of the function $y = f(x)$ that we considered in chapter 7. The only difference lies in the 'curly d' that we write to remind ourselves that each direct second order partial derivative is evaluated with another variable, either x or y, being held constant throughout.

Evaluating direct second order partial derivatives

This is best shown by an example.

EXAMPLE 14.5

We will continue with the function used in example 14.4, which was

$z = x^{0.5}y^{0.5} - 10$ where we now want to find $\frac{\partial^2 z}{\partial x^2}$ and $\frac{\partial^2 z}{\partial y^2}$.

(1) Finding $\frac{\partial^2 z}{\partial x^2}$. From example 14.4 we know that the partial derivative $\frac{\partial z}{\partial x}$ for this function is

$$\frac{\partial z}{\partial x} = 0.5x^{-0.5}(y^{0.5}) \quad \text{(the brackets are not strictly necessary)}$$

The direct second partial derivative is therefore the partial derivative of this: that is

$$\frac{\partial^2 z}{\partial x^2} \equiv \frac{\partial}{\partial x} \text{ of } 0.5x^{-0.5}(y^{0.5})$$

To find this derivative we therefore have to apply the standard rules of differentiation to $0.5x^{-0.5}(y^{0.5})$, treating x as the variable and y as a constant. Applying the power rule of differentiation to $x^{-0.5}$, and treating 0.5 and $y^{0.5}$ as multiplicative constants (which therefore reappear in the derivative), we get

$$\frac{\partial^2 z}{\partial x^2} = \underbrace{-0.5x^{-1.5}}_{\text{derivative of } x^{-0.5}}\underbrace{(0.5)(y^{0.5})}_{\substack{\text{multiplicative} \\ \text{constants (reappear)}}} = -0.25x^{-1.5}y^{0.5}$$

(14.2)

Given our earlier assumption that x and y are positive, and since we know that any positive number raised to any power is positive (see section 2.12), we can say that $y^{0.5}$ and $x^{-1.5}$ are both positive. Therefore the -0.25 in equation (14.2) tells us that $\frac{\partial^2 z}{\partial x^2}$ is negative, meaning that the slope in the x direction decreases as x increases with y held constant. This confirms our expectation based on inspection of figure 14.8 or figure 14.9(a).

(2) Finding $\frac{\partial^2 z}{\partial y^2}$. From example 14.4 we know that the partial derivative $\frac{\partial z}{\partial y}$ for this function is

$$\frac{\partial z}{\partial y} = 0.5y^{-0.5}(x^{0.5})$$

The direct second partial derivative is therefore the partial derivative of this; that is

$$\frac{\partial^2 z}{\partial y^2} \equiv \frac{\partial}{\partial y} \text{ of } 0.5y^{-0.5}(x^{0.5})$$

To find this derivative we have to apply the standard rules of differentiation to $0.5y^{-0.5}(x^{0.5})$, this time treating y as the variable and x as a constant. Applying the power rule of differentiation and treating 0.5 and $x^{0.5}$ as multiplicative constants, we get

$$\frac{\partial^2 z}{\partial y^2} = -0.5y^{-1.5}(0.5)x^{0.5} = -0.25x^{0.5}y^{-1.5}$$

(14.3)

As with equation (14.2) above, this is negative, confirming our expectation based on figure 14.8 or figure 14.9(b).

To summarize the example above, we found that at a point such as P in figures 14.8 and 14.9:

(1) $\frac{\partial z}{\partial x} > 0$ and $\frac{\partial^2 z}{\partial x^2} < 0$. Thus in the x direction the surface is positively sloped but the slope diminishes as x increases.

(2) $\frac{\partial z}{\partial y} > 0$ and $\frac{\partial^2 z}{\partial y^2} < 0$. Thus in the y direction the surface is also positively sloped but the slope diminishes as y increases.

But this is only one of no fewer than 16 possible permutations of signs for the four derivatives. You should make sure that you can sketch or at least describe the shape of the surface at any point P for each of the other 15 permutations.

Cross partial derivatives

Suppose the function $z = f(x, y)$ describes a surface from which two sections have been cut and are shown in figure 14.11. The tangent drawn at P shows the slope in the x direction when $x = x_0$ and $y = y_0$, and the tangent drawn at Q shows the slope in the x direction when $x = x_0$ and $y = y_0 + dy$. (The distance dy is supposed to be very small, as otherwise the algebra doesn't work, but we've shown dy as quite large in order to make the diagram as clear as possible.)

The cross partial derivative $\frac{\partial^2 z}{\partial y \partial x}$

Suppose we now ask: 'Starting from P, how does the slope in the x direction change when y increases by a small amount dy, with x constant?' In figure 14.11 we can see that an increase in y by a small amount dy, with x constant, takes us from P to Q. And it is clear that the slope of the tangent at Q (drawn in the x direction: that is, parallel to the $0xz$ plane) is steeper than the slope of the corresponding tangent at P. So the answer is: 'It increases.'

It might help to make this more concrete if you imagine yourself standing at P on a hillside, facing uphill: that is, in the x direction. You examine the steepness of the slope carefully. Then you take one small step to the left, and arrive at Q. Once again you carefully examine the steepness of the slope in the x direction, and then ask yourself 'is the slope steeper at Q than at P?'

The key point now is that, provided the increase in y, dy, is sufficiently small, we can answer the above question algebraically by evaluating $\frac{\partial}{\partial y}$ of $\frac{\partial z}{\partial x}$. That is, we take $\frac{\partial z}{\partial x}$ and find its derivative with respect to y. The resulting expression measures how the slope in the x direction changes when y increases by a small amount dy, with x constant. It is called a **cross partial derivative**. At P in figure 14.11, this cross partial derivative will be positive because the tangent at Q is steeper than the tangent at P. A small increase in y causes the slope in the x direction to increase.

For notational compactness, this cross partial derivative $\frac{\partial}{\partial y}$ of $\frac{\partial z}{\partial x}$ is written as $\frac{\partial^2 z}{\partial y \partial x}$. Notice that the order in which x and y are written in the denominator shows the order of differentiation; that is, first with respect to x to get $\frac{\partial z}{\partial x}$, then with respect to y to get $\frac{\partial}{\partial y}$ of $\frac{\partial z}{\partial x}$.

As a matter of algebraic technique, we evaluate this derivative in any particular case by simply

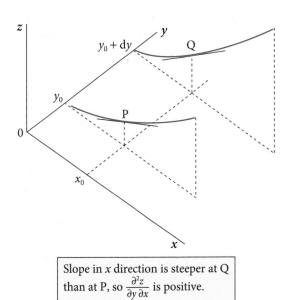

Slope in x direction is steeper at Q than at P, so $\frac{\partial^2 z}{\partial y \partial x}$ is positive.

Figure 14.11 Interpreting the cross partial derivative $\frac{\partial^2 z}{\partial y \partial x}$.

applying the normal rules of differentiation. Since we are differentiating with respect to y, we treat x as a constant. Once again this is best illustrated by a concrete example. The function

$$z = x^3 y^3$$

gives a shape broadly consistent with the sections drawn in figure 14.11. If we first differentiate this with respect to x, treating y^3 as a multiplicative constant, we get

$$\frac{\partial z}{\partial x} = \frac{\partial}{\partial x} \text{ of } x^3 y^3 = 3x^2 y^3$$

This measures the slope in the x direction and is positive when x and y are both positive. As x and y are both positive at P, the slope is therefore positive at P, confirming that figure 14.11 is correctly drawn.

If we now differentiate this expression with respect to y, treating $3x^2$ as a multiplicative constant, we get

$$\frac{\partial^2 z}{\partial y \partial x} = \frac{\partial}{\partial y} \text{ of } 3x^2 y^3 = 3x^2 3y^2 = 9x^2 y^2 \tag{14.4}$$

As this expression is positive when x and y are positive, it tells us that the slope in the x direction increases when there is a small increase in y, as in figure 14.11.

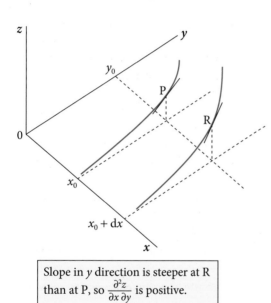

Slope in y direction is steeper at R than at P, so $\frac{\partial^2 z}{\partial x \, \partial y}$ is positive.

Figure 14.12 Interpreting the cross partial derivative $\frac{\partial^2 z}{\partial x \partial y}$.

The cross partial derivative $\frac{\partial^2 z}{\partial x \partial y}$

In exactly the same way, we could also ask how the slope in the y direction changes when there is a small increase in x. This is illustrated in figure 14.12. The slope of the tangent at P measures the slope in the y direction and is given by $\frac{\partial z}{\partial y}$ evaluated with $x = x_0$ and $y = y_0$. The slope of the tangent at R also measures the slope in the y direction and is given by $\frac{\partial z}{\partial y}$ evaluated with $x = x_0 + dx$ and $y = y_0$. The question is whether the slope at R is greater or less than the slope at P.

We can answer this question algebraically by evaluating the cross partial derivative $\frac{\partial}{\partial x}$ of $\frac{\partial z}{\partial y}$, which we write for compactness as $\frac{\partial^2 z}{\partial x \partial y}$. To illustrate the algebra, we continue with the function

$$z = x^3 y^3$$

If we first differentiate this with respect to y, treating x^3 as a multiplicative constant, we get

$$\frac{\partial z}{\partial y} = \frac{\partial}{\partial y} \text{of } z = x^3 y^3 = x^3 3y^2$$

If we now differentiate this expression with respect to x, treating $3y^2$ as a multiplicative constant, we get

$$\frac{\partial^2 z}{\partial x \partial y} = \frac{\partial}{\partial x} \text{of } x^3 3y^2 = 3x^2 3y^2 = 9x^2 y^2 \tag{14.5}$$

The fact that this is positive tells us that in figure 14.12 the slope in the y direction is steeper at R than at P, as drawn.

Thus we have the following rule:

RULE 14.3 Second order cross partial derivatives

Any function $z = f(x, y)$ has two second order cross partial derivatives:

(1) $\frac{\partial}{\partial y}$ of $\frac{\partial z}{\partial x}$, which we write more compactly as $\frac{\partial^2 z}{\partial y \partial x}$. The sign (+/−) of this tells us whether the slope in the x direction increases or decreases as y increases.

(2) $\frac{\partial}{\partial x}$ of $\frac{\partial z}{\partial y}$, which we write more compactly as $\frac{\partial^2 z}{\partial x \partial y}$. The sign (+/−) of this tells us whether the slope in the y direction increases or decreases as x increases.

Young's theorem

If we compare equations (14.4) and (14.5), we see that we have

$$\frac{\partial^2 z}{\partial x \partial y} = 9x^2 y^2 \quad \text{and} \quad \frac{\partial^2 z}{\partial x \partial y} = 9x^2 y^2$$

Thus the two cross partial derivatives are equal. This is not merely a quirk of the particular function $z = x^3 y^3$ that we have chosen to use here. In fact, it is a property of most functions. This equality between the two cross partial derivatives is known as Young's theorem. It's a rather surprising fact, since there seems to be no obvious reason why, at any point, the effect of an increase in y on the slope in the x direction should be the same as the effect of an increase in x on the slope in the y direction. This is apparent when we compare figures 14.12 and 14.13. However, we must keep in mind that the algebra only works when dx and dy are very small.

Summary of partial derivatives

Summarizing sections 14.4–14.6, for any function $z = f(x, y)$ we have two first order partial derivatives and four second order partial derivatives, with the interpretations shown in table 14.2.

Table 14.2 Interpreting the first and second order partial derivatives.

Partial derivative	What it tells us about any point P on the surface $z = f(x, y)$
$\frac{\partial z}{\partial x}$	sign (+/−) tells us whether slope in x direction is positive or negative
$\frac{\partial^2 z}{\partial x^2}$	sign (+/−) tells us whether slope in x direction increases/decreases as x increases (y constant)
$\frac{\partial^2 z}{\partial y \partial x}$	sign (+/−) tells us whether slope in x direction increases/decreases as y increases (x constant)
$\frac{\partial z}{\partial y}$	sign (+/−) tells us whether slope in y direction is positive or negative
$\frac{\partial^2 z}{\partial y^2}$	sign (+/−) tells us whether slope in y direction increases/decreases as y increases (x constant)
$\frac{\partial^2 z}{\partial x \partial y}$	sign (+/−) tells us whether slope in y direction increases/decreases as x increases (y constant)
Young's theorem: $\frac{\partial^2 z}{\partial x \partial y} = \frac{\partial^2 z}{\partial y \partial x}$	

The example below shows all of these partial derivatives in action.

EXAMPLE 14.6

Given the function $z = x^3 + x^4y^2 + y^3$ we can find the two first order partial derivatives and the four second order partial derivatives as follows:

$\dfrac{\partial z}{\partial x} = \dfrac{\partial}{\partial x}$ of $x^3 + x^4y^2 + y^3 = 3x^2 + 4x^3y^2$ In $x^3 + x^4y^2 + y^3$, y^3 is an additive constant, hence disappears. But y^2 is a multiplicative constant, hence retained.

$\dfrac{\partial^2 z}{\partial x^2} = \dfrac{\partial}{\partial x}$ of $3x^2 + 4x^3y^2 = 6x + 12x^2y^2$ In $3x^2 + 4x^3y^2$, y^2 is a multiplicative constant, hence retained.

$\dfrac{\partial^2 z}{\partial y \partial x} = \dfrac{\partial}{\partial y}$ of $3x^2 + 4x^3y^2 = 8x^3y$ In $3x^2 + 4x^3y^2$, $3x^2$ is additive, hence disappears. But $4x^3$ is multiplicative, hence retained.

$\dfrac{\partial z}{\partial y} = \dfrac{\partial}{\partial y}$ of $x^3 + x^4y^2 + y^3 = 2x^4y + 3y^2$ In $x^3 + x^4y^2 + y^3$, x^3 is additive, hence disappears. But x^4 is multiplicative, hence retained.

$\dfrac{\partial^2 z}{\partial y^2} = \dfrac{\partial}{\partial y}$ of $2x^4y + 3y^2 = 2x^4 + 6y$ In $2x^4y + 3y^2$, $2x^4$ is multiplicative, hence retained, multiplied by derivative of y, which $= 1$.

$\dfrac{\partial^2 z}{\partial x \partial y} = \dfrac{\partial}{\partial x}$ of $2x^4y + 3y^2 = 8x^3y$ In $2x^4y + 3y^2$, $3y^2$ is additive, hence disappears. But y is multiplicative, hence retained.

Young's theorem: $\dfrac{\partial^2 z}{\partial x \partial y} = \dfrac{\partial^2 z}{\partial y \partial x} = 8x^3y$

Generalization to functions of many variables

These concepts of partial derivatives and techniques for evaluating them extend straightforwardly to functions with many independent (right-hand side) variables. For example, if the function is

$z = \mathrm{F}(x, y, q)$ where x, y, and q are three variables,

we have three first order partial derivatives, $\frac{\partial z}{\partial x}$, $\frac{\partial z}{\partial y}$, and $\frac{\partial z}{\partial q}$, and three direct second order partial derivatives $\frac{\partial^2 z}{\partial x^2}$, $\frac{\partial^2 z}{\partial y^2}$, and $\frac{\partial^2 z}{\partial q^2}$. We now have a total of six second order cross partial derivatives. These comprise $\frac{\partial^2 z}{\partial y \partial x}$ and $\frac{\partial^2 z}{\partial q \partial x}$, which measure the effect of an increase in y and q on the slope in the x direction; $\frac{\partial^2 z}{\partial x \partial y}$ and $\frac{\partial^2 z}{\partial q \partial y}$, which measure the effect of an increase in x and q on the slope in the y direction; and $\frac{\partial^2 z}{\partial x \partial q}$ and $\frac{\partial^2 z}{\partial y \partial q}$, which measure the effect of an increase in x and y on the slope in the q direction. Note that in evaluating any of these partial derivatives, we must now hold two variables constant. In general, however many independent variables there may be, we hold all of them constant except one in evaluating any partial derivative.

An alternative notation

The format $\frac{\partial z}{\partial x}, \frac{\partial^2 z}{\partial x \partial y}$, and so on which we have used to write partial derivatives up to this point is cumbersome and wastes space on the page. There is an alternative, more compact notation which parallels that mentioned in section 7.14.

For the function $z = f(x, y)$, an alternative for the various partial derivatives is to write

$$f_x \equiv \frac{\partial z}{\partial x}, \quad f_{xx} \equiv \frac{\partial^2 z}{\partial x^2}, \quad f_{yx} \equiv \frac{\partial^2 z}{\partial y \partial x}, \quad f_{xy} \equiv \frac{\partial^2 z}{\partial x \partial y}, \quad \text{and so on}$$

If the function is $z = g(x, y)$ or $z = \phi(x, y)$, we can simply write, for example, g_x or ϕ_x to denote $\frac{\partial z}{\partial x}$, and so on. This is a better notation because if we wrote, say, $\frac{\partial z}{\partial x}$, it might sometimes be unclear whether this derivative was from $z = g(x, y)$ or $z = \phi(x, y)$.

We shall use this notation increasingly in the remainder of this book.

The Online Resource Centre shows how to use Maple to evaluate partial derivatives.
www.oxfordtextbooks.co.uk/orc/renshaw3e/

Progress exercise 14.3

1. (a) Given $z = x^3 + 5xy + 4y^2$, find

   $$\frac{\partial z}{\partial x}; \frac{\partial^2 z}{\partial x^2}; \frac{\partial^2 z}{\partial y \partial x}; \frac{\partial z}{\partial y}; \frac{\partial^2 z}{\partial y^2}; \text{ and } \frac{\partial^2 z}{\partial x \partial y}$$

   (b) Explain *briefly* what, in general, each of these partial derivatives measures.

2. For each of the following functions, find

   $$\frac{\partial z}{\partial x}; \frac{\partial^2 z}{\partial x^2}; \frac{\partial z}{\partial y}; \text{ and } \frac{\partial^2 z}{\partial x \partial y}$$

   (a) $z = x^2 + 2x - 3y + y^2$ (b) $z = x^2 + xy + y^2$ (c) $z = x^3 + 3x^2 - 2xy - xy^2 + 3y^2 + x^2 y$

   (d) $z = 15x + 3x^2 - 2xy - 2y^2 + 12y$ (e) $z = 10x^{0.4} y^{1.5}$

   (f) $z = x^\alpha y^\beta$, (where α and β are parameters)

   (g) $z = 0.25 \ln x + 0.5 \ln y$ (h) $z = \alpha \ln x + \beta \ln y$ (α, β, parameters)

   (i) $z = \ln(x^2 + 3y)$ (j) $z = e^{2x + 3y}$

3. For each of the following functions, find

   $$\frac{\partial z}{\partial x}; \frac{\partial^2 z}{\partial x^2}; \text{ and } \frac{\partial z}{\partial y}$$

   (a) $z = (2x - 3y)^{0.5}$ (b) $z = \frac{3x^2}{4y^2}$ (c) $z = \frac{x^2 + y^2}{x^2 - y}$ (d) $z = \left(\frac{x^2}{1 - y^3}\right)^{0.5}$

   (e) $z = x^2 e^{3y}$ (f) $z = \ln\left(\frac{x^2 + 1}{y^2 - 1}\right)$ (g) $z = \ln(x^{0.5} y^{0.25})$

4. Suppose you are given the following information about a function $z = f(x, y)$:

   At a point P, $\frac{\partial z}{\partial x} > 0$; $\frac{\partial^2 z}{\partial x^2} > 0$; $\frac{\partial^2 z}{\partial y \partial x} < 0$; $\frac{\partial z}{\partial y} > 0$; $\frac{\partial^2 z}{\partial y^2} < 0$

   Use this information to sketch the shapes of the iso-x and iso-y sections at P and thus indicate the shape of the surface in the vicinity of P. (This isn't easy, but don't be afraid to give it a try.) Figures 14.11 and 14.12 are relevant.

5. Repeat question 4 for a function $z = g(x, y)$, where at a point Q:

   $$\frac{\partial z}{\partial x} < 0; \frac{\partial^2 z}{\partial x^2} > 0; \frac{\partial^2 z}{\partial y \partial x} < 0; \frac{\partial z}{\partial y} < 0; \frac{\partial^2 z}{\partial y^2} < 0$$

14.7 Economic applications 1: the production function

The production function is an important part of the economist's toolkit. In general a production function gives the relationship between *inputs* of capital (K) and labour (L) and resulting *output* of some product (Q). This relationship may exist at the level of an individual firm, an industry or the whole economy. Here we will consider only the first case, the production function of a firm. (We have chosen to use capital letters for the variables, reflecting common but by no means universal practice among economists.)

Note that K and L are *flows* of physical input and Q is the resulting *flow* of physical output. We need to specify the units of measurement for these physical quantities. Output can be measured in its own units, such as litres per week in the case of beer production. The flow of labour input can also be measured in worker-hours per week: that is, the aggregate hours worked by all workers in that week. Measuring the capital input is less straightforward because capital equipment is embodied in many different types of machine, buildings, and so on. Economists often evade these complications by assuming that all capital is embodied in one type of machine, so that the flow of capital input can be measured in machine-hours per week: that is, the aggregate hours of operation of all the firm's machines in that week. Economists also simplify by assuming that capital and labour are the only inputs required to produce output, ignoring the fact that in the real world firms typically buy in raw materials and many components: for example, car producers do not themselves make the cars' tyres.

Another simplification is to assume that the production function, being a relationship between physical quantities of inputs and the resulting physical output, is a purely technological relationship. This means that to find out how much output can be produced from given quantities of labour and capital input, we should ask an engineer, not an economist.

The above assumptions are very general. The *neoclassical school* of economic thought (which includes the majority of economists) makes some further assumptions that result in what is known as the neoclassical production function. If we write the production function in mathematical form as

$$Q = f(K, L)$$

then neoclassical production theory assumes the following:

(1) Q, K, and L are infinitely divisible, and the production function $f(K, L)$ is smooth and continuous.

(2) If either K or L is zero, output is also zero. In other words, you need at least some of *both* inputs to produce any output. Workers can produce nothing unless they have at least some machinery to help them, and machines can produce nothing without at least some workers to operate them. Within these units there are many combinations of L and K that can produce a given output.

(3) Provided K and L are both positive, an increase in either L or K will always increase Q.

(4) The 'law' of diminishing marginal productivity holds, either at all levels or at least at sufficiently high levels of output. We will explain this shortly.

14.8 The shape of the production function

Given these assumptions, what shape will the production function $Q = f(K, L)$ have? A three-dimensional perspective drawing of the general shape is shown in figure 14.13, where the inputs of K and L are measured along two axes, and the resulting output, Q, is measured along the

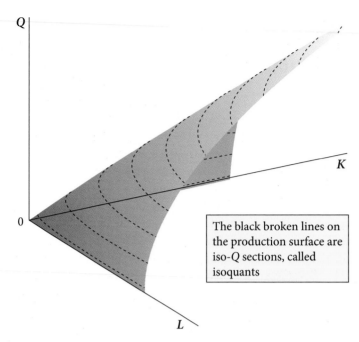

The black broken lines on the production surface are iso-Q sections, called isoquants

Figure 14.13 A plausible shape for a production function $Q = f(K, L)$.

third axis. We see that at the origin, we have $K = L = 0$ and therefore $Q = 0$ because with no input there is no output. As we move away from the origin along the L-axis, we have $K = 0$ and hence $Q = 0$ because of neoclassical assumption (2) above, that neither capital nor labour alone can produce anything. For the same reason, along the K-axis we have $L = 0$ and hence $Q = 0$. As we move away from the origin in any direction such that both K and L are positive, Q is positive and increases as either K or L, or both, increase. This reflects neoclassical assumption (3).

Isoquants

The production function surface in figure 14.13 is marked with black broken lines. Each of these identifies a horizontal section through the surface of the production function, parallel to the $0KL$ plane, such that Q is constant. They are thus iso-Q sections. In the language of economics, an iso-Q section is a locus of various combinations of K and L that are capable of producing a given output. The iso-Q sections are called **isoquants** (from iso, meaning equal, and quantity).

Figure 14.14 shows two isoquants from figure 14.13, again in a three-dimensional perspective drawing. The isoquants are the lines ABC and DEF, for constant output levels of $Q = 100$ and $Q = 60$, respectively.

By looking along the $Q = 100$ isoquant in figure 14.14 we can see some of the combinations of K and L that are required to produce 100 units of output. For example, at point A, 100 units of output are produced by L_1 units of labour and K_3 units of capital. At point B, 100 units of output are produced by L_2 units of labour and K_2 units of capital. Because of assumption (3) above, in moving along the isoquant from A to B, the capital input has to be reduced to compensate for the increase in the labour input, since otherwise output would not remain constant. Thus in moving from A to B, a *substitution* of labour for capital occurs. The distances AA′, BB′, and CC′ are equal to one another and equal 100, which also equals the distance up the Q-axis from 0 to 100. The equation of this isoquant is found by setting $Q = 100$ in the production function $Q = f(K, L)$, giving $100 = f(K, L)$, an implicit function relating K and L.

In the same way, the isoquant DEF shows the combinations of K and L required to produce 60 units of output, such as L_0 units of labour and K_3 units of capital at point D, or L_2 units of labour and K_1 units of capital at point E. The distances DD′, EE′, and FF′ are equal to one another and equal 60, which is also the distance up the Q-axis from 0 to 60. The equation of

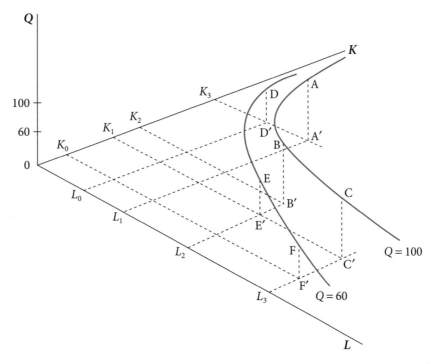

Figure 14.14 Production function isoquants showing combinations of K and L that can produce $Q = 60$ and $Q = 100$ units of output.

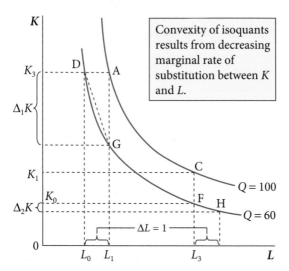

Figure 14.15 Isoquants for $Q = 60$ and $Q = 100$ units of output, projected on to $0KL$ plane.

Convexity of isoquants results from decreasing marginal rate of substitution between K and L.

the isoquant DEF is found by setting $Q = 60$ in the production function $Q = f(K, L)$, giving the implicit function $60 = f(K, L)$.

Figure 14.14 contains only two isoquants, which are themselves incomplete, because each can be extended indefinitely in both directions. But obviously the whole of the three-dimensional space is packed with isoquants for different levels of output. No isoquant can cut an axis, due to assumption (2) in section 14.7 above; and no isoquant can cut another isoquant, since this would violate assumption (3). (You may need a pause for thought, with the aid of pencil and paper, to understand why this last point is true.)

Figure 14.14 attempts a perspective drawing of the isoquants relative to the three axes. However, we can also project the isoquants on to the K, L plane, as in figure 14.15. This conveys essentially the same information as figure 14.14. The labelling of the points A, D, C, and F in figure 14.15 is the same as in figure 14.14.

The marginal rate of substitution

In figure 14.15 we see that the isoquants are negatively sloped in the $0KL$ plane, reflecting the fact that if one input is reduced the other must be increased in order to maintain output constant. A distinctive feature of the isoquants is that they are convex from below (see section 7.13 for revision of convexity). This convexity reflects an assumption that substitution of labour for capital becomes technically more difficult as we move down an isoquant. Economists describe this as a **decreasing marginal rate of substitution** (*MRS*) in production. (The *MRS* in production is also widely known as the marginal rate of technical substitution.)

The *MRS* in production is often defined in introductory textbooks as the amount of capital released when one more unit of labour is employed while keeping output constant. In figure 14.15, suppose production is initially at D where output of $Q = 60$ is produced using K_3 units of capital and L_0 units of labour. Now suppose the firm decides to employ one more unit of labour while maintaining output constant. The production point therefore moves to G and the amount of capital withdrawn or released is given by the distance $\Delta_1 K$.

However, if the same experiment is conducted from an initial position at F, production moves to H and only $\Delta_2 K$ units of capital are released. As $\Delta_2 K$ is clearly much smaller than $\Delta_1 K$, we see that the *MRS* decreases as *L* increases. It is easy to see that the *MRS* decreases because the isoquants are convex. If the isoquants were linear, the *MRS* would be constant. We then say that *K* and *L* are perfect substitutes.

Let us examine more closely the link between the *MRS* and the slope of the isoquant. Looking at point D in figure 14.15, for example, the *MRS* is $\Delta_1 K$. But since by definition $\Delta L = 1$, we can write the *MRS* as $\frac{\Delta_1 K}{\Delta L}$. We can do this anywhere on the curve, so the subscript 1 can be dropped, giving us, in general,

$$MRS = \frac{\Delta K}{\Delta L} \quad \text{(evaluated at any point on any isoquant; that is, with } Q \text{ held constant)}$$

The right-hand side of this equation is the difference quotient, with ΔL set equal to 1. At D, it measures the slope of the chord DG, which equals the average slope of the isoquant between D and G. As we have already seen in several contexts, the fact that the difference quotient measures only the average slope between two points on the curve in question makes it somewhat imprecise as a measure of slope. Therefore it is standard practice in theoretical economic analysis to replace the difference quotient by the derivative of the isoquant. The definition of the *MRS* is then as follows:

RULE 14.4 **The marginal rate of substitution in production**

For any production function $Q = f(K, L)$, the marginal rate of substitution (*MRS*) in production is defined as:

$MRS = \frac{dK}{dL}$ (evaluated with Q held constant)

This measures the slope of the isoquant at any point (with *K* on the vertical and *L* on the horizontal axis).

We shall see in chapter 16 how the *MRS* is linked to the **marginal products** of capital and labour. We will define the marginal products shortly. (Note that, in rule 14.4, $\frac{dK}{dL}$ is not, mathematically, a partial derivative, even though it is evaluated with another variable, *Q*, held constant. The reason for this will become clear in section 15.8.)

There remains a small but troublesome wrinkle to be ironed out. Clearly $\frac{dK}{dL}$ is negative, since the isoquant is downward sloping in the 0*KL* plane. In the language of mathematics, therefore, convexity means that the slope increases (becomes less negative) as *L* increases. For example, in figure 14.15 the slope is about −3 at D but *increases* to about −0.25 at F (an increase because −0.25 lies to the right of −3 on the number line).

However, economists invariably talk of the *MRS* as if it were positive, and therefore would say that the *MRS decreases* from about 3 to about 0.25 between D and F. In effect, despite having defined the *MRS* as $\frac{dK}{dL}$, which is always negative, economists speak and write about the absolute value of the *MRS*, $\left|\frac{dK}{dL}\right|$, which is always positive.

We follow this somewhat untidy economists' convention here: we define the *MRS* as the slope of the isoquant, which means that it is always negative, but in discussion we refer to its absolute value, so we describe the *MRS* as decreasing as we move down the isoquant. (Some economists who want to avoid this sloppiness define the *MRS* as the absolute value $\left|\frac{dK}{dL}\right|$, or as $-\frac{dK}{dL}$, in both cases thereby ensuring that the *MRS* is positive.)

Thus discussion of a decreasing *MRS* in production can be somewhat confusing. To minimize confusion, it helps to hold tight to the basic idea, which is that the isoquants are convex. Further, we know that when a curve is convex from below, its second derivative is positive (see section 7.13). So we expect $\frac{d^2K}{dL^2}$ to be positive.

Finally, this is a good point to introduce the concept of **capital intensity**. This is of great interest to both theoretical and applied economists, though we shall not use it much in this book. It can be defined either as capital per unit of labour, $\frac{K}{L}$, or as capital per unit of output, $\frac{K}{Q}$. In figure 14.15, $\frac{K}{L}$ at any point on an isoquant is measured by the slope of a ray from the origin to the point in question. Clearly, as we move down any isoquant $\frac{K}{L}$ is falling; the technique of production is becoming less capital intensive, or more labour intensive.

The short-run production function

It is also of interest to take sections through the production function for fixed values of K: that is, iso-K sections. Figure 14.16 reproduces the two isoquants of figure 14.14 for $Q = 100$ and $Q = 60$. Now suppose we take a section through the surface with the capital input fixed at $K = K_3$. The result is the section K_3EG, which is parallel to the $0QL$ plane. This section measures how output varies as the labour input varies with K constant at the level K_3. The equation of the section K_3EG is found by substituting $K = K_3$ into the production function $Q = f(K, L)$, giving $Q = f(K_3, L)$, where the variables are Q and L, and K_3 is the fixed capital input. We can repeat this with K fixed at $K = K_4$ and obtain the section K_4DA. We can derive as many of these curves as we wish by taking different fixed values of K.

The two iso-K sections K_3EG and K_4DA are projected on to the $0QL$ plane in figure 14.17a. Iso-K sections such as K_3EG and K_4DA are known as **short-run production functions**. They are so named because it is reasonable to assume that a firm cannot increase its capital input quickly, as to do so new machines have to be ordered, more office and workshop space has to be bought

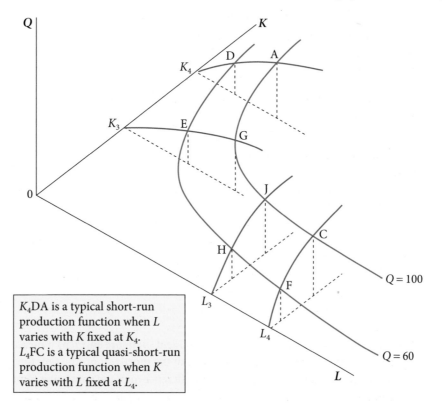

K_4DA is a typical short-run production function when L varies with K fixed at K_4.
L_4FC is a typical quasi-short-run production function when K varies with L fixed at L_4.

Figure 14.16 Short-run and quasi-short-run production functions.

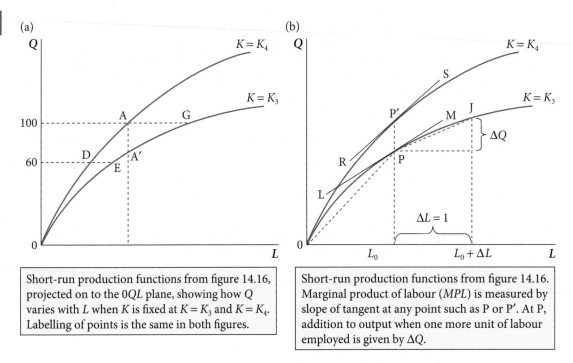

Short-run production functions from figure 14.16, projected on to the 0QL plane, showing how Q varies with L when K is fixed at K = K₃ and K = K₄. Labelling of points is the same in both figures.	Short-run production functions from figure 14.16. Marginal product of labour (MPL) is measured by slope of tangent at any point such as P or P′. At P, addition to output when one more unit of labour employed is given by ΔQ.

Figure 14.17 (a) Short-run production functions from figure 14.16; (b) measuring the marginal product of labour.

or rented, and so on—all of which inevitably takes time. For similar reasons the firm cannot reduce its capital input quickly either. Thus in the short run a firm might find itself with a capital input of, say, K_3 in figure 14.16 which it cannot vary. The firm's production possibilities are then restricted to the iso-K section K_3EG. Similarly if the firm's capital input were K_4, then its short-run production function would be the iso-K section K_4DA. It remains an open question how long the 'short run' is.

Note that in figure 14.17(a) the curve 0DA lies entirely above 0EG. This is because $K_4 > K_3$ (see figure 14.16). Thus the fixed capital input is higher along 0DA than along 0EG. This means that if we compare any two points with the same labour input, such as A and A′ in figure 14.17(a), output is larger at A because the capital input is larger. The same labour input is able to produce more output because it has more capital equipment to assist it. This reflects neoclassical assumption (3) in section 14.7, that having more K will always increase output (for any given L, that is).

The marginal product of labour

The *marginal product of labour* (MPL) is often defined as the increase in output that results when one more unit of labour is employed, with the capital input constant. For example, consider figure 14.17(b), which repeats the short-run production functions from figure 14.17(a). From an initial position at P, if the labour input is increased by $\Delta L = 1$ unit, production moves up the short-run production function to J and thus $MPL = \Delta Q$. However since $\Delta L = 1$, we can write this as

$$MPL = \frac{\Delta Q}{\Delta L}$$

Here, the right-hand side is the difference quotient. In figure 14.17(b) it measures the slope of the chord PJ, which equals the average slope of the short-run production function between P and J. As in earlier discussions, we find the difference quotient somewhat imprecise because it measures only the average slope of the curve, and because its value depends on the size of ΔL (when we hire 1 more unit of labour, is this one worker for 1 hour or one worker for 1 year?).

For these reasons, as we have already seen in other situations, economists almost invariably replace the difference quotient with the derivative of the function in question. In this case, since we are looking at an iso-K section through the production function, along which K is of course constant, the relevant derivative is the partial derivative $\frac{\partial Q}{\partial L}$. Thus we have the rule:

RULE 14.5a The marginal product of labour

For any production function $Q = f(K, L)$, the marginal product of labour (MPL) is defined as:

$$MPL = \frac{\partial Q}{\partial L}$$

It measures the slope of the short-run production function (that is, with K fixed).

Thus the MPL is given by the slope of the tangent to the short-run production function. At P in figure 14.17(b) the slope of the tangent LPM gives us the MPL at that point.

Because of assumption (3) in section 14.7 above, that an increase in the labour input always increases output, $\frac{\partial Q}{\partial L}$ is always positive: that is, the short-run production function is always upward sloping, as drawn in figure 14.17.

The 'law' of diminishing marginal productivity

We are now ready to explain assumption (4) in section 14.7 above, the law of **diminishing marginal productivity**. (Note that 'marginal productivity' and 'marginal product' are used interchangeably.) As applied to labour, the law states that, as we repeatedly increase the labour input by one (small) unit with the capital input held constant, then the successive increases in output become smaller and smaller. Thus we have a diminishing or decreasing marginal product of labour ($DMPL$).

The underlying reason for this is as follows. The capital input is constant, by assumption. Suppose an additional worker is hired. Then, given neoclassical assumption (2) above, that a worker can produce nothing without some capital to help her, some capital equipment must be taken away from existing workers and given to her. As existing workers now have less capital to assist their efforts, their output must fall, and this loss must be subtracted from the extra output produced by the newly hired worker to arrive at her (net) marginal product. As more and more workers are hired, this effect becomes stronger because capital per worker becomes smaller and smaller. However, $DMPL$ may not be present at low levels of output. This possibility is examined in section 14.10.

In figure 14.17, $DMPL$ is shown by the fact that the slope of the short-run production function decreases as L increases. For example, in figure 14.17(b), if we drew a tangent at J it would be flatter than the tangent drawn at P. In terms of partial derivatives, $DMPL$ means that $\frac{\partial Q}{\partial L}$ decreases as L increases (with K constant). And if $\frac{\partial Q}{\partial L}$ decreases as L increases, this implies that $\frac{\partial^2 Q}{\partial L^2}$ is negative. (See section 14.6 if you need to revise second derivatives.)

Finally, a feature of figure 14.17(b) that may have caught your eye is that the tangent RS at P′ is steeper than tangent LM at P; that is, $\frac{\partial Q}{\partial L}$ is higher at P′ than at P. Assuming P′ is very close to P, this means that a small increase in K (with L constant) increases $\frac{\partial Q}{\partial L}$. This implies that the cross partial derivative, $\frac{\partial^2 Q}{\partial K \partial L}$, is positive. It is not strictly necessary that the tangent at P′ must be steeper than at P, whatever the position of P. But it must be true at some points at least; otherwise the short-run production function for $K = K_4$ could not lie above that for $K = K_3$ (since they both start from the origin).

Confusion is possible over two points. We have found that, comparing P′ with P in figure 14.17(b), (1) Q is larger and (2) the MPL is higher. These two are logically distinct. The first is true simply because we assume that the marginal product of K is positive, so that the increase in K (with L constant) that causes the movement from P to P′ necessarily increases output. The second depends on the sign of the cross partial derivative, as just discussed, and economic theory gives us no strong reasons for expecting this sign to be invariably positive at any point on any short-run production function. But, as noted in the previous paragraph, it must be true at *some* points; otherwise one curve could not lie above the other, as we know it does.

The average product of labour

The average product of labour (*APL*) is also important. This is defined, quite straightforwardly, as $\frac{Q}{L}$ and is also known as output per worker or labour productivity. Note that, since Q and L are measured in physical units, $\frac{Q}{L}$ is also in physical units. For example, labour productivity in coal mining is commonly measured in tonnes per worker per shift. (Measuring output per worker is more difficult if a firm or industry produces many different types of output and employs many different types of labour. We avoid these complications here, by assuming that there is only one type of output and only one type of labour.)

In figure 14.17(b), $\frac{Q}{L}$ at P is measured by the slope of the straight line 0P from the origin (known as a ray). This is true because the slope of this ray is given by $\frac{L_0P}{0L_0}$. Since the distance L_0P measures the level of output at P, and the distance $0L_0$ measures the labour input, we have $\frac{L_0P}{0L_0} = \frac{Q}{L}$, the *APL*. (Note that this way of measuring the average product of labour parallels exactly the way in which we measured average cost in section 8.3, figure 8.1.)

From figure 14.17(b), two key characteristics of the *APL* may be seen. First, the *MPL* is always less than the *APL*. This may be seen at point P, where the slope of the tangent LM (*MPL*) is flatter than the slope of the ray 0P (*APL*). It is easy to see that this remains true wherever we place point P.

The second characteristic of the *APL* is that it falls as output increases. This may also be seen in figure 14.17(b), where the slope of the ray 0P, which measures *APL*, becomes flatter (less steep) as P moves up the short-run production function.

However, note that these properties, of *MPL* < *APL* at every output level, and *APL* falling continuously as output increases, are not found in every production function that conforms to the neoclassical assumptions listed in section 14.7. Shortly we will examine a production function in which at low output levels *MPL* > *APL*, and hence *APL* rises as output increases.

The quasi-short-run production function

As we have just seen, a short-run production function shows how Q varies with L when K is constant. In a completely symmetrical way, we can also examine how Q varies with K when L is constant. For example, in figure 14.16 we can take a section through the surface with the labour input fixed at, say, $L = L_3$. The result is the iso-L section L_3HJ, which is parallel to the 0QK plane. Along this curve, labour input is constant so the curve measures how output, Q, varies as capital input, K, alone varies. The equation of the section L_3HJ is found by substituting $L = L_3$ into the production function $Q = f(K, L)$, giving $Q = f(K, L_3)$, where the variables are Q and K, and L_3 is the fixed labour input. Similarly, when we fix the labour input at $L = L_4$ we derive the section L_4FC as the relationship between Q and K with L constant at the level L_4. Obviously we can draw as many curves like L_3HJ and L_4FC as we wish by taking different fixed values of L. We can project the iso-L sections L_3HJ and L_4FC on to the Q,K plane as in figure 14.18.

We will describe these iso-L sections as quasi-short-run production functions. The word 'quasi' means 'as if, but not really'. A quasi-short-run production function thus describes the firm's production possibilities when L is fixed and K variable. We have used the word 'quasi' because it seems improbable that in the real world any firm in any industry could find itself in a short-run situation in which it could vary its capital input but at the same time its labour input was fixed. However, although improbable it is not impossible. It could happen if, say, all the firm's machines are rented on a daily basis from the local tool-hire shop (and therefore their number can quickly be increased or decreased), while employment protection laws require lengthy negotiations or advance notice before the number of workers or their hours can be changed.

The improbability of a firm finding itself, in the short run, able to vary its capital input but not its labour input does not mean that the quasi-short-run production function is irrelevant

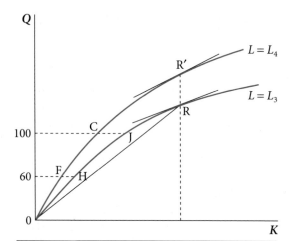

The quasi-short-run production function gives Q as a function of K, with L fixed. At R the marginal product of capital, MPK, is given by the slope of the tangent. The average product of capital, Q/L, is given by the slope of the ray 0R. Points F, C, H, and J link with figure 14.16.

Figure 14.18 Quasi-short-run production functions projected on to the 0QK plane.

to the real world. The quasi-short-run production function is highly relevant to a firm that is considering whether to install more machinery without changing the size of its workforce, as this function tells the firm how much extra output will result. Here the firm is choosing to treat its labour *as if* it were fixed, in order to focus on the effect on output of possible new machinery. However, as noted earlier, in the real world the installation of new machinery takes time, so the iso-L curves relate not to a truly short run but to an 'as if, but not really' short run; that is, a quasi short run.

The neoclassical production function assumes that labour and capital are completely symmetrical in their contributions to production. Thus the quasi-short-run production function has properties identical to those of the short-run production function. Therefore, rather than repeat the whole of our discussion above of the short-run production function, with the roles of K and L interchanged, we will simply state the key conclusions.

First, in figure 14.18, the 0FC curve lies entirely above the 0HJ curve. This is because the fixed labour input is L_4 along the 0FC curve, but L_3 along the 0HJ curve, and $L_4 > L_3$. Therefore any given capital input produces more output because there is more labour to work with it; for example, compare points R and R'.

Second, we can define the *marginal product of capital*, *MPK*, as $MPK = \frac{\partial Q}{\partial K}$. We state this as a new rule:

RULE 14.5b **The marginal product of capital.**

For any production function $Q = f(K, L)$, the marginal product of capital (MPK) is defined as

$$MPK = \frac{\partial Q}{\partial K}$$

It measures the slope of the quasi-short-run production function (that is, with L fixed).

Thus the MPK is given by the slope of a tangent drawn to any quasi-short-run production function, such as those in figure 14.18. The analogue of the diminishing marginal productivity of labour (*DMPL*) discussed above is diminishing marginal productivity of capital, *DMPK*. The underlying reason for *DMPK* is the same as that for *DMPL*: every time a new machine is installed, workers have to be taken away from other machines to operate it. Because of this, $MPK = \frac{\partial Q}{\partial K}$ decreases as we move up any quasi-short-run production function, such as those in figure 14.18, implying that $\frac{\partial^2 Q}{\partial K^2}$ is negative.

Third, a small increase in the fixed labour input, which takes production from, say, R to R' in figure 14.18, increases the *MPK*. Therefore the cross partial derivative $\frac{\partial^2 Q}{\partial L \partial K}$ is positive, at some points on the curve at least.

Finally, we can also define the average product of capital as $APK = \frac{Q}{K}$. This has the same properties as the average product of labour examined above. In figure 14.18 it is measured at R by the slope of the ray 0R from the origin. At any point such as R, we have $MPK < APK$, as is shown by the fact that the tangent at R has a flatter slope than the ray 0R.

Because we have skipped much of the supporting discussion, it is advisable to check that you understand why these properties hold, and why the various partial derivatives have their signs, before moving on. Rework fully the explanation above of the short-run production function, with capital and labour interchanged.

14.9 The Cobb–Douglas production function

A simple function that has a three-dimensional shape identical to that examined in the previous section has the general form

$$Q = AK^{\alpha}L^{\beta}$$

where the parameters A, α, and β are assumed positive. We shall see later in this chapter that it is also very likely, and arguably a logically necessity, that α and β are less than 1. This is called a Cobb–Douglas production function, named after two American economists, Cobb and Douglas, who discovered in the 1930s that a function of this form fitted actual data on firms' labour inputs, output, and other variables remarkably well.

Let's take a specific example by setting $A = 10$ and $\alpha = \beta = 0.5$, so we have

$$Q = 10K^{0.5}L^{0.5}$$

Isoquants and the *MRS*

Considering first the isoquants of this production function, the isoquant for, say, 100 units of output is found by substituting $Q = 100$ into the production function, giving

$$100 = 10K^{0.5}L^{0.5} \quad \text{which simplifies to } 10 = K^{0.5}L^{0.5}$$

This is an implicit function relating K and L. To rearrange it in explicit form, we can square both sides and get

$$10^2 = (K^{0.5}L^{0.5})^2 \quad \text{which simplifies to } 100 = KL, \text{ from which } K = \frac{100}{L} \text{ or } K = 100L^{-1}$$

This is the equation of the isoquant for $Q = 100$. The first derivative is $\frac{dK}{dL} = -100L^{-2} = -\frac{100}{L^2}$, which must be negative since L^2 is positive. Therefore the isoquant is negatively sloped. The second derivative is $\frac{d^2K}{dL^2} = \frac{200}{L^3}$, which is positive since L and therefore L^3 are positive. Therefore the isoquant is convex from below in the $0KL$ plane, reflecting a decreasing marginal rate of substitution.

Similarly, setting $Q = 60$ and following the steps above gives the isoquant for 60 units of output as

$$K = \frac{36}{L}$$

Hopefully you will recognize $K = \frac{100}{L}$ and $K = \frac{36}{L}$ as rectangular hyperbolas. When graphed, their shapes are similar to the isoquants in figure 14.16. Isoquants for other output levels can be obtained in the same way.

The short-run Cobb–Douglas production function

To obtain the short-run production function (that is, when the capital input is fixed) we set $K = 100$, say. In mathematical language we are taking an iso-K section through the production surface. The production function then becomes

$$Q = 10(100)^{0.5}L^{0.5} \quad \text{which simplifies to} \quad Q = 100L^{0.5}$$

Similarly, the short-run production function when K is fixed at, say, 49 is $Q = 70L^{0.5}$. The graphs of these two short-run production functions are shown in figure 14.19(a). (Check this for yourself, by sketching their graphs. For the graph of $Q = L^{0.5}$, see section 4.9.)

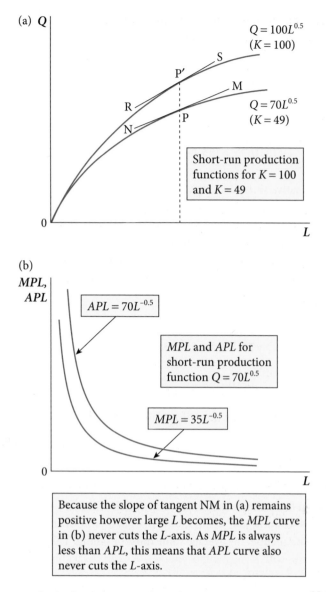

Figure 14.19 Short-run production functions, *MPL* and *APL*, for production function $Q = 10K^{0.5}L^{0.5}$.

The marginal product of labour

From the previous section we know that the marginal product of labour (*MPL*) is given by the slope of the tangent drawn to any short-run production function at any point; that is, $MPL = \frac{\partial Q}{\partial L}$. We can check this algebraically for the short-run production functions derived above. For example, we saw above that when the capital input is fixed at $K = 49$, the short-run production function is $Q = 70L^{0.5}$. So the MPL is given by

$$MPL = \frac{\partial Q}{\partial L} = \frac{\partial}{\partial L} \text{ of } 70L^{0.5} = 70(0.5)L^{-0.5} = \frac{35}{L^{0.5}} \text{ or } 35L^{-0.5} \tag{14.6}$$

This is, of course, a partial derivative because K is constant. The *MPL* is positive because L is positive and we know that a positive number raised to any power is positive (see section 2.12). Equation (14.6) gives us the *MPL* at any point on the short-run production function $Q = 70L^{0.5}$ when the value of L at that point is substituted in. (We already know that $K = 49$, of course, in this example.) Equation (14.6) is graphed in figure 14.19(b).

Diminishing marginal productivity

In figure 14.19(a), $DMPL$ is shown by the fact that the slope of the short-run production function decreases as L increases. In terms of partial derivatives, $DMPL$ means that $\frac{\partial Q}{\partial L}$ decreases as L increases (with K constant), which in turn means that $\frac{\partial^2 Q}{\partial L^2}$ is negative.

We can check this for the short-run production function $Q = 70L^{0.5}$ by finding the second derivative. This is

$$\frac{\partial^2 Q}{\partial L^2} = \frac{\partial}{\partial L} \text{ of } (35)L^{-0.5} = (35)(-0.5)L^{-1.5} = (-17.5)L^{-1.5}$$

which is negative since $L^{-1.5}$ is positive, for the same reason that $L^{-0.5}$ is positive in equation (14.6) above. (It may help to recall that $L^{-1.5} = (L^{-3})^{0.5} = 1/\sqrt{L^3}$.)

Thus in this example of the Cobb–Douglas production function, $\frac{\partial^2 Q}{\partial L^2}$ is negative (and therefore $DMPL$ operates) at *all* levels of output. This means that the MPL falls monotonically as L increases, confirming that the graph of the MPL in figure 14.19(b) is correctly drawn.

The average product of labour

The average product of labour (APL) is defined as $\frac{Q}{L}$. For the short-run production function we are considering, $Q = 70L^{0.5}$, the APL is

$$APL \equiv \frac{Q}{L} = \frac{70L^{0.5}}{L} = 70L^{-0.5} \tag{14.7}$$

It is interesting to compare the APL and the MPL. From equations (14.6) and (14.7) we have

$$MPL = \frac{\partial Q}{\partial L} = 35L^{-0.5} \quad \text{and} \quad APL \equiv \frac{Q}{L} = \frac{70L^{0.5}}{L} = 70L^{-0.5}$$

Since $35 < 70$, we have $MPL < APL$ for any value of L. More specifically, we have $\frac{MPL}{APL} = \frac{35L^{-0.5}}{70L^{-0.5}} = \frac{1}{2}$. Thus in this example the MPL is exactly half of the APL, for any given labour input. From section 9.24 we know that whenever the marginal value of any function is less than the average value of that function, the average value must be falling. This is indeed the case here: APL falls as L increases (with K constant, of course). We can check this by taking the partial derivative of APL; that is, $\frac{\partial}{\partial L}$ of APL. This is

$$\frac{\partial}{\partial L} \text{ of } 70L^{-0.5} = -35L^{-1.5}$$

which is negative (as $L^{-1.5}$ is positive, for reasons discussed immediately above). The APL for the short-run production function $Q = 70L^{0.5}$ is graphed in figure 14.19(b). It lies above the graph of MPL because $MPL < APL$ for any L.

Note that although MPL falls continuously, it remains positive however large L becomes. Falling MPL does not necessitate that it eventually become negative. Instead, it could approach the L-axis asymptotically, as here.

As a final feature of the Cobb–Douglas production function, we noticed in the previous section when discussing the short-run production function that in figure 14.17(b) the MPL, $\frac{\partial Q}{\partial L}$, was higher at P′ than at P. This implied that the cross partial derivative, $\frac{\partial^2 Q}{\partial K \partial L}$, which measures how the slope in the L direction changes when there is a small change in K, was positive.

We will now check that this is true for our example Cobb–Douglas production function $Q = 10K^{0.5}L^{0.5}$. We first find $MPL = \frac{\partial Q}{\partial L}$, treating K as a (multiplicative) constant. Given $Q = 10K^{0.5}L^{0.5}$, we have

$$\frac{\partial Q}{\partial L} = 10K^{0.5}(0.5)L^{-0.5} = 5K^{0.5}L^{-0.5}$$

The cross partial derivative is then found by differentiating this with respect to K, this time treating L as a (multiplicative) constant. Thus

$$\frac{\partial^2 Q}{\partial K \partial L} = \frac{\partial}{\partial K} \text{ of } 5K^{0.5}L^{-0.5} = 5(0.5)K^{-0.5}L^{-0.5}$$

$$= \frac{2.5}{K^{0.5}L^{0.5}}$$

which is positive since K and L are positive and a positive number raised to any power is positive (see section 2.12).

The quasi-short-run Cobb–Douglas production function

As discussed in the previous section, what we have called the quasi-short-run Cobb–Douglas production function consists of iso-L sections through the production function. These show how output varies as the capital input varies, with the labour input constant. The shape and slope of such sections would be of interest to a firm that was considering adding to its capital input without changing its labour input—though this is not a truly short-run choice.

In the case of our example, $Q = 10K^{0.5}L^{0.5}$, we can find a typical quasi-short-run production function by assigning a fixed value to L, say, $L = 64$. We then have $L^{0.5} = 8$, giving us $Q = 80K^{0.5}$ as the quasi-short-run production function. Similarly, if we set $L = 81$, we obtain $Q = 90K^{0.5}$. These curves when graphed have the same shape as the short-run production functions in figure 14.19a (but with K on the horizontal axis).

For $L = 64$, we have $Q = 80K^{0.5}$, so the marginal product of capital is found as

$$MPK \equiv \frac{\partial Q}{\partial K} = 40K^{-0.5}$$

The sign of the second derivative, $\frac{\partial^2 Q}{\partial K^2}$, tells us how MPK varies as K increases and we move up the quasi-short-run production function. We find that $\frac{\partial^2 Q}{\partial K^2} = -20K^{-1.5} = -\frac{20}{K^{1.5}}$, which is negative because $K^{-1.5}$ is positive. Therefore the quasi-short-run production function is concave from below: MPK decreases as K increases (with L constant). Thus we have diminishing marginal productivity of capital ($DMPK$).

Again for $L = 64$, we have $Q = 80K^{0.5}$, so the average product of capital is found as

$$APK \equiv \frac{Q}{K} = \frac{80K^{0.5}}{K} = 80K^{-0.5}$$

Comparing $MPK = 40K^{-0.5}$ and $APK = 80K^{-0.5}$, we see that MPK is always less than APK for any capital input, so APK is always decreasing as the capital input increases (with labour input fixed, of course). The marginal and average product curves have the same shape as the MPL and APL curves in figure 14.19(b).

The cross partial derivative $\frac{\partial^2 Q}{\partial L \partial K}$ tells us how the MPK changes when there is a small increase in the labour input, with capital input constant. To find this derivative for our example Cobb–Douglas production function $Q = 10K^{0.5}L^{0.5}$, we first find $\frac{\partial Q}{\partial K} = 10(0.5)K^{-0.5}L^{0.5} = 5K^{-0.5}L^{0.5}$. Note that we treated $L^{0.5}$ as a multiplicative constant in finding this derivative. Then we differentiate again, this time with respect to L, now with $K^{-0.5}$ as a multiplicative constant, to get

$$\frac{\partial^2 Q}{\partial L \partial K} = \frac{\partial}{\partial L} \text{ of } 5K^{-0.5}L^{0.5} = (2.5)K^{-0.5}L^{-0.5} = \frac{2.5}{K^{0.5}L^{0.5}}$$

This is positive, telling us that a small increase in the labour input, with capital input constant, increases the MPK. Earlier we saw this in figure 14.18 by comparing the slopes of the tangents at R′ and R.

The general Cobb–Douglas production function

It is worth briefly repeating the analysis above for the general form of the Cobb–Douglas production function, $Q = AK^\alpha L^\beta$, with A, α, and β positive. By setting Q equal to some fixed value, Q_0, we obtain the equation of an isoquant as

$$Q_0 = AK^\alpha L^\beta$$

Since Q is now a constant, this is an implicit function relating K and L. To rearrange this so as to make K the dependent variable, we divide both sides by AL^β and then raise both sides to the power $1/\alpha$, resulting in

$$(K^\alpha)^{1/\alpha} = \left(\frac{Q_0}{AL^\beta}\right)^{1/\alpha}$$

and, since the left-hand side equals simply K, we have

$$K = \left(\frac{Q_0}{AL^\beta}\right)^{1/\alpha} = \left(\frac{Q_0}{A}\right)^{1/\alpha} L^{-\beta/\alpha} \tag{14.8}$$

so K is now an explicit function of L. This is the equation of the isoquant for $Q = Q_0$.

We can differentiate equation (14.8) to find the slope of the isoquant as

$$\frac{dK}{dL} = -\frac{\beta}{\alpha}\left(\frac{Q_0}{A}\right)^{1/\alpha} L^{-(\beta/\alpha)-1}$$

This is negative because α and β are positive, and so are $(\frac{Q_0}{A})^{1/\alpha}$ and $L^{-(\beta/\alpha)-1}$ because Q, A, and L are assumed positive and a positive number raised to any power is positive (see section 2.12).

The second derivative is

$$\frac{d^2K}{dL^2} = \left(-\frac{\beta}{\alpha} - 1\right)\left(-\frac{\beta}{\alpha}\right)\left(\frac{Q_0}{A}\right)^{1/\alpha} L^{-(\beta/\alpha)-2}$$

Here $(\frac{Q_0}{A})^{1/\alpha}$ and $L^{-(\beta/\alpha)-2}$ are positive, as explained immediately above, while $(-(\beta/\alpha) - 1)$ and $(-\beta/\alpha)$ are both negative, hence their product is positive. So $\frac{d^2K}{dL^2}$ is positive.

We conclude that since the first derivative is negative and the second positive, the isoquant for any output Q_0 is downward sloping and convex from below.

Incidentally, you might think that all of the derivatives in this sub-section are partial derivatives (since Q is being held constant) and therefore $\frac{dK}{dL}$ should be written as $\frac{\partial K}{\partial L}$, and so on. However, Q is the *dependent* variable and a partial derivative arises only when there is one or more *independent* variables being held constant. This will become clearer in section 15.8 (see also rule 14.4).

The short-run production function

The equation of an iso-K section is found by setting K equal to some fixed value K_0. This gives $Q = AK_0^\alpha L^\beta$. Here output, Q, is a function of L alone, and this is therefore the equation of the short-run production function. The marginal product of labour is given by the partial derivative

$$\frac{\partial Q}{\partial L} = \beta AK_0^\alpha L^{\beta-1}$$

This is positive because β and A are positive, and K_0^α and $L^{\beta-1}$ are positive because K and L are positive (see section 2.12). The second derivative is

$$\frac{\partial^2 Q}{\partial L^2} = (\beta - 1)\beta AK_0^\alpha L^{\beta-2} \tag{14.9}$$

Here, β, A, K_0^α, and $L^{\beta-2}$ are all positive, so the sign of $\frac{\partial^2 Q}{\partial L^2}$ depends on the first term, $\beta - 1$. If β is less than 1, then $\beta - 1$ and therefore $\frac{\partial^2 Q}{\partial L^2}$ are negative. We may consider this to be the normal case because $\frac{\partial^2 Q}{\partial L^2}$ negative means that the law of diminishing marginal productivity is in operation.

If instead β is greater than 1, then $\beta - 1$ and therefore $\frac{\partial^2 Q}{\partial L^2}$ are positive at all levels of output. This means that $\frac{\partial Q}{\partial L}$, the marginal product of labour, increases as the labour input increases with the capital input held constant. This case must be ruled out because, as discussed in section 14.8, if capital and labour contribute *jointly* or *cooperatively* to output, then *MPL* must decline as the fixed quantity of capital has to be shared among an ever-increasing number of workers. The boundary case of $\beta = 1$ implies constant marginal productivity at all levels of output, which should be ruled out also for the same reason. It therefore seems reasonable to impose the restriction that $\beta < 1$.

However, a productive process in which *MPL* is constant or increasing over a certain range of output, but eventually decreases, is perfectly possible, and we discuss such a case in the next section. The Cobb–Douglas is too simple a functional form to be able to accommodate this subtlety.

The quasi-short-run production function

The results above for the iso-K sections hold equally for the iso-L sections of the Cobb–Douglas production function. By setting L equal to some fixed value L_0, we get $Q = AK^\alpha L_0^\beta$, which gives output as a function of capital input alone. The marginal product of capital is given by the partial derivative

$$\frac{\partial Q}{\partial K} = \alpha AK^{\alpha-1}L_0^\beta$$

This is positive for the same reasons that $\frac{\partial Q}{\partial L}$ was positive above. The second derivative is

$$\frac{\partial^2 Q}{\partial K^2} = (\alpha - 1)(\alpha)AK^{\alpha-2}L_0^\beta \tag{14.10}$$

By the same reasoning that we applied to $\frac{\partial^2 Q}{\partial L^2}$ above, this is positive or negative according to whether α is less or greater than 1. To make the Cobb–Douglas production function a plausible representation of the real world, we must impose the restriction that $\alpha < 1$. The reason for this was explained in our discussion of the sign of equation (14.9) above.

14.10 Alternatives to the Cobb–Douglas form

The Cobb–Douglas production function is a very simple functional form. The real world may be more complex. One possibility is that the short-run production function could have the shape shown in figure 14.20, where we have drawn the isoquants for three levels of output, Q_0, Q_1, and Q_2. By fixing the capital input at $K = K_0$ we obtain the short-run production function K_0PJ. Looking at this curve, we see that at low levels of labour input and output the slope (= *MPL*) is increasing rather than decreasing. The slope reaches a maximum at P with labour input L_0, and thereafter decreases. Thus *DMPL* begins to take effect at only relatively high levels of labour input and output. The mathematics of this short-run production function are that there is a point of inflection at point P, where the slope reaches its maximum and the sign of the second derivative changes from positive to negative as L increases.

One reason for this could be that, for technological reasons, the fixed stock of capital equipment cannot be operated efficiently when the labour input is very small. A machine may have been designed to be operated by a crew of three, say, and with only two operators it is very slow. Hiring extra workers then raises marginal productivity instead of lowering it. As more and

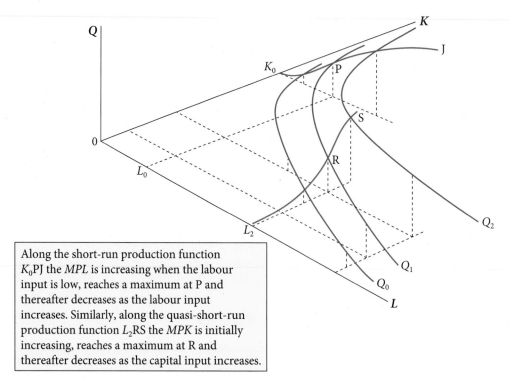

Along the short-run production function K_0PJ the MPL is increasing when the labour input is low, reaches a maximum at P and thereafter decreases as the labour input increases. Similarly, along the quasi-short-run production function L_2RS the MPK is initially increasing, reaches a maximum at R and thereafter decreases as the capital input increases.

Figure 14.20 A production function in which MPL and MPK are increasing when output is low.

more workers are hired, MPL continues to rise until it reaches a maximum at P in figure 14.20. Beyond this point $DMPL$ begins to take effect and the slope starts to decrease. Every machine now has its full crew and additional workers can contribute relatively little to increasing output.

The algebra of a production function with these characteristics is inevitably complex, so we will content ourselves with a purely graphical depiction, in figure 14.21, of the shapes of the MPL and APL curves implied by the production function in figure 14.20. Figure 14.21(a) shows the short-run production function, which is the section K_0PJ in figure 14.20. As already discussed, in figure 14.21(a) the MPL, measured by the slope of the tangent, is initially rising, reaches its maximum at P with labour input L_0 and thereafter decreases as L increases. The APL, measured by the slope of a line (ray) from the origin in figure 14.21(a), is also initially rising, reaches a maximum at N with labour input L_1 and thereafter decreases at L increases. Because the ray from the origin is tangent to the curve at N, MPL and APL are equal at this point. The MPL and APL are graphed in figure 14.21(b). Note that between P and N, the MPL is falling but the APL is rising, which seems at first sight counter-intuitive.

Another connection we can make here is between the firm's short-run production function and its short-run total cost function. Although we are not yet ready to make this link rigorously, you can probably see in general terms the link between the short-run production function in figure 14.21(a) and the slightly S-shaped total cost curve in section 8.20, figure 8.18. It is the fact that the MPL is rising initially in figure 14.21(b) that causes marginal cost to fall initially in figure 8.18(b). At higher levels of output, $DMPL$ starts to operate and marginal cost starts to rise.

This example illustrates that the law of $DMPL$ need not hold at all levels of output. However, it must eventually start to operate at sufficiently high levels of output, as more and more labour is employed to operate a fixed stock of capital equipment and the equipment has to be shared between more and more workers. For example, we can imagine workers having to queue up to use a certain machine. To put the point slightly differently, if the marginal product of labour (and therefore the average product) were constant irrespective of the amount of capital equipment per worker, then this would imply that capital contributed nothing to production, which is a logical contradiction if K appears in the specification of the production function. So the law

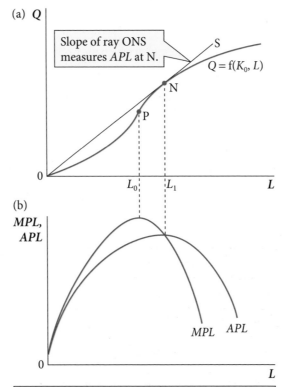

On the short-run production function in (a), $MPL \equiv \partial Q/\partial L$ initially increases, reaches its maximum at P (a point of inflection), and then decreases. $APL \equiv Q/L$ is measured at any point on the short-run production function by the slope of a ray from the origin to that point. Ray 0NS is the steepest such ray that can be drawn to the production function; therefore APL is at its maximum at N.

Because 0NS is tangent to the curve at N, the slope of the tangent (= MPL) equals the slope of a ray from the origin (= APL), so $MPL = APL$ at N.

Figure 14.21 Short-run production function, MPL and APL from figure 14.20.

of $DMPL$ is a law in the sense of being a logical necessity; although, because it need not hold at all levels of output, the use of the word 'law' is perhaps slightly misleading.

Finally, referring once again to figure 14.20, by fixing the labour input at $L = L_2$ we obtain the quasi-short-run production function L_2RS. This has a shape very similar to the short-run production function K_0PJ, and thus we can immediately infer that the marginal product of capital is initially increasing as capital input and output increase, reaching a maximum at R and thereafter declining. Thus the graphs of the marginal and average products of capital (MPK and APK) have shapes like those of the MPL and APL in figure 14.21(b). (In figure 14.20, it is purely for diagrammatic simplicity that we have placed P and R on the same isoquant.)

 Hint The concept of diminishing marginal productivity (of labour, or of capital), in the sense discussed above, is also sometimes referred to as 'diminishing returns'. This terminology is best avoided as it invites confusion with diminishing returns to scale, a very different idea which is discussed in chapter 17 below.

We return to the production function in chapter 15.

Summary of sections 14.7–14.10

The neoclassical production function, of which the Cobb–Douglas form is the leading example, assumes the following:

(1) Isoquants are downward sloping and, due to decreasing *MRS* in production, are convex.

(2) $\frac{\partial Q}{\partial L}$, the marginal product of labour (MPL), and $\frac{\partial Q}{\partial K}$, the marginal product of capital (*MPK*), are always positive. In words, an increase in the capital input with the labour input constant always increases output, and an increase in the labour input with the capital input constant also always increases output.

(3) The direct second derivatives $\frac{\partial^2 Q}{\partial L^2}$ and $\frac{\partial^2 Q}{\partial K^2}$ are negative, at least at sufficiently high levels of output. When $\frac{\partial^2 Q}{\partial L^2}$ is negative this means that an increase in the labour input (with K constant) reduces the *MPL*. We call this reduction diminishing marginal product of labour (*DMPL*). This case is of interest because it relates to the firm's short-run production possibilities. When $\frac{\partial^2 Q}{\partial K^2}$ is negative, an increase in the capital input (with L constant) reduces the *MPK* and the law of *DMPK* operates. This case is of interest to a firm considering whether to invest; that is, increase its capital input without increasing its labour force. In the case of the Cobb–Douglas production function $\frac{\partial^2 Q}{\partial L^2}$ and $\frac{\partial^2 Q}{\partial K^2}$ are negative at all levels of output provided α and β are each less than 1. We argued that diminishing marginal products of K and L, and therefore α and β being less than 1, was a reasonable and even logically necessary assumption.

(4) The cross partial derivatives $\frac{\partial^2 Q}{\partial K \partial L}$ and $\frac{\partial^2 Q}{\partial L \partial K}$ may be positive or negative, but in the case of the Cobb–Douglas production function $Q = AK^{\alpha}L^{\beta}$ they are positive for all values of K and L provided α and β are each less than 1. We argued that α and β are less than 1 was a reasonable and even logically necessary assumption. When both cross partials are positive, this means that an increase in the capital input (with L constant) increases the *MPL*, and an increase in the labour input (with K constant) increases the *MPK*.

In section 14.10 we considered a production function that was more complex than the simple Cobb–Douglas form, and which had marginal products that were increasing at low levels of output but decreasing at higher levels.

14.11 Economic applications 2: the utility function

A utility function is a relationship between the quantities of various goods and services that a person consumes, and the resulting satisfaction or utility that the person enjoys.

Assuming for simplicity that there are only two goods, we can write the consumer's utility function in mathematical form as

$U = f(X, Y)$

where U = some measure of the consumer's satisfaction or utility, X and Y are the quantities of the two goods consumed, and f(X, Y) is some functional form. Note that U, X, and Y are all flows through time, so for example if good X is beer, then it is measured in pints consumed per week.

An immediate problem is that the units in which utility is measured, and the zero point from which it is measured, cannot be uniquely specified. A consumer may be able to *compare* utilities because such comparisons do not require specifying any units for U. Thus a consumer could say that she derives equal satisfaction from eating either three bananas and two apples, or four bananas and one apple. She might also say that she prefers four bananas and two apples to three bananas and three apples. These comparisons of bundles (combinations) of goods require only what is called an *ordinal* measure of utility: that is, the ability to rank bundles of goods according to their relative desirability. But a consumer would be met with puzzled looks if she said that she derived two units of satisfaction from eating one apple and two bananas, but only 1.5 units from eating two apples and one banana. To be able to make such a statement requires what is called a *cardinal* measure of utility. There exist some advanced theories that attempt to measure utility on a cardinal scale, but we will not consider them here.

14.12 The shape of the utility function

The neoclassical version of the utility function makes assumptions analogous to those of the neoclassical production function (see section 14.7 above):

(1) U, X, and Y are infinitely divisible, and the utility function is smooth and continuous.

(2) An increase in either X or Y will always increase U. In other words, a little more of either good always increases utility. This property is called **non-satiation**.

(3) The utility function is characterized by a decreasing marginal rate of substitution (*MRS*) in consumption. We explain this below.

Because these assumptions are so closely analogous to those underlying the neoclassical production function, they result in a utility function that has the same general shape and mathematical properties as the production function we have just considered (though of course the two address very different areas of economic analysis). Consequently, with suitable relabelling of axes (replacing Q with U, and K and L with X and Y), the surface that we used in figure 14.13 to represent the production function can be used to depict the utility function: see figure 14.22.

However, we must keep in mind that the scale on the U-axis measures only the consumer's *ranking* of utilities. Thus a utility level of 40 is higher than a utility level of 20, but cannot meaningfully be said to be twice as high. Such a scale is called an ordinal scale.

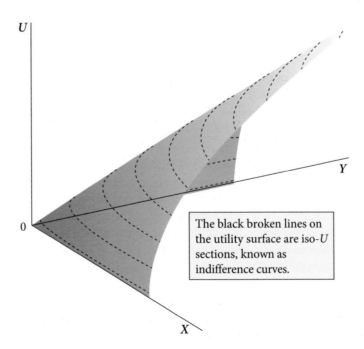

The black broken lines on the utility surface are iso-U sections, known as indifference curves.

Figure 14.22 A plausible shape for the utility function $U = f(X, Y)$.

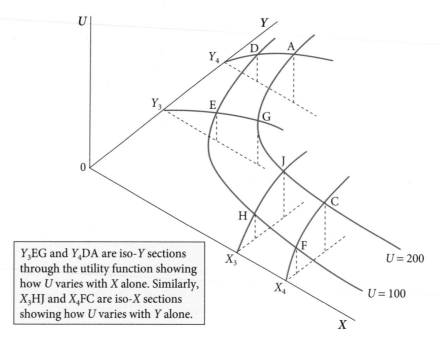

Y_3EG and Y_4DA are iso-Y sections through the utility function showing how U varies with X alone. Similarly, X_3HJ and X_4FC are iso-X sections showing how U varies with Y alone.

Figure 14.23 Iso-X and iso-Y sections through the utility function.

Indifference curves

The black broken lines on the utility surface in figure 14.22 are parallel with the $0XY$ plane and are therefore iso-U sections. An iso-U section identifies combinations of the goods X and Y such that the consumer's utility ranking is the same for all combinations. These iso-U sections are called **indifference curves**. The equation of a typical indifference curve is $U_0 = f(X, Y)$, where U_0 is the fixed level of utility, and is thus an implicit function relating Y and X.

Two indifference curves from figure 14.22, for utility levels $U = 100$ and $U = 200$, are shown as a three-dimensional sketch in figure 14.23, and have been projected on to the $0XY$ plane in figure 14.24. The consumer constructs an indifference curve in her mind by identifying combinations of goods X and Y which yield equal satisfaction and which she is therefore indifferent between. As noted above, the fact that no meaningful units exist with which to measure satisfaction is not a barrier to *comparing* the satisfaction derived from different bundles of X and Y, as a comparison requires only a ranking of utilities. An important assumption about consumer preferences is that they are **transitive**. This means that if the consumer prefers combination of goods X_1, Y_1 to combination X_2, Y_2 (or is indifferent between them), and also prefers combination X_2, Y_2 to combination X_3, Y_3 (or is indifferent between them), then the consumer must prefer combination X_1, Y_1 to combination X_3, Y_3 (or must be indifferent between them). For more detail, see any introductory microeconomics textbook.

Because there are no unique units with which to measure utility, the calibration of the U-axis is arbitrary. In figure 14.24 we have arbitrarily assigned the number 200 to the level of utility on the indifference curve AGJC, and 100 to indifference curve DEHF. These numbers have no meaning in themselves; they could equally be 2.5 and 1.9 respectively. However, we can be sure that the consumer's utility is higher along AGJC than along DEHF. This is because, if we pick any point on DEHF, we can always find a point on AGJC where the quantities consumed of *both* goods are higher, which ensures that utility is higher by virtue of assumption (2) above. And having shown that there is one point on AGJC that is better than one point on DEHF, it immediately follows that every point on AGJC is better than every point on DEHF, given transitivity of preferences and because every point on AGJC has the same utility, as does every point on DEHF. But we can't say *how much* higher utility is on AGJC compared with DEHF because we cannot measure utility on a cardinal scale.

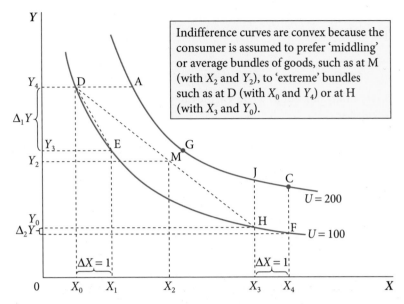

Figure 14.24 Indifference curves for $U = 100$ and $U = 200$, from figure 14.23, projected on to $0XY$ plane.

Assumption (2) also ensures that the indifference curves have a negative slope. For example, in moving along indifference curve AGJC from A to C, consumption of good Y has to be reduced to compensate for the increase in consumption of good X, since otherwise utility would rise. Thus we have the idea of *substitution* in consumption between the two goods. Note also that we can create an indefinitely large number of indifference curves because the quantities of X and Y are assumed infinitely divisible and we assume that the consumer can rank all conceivable bundles in order of **preference**. Assumption (2) above also means that indifference curves cannot cut one another (though this is not immediately obvious and you may need to think it through).

Finally, an important feature of the utility function shown in figures 14.22–14.24 is that the indifference curves are convex from below (see section 7.13 if you need to refresh convexity). We shall see in section 16.11 that some problems arise if indifference curves are not convex, and for this reason a utility function with convex indifference curves is often described as 'well behaved'. Convexity results from assumption (3) above, of a decreasing marginal rate of substitution (*MRS*) in consumption, as we will now show.

In a way that parallels the *MRS* in production discussed in the previous section, the *MRS* in consumption is usually defined as the amount of additional Y that a consumer is willing to give up in exchange for one additional unit of X, while maintaining a constant level of utility. In figure 14.24, we see that from an initial position at D, if the consumer is given one more unit of X she is willing to give up $\Delta_1 Y$ units of Y in exchange and thus moves to E with her utility unchanged. Her *MRS* is therefore given by the distance $\Delta_1 Y$. However, if the same experiment is conducted from an initial position at H, the consumer moves to F and is willing to give up only $\Delta_2 Y$ units of Y. As this is clearly much smaller than $\Delta_1 Y$, we see that her *MRS* decreases as X increases.

Let us examine more closely the link between the *MRS* and the slope of the indifference curve. Looking at point D in figure 14.24, for example, the *MRS* is $\Delta_1 Y$. But since by definition $\Delta X = 1$, we can write the *MRS* as $\frac{\Delta_1 Y}{\Delta X}$. We can do this anywhere on the curve, so the subscript 1 can be dropped, giving us, in general,

$$MRS = \frac{\Delta Y}{\Delta X} \quad \text{(evaluated along any indifference curve; that is, with } U \text{ held constant)}$$

We immediately recognize the right-hand side of this equation as the difference quotient, with ΔX set equal to 1. At D, it measures the slope of the chord DE, which equals the average slope of the indifference curve between D and E. As we have already seen in several contexts, the difference quotient is imprecise measure; hence in theoretical economic analysis it is normally replaced by the derivative of the indifference curve. Thus we have a new rule:

RULE 14.6 The marginal rate of substitution in consumption

For any utility function $U = f(X, Y)$, the marginal rate of substitution (MRS) in consumption is defined as

$$MRS = \frac{dY}{dX} \text{ (evaluated with } U \text{ held constant)}$$

The MRS measures the slope of the indifference curve at any point.

Note that, in rule 14.6, $\frac{dY}{dX}$ is not, mathematically, a partial derivative, even though it is evaluated with another variable, Q, held constant. This is because a partial derivative is a relationship between the *dependent* variable and one of the independent variables, whereas $\frac{dY}{dX}$ is a relationship between two independent variables. This will become clearer in section 15.8.

In examining the production function in section 14.8 we noted that confusion can arise in discussing the MRS in production. The same problem arises here concerning the MRS in consumption. Since the indifference curve is downward sloping, its convexity means that the slope increases (becomes less negative) as X increases. For example, in figure 14.24 the slope is about -3 at D but *increases* to about -0.25 at H. However, as with the slope of the isoquant discussed in section 14.8 above, economists invariably ignore the minus sign and consider only the absolute value of the slope. We will follow this convention, defining the MRS as the slope of the indifference curve, but in discussion referring to its absolute value. So we will say that the MRS *decreases* from about 3 to about 0.25 between D and H.

What pattern of consumer tastes is implied by a decreasing MRS? What is implied is that the consumer derives relatively little satisfaction from 'extreme' bundles: that is, combinations that comprise relatively little of one good and a relatively large amount of the other. To say this in a different way, the consumer prefers average or 'middling' bundles (combinations) of X and Y to extreme bundles. This can be seen in figure 14.24. If the consumer were given Y_2 units of good Y, where Y_2 is the average of Y_0 and Y_4; together with X_2 units of good X, where X_2 is the average of X_0 and X_4, (that is, a 'middling' bundle), she would then be at point M, which puts her on a higher indifference curve (not shown in figure 14.24) than she reached with the 'extreme' bundles at D and H. Only if the indifference curve were linear rather than convex would points D, M, and H all lie on the same indifference curve.

Iso-X and iso-Y sections

We can also take sections through the utility function for fixed values of X or Y, as in figure 14.23. The sections X_3HJ and X_4FC are iso-X sections, and show how utility varies with consumption of good Y when the quantity of good X consumed is fixed at X_3 and X_4 respectively. Similarly, the sections Y_3EG and Y_4DA are iso-Y sections, and show how utility varies with consumption of good X when the quantity of good Y consumed is fixed at Y_3 and Y_4 respectively. These iso-X and iso-Y sections can be projected on to the $0UY$ and $0UX$ planes, as in figure 14.25. The labelling of points in figures 14.23 and 14.25 is the same.

Note that the section for $X = X_4$ in figure 14.25(a) lies entirely above the section for $X = X_3$. This is because $X_4 > X_3$, so for any given quantity of Y the consumer has more X on the $X = X_4$ section than on the $X = X_3$ section and is therefore better off. For the equivalent reason, in figure 14.25(b) the $Y = Y_4$ section lies entirely above the $Y = Y_3$ section.

Marginal utility

This is often defined in introductory textbooks as the increase in utility that occurs when the consumer is given one more unit of one of the goods, with the quantity of the other good held constant. For example, at M in figure 14.25(a) the marginal utility of Y equals the distance ΔU.

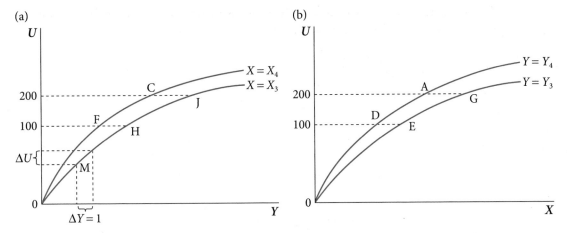

Figure 14.25 Iso-X and iso-Y sections from the utility function in figure 14.23, projected on to the $0UX$ and $0UY$ planes (not to scale).

However, since $\Delta Y = 1$ by definition, this equals the difference quotient $\frac{\Delta U}{\Delta Y}$. In the same way, we can measure the marginal utility of X by the difference quotient $\frac{\Delta U}{\Delta Y}$.

Because the difference quotient lacks precision (as already discussed at several points in this book), economists normally use the partial derivatives $\frac{\partial U}{\partial X}$ and $\frac{\partial U}{\partial Y}$ to measure the **marginal utilities** of X and Y. We will state this as a rule:

>
>
> **RULE 14.7 Marginal utilities**
>
> For any utility function $U = \text{f}(X, Y)$, the marginal utilities of goods X and Y, MU_X and MU_Y, are defined as the partial derivatives $\frac{\partial U}{\partial X}$ and $\frac{\partial U}{\partial Y}$ respectively.
>
> Therefore MU_X is measured by the slope of the tangent to any iso-Y section, and MU_Y is measured by the slope of the tangent to any iso-X section.

So for example at M in figure 14.25(a) the slope of the tangent (not drawn) is the measure of the marginal utility of Y at that point.

It's important to note that, because the units of measurement of utility are arbitrary, no significance can be attached to the numerical values of the marginal utilities, whether these be defined as difference quotients or partial derivatives. We know only that the marginal utilities are positive, reflecting assumption (2) above, non-satiation.

Diminishing marginal utility

As figure 14.25 is drawn, the slopes of the iso-X and iso-Y sections decrease as X and Y respectively increase. Since these slopes are measured by the derivatives $\frac{\partial U}{\partial X}$ and $\frac{\partial U}{\partial Y}$, a decreasing slope means that the direct second order derivatives, $\frac{\partial^2 U}{\partial X^2}$ and $\frac{\partial^2 U}{\partial Y^2}$, are negative. This decreasing slope is linked to the concept of **diminishing marginal utility** and is an idea almost as old as the study of economics itself. Note that in a mathematical sense it is directly analogous to diminishing marginal productivity in the production function (see section 14.8 above).

The underlying idea is that as the individual increases her consumption of one good, with all others held constant, then each successive unit consumed yields a smaller increase in utility than did the preceding unit. An example might be Imelda Marcos, a politician's wife who was widely reported to have owned 3000 pairs of shoes (though she only admitted to 1000). It may seem intuitively plausible that when her stock of shoes increased from 2999 to 3000 pairs, the increase in her utility was smaller than when her stock increased from 1 pair to 2. But this cannot be proved rigorously, and it is perfectly possible that Mrs Marcos derived equal satisfaction from both increases. It is even possible that the marginal utility of the 3000th pair of shoes was actually greater than the marginal utility of the second pair—a case of an 'appetite that grows with eating'.

Thus it remains an open question whether people do in fact experience diminishing marginal utility from a good as their consumption of it increases (with consumption of other goods constant). Moreover, whether they do or do not, the fact that utility can only be measured in arbitrary units means that no significance can be attached to the concavity of the iso-X and iso-Y sections in figure 14.25. We could easily choose a different set of units of measurement on the U-axis which would make these sections linear, concave, or convex. Therefore the signs of the direct second derivatives $\frac{\partial^2 U}{\partial X^2}$ and $\frac{\partial^2 U}{\partial Y^2}$ have, in themselves, no economic significance.

We will return to marginal utilities in section 15.12.

14.13 The Cobb–Douglas utility function

Because in neoclassical economics the utility function is assumed to have the same general shape as the production function, the Cobb–Douglas functional form that we used earlier as a production function can serve as a utility function too (though the original research by Cobb and Douglas was concerned solely with production functions). The general form of the Cobb–Douglas utility function is

$$U = X^\alpha Y^\beta$$

where α and β are positive parameters. Notice that the multiplicative constant that we saw in the Cobb–Douglas production function is redundant here, because the units in which U is measured are arbitrary.

Let us consider a specific case, by setting $\alpha = \beta = 0.5$. Then we have

$$U = X^{0.5} Y^{0.5}$$

Indifference curves

To find the equation of the indifference curve for, say, 10 units of utility we simply substitute $U = 10$ into the utility function, which gives

$$10 = X^{0.5} Y^{0.5}$$

If we then square both sides of this equation, it becomes

$$100 = (X^{0.5} Y^{0.5})^2$$

and after multiplying out the brackets this becomes

$$100 = XY \quad \text{from which} \quad Y = \frac{100}{X}$$

In the same way we can show that the equation of the indifference curve for $U = 20$ is

$$Y = \frac{400}{X}$$

So these are the equations of the indifference curves for $U = 10$ and $U = 20$, respectively. Hopefully you recognize them as rectangular hyperbolas (see section 5.5 if you don't). When graphed, they have a shape broadly the same as the indifference curves in figures 14.23 and 14.24 (but check this for yourself).

As discussed earlier, a 'well-behaved' utility function has indifference curves that are downward sloping and convex from below. This means that the first derivative must be negative and the second derivative positive. Let us check that this is the case for these two indifference curves. Taking the indifference curve for $U = 20$ (that is, $Y = \frac{400}{X}$), we can write this as $Y = 400X^{-1}$, so

$$\frac{dY}{dX} = -400X^{-2} = -\frac{400}{X^2}$$

which is negative since X^2 is positive. This derivative, of course, measures the marginal rate of substitution (see rule 14.6).

Also, by differentiating again, we have

$$\frac{d^2Y}{dX^2} = 800X^{-3} = \frac{800}{X^3}$$

which is positive since X and therefore X^3 are positive. Therefore the indifference curve is convex from below, as drawn in figure 14.24.

Iso-X and iso-Y sections

The equation of an iso-X or iso-Y section like those in figure 14.25 can be found by fixing X or Y equal to some constant value. For example, the iso-Y section for $Y = 100$ is found by substituting $Y = 100$ into the utility function $U = X^{0.5}Y^{0.5}$. This gives

$$U = X^{0.5}100^{0.5} = 10X^{0.5}$$

This equation shows how utility, U, varies with consumption of X with Y held constant at $Y = 100$.

To find the slope of this iso-Y section, $U = 10X^{0.5}$, we take the partial derivative $\frac{\partial U}{\partial X}$, which is

$$\frac{\partial U}{\partial X} = \frac{\partial}{\partial X} \text{ of } 10X^{0.5} = (0.5)10X^{-0.5} = \frac{5}{X^{0.5}}$$

This is positive as required by the assumption of non-satiation. (It is positive because X is positive and any positive number raised to any power is also positive.) Note that it is a partial derivative because Y is held constant at $Y = 100$. To see whether there is diminishing marginal utility we take the direct second order partial derivative, $\frac{\partial^2 U}{\partial X^2}$. This is

$$\frac{\partial^2 U}{\partial X^2} = \frac{\partial}{\partial X} \text{ of } \frac{5}{X^{0.5}} = \frac{\partial}{\partial X} \text{ of } 5X^{-0.5} = (-2.5)X^{-1.5} = -\frac{2.5}{X^{1.5}}$$

This is negative because $X^{1.5}$ is positive. So this particular utility function has the property of diminishing marginal utility with respect to X. However, as explained above, this is an illusion as the signs of the second derivatives depend on the calibration of the U-axis, which is inherently arbitrary.

Repeating the steps above with X fixed and Y variable gives the same results: $\frac{\partial U}{\partial Y}$ positive and $\frac{\partial^2 U}{\partial Y^2}$ negative.

In the same way, we can get the iso-X section for $X = 64$ (say) by setting $X = 64$ in the utility function. We get

$$U = 8Y^{0.5} \quad \text{(Check this for yourself.)}$$

The graphs of these iso-Y and iso-X sections have the same general shape as those in figure 14.25. (See section 7.13 if you are uncertain about how the shapes of these curves can be inferred from the signs of the first and second derivatives.)

The general Cobb–Douglas utility function

It is worth repeating the analysis above for the general form of the Cobb–Douglas utility function, $U = X^{\alpha}Y^{\beta}$ (with α and β both positive). We can go quickly through this because it closely parallels our discussion of the general Cobb–Douglas production function in the previous section. First, by fixing utility at some level U_0 we obtain the equation of an indifference curve as $U_0 = X^{\alpha}Y^{\beta}$. This is an implicit function relating Y and X, and can be rearranged as the explicit function

$$Y = \left(\frac{U_0}{X^{\alpha}}\right)^{1/\beta} = U_0^{1/\beta}X^{-\alpha/\beta}$$

The slope of the indifference curve is therefore

$$\frac{dY}{dX} = -\frac{\alpha}{\beta}U_0^{1/\beta}X^{-(\alpha/\beta)-1} = -\frac{\alpha}{\beta}\frac{Y}{X} \quad \text{(since } Y = U_0^{1/\beta}X^{-\alpha/\beta})$$

which is negative since α and β are assumed positive, so the indifference curves are downward sloping. The second derivative is

$$\frac{d^2Y}{dX^2} = \left(-\frac{\alpha}{\beta}-1\right)\left(-\frac{\alpha}{\beta}\right)(U_0)^{1/\beta}X^{-(\alpha/\beta)-2} = \left(-\frac{\alpha}{\beta}-1\right)\left(-\frac{\alpha}{\beta}\right)\frac{Y}{X^2}$$

which is positive since α and β are assumed positive, so the indifference curves are convex.

From $U = X^\alpha Y^\beta$ we get the marginal utility of good X as $\frac{\partial U}{\partial X} = \alpha X^{\alpha-1}Y^\beta$, which is positive. The second derivative is

$$\frac{\partial^2 U}{\partial X^2} = (\alpha-1)\alpha X^{\alpha-1}Y^\beta$$

which is negative if α is less than one, and thus we have diminishing marginal utility of good X. Similarly, there is diminishing marginal utility of good Y if β is less than 1; the demonstration of this is left to you. However, as previously noted, this is not a matter of any real analytical significance because of the arbitrary scale on the U-axis. Note particularly that the convexity of an indifference curve is not dependent on diminishing marginal utilities, as we will demonstrate in section 15.12.

The appendix to this chapter demonstrates a convenient and frequently used manipulation of the partial derivatives of the Cobb–Douglas production function. The same tricks can be performed on the Cobb–Douglas utility function that we have just been examining.

Summary of sections 14.11–14.13

(1) The neoclassical version of the utility function, of which the Cobb–Douglas form is the most widely used, makes assumptions analogous to those of the neoclassical production function (see section 14.7). A key difference, however, is that utility cannot be directly measured.

(2) One of the most important assumptions is non-satiation (meaning that a little more of either good always increases utility), which means that the marginal utilities (defined as the partial derivatives $\frac{\partial U}{\partial X}$ and $\frac{\partial U}{\partial Y}$ of the utility function) are always positive.

(3) A second important assumption is that the utility function is characterized by a decreasing marginal rate of substitution (MRS) in consumption. In words, this means that the consumer is assumed to prefer average or 'middling' bundles of goods rather than extreme bundles—those that contain large quantities of some goods but small quantities of others. The effect of decreasing MRS is that the indifference curves are convex. This is a very important property, as we shall see in chapters 15 and 16.

(4) Decreasing MRS is sometimes wrongly equated with diminishing marginal utility, the latter meaning that the direct second derivatives $\frac{\partial^2 U}{\partial X^2}$ and $\frac{\partial^2 U}{\partial Y^2}$ are negative. However, decreasing MRS occurs if, as we move down the indifference curve and consumption of X increases and of Y decreases, the marginal utility of X falls *relative* to the marginal utility of Y. This does not require the marginal utility of X to fall *absolutely*. We illustrated this for the general Cobb–Douglas utility function.

Moreover, although diminishing marginal utility is an idea that many throughout the history of economic thought have found intuitively appealing, no way has yet been found of verifying it due to the impossibility of directly observing utility and the so far insuperable difficulties of measuring it.

The Online Resource Centre contains more discussion of the economic applications of this chapter, including how to graph isoquants and indifference curves in Excel® and Maple.

Progress exercise 14.4

1. A firm finds that its production function is of the Cobb–Douglas form:
   $$Q = 100K^{0.25}L^{0.75}$$
   where Q is weekly output (in tonnes) and K and L are weekly inputs of machine-hours and worker-hours respectively.

   (a) (i) Show that the equation of the isoquant for $Q = 100$ is given by $K = \dfrac{1}{L^3}$.

       (ii) By differentiation, show that the isoquant is negatively sloped and convex.

       (iii) Sketch the graph of the isoquant.

       (iv) Give an economic interpretation to the slope and curvature of the isoquant.

       (v) Would other isoquants for this production function have a similar shape?

   (b) Treating output now as variable:

       (i) By differentiation, find the marginal products of labour and capital.

       (ii) Show that they are positive at every level of output.

       (iii) Would you expect this to be true of every production function?

   (c) By examining the signs of the direct second derivatives, show that the marginal product of labour diminishes as the labour input increases with the capital input held constant; and similarly that the marginal product of capital diminishes as the capital input increases with the labour input held constant.

   (d) Use the information from (b) and (c) above to sketch the graphs of the marginal products of labour and capital.

   (e) (i) How would you explain the 'law' of diminishing marginal productivity (DMP) implied by (b) and (c) above? Would you expect it to be true for all production functions and at every level of output?

       (ii) What does DMP imply about the graph of the average product of labour (APL) and of capital (APK)?

       (iii) Check your answer to (ii) by deriving the equations of the APL and APK and sketching their graphs.

   (f) (i) By examining the second order cross partial derivatives of the production function, show that an increase in capital input increases the marginal product of labour, and vice versa.

       (ii) Give an economic interpretation. Would you expect this to be generally true?

   (g) (i) Derive the equation of the quasi-short-run production function when the capital input is fixed at 160,000, and sketch its graph.

       (ii) How does the shape of this graph relate to your findings in (b) and (c) above?

       (iii) Repeat for a capital input fixed at 810,000, sketching the graph on the same axes.

       (iv) How does your answer to (f) above reveal itself in these graphs?

   (h) (i) With the labour input fixed at 160,000, derive the relationship between capital input and output, and sketch its graph. (This is the quasi-short-run production function and is sometimes called the total product of capital curve.)

       (ii) How does the shape of this graph relate to your findings in (b) and (c) above?

2. Consider the general Cobb–Douglas production function:
   $$Q = AK^{\alpha}L^{\beta} \quad \text{where } A, \alpha \text{ and } \beta \text{ are positive parameters.}$$

   (a) Show by partial differentiation that the production function has the property of increasing marginal productivity of capital (if $\alpha > 1$) and of labour (if $\beta > 1$).

(b) Explain the economic significance of this. Does this explain why we normally assume that α and β are less than 1?

3. An individual's utility function has the Cobb–Douglas form:

$$U = X^{0.75}Y^{1.5}$$

where U is an index of her utility and X and Y are the weekly quantities of the two goods consumed.

(a) (i) Show that the equation of the indifference curve for $U = 27$ is

$$Y = \frac{9}{X^{0.5}}$$

(ii) By differentiation, show that the indifference curve is negatively sloped and convex.

(iii) Give an economic interpretation to the slope and curvature of the indifference curve.

(iv) Sketch the graph of the indifference curve.

(b) (i) Repeat (a) for the indifference curve for $U = 8$.

(ii) What meaning can be attached to the levels of utility experienced on these two indifference curves? Can the levels be compared?

(iii) How does the shape of other indifference curves of this utility function compare with the two examined here?

Treating the level of U now as variable:

(c) (i) By differentiation, find the marginal utilities of the two goods.

(ii) Show that they are positive whatever the levels of consumption of the two goods.

(iii) Would you expect this to be true of every good and every utility function?

(d) (i) By examining the signs of the direct second derivatives, determine whether the marginal utility of each good increases, decreases, or remains constant as consumption of it increases with consumption of the other good held constant.

(ii) What economic significance (if any) can be attached to the signs of these partial derivatives?

(e) Use the information obtained from (c) and (d) above to sketch the graph of a typical iso-Y section: that is, the curve showing how U varies as consumption of X increases with consumption of Y held constant. Similarly, sketch the graph of an iso-X section.

(f) (i) By examining the signs of the cross partial derivatives, determine whether an increase in consumption of X increases the marginal utility of Y, and vice versa.

(ii) What economic significance (if any) can be attached to the signs of these cross partial derivatives?

4. (a) By means of a numerical example, show that a consumer with the utility function of question 3 will always prefer an average or 'middling' combination of X and Y to extreme combinations. (Hint: you could start by considering the bundle of goods $X = 256$, $Y = 9$ and the bundle $X = 81$, $Y = 16$. Your answer does not need to be absolutely precise in its numerical values.)

(b) Illustrate your answer with a sketch graph, including the relevant indifference curve.

(c) Prove algebraically that the result in (a) necessarily holds for any utility function if the indifference curves are convex. (Hint: this is a lot easier than you might think. Start by considering the bundle of goods $X = X_0$, $Y = Y_0$ and the bundle $X = X_1$, $Y = Y_1$.)

5. (a) Repeat question 3 for the utility function $U = XY$.

(b) Repeat question 4 for the utility function $U = XY$.

(c) Comment on the suitability of this functional form for use as a utility function.

Checklist

Be sure to test your understanding of this chapter by attempting the progress exercises (answers are at the end of the book). The Online Resource Centre contains further exercises and materials relevant to this chapter www.oxfordtextbooks.co.uk/orc/renshaw3e/.

This has been a long and difficult chapter because we have extended our analysis to functions with two or more independent variables, including the concepts of first and second order partial derivatives of these functions. We also looked at some important economic applications. The key points were:

✔ **Functions of two or more independent variables.** How a function with two independent variables corresponds to a surface in three-dimensional space, and with n variables to a hyper-surface in $(n + 1)$-dimensional space.

✔ **Sections through a surface.** How to take sections through a three-dimensional surface in order to analyse how any two variables are related.

✔ **Partial derivatives.** Understanding what first and second order partial derivatives measure, and how to obtain them.

✔ **The production function.** Understanding the assumptions behind the neoclassical production function and its resulting shape.

✔ **Marginal productivities.** How the partial derivatives from the production function measure the marginal products of the inputs. Finding the slope of an isoquant, and testing for a decreasing marginal rate of substitution in production.

✔ **The utility function.** Understanding the assumptions behind the neoclassical utility function and its resulting shape.

✔ **Marginal utilities.** How the partial derivatives from the utility function measure the marginal utilities of the goods. The problem of units of measurement. Finding the slope of an indifference curve, and checking for a decreasing marginal rate of substitution in consumption.

✔ **The Cobb–Douglas form of production and utility functions.** Applying the above concepts and techniques to the Cobb–Douglas production and utility functions.

The next chapter builds on the ideas developed in this chapter, so it is advisable to be sure that you fully understand this chapter before moving on.

Appendix 14.1 A variant of the partial derivatives of the Cobb–Douglas function

This appendix may be skipped without loss of continuity. Its purpose is to demonstrate a convenient and frequently used manipulation of the partial derivatives of the Cobb–Douglas function. We will demonstrate it for the general Cobb–Douglas production function

$$Q = AK^{\alpha}L^{\beta}$$

We know from section 14.9 above that the *MPL* is $\frac{\partial Q}{\partial L} = \beta AK^{\alpha}L^{\beta-1}$. If we simultaneously multiply and divide the right-hand side by L, this becomes

$$\frac{\partial Q}{\partial L} = \frac{\beta AK^{\alpha}L^{\beta-1}L}{L} = \beta \frac{AK^{\alpha}L^{\beta}}{L} = \beta \frac{Q}{L}$$

since $Q = AK^{\alpha}L^{\beta}$. Thus we have

$$\frac{\partial Q}{\partial L} = \beta \frac{Q}{L} \quad \text{that is:} \quad MPL = \beta \times APL$$

Thus in the Cobb–Douglas production function, the marginal product of labour, $\frac{\partial Q}{\partial L}$, and the average product of labour, $\frac{Q}{L}$, are connected to one another by the parameter β. If we restrict β to being less than 1, for the reasons discussed in section 14.9, it follows that the marginal product of labour is always less than the average product of labour. In turn this means that the average product of labour (that is, labour productivity) must be falling, since we know from section 9.24 that when the marginal value of any function is below its average value, the average value must be falling.

We can perform a similar manipulation on the second derivatives. From section 14.9 above:

$$\frac{\partial^2 Q}{\partial L^2} = (\beta - 1)(\beta)AK^{\alpha}L^{\beta-2}$$

If we simultaneously multiply and divide the right-hand side by L^2, this becomes

$$\frac{\partial^2 Q}{\partial L^2} = \frac{(\beta-1)(\beta)AK^{\alpha}L^{\beta-2}L^2}{L^2} = (\beta-1)\beta\frac{Q}{L^2}$$

which is negative because we have restricted β to being less than 1.

Finally, by a similar manipulation (left to you as an exercise) we can show that the cross partial derivative is

$$\frac{\partial^2 Q}{\partial K \partial L} = \alpha\beta AK^{\alpha-1}L^{\beta-1} = \alpha\beta\frac{Q}{KL}$$

which is positive.

The same trick can be performed on the partial derivatives with respect to K. This is left to you as an exercise. All of the above is equally valid for any other application of the Cobb–Douglas function, such as the utility function (see section 14.13 above).

Chapter 15
Maximum and minimum values, the total differential, and applications

OBJECTIVES

Having completed this chapter you should be able to:

- Use partial derivatives to identify maxima, minima, and saddle points of a function of two or more independent variables.
- Understand how the differential of a function is obtained and what it measures.
- Use the differential to find the derivative of an implicit function or a function of a function.
- Use the differential to show that the slope of an isoquant is given by the ratio of marginal products, and that the slope of an indifference curve is given by the ratio of marginal utilities.

15.1 Introduction

In the previous chapter we developed partial derivatives. We begin this chapter by putting partial derivatives to work to find the maximum and minimum values of any function $z = f(x, y)$. We shall find that this requires only a small extension of the rules for finding maximum and minimum values that we developed earlier for functions with only one independent variable (see section 7.8). However, in a surface in three dimensions we find a new phenomenon, the saddle point, which wasn't possible in the two-dimensional relationships analysed in chapter 7.

Next we develop a new concept, the total differential. For a function $z = f(x, y)$, this formula measures the impact on z of small changes in x and y. The total differential proves to be a very flexible tool with important applications. We can use it to find the derivative of an implicit function, where standard techniques of differentiation from chapters 6 and 13 will not work. We can also use it to find the derivative of a function such as $z = f(x, y)$, where x and y are not independent of one another. In the last part of the chapter we show how these techniques can be applied to the production function and the utility function. By using implicit differentiation we can show how the slope of a production function isoquant is dependent on the marginal productivities of labour and capital. Similarly, in the case of the utility function we show that the slope of an indifference curve is determined by the relative marginal utilities of the two goods. Finally, we look at a simple macroeconomic model and how the total differential can be used to analyse the effects of changes in both endogenous and exogenous variables.

15.2 Maximum and minimum values

Conditions for a maximum

In figure 15.1(a) we have created a three-dimensional perspective drawing of a function $z = f(x, y)$ which has the shape of a cone or, if you prefer, a pudding basin. It is, in a sense, upside down and lying on an imaginary 'floor' formed by the $0xy$ plane. Its peak, or maximum value of z, is at point P with coordinates $x = x_0$, $y = y_0$, and this is the point that we want to find.

We first take an iso-y section through the cone for $y = y_0$. This gives us the section APB, which we have projected onto the $0zx$ plane in figure 15.1(b). As usual, since we are holding y constant, z varies with x alone.

We see that any tangent drawn to APB has a positive slope to the left of P, a slope of zero at P, and a negative slope to the right of P. Since from chapter 14 we know that the slope of this tangent is measured by the partial derivative $\frac{\partial z}{\partial x}$, it follows that $\frac{\partial z}{\partial x}$ is positive to the left of P, equals zero at P, and is negative to the right of P. Therefore the graph of $\frac{\partial z}{\partial x}$ must have broadly the shape shown in figure 15.1(c); that is, it is negatively sloped. And since we know that the slope of this graph is measured by the second derivative, $\frac{\partial^2 z}{\partial x^2}$, it follows that this second derivative is negative throughout the section APB, and in particular is negative at P. Thus we conclude that, at P,

$$\frac{\partial z}{\partial x} = 0 \text{ and } \frac{\partial^2 z}{\partial x^2} < 0$$

You can probably guess what's coming next. We simply repeat the analysis above, but this time with x constant instead of y. Looking now at figure 15.2(a), we take an iso-x section through the cone for $x = x_0$. This gives us the section CPD, which we have projected onto the $0yz$ plane in figure 15.2(b). Along this section, z depends only on y. The slope of CPD is measured by the partial derivative $\frac{\partial z}{\partial y}$, which therefore is positive to the left of P, equal to zero at P, and negative to the right of P. Consequently, the graph of $\frac{\partial z}{\partial y}$ is negatively sloped as in figure 15.2c, and the second derivative, $\frac{\partial^2 z}{\partial y^2}$, is therefore negative throughout the section CPD, and in particular is negative at P. Thus we conclude that, at P,

$$\frac{\partial z}{\partial y} = 0 \text{ and } \frac{\partial^2 z}{\partial y^2} < 0$$

Collecting these two results, we can conclude as follows:

RULE 15.1 A maximum of $z = f(x, y)$

At a maximum of any function $z = f(x, y)$, such as P in figures 15.1 and 15.2,

(1) $\dfrac{\partial z}{\partial x} = 0$ and $\dfrac{\partial z}{\partial y} = 0$ and

(2) $\dfrac{\partial^2 z}{\partial x^2} < 0$ and $\dfrac{\partial^2 z}{\partial y^2} < 0$

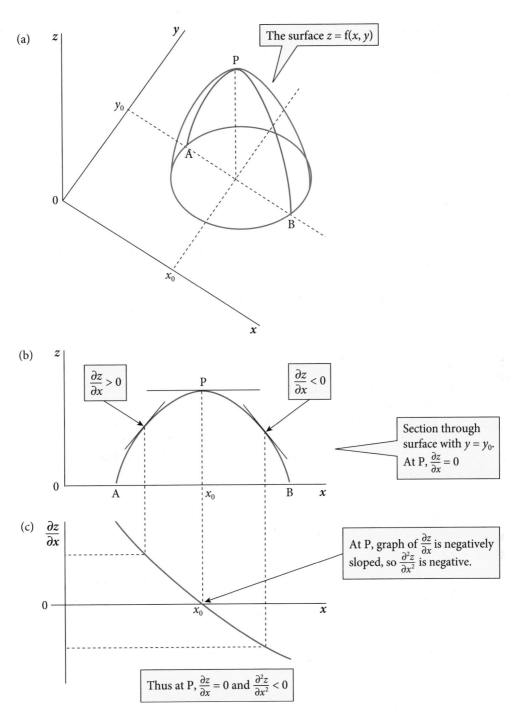

Figure 15.1 Examining an iso-y section to find necessary conditions for a maximum of $z = \mathrm{f}(x, y)$.

Conditions for a minimum

In figure 15.3(a) we have a perspective drawing of a function $z = \mathrm{f}(x, y)$, a cone floating mysteriously above an imaginary 'floor' formed by the $0xy$ plane. Its minimum is at point R with coordinates $x = x_0$, $y = y_0$, and this is the point that we want to find.

We begin by taking an iso-y section through the cone for $y = y_0$. This gives us the section ERF, which we have projected on to the $0xz$ plane in figure 15.3(b). As usual this section shows how z varies as x varies, with y constant. In this case we see that $\frac{\partial z}{\partial x}$ is negative to the left of R, equals zero at R, and is positive to the right of R. Therefore the graph of $\frac{\partial z}{\partial x}$ must have broadly the shape shown in figure 15.3(c); that is, it is positively sloped. So it follows in this case that the second

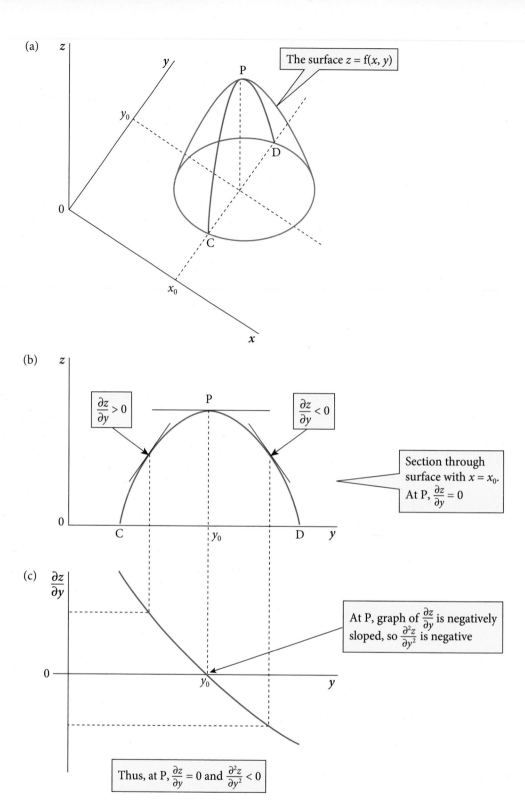

Figure 15.2 Examining an iso-x section to find necessary conditions for a maximum of $z = f(x, y)$.

derivative, $\frac{\partial^2 z}{\partial x^2}$, is positive throughout the section ERF, and in particular is positive at R. Thus we conclude that, at R,

$$\frac{\partial z}{\partial x} = 0 \text{ and } \frac{\partial^2 z}{\partial x^2} > 0$$

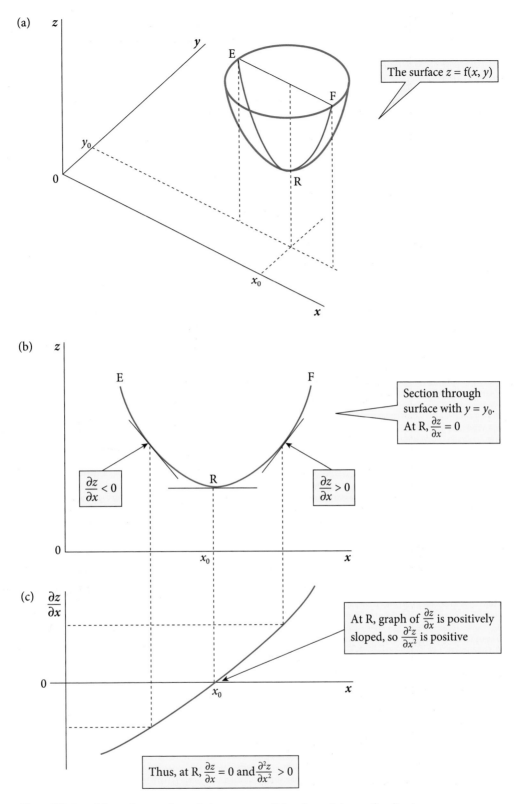

Figure 15.3 Examining an iso-y section to find necessary conditions for a minimum of $z = f(x, y)$.

Finally, in figure 15.4(a) we take an iso-x section through the cone for $x = x_0$. This gives us the section GRH, which we have projected onto the $0yz$ plane in figure 15.4b. The slope of GRH is measured by the partial derivative $\frac{\partial z}{\partial y}$, which therefore is negative to the left of R, equal to zero at R, and positive to the right of R. Consequently, the graph of $\frac{\partial z}{\partial y}$ is positively sloped as in

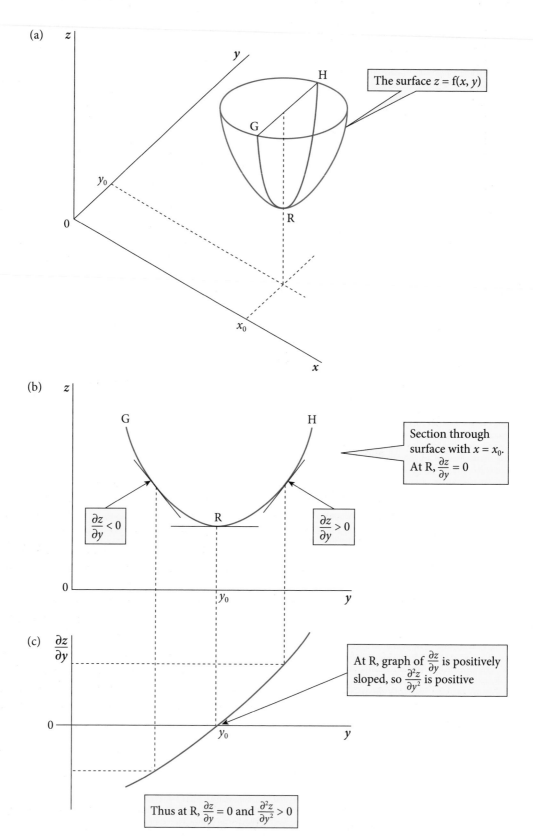

Figure 15.4 Examining an iso-x section to find necessary conditions for a minimum of $z = f(x, y)$.

figure 15.4c, and the second derivative, $\frac{\partial^2 z}{\partial y^2}$ is therefore positively sloped throughout the section GRH, and in particular is positive at R. Thus we conclude that, at R,

$$\frac{\partial z}{\partial y} = 0 \text{ and } \frac{\partial^2 z}{\partial y^2} > 0$$

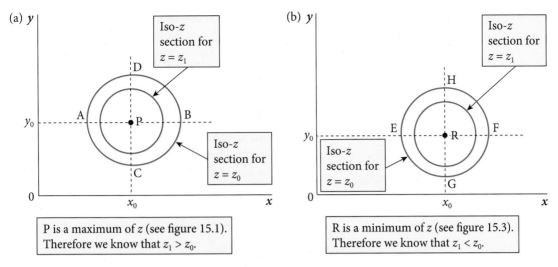

Figure 15.5 Iso-z sections near to a maximum or minimum.

Collecting these two results, we can conclude:

> **RULE 15.2 A minimum of $z = f(x, y)$**
>
> At a minimum of any function $z = f(x, y)$, such as R in figures 15.3 and 15.4,
>
> (1) $\dfrac{\partial z}{\partial x} = 0$ and $\dfrac{\partial z}{\partial y} = 0$ and
>
> (2) $\dfrac{\partial^2 z}{\partial x^2} > 0$ and $\dfrac{\partial^2 z}{\partial y^2} > 0$

It is also worth thinking about the shape of the iso-z sections near to a maximum or minimum. If we take some iso-z sections of the cone in figures 15.1 and 15.2 and project them on to the $0xy$ plane, we will get a series of circles (not necessarily regular in shape) surrounding P as shown in figure 15.5(a). The points A, B, C, and D from figures 15.1 and 15.2 have been put in to help you see the relationship between the two figures. Note that because P is a maximum of z, the iso-z values for each section increase as we move towards P.

If we think of z as measuring height, these iso-z sections are the exact equivalents of the contour lines that we see on a relief map, where the labelling of the contour lines gives their height above sea-level. Point P would then be a mountain or hill peak.

If we repeat this exercise for figures 15.3 and 15.4, again we get a set of iso-z sections that form circles around R, as in figure 15.5(b). Note the similarity to figure 15.5(a). The only difference is in the labelling of the iso-z sections. Because R is a minimum of z, the iso-z values for each section decrease as we move towards R. In terms of the relief map analogy, R would be the lowest point of a hole or crater in the ground. Note that in neither the maximum nor the minimum case do the iso-z sections cut one another.

An additional condition

The conditions for a maximum or minimum value described in rules 15.1 and 15.2 may appear comprehensive, but in fact they only tell us what happens to z when x varies with y constant, or vice versa. So even when these conditions are satisfied, there remains the possibility that, for example, we could be at a point such as S in figure 15.6. From S, a variation in x alone takes us downhill, towards either point A or point B. And a variation in y alone takes us downhill, towards either point C or point D. Thus S is a maximum for separate variation in either x or y, and the conditions for a maximum (rule 15.1) are satisfied. But if at S we increase or decrease x and y together, by roughly equal amounts, we move *uphill* towards either E or F. So S is a maximum

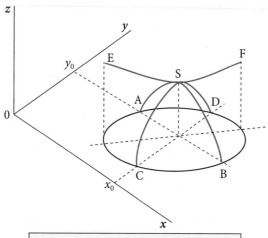

From S, any variation in x or y separately reduces z, but if x and y both increase or decrease together, z may increase.

Figure 15.6 How separate and simultaneous variations in x and y can have different effects on z.

for variation in x or y separately, but a minimum for at least some combinations of simultaneous variation in both x and y.

Point S is a **saddle point**, so called because the surface has roughly the shape of a saddle (of a horse, not a bike!). Even if you have never ridden a horse, hopefully you can see that in figure 15.6 you would sit with your butt at S, with one leg dangling somewhere between A and D, the other leg between C and B, and facing towards either E or F (depending on where the horse's head is). A saddle point is defined as a stationary point (at which therefore $\frac{\partial z}{\partial x} = \frac{\partial z}{\partial y} = 0$) such that some small variations in x and y take us uphill, and others downhill. To preclude the possibility that the point we are examining is a saddle point, or in other words to ensure that a point is a maximum (or minimum) not just for separate variation in either x or y but also for any simultaneous variation in both x and y, the following condition, a *sufficient* condition, additional to rules 15.1 or 15.2, is:

RULE 15.3 Additional second order condition for a maximum or a minimum of the function $z = f(x, y)$.

In addition to rules 15.1 and 15.2, the additional condition which is sufficient to ensure that the point is either a maximum or minimum, and not a saddle point, is:

$$\frac{\partial^2 z}{\partial x^2} \cdot \frac{\partial^2 z}{\partial y^2} > \frac{\partial^2 z}{\partial x \partial y} \cdot \frac{\partial^2 z}{\partial y \partial x}$$

In words, the rule says that the product of the direct second derivatives must exceed the product of the cross partial derivatives. Note that Young's theorem (see section 14.6) implies that the right-hand side of rule 15.3 can be written as $(\frac{\partial^2 z}{\partial x \partial y})^2$ or $(\frac{\partial^2 z}{\partial y \partial x})^2$. Note too that rule 15.3 is a sufficient but not a necessary condition; that is, some functions exist with a maximum or minimum at which rule 15.3 is not satisfied. The maths underlying this condition is explored in chapter W21, which can be downloaded from the Online Resource Centre (www.oxfordtextbooks.co.uk/orc/renshaw3e/).

15.3 Saddle points

As noted above, a saddle point is defined as a stationary point (at which therefore $\frac{\partial z}{\partial x} = \frac{\partial z}{\partial y} = 0$) such that some small variations in x and y take us uphill, and others downhill. A sufficient condition to ensure that the point in question is a saddle point, and not a maximum or minimum, is given in rule 15.4, which is simply the reverse of the inequality in rule 15.3:

RULE 15.4 A sufficient condition for a saddle point of the function $z = f(x, y)$.

In addition to the condition $\frac{\partial z}{\partial x} = \frac{\partial z}{\partial y} = 0$, the additional condition which is sufficient to ensure that the point is a saddle point, and not a maximum or minimum, is:

$$\frac{\partial^2 z}{\partial x^2} \cdot \frac{\partial^2 z}{\partial y^2} < \frac{\partial^2 z}{\partial x \partial y} \cdot \frac{\partial^2 z}{\partial y \partial x}$$

Note that, like rule 15.3, rule 15.4 is a sufficient but not a necessary condition; that is, some functions exist with a saddle point at which rule 15.4 is not satisfied.

Note too that rules 15.3 and 15.4 are incomplete: that is, they do not cover every logical possibility. For we do not know what to infer if we find that:

$$\frac{\partial^2 z}{\partial x^2} \cdot \frac{\partial^2 z}{\partial y^2} = \frac{\partial^2 z}{\partial x \partial y} \cdot \frac{\partial^2 z}{\partial y \partial x}$$

A stationary point at which this equality is satisfied may be either a maximum, minimum or saddle point. This case is addressed in section W21.2, which can be downloaded from the Online Resource Centre (www.oxfordtextbooks.co.uk/orc/renshaw3e/).

Saddle points can occur in several different ways. One type of saddle point that is found quite frequently in economics is shown in figure 15.7.

The key feature of this shape of surface is that, from point S, any variation in x alone takes us *uphill* towards either C or D. So point S is a minimum for variation in x alone. For variation in x, it will therefore satisfy the conditions for a minimum (see rule 15.2 above): that is, $\frac{\partial z}{\partial x} = 0$ and $\frac{\partial^2 z}{\partial x^2} > 0$. In contrast, from S any variation in y alone takes us *downhill* towards either A or B. So point S is a maximum for variation in y alone. For variation in y, it will therefore satisfy the conditions for a maximum (see rule 15.1 above): that is, $\frac{\partial z}{\partial y} = 0$ and $\frac{\partial^2 z}{\partial y^2} < 0$.

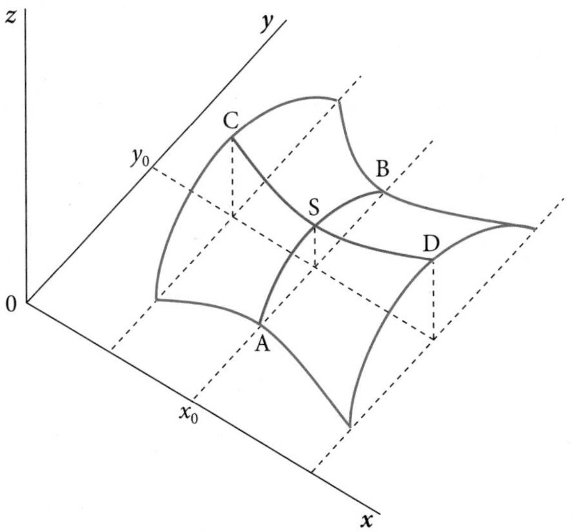

Figure 15.7 shows one possible orientation of the saddle, in which the rider faces in the x direction (towards either C or D). But we could easily turn the saddle through 90 degrees relative to the x-axis so that the rider faced in the y direction. (We could achieve the same result by swapping the labels on the x- and y-axes.) In that case, S would be a maximum for variation in x alone and a minimum for variation in y alone. So then we would have $\frac{\partial z}{\partial x} = 0$ and $\frac{\partial^2 z}{\partial x^2} < 0$ together with $\frac{\partial z}{\partial y} = 0$ and $\frac{\partial^2 z}{\partial y^2} > 0$.

In both cases we have a saddle point and therefore, combining these results, we have:

Figure 15.7 A saddle point at S that is a maximum for variation in y alone and a minimum for variation in x alone.

RULE 15.5 Conditions for a saddle point of the type shown in figure 15.7.

(1) $\frac{\partial z}{\partial x} = 0$ and $\frac{\partial z}{\partial y} = 0$;

(2) $\frac{\partial^2 z}{\partial x^2}$ and $\frac{\partial^2 z}{\partial y^2}$ have opposite signs (that is, one positive and the other negative);

(3) As a sufficient condition, rule 15.4 is satisfied.

As we did with the maximum and minimum in figure 15.5, it is interesting to think about the shape of the iso-z sections near to a saddle point. Figure 15.8 repeats figure 15.7, but now shows some iso-z sections through the surface. The key feature is that the iso-z section for $z = 20$ (shown in red) is in two branches, which are tangent to one another at S, forming roughly an X-shape. This contrasts with the maximum or minimum cases where, as we saw above, the iso-z sections form closed loops around the maximum or minimum point.

Using the analogy of the relief map once again, with the iso-z sections as the contour lines, an experienced map-reader would recognize point S, where the contour line intersects itself, as a saddle point; for this is a geographer's term too. If you were walking on this surface and your

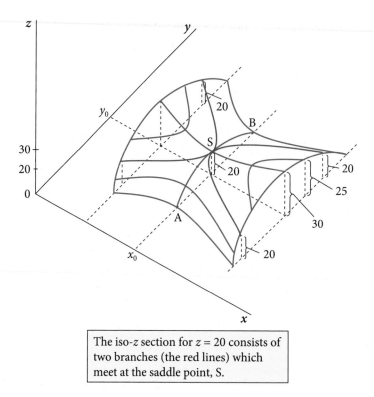

The iso-z section for $z = 20$ consists of two branches (the red lines) which meet at the saddle point, S.

Figure 15.8 Iso-z sections at and near a saddle point.

destination was in the y-direction, S would be a 'pass'—the lowest point at which you can pass over from A, which is in one valley, to B, which is in another.

Another possible way in which a saddle point may arise is shown in figure 15.9. Point P is a stationary point because we have $\frac{\partial z}{\partial x} = \frac{\partial z}{\partial y} = 0$. The surface inflects at point P, in the sense that, as we pass through P with x and y increasing either separately or together, the surface switches from concave to convex. Such a point is classified as a saddle point because it satisfies the definition given above: that is, it is a stationary point at which some variations in x and y take us downhill, and others uphill. The function $z = x^3 + y^3$ has a shape like that of figure 15.9, with P at the origin. There are other possible shapes for saddle points and you may be able to think of some yourself.

Local versus global maxima and minima

As in the two-variable case considered in chapter 7, the maximum and minimum values that we find using the above technique may be merely *local* maxima and minima. For example, in figure 15.1 the fact that there is a maximum at P does not preclude the possibility that z reaches higher values at some other point in three-dimensional space that does not appear in our drawing.

Extension to a function of n variables

This analysis of maximum and minimum values, and other possible shapes for the function, can be extended to a function with an unspecified number of independent variables: that is, a function of the form $z = f(x_1, x_2, x_3, \ldots, x_n)$. Some additional complications arise, however; these are taken up in chapter W21, which can be downloaded from the Online Resource Centre.

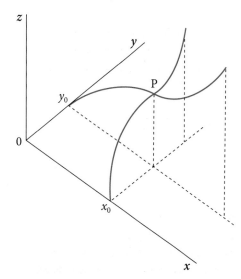

Figure 15.9 A saddle point that is a point of inflection.

The example below shows how to apply rules 15.1–15.5.

EXAMPLE 15.1

Suppose we are asked to examine the function

$$z = 2x + y - x^2 + xy - y^2$$

for maximum and minimum values, and saddle point(s). From rules 15.1, 15.2, and 15.4 we know that in all three cases (maximum, minimum, or saddle point), $\frac{\partial z}{\partial x} = 0$ and $\frac{\partial z}{\partial y} = 0$. So our method of solution is, first, to find these two partial derivatives. They are

$$\frac{\partial z}{\partial x} = 2 - 2x + y \quad \text{and} \quad \frac{\partial z}{\partial y} = 1 + x - 2y$$

The key insight is that at the point(s) we are looking for, both of these equations are equal to zero. So, if we set each of them equal to zero and solve them as a pair of simultaneous equations, the solution(s) will give us the values of x and y that we are looking for. Thus our simultaneous equations are

$$2 - 2x + y = 0 \quad \text{and} \quad 1 + x - 2y = 0$$

As both are linear functions, there is one solution: $x = \frac{5}{3}$, $y = \frac{4}{3}$, with $z = \frac{7}{3}$. (See section 3.9 if you need to revise simultaneous linear equations.)

This point could be a maximum, a minimum, or a saddle point. To find which of these three we have, we take the direct second derivatives and examine their signs at the point $x = \frac{5}{3}$, $y = \frac{4}{3}$. The second derivatives are

$$\frac{\partial^2 z}{\partial x^2} = \frac{\partial}{\partial x} \text{ of } 2 - 2x + y = -2 \quad \text{and}$$

$$\frac{\partial^2 z}{\partial y^2} = \frac{\partial}{\partial y} \text{ of } 1 + x - 2y = -2$$

These are both negative when $x = \frac{5}{3}$, $y = \frac{4}{3}$ (and for all values of x and y, as it happens, in this case). So when $x = \frac{5}{3}$, $y = \frac{4}{3}$:

$$\frac{\partial z}{\partial x} = 0 \text{ and } \frac{\partial z}{\partial y} = 0 \quad \text{and}$$

$$\frac{\partial^2 z}{\partial x^2} < 0 \text{ and } \frac{\partial^2 z}{\partial y^2} < 0$$

So this point is probably a maximum. To be quite sure, we must check the additional condition, which requires (rule 15.3)

$$\frac{\partial^2 z}{\partial x^2} \cdot \frac{\partial^2 z}{\partial y^2} > \frac{\partial^2 z}{\partial x \partial y} \cdot \frac{\partial^2 z}{\partial y \partial x}$$

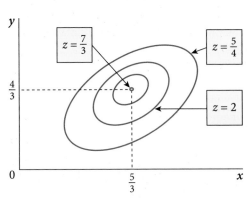

Figure 15.10 Iso-z sections of $z = 2x + y - x^2 + xy - y^2$.

In this case:

$$\frac{\partial^2 z}{\partial x \partial y} = \frac{\partial}{\partial x} \text{ of } \frac{\partial z}{\partial y} = 1; \text{ and } \frac{\partial^2 z}{\partial y \partial x} = \frac{\partial}{\partial y} \text{ of } \frac{\partial z}{\partial x} = 1$$

So the additional condition is satisfied since $(-2)(-2) > (1)(1)$.

Some iso-z sections of the surface are shown in figure 15.10. The labels on the iso-z sections show that z rises as we move across the surface towards J, where z reaches its maximum of $\frac{7}{3}$. A three-dimensional perspective drawing of this function would be very similar to figure 15.1, with $x_0 = \frac{5}{3}$, $y_0 = \frac{4}{3}$.

Summary of sections 15.1–15.3

In sections 15.1–15.3 we showed that the conditions for a maximum, minimum, or saddle point of a function $z = f(x, y)$ at a point P were:

	First order necessary conditions	Second order sufficient conditions		
Maximum:	$\frac{\partial z}{\partial x} = 0$ and $\frac{\partial z}{\partial y} = 0$	$\frac{\partial^2 z}{\partial x^2} < 0$ and $\frac{\partial^2 z}{\partial y^2} < 0$	;	$\frac{\partial^2 z}{\partial x^2} \cdot \frac{\partial^2 z}{\partial y^2} > \frac{\partial^2 z}{\partial x \partial y} \cdot \frac{\partial^2 z}{\partial y \partial x}$
Minimum:	$\frac{\partial z}{\partial x} = 0$ and $\frac{\partial z}{\partial y} = 0$	$\frac{\partial^2 z}{\partial x^2} > 0$ and $\frac{\partial^2 z}{\partial y^2} > 0$	;	$\frac{\partial^2 z}{\partial x^2} \cdot \frac{\partial^2 z}{\partial y^2} > \frac{\partial^2 z}{\partial x \partial y} \cdot \frac{\partial^2 z}{\partial y \partial x}$
Saddle point:	$\frac{\partial z}{\partial x} = 0$ and $\frac{\partial z}{\partial y} = 0$	$\frac{\partial^2 z}{\partial x^2}$ and $\frac{\partial^2 z}{\partial y^2}$ have any values;		$\frac{\partial^2 z}{\partial x^2} \cdot \frac{\partial^2 z}{\partial y^2} < \frac{\partial^2 z}{\partial x \partial y} \cdot \frac{\partial^2 z}{\partial y \partial x}$

If $\frac{\partial^2 z}{\partial x^2} \cdot \frac{\partial^2 z}{\partial y^2} = \frac{\partial^2 z}{\partial x \partial y} \cdot \frac{\partial^2 z}{\partial y \partial x}$, the point may be a maximum, minimum, or saddle point.

The Online Resource Centre gives more examples of functions with maximum, minimum, and saddle points, and uses Maple to plot their graphs.
www.oxfordtextbooks.co.uk/orc/renshaw3e/

Progress exercise 15.1

1. Show that the function $z = 60x + 34y - 6x^2 - 4xy - 3y^2 + 5$ has a single stationary point at $x = 4$, $y = 3$ and that this is a maximum.

2. Show that the function $z = 4x^2 - xy + y^2 - x^3$ has a minimum at $x = y = 0$ and a saddle point at $x = 2.5$, $y = 1.25$.

3. Find the stationary points of the following functions, and determine whether each is a maximum, minimum, or saddle point.

   (a) $z = x^2 + y^2 - 10x - 12y$

   (b) $z = 20x - x^2 - y^2 + 8y$

   (c) $F(X, Y) = X^2 + 2XY + 2Y^2 - 16X - 20Y$

   (d) $f(x, y) = x^3 + y^3 - 3x - 12y + 20$

15.4 The total differential of $z = f(x, y)$

In section 7.15 we saw that the derivative $\frac{dy}{dx}$ of the function $y = f(x)$ could be used as an approximation in calculating the effect on y of a change in x. If x changes by an amount dx, the resulting change in y is given by the differential formula

$$dy = \frac{dy}{dx}dx \quad \text{which can also be written as } dy = f'(x)dx \quad \text{(see rule 7.4)}$$

The differential involves an error, but we can make this error as small as we wish by making dx sufficiently small. We also saw that this formula can be viewed as a linear approximation to the function itself, because there would be no error in the formula if the function $f(x)$ were a straight line. By ignoring the error we are effectively treating the function as if it were linear over the distance set by the size of dx.

Extending this idea to $z = f(x, y)$

This idea of a differential can be extended in a quite straightforward way to a function with two independent variables, $z = f(x, y)$. Suppose we are at some initial position, P, on the surface $z = f(x, y)$ (see figure 15.11(a)). Now suppose x increases by a small amount dx (with y constant). The resulting change in z is given by the distance J′K, which approximately equals the distance JK. We can write JK as $JK = \frac{JK}{PK}PK$. Here, $\frac{JK}{PK}$ is the slope of the tangent at P, which is given by the partial derivative $\frac{\partial z}{\partial x}$; while PK = d$x$. Thus we have, as an approximation,

$$J'K = \frac{\partial z}{\partial x}dx \quad (= \text{the change in } z \text{ when only } x \text{ changes}) \qquad (15.1)$$

As in the case of $y = f(x)$, we can make the error in this approximation as small as we wish by making dx sufficiently small, for then J′ and J in figure 15.11(a) approach one another.

Similarly, and again starting from P, suppose now that y increases by a small amount dy (with x constant). The resulting change in z is given by the distance L′M, which approximately equals the distance LM. We can write LM as $LM = \frac{LM}{PM}PM$, where $\frac{LM}{PM}$ is the slope of the tangent at P, which is given by the partial derivative $\frac{\partial z}{\partial y}$; and PM = d$y$. Thus we have the approximation

$$L'M = \frac{\partial z}{\partial y}dy \quad (= \text{the change in } z \text{ when only } y \text{ changes}) \qquad (15.2)$$

Again we can make the error in this approximation as small as we wish by making dy sufficiently small, for then L′ and L in figure 15.11(a) approach one another.

Thus equation (15.1) gives us the change in z when x alone changes by a small amount. And equation (15.2) gives us the change in z when y alone changes by a small amount. But what if both x and y change simultaneously? We can answer this by looking at figure 15.11(b). If x changes by dx, we move from P to J′ and the change in z is the distance J′K. If y then changes by dy, we move from J′ to Q and the change in z is the distance QR. Therefore the total change in z when both x and y change, which we will call dz, is J′K + QR. But since dx is small, the distance QR cannot differ much from the distance L′M in figure 15.11(a). So we can write, as an approximation,

$$dz = J'K + L'M$$

Substituting from equations (15.1) and (15.2) into this expression, we obtain

$$dz = \frac{\partial z}{\partial x}dx + \frac{\partial z}{\partial y}dy$$

This expression is called the **total differential**. It tells us that the total change in z equals the change in z due to the change in x, plus the change in z due to the change in y. This simplicity is very convenient. We will state this result as a rule:

RULE 15.6 **The total differential of the function $z = f(x, y)$**

For any function $z = f(x, y)$, if x changes by a small amount dx and y changes by a small amount dy, the resulting change in z, dz, is called the total differential, defined as

$$dz = \frac{\partial z}{\partial x}dx + \frac{\partial z}{\partial y}dy \qquad (15.3)$$

The term 'total differential' is often shortened to 'differential' and we will follow this practice.

The word 'differential' simply means 'difference', and the total differential, dz, simply measures (approximately) the difference between the new value of z (at Q in figure 15.11(b)), and the initial value (at P), when both x and y change by small amounts. There is a small error, but this can be made as small as we wish by making dx and dy sufficiently small.

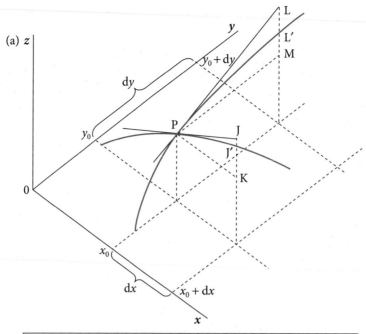

From P, when x changes by dx the true change in z is distance J′K. We approximate this with the distance $JK = \frac{\partial z}{\partial x}dx$ (a partial differential). The error is distance J′J. Similarly, when y changes by dy the true change in z is distance L′M. We approximate this with the distance $LM = \frac{\partial z}{\partial y}dy$ (a partial differential). The error is distance L′L. Finally, we approximate the total change in z when x changes by dx and y changes by dy as $JK + LM = \frac{\partial z}{\partial x}dx + \frac{\partial z}{\partial y}dy$ (the total differential). Total error is therefore J′J + L′L, plus whatever is the difference between L′M in (a) and QR in (b).

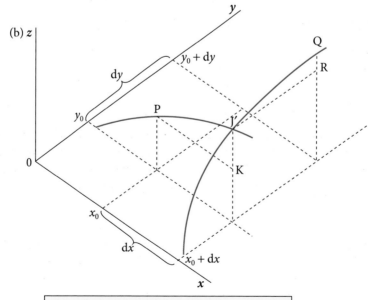

From P, when x changes by dx and y changes by dy, the true change in z is J′K + QR in figure 15.11(b). The total differential formula approximates this as JK + LM in figure 15.11(a).

Figure 15.11 The nature of the approximation in the total differential formula.

Why use an approximation rather than the real thing?

As we have seen, in deriving the total differential we chose to ignore the distances J′J and L′L in figure 15.11(a). We also treat the distances L′M and QR in figure 15.11(b) as equal. This is equivalent to treating the function in question as if it were linear (that is, a plane if there are three variables or a hyper-plane if there are more than three). This simplification permits us to answer questions and solve problems that otherwise would have no solution or, at best, a complicated solution. The slight loss of accuracy is considered a price worth paying to achieve these benefits. At a more mundane level, the differential is easier to calculate than the true change in z that follows small changes in x and y.

The partial differential

The differential formula (rule 15.6 above) gives the change in z when both x and y change. If, say, x changes while y remains constant, we have $dy = 0$ and equation (15.3) collapses to $dz = \frac{\partial z}{\partial x} dx$. This is called the **partial differential** with respect to x. Similarly, if y alone changes by a small amount with x constant, the total differential collapses to $dz = \frac{\partial z}{\partial y} dy$. This is the partial differential with respect to y.

> **RULE 15.7 The partial differentials of $z = f(x, y)$**
>
> The partial differentials of $z = f(x, y)$ with respect to x and y are
>
> $$dz = \frac{\partial z}{\partial x} dx \quad \text{and} \quad dz = \frac{\partial z}{\partial y} dy$$
>
> Each measures (with a small error) the change in z that occurs when either x or y changes by a small amount.

Equations (15.1) and (15.2) were in fact partial differentials.

Generalization to many variables

For any function of n independent variables

$$z = F(x_1, x_2, x_3, \ldots, x_n)$$

the differential of the function is

$$dz = \frac{\partial z}{\partial x_1} dx_1 + \frac{\partial z}{\partial x_2} dx_2 + \frac{\partial z}{\partial x_3} dx_3 + \cdots + \frac{\partial z}{\partial x_n} dx_n$$

This measures the change in z, dz, that follows small changes in the independent variables, dx_1, dx_2, dx_3, $\ldots$, dx_n. It is equally valid if some variables change while others remain constant. If, for example, variable x_3 did not change, we simply set $dx_3 = 0$ in the formula above. The resulting expression would then be a partial differential.

EXAMPLE 15.2

This very simple example gives us some practice at evaluating differentials and also helps to bring out the nature of the approximation involved. Consider the function

$z = xy$

We can think of z as the area of a rectangle with sides of length x and y (see figure 15.12). Initially the sides have length x_0 and y_0, so the initial area is z_0 (which equals $x_0 y_0$).

If side x increases by dx and side y by dy, the resulting change in area, dz, is given by the sum of the areas of the three rectangles A, B, and C. So we have

$dz = $ areas $A + B + C$

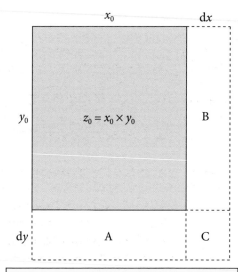

When x changes by dx and y by dy, change in area is $A + B + C = x_0 dy + y_0 dx + dx dy$. Differential is $x_0 dy + y_0 dx = A + B$. So in this example error in differential is $C = dx dy$ which is 'super-small'.

Figure 15.12 An example of the approximation in the total differential formula.

But rectangle A has one side of length x_0 and one of length dy, so its area is $x_0 dy$. Similarly, rectangle B has area $y_0 dx$; and rectangle C has area $dx dy$. So

$$dz = x_0 dy + y_0 dx + dx dy \qquad (15.4)$$

This is the *true* change in area; no approximation is involved.

Now we compare this with the value of dz given by the differential (rule 15.6). In this example, $z = xy$, so $\frac{\partial z}{\partial x} = y$ and $\frac{\partial z}{\partial y} = x$. Substituting these values into rule 15.6 we get

$$dz = \frac{\partial z}{\partial x} dx + \frac{\partial z}{\partial y} dy = y dx + x dy \quad (15.5)$$

(where it is understood that x and y are the initial values).

If we compare equations (15.4) and (15.5) we see that the error in the approximation is $dx dy$. From figure 15.12 we see that this is the area of the small rectangle, C. When dx and dy are both small, this area is the product of two small numbers and hence is 'super-small'. For example, if $dx = 0.1$ and $dy = 0.1$, their product $dx dy$ is $(0.1)(0.1) = 0.01$.

This example illustrates that there is always an error in the differential formula but it is small relative to the sizes of dx and dy and can be made as small as we wish by making dx and dy sufficiently small.

Progress exercise 15.2

1. Find the total differential (dz) of the following functions:

   (a) $z = x^2 + xy + y^2 - 1$

   (c) $z = x^{0.25} y^{0.5}$

   (b) $z = x^\alpha y^\beta$

   (d) $z = (x^2 + y^3)^{0.5}$

2. Suppose I have a rectangular lawn in my garden with one side of length 10 metres and the other of length 5 metres. (a) If I plan to buy additional turf in order to enlarge the lawn to 11 metres by 6 metres, what error will result if I use the total differential to calculate the additional turf needed? (b) Illustrate diagrammatically.

3. If $z = x^2 + y^2$, suppose x increases from 5 to 5.1 and y increases from 10 to 10.1. What percentage error occurs when we use the total differential, dz, of the function to calculate the resulting change in z?

(Note that you can illustrate this question diagrammatically if you draw two squares, one with sides of length x, and the other with sides of length y. The dependent variable z is then given by the sum of the areas.)

15.5 Differentiating a function of a function

The differential is a very flexible tool that is useful in a wide variety of situations. In this and the following sections we will look at the most important applications.

EXAMPLE 15.3

Suppose we are given

$$z = x^2 + y^3 \qquad (15.6)$$

Now suppose we are given the additional information that x and y are not independent of one another, but instead are related by the function $y = 2x$. This is analogous to the 'function of a function' relationship that we met in chapter 6; see rule D5 and example 6.13. It means that the variable y is not an independent variable: it has no independent life of its own, but depends for its value on x. Therefore z varies only if x varies. A change in x affects z both directly and also indirectly via its effect on y. The direct and indirect effects can be seen schematically as follows:

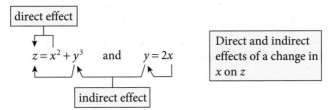

Because z depends only on x, the derivative $\frac{dz}{dx}$ must exist. Note that this is not a partial derivative because no other variable is being held constant. One way of finding this derivative would be to go back to equation (15.6) and substitute the additional information (that $y = 2x$) into it, giving

$$z = x^2 + y^3 = x^2 + (2x)^3$$

After multiplying out the brackets this becomes

$$z = x^2 + 8x^3$$

Thus y has been eliminated and z depends only on x. So the derivative is

$$\frac{dz}{dx} = 2x + 24x^2$$

That would seem to be the end of the story. However, there is another way of finding the derivative. Given $z = x^2 + y^3$, the partial derivatives are

$$\frac{\partial z}{\partial x} = 2x \quad \text{and} \quad \frac{\partial z}{\partial y} = 3y^2$$

So the differential is

$$dz = \frac{\partial z}{\partial x}dx + \frac{\partial z}{\partial y}dy = (2x)dx + (3y^2)dy \qquad (15.7)$$

If we divide both sides of equation (15.7) by dx, we get

$$\frac{dz}{dx} = \frac{\partial z}{\partial x}\cdot\frac{dx}{dx} + \frac{\partial z}{\partial y}\cdot\frac{dy}{dx} = 2x + (3y^2)\frac{dy}{dx} \qquad \left(\text{since}\frac{dx}{dx} \equiv 1\right) \qquad (15.8)$$

But, since we are given that $y = 2x$, we know that $\frac{dy}{dx} = 2$. We can substitute both of these pieces of information into equation (15.8) and get

$$\frac{dz}{dx} = 2x + (3[2x]^2)(2) = 2x + 24x^2$$

Thus we have found the derivative $\frac{dz}{dx}$ by a new route, in which we take the differential, dz, divide it by dx, and then substitute to eliminate y and $\frac{dy}{dx}$. You may ask what is the point of this. The answer is that in this particular example there is not much point, as we found the derivative equally easily by direct substitution—that is, replacing y with $2x$ in the original function. But in more complex cases this direct substitution may be difficult, in which case the alternative method using the differential is very useful.

Another situation where it can be beneficial to use the differential in this way is when the functional form is unspecified. We will illustrate this shortly.

Finally, note that the relationship $y = 2x$ means that only one of these variables is truly independent. In the above example we chose to eliminate y and thus focus on the relationship between z and x. But we could equally have chosen to eliminate x and thence focused on the relationship between z and y. Try working this through as an exercise; that is, eliminate x and find $\frac{dz}{dy}$.

Generalization

We can generalize the previous example. Given any function $z = f(x, y)$ where x and y are not independent of one another but instead are linked by the function $y = g(x)$, we have a 'function of a function' situation, in the sense that z depends only on x. This implies that the derivative $\frac{dz}{dx}$ exists. We can find this derivative as follows. First, we take the differential of $z = f(x, y)$, which is

$$dz = \frac{\partial z}{\partial x}dx + \frac{\partial z}{\partial y}dy$$

where, of course, the partial derivatives are obtained from the function f(x, y). Then, we divide through by dx, and get

$$\frac{dz}{dx} = \left(\frac{\partial z}{\partial x}\right) + \left(\frac{\partial z}{\partial y}\frac{dy}{dx}\right)$$

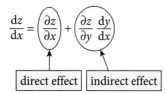

direct effect | indirect effect

Direct and indirect effects of a change in x on z

This expression is often called a **total derivative**. We will state is as a rule:

> **RULE 15.8** The total derivative of $z = f(x, y)$
>
> For any function $z = f(x, y)$ where x and y are themselves linked by the function $y = g(x)$, the total derivative is given by
>
> $$\frac{dz}{dx} = \frac{\partial z}{\partial x} + \frac{\partial z}{\partial y}\frac{dy}{dx}$$
>
> It is 'total' in the sense that it captures the total effect on z of a change in x, comprising both the direct effect $(\frac{\partial z}{\partial x})$ and the indirect effect $(\frac{\partial z}{\partial y}\frac{dy}{dx})$ that operates via the change in y that is induced by the change in x.

15.6 Marginal revenue as a total derivative

In this section we will show how in general three basic concepts in microeconomics—price, marginal revenue, and the elasticity of demand—are related. We derived the same result by a slightly different method in appendix 8.2. By definition, total revenue (TR) is given by

$$TR = pq$$

where p = price, q = quantity. The partial derivatives are

$$\frac{\partial TR}{\partial p} = q; \quad \text{and} \quad \frac{\partial TR}{\partial q} = p \quad \text{So the differential is}$$

$$dTR = \frac{\partial TR}{\partial p}dp + \frac{\partial TR}{\partial q}dq = qdp + pdq \tag{15.9}$$

However, we know that in general p and q are not independent of one another. Instead, they are linked through the *inverse* demand function (see section 3.15 to review this), which we can write as

$$p = f(q) \quad \text{with derivative} \quad \frac{dp}{dq}$$

Thus we have a typical 'function of a function' situation, in which TR depends only on q via a direct and an indirect effect. This may be seen below:

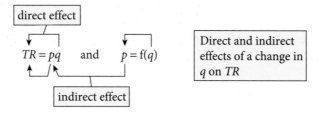

Direct and indirect effects of a change in q on TR

The direct effect arises because since $TR = pq$, any change in q (with p held constant) changes TR. The indirect effect arises from the fact that if q changes, p must change too in order to satisfy the inverse demand function $p = f(q)$; and since $TR = pq$, any change in p changes TR.

Therefore there is a total derivative $\frac{dTR}{dq}$ which we can find by proceeding as we did in example 15.3. We divide equation (15.9) through by dq, which gives

$$\frac{dTR}{dq} = q\frac{dp}{dq} + p \quad \left(\text{since} \frac{dq}{dq} = 1\right)$$

This is the total derivative with respect to q. You will immediately recognize the left-hand side as marginal revenue (MR) (see sections 8.7–8.10, and appendix 8.2) to review this concept). So we have

$$MR = \left(q\frac{dp}{dq}\right) + \left(p\right) \tag{15.10}$$

indirect effect   direct effect

Let's translate into words what equation (15.10) is telling us. If we think of MR as the additional revenue from selling one more (very small) unit, then equation (15.10) is saying that selling one more unit has a direct and an indirect effect on TR. The direct effect is that selling one more unit brings in extra revenue equal to its price, p. The indirect effect is the loss of revenue on the initial quantity sold, due to the fact that the price has to be reduced in order to sell one more unit. The revenue loss equals the price reduction multiplied by the initial quantity, q. The price reduction is given by the partial differential $dp = \frac{dp}{dq}dq = \frac{dp}{dq}$ (since $dq = 1$). Therefore the revenue loss is $q\frac{dp}{dq}$. (Note that $\frac{dp}{dq}$ is negative, since we assume that the demand function is negatively sloped.)

Thus equation (15.10) gives us a more precise picture than we previously had about the relationship between marginal revenue, price, and the slope of the demand function.

Moreover, we can also bring the elasticity of demand into the story. By simple algebra, the right-hand side of equation (15.10) can be rearranged to give

$$MR = p\left(\frac{q}{p}\frac{dp}{dq} + 1\right) \quad \text{(check for yourself)}$$

Also, we know that the elasticity of the demand function (E^D) is defined as

$$E^D = \frac{p}{q}\frac{dq}{dp} \quad \text{(see section 9.14).}$$

Substituting this into the right-hand side of the MR expression gives our next rule:

RULE 15.9 Marginal revenue, price, and the elasticity of demand

The relationship between marginal revenue, price, and the elasticity of demand is

$$MR = p\left(\frac{1}{E^D} + 1\right)$$

This is the same formula as we derived as rule 9.6 in section 9.18 by direct substitution of the inverse demand function into the total revenue function. (See also appendix 8.2, rule 8.4.)

15.7 Differentiating an implicit function

Suppose we are given the implicit function

$$x^2 + y^2 - 10 = 0$$

and asked to find $\frac{dy}{dx}$. This equation quickly rearranges as $y^2 = 10 - x^2$ and by raising both sides to the power 0.5:

$$y = (10 - x^2)^{0.5}$$

This can now be differentiated using the function of a function rule. We write

$$y = u^{0.5} \text{ where } u = 10 - x^2 \text{ from which}$$

$$\frac{dy}{dx} = \frac{dy}{du}\frac{du}{dx} = 0.5u^{-0.5}(-2x) = 0.5(10 - x^2)^{-0.5}(-2x) = -x(10 - x^2)^{-0.5}$$

$$= -\frac{x}{(10 - x^2)^{0.5}} = -\frac{x}{y} \tag{15.11}$$

However, this method won't always work because it's sometimes impossible to rearrange an implicit function into explicit form, which was our first step above.

In such a case, we can proceed as follows. The equation we were given at the beginning was

$$0 = x^2 + y^2 - 10 \tag{15.12}$$

We now create a new variable, z, defined as

$$z = x^2 + y^2 - 10$$

The partial derivatives are therefore

$$\frac{\partial z}{\partial x} = 2x \quad \text{and} \quad \frac{\partial z}{\partial y} = 2y$$

The differential of the function is therefore

$$dz = \frac{\partial z}{\partial x}dx + \frac{\partial z}{\partial y}dy = (2x)dx + (2y)dy \tag{15.13}$$

But to satisfy equation (15.12) we must have $z = 0$, which in turn implies $dz = 0$. In words, if equation (15.12) is to be satisfied, z must be a constant (and equal to zero). And if z is a constant, then the change in z, dz, must be zero.

So, in order to satisfy equation (15.12), equation (15.13) must equal zero. Setting equation (15.13) equal to zero gives

$(2x)dx + (2y)dy = 0$ which rearranges as

$(2y)dy = -(2x)dx$ from which

$$\frac{dy}{dx} = -\frac{2x}{2y} = -\frac{x}{y}$$ (the same answer as equation (15.11) above)

Generalization

Given any implicit function $f(x, y) = 0$, we can find the derivative, $\frac{dy}{dx}$, as follows. We define a new variable, z, such that $z = f(x, y)$ with differential

$$dz = \frac{\partial z}{\partial x}dx + \frac{\partial z}{\partial y}dy$$

We then set $dz = 0$ to reflect the fact that z is a constant (and equal to zero) when the given equation is satisfied. This gives us

$$0 = \frac{\partial z}{\partial x}dx + \frac{\partial z}{\partial y}dy \tag{15.14}$$

Equation (15.14) is the differential of the implicit function $f(x, y) = 0$. By subtracting $\frac{\partial z}{\partial x}dx$ from both sides, then dividing by $\frac{\partial z}{\partial y}$ and by dx, it rearranges as

$$\frac{dy}{dx} = -\frac{\frac{\partial z}{\partial x}}{\frac{\partial z}{\partial y}}$$

This is an untidy expression which is sometimes written as $\frac{dy}{dx} = -\frac{\partial z}{\partial x}/\frac{\partial z}{\partial y}$, but this is not much better. However, we can use the alternative notation (see final paragraph of section 14.6), in which we write

$$\frac{\partial z}{\partial x} \equiv f_x \quad \text{and} \quad \frac{\partial z}{\partial y} \equiv f_y$$

Then we can rewrite the previous expression as

$$\frac{dy}{dx} = -\frac{f_x}{f_y}$$

This is a better notation because it's not only neater but gets rid of the fictional z. Also the f notation reminds us that the partial derivatives are those of the function $f(x, y)$.

We can summarize this as the following rule:

RULE 15.10 The derivative of an implicit function (with two variables)

The derivative of the implicit function $f(x, y) = 0$ is given by

$$\frac{dy}{dx} = -\frac{f_x}{f_y}$$

where f_x and f_y are the partial derivatives of the function $z = f(x, y)$. This technique for finding the derivative is known as implicit differentiation.

Implicit differentiation is a vital tool when we are dealing with an implicit function $f(x, y) = 0$, as it is often difficult or impossible to find the corresponding explicit function giving y as a function of x or x as a function of y.

Three or more variables

Suppose we have an implicit function with three variables, $F(x, y, v) = 0$. Then equation (15.14) above becomes

$$0 = \frac{\partial z}{\partial x} dx + \frac{\partial z}{\partial y} dy + \frac{\partial z}{\partial v} dv \qquad (15.14a)$$

In order to get a derivative out of this, we have to hold one of the variables constant. If we hold v constant, we have $dv = 0$, and the equation above becomes

$$0 = \frac{\partial z}{\partial z} dx + \frac{\partial z}{\partial y} dy \quad \text{which rearranges as}$$

$$\frac{dy}{dx} = -\frac{\frac{\partial z}{\partial x}}{\frac{\partial z}{\partial y}}$$

as in the two-variable case, $f(x, y)$, considered above.

However, there is a crucial change we must make. We must take account of the fact that there is another variable, v, being held constant. This means that what we have found here is not a derivative, but a partial derivative. So we must change the notation on the left-hand side, and write instead

$$\frac{\partial y}{\partial x} = -\frac{\frac{\partial z}{\partial x}}{\frac{\partial z}{\partial y}}$$

which, using the alternative notation (see section 14.6), we can write more compactly as

$$\frac{\partial y}{\partial x} = -\frac{F_x}{F_y}$$

Thus we obtain the partial derivative $\frac{\partial y}{\partial x}$ as the ratio of two differentials, dy and dx. In the same way, by holding y constant we can obtain $\frac{\partial v}{\partial x} = -\frac{F_x}{F_v}$, and by holding x constant, $\frac{\partial y}{\partial v} = -\frac{F_v}{F_y}$. We could deal with additional variables in the same way.

We summarize this as a new rule:

> **RULE 15.11 The partial derivatives of an implicit function (with three or more variables)**
>
> The partial derivatives of an implicit function $F(x, y, v) = 0$ are given by
>
> $$\frac{\partial y}{\partial x} = -\frac{F_x}{F_y}, \quad \frac{\partial v}{\partial x} = -\frac{F_x}{F_v}, \quad \text{and} \quad \frac{\partial y}{\partial v} = -\frac{F_v}{F_y}$$
>
> where F_x, F_y, and F_v are the partial derivatives of the function $z = F(x, y, v)$. This technique is known as implicit partial differentiation. The technique extends straightforwardly to functions with more than three variables.

This rule illustrates the flexibility and power of the differential concept. We will use this in an economic application later in this chapter.

15.8 Finding the slope of an iso-z section

This is really no more than a small extension of the idea of implicit differentiation, which we have just considered.

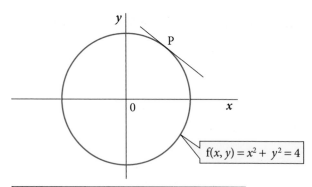

The slope at any point such as P is $\dfrac{dy}{dx} = -\dfrac{f_x}{f_y}$.

Figure 15.13 Finding the slope of an iso-z section by implicit differentiation.

Suppose we are looking at the function $z = x^2 + y^2$. We know that this function corresponds to a cone in three-dimensional space (see section 14.3, example 14.2). We can define an iso-z section by fixing the value of z at, say, $z = 4$. The equation of the iso-z section is then $x^2 + y^2 = 4$. This defines a circle with radius 2 which is projected onto the $0xy$ plane in figure 15.13.

Suppose now we wish to find the slope of this iso-z section at some point P. The equation of the contour is $x^2 + y^2 = 4$, which can of course be written as

$$x^2 + y^2 - 4 = 0$$

The key point is that this is an implicit function relating y and x. Its partial derivatives are

$$f_x = 2x \quad \text{and} \quad f_y = 2y$$

So applying rule 15.10 for implicit differentiation we can say straight away that the derivative is

$$\frac{dy}{dx} = -\frac{f_x}{f_y} = -\frac{2x}{2y} = -\frac{x}{y}$$

This means that we can find the slope of the tangent at any point such as P in figure 15.13 by evaluating the partial derivatives at P and substituting them into the expression above. For example, if at P we have $x = 2$, $y = 3$, the slope would be $-\frac{2}{3}$. Thus we have a new rule:

> **RULE 15.12 The slope of an iso-z section**
>
> Given any function $z = f(x, y)$ with partial derivatives f_x and f_y, the equation of an iso-z section when $z = z_0$ is
>
> $$f(x, y) - z_0 = 0$$
>
> Then by implicit differentiation (rule 15.10) the derivative $\frac{dy}{dx}$ is given by
>
> $$\frac{dy}{dx} = -\frac{f_x}{f_y}$$
>
> which measures the slope of the iso-z section at any point.

15.9 A shift from one iso-z section to another

Suppose $z = f(x, y)$ and initially we are on the iso-z contour for $z = z_0$. Now suppose z changes to a new fixed value, z_1, so we move onto the iso-z contour for $z = z_1$. How does this affect the values of x and y? Since the differential, dz, denotes a small change in z, we can use the differential to answer this question, provided the change in z is small. Thus when the jump in z from the z_0 to the z_1 iso-z contour occurs, we can write the change in z as

$$z_1 - z_0 = dz \quad \text{where} \quad dz = f_x dx + f_y dy$$

and dx and dy are the changes in x and y that accompany the change in z.

In words, the differential formula, which measures the effect of a small change in z when z is viewed as a variable, is equally valid for measuring the effect of a small change in z from one fixed value to a new fixed value. Any change in z is as valid as any other change in z, as far as the differential formula is concerned. The difference between this case and the previous cases we have considered is that the change in z now occurs spontaneously (or exogenously, to use a more precise term), and x and y have to change in response, rather than the other way round.

As an example, if $z = x^2 + y^2$, and z changes by a small amount dz from a fixed value z_0 to a new fixed value z_1, the associated changes in x and y are given by

$$dz = 2x\,dx + 2y\,dy$$

Of course, for the given dz there are many combinations of dx and dy that satisfy this equation. An economic application of this is given in section 15.17 below.

Summary of sections 15.4–15.9

We first derived the formula for the total differential (rule 15.6). For any function $z = f(x, y)$, if x changes by a small amount dx and y changes by a small amount dy, the resulting change in z, dz, is given by

$$dz = \frac{\partial z}{\partial x}dx + \frac{\partial z}{\partial y}dy$$

We generalized this to the case of many variables, and also showed that when either dx or dy is zero we obtain the partial differential (rule 15.7).

We also examined two important extensions of the total differential:

(1) *The derivative of a function of a function* (rule 15.8). If we have a function $z = f(x, y)$ where x and y are linked by another function, $y = g(x)$, we have a 'function of a function', in the sense that z depends only on x. The total derivative, which captures both the direct and the indirect effect on z of a change in x, is then

$$\frac{dz}{dx} = \frac{\partial z}{\partial x} + \frac{\partial z}{\partial y}\frac{dy}{dx}$$

We applied rule 15.8 to show the relationship between marginal revenue, price, and the elasticity of demand, resulting in rule 15.9: $MR = P(\frac{1}{E^D} + 1)$

(2) *The derivative of an implicit function* (rules 15.10 and 15.11). Rule 15.10 enables us to find the derivative, $\frac{dy}{dx}$, of any implicit function $f(x, y) = 0$. We found that the derivative is

$$\frac{dy}{dx} = -\frac{f_x}{f_y}$$

where $f_x \equiv \frac{\partial z}{\partial x}$ and $f_y \equiv \frac{\partial z}{\partial y}$ are the partial derivatives of $z = f(x, y)$.

We also extended rule 15.10 to obtain rule 15.12, for finding the slope of an iso-z section of any function $z = f(x, y)$ (rule 15.12). If $z = z_0$, the equation of the iso-z contour is $f(x, y) - z_0 = 0$, and its slope is again $\frac{dy}{dx} = -\frac{f_x}{f_y}$.

Finally, we showed (section 15.9) that, given $z = f(x, y)$, the differential formula (rule 15.6) is also valid for finding the resulting changes in x and y, following a given change in z, dz. This is a shift from one iso-z contour to another.

The Online Resource Centre gives more explanation and examples of total differential and implicit differentiation. www.oxfordtextbooks.co.uk/orc/renshaw3e/

Progress exercise 15.3

1. Use the differential (dz) to find the total derivative ($\frac{dz}{dx}$), given the following functions (in which the variables on the right-hand side are not independent of one another). Then check your answers by direct substitution.

    (a) $z = x^3 + y^2$ where $y = x^2$

    (b) $z = 3x^2 + 4xy^3$ where $y = (x + 1)^2$

    (c) $z = uv$ where $u = 3x + 2$ and $v = 3x^2$

2. (a) Use the differential to find the total derivative $\frac{dTR}{dq}$, given the definition of total revenue as $TR = pq$ (where p = price, q = quantity), and given also the inverse demand function $p = 50e^{-0.5q}$.

    (b) Check your answer by direct substitution.

    (c) Give a verbal interpretation to your answer to (a).

    (d) What is the economist's term for $\frac{dTR}{dq}$?

3. Use the differential (dz) to find the derivative ($\frac{dy}{dx}$) of each of the following implicit functions:

    (a) $2x - 3y = 0$ (b) $0 = x^2 + xy + y^2$ (c) $100 = x^\alpha y^\beta$ (d) $(x^2 + 3y)^{0.5} = 0$

4. For each of the functions involving z in question 1 above, use the differential to find the slope of an iso-z section. (Treat x and y as independent of one another; that is, ignore the functions relating y to x and u and v to x.)

15.10 Economic applications 1: the production function

The slope of an isoquant

In section 14.8 we looked at the production function $Q = f(K, L)$. We obtained an iso-Q section (an isoquant) by setting Q equal to some fixed value Q_0. Figure 14.15, repeated here as figure 15.14, shows two such isoquants for $Q = 60$ and $Q = 100$.

We can find the slope of any isoquant at any point by implicit differentiation. For example, the equation of the $Q = 60$ isoquant is

$$f(K, L) - 60 = 0$$

So, using rule 15.12 above, its slope at any point such as R in figure 15.14 is given by

$$\frac{dK}{dL} = -\frac{f_L}{f_K}$$

The economic interpretation of this result is interesting. By definition, f_K and f_L are the partial derivatives of the production function $Q = f(K, L)$. But, from section 14.8 we know that these partial derivatives are the marginal products of capital and labour. In section 14.8 we also defined the marginal rate of substitution in production (MRS) as the slope of the isoquant. Collecting all this together, we have rule 15.13:

RULE 15.13 The slope of an isoquant

Given any production function $Q = f(K, L)$, the slope of any isoquant, which by definition equals the MRS, at any point is given by

$$MRS \equiv \frac{dK}{dL} = -\frac{f_L}{f_K} = -\frac{MPL}{MPK}$$

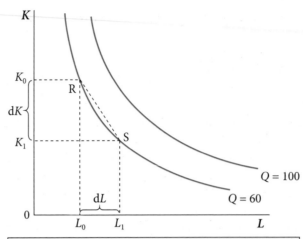

When production moves from R to S, there is a loss in output measured approximately by the partial differential $f_K dK$ and a gain in output measured approximately by the partial differential $f_L dL$. As output is unchanged, we must have $f_K dK + f_L dL = 0$, which rearranges as $\frac{dK}{dL} = -\frac{f_L}{f_K}$ = slope of the indifference curve at R. (This assumes S very close to R, otherwise error in approximation becomes unacceptably large.)

Figure 15.14 Economic interpretation of the slope of an isoquant.

This relationship between the slope of the isoquant and the marginal products of labour and capital can be explained in words as follows. Suppose production is initially at R in figure 15.14, using K_0 units of capital and L_0 units of labour to produce 60 units of output. Now suppose the production point moves to S as the result of a small reduction, dK, in the capital input which is offset by a small increase, dL, in the labour input so as to leave output unchanged.

Using the formula for the partial differential (see section 15.4, rule 15.7 above), we can say that the loss of output due to the reduced capital input, which we will label dQ_1, is given by

$$dQ_1 = f_K \times dK = MPK \times dK$$

Similarly, the gain in output due to the increased labour input, which we will label dQ_2, is given by

$$dQ_2 = f_L \times dL = MPK \times dL$$

Since output is left unchanged, the loss of output dQ_1 and the gain in output dQ_2 must be equal in absolute magnitude but of opposite sign. Therefore

$$MPK \times dK = -MPL \times dL$$

Dividing both sides of this by MPK and by dL gives rule 15.13 above. As may be seen in figure 15.14, $\frac{dK}{dL}$ actually measures the slope of the chord RS, not the slope of the isoquant at R. However, when dL is sufficiently small, this error can be ignored.

The convexity of an isoquant

In section 14.8 we said that the isoquants of the neoclassical production function were convex, reflecting the assumption of a decreasing marginal rate of substitution in production. We can now examine the conditions that must be satisfied if the isoquant for, say, $Q = 100$ is to be convex.

From above, the equation of the isoquant is the implicit function $f(K, L) - 100 = 0$ with slope given by the first derivative, $\frac{dK}{dL} = -\frac{f_L}{f_K}$. For convexity, the second derivative $\frac{d^2K}{dL^2}$ must be positive (see section 7.13 to review this). To find this second derivative, we have to find $\frac{d}{dL}$ of $-\frac{f_L}{f_K}$.

Using the quotient rule of differentiation to find the derivative of $-\frac{f_L}{f_K}$, with $u = f_L$ and $v = f_K$, we get

$$\frac{d^2K}{dL^2} = -\frac{f_K \frac{d(f_L)}{dL} - f_L \frac{d(f_K)}{dL}}{(f_K)^2} \tag{15.15}$$

To evaluate $\frac{d(f_L)}{dL}$ and $\frac{d(f_K)}{dL}$ we must take into account the fact that f_L and f_K are functions of both K and L. Moreover, along any isoquant any change in L necessitates an accompanying change in K so as to maintain the constant level of output. The derivative $\frac{d(f_L)}{dL}$ is therefore a total derivative, consisting of the direct effect on f_L of the change in L, plus the indirect effect on f_L of the change in L that works through the accompanying change in K. In the same way, $\frac{d(f_K)}{dL}$ is a total derivative too.

We can find the total derivative $\frac{d(f_L)}{dL}$ using the method described in section 15.5 above. Suppose the function relating f_L to K and L is $f_L = g(K, L)$, with differential

$$d(f_L) = \frac{\partial(f_L)}{\partial K}dK + \frac{\partial(f_L)}{\partial L}dL$$

$$= f_{KL}dK + f_{LL}dL$$

To obtain the total derivative we simply divide through by dL, giving

$$\frac{d(f_L)}{dL} = f_{KL}\frac{dK}{dL} + f_{LL} \qquad \left(\text{since } \frac{dL}{dL} \equiv 1\right)$$

In the above equation, $\frac{dK}{dL}$ is the slope of the isoquant, which we know is $-\frac{f_L}{f_K}$. Substituting this in, we get

$$\frac{d(f_L)}{dL} = f_{KL}\left(-\frac{f_L}{f_K}\right) + f_{LL} \tag{15.16}$$

In the same way, we can suppose the function relating f_K to K and L is $f_K = h(K, L)$. Then by repeating the steps above we can arrive at

$$\frac{d(f_K)}{dL} = f_{KK}\left(-\frac{f_L}{f_K}\right) + f_{LK} \tag{15.17}$$

We can then substitute equations (15.16) and (15.17) into equation (15.15), and, after some manipulation, which includes using Young's theorem (see section 14.6), arrive at

$$\frac{d^2K}{dL^2} = -\frac{1}{(f_K)^3}[(f_K)^2 f_{LL} - 2f_L f_K f_{KL} + (f_L)^2 f_{KK}] \tag{15.18}$$

Convexity of the isoquant requires equation (15.18) to be positive, which (since we have assumed $f_K > 0$) will be true when the expression in square brackets is negative. We can see that this expression is quite complex, involving the marginal products f_K and f_L and their squared values, the direct second derivatives f_{LL} and f_{KK}, and the cross partial derivative f_{KL}. Clearly therefore there are many possible combinations of values for these first and second derivatives of the production function which would cause the isoquant to be convex.

However, one relatively simple (and therefore appealing) case is worth noting. We have assumed that both marginal products, f_K and f_L, are positive. Suppose also that capital and labour inputs are both subject to diminishing marginal productivity, so that f_{LL} and f_{KK} are both negative. Suppose finally that the cross-partial derivative f_{KL} is positive, indicating that an increase in the capital input raises the marginal product of labour (and vice versa, due to Young's theorem). Then all three terms in the square brackets are negative, and $\frac{d^2K}{dL^2}$ is therefore positive, indicating convexity of the isoquants. Note, though, that while these assumptions are *sufficient* to give convexity, they are not *necessary*, as other assumptions could also make (15.18) positive. In the next section we will give an example where these assumptions are not fulfilled, but the isoquants are nevertheless convex. (On necessary and sufficient conditions, see section 2.13.)

More than two inputs

All of the above analysis of the production function can be quite easily extended to cover cases where there are three or more inputs. The production function then becomes a 'hyper-surface' in multi-dimensional space, with equation $Q = f(X_1, X_2, \ldots, X_n)$ where n is the number of inputs. We can take an iso-section through this hyper-surface by holding all inputs constant except, say, the ith (where i refers to any one of the n inputs). The partial derivative $\frac{dQ}{dX_i}$ (which we can alternatively write as f_i) is then the marginal product of the ith input and measures the slope of this iso-section in the X_i direction. We can also define an isoquant by fixing the level of output and then holding all inputs constant except two, say input i and input j. The isoquant would then show all the combinations of inputs i and j consistent with this fixed output, and subject to the fixed levels of all other inputs. The slope of this isoquant is given by $\frac{\partial X_i}{\partial X_j} = -\frac{\partial Q/\partial X_j}{\partial Q/\partial X_i} \equiv -\frac{f_j}{f_i}$ (see equation (15.14a) above). The only difference is in the number of variables being held constant.

15.11 Isoquants of the Cobb–Douglas production function

The Cobb–Douglas production function was introduced in section 14.9. It has the general form

$$Q = AK^\alpha L^\beta \quad \text{with } A, \alpha, \text{ and } \beta \text{ positive}$$

First we will link with the specific example considered in section 14.9, which was

$$Q = K^{0.5}L^{0.5}$$

The marginal products of labour and capital are given by their respective partial derivatives. Thus

$$MPL \equiv \frac{\partial Q}{\partial L} = (K^{0.5})(0.5)L^{-0.5} \quad \text{and}$$

$$MPK \equiv \frac{\partial Q}{\partial K} = (L^{0.5})(0.5)K^{-0.5}$$

So using rule 15.13, the slope of an isoquant for this production function is

$$\frac{dK}{dL} = -\frac{MPL}{MPK} = -\frac{K^{0.5}(0.5)L^{-0.5}}{L^{0.5}(0.5)K^{-0.5}} \quad \text{which simplifies to}$$

$$\frac{dK}{dL} = -\frac{K}{L}$$

Thus the slope is negative and is steeper, the higher is the ratio of capital to labour, $\frac{K}{L}$.

Now let us generalize this. In section 14.9 we also considered the general form of the Cobb–Douglas production function, $Q = AK^\alpha L^\beta$. Repeating the analysis above for this case, we obtain

$$MPL \equiv \frac{\partial Q}{\partial L} = AK^\alpha \beta L^{\beta-1} \quad \text{and}$$

$$MPK \equiv \frac{\partial Q}{\partial K} = A\alpha K^{\alpha-1}L^\beta$$

Again using rule 15.13, the slope of an isoquant is

$$\frac{dK}{dL} = -\frac{\beta}{\alpha}\frac{K}{L} \tag{15.19}$$

Again the slope of the isoquant is negative and becomes steeper as $\frac{K}{L}$ rises. Thus the *MPL* rises, relative to the *MPK*, as $\frac{K}{L}$ rises. (The ratio $\frac{K}{L}$ is one way of measuring *capital intensity*, though capital intensity can also be measured by $\frac{K}{Q}$, capital per unit of output or the capital/output ratio.)

Convexity of the Cobb–Douglas isoquant

We could check whether the isoquants of the Cobb–Douglas function are convex by finding all the first and second partial derivatives and substituting them into equation (15.18). This is a laborious task. Given that we have already found the slope, $\frac{dK}{dL}$ (equation 15.19), a simpler route is to differentiate for a second time to find $\frac{d^2K}{dL^2}$. As explained in the previous section, in finding this derivative we must take into account not just the direct effect of a change in L but also the indirect effect via the accompanying change in K that must occur to maintain the given level of output. Therefore the derivative we are looking for is a total derivative. To find it, we first find the differential of $\frac{dK}{dL} = -\frac{\beta}{\alpha}\frac{K}{L}$. As the left-hand and right-hand sides of this equation are always equal, the differential of the left-hand side must equal the differential of the right-hand side; that is:

$$d\left[\frac{dK}{dL}\right] = d\left[-\frac{\beta}{\alpha}\frac{K}{L}\right] \tag{15.20}$$

Applying the definition of the differential (see rule 15.6) to the right-hand side, we get

$$d\left[\frac{dK}{dL}\right] = d\left[-\frac{\beta}{\alpha}\frac{K}{L}\right] = \frac{\partial}{\partial L}\left[-\frac{\beta}{\alpha}\frac{K}{L}\right]dL + \frac{\partial}{\partial K}\left[-\frac{\beta}{\alpha}\frac{K}{L}\right]dK \tag{15.21}$$

We can evaluate the two partial derivatives in this expression as

$$\frac{\partial}{\partial L}\left[-\frac{\beta}{\alpha}\frac{K}{L}\right] = \frac{\partial}{\partial L}\left[-\frac{\beta}{\alpha}KL^{-1}\right] = \frac{\beta}{\alpha}\frac{K}{L^2} \quad \text{and} \quad \frac{\partial}{\partial K}\left[-\frac{\beta}{\alpha}\frac{K}{L}\right] = -\frac{\beta}{\alpha}\frac{1}{L}$$

Substituting these into the right-hand side of (15.21) gives

$$d\left[\frac{dK}{dL}\right] = \left[\frac{\beta}{\alpha}\frac{K}{L^2}\right]dL + \left[-\frac{\beta}{\alpha}\frac{1}{L}\right]dK \tag{15.22}$$

If we now divide through by dL, the left-hand side of this expression becomes

$$\frac{d}{dL}\left[\frac{dK}{dL}\right] \equiv \frac{d^2K}{dL^2} \quad \text{and the right-hand side becomes} \quad \frac{\beta}{\alpha}\frac{K}{L^2} + \left[-\frac{\beta}{\alpha}\frac{1}{L}\right]\frac{dK}{dL}$$

Thus (15.22) is now

$$\frac{d^2K}{dL^2} = \frac{\beta}{\alpha}\frac{K}{L^2} + \left[-\frac{\beta}{\alpha}\frac{1}{L}\right]\frac{dK}{dL} \tag{15.23}$$

From equation (15.19), we have $\frac{dK}{dL} = -\frac{\beta}{\alpha}\frac{K}{L}$ (the slope of the isoquant). Using this to substitute for $\frac{dK}{dL}$ in (15.23), we get

$$\frac{d^2K}{dL^2} = \frac{\beta}{\alpha}\frac{K}{L^2} + \left[-\frac{\beta}{\alpha}\frac{1}{L}\right]\left[-\frac{\beta}{\alpha}\frac{K}{L}\right] \quad \text{which simplifies to}$$

$$\frac{d^2K}{dL^2} = \frac{\beta}{\alpha}\frac{K}{L^2}\left[1 + \frac{\beta}{\alpha}\right] \tag{15.24}$$

This is positive, as we assume α, β, and K positive, so we conclude that the isoquants of the Cobb–Douglas production function are convex.

There is an important general conclusion to be drawn here. In section 14.9 we showed that there is diminishing marginal productivity in the Cobb–Douglas production function only if α and β are positive but less than 1 (see equations (14.9) and (14.10)). Equation (15.24) however tells us that the isoquants are convex, provided α and β are positive, regardless of whether they are greater or less than 1. This illustrates the point we made when discussing equation (15.18), that diminishing marginal productivity (together with $\frac{d^2K}{dKdL}$ positive) is sufficient, but not necessary, for convexity of Cobb–Douglas isoquants.

For example, if $A = 1$ and $\alpha = \beta = 2$, the production function is $Q = AK^{\alpha}L^{\beta} = K^2L^2$. Then the marginal products are $f_K = 2KL^2$ and $f_L = 2LK^2$, so we see immediately that f_K increases as K increases, and f_L increases as L increases; that is, we have increasing marginal productivity. This is confirmed when we take the direct second derivatives; these are $f_{KK} = 2L^2$ and $f_{LL} = 2K^2$, which are positive. However, when we substitute $\alpha = \beta = 2$ into equation (15.24) we find $\frac{d^2K}{dL^2} = 2\frac{K}{L^2}$, which is positive, so the isoquants of $Q = K^2L^2$ are convex.

15.12 Economic applications 2: the utility function

In section 14.11 we examined the utility function $U = f(X, Y)$, where X and Y are the quantities of two goods consumed and U is the individual's utility or satisfaction. We obtained an iso-U section, or indifference curve, by setting U equal to any fixed value U_0. Two such indifference curves, for $U = 100$ and $U = 200$, were shown in figure 14.24 and are repeated here as figure 15.15.

As we did earlier with the production function isoquant, we can find the slope of any indifference curve at any point by implicit differentiation. For example, the equation of the $U = 100$ indifference curve in figure 15.15 is found by setting $U = 100$ in the utility function $U = f(X, Y)$, giving us the implicit function

$$f(X, Y) - 100 = 0$$

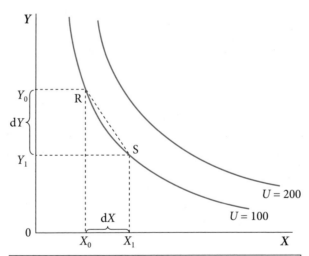

When consumption moves from R to S, there is a loss in utility measured approximately by the partial differential $f_Y dY$ and a gain in utility measured approximately by the partial differential $f_X dX$. As utility is unchanged, we must have $f_Y dY + f_X dX = 0$, which rearranges as $\frac{dY}{dX} = -\frac{f_X}{f_Y}$ = slope of the isoquant at R. (This assumes S very close to R.)

Figure 15.15 Economic interpretation of the slope of an indifference curve.

So, using rule 15.12 above, its slope at any point is given by

$$\frac{dY}{dX} = -\frac{f_X}{f_Y}$$

where f_X and f_Y are the partial derivatives of the utility function $U = f(X, Y)$. These partial derivatives are of course the marginal utilities of goods X and Y (see section 14.12). Thus:

> **RULE 15.14** **The slope of an indifference curve**
>
> Given any utility function $U = f(X, Y)$, the slope of any indifference curve, which by definition is the marginal rate of substitution (MRS) in consumption, is given by
>
> $$MRS \equiv \frac{dY}{dX} = -\frac{f_X}{f_Y} = -\frac{MU_X}{MU_Y}$$
>
> where MU_X and MU_Y are the marginal utilities of the two goods.

In section 14.12 we also saw that the neoclassical assumption that indifference curves are convex means that the MRS decreases in absolute value (that is, the curve becomes flatter) as we move down an indifference curve.

The convexity of an indifference curve

In section 15.10 above we derived equation (15.18) as the second derivative of an isoquant of the production function $Q = f(K, L)$. We now use the same technique to examine for convexity of the indifference curve for, say, $U = 100$. Starting with the expression for the slope in rule 15.14, $\frac{dY}{dX} = -\frac{f_X}{f_Y}$, we can follow exactly the same steps as we did with the production function isoquant in section 15.10 and obtain the second derivative as

$$\frac{d^2Y}{dX^2} = -\frac{1}{(f_Y)^3}[(f_Y)^2 f_{XX} - 2f_X f_Y f_{YX} + (f_X)^2 f_{YY}] \tag{15.25}$$

This equation is identical to equation (15.18) except that f_X, f_Y, f_{YX}, and so on are now the various partial derivatives of the utility function $U = f(X, Y)$. The condition for convexity of an indifference curve is thus formally identical to that for convexity of an isoquant: the expression in square brackets in equation (15.25) must be negative.

As with the production function, there are many combinations of values for the various partial derivatives that can result in convexity. As with the production function, there is one case that appeals by virtue of its simplicity. Assume non-satiation, so that the marginal utilities f_X and f_Y are positive; diminishing marginal utility so that f_{XX} and f_{YY} are negative; and finally that f_{YX} is positive, meaning that an increase in the quantity of good Y increases the marginal utility of good X. Then the expression in square brackets in equation (15.25) is negative and the isoquant is convex.

However, as we saw in section 14.12, the notion of diminishing marginal utility is elusive. While an individual may well experience diminishing marginal utility as a mental state, this can only ever be observed indirectly, in her preference ranking of alternative bundles of goods. This ranking can only tell us that one bundle of goods is preferred to another. It cannot tell us how much utility is derived from either bundle. To assume, as we have done for mathematical convenience, that an individual's utility function has the form $U = f(X, Y)$ requires us to choose some unit of measurement for utility. Because the choice of unit of measurement is inherently arbitrary, whether or not there is diminishing marginal utility depends simply on the units of measurement chosen, and no significance can be attached to the presence or absence of diminishing marginal utility in any particular utility function.

Fortunately, diminishing marginal utility is not a necessary condition for indifference curves to be convex. Since the slope of an indifference curve is given by $-\frac{MU_X}{MU_Y}$, the ratio of marginal utilities (rule 15.14), convexity of indifference curves requires only that MU_X must fall *relative* to MU_Y as we move down the indifference curve. This is consistent with both *absolute* marginal utilities increasing, both decreasing, or one increasing and the other decreasing. Thus while no meaningful statement about how *absolute* marginal utilities change as we move along an indifference curve is possible, no such statement is necessary.

15.13 The Cobb–Douglas utility function

In section 14.13 we considered the Cobb–Douglas utility function as a specific example widely used in economics. We saw that the general form of this function is

$$U = X^\alpha Y^\beta \quad \text{(where } \alpha \text{ and } \beta \text{ are positive)}$$

and took as a specific example

$$U = X^{0.5} Y^{0.5}$$

The marginal utilities of the goods X and Y are given by their respective partial derivatives. Thus

$$MU_Y \equiv \frac{\partial U}{\partial Y} = (X^{0.5})(0.5)Y^{-0.5} \quad \text{and}$$

$$MU_X \equiv \frac{\partial U}{\partial X} = (Y^{0.5})(0.5)X^{-0.5}$$

So using implicit differentiation (rule 15.12) the slope of an indifference curve (an iso-U section) is

$$\frac{dY}{dX} = -\frac{MU_X}{MU_Y} = -\frac{Y^{0.5}(0.5)X^{-0.5}}{X^{0.5}(0.5)Y^{-0.5}} \quad \text{which simplifies to} \quad \frac{dY}{dX} = -\frac{Y}{X}$$

Note that this is the same result as we obtained in chapter 14, but here we have reached it by our newly acquired technique of implicit differentiation. In the case of this particular Cobb–Douglas utility function, we get a very simple result. It says that the slope of the indifference curve depends solely on the relative quantities of the two goods consumed. If we take the more general form of the Cobb–Douglas utility function, $U = AX^\alpha Y^\beta$, then by following the steps above we arrive at

$$\frac{dY}{dX} = -\frac{\alpha}{\beta}\frac{Y}{X}$$

This result has some further significance which we examine in sections 16.11, 16.12, and example 17.6.

Convexity of an indifference curve

By following the steps laid out in section 15.11 for the Cobb–Douglas production function, which led to equation (15.24), we can obtain the second derivative of a Cobb–Douglas indifference curve as

$$\frac{d^2Y}{dX^2} = \frac{\alpha}{\beta}\frac{Y}{X^2}\left[1 + \frac{\alpha}{\beta}\right] \tag{15.26}$$

This is always positive, so we conclude that the indifference curves for any Cobb–Douglas utility function are convex.

15.14 Economic application 3: macroeconomic equilibrium

In section 3.17 we considered a very simple macroeconomic model from which we derived the equilibrium condition

$$Y = C + I$$

where Y is aggregate output and real income, C is planned and actual consumption, and I is spending on investment goods by firms. When this equation is satisfied, there is equilibrium in the market for all goods produced in the economy, in the sense that the plans or desires of households to spend on consumption goods, and of firms to spend on investment goods, are consistent with the economy's actual production of these goods.

We will now develop this model a little further. First, we add some additional types of expenditure: government expenditure, G, and expenditure by foreigners on the country's exports, X. We treat these, as well as investment, as exogenous variables (see section 3.17 for explanation of this term). Our new equilibrium condition is then

$$Y = C + I + G + X \tag{15.27}$$

After subtracting Y from both sides this becomes the implicit function

$$C - Y + I + G + X = 0$$

involving the five variables $Y, C, I, G,$ and X. If we write this implicit function as $F(Y, C, I, G, X)$, then we have

$$F(Y, C, I, G, X) = 0$$

The differential of this implicit function (see equation (15.14) above) is

$$F_Y dY + F_C dC + F_I dI + F_G dG + F_X dX = 0$$

But since in this case $F(Y, C, I, G, X) = C - Y + I + G + X$, we can see that the partial derivatives are

$$F_C = F_I = F_G = F_X = 1 \text{ and } F_Y = -1; \text{ therefore}$$

$$F_Y dY + F_C dC + F_I dI + F_G dG + F_X dX = -dY + dC + dI + dG + dX = 0$$

This rearranges as

$$dY = dC + dI + dG + dX \tag{15.28}$$

This is the total differential of the equilibrium condition (equation (15.27)) above. Though very simple, equation (15.28) is a very powerful analytical tool in macroeconomics, as are other similar formulae based on the total differential.

15.15 The Keynesian multiplier

In section 3.17 we assumed a consumption function in which planned consumption, C, was a linear function of Y. In section 9.23 we defined the marginal propensity to consume (MPC) as the derivative, $\frac{dC}{dY}$, of the consumption function. We will now generalize this by assuming that planned consumption is related to income in an unspecified way, and further that consumption also depends on other variables that we will not try to identify here. Thus we write

$$C = f(Y, \ldots) \tag{15.29}$$

where the dots indicate unspecified, omitted variables. We assume that $\frac{\partial C}{\partial Y}$ (the MPC, which we can also write as f_Y, and which is now a partial derivative) is positive but less than 1. Note that the *MPC* is a partial derivative because there are other (unspecified) variables in the consumption function. However, in what follows we will assume that these unspecified variables remain constant.

The partial differential (see rule 15.7) of equation (15.29) is

$$dC = f_Y dY \quad \text{or} \quad dC = \frac{\partial C}{\partial Y} dY \tag{15.30}$$

If we next substitute this partial differential into equation (15.28) we get

$$dY = f_Y dY + dI + dG + dX \tag{15.31}$$

After subtracting $f_Y dY$ from both sides, and a little rearrangement, this becomes

$$dY = \frac{1}{1 - f_Y}[dI + dG + dX] \tag{15.32}$$

Equation (15.32) tells us the resulting change in Y when I, G, and X vary separately or together. If only I, say, varies, we have $dG = dX = 0$. Equation (15.32) then becomes the partial differential

$$dY = \frac{1}{1 - f_Y} dI$$

Then we can divide both sides by dI and get

$$\frac{\partial Y}{\partial I} = \frac{1}{1 - f_Y} \tag{15.33}$$

Note that dividing by dI turns the left-hand side into a partial derivative. This is because when we divide dY by dI, with other variables held constant, we are measuring the slope of the surface in the I direction, and this slope is measured by the partial derivative. Thus a partial derivative can be viewed as a ratio of differentials.

Equation (15.33) is the Keynesian investment multiplier, first developed by the economist J.M. Keynes in the 1930s. It tells us the size of the change in Y necessary to restore equilibrium in the goods market following a change in investment, I. A key point is that the multiplier, $\frac{1}{1-f_Y}$, is greater than 1 due to our assumption that f_Y is positive but less than 1. Note that we found a similar formula, without using calculus, at the end of section 3.17.

15.16 The *IS* curve and its slope

Up to this point we have treated investment as exogenous. Now, we add a new equation to the model above: the behavioural relationship

$$I = g(r, \ldots) \quad \text{(where } r \text{ is the interest rate)} \tag{15.34}$$

This is called the investment function. It assumes that investment decisions by firms are influenced negatively by the interest rate, since this measures the cost of borrowing money to buy new capital equipment, or measures the income forgone if the firm buys the equipment with its own money. Thus we assume that I falls when r rises, so the partial derivative $\frac{\partial I}{\partial r} \equiv g_r$ is negative. As with the consumption function, this derivative is a partial derivative because investment is also influenced by other unspecified variables, as indicated by the dots in the function. For this analysis we will assume these unspecified variables to be constant.

Shortly we will need the partial differential of the investment function, which is

$$dI = g_r dr \quad \text{or} \quad dI = \frac{\partial I}{\partial r} dr \tag{15.35}$$

To obtain the *IS* curve we substitute the investment function (equation (15.34)) and the consumption function (equation (15.29)) into equation (15.27), resulting in

$$Y = f(Y, \ldots) + g(r, \ldots) + G + X \quad \text{which we can write as}$$

$$0 = f(Y, \ldots) - Y + g(r, \ldots) + G + X \tag{15.36}$$

For given values of G, X, and the unspecified variables, equation (15.36) defines an implicit functional relationship between the two variables Y and r. *This implicit function is known as the IS curve.* The *IS* curve gives us the combinations of Y and r consistent with equilibrium in the goods market, in the sense that the demand for goods by households (C), firms (I), government (G), and foreigners (X) equals actual production (Y).

As we have assumed that the unspecified variables in the consumption function $f(Y, \ldots)$ and investment function $g(r, \ldots)$ are constant, the differential of equation (15.36) (see equation 15.14) is

$$0 = f_Y dY - dY + g_r dr + dG + dX \tag{15.37}$$

where $f_Y dY$ and $g_r dr$ are from equations (15.30) and (15.35) respectively. This is a very useful equation. It tells us how changes in Y, r, G, and X must be related to one another if the economy is to remain on its *IS* curve.

The slope of the *IS* curve

With G and X held constant, so that $dG = dX = 0$, equation (15.37) becomes

$$0 = f_Y dY - dY + g_r dr \tag{15.38}$$

Suppose now the interest rate changes by an amount dr. Then from equation (15.38), the change in Y necessary to keep the economy on its *IS* curve is

$$dY = \frac{g_r}{1 - f_Y} dr \tag{15.39}$$

Since g_r is assumed negative, while $1 - f_Y$ is positive due to our assumption that f_Y is less than 1, $\frac{g_r}{1 - f_Y}$ is negative. Therefore a rise in the interest rate reduces income, and the size of the reduction is greater, the greater is g_r in absolute value and the greater is f_Y.

When graphing the *IS* curve it is customary to put r on the vertical axis and Y on the horizontal. The slope is then $\frac{\partial r}{\partial Y}$, which, by dividing both sides of (15.39) by dY and rearranging, we can obtain as

$$\frac{\partial r}{\partial Y} = \frac{1 - f_Y}{g_r} \tag{15.40}$$

The slope, $\frac{\partial r}{\partial Y}$, is a partial derivative because there are other variables (G and X) in the equation of the *IS* curve which are being held constant. Note that $\frac{\partial r}{\partial Y}$ is obtained as the ratio of two differentials (see also the remarks following equation (15.33) above).

The sign of the slope, and its absolute magnitude, are important. The slope is negative because, as just discussed, g_r is negative and f_Y is less than 1. In absolute terms the slope is greater (that is, more steeply negative) the smaller is g_r (in absolute magnitude) and the smaller is f_Y. (Note that equations (15.39) and (15.40) are easily confused.)

15.17 Comparative statics: shifts in the *IS* curve

If G or X varies, then the *IS* curve shifts in the $0Yr$ plane. We can analyse this using equation (15.37). Assume we have $dG > 0$ (an increase in government expenditure) with $dX = 0$. Then equation (15.37) becomes

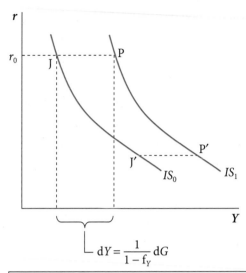

$$dY = \frac{1}{1 - f_Y} dG$$

dY (= distance JP) is rightward shift in *IS* curve that follows increase dG in government expenditure. As dY in equation (15.42) is independent of Y and r, the distance JP is the same whatever the position of J on *IS*.

Thus JP = J'P', though in the figure, our eye is deceived into thinking this is not the case.

Figure 15.16 Shift in the *IS* curve following an increase in government spending.

$$0 = f_Y dY - dY - g_r dr + dG \qquad (15.41)$$

This equation tells us how the changes in Y, r, and G must be related if the economy is to remain on its IS curve. As there are three variables in this equation (dY, dr, and dG) we can't take the analysis any further unless we arbitrarily hold one of them constant. If we decide to hold r constant so that dr = 0, the equation above reduces to $0 = f_Y dY - dY + dG$, which rearranges as

$$dY = \frac{1}{1 - f_Y} dG \qquad (15.42)$$

This is positive and gives us the change in Y that follows a change in G, with X and r held constant. We can see this graphically in figure 15.16. We assume an arbitrary initial position at J on the *IS* curve IS_0. The rise in government expenditure shifts the curve to IS_1. With the interest rate held constant at r_0 by assumption, the new equilibrium is at P. The horizontal distance from IS_0 to IS_1 is given by equation (15.42). As this equation does not contain either Y or r, the rightward shift of the *IS* curve is uniform; that is, the distance from J to P is the same, wherever J may be on IS_0. Of course, this analysis merely establishes the position of the new *IS* curve; whether the economy actually finds its new equilibrium at P, with an unchanged interest rate, is a matter for further analysis.

 Summary of sections 15.10–15.17

In sections 15.10 and 15.11 we used implicit differentiation to find the slope and convexity of a production function isoquant as functions of the marginal products of labour and capital, f_L and f_K, and the second derivatives of the production function, f_{LL}, f_{KK}, and f_{KL}. We applied these results to the specific case of the Cobb–Douglas production function, and found that this function has convex isoquants for all values of the parameters α and β.

In sections 15.12 and 15.13 we used the same technique to find the slope and convexity of an indifference curve as functions of the marginal utilities of the two goods, f_X and f_Y, and the second derivatives of the utility function, f_{XX}, f_{YY}, and f_{XY}. We saw that diminishing marginal utility was not a necessary condition for convexity of indifference curves.

Finally in sections 15.14–15.17 we showed how the differential was a powerful tool for analysing the effects of disturbances to equilibrium in a simple macroeconomic model, and in particular how the Keynesian multiplier could be derived as a partial differential. We also used the differential to find the slope of the *IS* curve, and the size and direction of its shift following a change in an exogenous variable such as government expenditure.

Progress exercise 15.4

1. A firm's production function is

$$Q = K^{0.5}L^{0.5}$$

(a) Find the marginal products of capital and labour. Are they always positive? Sketch their graphs.

(b) Find the equations of the isoquants for (i) 10 units and (ii) 100 units of output.

(c) By implicit differentiation, find the slope of any isoquant and show that this slope is given by the ratio of the marginal products of capital and labour. Is the slope always negative?

(d) Use the information from (b) and (c) above to sketch the isoquants.

(e) Suppose that, from any initial position, the labour input increases by a small amount dL. What change in the capital input is necessary to restore output to its initial level? Explain your answer in words.

2. A firm's production function is

$$Q = 4KL - K^2 - L^2$$

(a) Find the marginal products of capital and labour. Are they always positive? Sketch their graphs.

(b) Use the differential of the production function to find the slope of any isoquant and show that this slope is given by the ratio of the marginal products of capital and labour. Is the slope always negative?

(c) Use the information from (a) and (b) above to sketch the isoquants.

(d) Write down the equation of a typical short-run production function. What determines its slope? What is true at its maximum value? Sketch its graph.

3. Consider the utility function of exercise 14.4, question 3, which was

$$U = X^{0.75}Y^{1.5}$$

(a) By implicit differentiation, find the slope of any indifference curve and show that it is given by the ratio of the marginal utilities of the two goods.

(b) Compare this with your answer to exercise 14.4, question 3, part (a)(ii). Are the two answers different? If so, why?

4. An individual's utility function is

$$U = (X+1)^{0.5}(Y+2)^{0.5}$$

(a) By implicit differentiation, find the slope of any indifference curve and show that it is given by the ratio of marginal utilities of the two goods.

(b) Find the indifference curve for $U = 6$ and show that it is negatively sloped and convex. Does the indifference curve cut the axes? What does this imply about this individual's tastes?

(c) Repeat (a) for the generalized form of the utility function above, $U = (X+a)^{\alpha}(Y+b)^{\beta}$ (where a, b, α, and β are all positive).

5. Suppose there are two individuals, Larry and Milly. Larry's utility function is $U_L = (X+a)^{\alpha}(Y+b)^{\beta}$, while Milly's is $U_M = [(X+a)^{\alpha}(Y+b)^{\beta}]^2$, where a, b, α, and β are all positive. Show by implicit differentiation that, for any given combination of X and Y, Larry's indifference curve has the same slope as Milly's. What does this imply about their tastes?

Checklist

Be sure to test your understanding of this chapter by attempting the progress exercises (answers are at the end of the book). The Online Resource Centre contains further exercises and materials relevant to this chapter www.oxfordtextbooks.co.uk/orc/renshaw3e/.

In this chapter we have put partial derivatives to work to find optimum values of a function. We have also introduced a new concept, the total differential, and applied it. The key points were:

✔ **Optimization.** Using partial derivatives to identify maxima, minima, and saddle points of a function of two or more independent variables.

✔ **The total differential.** How the differential of a function is obtained and what it measures.

✔ **Implicit differentiation.** Using the differential to find the derivative of an implicit function or a function of a function.

✔ **Slope and curvature of an isoquant.** Using the differential to show that the slope of an isoquant is given by the ratio of marginal productivities, and the conditions required for convexity of an isoquant.

✔ **Slope and curvature of an indifference curve.** Using the differential to show that the slope of an indifference curve is given by the ratio of marginal utilities, and the conditions required for convexity of an indifference curve.

✔ **Comparative statics.** Using the differential to analyse the slope and effect of shifts of the *IS* curve in a simple macroeconomic model.

In the next chapter we introduce a powerful new tool, the Lagrange multiplier, which enables us to find solutions to problems such as cost minimization by a firm.

Chapter 16
Constrained maximum and minimum values

OBJECTIVES

Having completed this chapter you should be able to:

- Understand the concept of a constrained maximum or minimum value.

- Understand the characteristics of the constrained optimization problem and the conditions that enable a constrained optimum to be identified.

- Understand how to use the Lagrange multiplier method to solve constrained optimization problems.

- Use the Lagrange multiplier method to solve the producer's problem of minimizing costs and maximizing profit.

- Understand the nature of the solution to the producer's problem and how the solution may fail on certain assumptions.

- Use the Lagrange multiplier method to solve the consumer's utility maximization problem.

- Derive the consumer's demand functions and price and income elasticities from the solution to the consumer's optimizing problem in simple cases.

16.1 Introduction

The process of finding maximum and minimum values is known as optimization. In chapter 7 we examined the technique of optimization for a function with one independent variable, of the form $y = f(x)$. Then in chapter 15 we extended the technique to cover functions with two or more independent variables, such as $z = f(x, y)$. In this chapter we develop the optimization technique in an important new direction: to cover cases where the independent variables are constrained or restricted in the values they can take on.

In this scenario, the point or points we are looking for are called constrained optima or constrained stationary points. As a preliminary, we first show the characteristics of such points by purely diagrammatic methods. From this we derive a simple procedure for finding these points by setting up and solving a set of simultaneous equations. Then we explain the method of

Lagrange multipliers, which provides a more powerful but less intuitive method of locating constrained stationary points. The Lagrange method gives us a very powerful tool for deriving many key results in economic theory, and we illustrate three of these in the second half of the chapter. We explain first how a firm finds the method of production that minimizes the total cost of any given output; second, how a firm achieves maximum profit; and finally, how a consumer with a given budget achieves maximum utility.

16.2 The problem, with a graphical solution

The meaning of a constraint

We will now develop the idea of constrained optimization of some function $z = f(x, y)$. Although the technique can handle any number of independent variables, for simplicity we assume that there are three variables, z, x, and y, which take only positive values.

Up to now, when looking at the function $z = f(x, y)$ we have mostly assumed that x and y are independent variables, meaning that x can take any value, regardless of the value of y, and vice versa. The exception was when we considered cases where x and y where related by some function (see section 15.5), and showed how to derive the total derivative.

Here we take up again that idea of a functional relationship between y and x. We suppose there is some function $z = f(x, y)$, where x and y are not independent of one another. Instead, we assume that x and y can take only the values that satisfy some equation: say, the equation $y = 100 - 0.5x$. This equation is called a constraint, since it constrains or restricts the permitted values of x and y. It means, for example, that if x takes the value 2, y must take the value $100 - 0.5(2) = 99$.

In figure 16.1 the graph of the constraint $y = 100 - 0.5x$ is plotted. Since this is a linear function, its graph is a straight line in the $0xy$ plane. This gives us the locus of points where $y = 100 - 0.5x$. At all of these points we say that the constraint is satisfied. For example, point A where $y = 40$, $x = 120$ satisfies the constraint. Notice that we haven't yet mentioned z, because we are considering only the relationship between x and y.

It now occurs to us that the constraint is not only satisfied at point A, but at any point vertically above A: that is, anywhere on the dotted line AB in figure 16.1. This is true because, as we rise vertically from A, only z is varying, not x and y; and therefore the constraint remains satisfied. Moreover, this is true of *any* point (not merely A) on the line CF. Therefore in figure 16.2 any point on the *surface* CDEF satisfies the constraint. And this surface can be extended indefinitely upwards (in the $+z$ direction).

Thus in three dimensions, a constraint on the permissible combinations of x and y is represented graphically by a surface parallel to the z-axis. In our example, the constraint is a linear function, so the surface CDEF in figure 16.2 is a plane.

Introducing the objective function

Next, suppose we are given a function $z = f(x, y)$. We will assume its shape is conical, as in figure 16.3, with a maximum of z at R. This function is called the **objective function**.

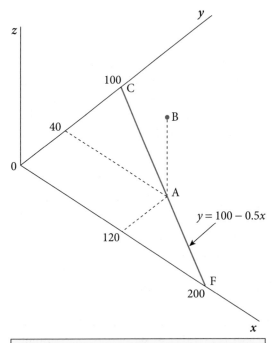

All points such as A on the line $y = 100 - 0.5x$ satisfy the constraint, but so do all points that lie vertically above it. Thus B satisfies the constraint because it has the same x- and y-coordinates as A.

Figure 16.1 The constraint $y = 100 - 0.5x$.

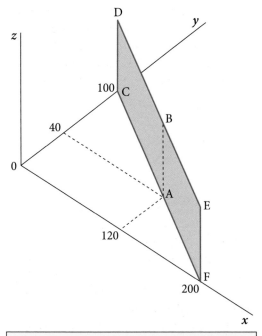

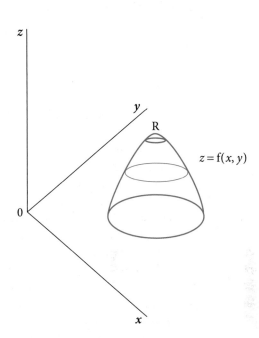

All points on the vertical plane CDEF satisfy the constraint $y = 100 - 0.5x$. The plane extends indefinitely upwards, as satisfying the constraint depends only on the values of y and x. The value of z is immaterial.

Figure 16.2 Vertical plane showing points satisfying the constraint $y = 100 - 0.5x$.

Figure 16.3 The objective function.

Now we come to the crux of our problem. We suppose we are asked to find the maximum value of the objective function $z = f(x, y)$, while at the same time satisfying the constraint $y = 100 - 0.5x$. Solving this problem is called *constrained optimization*.

So, we have to find the values of x and y (let's call them x^* and y^*) that (1) satisfy the constraint *and* (2), when substituted into $z = f(x, y)$, give us the largest possible value of z.

Graphical solution

We can find x^* and y^* graphically by combining figures 16.2 and 16.3 to produce figure 16.4.

In figure 16.4 we imagine the constraint CDEF to be slicing vertically through the surface of the cone, like the blade of a knife cutting through an apple. The slice of the cone that is thereby cut off is moved to one side. This reveals the line JPT as the locus, or collection, of points that lie on the surface of the cone and also on the constraint (that is, all the points on the surface of the apple that were in contact with the knife blade). Of these points, the point with the highest value of z is point P.

Point P is the solution to our problem. It is true that there are other points on the objective function (such as S) where the value of z is higher than at P, but at S the constraint is not satisfied. There are also points on the objective function (such as V) where the constraint is satisfied, but where the value of z is lower than at P.

So point P, with coordinates x^* and y^*, is the point we are looking for. At P, the value of z is given by the vertical distance PA. This value is also given by the label on the iso-z section upon which P lies. We will label this value z^*.

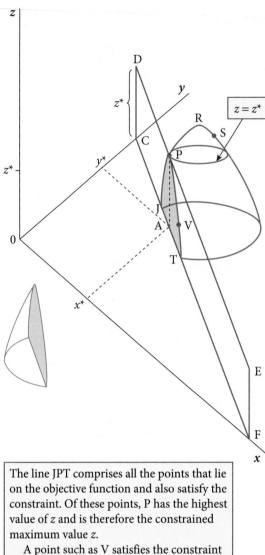

The line JPT comprises all the points that lie on the objective function and also satisfy the constraint. Of these points, P has the highest value of z and is therefore the constrained maximum value z.

A point such as V satisfies the constraint but the value of z is lower than at P. A point such as S has a higher value of z than at P but does not satisfy the constraint.

Figure 16.4 Graphical solution to the constrained maximization problem.

16.3 Solution by implicit differentiation

What we need now is an algebraic technique for locating point P (that is, finding the values of x^* and y^*) in any given case.

Figure 16.5 essentially repeats figure 16.4 but from a bird's-eye view: that is, with the constraint and some iso-z sections of the function $z = f(x, y)$ projected on to the $0xy$ plane. To help you to link the two diagrams, we have included points T, J, and S, and the iso-z contours on which each lies. But our interest is focused on point P.

The key point is that at P (and only at P), two things are simultaneously true:

(1) An iso-z section of the objective function is tangent to the constraint (other iso-z sections merely cut the constraint, for example at T). This tangency means that at P, the iso-z section and the constraint have the same slope in the $0xy$ plane.

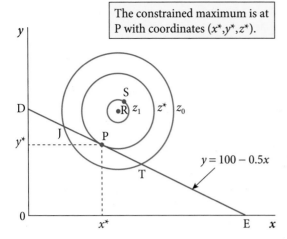

The constrained maximum is at P with coordinates (x^*, y^*, z^*).

Figure 16.5 Bird's-eye view of the objective function and constraint.

(2) The constraint is satisfied because P lies on the plane CDEF (vertically above A in figure 16.4).

Our next task is translate these two conditions into algebraic form.

Condition (1)

This condition says that the iso-z contour and the constraint have the same slope. We can find both of these slopes, as follows:

(a) Slope of constraint: the equation of the constraint is $y = 100 - 0.5x$. So its slope is given by

$$\frac{dy}{dx} = -0.5$$

(b) Slope of iso-z section: the equation of the iso-z section is $z^* = f(x, y)$, where z^* is constant by definition. We can find the slope of this by implicit differentiation (see section 15.7). The derivative is

$$\frac{dy}{dx} = -\frac{f_x}{f_y} \quad \text{where } f_x \text{ and } f_y \text{ are the partial derivatives of } z = f(x, y)$$

So to satisfy condition (1) we set these two slopes equal to one another, which gives

$$-0.5 = -\frac{f_x}{f_y} \tag{16.1}$$

Condition (2)

This condition simply says that the constraint is satisfied, that is

$$y = 100 - 0.5x \tag{16.2}$$

Thus at P, equations (16.1) and (16.2) both hold. Therefore we can find P by solving equations (16.1) and (16.2) as simultaneous equations. In this example these simultaneous equations have only one solution, which therefore gives us the values of x^* and y^*. These values can then be substituted into the objective function $z = f(x, y)$ to find z^*. In our example, this will be the highest value of z which is compatible with the assumed constraint, $y = 100 - 0.5x$.

Distinguishing a maximum from a minimum

The algebraic technique that we have just developed gives us only first order conditions. It does not tell us that our solution for z is a maximum; it could equally be a minimum. Figure 16.6 shows such a case. The objective function is cone-shaped as before, but is now inverted. The locus WJT is the locus of points that lie on the surface of the cone and also on the constraint. Point W is now the *minimum* value of z compatible with the constraint. At a point such as S the value of z is lower but the constraint is not satisfied. At a point such as T the constraint is satisfied but the value of z is higher than at W.

In this case, solving simultaneous equations (16.1) and (16.2) above would identify point W. Thus the technique we have just developed gives us only the first order conditions for a maximum or minimum (see section 7.8). The technique does not permit us to say whether the point we have found is a maximum (like P in figure 16.4) or a minimum (like W in figure 16.6) of the objective function. There exist second order conditions that enable us to determine whether we have found a maximum or a minimum, but we won't cover them in this book.

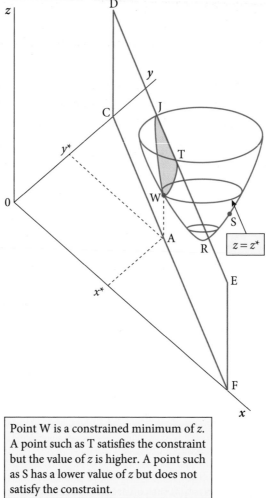

Point W is a constrained minimum of z. A point such as T satisfies the constraint but the value of z is higher. A point such as S has a lower value of z but does not satisfy the constraint.

Figure 16.6 The solution as a constrained minimum of the objective function.

Fortunately when applying this method to an economic problem, it is usually reasonably clear from the context whether the point we have found is a maximum or a minimum. Let us now see how the method we have just developed works out in an example.

EXAMPLE 16.1

We are asked to find the stationary point (that is, maximum or minimum) of $z = x^2 + y^2$ when x and y are subject to the constraint $y = 10 - x$.

Considering first condition (1), the objective function is $z = f(x, y) = x^2 + y^2$, so the partial derivatives are

$$f_x = 2x \quad \text{and} \quad f_y = 2y$$

By implicit differentiation (see section 15.7) the slope of any iso-z section at any point is given by

$$-\frac{f_x}{f_y} = -\frac{2x}{2y} = -\frac{x}{y}$$

The constraint is $y = 10 - x$, so its slope is $\frac{dy}{dx} = -1$.

Setting the two slopes equal to one another gives us

$$-\frac{x}{y} = -1 \tag{16.3}$$

(that is, slope of iso-z contour = slope of constraint). Equation (16.3) is the equivalent of equation (16.1) above.

Condition (2) is simply that the constraint must be satisfied, that is

$$y = 10 - x \tag{16.4}$$

Equation (16.4) is the equivalent of equation (16.2) above. We now have to solve equations (16.3) and (16.4) as a pair of linear simultaneous equations. The solution is $x = y = 5$. (Check this for yourself.) So the solution value of z is

$$z = x^2 + y^2 = 5^2 + 5^2 = 50$$

As we know from section 14.3, the objective function $z = x^2 + y^2$ is a cone with its vertex (point) at the origin. Some iso-z sections of this cone are shown in figure 16.7. In the $0xy$ plane they are circles centred at the origin. The solution is at P where the iso-z section for $z^\star = 50$ is tangent to the linear constraint, $y = 10 - x$.

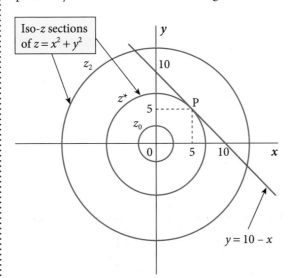

The objective function $z = x^2 + y^2$ is a cone with its point at the origin. P is the constrained minimum of z: the smallest value of z that satisfies the constraint.

Figure 16.7 The objective function $z = x^2 + y^2$ with constraint $y = 10 - x$.

As explained above, we don't know whether $z = 50$ is a maximum or minimum of z. A crude way of answering this is to examine points near P in figure 16.7 that also lie on the constraint. For example, we could try $x = 4.9$, which implies $y = 5.1$ in order to satisfy the constraint $y = 10 - x$. These values give $z = (4.9)^2 + (5.1)^2 = 50.02 > 50$. Thus we have found another solution in which z is greater than our solution value of $z = 50$. So it appears that $z = 50$ is the minimum of z, subject to the constraint. However, this method is not rigorous, as it remains logically possible that we could try another point on the constraint and find z was lower there. (In fact, in this example we know perfectly well that P is minimum because $z = x^2 + y^2$ is a very simple function and we know its shape from section 14.3. In many cases, though, we don't know the shape of the objective function.)

16.4 Solution by direct substitution

It may have occurred to you that an alternative way of solving the previous example would be to proceed as follows. When the constraint is satisfied, we know that $y = 10 - x$ (by definition). So, if we set y equal to $10 - x$ in the objective function, we can be sure that the constraint is satisfied. To do this, we simply replace y with $10 - x$ in the objective function, and get

$$z = x^2 + y^2 = x^2 + (10 - x)^2$$

Multiplying out the brackets gives

$$z = 2x^2 - 20x + 100$$

The effect of this substitution is that instead of looking at $z = x^2 + y^2$, which is all of the points on the surface of the cone, we are now looking at $z = 2x^2 - 20x + 100$, which is only those points on the surface of the cone that also satisfy the constraint. These points are the equivalent of the locus of points WJT in figure 16.6. (Note that $z = 2x^2 - 20x + 100$ is a quadratic function and has a shape similar to the locus WJT in figure 16.6.)

Thus $z = 2x^2 - 20x + 100$ gives us all the values of z that also lie on the constraint. From among these points we now want to find the largest or smallest value of z. Because we now know that the constraint is satisfied at all of these points, we can simply go ahead and find the *unconstrained* maximum or minimum of the function $z = 2x^2 - 20x + 100$ in the usual way, by taking the first derivative and setting it equal to zero. This gives

$$\frac{dz}{dx} = 4x - 20 = 0$$

with solution $x = 5$. Since we know that $y = 10 - x$, it follows that $y = 5$ also. Finally, $\frac{d^2z}{dx^2} = 4$ (that is, positive), so this is a minimum of z.

This method of solution, by direct substitution of the constraint into the objective function, is obviously attractive. It uses a technique that we are already familiar with and moreover gives us a second order condition for distinguishing a maximum from a minimum. Why then do we bother with the previous method of implicit differentiation?

The answer is: (1) the method of direct substitution is not always possible when the constraint is an implicit function; (2) more importantly, the earlier method brings out more clearly the structure of the problem and the nature of its solution. We are more interested in these broader aspects than in merely finding the answer.

Progress exercise 16.1

1. For each of the following functions, find the optimum value of z (that is, maximum or minimum) subject to the given constraint (a) by the method of direct substitution, and (b) by setting the slope of the iso-z section equal to the slope of the constraint. (c) In each case, attempt to assess informally the shapes of the two surfaces and whether the optimum is a maximum or minimum of z.

(i) $z = x^2 - xy + 2y^2$ subject to the constraint $y = 40 - x$

(ii) $z = 4x^2 + 3xy + 6y^2$ subject to the constraint $x + y = 56$

(iii) $z = (x - 1)^2 + (y - 1)^2$ subject to the constraint $2x + y = 38$

16.5 The Lagrange multiplier method

There is an alternative method to the two methods we have just looked at for finding a constrained maximum or minimum. It is called the Lagrange multiplier method, invented in 1797. Why are we looking at yet another way of achieving the same result? The answer is that while the Lagrange multiplier method is less intuitive than the methods above, it is better suited to dealing with more than two variables, more than one constraint, and also with cases where the constraint is an implicit rather than explicit function. We won't attempt to prove how or why the method works; we will just lay out the procedural rules and then do some worked examples.

Suppose we wish to find the maximum or minimum of $z = f(x, y)$, subject to the constraint $g(x, y) = 0$. Notice that the constraint here is an implicit function: it has zero on one side of the equals sign and everything else on the other side.

Method of solution

Step 1

We create a new variable, V, defined as

$$V = f(x, y) + \lambda[g(x, y)] \tag{16.5}$$

Note that V is obtained simply by adding the constraint to the objective function, and multiplying the constraint by a new variable, λ, that we have just produced out of thin air. (Well, we said the method wasn't very intuitive!) (λ is the Greek letter 'lambda'. The square brackets are not strictly necessary.)

This new variable, λ, is called a Lagrange multiplier, and equation (16.5) is called the Lagrangean expression.

Step 2

We find the *unconstrained* maximum or minimum of V in equation (16.5). To do this we proceed in the usual way. We take all the partial derivatives of the function, set them all equal to zero, and solve as simultaneous equations.

In equation (16.5) we have three variables on the right-hand side: x, y, and the new variable, λ, that we have just introduced. Therefore there are three partial derivatives: $\frac{\partial V}{\partial x}$, $\frac{\partial V}{\partial y}$, and $\frac{\partial V}{\partial \lambda}$. So, to implement step 2, we first evaluate $\frac{\partial V}{\partial x}$ (using the normal rules of differentiation) and set it equal to zero. We get

$$\frac{\partial V}{\partial x} = f_x + \lambda g_x = 0 \tag{16.6}$$

How did we get this? To find $\frac{\partial V}{\partial x}$, we have to go through the right-hand side of equation (16.5), differentiating each term with respect to x, treating the other two variables, y and λ, as constants. So, the first term on the right-hand side of equation (16.5) is $f(x, y)$, and the partial derivative of this with respect to x is f_x. Next, we come to λ, which is a multiplicative constant and therefore reappears in the partial derivative. Finally we come to $g(x, y)$, and the partial derivative of this with respect to x is g_x.

Repeating the steps of the previous paragraph, but this time differentiating with respect to y, we get

$$\frac{\partial V}{\partial y} = f_y + \lambda g_y = 0 \tag{16.7}$$

Finally, we have to differentiate equation (16.5) with respect to λ with x and y held constant. We see that $f(x, y)$ is now an additive constant, which therefore disappears. However $g(x, y)$ is now a multiplicative constant, which therefore reappears in the derivative. It multiplies the derivative of λ, with respect to λ, which is simply 1. So we get

$$\frac{\partial V}{\partial \lambda} = g(x, y) = 0 \tag{16.8}$$

We now have to solve the three equations (16.6)–(16.8) simultaneously. The solutions for x and y will give us the constrained maximum or minimum of z. This is as far as we can usefully take the analysis, because the functional forms $f(x, y)$ and $g(x, y)$ have not been given to us.

Note that, as with the method of implicit differentiation in section 16.3, this method, as presented here, gives us only the first order conditions for a maximum or minimum. We lack the second order conditions necessary to distinguish a maximum from a minimum (see section 7.8). In an economic application, the context of the problem often tells us whether we have found a maximum or a minimum. The Lagrange multiplier method can be extended to provide second order conditions, but we won't cover this in this book.

Hint Here is an intuitive explanation of why the Lagrange method works (but don't worry if you don't find this explanation helpful):

(1) Notice that equation (16.8) is simply the constraint. So by solving equations (16.6)–(16.8) as simultaneous equations, we automatically ensure that the constraint is satisfied.

(2) When the constraint is satisfied (as it must be, in the solution) then we will have, by definition, $g(x, y) = 0$. In that case, equation (16.5) will become $V = f(x, y)$. So V and z will be identical, and finding the unconstrained maximum or minimum of V is the same as finding the maximum or minimum of z, with the constraint satisfied.

EXAMPLE 16.2

We are asked to find the maximum or minimum of $z = x^2 + y^2$ subject to the constraint $y = 10 - x$. This is the same as example 16.1 above, but this time we will solve it by the Lagrange multiplier method.

As a preliminary, we have to rearrange the constraint with zero on one side of the equation because the Lagrange method only works if the constraint is in this form. The constraint thus becomes $10 - x - y = 0$ (though $x + y - 10 = 0$ would serve equally well).

Step 1. We form the Lagrangean expression

$$V = x^2 + y^2 + \lambda[10 - x - y] \tag{16.9}$$

Step 2. We take the three partial derivatives of V, set each of them equal to zero, and solve the resulting simultaneous equations for the three unknowns, x, y, and λ. This gives

$$\frac{\partial V}{\partial x} = 2x + \lambda(-1) = 0 \tag{16.10}$$

$$\frac{\partial V}{\partial y} = 2y + \lambda(-1) = 0 \tag{16.11}$$

$$\frac{\partial V}{\partial \lambda} = 10 - x - y = 0 \tag{16.12}$$

Solving three simultaneous equations in three unknowns may seem a difficult task, but the methods are no different from the case of two equations and unknowns. From equation (16.10), $2x = \lambda$. From equation (16.11), $2y = \lambda$. Combining these, $2x = 2y$, so $x = y$. Using this to substitute for y in equation (16.12) gives

$$10 - x - x = 0$$

from which $x = 5$. Therefore $y = 5$ also.

Therefore

$$z = x^2 + y^2 = 5^2 + 5^2 = 50$$

This, of course, is the same answer as we got in example 16.1 above.

Note incidentally that the solution for the third variable, λ, is found by substituting $x = 5$ into equation (16.10), or $y = 5$ into equation (16.11). Either gives $\lambda = 10$. We will explain the significance of λ immediately after the next example.

EXAMPLE 16.3

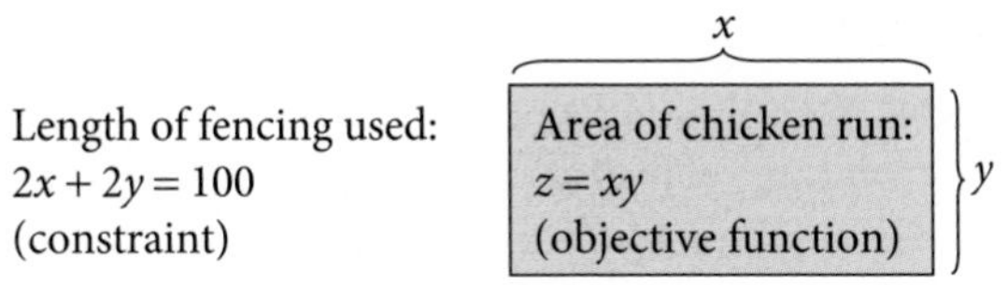

Length of fencing used:
$2x + 2y = 100$
(constraint)

x

Area of chicken run:
$z = xy$
(objective function)

y

Figure 16.8 Maximizing area subject to constraint on length of perimeter.

Suppose a farmer plans to fence off a rectangular area in a field to make a chicken run. He has 100 metres of fencing. What is the largest rectangular area, z, that the fencing can enclose?

To answer this, let x and y denote the lengths of the adjacent sides of the rectangle (see figure 16.8). Then the area fenced off is given by $z = xy$. This is the objective function. The distance round the perimeter of the rectangle is $2x + 2y$ and this must equal 100. So the constraint is $2x + 2y = 100$.

Thus we want to find the maximum of $z = xy$, subject to the constraint $2x + 2y = 100$. We rewrite the constraint as $100 - 2x - 2y = 0$ to make it equal zero.

The Lagrangean expression is therefore

$$V = xy + \lambda[100 - 2x - 2y]$$

The three partial derivatives, set equal to zero, are therefore

$$\frac{\partial V}{\partial x} = y + \lambda(-2) = 0 \tag{16.13}$$

$$\frac{\partial V}{\partial y} = x + \lambda(-2) = 0 \tag{16.14}$$

$$\frac{\partial V}{\partial \lambda} = 100 - 2x - 2y = 0 \tag{16.15}$$

From equation (16.13), $y = 2\lambda$ and from equation (16.14), $x = 2\lambda$. Combining these gives $x = y$. Substituting this into (16.15) gives

$$100 - 2x - 2x = 0$$

from which $x = 25$. Therefore $y = 25$. So $z = xy = 25 \times 25 = 625$ (square metres). Thus the largest rectangle is a square with sides of 25 and area $25^2 = 625$.

As already noted, the Lagrange method as presented here doesn't tell us whether this is the maximum or minimum area but it's intuitively fairly obvious that it's a maximum. To check non-rigorously, we could try a rectangle with sides of 24 and 26 (hence satisfying the constraint that the perimeter equals 100). Then the area is $z = 24 \times 26 = 624$. As this is smaller than our answer above, it seems that we have found the maximum area.

The interpretation of λ, the Lagrange multiplier

In example 16.3 we can easily find λ by substituting $y = 25$ into equation (16.13), giving $\lambda = 12.5$. What does this mean?

Suppose we were given one extra metre of fencing. We describe this as a marginal 'relaxation' of the constraint. If we re-solve the problem with 101 metres of fencing instead of 100, equations (16.13) and (16.14) above are unchanged, so our solution is $x = y$ as before. But equation (16.15) now becomes

$$101 - 2x - 2y = 0$$

and with $x = y$ this gives

$$101 - 2x - 2x = 0$$

from which $x = y = 25.25$ and therefore $z = xy = (25.25)(25.25) = 637.5625$.

The answer in example 16.3 was $z = 625$. So the *increase* in area, Δz, that is made possible by 1 extra metre of fencing is 12.5625 (square metres). This is approximately equal to the solution value of λ, 12.5 (see above).

This is no coincidence: in general, the solution value of λ (12.5 in this example) gives us (approximately) the increase in the solution value of z, the objective function, resulting from a 1-unit relaxation of the constraint. The approximation arises for the reason that we have already met at many points in this book: results obtained using derivatives assume an indefinitely small increase in the fencing available, not an increase as large as in this example, 1 metre.

To summarize, in any problem solved by the Lagrange multiplier method, the solution value for λ gives us the 'value' (that is, the increase/decrease in z that it makes possible) of a small relaxation of the constraint. This has some important uses in economics.

Summary of sections 16.1–16.5

In section 16.2 we explained that a constrained optimization problem is a situation in which we are seeking the maximum or minimum of a function $z = f(x, y)$ (the objective function), subject to some constraint on the values of x and y, such as the function $g(x, y) = 0$ (the constraint). The problem can be solved diagrammatically (section 16.2), by implicit differentiation (16.3) or by direct substitution (16.4), but the Lagrange multiplier method is a more powerful method of solution, though it gives us only first order conditions (section 16.5).

Progress exercise 16.2

Solve each of the problems in exercise 16.1 using the Lagrange multiplier method. In question (i), find the value of λ and verify by recalculating the solution that the value of λ gives (with a small error) the change in z when the constraint is relaxed by one unit.

16.6 Economic applications 1: cost minimization

Cost minimization with perfect competition

In this section we consider a key optimizing choice that every firm has to make: how to produce a given output at minimum total cost. Throughout this section we will assume perfect competition. We begin with a specific example and then generalize it.

EXAMPLE 16.4

First we examine how the firm's costs vary with the quantities of labour and capital that it hires. As usual, we assume that these are the firm's only inputs.

Isocost lines

We assume the firm operates under conditions of perfect competition in the markets for labour and capital, meaning that the price that it pays for each is a given constant, set in the market place. To make the problem more concrete we assume it hires labour inputs (measured in worker-hours) at a constant hourly wage, $w = 4$ (euros). So its total wage costs equal the number of worker-hours hired, L, multiplied by the hourly wage rate, w: that is, $wL = 4L$.

Similarly, the firm hires capital inputs (measured in machine-hours, say) at a constant hourly rental, $r = 2$ (euros). So its total capital costs equal the number of machine-hours hired, K, multiplied by the hourly rental rate, r: that is, $rK = 2K$. This may seem a bit unrealistic because in the real world firms usually buy their machines rather than renting them, but nevertheless there is a large rental (or leasing) market for machinery. Moreover, even when the firm owns its machines, it could be considered to be renting them from itself. So we are perhaps not so unrealistic after all.

Therefore the firm's total cost, TC, (= the cost of its labour and capital inputs) is given by

$$TC = wL + rK = 4L + 2K \tag{16.16}$$

If we now take a given level of total cost, say $TC = 150$, then we have

$$150 = 4L + 2K \tag{16.17}$$

This is an implicit linear function which specifies the various combinations of L and K that the firm can hire for the given level of TC. We call this an **isocost line**, since everywhere along it total cost is constant at 150. It is also known as a **budget line**, since we can think of the given level of total cost as being a budget that the firm's managers are not allowed to exceed. The point to keep in mind is that total cost, or total expenditure, is by definition constant along any isocost or budget line.

The implicit function in equation (16.17) can easily be rearranged to give K as an explicit function of L. This gives

$$K = 75 - 2L \tag{16.17a}$$

We see that with K on the vertical and L on the horizontal axis, this is a downward-sloping straight line with a K-intercept of 75 and a slope of -2. The intercept tells us that if the firm spends its whole budget on capital, it can hire 75 machines. The slope of -2 tells us that the firm must hire 2 fewer machine-hours (saving €4) if it wishes to hire one more worker-hour (costing €4).

If we repeat this process with a new level of total cost, say $TC = 120$, we obtain a second isocost line

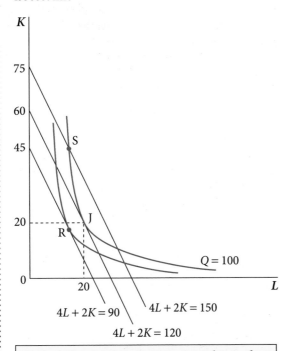

$$120 = 4L + 2K \tag{16.18}$$

which, when rearranged in explicit form, becomes

$$K = 60 - 2L \tag{16.18a}$$

Similarly for $TC = 90$, we obtain a third isocost line in implicit form as

$$90 = 4L + 2K \tag{16.19}$$

with explicit form $K = 45 - 2L$.

In figure 16.9 we have plotted these three isocost lines (equations (16.17), (16.18), and (16.19)). They are parallel lines because their slope depends on the values of w and r, as we will show more clearly in the next example. The higher is the fixed level of TC, the further away is the isocost line from the origin. For example, total cost is higher at J than at R because the budget line that J lies on is further away from the origin. Clearly we can derive as many of these isocost lines as we wish by assigning different values to total cost, so we should think of *every* point in figure 16.9 as lying on an isocost line even though we cannot draw them all.

Given the input prices $w = 4$, $r = 2$, total cost of producing 100 units of output is minimized at J where the $Q = 100$ isoquant is tangent to an isocost line. A point such as R is inferior to J because although it is on a lower isocost line it is also on a lower isoquant. A point such as S is inferior to J because although it is on the same isoquant it is on a higher isocost line.

Figure 16.9 Cost minimization subject to an output constraint.

Turning now to the firm's production, we assume that the firm's production function is

$$Q = 5K^{1/3}L^{2/3} \tag{16.20}$$

where Q = output, and K and L are inputs of capital and labour. (Note that this is a Cobb–Douglas production function, but this is chosen for its algebraic convenience and is not a necessary assumption.) We assume that the firm wishes to produce a given output, $Q_0 = 100$. When this is substituted into the production function, it becomes

$$100 = 5K^{1/3}L^{2/3} \tag{16.20a}$$

Thus the problem facing the firm is that it wishes to minimize its total cost (16.16), subject to the constraint of producing 100 units of output (16.20a). The Lagrangean expression is therefore

$$V = 4L + 2K + \lambda[5K^{1/3}L^{2/3} - 100]$$

(Note that we have rearranged the constraint so that it is equal to zero inside the square brackets, as required for this method to work.)

The partial derivatives, set equal to zero, are

$$\frac{\partial V}{\partial L} = 4 + \lambda 5 K^{1/3}(\tfrac{2}{3})L^{-1/3} = 0 \tag{16.21}$$

$$\frac{\partial V}{\partial K} = 2 + \lambda 5 L^{2/3}(\tfrac{1}{3})K^{-2/3} = 0 \tag{16.22}$$

$$\frac{\partial V}{\partial \lambda} = 5K^{1/3}L^{2/3} - 100 = 0 \tag{16.23}$$

Solving these three simultaneous equations for L, K, and λ is easier than appears at first sight.

From (16.21):

$$5K^{1/3}(\tfrac{2}{3})L^{-1/3} = \frac{-4}{\lambda} \tag{16.21a}$$

From (16.22), after multiplying both sides by 2:

$$10L^{2/3}(\tfrac{1}{3})K^{-2/3} = \frac{-4}{\lambda} \tag{16.22a}$$

Combining (16.21a) and (16.22a) gives

$$5K^{1/3}(\tfrac{2}{3})L^{-1/3} = 10L^{2/3}(\tfrac{1}{3})K^{-2/3}$$

Dividing both sides by the right-hand side, and tidying up, gives

$$\frac{(\tfrac{10}{3})K^{1/3}L^{-1/3}}{(\tfrac{10}{3})L^{2/3}K^{-2/3}} = 1$$

By using the rule of powers that says that $\dfrac{X^m}{X^n} = X^{m-n}$, this becomes

$$\frac{(\tfrac{10}{3})K^{1/3-(-2/3)}L^{-(1/3)-(2/3)}}{\tfrac{10}{3}} = K^{+1}L^{-1} = 1$$

that is, $\dfrac{K}{L} = 1$, or $K = L$.

Using this to substitute for L in equation (16.23) gives

$$5K^{1/3}K^{2/3} - 100 = 0 \quad \text{from which}$$

$$K = \frac{100}{5}$$

that is, $K = 20$ and therefore $L = 20$ (since we know $K = L$).

So the cost-minimizing production technique for producing 100 units of output is to use 20 worker-hours of labour and 20 machine-hours of capital. (The fact that these two quantities are equal is purely an accident of the numbers in this example.)

We should, of course, check that these quantities do in fact produce the given level of output. We do this by substituting $L = 20$ and $K = 20$ into the production function. This gives

$$Q = 5K^{1/3}L^{2/3} = 5(20)^{1/3}(20)^{2/3} = 5(20)^{1/3+2/3} = 100 \quad \text{(as required)}$$

The minimized level of total cost is found by substituting $L = 20$ and $K = 20$ into the TC function, giving

$$TC = 4L + 2K = 4(20) + 2(20) = 120$$

The solution is seen graphically in figure 16.9. The isoquant for $Q = 100$ is tangent to the isocost line for $TC = 120$ at J, where $L = 20$ and $K = 20$. This combination of L and K puts the firm on the isocost line that is the closest possible to the origin (and thus minimizes total cost) while satisfying the output constraint. There are other points, such as S, that satisfy the output constraint but lie on a higher isocost line; and there are other points, such as R, that lie on a lower isocost line but do not satisfy the output constraint.

Minimum or maximum TC?

Although it is clear from figure 16.9 that our solution at J is a point of minimum total cost for the given output, rather than maximum cost, you may be healthily sceptical and demand some algebraic demonstration of this. As mentioned earlier, this can be settled by examining second order conditions, but this is a more complex task than we wish to take on in this book. However, as a rough check, we can try some neighbouring values of K and L (which also produce 100 units of output) to see whether TC is then higher or lower. The production constraint (equation (16.20a) above) is $100 = 5K^{1/3}L^{2/3}$, and this can be rearranged (by dividing both sides by $5L^{2/3}$ and raising both sides to the power 3) as

$$K = \frac{20^3}{L^2}$$

Using this equation we can calculate that if the firm used a little less labour, say $L = 19$, it would need to use 22.16066 units of capital in order to satisfy its output constraint. (We assume infinite divisibility of labour and capital units.) If we then substitute $L = 19$, $K = 22.16066$ into the function $TC = 4L + 2K$, we get $TC = 120.3213$, which is higher than our solution value of $TC = 120$.

Similarly, if the firm used a little more labour, say $L = 21$, it would then need to use 18.14059 units of capital to produce the required output, and then we get $TC = 120.2812$, again higher than our solution value. Although not rigorous, since it's logically possible that some other trial, say $L = 19.5$, might give lower TC, this strongly suggests that we have found a minimum of TC subject to the output constraint.

EXAMPLE 16.5

We now reconsider the cost-minimization problem of the previous example in its most general form. Thus we will not specify any values for the input prices, w for labour and r for capital. Nor will we assign any values to the isocost lines. So we now have a general equation for the whole family of isocost lines, which is

$$TC = wL + rK \tag{16.24}$$

When we rearrange this as an explicit function with K as the dependent variable, it becomes

$$K = \frac{TC}{r} - \frac{w}{r}L \tag{16.24a}$$

which is a linear function with an intercept of $\frac{TC}{r}$ on the K-axis, and a slope of $-\frac{w}{r}$. Note that the slope is thus given by the ratio of input prices. (Be sure that you are clear about the slope and intercept of equation (16.24a) before you move on.)

For maximum generality we will also leave the production function in the general form $Q = f(K, L)$. Therefore the partial derivatives (the marginal products of L and K) are

$$\frac{\partial Q}{\partial L} = f_L \quad \text{and} \quad \frac{\partial Q}{\partial K} = f_K$$

With the required level of output at some unspecified, but fixed, level Q_0, the firm's production constraint is

$$Q_0 = f(K, L) \tag{16.25}$$

So the firm's optimization problem is to minimize equation (16.24), subject to the constraint of equation (16.25). The Lagrangean expression is therefore

$$V = wL + rK + \lambda[f(K, L) - Q_0] \tag{16.26}$$

The partial derivatives, set equal to zero, are

$$\frac{\partial V}{\partial L} = w + \lambda f_L = 0 \tag{16.27}$$

$$\frac{\partial V}{\partial K} = r + \lambda f_K = 0 \tag{16.28}$$

$$\frac{\partial V}{\partial \lambda} = f(K, L) - Q_0 = 0 \tag{16.29}$$

Now, we obviously can't solve this set of simultaneous equations until we are given the functional form of the production function and the numerical values of Q_0, w, and r. But we can take the algebra a little further and thereby derive an interesting and important general result.

From equation (16.27):

$$\frac{w}{f_L} = -\lambda \quad \text{and similarly, from equation (16.28):} \quad \frac{r}{f_K} = -\lambda$$

Combining these two gives

$$\frac{w}{f_L} = \frac{r}{f_K} \quad \text{which rearranges as}$$

$$\frac{f_L}{f_K} = \frac{w}{r} \tag{16.30}$$

The left-hand side of this equation, $\frac{f_L}{f_K}$, is the ratio of the marginal products of L and K. From section 15.10, rule 15.13 we know that this ratio gives the absolute slope of any production function isoquant.

The right-hand side, $\frac{w}{r}$, is the ratio of the input prices, which we saw in equation (16.24a) above gives the absolute slope of the isocost line.

Equation (16.30) therefore says that, in the solution to the cost-minimization problem, the slope of the production function isoquant must equal the slope of the isocost line. This is exactly as we found in the previous example, as shown in figure 16.9, where the isoquant was tangent to the isocost line at J and their slopes were therefore equal.

As well as this graphical interpretation, it is of interest to put equation (16.30) into words. It says that, in general, to minimize the cost of producing a given output, the firm must combine its inputs of L and K in such a way as to make their relative marginal products equal to their relative price. For example, if w is $2 \times r$, the firm must produce where $f_L = 2 \times f_K$. If it costs the firm twice as much to hire one more worker as to rent one more machine, an additional worker must add twice as much to output as does an additional machine.

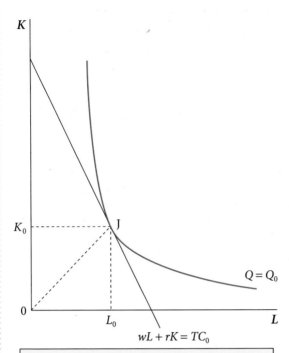

$$wL + rK = TC_0$$

Total cost of Q_0 units of output is minimized at J where the isoquant is tangent to an isocost line. At the tangency the slope of the isoquant, $-\frac{MPL}{MPK}$, equals the slope of the isocost line, $-\frac{w}{r}$.

Figure 16.10 Generalization of the solution to the firm's cost-minimization problem.

Equation (16.30) is a very powerful result because it is very general. However, there are two key assumptions. First, it assumes that the firm is perfectly competitive in the markets for capital and labour, so that w and r are given constants. Second, it assumes that the production function isoquants are convex. Without this assumption we cannot be sure that a tangency with an isocost line exists. From section 14.8 you will hopefully recall that it was the assumption of a decreasing marginal rate of substitution in production (MRS) that guaranteed convexity of the isoquants.

The diagrammatic treatment of this generalized example in figure 16.10 is essentially the same as figure 16.9. However, the specific isocost lines of figure 16.9 are replaced by the single, generic form of equation (16.24): $TC_0 = wL + rK$, where TC_0 is the minimized level of total cost. Also, the position of the production function isoquant is now fixed by the output constraint, Q_0. The cost-minimizing capital and labour inputs are K_0 and L_0. The cost-minimizing capital/labour ratio, or capital intensity, K_0/L_0, is given by the slope of the ray from the origin to J.

Summary

Our findings from examples 16.4 and 16.5 can be summarized in the following rule:

RULE 16.1 Cost minimization by a perfectly competitive firm

The condition $\frac{f_L}{f_K} = \frac{w}{r}$, where f_L and f_K are the marginal products of labour and capital and w and r are the prices of labour and capital, is the *necessary* condition for minimizing the total cost of any given output, for a firm that is perfectly competitive in the markets for labour and capital. Since by assumption the firm cannot influence w and r, it achieves this equality by adjusting its inputs of K and L, thereby changing f_L and f_K.

We will refer to rule 16.1 as the cost-minimizing condition. Note that a point where $\frac{f_L}{f_K} = \frac{w}{r}$ could, logically, be a point where the cost of a given output is maximized rather than min-imized. This is because $\frac{f_L}{f_K} = \frac{w}{r}$ is merely the first order condition for minimizing cost. We have not considered the second order conditions that are needed to distinguish a maximum from a minimum or a point of inflection (see section 7.8).

Progress exercise 16.3

1. A firm's production function is $Q = K^{0.5}L^{0.5}$, where K is capital input measured in machine-hours and L is labour input measured in worker-hours. The firm is perfectly competitive and hires its machines at a constant rental rate of $r = €4$ per hour and its workers at a constant wage rate of $w = €1$ per hour.

(a) Show that the minimum total cost of producing 100 units of output is €400, and that this is achieved by employing 50 machines and 200 workers. (Hint: this can be done using the Lagrange multiplier method, but the other methods discussed earlier in this chapter will also work.)

(b) Show that cost minimization requires employing capital and labour so as to make the ratio of marginal products, $\frac{MPL}{MPK}$, equal to the ratio of input prices, $\frac{w}{r}$.

(c) How does the minimum total cost of producing 200 units compare with that of producing 100 units? What does your answer tell us about the firm's total and average cost curves, as a function of Q?

(d) In your solution to (a) above, you may have noticed that of the minimized total cost, €400, half of this is spent on hiring workers and half on hiring machines. Would this 50:50 split also occur if we had, say, $w = €9$ and $r = €1$? (Hint: the answer to (b) is relevant here.)

2. A firm's production function is $Q = K^{0.75}L^{0.5}$, where K and L are as in question 1. The firm is perfectly competitive and the factor prices are $r = €7.50$ per hour and $w = €3.20$ per hour.

(a) Using the Lagrange multiplier method, show that the minimum cost of producing 40 units of output is €200.

(b) Show that to achieve minimum cost the firm must choose values for L and K such that the isoquant for $Q = 40$ is tangent to an isocost line.

(c) Show that, whatever level of output the firm produces, it will always employ capital and labour in the proportion $\frac{K}{L} = \frac{16}{25}$ unless the relative input prices change.

3. A firm's production function is $Q = K^{0.4}L^{0.5}$, where K and L are as in question 1. The firm is perfectly competitive and the factor prices are $r = €4$ per hour and $w = €3$ per hour. The firm is run by accountants who have imposed a fixed budget on the production managers of €270 per hour.

(a) Using the Lagrange multiplier method, find the maximum hourly output that the firm can produce, given its budget.

(b) Show that this requires employing 30 machines and 50 workers.

(c) How does the solution to this problem, of maximizing output with a given budget, differ from the solutions to questions above in which the problem was to minimize the cost of a given output?

16.7 Economic applications 2: profit maximization

Profit maximization with perfect competition

In the cost-minimization problem examined in the previous section, we simplified by assuming that the firm's output was given. The optimization problem for the firm is then to choose values for K and L that minimize the total cost of the given output. We will now extend the problem facing the firm by treating output as an additional variable. The optimization problem for the firm is now to choose values for Q, K, and L that maximize the firm's profits. For most of this section we assume, as in section 16.6, that the firm is a perfect competitor in the markets for labour and capital, so their prices, w and r, are given constants as far as the firm is concerned. We add the assumption that the firm is a perfect competitor in the product market too, so that the product price, p, is also a given constant for the firm.

Our first task is to develop an equation that gives the firm's profits. We know from section 8.15 that profit, Π, equals total revenue (TR) minus total cost (TC). We also know that under perfect competition, $TR = pQ$, where p is the ruling market price and Q is the chosen level of output. From the previous section we know that $TC = wL + rK$. Assembling all these pieces, we have

$$\Pi = TR - TC = pQ - (wL + rK) \qquad (16.31)$$

This is the firm's objective function, with the choice variables being Q, L, and K. The prices p, w, and r are given constants by virtue of our assumption of perfect competition in the product market and in the factor markets (the markets for factors of production K and L).

In seeking to maximize equation (16.31), the firm is constrained by the fact that Q, L, and K are not independent of one another, but are linked through the production function

$$Q = f(K, L) \qquad (16.32)$$

Therefore the optimization problem for the firm is to maximize profits (equation (16.31)) subject to the constraint of the production function (equation (16.32)).

Solution by substitution

Using the method of direct substitution (see section 16.4), we use equation (16.32) to substitute for Q in equation (16.31). The latter equation then becomes

$$\Pi = p[f(K, L)] - (wL + rK) \qquad (16.31a)$$

The production function (the constraint) is now 'built in' to the profit function, thus automatically ensuring that the constraint is satisfied. Consequently we can now go ahead and seek the *unconstrained* maximum of equation (16.31a) using the technique we learned in section 15.2. On the right-hand side of (16.31a) we have only K and L, because Q has been 'substituted out'. Therefore the partial derivatives, set equal to zero, are

$$\frac{\partial \Pi}{\partial L} = pf_L - w = 0 \qquad (16.33)$$

$$\frac{\partial \Pi}{\partial K} = pf_K - r = 0 \qquad (16.34)$$

If we had actual numerical values for the prices p, w, and r, and a specific functional form for the production function, we could solve this pair of simultaneous equations and thereby find the values of K and L that maximized profit. Substituting these values of K and L into the production function would then give us the optimum (profit-maximizing) output. We would also, of course, have to examine the second derivatives ($\frac{\partial^2 \Pi}{\partial L^2}$, $\frac{\partial^2 \Pi}{\partial K^2}$, and $\frac{\partial^2 \Pi}{\partial K \partial L}$) to check whether the second order conditions for a maximum were also fulfilled (see section 15.2).

Solution by the Lagrange multiplier method

A second, alternative, way of solving the firm's optimization problem is to use the Lagrange multiplier method. Using this method, with equation (16.31) as the objective function and equation (16.32) as the constraint, the Lagrangean expression is

$$V = pQ - (wL + rK) + \lambda[f(K, L) - Q] \qquad (16.35)$$

(Note that as usual we have rearranged the constraint so that it is equal to zero, inside the square brackets.)

There are now four unknowns; Q, L, K, and λ. The four partial derivatives, set equal to zero, are

$$\frac{\partial V}{\partial Q} = p - \lambda = 0 \tag{16.36}$$

$$\frac{\partial V}{\partial L} = -w + \lambda f_L = 0 \tag{16.37}$$

$$\frac{\partial V}{\partial K} = -r + \lambda f_K = 0 \tag{16.38}$$

$$\frac{\partial V}{\partial \lambda} = f(K, L) - Q = 0 \tag{16.39}$$

Again, if we had actual numerical values for the prices p, w, and r, and a specific functional form for the production function, we could solve this set of four simultaneous equations and thereby find the values of Q, K, and L that maximized profit. We will illustrate this in some examples later. Meanwhile, we can extract some very significant general conclusions about the profit-maximizing firm's behaviour from these four equations.

From equation (16.36), we see immediately that $p = \lambda$. Substituting this into equations (16.37) and (16.38), and rearranging a little, we obtain

$$w = pf_L \tag{16.37a}$$

$$r = pf_K \tag{16.38a}$$

Note that these two equations are identical to equations (16.33) and (16.34) above, confirming that solution by direct substitution and solution by the Lagrange multiplier method give the same answer. These two equations, known as the **marginal productivity conditions**, are key optimizing conditions, and we will now discuss their economic interpretation.

The optimum labour input

We first consider equation (16.37a). On the right-hand side we have f_L, the marginal product of labour (*MPL*), multiplied by p, the product price. We know that *MPL* measures (approximately) the addition to physical output that results from hiring one more worker/hour of labour. For example, if the firm produces beer, we might have $f_L = 3$ litres. By multiplying this by p, the product price, we get the addition to total revenue that results from hiring one more worker/hour of labour. This is called the **marginal revenue product** of labour, or the **value of the marginal product** of labour (*VMPL*). For example, if $f_L = 3$ litres and $p = 5$ euros per litre, then $VMPL = p \times f_L = 15$ euros.

On the left-hand side of equation (16.37a) we have the wage rate, w, which under perfect competition is the marginal cost of hiring one more worker/hour of labour. So equation (16.37a) says that, in order to maximize profits, a firm must adjust its labour input to reach a position where hiring one more worker/hour of labour adds the same amount to total revenue ($p \times f_L$) as it adds to total cost (w). Thus the wage rate must equal the value of the marginal product of labour.

A second way of analysing equation (16.37a) is to divide both sides by p, giving

$$\frac{w}{p} = f_L \tag{16.37b}$$

Here the left-hand side, $\frac{w}{p}$, is the **real wage**. This is a key concept in both macro- and micro-economics. To grasp its meaning, suppose $w = 5$ euros per hour and the product price, p, is 2.5 euros. Then the real wage is $\frac{w}{p} = \frac{5}{2.5} = 2$, meaning that the amount of money earned in 1 hour is enough to buy 2 units of output. This is a real magnitude—an amount of output—hence the term 'real wage'. From the point of view of the worker, it is the real reward, in terms of output, for an additional hour's work. From the point of view of the firm, it is the amount of additional output it must *sell* to generate the money receipts sufficient to pay for one additional hour's labour. It is thus the real cost to the firm (in units of output) of one additional hour of labour. Another name for the real wage defined in this way is the **product wage**.

On the right-hand side of equation (16.37b) we have f_L, the marginal product of labour. So equation (16.37b) says that, in order to maximize profits, a firm must adjust its labour input to reach a position where the additional output from one additional worker/hour of labour input (f_L) equals the real cost of one additional worker/hour of labour input ($\frac{w}{p}$).

The optimum capital input

Looking now at equation (16.38a), our discussion of the optimum capital input exactly parallels that of the optimum labour input above. On the right-hand side of equation (16.38a) we have f_K, the marginal product of capital (*MPK*), multiplied by p, the product price. Thus the right-hand side is the addition to total revenue that results from hiring one more machine/hour of capital. This is called the *marginal revenue product* of capital, or the *value of the marginal product* of capital (*VMPK*).

On the left-hand side of equation (16.38a) we have the cost of renting one machine/hour of capital, r, which under perfect competition is the marginal cost of hiring one more machine/hour of capital input. So equation (16.38a) says that, in order to maximize profits, a firm must adjust its capital input to reach a position where hiring one more machine/hour of capital adds the same amount to total revenue ($p \times f_K$) as it adds to total cost (r).

We can also divide both sides of (16.38a) by p, giving

$$\frac{r}{p} = f_K \tag{16.38b}$$

Here the left-hand side, $\frac{r}{p}$, is the analogue of the real wage that we discussed immediately above. From the point of view of the owner of the capital, it is the **real reward of capital**, the compensation for making it available to the firm. From the point of view of the firm, it is the **real cost of capital** (measured in units of output). So equation (16.38b) says that, in equilibrium, hiring one more unit of capital incurs a real cost ($\frac{r}{p}$) that exactly equals the resulting additional output (f_K).

The optimum combination of *L* and *K*

Since we have established above the conditions for optimizing the labour input and also the capital input, there is some redundancy in examining the optimum combination of L and K, since if each is separately optimized their combination should be automatically optimized too. However, their optimum combination is worth at least a brief examination. By dividing both sides of equation (16.37a) by f_L we get

$$\frac{w}{f_L} = p \tag{16.40}$$

Similarly, by dividing both sides of equation (16.38a) by f_K we get

$$\frac{r}{f_K} = p \tag{16.41}$$

Since these two equations have the same right-hand side, their left-hand sides must be equal, so we have $\frac{w}{f_L} = \frac{r}{f_K}$, which quickly rearranges as

$$\frac{w}{r} = \frac{f_L}{f_K} \tag{16.42}$$

This equation should be familiar, as we met it in the previous section (see rule 16.1). It is the cost-minimizing condition: that is, the condition for minimizing the total cost of a given output. The fact that it has popped up again here confirms something that we would intuitively expect: that in order to maximize profits, a necessary condition is that costs be minimized for any given output. Logically, we can think of profit maximization as being broken down into two sub-tasks: first, minimizing the cost of any given output; second, deciding on the optimum level of output.

The optimum output

Up to this point we have examined only the optimum inputs of K and L, without explicit reference to output, Q. What can we infer about the optimization of Q from this discussion? The answer is that once the firm has determined its optimal labour input and its optimal capital input by solving equations (16.36), (16.37), and (16.38), in the way we have just discussed, then there is in principle nothing left to be done except to feed these optimized values for K and L into the production function (equation (16.39) above) and thereby find the optimum level of output. However, finding the optimum values of K and L, and from them the optimum value of Q, is not always as straightforward as we might expect. Some difficulties that may arise are discussed in sections 16.9 and 17.4.

Summary

We can collect our results from this section and express them as the following rule:

RULE 16.2 **Profit maximization: the marginal productivity conditions**

As a necessary condition for profit maximization under perfect competition, the firm must adjust its capital and labour inputs so as to achieve

(1) equality between the real wage and the *MPL*, that is

$$\frac{w}{p} = f_L \tag{16.37b}$$

(2) equality between the real cost of capital and the *MPK*, that is

$$\frac{r}{p} = f_K \tag{16.38b}$$

These two equations, known as the marginal productivity conditions, are solved simultaneously to find the profit-maximizing values of K and L. Substitution of these values into the production function then gives the optimum, or profit-maximizing, output.

Note that this rule gives only the first order conditions, or necessary conditions, for maximum profit. The second order conditions for distinguishing a maximum from a minimum or saddle point, when using the Lagrange multiplier method, are not covered in this book, though we examine second order conditions in a slightly different way in section 17.4.

Note also that if equations (16.37b) and (16.38b) are satisfied, then we have (after a little rearrangement)

$$\frac{w}{r} = \frac{f_L}{f_K}$$

which tells us that when rule 16.2 is satisfied, the condition for cost minimization is also satisfied (see rule 16.1 and equation (16.42) above).

16.8 A worked example

Next we will work through a straightforward example of the algebraic model of profit maximization that we have just discussed, using the Lagrange multiplier method. Our starting point is exactly the same as in the previous section. The firm wishes to maximize its profits, given by

$$\Pi = TR - TC = pQ - (wL + rK) \tag{16.43}$$

subject to the constraint of the production function

$$Q = f(K, L) \tag{16.44}$$

EXAMPLE 16.6

In this example we assume that the prices are $p = 20$, $r = 8$, and $w = 2$, and that the production function is the Cobb–Douglas form $Q = K^{0.2}L^{0.6}$.

Thus equations (16.43) and (16.44) become

$$\Pi = 20Q - (2L + 8K) \tag{16.43a}$$

$$Q = K^{0.2}L^{0.6} \tag{16.44a}$$

The Lagrangean expression is therefore

$$V = 20Q - (2L + 8K) + \lambda[K^{0.2}L^{0.6} - Q]$$

The partial derivatives, set equal to zero, are

$$\frac{\partial V}{\partial Q} = 20 - \lambda = 0 \tag{16.45}$$

$$\frac{\partial V}{\partial L} = -2 + \lambda K^{0.2}(0.6)L^{-0.4} = 0 \tag{16.46}$$

$$\frac{\partial V}{\partial K} = -8 + \lambda(0.2)K^{-0.8}L^{0.6} = 0 \tag{16.47}$$

$$\frac{\partial V}{\partial \lambda} = K^{0.2}L^{0.6} - Q = 0 \tag{16.48}$$

We have here a set of four simultaneous equations with four unknowns: λ, Q, K, and L. They are actually quite easy to solve. From equation (16.45), we see immediately that $\lambda = 20$, and we can use this to substitute for λ in equations (16.46) and (16.47). After this substitution, and a little rearrangement, (16.46) and (16.47) become

$$(0.6)K^{0.2}L^{-0.4} = \frac{2}{20} \tag{16.46a}$$

$$(0.2)K^{-0.8}L^{0.6} = \frac{8}{20} \tag{16.47a}$$

Note that these two equations are in fact the marginal productivity conditions (see rule 16.2). For example, the left-hand side of (16.46a) is the marginal product of labour, and the right-hand side is the real wage.

From equation (16.46a), after a little rearrangement, we get $K^{0.2} = \frac{1}{6}L^{0.4}$. Raising both sides to the power 5, this becomes

$$K = \left(\frac{1}{6}\right)^5 L^2 \tag{16.46b}$$

From equation (16.47a), after similar manipulation, we get $K^{0.8} = \frac{1}{2}L^{0.6}$. Raising both sides to the power 1.25, this becomes

$$K = \left(\frac{1}{2}\right)^{1.25} L^{0.75} \qquad (16.47b)$$

Since equations (16.46b) and (16.47b) have the same left-hand side, their right-hand sides must be equal, therefore $(\frac{1}{6})^5 L^2 = (\frac{1}{2})^{1.25} L^{0.75}$. Dividing both sides by $L^{0.75}$ and by $(\frac{1}{6})^5$,

we get $L^{1.25} = \dfrac{(\frac{1}{2})^{1.25}}{(\frac{1}{6})^5}$. Raising both sides to the power $\frac{1}{1.25}$ gives

$$L = \frac{\frac{1}{2}}{\left[(\frac{1}{6})^5\right]^{1/1.25}} = \frac{\frac{1}{2}}{(\frac{1}{6})^4} = \frac{1}{2}6^4 = 648$$

Therefore, using (16.46b), $K = (\frac{1}{6})^5 L^2 = (\frac{1}{6})^5 (648)^2 = \dfrac{648^2}{6^5} = 54$

Thus we have obtained $L = 648$ and $K = 54$ from equations (16.45)–(16.47). Substituting these values into equation (16.48) gives

$$Q = 54^{0.2} 648^{0.6} = 108$$

We can now substitute $L = 648$, $K = 54$, and $Q = 108$ into the profit function (equation (16.43a)) and get

$$\Pi = 20Q - (2L + 8K) = 20(108) - 2(648) - 8(54) = 432$$

So the first order conditions for maximum profit are satisfied when the firm hires 648 units of labour and 54 units of capital, with which it produces 108 units of output. Profit is then 432.

However, as explained earlier, this could be a minimum of profits rather than a maximum, because only the first order conditions are satisfied. As a non-rigorous check on this point, we can examine profits when a slightly different combination of K and L is used. Let's assume the firm decides to use 660 units of labour instead of 648. Then it seems reasonable to assume that the firm will continue to minimize costs, by using a capital/labour ratio such that $\frac{f_L}{f_K} = \frac{w}{r}$ (see rule 16.1). In this example we have $w = 2$, $r = 8$, and $f(K, L) = Q = K^{0.2}L^{0.6}$, so $f_L = 0.6K^{0.2}L^{-0.4}$ and $f_K = 0.2K^{-0.8}L^{0.6}$. Substituting these values into rule 16.1 gives

$$\frac{0.6K^{0.2}L^{-0.4}}{0.2K^{-0.8}L^{0.6}} = \frac{2}{8}$$

After simplifying the left-hand side, this becomes $3\frac{K}{L} = \frac{2}{8}$, from which $K = \frac{1}{12}L$. Therefore, if the firm uses 660 units of labour, it will use $\frac{1}{12}660 = 55$ units of capital. Then, from the production function, output will be $Q = K^{0.2}L^{0.6} = 55^{0.2}660^{0.6} = 109.6$. Profit is therefore $\Pi = 20Q - (2L + 8K) = 20(109.6) - 2(660) - 8(55) = 413.9$. Thus the slightly larger inputs of capital and labour result in slightly larger output but slightly lower profits. Though not conclusive, this suggests that our solution is probably a maximum of profit. It is certainly not a minimum, but a saddle point is a logical possibility that we will not pursue here. We examine this finding further in the next section.

16.9 Some problems with profit maximization

In equations (16.37b) and (16.38b) above, the variables are K and L since they determine the marginal products, f_L and f_K. The firm achieves equilibrium by varying K and L both relatively and absolutely, thereby causing f_K and f_L to vary. However, we know from section 3.11 that a pair of simultaneous equations may not have a unique solution. Instead, the pair may have no solution or an infinite number of solutions. This is one problem that can arise when we try to solve equations (16.37b) and (16.38b). A second problem that can arise is that the marginal productivity conditions (equations (16.37b) and (16.38b)) are satisfied, but the solution gives a minimum of profits rather than a maximum.

We will now illustrate both of these problems by means of a worked example. We also analyse these problems further in section 17.4.

EXAMPLE 16.7

In this example we will continue to assume perfect competition, but to make our analysis as general as possible we will leave the values of p, w, and r unspecified. We will also continue to assume that the production function is of Cobb–Douglas form but, again to make the analysis as general as possible we will not assign specific values to the parameters α and β. Thus the firm wishes to maximize its profits, given by

$$\Pi = TR - TC = pQ - (wL + rK)$$

subject to the constraint of its production function

$$Q = K^{\alpha}L^{\beta}$$

The Lagrangean expression is therefore

$$V = pQ - (wL + rK) + \lambda[K^{\alpha}L^{\beta} - Q]$$

with partial derivatives set equal to zero:

$$\frac{\partial V}{\partial Q} = p - \lambda = 0 \tag{16.49}$$

$$\frac{\partial V}{\partial L} = -w + \lambda K^{\alpha}\beta L^{\beta-1} = 0 \tag{16.50}$$

$$\frac{\partial V}{\partial L} = -r + \lambda \alpha K^{\alpha-1}L^{\beta} = 0 \tag{16.51}$$

$$\frac{\partial V}{\partial \lambda} = K^{\alpha}L^{\beta} - Q = 0 \tag{16.52}$$

As in the previous example we see from equation (16.49) that $\lambda = p$, and this can be immediately substituted into equations (16.50) and (16.51). After this substitution, and some rearrangement, equations (16.50) and (16.51) become

$$\frac{w}{K^{\alpha}\beta L^{\beta-1}} = p \tag{16.50a}$$

$$\frac{r}{\alpha K^{\alpha-1}L^{\beta}} = p \tag{16.51a}$$

As these two equations have the same right-hand side, their left-hand sides must be equal, so we have

$$\frac{r}{\alpha K^{\alpha-1}L^{\beta}} = \frac{w}{K^{\alpha}\beta L^{\beta-1}} \quad \text{This rearranges as } \frac{K^{\alpha}\beta L^{\beta-1}}{\alpha K^{\alpha-1}L^{\beta}} = \frac{w}{r}, \text{ which simplifies to}$$

$$K = \frac{\alpha}{\beta}\frac{w}{r}L \tag{16.53}$$

Note that (16.53) is the cost-minimizing condition (see rule 16.1).

Next, we use (16.53) to substitute for K in equation (16.50) (though we could equally have used equation (16.51)). Equation (16.50), with λ replaced by p, then becomes

$$-w + p\left[\frac{\alpha}{\beta}\frac{w}{r}L\right]^{\alpha}\beta L^{\beta-1} = 0$$

After some manipulation, we obtain

$$
L = \left[\frac{w}{\beta p(\frac{\alpha}{\beta}\frac{w}{r})^{\alpha}} \right]^{\frac{1}{\alpha+\beta-1}}
\tag{16.54}
$$

This gives us a general solution for L in terms of the parameters, p, w, r, α, and β. As a check on the accuracy of our algebra we can substitute into (16.54) the parameter values of example 16.6: $p = 20$, $w = 2$, $r = 8$, $\alpha = 0.2$, and $\beta = 0.6$. This gives

$$
L = \left[\frac{2}{0.6(20)(\frac{0.2}{0.6}\frac{2}{8})^{0.2}} \right]^{\frac{1}{0.8-1}} = 648 \quad \text{(the same answer as we found in example 16.6)}
$$

Having found L, we can quickly find general solutions for K and Q. By using (16.54) to substitute for L in (16.53), we obtain

$$
K = \frac{\alpha}{\beta}\frac{w}{r} \left[\frac{w}{\beta p(\frac{\alpha}{\beta}\frac{w}{r})^{\alpha}} \right]^{\frac{1}{\alpha+\beta-1}}
\tag{16.55}
$$

Finally, using (16.54) and (16.55) to substitute for L and K in the production function (equation (16.52)), we obtain Q as

$$
Q = \left(\frac{\alpha}{\beta}\frac{w}{r} \right)^{\alpha} \left[\frac{w}{\beta p(\frac{\alpha}{\beta}\frac{w}{r})^{\alpha}} \right]^{\frac{\alpha+\beta}{\alpha+\beta-1}}
\tag{16.56}
$$

Deriving the firm's cost curves

The solutions for L, K, and Q given by equations (16.54)–(16.56) fail if $\alpha + \beta = 1$ (the case of constant returns to scale), for then we find ourselves dividing by zero in the exponents of all three equations, and dividing by zero has an undefined result (see section 1.3). Moreover, even when $\alpha + \beta \neq 1$ so that solutions exist, it remains unclear whether the solutions give us a maximum of profits or a minimum. We can achieve some insight into these questions in the following way.

First, we take the cost-minimizing condition (equation (16.53)) and use it to substitute for K in the firm's total cost function. We get

$$
TC = wL + rK = wL + r\left(\frac{\alpha}{\beta}\frac{w}{r}L \right) = wL\left(1 + \frac{\alpha}{\beta} \right)
\tag{16.57}
$$

This gives total cost as a function of labour input only, when costs are minimized.

Next, we again take the cost-minimizing condition (equation (16.53)) and this time use it to substitute for K in the firm's production function. We get

$$
Q = K^{\alpha}L^{\beta} = \left(\frac{\alpha}{\beta}\frac{w}{r}L \right)^{\alpha} L^{\beta} = \left(\frac{\alpha}{\beta}\frac{w}{r} \right)^{\alpha} L^{\alpha+\beta}
$$

Then we invert this function to give L as a function of Q. We get

$$
L = \left[Q\left(\frac{\beta}{\alpha}\frac{r}{w} \right)^{\alpha} \right]^{\frac{1}{\alpha+\beta}}
\tag{16.58}
$$

Finally, we use equation (16.58) to substitute for L in the total cost function (equation (16.57)). The result is

$$TC = wL\left(1 + \frac{\alpha}{\beta}\right) = w\left[Q\left(\frac{\beta}{\alpha}\frac{r}{w}\right)^{\alpha}\right]^{\frac{1}{\alpha+\beta}}\left(1 + \frac{\alpha}{\beta}\right)$$

We can rearrange this more tidily as

$$TC = w\left(1 + \frac{\alpha}{\beta}\right)\left(\frac{\beta}{\alpha}\frac{r}{w}\right)^{\frac{\alpha}{\alpha+\beta}}Q^{\frac{1}{\alpha+\beta}}$$

For compactness, let us define a new parameter, Z, as $Z = w(1 + \frac{\alpha}{\beta})(\frac{\beta}{\alpha}\frac{r}{w})^{\frac{\alpha}{\alpha+\beta}}$. Note that Z is constant unless one or more of the parameters w, r, α, and β change. Then we have

$$TC = ZQ^{\frac{1}{\alpha+\beta}} \tag{16.59}$$

This equation gives the firm's total cost as a function of output. Recall that we derived this function using the cost-minimizing condition (equation (16.53)), so we know this gives us the *minimum* total cost of any output. Note that this is long-run total cost since we assume both K and L to be variable.

From the total cost curve we can derive marginal cost (MC) and average cost (AC) in the usual way as

$$MC = \frac{dTC}{dQ} = \frac{1}{\alpha+\beta}ZQ^{\frac{1}{\alpha+\beta}-1} \quad \text{and} \quad AC = \frac{TC}{Q} = \frac{ZQ^{\frac{1}{\alpha+\beta}}}{Q} = ZQ^{\frac{1}{\alpha+\beta}-1}$$

Note that $MC = \frac{1}{\alpha+\beta}AC$.

Conclusions

In this section we have analysed profit maximization by a firm in conditions of perfect competition with a Cobb–Douglas production function. There are three cases, depending on the values of α and β:

(1) $\alpha + \beta$ is less than 1. For example, suppose $\alpha + \beta = 0.5$. Then in equation (16.59) we have $TC = ZQ^2$. Therefore $MC = 2ZQ$, and $AC = ZQ$. These total, marginal, and average cost curves are graphed in figure 16.11. Both MC and AC increase as output increases, and $MC > AC$. (To be precise, $MC = 2AC$.)

With perfect competition, price is constant, and equals marginal revenue. Therefore in figure 16.11(b) we can draw the $p = MR$ function as a horizontal straight line. The first order condition for a maximum of profit is satisfied at Q_0, where $MC = MR$. Moreover, because at Q_0 the slope of the MC curve is greater than the slope of the MR curve, the second order condition for a maximum of profits is also satisfied (see section 8.20).

We can also read off the maximized profits in figure 16.11. We know that profits, Π, are given by $\Pi = TR - TC$. Under perfect competition, $TR = pQ$; and, by definition, $TC \equiv AC \times Q$. Therefore we can write profits as $\Pi = (p - AC)Q$. In figure 16.11, the shaded rectangle has a height of $p - AC$ and a base of Q_0. Its area thus gives us profits when $Q = Q_0$. (Example 16.6 above, where $\alpha + \beta = 0.8$, falls into this category.)

(2) $\alpha + \beta$ is greater than 1. For example, suppose $\alpha + \beta = 2$. Then we have $TC = ZQ^{0.5}$, $MC = (0.5)ZQ^{-0.5}$, and $AC = ZQ^{-0.5}$. These total, marginal, and average cost curves are graphed in figure 16.12. Both MC and AC decrease as output increases, and $MC < AC$. (To be precise, $MC = 0.5AC$.) As in figure 16.11, we can draw the $p = MR$ function as a horizontal straight line. The first order condition for a maximum of profit is satisfied at Q_0, where $MC = MR$. However, because at Q_0 the slope of the MC curve is less than the slope of the MR curve, the second order condition for a maximum of profits is not satisfied (see section 8.20). This in fact is a point of

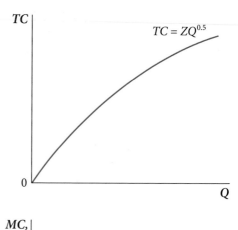

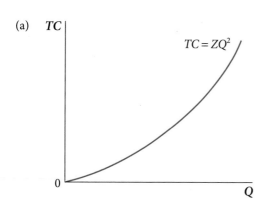

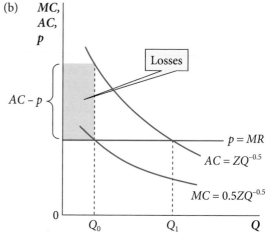

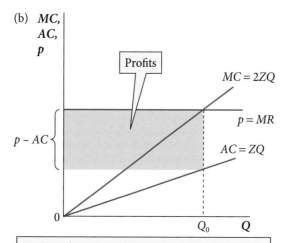

With perfect competition and a Cobb–Douglas production function $Q = K^\alpha L^\beta$, if $\alpha + \beta = 0.5$, the firm's AC increases with output, and therefore $MC > AC$. The MC curve lies above the AC curve. There is a unique profit-maximizing output, Q_0.

With perfect competition and the Cobb–Douglas production function $Q = K^\alpha L^\beta$, if $\alpha + \beta = 2$, the firm's AC decreases with output, and therefore $MC < AC$. The MC curve lies below the AC curve. There is a unique output, Q_0, at which the first order conditions for a maximum of profit are satisfied, but the second order conditions are not. This in fact is a minimum of profits, or a maximum of losses. To the right of Q_1, profits are positive and increase with output.

Figure 16.11 Profit maximization with production function $Q = K^\alpha L^\beta$, $\alpha + \beta = 0.5$.

Figure 16.12 Profit maximization with production function $Q = K^\alpha L^\beta$, $\alpha + \beta = 2$.

minimum profit, or maximum loss. The maximized loss is measured by the area of the shaded rectangle, which has a height of $AC - p$ and a base of Q_0.

Although the firm might find itself briefly producing at $Q = Q_0$ (perhaps on the advice of an economist who looked only at the first order condition for maximum profit, forgetting about the second order condition), it will not remain there. For it will notice that to the right of Q_0, marginal revenue is greater than marginal cost, and the firm can therefore increase its profits (that is, reduce its losses) by increasing output. Because $MR > MC$ everywhere to the right of Q_0 (see section 8.20), the firm will wish to expand its output without limit. As output increases from Q_0, losses fall and eventually, to the right of Q_1, profits become positive and increase without limit as output increases.

(3) The boundary between the two cases above is where $\alpha + \beta = 1$. Then we have $TC = ZQ$, and $MC = AC = Z$. Thus total cost is a linear function of Q, with a slope given by the

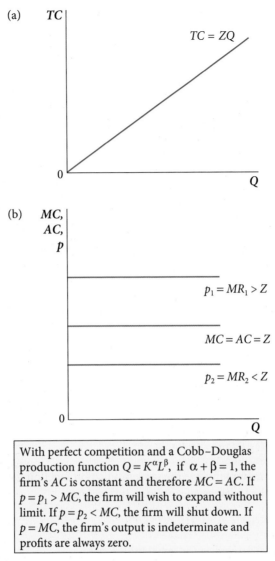

Figure 16.13 Profit maximization with production function $Q = K^{\alpha}L^{\beta}$, $\alpha + \beta = 1$.

parameter Z. With a linear total cost curve, it is no surprise to see that marginal cost is constant. Nor is it a surprise that when marginal cost is constant, marginal and average cost are equal (see section 8.5 if you need to revise this point).

These total, marginal, and average cost curves are graphed in figure 16.13. With $MC = AC = Z$, the marginal cost curve is a horizontal straight line with the average cost curve lying on top of it, and both equal to Z. The key question now is whether the ruling market price of the product, p, is greater, equal to or less than Z. In the figure we show two possible prices: $p_1 > Z$ and $p_2 < Z$.

With price p_1, the $p_1 = MR_1$ function lies always above the MC curve, so there is no output at which $MC = MR$, and therefore no output at which the first order conditions for maximum profit are fulfilled. However, since $p_1 > AC$, profits, given by $\Pi = (p_1 - AC)Q$, are positive and increase as Q increases. The firm will therefore wish to increase output without limit. Another way of reaching the same conclusion is by noting that, because MR is always greater than MC, the firm can always increase its profits by increasing output (see section 8.20).

With price p_2, the $p_2 = MR_2$ function lies always below the MC curve, so again there is no output at which $MC = MR$, and therefore no output at which the first order conditions for maximum profit are fulfilled. In this case, since $p_2 < AC$, profits, given by $\Pi = (p_2 - AC)Q$, are always negative, and losses increase as Q increases. Therefore the firm will wish to reduce its output without limit; that is, it will cease production. Another way of reaching the same conclusion is by noting that, because MC is always greater than MR, the firm can always increase its profits (that is, reduce its losses) by reducing output (again, see section 8.20).

The final possibility, not shown in figure 16.13, is that $p = MC$. In this case, the $p = MR$ function lies on top of the MC curve, so that $MC = MR$ at *every* output. Since the first order condition for maximum profit is fulfilled at every output, the firm has no criterion for choosing one level of output in preference to another; so the output it will choose is indeterminate. Furthermore, since we now have $p = AC$, profits, given by $\Pi = (p - AC)Q$, are zero at every output.

From this section we have some interesting results concerning the relationship between production function parameters and the resulting cost curves. Although we have derived these results assuming the production function to be of Cobb–Douglas form, they are not restricted to the Cobb–Douglas case, as we will show in section 17.4.

16.10 Profit maximization by a monopolist

In sections 16.7–16.9 we examined profit maximization by a perfectly competitive firm. In this section we will briefly analyse how our conclusions are changed when the firm is a monopolist in the market for the product it sells. The key point is that the firm now faces a downward sloping inverse demand function for its product, $p = f(Q)$, with slope $\frac{dp}{dQ} = f'(Q)$, rather than the exogenously determined ruling market price, p, of the perfectly competitive firm. When this inverse demand function is substituted into the profit function, it becomes

$$\Pi = TR - TC = f(Q)Q - (wL + rK) \tag{16.31b}$$

Assuming the monopolist has the same production function $Q = f(K, L)$ as the perfectly competitive firm, the Lagrangean expression is therefore

$$V = f(Q)Q - (wL + rK) + \lambda[f(K, L) - Q]$$

The four partial derivatives, set equal to zero, are

$$\frac{\partial V}{\partial Q} = f(Q) + f'(Q)Q - \lambda = 0 \tag{16.60}$$

$$\frac{\partial V}{\partial L} = -w + \lambda f_L = 0 \tag{16.61}$$

$$\frac{\partial V}{\partial K} = -r + \lambda f_K = 0 \tag{16.62}$$

$$\frac{\partial V}{\partial \lambda} = f(K, L) - Q = 0 \tag{16.63}$$

Note that in (16.60), $f(Q) + f'(Q)Q$ is obtained by differentiating $f(Q)Q$ using the product rule. Since we have $p = f(Q)$ and $\frac{dp}{dQ} = f'(Q)$, equation (16.60) can be written as

$$p + \frac{dp}{dQ}Q = \lambda$$

Hopefully you will recognize the left-hand side as marginal revenue (MR). (See section 15.6 if you don't.) We can therefore replace λ with MR in equations (16.61) and (16.62) and obtain the marginal productivity conditions for a monopoly firm:

$$\frac{w}{MR} = f_L \tag{16.61a}$$

$$\frac{r}{MR} = f_K \tag{16.62a}$$

Thus, compared to the marginal productivity conditions in rule 16.2, we have MR instead of p on the left-hand side of both equations. Since $MR < p$, for a given w the monopolist will choose a higher value of f_L than a perfectly competitive firm, which, assuming diminishing marginal productivity, will require a *lower* labour input. By the same reasoning, the capital input will be lower too. With both inputs lower, output must also be lower, corroborating our finding in section 8.18 that the monopolist produces a smaller output than would be produced under perfect competition with the same market price.

Another possibility is that a firm with a monopoly in the product that it sells is also, in the economic sense, sufficiently large that as it hires more workers this increase in demand for labour drives up the market wage rate; and similarly as it rents more machines this drives up their rental rate. We then say that the firm faces upward sloping factor supply curves. This changes the firm's cost-minimizing condition compared with the perfectly competitive case (rule 16.1). Regrettably, space constraints preclude consideration of this case, but you may wish to pursue this on your own. (Hint: write these upward sloping factor supply curves as $w = g(L)$ and $r = h(K)$, and substitute them into the total cost function (equation 16.24).)

Summary of sections 16.6–16.10

In section 16.6 we used the Lagrange multiplier method to find the minimum cost of producing a given output for a firm operating in conditions of perfect competition. We found that this required the firm, as a necessary (first order) condition, to adjust its inputs of K and L so as to satisfy the *cost-minimizing condition* (see rule 16.1). We did not consider the second order conditions that are needed to distinguish a maximum from a minimum or a point of inflection in total cost.

In section 16.7 we used the Lagrange multiplier method to find the profit-maximizing levels of capital and labour inputs, and resulting output, for a perfectly competitive firm. We found that profit maximization required the firm to adjust its inputs of K and L (and, consequently, the level of output) so as to satisfy the *marginal productivity conditions* (see rule 16.2). Again, this rule gives only the first order conditions, or necessary conditions, for maximum profit.

In section 16.8 we applied the methods of section 16.7 in a worked example using a Cobb–Douglas production function and found the profit-maximizing values of K, L, and Q, and maximum profit.

In section 16.9 we explored our solution from section 16.8 in greater depth. We found that under perfect competition, with a Cobb–Douglas production function and given input prices, w and r, the shapes of the firm's total, average, and marginal cost curves depend on the production function parameters α and β. If $\alpha + \beta$ is less than 1, there is a unique profit-maximizing output. If $\alpha + \beta$ is greater than 1, the firm will wish to expand its output without limit. In the boundary case, where $\alpha + \beta = 1$, then depending on the product price there are three possibilities: the firm will wish to expand production without limit, will shut down, or will find that one output is as good as any other because profits are always zero. (See figures 16.11–16.13.)

In section 16.10 we found the equivalent of rule 16.2 for profit maximization by a monopoly firm (equation (16.61a) and (16.62a)).

At the Online Resource Centre you will find more examples of constrained optimization and its application to cost minimization and profit maximization by the firm.

?

Progress exercise 16.4

1. A firm's production function is $Q = K^{0.4}L^{0.5}$ (the same production function as in exercise 16.3, question 3). The firm is perfectly competitive and the factor prices are $r = €4$ per hour and $w = €5$ per hour. The ruling market price is $p = €20$.

(a) Show that the most profitable output is $Q = 512$. Find the profits at this output.

(b) How would the most profitable output change if p rose from 20 to, say, 22 (with input prices unchanged)? What does this tell us about the firm's supply function (that is, Q as a function of p)? Does supply appear to be elastic or inelastic?

(c) Show that cost minimization requires $K = L$. Illustrate graphically, showing the isoquants and isocost lines.

(d) Using (b), find total cost as a function of output, and sketch the graph of this function. (Hint: eliminate either K or L from the TC function, then substitute Q in place of K or L. See example 16.7.) What can you deduce about marginal cost and average cost, as functions of output? How does marginal cost compare with marginal revenue?

16.11 Economic applications 3: utility maximization by the consumer

In sections 14.11 and 15.12 we introduced the utility function $U = f(X, Y)$, where X and Y are the quantities of two goods consumed and U is the individual's utility or satisfaction. We saw that the equation of an indifference curve, or iso-U section, is defined by setting U equal to any fixed value U_0, giving the implicit function

$$f(X, Y) - U_0 = 0$$

In section 15.12 (rule 15.14) we showed by implicit differentiation that the slope of an indifference curve at any point is given by

$$\frac{dY}{dX} = -\frac{f_X}{f_Y}$$

where f_X and f_Y are the partial derivatives of the utility function $U = f(X, Y)$. These partial derivatives are the marginal utilities of goods X and Y. Thus the slope of an indifference curve (which by definition equals the *MRS*) depends solely on the *ratio* of marginal utilities of the two goods, and not on their absolute magnitudes, which is good news since there is no reliable and generally accepted way of measuring absolute utility (though the search goes on).

We will now analyse the optimizing choice that every consumer has to make: how to achieve maximum utility within the constraint of a limited budget. In this case we will begin with a generalized model, then follow this up with a specific example.

First we consider the budget or income constraint that the consumer faces. We assume there are two goods, X and Y, which the consumer can buy at prices p_X and p_Y. We assume that the consumer is perfectly competitive in the goods markets, and thus cannot influence these prices by buying more or less of the goods.

The consumer's expenditure on good X is, in the usual way, price × quantity, that is $p_X X$. Similarly her expenditure on good Y is $p_Y Y$. So total expenditure on the two goods is $p_X X + p_Y Y$.

We assume that the consumer has a given money income or budget, B_0, and spends all of it on the two goods. Thus the consumer's *budget constraint* is

$$p_X X + p_Y Y = B_0 \qquad (16.64)$$

Given the budget B_0 and the prices p_X and p_Y, this equation defines an implicit function relating X and Y. We can visualize the shape of this implicit function more easily if we convert it into an explicit function with Y as the dependent variable. We can do this by subtracting $p_X X$ from both sides of equation (16.64), then dividing through by $p_Y Y$, giving

$$Y = \frac{B_0}{p_Y} - \frac{p_X}{p_Y} X \qquad (16.64a)$$

We recognize equation (16.64a) as a linear function with an intercept on the Y-axis of $\frac{B_0}{p_Y}$ and a slope of $-\frac{p_X}{p_Y}$. The Y intercept, $\frac{B_0}{p_Y}$, gives us the maximum amount of Y that the consumer can buy if she spends nothing on X. For example, if $B_0 = 100$ and $p_Y = 4$, the consumer can buy $\frac{B_0}{p_Y} = \frac{100}{4} = 25$ units of Y if she spends nothing on X. The absolute value of the slope, $\frac{p_X}{p_Y}$, tells us by how much the consumer must reduce her consumption of Y in order to buy one more unit of X. For example, if $p_X = 8$ and $p_Y = 4$, the consumer must give up $\frac{p_X}{p_Y} = \frac{8}{4} = 2$ units of Y in order to buy one more unit of X. A key point to remember about any budget line or budget constraint is that, by definition, total expenditure is the same at every point on it.

Equation (16.64) (or equally, equation (16.64a) if you feel more comfortable with the explicit form) is graphed in figure 16.14. We have also graphed three indifference curves: one for a utility level of U_0; one for a utility level higher than U_0; and a third for a utility level lower than U_0. As discussed in section 14.12, these indifference curves are convex and non-intersecting.

We are now ready to examine the consumer's optimization problem, which is to maximize her utility, given by $U = f(X, Y)$, subject to the constraint of her budget, $p_X X + p_Y Y = B_0$. The Lagrangean expression is therefore

$$V = f(X, Y) + \lambda[p_X X + p_Y Y - B_0]$$

Note that, inside the square brackets, we have as usual rearranged the constraint so that it equals zero.

In the Lagrangean expression, the variables are X, Y, and λ, with the prices of the goods and the consumer's budget (income) as parameters. The partial derivatives, set equal to zero, are

$$\frac{\partial V}{\partial X} = f_X + \lambda p_X = 0 \qquad (16.65)$$

$$\frac{\partial V}{\partial Y} = f_Y + \lambda p_Y = 0 \qquad (16.66)$$

$$\frac{\partial V}{\partial \lambda} = p_X X + p_Y Y - B_0 = 0 \qquad (16.67)$$

A little manipulation of equation (16.65) gives

$$\frac{f_X}{p_X} = -\lambda \qquad (16.65a)$$

The same rearrangement of equation (16.66) gives

$$\frac{f_Y}{p_Y} = -\lambda \qquad (16.66a)$$

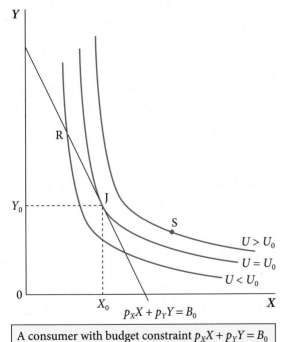

A consumer with budget constraint $p_X X + p_Y Y = B_0$ achieves maximum utility at J where the slope of the indifference curve, $-\frac{MU_X}{MU_Y}$, equals the slope of the budget constraint, $-\frac{p_X}{p_Y}$

Figure 16.14 Utility maximization by the consumer.

Combining equations (16.65a) and (16.66a), we get

$$\frac{f_X}{p_X} = \frac{f_Y}{p_Y}$$

which easily rearranges to give us the following rule:

RULE 16.3 Utility maximization by the consumer

To maximize utility requires, as a necessary condition

$$\frac{f_X}{f_Y} = \frac{p_X}{p_Y} \tag{16.68}$$

where f_X and f_Y are the marginal utilities of the two goods, and p_X and p_Y are their prices. Note that this rule gives only the necessary, or first order, condition for maximum utility (see section 7.8).

Equation (16.68) is a core finding in the theory of consumer behaviour. The left-hand side is the ratio of marginal utilities, which is known as the marginal rate of substitution, and which we know from section 15.12, rule 15.14 equals the absolute slope (that is, ignoring the minus sign) of an indifference curve. The right-hand side is the ratio of prices, which we know from the discussion immediately above is the absolute slope of the budget constraint. So equation (16.68) tells us that the consumer will maximize her utility by buying quantities of X_0 and Y_0 that place her on an indifference curve at a point that has the same slope as the budget constraint. In other words, to maximize utility the consumer must choose quantities such that the relative marginal utilities of the two goods (the consumer's *MRS*) equals their relative price. For example, if $p_X = 2p_Y$, in equilibrium the marginal utility of good X must be twice that of good Y.

Equation (16.68) does not have a unique solution, however, since every indifference curve has a point on it where its slope equals the slope of the budget constraint. But when we bring in the additional condition (equation (16.67)) that the point chosen must lie on the budget constraint, we have a unique solution, at the point of tangency of the budget constraint with an indifference curve.

This is shown graphically in figure 16.14, where we assume that the consumer's budget is B_0. The indifference curve for $U = U_0$ is tangent to the budget constraint at J, where the consumer buys quantities X_0 and Y_0. There are other combinations of X and Y that the consumer could buy, such as at R, that would satisfy the budget constraint; but all such points lie on a lower indifference curve, as the example of R illustrates. There are other combinations of X and Y, such as at S, that put the consumer on a higher indifference curve, but as the case of S illustrates, they do not satisfy the budget constraint. Thus any combination of X and Y other than X_0 and Y_0 either gives a lower level of utility or violates the budget constraint, B_0. Note that the existence and uniqueness of the tangency at J results from the convexity of the indifference curves, which in turn results from the assumption of decreasing marginal rate of substitution in consumption (*MRS*) discussed in section 14.12.

16.12 Deriving the consumer's demand functions

Equation (16.68) above, together with the consumer's budget constraint, can be used to find the consumer's demand functions for the two goods. This is true quite generally, but here we will demonstrate it on the simplifying assumption that the consumer's utility function is the Cobb–Douglas form (discussed in sections 14.13 and 15.13)

$$U = X^\alpha Y^\beta$$

Let us write the marginal utilities of the two goods as MU_X and MU_Y. The equivalent of equation (16.68) above is then

$$\frac{MU_X}{MU_Y} = \frac{p_X}{p_Y} \tag{16.68a}$$

From the utility function $U = X^{\alpha} Y^{\beta}$, we have the marginal utilities

$$MU_X = \alpha X^{\alpha-1} Y^{\beta} \quad \text{and} \quad MU_Y = \beta X^{\alpha} Y^{\beta-1}$$

Substituting these into equation (16.68a) gives

$$\frac{\alpha X^{\alpha-1} Y^{\beta}}{\beta X^{\alpha} Y^{\beta-1}} = \frac{p_X}{p_Y}$$

Collecting together the exponents on the left-hand side, this becomes

$$\frac{\alpha X^{\alpha-1-\alpha} Y^{\beta-(\beta-1)}}{\beta} = \frac{p_X}{p_Y} \quad \text{which simplifies to}$$

$$\frac{\alpha Y}{\beta X} = \frac{p_X}{p_Y}$$

If we multiply both sides by $\frac{\beta X p_Y}{\alpha}$, this becomes

$$p_Y Y = \frac{\beta}{\alpha} p_X X \tag{16.69}$$

which we can use to substitute for $p_Y Y$ in the consumer's budget constraint, $p_X X + p_Y Y = B_0$ (equation (16.64) above). We then get

$$p_X X + \frac{\beta}{\alpha} p_X X = B_0 \quad \text{which rearranges as}$$

$$X = \frac{\alpha}{\alpha+\beta} \frac{B_0}{p_X} \tag{16.70}$$

For given values of B_0 and p_X (and given also, of course, α and β, the parameters of the individual's utility function) we can solve (16.70) to find the utility-maximizing value of X, which we labelled X_0 in figure 16.14 above.

However, if we now treat B_0 and p_X as variables, equation (16.70) becomes the consumer's demand function for good X as a function of these two variables. Thus if either B_0 or p_X changes, the consumer re-solves equation (16.70) in order to re-establish the maximum level of utility, and thereby arrives at a new equilibrium quantity of X. We might have expected to see the price of good Y, p_Y, on the right-hand side of equation (16.70), but it is a peculiarity of the Cobb–Douglas form of utility function that p_Y does not influence the consumer's demand for X (nor does p_X influence her demand for Y).

We can similarly derive the consumer's demand function for good Y. Equation (16.69) easily rearranges as $p_X X = \frac{\alpha}{\beta} p_Y Y$. We then substitute this into equation (16.64) and obtain

$$Y = \frac{\beta}{\alpha+\beta} \frac{B_0}{p_Y} \tag{16.71}$$

Although we have derived the demand functions only for the specific case of a Cobb–Douglas utility function and with only two goods, in principle the demand functions can be derived whatever the utility function may be and with any number of goods. Thus in general the quantity of any good demanded by the consumer is the quantity that maximizes her utility, given her budget (income) and whatever prices are relevant.

Expenditure shares

If we multiply both sides of equation (16.70) by $\frac{p_X}{B_0}$, it becomes

$$\frac{p_X X}{B_0} = \frac{\alpha}{\alpha + \beta} \tag{16.72}$$

The left-hand side of this expression is expenditure on good X as a proportion of the consumer's budget: that is, the share of total expenditure that is devoted to good X. In the same way, by multiplying both sides of equation (16.71) by $\frac{p_Y}{B_0}$, we obtain

$$\frac{p_Y Y}{B_0} = \frac{\beta}{\alpha + \beta} \tag{16.73}$$

which gives the share of total expenditure devoted to good Y. (As a consistency check, note that the sum of the shares is $\frac{\alpha}{\alpha+\beta} + \frac{\beta}{\alpha+\beta}$, which equals 1, as it should.)

Thus the distribution of total spending between the two goods is determined entirely by the relative values of α and β. For example if $\alpha = 0.5$ and $\beta = 0.75$, then the share of expenditure devoted to X is $\frac{\alpha}{\alpha+\beta} = \frac{0.5}{1.25} = 0.4$, or 40 per cent.

Thus, contrary to what we might expect, the distribution of spending between the two goods does *not* depend on the prices of the goods. If the price of good X, say, rises, then the utility-maximizing quantity of X goes down, but the amount spent on X remains unchanged. This, of course implies that the price elasticity of demand for X is constant and equals -1 (see below). (See chapter 9 if you need to revise basics of elasticity.)

However, we must emphasize once again that these somewhat counter-intuitive findings result from the Cobb–Douglas utility function and therefore are not generally true for real-world consumers whose tastes are undoubtedly more complex than those implied by the Cobb–Douglas form.

Price and income elasticities

By differentiating the demand functions (equations (16.70) and (16.71)), the price and income elasticities of demand for X and Y can be obtained. We will do this in section 17.6, below.

EXAMPLE 16.10

We finish this chapter with a worked example of utility maximization by a consumer who faces given prices and is subject to a budget constraint. We suppose the consumer's utility function is the Cobb–Douglas form

$$U = X^{1/3} Y^{2/3}$$

We assume the prices are $p_X = 4$ and $p_Y = 2$, and the consumer's income (budget) is $B = 120$. So the consumer's optimization problem is to maximize $U = X^{1/3} Y^{2/3}$ subject to the constraint of her budget, $4X + 2Y = 120$. The Lagrangean expression is therefore

$$V = X^{1/3} Y^{2/3} + \lambda[4X + 2Y - 120]$$

The partial derivatives, set equal to zero, are

$$\frac{\partial V}{\partial X} = \frac{1}{3} Y^{2/3} X^{-2/3} + \lambda 4 = 0 \tag{16.74}$$

$$\frac{\partial V}{\partial Y} = \frac{2}{3} Y^{-1/3} X^{1/3} + \lambda 2 = 0 \tag{16.75}$$

$$\frac{\partial V}{\partial \lambda} = 4X + 2Y - 120 = 0 \tag{16.76}$$

Equation (16.74), after rearranging slightly, becomes

$$\tfrac{1}{3}Y^{2/3}X^{-2/3} = -\lambda 4$$

Equation (16.75), after slight rearrangement and multiplying by 2, becomes

$$\tfrac{4}{3}Y^{-1/3}X^{1/3} = -\lambda 4$$

As their right-hand sides are equal, we can combine these two equations and get

$$\tfrac{4}{3}Y^{-1/3}X^{1/3} = \tfrac{1}{3}Y^{2/3}X^{-2/3}$$

Dividing both sides by the right-hand side, and collecting powers of X and Y together results in

$$\frac{\tfrac{4}{3}Y^{-1/3}Y^{-2/3}X^{1/3}X^{2/3}}{\tfrac{1}{3}} = 1$$

from which $\tfrac{4X}{Y} = 1$, hence $Y = 4X$.

$Y = 4X$ can then be substituted into equation (16.76) to give

$$4X + 2(4X) - 120 = 0$$

from which $X = 10$, and therefore $Y = 40$.

So this consumer maximizes her utility by buying 10 units of X and 40 units of Y. We can check by substitution that these purchases satisfy the budget constraint $4X + 2Y = 120$. We get $4(10) + 40(2) = 120$, as required. The solution is shown graphically in figure 16.15.

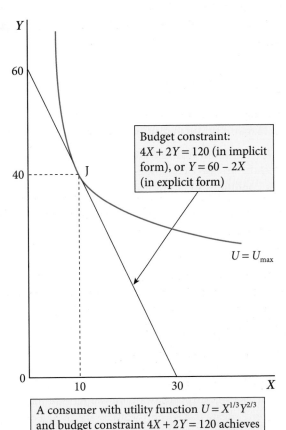

Budget constraint:
$4X + 2Y = 120$ (in implicit form), or $Y = 60 - 2X$ (in explicit form)

$U = U_{max}$

A consumer with utility function $U = X^{1/3}Y^{2/3}$ and budget constraint $4X + 2Y = 120$ achieves maximum utility at J, consuming 40 units of Y and 10 units of X.

Figure 16.15 Maximum utility for a consumer.

Note that the solution conforms with equations (16.72) and (16.73) above, which give the expenditure shares. For example, according to equation (16.72) the share of expenditure on X should be

$$\frac{\alpha}{\alpha + \beta} = \frac{\frac{1}{3}}{\frac{1}{3} + \frac{2}{3}} = \frac{1}{3}$$

Our answer has $p_X X = 4(10) = 40$, so $\frac{p_X X}{B} = \frac{40}{120} = \frac{1}{3}$, confirming equation (16.72). Thus the consumer spends one-third of her income on X and two-thirds on Y.

Summary of sections 16.11–16.12

In section 16.11 we considered the constrained optimization problem faced by the consumer who wishes to maximize her utility subject to a budget constraint, when product prices are given. We found that the necessary, or first order, condition for utility maximization is $\frac{f_X}{f_Y} = \frac{p_X}{p_Y}$ (see rule 16.3). In words, this says that the relative marginal utilities of the two goods must equal their relative price. This is a core finding in the theory of consumer behaviour.

From section 16.12, the key analytical conclusion is that an individual's demand functions for the two goods, equations (16.70) and (16.71), are obtained as the result of the utility-maximizing process by that individual, as analysed in section 16.11. This means that at any point on the individual's demand function for each good, her utility is maximized for the given prices and income (budget).

Chapter summary

This has been a long and important chapter, and therefore it merits an overall summary in addition to the section summaries. In sections 16.1–16.5 we developed the idea of a constrained optimization problem, and the Lagrange multiplier method of solution. In the remainder of the chapter we showed how this technique could be applied to the firm's goals of minimizing the cost of any given output; profit maximization; and the consumer's goal of utility maximization subject to a budget constraint.

In section 16.6 we obtained the *cost-minimizing condition* for a perfectly competitive firm: $\frac{f_L}{f_K} = \frac{w}{r}$ (rule 16.1). In words, this says that relative marginal productivities must equal relative input prices. This is the *necessary* condition for minimizing the total cost of any given output; the second order conditions were not explored. Since by assumption the firm cannot influence w and r, it achieves this cost minimization by adjusting its inputs of K and L, thereby changing f_L and f_K.

In section 16.7, with output now treated as a variable, we obtained the *marginal productivity conditions*: $\frac{w}{p} = f_L$ and $\frac{r}{p} = f_K$, for a perfectly competitive firm seeking maximum profit (rule 16.2). In words, this says that the real cost of each input must equal its marginal productivity. Again, this is the *necessary*, or first order, condition; the second order condition was not considered. In sections 16.8 and 16.9 we examined the conditions for the existence of a unique solution to the optimization problem of a perfectly competitive firm with a Cobb–Douglas production function. In section 16.10 we derived the analogue of rule 16.2 for a monopoly firm.

In section 16.11 we found the necessary utility-maximizing condition for a consumer facing given product prices and subject to a budget constraint: $\frac{f_x}{f_y} = \frac{p_x}{p_y}$ (rule 16.3). This says that the consumer must adjust her purchases until the relative marginal utilities of the goods equal their relative price. Again this is only the necessary, or first order, condition; second order conditions were not examined. Using this condition, we showed in the case of the Cobb–Douglas utility function how the consumer's demand functions for the two goods could be derived (section 16.12).

At the Online Resource Centre you will find more examples of constrained optimization and its application to utility maximization by the individual.

Progress exercise 16.5

1. An individual's utility function is $U = X^2Y^3$, where X and Y are weekly consumption levels of goods X and Y. The market prices are $p_X = €4$ and $p_Y = €3$, and the individual's weekly budget is $B = €120$.

   (a) Show that the individual will maximize her utility by buying 12 units of X and 24 units of Y per week.

   (b) Show graphically the individual's equilibrium as a tangency between an indifference curve and her budget line.

   (c) What is her weekly expenditure on each good (i) in money terms, and (ii) as a proportion of her budget?

   (d) Suppose p_X rises from €4 to €6. Find the new utility-maximizing quantities.

   (e) Following this price increase, what is the individual's new weekly expenditure on each good (i) in money terms, and (ii) as a proportion of her budget?

   (f) From your answer to (e), can you infer anything about the individual's price elasticity of demand for good X?

2. Generalize the previous example by considering an individual whose utility function is $U = X^\alpha Y^\beta$ with a budget of B and market prices of p_x and p_y.

   (a) Show that the individual will maximize her utility by adjusting the quantities bought until her marginal rate of substitution between the two goods equals the ratio of their prices.

   (b) Show this equilibrium graphically.

   (c) Show that this individual will consume the two goods in a fixed proportion, irrespective of her income, as long as the price ratio remains constant. Illustrate this graphically.

   (d) Show that her demand functions for the two goods are

$$X = \frac{\alpha}{\alpha + \beta} \frac{B}{p_X} \text{ and } Y = \frac{\beta}{\alpha + \beta} \frac{B}{p_Y}$$

   (Hint: you need to substitute the equilibrium condition from (a) above into the budget constraint.)

   (e) Using either (c) or (d), show that this individual will spend a fixed proportion of her income on each of the two goods, irrespective of their prices or her budget.

   (f) How would her demand for the two goods change if (i) one of the prices doubled or halved, and (ii) both prices doubled or halved?

   (g) How would her demand for the two goods change if her budget, B, doubled or halved?

   (h) How would her demand for the two goods change if the prices of the two goods and her budget all doubled? Is this answer surprising?

Checklist

Be sure to test your understanding of this chapter by attempting the progress exercises (answers are at the end of the book). The Online Resource Centre contains further exercises and materials relevant to this chapter www.oxfordtextbooks.co.uk/orc/renshaw3e/.

In this chapter we have identified the problem of optimization of a function subject to a constraint on the values of the independent variables. We showed that this problem could be solved in three ways, of which the Lagrange multiplier method was the most powerful. Specific topics were:

✔ **Constrained optimum values.** Understand the nature of a constrained maximum or minimum value and the characteristics of the solution.

✔ **The Lagrange multiplier method.** How to use the Lagrange multiplier method to solve a constrained optimization problem.

✔ **Optimization by a producer.** Using the Lagrange multiplier method to solve the producer's problems of minimizing costs and maximizing profit, assuming perfect competition. Depending on parameter values, a unique profit-maximizing output may not exist.

✔ **Optimization by a consumer.** Using the Lagrange multiplier method to solve the consumer's problem of optimizing purchases subject to a budget constraint.

This chapter should be studied with great care as the mathematical methods and results of the economic applications go the heart of economics. In the next chapter we will examine several equally important areas concerned with production and economic growth theory.

Chapter 17

Returns to scale and homogeneous functions; partial elasticities; growth accounting; logarithmic scales

OBJECTIVES

Having completed this chapter you should be able to:

- Define precisely the concepts of constant, increasing, and decreasing returns to scale in the production function.

- Demonstrate algebraically and graphically that a production function has constant, increasing, or decreasing returns to scale.

- Understand the meaning of homogeneity and how to test a function for homogeneity.

- Understand the main properties of homogeneous functions and their significance for economic analysis.

- Understand the concept of partial elasticities, how to evaluate them and their applications in economic analysis.

- Understand the concept of the proportionate differential.

- Understand and apply the technique of growth accounting.

- Understand the relationship between partial elasticities and logarithmic scales.

17.1 Introduction

In this concluding chapter of part 4 of the book, we examine several economic concepts that can be explained and analysed clearly using simple mathematical techniques. First, we consider the concept of economies of scale, also known as increasing returns to scale. We demonstrate diagrammatically the implied shape and position of the isoquants of the production function when economies or diseconomies of scale are present. We then look at the algebraic characteristics of the production function in these cases and how these relate to the mathematician's concept of a homogeneous function.

We go on from there to examine several other characteristics of a homogeneous production function of interest to economists, most importantly the relationship between returns to scale and the shape of the long-run average and marginal cost curves. Next we develop the concept of a partial elasticity as a necessary extension to the basic ideas on elasticity that we introduced in chapter 9. For example, the partial elasticities of a demand function measure the proportionate rate of change of quantity demanded resulting from proportionate changes in not only the product price but also other variables such as the prices of other goods, consumers' income, and so on.

Then we demonstrate another useful tool, the proportionate differential of a function, which embodies the partial elasticities. The most common use for the proportionate differential is for 'growth accounting', which plays a major role in both theoretical and empirical analysis of economic growth. Finally, we return once more to the use of logarithmic scales or logarithmic transformation. In section 13.10 we showed by means of an example that, for any function $y = f(x)$, when the function is graphed with logarithmic scales on both axes the slope measures the elasticity of the function. We now prove this result and extend it to any function $z = f(x, y)$. With logarithmic scales on all axes, the slope in, say, the $\ln x$ direction measures the partial elasticity of the function with respect to x. This applies immediately to economic relationships such as the demand function for a good.

17.2 The production function and returns to scale

In economics we use the term 'returns to scale' to describe an important characteristic of the production function. Suppose we are looking at some production function $Q = f(K, L)$. Then it is interesting and important to know what happens to output when both inputs are, say, doubled. When both inputs are doubled, we say that the *scale* of production is doubled. Note that a change in scale can occur only in the long run, because in the short run the capital input is assumed fixed.

If the effect of this doubling of both inputs is that output is also doubled, then we say that there are *constant* returns to scale. If the effect is that output is more than doubled, we say that there are *increasing* returns to scale; or alternatively that there are economies of scale. If the effect is that output is less than doubled, we say that there are *decreasing* returns to scale, or alternatively that there are diseconomies of scale.

Let us pin down this idea more precisely by means of a diagram and a little algebra. Figure 17.1 shows two isoquants, for output levels Q_0 and Q_1, from some unspecified production function $Q = f(K, L)$.

Suppose initially production is taking place at a randomly chosen point, J, with labour input L_0, capital input K_0, and resulting output Q_0.

Now suppose that capital and labour inputs are both increased by the factor λ, where λ is an arbitrary constant (assumed greater than 1). To make the discussion as concrete as possible, let us choose $\lambda = 2$. Thus the new labour input is $L_1 = \lambda L_0 = 2L_0$ and the new capital input is $K_1 = \lambda K_0 = 2K_0$. In other words, we have doubled both

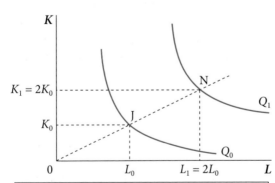

If both inputs are doubled, production moves from J to N. If $Q_1 = 2Q_0$ there are constant returns to scale. If Q_1 is greater (less) than $2Q_0$ there are increasing (decreasing) returns to scale.

Figure 17.1 Constant, increasing, and decreasing returns to scale.

inputs and thus, by definition, doubled the *scale* of production. (If we chose $\lambda = 1.5$, both inputs would be increased by 50%; if $\lambda = 1.2$, by 20%, and so on.)

This doubling of both inputs takes production to point N where the output level is now Q_1. The crucial question is the relationship between the new level of output (Q_1) and the initial level (Q_0). There are three possibilities:

(1) If $Q_1 = 2Q_0$, then output has exactly doubled. Thus output has increased in the same proportion as the increase in inputs. This is the case of *constant* returns to scale.

(2) If $Q_1 > 2Q_0$, then output has more than doubled. Thus output has increased in a greater proportion than the increase in inputs. This is the case of *increasing* returns to scale (also known as economies of scale).

(3) If $Q_1 < 2Q_0$, then output has less than doubled. Thus output has increased in a smaller proportion than the increase in inputs. This is the case of *decreasing* returns to scale (also known as diseconomies of scale).

Note that it is essential that both inputs increase in the same proportion. If one input were increased by a larger proportion than the other, we would be changing both the scale and the capital/labour ratio at the same time, and it would then be very difficult to analyse how much of the change in output was due to the former and how much to the latter.

Note also that we have set $\lambda = 2$ purely to make the discussion more concrete and in order to draw figure 17.1. In general, if both inputs increase (decrease) in some unspecified proportion, λ, then the crucial question is whether output increases (decreases) in a proportion that is equal to, greater than, or less than λ. If $Q_1 = \lambda Q_0$ we have constant returns to scale. If $Q_1 > \lambda Q_0$ we have increasing returns to scale. If $Q_1 < \lambda Q_0$ we have decreasing returns to scale.

Further, if a production function has increasing or decreasing returns to scale, we need to think about the *strength* of those increasing or decreasing returns. For example, if doubling both inputs results in a ten-fold increase in output, this is obviously a case of very strongly increasing returns $(Q_1 = 10Q_0)$. If the result of doubling both inputs (an increase of 100 per cent) is that output increases by only 90 per cent, this is a case of weakly decreasing returns $(Q_1 = 1.9Q_0)$.

Uniform versus variable returns to scale

To keep the analysis of returns to scale manageably simple, in economic theory it is common to choose very simple functional forms for the production function. These production functions have the mathematical property that when we carry out the experiment above, of choosing in figure 17.1 an initial position J and then a value for λ that moves us to N, the outcome is the same whatever the position of J and for any value of λ. This means that if in figure 17.1 we chose $L_0 = 50$, $K_0 = 75$, and $\lambda = 2$ (so that both inputs doubled) and then found that output increased four-fold, then output would also increase four-fold if we chose any other initial point such as $L_0 = 90$, $K_0 = 60$. It also means that if we chose another value for λ, say $\lambda = 3$, then output would increase six-fold (repeating the pattern that the increase in output is twice as large as the increase in inputs). A simple way of describing these characteristics is to say that returns to scale (whether increasing, decreasing, or constant) are *uniform*, and this is the label we will use.

In the real world, however, it is quite possible, and even probable, that whether we find constant, increasing, or decreasing returns to scale, and the strength of increasing or decreasing returns if they are present, will vary according to the initial position J and/or the value of λ. For example, in figure 17.2, if the initial position is J, there are constant returns to scale because a doubling of both inputs takes production to N where the new level of output Q_1 is exactly double the initial level Q_0. However, if the initial position is M, then an increase in both inputs of only about 40% is sufficient to move the production point to R, doubling output. So at M

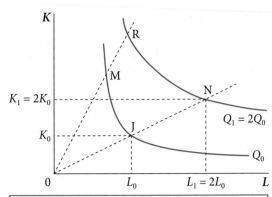

If production moves from J to N, there are constant returns to scale, but if production moves from M to R, there are increasing returns to scale. So this production function has variable returns to scale.

Figure 17.2 Variable returns to scale.

there are increasing returns to scale. Thus returns to scale are *variable*, depending on the initial position, but also possibly on the value of λ. This is because the isoquants in figure 17.2 have very different shapes from each other.

Another possibility is that there are constant returns to scale when output is small (the initial position, J, is close to the origin) but strongly increasing returns to scale when output is large (J is far from the origin). Thus again we have variable returns to scale. This seems a likely feature of the real world, where it is often found that 'economies of scale' (another term for increasing returns to scale) are achievable, as the terminology implies, only when the scale of production is large. Try sketching the isoquants for this case, to test your understanding.

 Hint Don't confuse decreasing returns to scale, which concerns simultaneous variation in both L and K, with diminishing marginal productivity, which concerns variation in one input with the other held constant. (See section 14.8 for discussion of diminishing marginal productivity.)

17.3 Homogeneous functions

In mathematics there exists the concept of a homogeneous function which is exactly equivalent to the economist's concept of uniform returns to scale. To explain this we refer back to the analysis above, but now treat the function $Q = f(K, L)$ not as a production function but as any function (in other words, K, L, and Q are just variables that don't denote anything in the real world).

Like the economist, the mathematician is interested in what happens to the dependent variable (Q) when the independent variables (K and L) are both increased by some factor λ from any randomly chosen initial position. If we take again the example of $\lambda = 2$ (a doubling of K and L), the mathematician would say that if the result is that Q also doubles, then the function is *homogeneous of degree 1*. But in this case we know from section 17.2 that if $Q = f(K, L)$ is a production function, an economist would say that there are *uniform constant returns to scale*. Thus the mathematician's concept of a function that is homogeneous of degree 1 corresponds exactly to the economist's concept of a production function that has uniform constant returns to scale.

Similarly, if in the above experiment Q more than doubles, the mathematician would say that the function is homogeneous of degree greater than 1, while the economist would say that there are uniform increasing returns to scale. Finally, if Q less than doubles, the mathematician would say that the function is homogeneous of degree less than 1, while the economist would say that there are uniform decreasing returns to scale. Table 17.1 summarizes the exact equivalence between uniform returns to scale and homogeneity of the function $Q = f(K, L)$, for the case $\lambda = 2$.

There is another possibility, namely that the change in Q does not correspond in any simple way to the changes in K and L. This is the case of variable returns to scale discussed at the end of section 17.2. In this case the mathematician would say that the function is not homogeneous.

Table 17.1 Returns to scale and homogeneity of $Q = f(K, L)$ when $\lambda = 2$.

Effect on Q when K and L are both doubled (from any initial position)	Economist's view (assuming $Q = f(K, L)$ is a production function)	Mathematician's view (assuming $Q = f(K, L)$ is a any function)
Q is exactly doubled	Function has uniform constant returns to scale	Function is homogeneous of degree 1
Q is more than doubled	Function has uniform increasing returns to scale	Function is homogeneous of degree greater than 1
Q is less than doubled	Function has uniform decreasing returns to scale	Function is homogeneous of degree less than 1

Testing for homogeneity

The question of homogeneity of a function has other applications in economics in addition to the production function. Therefore we need to understand how to determine whether a function is homogeneous, and if so, its degree of homogeneity.

The procedure for testing for homogeneity is as follows. Given any function $z = f(x, y)$:

Step 1. We suppose that we are initially at some point J on this surface, where $x = x_0, y = y_0$ and consequently the value of z is given by $z_0 = f(x_0, y_0)$.

Step 2. We increase x and y by the factor λ; their new values are therefore $x_1 = \lambda x_0$ and $y_1 = \lambda y_0$. The new value of z is given by $z_1 = f(x_1, y_1) = f(\lambda x_0, \lambda y_0)$.

Step 3. If we can show from the equation $z_1 = f(\lambda x_0, \lambda y_0)$ that $z_1 = \lambda^r z_0$, then the function is said to be homogeneous of degree r. If we cannot, the function is not homogeneous.

The following examples illustrate this procedure.

EXAMPLE 17.1

$z = x^2 + 2xy + y^2$

Step 1. $z_0 = x_0^2 + 2x_0 y_0 + y_0^2$

Step 2. $z_1 = (\lambda x_0)^2 + 2(\lambda x_0)(\lambda y_0) + (\lambda y_0)^2$

Step 3. Removing the brackets from step 2 gives

$$z_1 = \lambda^2 x_0^2 + 2\lambda x_0 \lambda y_0 + \lambda^2 y_0^2$$
$$= \lambda^2(x_0^2 + 2x_0 y_0 + y_0^2)$$
$$= \lambda^2 z_0$$

Thus the ratio $\frac{z_1}{z_0}$ depends only on λ, and therefore the function is homogeneous. It is homogeneous of degree 2, the power to which λ is raised in the last line.

EXAMPLE 17.2

$z = x^3 + 2xy + y^2$

Step 1. $z_0 = x_0^3 + 2x_0 y_0 + y_0^2$

Step 2. $z_1 = (\lambda x_0)^3 + 2(\lambda x_0)(\lambda y_0) + (\lambda y_0)^2$

Step 3. Removing the brackets from step 2 gives

$$z_1 = \lambda^3 x_0^3 + 2\lambda x_0 \lambda y_0 + \lambda^2 y_0^2$$
$$= \lambda^3 x_0^3 + 2\lambda^2 x_0 y_0 + \lambda^2 y_0^2 = \lambda^2(\lambda x_0^3 + 2x_0 y_0 + y_0^2)$$

The problem in this case is that we are unable to remove λ completely from the brackets on the right-hand side, unlike the previous example. Therefore the relationship between z_1 and z_0 does not depend solely on λ, but depends on x_0 and y_0 too. Thus the proportionate change in z that results from a given proportionate change in x and y will vary according to the initial values of x and y. Therefore the function is not homogeneous.

EXAMPLE 17.3

In this example we will examine the homogeneity of the Cobb–Douglas production function (see sections 14.9 and 15.11 if you want to review this function). The general form of the function is

$\qquad Q = AK^{\alpha}L^{\beta}$ where A, α, and β are positive constants.

Consider an example with $A = 1$, $\alpha = 0.75$, and $\beta = 0.5$, giving

$\qquad Q = K^{0.75}L^{0.5}$

Step 1. $Q_0 = K_0^{0.75}L_0^{0.5}$

Step 2. $Q_1 = (\lambda K_0)^{0.75}(\lambda L_0)^{0.5}$

Step 3. Removing the brackets from step 2 gives

$$Q_1 = \lambda^{0.75}K_0^{0.75}\lambda^{0.5}L_0^{0.5}$$
$$= \lambda^{1.25}K_0^{0.75}L_0^{0.5}$$
$$= \lambda^{1.25}Q_0$$

Therefore the function is homogeneous of degree 1.25, the power to which λ is raised in the last line. Referring to table 17.1, we see that this means that this production function has increasing returns to scale. The strength of the increasing returns is measured by degree of homogeneity. To illustrate the strength, suppose $Q_0 = 100$ and $\lambda = 2$ (that is, both inputs are doubled). Then $Q_1 = \lambda^{1.25}Q_0 = 2^{1.25}(100) = 237.8$. In words, an increase in inputs of 100% results in an increase in output of 137.8%. Try some other values for α and β to get a feel for this.

EXAMPLE 17.4

As a final example, let us consider the homogeneity of the general Cobb–Douglas production function in which the parameters A, α, and β are unspecified. Thus we start with

$\qquad Q = AK^{\alpha}L^{\beta}$

Step 1. $Q_0 = AK_0^{\alpha}L_0^{\beta}$

Step 2. $Q_1 = A(\lambda K_0)^{\alpha}(\lambda L_0)^{\beta}$

Step 3. Removing the brackets from step 2 gives

$$Q_1 = A\lambda^{\alpha}K_0^{\alpha}\lambda^{\beta}L_0^{\beta}$$
$$= A\lambda^{\alpha+\beta}K_0^{\alpha}L_0^{\beta}$$
$$= \lambda^{\alpha+\beta}[AK_0^{\alpha}L_0^{\beta}]$$
$$= \lambda^{\alpha+\beta}Q_0$$

Thus we see that the function is homogeneous of degree $\alpha + \beta$. Therefore we have three cases:

- If $\alpha + \beta = 1$, then $\lambda^{\alpha+\beta} = \lambda^1 = \lambda$, and the function is homogeneous of degree 1. Therefore the production function has constant returns to scale. For example, if $\lambda = 2$, $Q_1 = \lambda^{\alpha+\beta}Q_0 = 2^1Q_0 = 2Q_0$.

- If $\alpha+\beta>1$, then $\lambda^{\alpha+\beta}>\lambda$ (since we assumed $\lambda>1$), and the function is homogeneous of degree greater than 1. Therefore the production function has increasing returns to scale. For example, if $\lambda=2$ and $\alpha+\beta=1.5$, $Q_1=\lambda^{\alpha+\beta}Q_0=2^{1.5}Q_0=2.8Q_0>2Q_0$. Thus there are strongly increasing returns, since an increase of 100% in inputs leads to an increase of 180% in output.

- Finally, if $\alpha+\beta<1$, then $\lambda^{\alpha+\beta}<\lambda$, and the function is homogeneous of degree less than 1. The production function has decreasing returns to scale. For example, if $\lambda=2$ and $\alpha+\beta=0.75$, $Q_1=2^{0.75}Q_0=1.68Q_0<2Q_0$. An increase of 100% in inputs leads to an increase of only 68% in output.

Progress exercise 17.1

1. (a) Sketch and label some isoquants for a production function that has increasing returns to scale at low levels of output but decreasing returns to scale at high levels of output.

   (b) Sketch and label some isoquants for a production function that has increasing returns to scale when capital-intensive methods of production are used, and decreasing returns to scale when labour-intensive methods of production are used. Is either of the production functions in (a) and (b) homogeneous?

2. Which of the following functions are homogeneous, and if so, of what degree?

   (a) $Q=3LK+L^2$ (b) $Q=(3LK+L^2)^{1/2}$

   (c) $Q=4K^2+0.5LK+3L^2$ (d) $Q=3K^2+10LK^{0.5}+5L^2$

3. Find the degree of homogeneity of the following Cobb–Douglas production functions, and in each case say whether there are constant, increasing, or decreasing returns to scale. Draw and label some isoquants to illustrate each case.

   (a) $Q=100K^{0.3}L^{0.7}$ (b) $Q=50K^{0.6}L^{0.5}$ (c) $Q=25K^{0.3}L^{0.5}$

   (d) $Q=AK^\alpha L^\beta$ (e) $Q=AK^\alpha L^{1-\alpha}$

4. Show that the production function $Q=A(bK^\alpha+(1-b)L^\alpha)^{1/\alpha}$ is homogeneous of degree 1 and therefore has constant returns to scale. (This is called a constant elasticity of substitution (CES) production function and is examined further in chapter W21, to be found at the Online Resource Centre.)

17.4 Properties of homogeneous functions

We will now examine three properties of homogeneous functions that have important implications for economic analysis, perhaps most significantly in relation to the production function. We will not attempt to prove any of what follows.

Property 1: the 'parallel tangents' property

Property 1 says that if a function $z=f(x,y)$ is homogeneous (of any degree), then the ratio of its partial derivatives, $(\frac{\partial z}{\partial x})/(\frac{\partial z}{\partial y})$, which we can write more compactly as $\frac{f_x}{f_y}$, depends only on the ratio $\frac{y}{x}$ and not on the absolute magnitudes of y and x.

This has immediate economic application. Suppose we are looking at the production function $Q=f(K,L)$, which we assume is homogeneous of some unspecified degree.

Then according to property 1, $\frac{f_L}{f_K}$, depends only on $\frac{K}{L}$, capital per worker. Since by definition $\frac{f_L}{f_K}$ equals $\frac{MPL}{MPK}$, the ratio of marginal products, this means that the ratio of marginal products

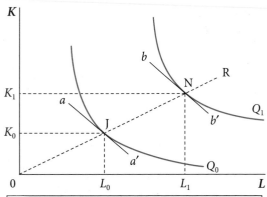

If the production function is homogeneous of any degree, then along any ray 0R the tangents such as aa' and bb' to any isoquants are parallel. Since the slope of the tangent equals $-\dfrac{MPL}{MPK}$, this ratio is constant along any ray.

Figure 17.3 The 'parallel tangents' property.

does not vary with the scale of output. This is because changing scale alone means changing K and L in the same proportion, which leaves their ratio $\dfrac{K}{L}$ unchanged.

Property 1 may be shown graphically. Figure 17.3 shows two isoquants from the production function $Q = f(K, L)$, assumed to be homogeneous of some degree. Since we have K and L on the axes, we know that the equation of any straight line (ray) from the origin is $K = aL$, for some constant a. Therefore at any point on this ray we have $\dfrac{K}{L} = a$ = constant. Therefore, because J and N both lie on the ray 0R, they have the same $\dfrac{K}{L}$.

Property 1 tells us that, because J and N have the same $\dfrac{K}{L}$, they have the same $\dfrac{MPL}{MPK}$. And we also know that the tangent drawn to any isoquant has a slope of $-\dfrac{MPL}{MPK}$ (see section 15.10, rule 15.13). Therefore the tangent aa' at J has the same slope as the tangent bb' at N. In other words, these two tangents are parallel to one another. This is the 'parallel tangents' property. The tangents aa' and bb' are parallel whatever the slope of 0R and wherever J and N are placed on 0R.

We can develop this analysis further. We know from section 16.6 (rule 16.1), that a firm that is perfectly competitive and pays w per unit for its labour and r per unit for its capital will minimize the total cost of any given output by producing at the point on the isoquant where $\dfrac{MPL}{MPK} = \dfrac{w}{r}$. Suppose the ruling market input prices, w and r, are such that $-\dfrac{w}{r}$ equals the slope of aa' and bb' in figure 17.3. Then the firm will produce at J if it produces Q_0 units of output and at N if it produces Q_1 units of output. The ray 0R is called the firm's **expansion path**, since with the given input prices the firm will always increase or decrease its output by moving along this ray.

In contrast, if the production function is not homogeneous, but instead has variable returns to scale (see figure 17.2), the firm's expansion path will not be a straight line. Thus in figure 17.2 we see that if tangents were drawn at J and N they would not be parallel, so the firm would not expand by moving from J to N unless input prices also changed.

Long-run average and marginal cost

This has important implications for the firm's costs. Let us now assume that the production function is homogeneous of degree 1 (constant returns to scale). At J in figure 17.3, total cost is $TC_0 = wL_0 + rK_0$. At N, total cost is $TC_1 = wL_1 + rK_1$. For simplicity of exposition, let us suppose that the movement from J to N in figure 17.3 results from a doubling of both inputs: that is, $K_1 = 2K_0$ and $L_1 = 2L_0$. Then we have

$$TC_1 = wL_1 + rK_1 = w2L_0 + r2K_0 = 2(wL_0 + rK_0) = 2TC_0$$

Thus the movement from J to N has doubled total cost. Moreover, since both inputs have doubled, and we have assumed constant returns to scale, output must have doubled also: that is, $Q_1 = 2Q_0$.

The significance of this is as follows. Suppose the firm was initially at J and then decided to double its output by moving production to some point on isoquant Q_1. Then with constant input prices w and r the firm would choose to produce at N as this would minimize total cost. But we have just shown that total cost at N is exactly twice as much as at J. Thus the movement from J to N doubles output (Q) but also doubles total cost (TC). Therefore *average* cost, $\dfrac{TC}{Q}$, must be the same at N as at J. In turn this implies that marginal cost (MC) equals

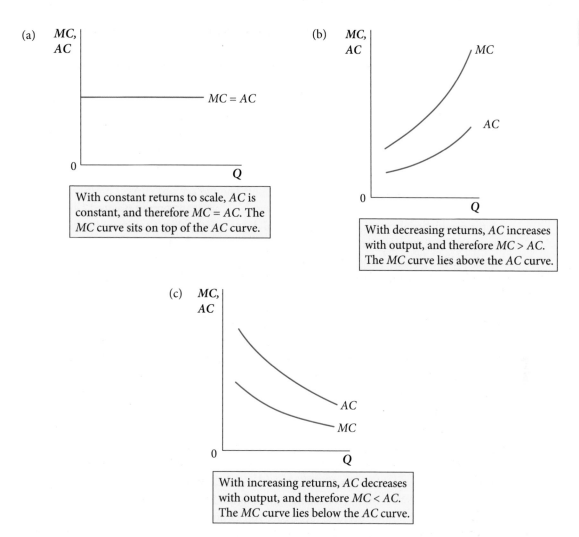

Figure 17.4 Marginal and average cost with constant, decreasing, and increasing returns to scale.

average cost (AC), since we know from section 8.5 that if MC is greater (less) than AC, AC is rising (falling). Because the function is homogeneous of degree 1, any doubling of output from any initial position will always double (minimized) total cost, so $MC = AC$ at every output level, so the AC and MC curves are as shown in figure 17.4(a), with the one curve lying on top of the other.

We can easily extend this result to cover the cases of decreasing and increasing returns to scale. If the production function is homogeneous of degree less than 1 (that is, there are decreasing returns to scale), then this means that in figure 17.3 $Q_1 < 2Q_0$, but nothing else changes. Therefore, comparing J and N, we have $Q_1 < 2Q_0$ and $TC_1 = 2TC_0$. Since total cost has doubled, but output has less than doubled, it follows that average cost at N is higher than at J. Thus the long-run average cost curve is positively sloped, and marginal cost must be greater than average cost at every output level (see figure 17.4(b)).

If the production function is homogeneous of degree greater than 1 (increasing returns to scale), then, comparing J and N, we have $Q_1 > 2Q_0$ and $TC_1 = 2TC_0$. Since total cost has doubled, but output has more than doubled, it follows that average cost at N is lower than at J. The long-run average cost curve is negatively sloped, and marginal cost must be less than average cost at every output level (see figure 17.4(c)).

We can summarize these results as a rule:

> **RULE 17.1** **Relationship between returns to scale and long-run average and marginal cost, given constant input prices**
>
> If the production function is homogeneous of degree 1 (uniform constant returns to scale), then average cost is the same at every level of output. The average cost curve, as a function of output, is therefore a horizontal straight line. Consequently $MC = AC = $ a constant, at every output level (figure 17.4(a)).
>
> If the production function is homogeneous of degree less than 1 (uniform decreasing returns to scale), average cost increases with output. The average cost curve is positively sloped, and marginal cost must be greater than average cost at any output (figure 17.4(b)).
>
> If the production function is homogeneous of degree greater than 1 (uniform increasing returns to scale), average cost decreases with output. The average cost curve is negatively sloped, and marginal cost must be less than average cost at any output (figure 17.4(c)).
>
> (Don't forget that this rule refers to *long-run* average and marginal cost, since both inputs are varying, which we assume is impossible in the short run.)

Thus property 1 enables us to establish an important link between returns to scale and the slope of the long-run average cost curve. This is useful in proving many other theorems: for example, in international trade theory and in the theory of economic growth. It is a powerful result because it is valid for any homogeneous production function.

Rule 17.1 in the Cobb–Douglas case

You have probably already noticed the similarity between figure 17.4 and figures 16.11–16.13. This is no accident. We saw in example 17.4 that the Cobb–Douglas production function $Q = AK^\alpha L^\beta$ has constant, increasing, or decreasing returns to scale according to whether the sum $\alpha + \beta$ is equal to, greater than, or less than 1 respectively. We also found in example 16.7 that when $\alpha + \beta = 1$, $MC = AC = Z$ (a constant) (see figure 16.13); that when $\alpha + \beta > 1$, $MC < AC$ and AC decreases as output increases (see figure 16.12); and that when $\alpha + \beta < 1$, $MC > AC$ and AC increases as output increases (see figure 16.11).

Thus rule 17.1 extends our findings in example 16.7, which were specific to the Cobb–Douglas production function, to any homogeneous production function.

Property 1 in the Cobb–Douglas case

Consider the Cobb–Douglas production function $Q = AK^\alpha L^\beta$. We know that this function is homogeneous of degree $\alpha + \beta$ (see example 17.4 above). So therefore property 1 should hold. Let's check whether this is so.

The partial derivatives (marginal products) are

$$\frac{\partial Q}{\partial L} = AK^\alpha \beta L^{\beta-1} \quad \text{and} \quad \frac{\partial Q}{\partial K} = A\alpha K^{\alpha-1}L^\beta$$

So the ratio of marginal products, $\frac{MPL}{MPK}$, is

$$\frac{\partial Q/\partial L}{\partial Q/\partial K} = \frac{AK^\alpha \beta L^{\beta-1}}{A\alpha K^{\alpha-1}L^\beta} = \frac{\beta}{\alpha}K^{\alpha-(\alpha-1)}L^{(\beta-1)-\beta} = \frac{\beta}{\alpha}K^1 L^{-1} = \frac{\beta}{\alpha}\frac{K}{L}$$

which is constant when $\frac{K}{L}$ is constant (for given values of the parameters α and β, of course). Thus property 1 holds.

Property 2: how marginal and average products vary with scale

Property 2 concerns how the absolute values of *MPL*, *APL*, *MPK*, and *APK* vary as the scale of production increases. To isolate the effect of increasing scale we hold capital intensity, $\frac{K}{L}$, constant while increasing output, thus moving from J to N in figure 17.3.

We can illustrate the effect of this on *MPL* and *APL* with the Cobb–Douglas production function $Q = AK^{\alpha}L^{\beta}$, which we know is homogeneous of degree $\alpha + \beta$ (see example 17.4). If we assume the function is homogeneous of degree v, we have $\alpha + \beta = v$, hence

$$Q = AK^{\alpha}L^{v-\alpha} = A\left(\frac{K}{L}\right)^{\alpha}L^{v}$$

As we are considering only points on the ray 0R in figure 17.3, $\frac{K}{L}$ is held constant. Given this constraint, the *MPL* is found by differentiating the above expression with respect to L, treating $\frac{K}{L}$ as a constant. This gives

$$MPL \equiv \frac{\partial Q}{\partial L} = A\left(\frac{K}{L}\right)^{\alpha}vL^{v-1}$$

Differentiating for a second time, again with $\frac{K}{L}$ constant, we get

$$\frac{\partial MPL}{\partial L} \equiv \frac{\partial Q^{2}}{\partial L^{2}} = A\left(\frac{K}{L}\right)^{\alpha}(v-1)vL^{v-2}$$

The sign of this expression depends on the sign of $v - 1$. If there are increasing (decreasing) returns, $v - 1$ is positive (negative), hence *MPL* increases (decreases) as L increases with $\frac{K}{L}$ constant, that is, as we move from J to N in figure 17.3. The boundary between these two cases is when there are constant returns to scale, for then $v - 1 = 0$, so $\frac{\partial MPL}{\partial L} = 0$ and *MPL* is constant as we move from J to N.

The effect of increasing scale on the *APL* is easily found. By dividing the expression for Q above by L, we get

$$APL \equiv \frac{Q}{L} = A\left(\frac{K}{L}\right)^{\alpha}L^{v-1} = \frac{1}{v}MPL$$

Thus, as scale increases, *APL* varies as *MPL* varies; and so increases, decreases, or remains constant according as there are increasing, decreasing, or constant returns to scale respectively. Note too from the equation above that $MPL > APL$ if $v > 1$. (Don't forget that these expressions for *MPL* and *APL* differ from those normally obtained because we are here holding $\frac{K}{L}$ constant, rather than K as in the normal case; see section 15.11.) Clearly *MPK* and *APK* vary with scale in the same way. The demonstration of this is left to you as an exercise.

As noted earlier, property 2 is valid for any homogeneous production function, not just the Cobb–Douglas. It is used to prove an important theorem concerning the effect of international trade on real wages.

Property 3: Euler's theorem and the adding-up problem

Euler's theorem states that if any function $z = \mathrm{f}(x, y)$ is homogeneous of degree v, then the following is true:

$$\frac{\partial z}{\partial x}x + \frac{\partial z}{\partial y}y = vz \qquad (17.1)$$

Note that this is true for all values of x, y, and z (provided they satisfy the function $z = \mathrm{f}(x, y)$, of course). Don't confuse equation (17.1) with the differential, $\frac{\partial z}{\partial x}\mathrm{d}x + \frac{\partial z}{\partial y}\mathrm{d}y = \mathrm{d}z$, which holds for *any* function (see section 15.4). ('Euler' may be pronounced 'Yew-ler' or 'Oil-er'.)

When Euler's theorem is applied to the production function, there are three interesting cases:

(1) Constant returns to scale. If the production function $Q = \mathrm{f}(K, L)$ is homogeneous of degree 1 (that is, there are constant returns to scale), then in equation (17.1) we have $v = 1$ and Euler's theorem says that

$$\frac{\partial Q}{\partial K}K + \frac{\partial Q}{\partial L}L = vQ = Q \qquad (17.1a)$$

Since $\frac{\partial Q}{\partial K}$ and $\frac{\partial Q}{\partial L}$ are the marginal products, we can read this equation as

'*MPK* times *K* plus *MPL* times *L* equals *Q*'

This equation has a number of uses in economics, but its most important theoretical application is called the 'adding-up problem'. From section 16.7, rule 16.2, we know that a firm under perfect competition will wish to adjust its inputs of L and K so as to make

$$\frac{\partial Q}{\partial K} = \frac{r}{p} \quad \text{and} \quad \frac{\partial Q}{\partial L} = \frac{w}{p}$$

thereby satisfying the first order condition for profit maximization, where, as usual, w, r, and p are all given constants. Recall that these two equations are known as the *marginal productivity conditions*.

If we substitute the marginal productivity conditions into equation (17.1a), and multiply both sides by p, it becomes

$$rK + wL = pQ \qquad (17.2)$$

However, profit, Π, is defined as $\Pi = pQ - (wL + rK)$ (see section 16.7, equation 16.31). Therefore when equation (17.2) is satisfied, $\Pi = 0$.

Let us pause to review what assumptions we made in arriving at this rather surprising result. We derived equation (17.1a) by assuming that the production function was homogeneous of degree 1 (constant returns to scale) and then applying Euler's theorem. We then derived equation (17.2) from equation (17.1a) by assuming that the marginal productivity conditions were fulfilled (see rule 16.2).

Therefore we have shown that, for a firm producing under constant returns to scale and satisfying the marginal productivity conditions, profits are zero. This is exactly as we found in section 16.9. We also found that if the marginal productivity conditions were fulfilled at one level of output, they were fulfilled at any level of output. There was therefore no unique solution to the profit-maximization problem.

(2) Decreasing returns to scale. In the case of decreasing returns to scale, the production function is homogeneous of degree v, with v less than 1. Equation (17.1a) then becomes

$$\frac{\partial X}{\partial K}K + \frac{\partial X}{\partial L}L = vQ \quad \text{with } vQ < Q \text{ (because } v < 1). \qquad (17.1b)$$

Because $vQ < Q$, when we substitute the marginal productivity conditions into equation (17.1b) we get, as the equivalent of equation (17.2) above,

$$rK + wL = pvQ < pQ \qquad (17.2a)$$

Equation (17.2a) says that total cost, $rK + wL$, is less than total revenue, pQ, and therefore profits are positive. Thus at the output at which the marginal productivity conditions are fulfilled (and the production function is of course satisfied), profits are positive. This is as we found in section 16.9, where the production function had decreasing returns to scale. We found that the output at which the marginal productivity conditions were fulfilled yielded positive profits. The second order condition for a maximum of profits was also satisfied at this point.

(3) Increasing returns to scale. In this case, the production function is homogeneous of degree v, with v greater than 1, so that $vQ > Q$. The equivalent of equation (17.2a) above is now

$$rK + wL = pvQ > pQ \qquad (17.2b)$$

Thus total cost, $rK + wL$, is greater than total revenue, pQ, and therefore profits are negative. Thus at the output at which the marginal productivity conditions are fulfilled (and the production function is of course satisfied), profits are negative. This is as we found in section 16.9, where the production function had increasing returns to scale. We found that the output at which the marginal productivity conditions were fulfilled yielded negative profits, and the second order condition for a maximum of profits was *not* satisfied at this point. Moreover, the firm could always increase its profits by increasing output.

To summarize, the adding-up problem concerns the question whether, when the marginal productivity conditions are fulfilled and we 'add up' the factor rewards wL and rK and thus arrive at total cost, this total will be equal to, less than, or greater than total revenue. We have shown that the answer depends on whether the production function has, respectively, constant, decreasing, or increasing returns to scale.

The Online Resource Centre contains more discussion and worked examples on homogeneous functions, returns to scale in the production function, and Euler's theorem.

Progress exercise 17.2

1. A firm's production function is $Q = AK^{\alpha}L^{\beta}$

(a) Show that there are constant, increasing, or decreasing returns to scale, depending on the values of α and β.

(b) Show that the cost-minimizing capital/labour ratio depends only on $\frac{w}{r}$, the ratio of input prices, and is independent of the level of output.

(c) From (b), show diagrammatically how the cost-minimizing levels of K and L vary as output varies, assuming $\frac{w}{r}$ constant.

(d) Assuming $\frac{w}{r}$ constant, analyse and explain the effect on total cost and on average cost of a doubling of output, assuming (1) $\alpha + \beta > 1$, (2) $\alpha + \beta = 1$, and (3) $\alpha + \beta < 1$. What inference can we draw about the slope of the marginal cost and average curves, in each case? Illustrate the three cases diagrammatically.

(e) Answer (d) more rigorously and transparently by deriving an equation giving total cost as a function of Q, with w, r, and $\alpha + \beta$ as parameters. (Hint: see chapter 16, equation (16.59).)

(f) Suppose this firm is perfectly competitive and can therefore sell as much output as it wishes at the ruling market price, p. Consider and explain how the firm will identify the most profitable level of output in each of the three cases: (1) $\alpha + \beta > 1$, (2) $\alpha + \beta = 1$, and (3) $\alpha + \beta < 1$. (Hint: how did the perfectly competitive firm in section 8.17 identify the most profitable output?)

2. In neoclassical economic theory it is argued that both the supply and demand for labour should be functions of the real wage, $\frac{w}{p}$, where w is the money wage and p the price level. For example, we might have a labour supply function $L^S = A(\frac{w}{p})^{\alpha}$ and a labour demand function $L^D = B(\frac{w}{p})^{-\beta}$ (where A, B, α, and β are positive parameters). Assess the degree of homogeneity of these two functions and comment on the effect on equilibrium employment of an equal proportionate rise or fall in w and p.

17.5 Partial elasticities

In chapter 9 we developed two elasticity concepts, arc elasticity and point elasticity, for any function $y = f(x)$ (that is, with one independent variable). We defined the arc elasticity as $\frac{x}{y}\frac{\Delta y}{\Delta x}$, where $\frac{\Delta y}{\Delta x}$ is the difference quotient, and the point elasticity as $\frac{x}{y}\frac{dy}{dx}$, where $\frac{dy}{dx}$ is the derivative of the function.

We saw that both elasticities measure the rate of *proportionate* change of y. This contrasts with the difference quotient $\frac{\Delta y}{\Delta x}$ and the derivative $\frac{dy}{dx}$, both of which measure the rate of *absolute* change. The difference between the two elasticity concepts is that the arc elasticity measures the average rate of proportionate change following a discrete change in x, while the point elasticity measures the rate of proportionate change at a point on the function.

These concepts are easily extended to a function with two or more independent variables. Given

$z = f(x, y)$

we can start from the difference quotient $\frac{\Delta z}{\Delta x}$ that we derived in section 14.4. Recall that this is evaluated with the other variable, y, held constant. It measures the average rate of change of z between two points on the function following a discrete change in x, Δx. If we multiply this by $\frac{x}{z}$, we obtain

$$\frac{x}{z}\frac{\Delta z}{\Delta x}$$

which we can define as the **arc partial elasticity** of z with respect to x. This measures the average rate of proportionate change in z following a discrete change in x, Δx.

We also have the partial derivative $\frac{\partial z}{\partial x}$ that we derived in section 14.4. Again this is evaluated with the other variable, y, held constant. If we multiply this by $\frac{x}{z}$ we obtain

$$\frac{x}{z}\frac{\partial z}{\partial x}$$

which we define as the **point partial elasticity** of z with respect to x. It measures the rate of proportionate change of z at a point on the function, as x changes with y held constant. Both elasticities are called partial elasticities because they measure the response of z to a change in x alone, with y held constant.

In exactly the same way, by allowing y to vary with x held constant, we derive

$$\frac{y}{z}\frac{\Delta z}{\Delta y} \quad \text{and} \quad \frac{y}{z}\frac{\partial z}{\partial y}$$

as, respectively. the arc and point partial elasticities of z with respect to y. The first measures the average rate of proportionate change in z following a discrete change in y, Δy. The second measures the rate of proportionate change of z at a point on the function, as y changes with x held constant.

Although the arc partial elasticities are useful for analysing discrete changes, they are seldom used in theoretical economics, so from now on we will focus exclusively on the point partial elasticities, and refer to them simply as partial elasticities. Thus we have rule 17.2:

RULE 17.2 Partial elasticities

The partial elasticities of the function $z = \mathrm{f}(x, y)$ are

$$\frac{x}{z}\frac{\partial z}{\partial x} \text{ and } \frac{y}{z}\frac{\partial z}{\partial y}$$

17.6 Partial elasticities of demand

Suppose there are two goods, a and b, and Y = consumers' income. Let q_a denote the total quantity of good a demanded by all consumers, and p_a and p_b denote the prices of goods a and b. Then the combined demands of all consumers, known as the market demand function for good a, might be of the form

$$q_a = \mathrm{f}(p_a, p_b, Y)$$

In words, this says that the quantity of good a demanded depends on the prices of goods a and b, and on consumers' income.

Then there is a partial elasticity associated with each of the three variables on the right-hand side of the demand function. These partial elasticities are:

(1) $\dfrac{p_a}{q_a}\dfrac{\partial q_a}{\partial p_a}$ = the partial elasticity of q_a with respect to p_a. This is known as the 'own-price' elasticity of demand because it measures the response of quantity of good a demanded to a change in the price of good a itself.

(2) $\dfrac{p_b}{q_a}\dfrac{\partial q_a}{\partial p_b}$ = the partial elasticity of q_a with respect to p_b. This is known as the 'cross-price' elasticity of demand because it measures the response of quantity of good a demanded to a change in the price of another good, good b.

(3) $\dfrac{Y}{q_a}\dfrac{\partial q_a}{\partial Y}$ = the partial elasticity of q_a with respect to Y. This is known as the income elasticity of demand.

Note that in this example we have *three* variables on the right-hand side. Therefore each partial derivative, and therefore each partial elasticity, is evaluated with the other *two* variables held constant. For example, $\dfrac{\partial q_a}{\partial p_a}$ is evaluated with p_b and Y held constant.

The following two examples illustrate how we derive these elasticities in particular cases.

EXAMPLE 17.5

Consider the demand function

$$q_a = 10Y - 0.05Yp_a + 0.02Yp_b$$

where q_a is the quantity demanded of good a; p_a is the price of good a; p_b is the price of good b; and Y is the individual's budget or income.

The partial derivatives are

$$\frac{\partial q_a}{\partial p_a} = -0.05Y; \quad \frac{\partial q_a}{\partial p_b} = 0.02Y; \quad \frac{\partial q_a}{\partial Y} = 10 - 0.05p_a + 0.02p_b$$

So the partial elasticities are:

(1) Own-price elasticity: $\dfrac{p_a}{q_a}\dfrac{\partial q_a}{\partial p_a} = \dfrac{-0.05p_a}{10 - 0.05p_a + 0.02p_b}$

As the numerator is negative, this elasticity will be negative unless the denominator is negative, which requires $p_a > 200 + 0.4p_b$. In other words, provided $p_a < 200 + 0.4p_b$, we have the 'normal' case in which the 'partial' demand curve, which relates q_a to p_a with Y and p_b held constant, is downward sloping.

(2) Cross-price elasticity: $\dfrac{p_b}{q_a}\dfrac{\partial q_a}{\partial p_b} = \dfrac{0.02p_b}{10 - 0.05p_a + 0.02p_b}$

As the numerator is positive, this elasticity will be positive unless the denominator is negative, which again requires $p_a > 200 + 0.4p_b$. In other words, provided $p_a < 200 + 0.4p_b$, a rise in the price of good b, with income and the price of good a constant, increases the demand for good a. The goods are substitutes.

(3) Income elasticity: $\dfrac{Y}{q_a}\dfrac{\partial q_a}{\partial Y} = \dfrac{Y(10 - 0.05p_a + 0.02p_b)}{10Y - 0.05Yp_a + 0.02Yp_b} = 1$

The fact that the income elasticity is constant and equal to 1 means that consumers respond to a 10% increase in their incomes by buying 10% more units of this good, provided of course that neither its price nor the price of good b has changed. (A point elasticity is not normally an accurate guide to the effect of a change as large as 10% in the independent variable, but this case is exceptional because the income elasticity is constant.)

It's also interesting to note that in this example the own-price and cross-price elasticities turn out to be independent of the level of income. Be sure to check these for yourself before moving on.

EXAMPLE 17.6

In this example we will use the demand functions for two goods, X and Y, derived in section 16.12 where we assumed the individual's utility function was Cobb–Douglas in form. The demand function for X (equation (16.70)) is

$$X = \frac{\alpha}{\alpha + \beta} \frac{B}{p_X} = \frac{\alpha}{\alpha + \beta} B(p_X)^{-1}$$

where B = consumer's budget or income and p_X and p_Y are the prices. Note that B is now treated as a variable, although in section 16.12 we treated it as a constant, B_0, in deriving the consumer's utility-maximizing quantities of X and Y.

The partial derivative with respect to p_x is

$$\frac{\partial X}{\partial p_X} = -\frac{\alpha}{\alpha + \beta} B(p_X)^{-2}$$

Multiplying this partial derivative by $\frac{p_x}{X}$ we arrive at the own-price elasticity, E^D, as

$$E^D \equiv \frac{p_X}{X} \frac{\partial X}{\partial p_X} = -\frac{\alpha}{\alpha + \beta} B(p_X)^{-2} \frac{p_X}{X} = -\frac{\alpha}{\alpha + \beta} B(p_X)^{-1} \frac{1}{X}$$

$$= -1 \quad \text{(since, from the demand function, } \tfrac{\alpha}{\alpha + \beta} B(p_X)^{-1} = X\text{)}$$

The partial derivative of the demand function, with respect to B, is

$$\frac{\partial X}{\partial B} = \frac{\alpha}{\alpha + \beta} (p_X)^{-1}$$

Multiplying this partial derivative by $\frac{B}{X}$ we arrive at the income elasticity, E^B, as

$$E^B \equiv \frac{B}{X} \frac{\partial X}{\partial B} = \frac{\alpha}{\alpha + \beta} (p_X)^{-1} \frac{B}{X}$$

$$= 1 \quad \text{(since, from the demand function, } \tfrac{\alpha}{\alpha + \beta} B(p_X)^{-1} = X\text{)}$$

Thus with a Cobb–Douglas utility function the price elasticity of demand is −1 and the income elasticity of demand is 1. Note that, because p_Y does not appear in the demand function for X (equation (16.70)), it follows automatically that the cross-price elasticity of demand for X, $\frac{p_Y}{X} \frac{\partial X}{\partial p_Y}$, is zero.

In the same way, from section 16.12 we can take the demand function for good Y, equation (16.71), which is

$$Y = \frac{\beta}{\alpha + \beta} \frac{B_0}{p_Y}$$

and derive the own-price and income elasticities of demand for good Y. This is left to you as an exercise. (See section 9.11 if you need to revise demand elasticities.)

17.7 The proportionate differential of a function

In section 15.4 we saw that the total differential of a function $z = f(x, y)$ is given by

$$dz = \frac{\partial z}{\partial x} dx + \frac{\partial z}{\partial y} dy \tag{17.3}$$

This formula gives us the change in z, dz, that results when x changes by a small amount dx and at the same time y changes by a small amount dy. The key point here is that these changes are *absolute* changes.

However, in economics, as we have already seen at several points in this book, we are often more interested in analysing proportionate changes than absolute changes. Given some function $z = f(x, y)$, we often want to ask the question: 'If there occurs a small proportionate change in x, and at the same time a small proportionate change in y, what is the resulting proportionate change in z?'

We can answer this question by modifying the differential of the function. Starting from equation (17.3) above, we want to get from the absolute changes, dz, dx, and dy, in equation (17.3), to the proportionate changes, $\frac{dz}{z}$, $\frac{dx}{x}$, and $\frac{dy}{y}$, that we are interested in. To do this, we proceed as follows:

Step 1: Divide both sides of equation (17.3) by z, giving

$$\frac{dz}{z} = \frac{1}{z}\frac{\partial z}{\partial x}dx + \frac{1}{z}\frac{\partial z}{\partial y}dy$$

We now have the proportionate change in z on the left-hand side.

Step 2: Simultaneously divide and multiply the first term on the right-hand side by x, and the second term by y. This gives

$$\frac{dz}{z} = \frac{x}{x}\frac{1}{z}\frac{\partial z}{\partial x}dx + \frac{y}{y}\frac{1}{z}\frac{\partial z}{\partial y}dy, \text{ which rearranges to give us rule 17.3:}$$

RULE 17.3 The proportionate differential

The proportionate differential of the function $z = f(x, y)$ is

$$\frac{dz}{z} = \left(\frac{x}{z}\frac{\partial z}{\partial x}\right)\frac{dx}{x} + \left(\frac{y}{z}\frac{\partial z}{\partial y}\right)\frac{dy}{y}$$

At first sight rule 17.3 seems a deeply mysterious expression, but it can be decoded. On the left-hand side we have $\frac{dz}{z}$, the proportionate change in z. On the right-hand side, the terms in brackets are the partial elasticities of the function. The partial elasticity with respect to x is multiplied by $\frac{dx}{x}$, the proportionate change in x. Similarly, the partial elasticity with respect to y is multiplied by $\frac{dy}{y}$, the proportionate change in y.

So, in words, rule 17.3 is answering our question above as follows: 'The resulting proportionate change in z, $\frac{dz}{z}$, is given by the sum of the proportionate changes in x and y, each multiplied by its partial elasticity.' In effect, the partial elasticities act as *weights* on the proportionate changes in x and y. If for example $\frac{x}{z}\frac{\partial z}{\partial x}$ is large, then this acts as a heavy weight on $\frac{dx}{x}$ so that the latter has a large effect on z. This will become clearer in the examples below.

EXAMPLE 17.7

Suppose we wish to find the proportionate differential of $z = x^3 + 2y$.

The partial derivatives are

$$\frac{\partial z}{\partial x} = 3x^2 \quad \text{and} \quad \frac{\partial z}{\partial y} = 2$$

So the partial elasticities are

$$\frac{x}{z}\frac{\partial z}{\partial x} = \frac{x}{z}3x^2 = \frac{3x^3}{z} \quad \text{and} \quad \frac{y}{z}\frac{\partial z}{\partial y} = \frac{2y}{z}$$

These values are then substituted into the proportionate differential formula (rule 17.3), giving

$$\frac{dz}{z} = \left(\frac{3x^3}{z}\right)\frac{dx}{x} + \left(\frac{2y}{z}\right)\frac{dy}{y}$$

This is not a very neat result, but that's because the function we started with was not very neat either.

Hint The results shown in the next three examples are used very frequently in economics and deserve to be studied carefully.

EXAMPLE 17.8

Suppose we wish to find the proportionate differential of the Cobb–Douglas function:
$z = Ax^{\alpha}y^{\beta}$ (where A, α and β are constants).

The partial derivative with respect to x is

$$\frac{\partial z}{\partial x} = \alpha A x^{\alpha-1} y^{\beta}$$

If we simultaneously multiply and divide the right-hand side by x (see appendix 14.1), we get

$$\frac{\partial z}{\partial x} = \frac{\alpha A x^{\alpha-1} y^{\beta} x}{x} = \frac{\alpha A x^{\alpha} y^{\beta}}{x} = \alpha \frac{z}{x} \quad \text{(since } z = Ax^{\alpha}y^{\beta})$$

So the partial elasticity with respect to x is

$$\frac{x}{z}\frac{\partial z}{\partial x} = \frac{x}{z}\alpha\frac{z}{x} = \alpha$$

In exactly the same way, the partial elasticity with respect to y is

$$\frac{y}{z}\frac{\partial z}{\partial y} = \frac{y}{z}\beta\frac{z}{y} = \beta \quad \text{(Check this for yourself.)}$$

These values, α and β, are then substituted into the proportionate differential formula

$$\frac{dz}{z} = \left(\frac{x}{z}\frac{\partial z}{\partial x}\right)\frac{dx}{x} + \left(\frac{y}{z}\frac{\partial z}{\partial y}\right)\frac{dy}{y}$$

giving an important result, which we state as a rule:

RULE 17.4 The proportionate differential of the Cobb–Douglas function

The Cobb–Douglas function, $z = Ax^{\alpha}y^{\beta}$, has the proportionate differential

$$\frac{dz}{z} = \alpha\frac{dx}{x} + \beta\frac{dy}{y}$$

In rule 17.4 we can see clearly how the partial elasticities, α and β, act as weights on the proportionate changes $\frac{dx}{x}$ and $\frac{dy}{y}$. For example, for given values of $\frac{dx}{x}$ and $\frac{dy}{y}$, if α is large and β small, then $\frac{dz}{z}$ will be more influenced by $\frac{dx}{x}$ than by $\frac{dy}{y}$. We will examine this in an economic example in the next section.

EXAMPLE 17.9

To find the proportionate differential of $z = xy$.

This is really just a specific case of the previous example, as the function is of Cobb–Douglas form with $\alpha = \beta = 1$. So applying rule 17.4 we get

$$\frac{dz}{z} = \frac{dx}{x} + \frac{dy}{y}$$

In words, this result says: suppose some variable, z, is given by the *product*, xy, of two variables x and y; then the proportionate change in z is given by the *sum* of the associated proportionate changes in x and y. (An obvious example in economics would be total revenue, which is given by the product of price and quantity.)

EXAMPLE 17.10

To find the proportionate differential of $z = \frac{x}{y}$.

We can rewrite this as

$$z = xy^{-1}$$

Then we see that this again is a specific case of the Cobb–Douglas form, with $\alpha = 1$ and $\beta = -1$. So applying rule 17.4 we get

$$\frac{dz}{z} = \frac{dx}{x} - \frac{dy}{y}$$

In words, this result says: suppose some variable, z, is given by the *ratio*, $\frac{x}{y}$, of two variables x and y; then the proportionate change in z is given by the *difference* of the associated proportionate changes in x and y. (An obvious example in economics would be average cost, which is given by the ratio of total cost to output.)

17.8 Growth accounting

In analysing the growth of total output in an economy over time, the standard neoclassical approach is to assume that there exists an aggregate production function for the whole economy of the form $Q = F(K, L)$, where Q = total output of goods and services (GDP) and K and L are the total inputs of capital and labour.

Then applying the proportionate differential formula (rule 17.3) to this aggregate production function gives

$$\frac{dQ}{Q} = \left(\frac{K}{Q} \frac{\partial Q}{\partial K} \right) \frac{dK}{K} + \left(\frac{L}{Q} \frac{\partial Q}{\partial L} \right) \frac{dL}{L}$$

If we then expected the supply of capital to grow (due to investment) by, say, 20% over the next 10 years, and the supply of labour to grow by, say, 10% over the same period, then we could set $\frac{dK}{K} = 20\% = 0.2$ and $\frac{dL}{L} = 10\% = 0.1$. (A significant error may result from using the differential to model a change as large as 20%, but this is often ignored.) Substituting these values into the proportionate differential formula gives

$$\frac{dQ}{Q} = \left(\frac{K}{Q} \frac{\partial Q}{\partial K} \right)(0.2) + \left(\frac{L}{Q} \frac{\partial Q}{\partial L} \right)(0.1) \tag{17.4}$$

but this does not take us very far unless we can somehow find some plausible values for the partial elasticities $\frac{K}{Q} \frac{\partial Q}{\partial K}$ and $\frac{L}{Q} \frac{\partial Q}{\partial L}$.

One way of obtaining these is to assume that the aggregate production function is of Cobb–Douglas form with constant returns to scale: that is,

$$Q = AK^{\alpha}L^{1-\alpha} \quad \text{(where A and α are constants)}$$

Then we know from rule 17.4 above that the partial elasticities are

$$\frac{K}{Q} \frac{\partial Q}{\partial K} = \alpha \quad \text{and} \quad \frac{L}{Q} \frac{\partial Q}{\partial L} = 1 - \alpha \tag{17.5}$$

Unfortunately we are still no further forward unless we can come up with a reliable estimate of α. One way of estimating α is to use the marginal productivity conditions (see section 16.7). The marginal productivity conditions state that with perfect competition and profit maximization, we will have

$$\frac{\partial Q}{\partial K} = \frac{r}{p} \quad \text{and} \quad \frac{\partial Q}{\partial L} = \frac{w}{p} \tag{17.6}$$

where as usual p is the product price and w and r are the input prices. (Note that the marginal productivity conditions can be satisfied, given our assumption of constant returns to scale, only if profits are zero at any level of output; see section 16.9 and rule 17.1. This obviously raises questions but we will not attempt to answer them here.)

If we substitute the marginal productivity conditions (equation (17.6)) into the partial elasticities (17.5), the partial elasticities become

$$\frac{K}{Q}\frac{r}{p} = \alpha \quad \text{and} \quad \frac{L}{Q}\frac{w}{p} = 1 - \alpha \tag{17.7}$$

Here $K \times r$ is the income received by owners of capital and $Q \times p$ is aggregate income (GDP), so $\frac{K}{Q}\frac{r}{p}$ is the *share* of aggregate income received by owners of capital, which we can measure by looking at the share of profits in GDP. Similarly $L \times w$ is wage income, so $\frac{L}{Q}\frac{w}{p}$ is the share of wage income in GDP. (These are known as **factor shares** because L and K are the factors of production.)

Thus equation (17.7) tells us that we can use the shares of profits and wages in GDP as estimates of the partial elasticities α and $1 - \alpha$ respectively. Rough estimates of these shares might be $\alpha = 0.3$ and $1 - \alpha = 0.7$. (For more discussion, see the chapters on economic growth in any good macroeconomics textbook.) If we then substitute these values into equation (17.4), together with the assumed values 0.2 and 0.1 for the growth of capital and labour inputs, we get

$$\frac{dQ}{Q} = \alpha\frac{dK}{K} + (1-\alpha)\frac{dL}{L} = (0.3)(0.2) + (0.7)(0.1) = 0.06 + 0.07 = 0.13 = 13\%$$

This says that the assumed growth of capital and labour inputs over the 10-year period (20% and 10% respectively) will result in a growth of total output of 13%. Note that the growth of output (13%) is a weighted sum of the growth of inputs (20% and 10%), the weights being the partial elasticities (0.3 and 0.7). Thus in general this methodology gives us a tool for predicting the future growth of output as a function of predicted growth of inputs.

Equally, the methodology gives us a tool for explaining past growth of output by reference to the past growth of inputs. For example, suppose the above growth rates of 20% for K and 10% for L were not predictions about the future but actual growth in the past 10 years. The equation above, which says that $\frac{dQ}{Q} = 13\%$, is now a 'prediction' of past growth of output. If it turns out that the actual growth of output was, say, 15%, then we have an unexplained 'residual' of 15% − 13% = 2%. This 2% is growth of output which, if our model is correct, is *not* attributable to the growth of capital and labour inputs. Instead, this 2% is perhaps due to technological innovation which has made production more efficient, equivalent in effect to an increase over time in the constant term, A, in the production function $Q = AK^\alpha L^\beta$. Thus the 2% unexplained residual implies that the production function has 'shifted' over time, in the sense that the parameter A has changed in value.

A third use of the growth accounting methodology is to predict the effects of an increase or decrease in the growth rate of inputs. Suppose, for example, that government policies to encourage investment caused the expected growth of capital over the next 10 years to increase, from the 20% assumed above, to 25%. (We assume no change in growth of L.) Then substituting this new value into the growth accounting equation, we get

$$\frac{dQ}{Q} = \alpha\frac{dK}{K} + (1-\alpha)\frac{dL}{L} = (0.3)(0.25) + (0.7)(0.1) = 0.075 + 0.07 = 0.145 = 14.5\%$$

Thus the 5% increase in capital growth (from 20% to 25%) adds only 1.5% to the growth of output (from 13% to 14.5%). This is because α, the partial elasticity with which $\frac{dK}{K}$ is weighted, is only 0.3. So the boost to the growth of GDP is only $0.3 \times 5\% = 1.5\%$.

17.9 Elasticity and logs

In chapter 13, rule 13.9, we showed that for any function $y = f(x)$

$$\frac{d(\ln y)}{dx} = \frac{1}{y}\frac{dy}{dx}$$

We will now extend this idea a little further. The following rule is proved in appendix W17.1, which you can find at the Online Resource Centre (www.oxfordtextbooks.co.uk/orc/renshaw3e/).

RULE 17.5 Elasticity and slope with log-log scales

For any function $y = f(x)$

$$\frac{d(\ln y)}{d(\ln x)} = \frac{x}{y}\frac{dy}{dx}$$

In rule 17.5, the left-hand side is the slope of the function when it is plotted with natural log scales on both axes. The right-hand side is the elasticity of the function. So the rule says that when we plot the graph of any function $y = f(x)$, with natural log scales on both axes, the slope of this graph at any point measures the elasticity of the function at that point.

EXAMPLE 17.11

Consider the demand function $q = 100p^{-0.5}$.

First, we will find the elasticity of demand, E^D. The derivative is

$$\frac{dq}{dp} = (-0.5)100p^{-1.5}$$

so the elasticity is

$$E^D \equiv \frac{p}{q}\frac{dq}{dp} = \frac{p(-0.5)100p^{-1.5}}{q} = \frac{(-0.5)100p^{-0.5}}{q}$$

$$= -0.5 \quad (\text{since } q = 100p^{-0.5})$$

Next, if we take (natural) logs on both sides of the demand function, we get

$$\ln q = \ln 100 - 0.5(\ln p)$$

Hopefully you can see straight away that this is a linear function in which the variables are $\ln q$ and $\ln p$, with a constant term of $\ln 100$ and a slope of -0.5. If you can't, see the hint below.

A function such as this, which becomes linear after taking logs on both sides, is said to be log linear. (For greater precision it would be better described as being double log linear, to distinguish it from the semi-log linear function that is derived from the exponential function $y = ae^{rx}$, and which we met in section 13.9.)

We next find the derivative of $\ln q = \ln 100 - 0.5(\ln p)$. Since the independent variable is $\ln p$ and dependent variable is $\ln q$, the derivative is, by definition, $\frac{d(\ln q)}{d(\ln p)}$. Using the standard rules of differentiation, this derivative is

$$\frac{d(\ln q)}{d(\ln p)} = \frac{d}{d(\ln p)} \text{ of: } \ln 100 - 0.5(\ln p) = -0.5$$

and from above we see that this equals the elasticity, E^D. Thus rule 17.5 above is confirmed in this case. The graphs of the original demand function, $q = 100p^{-0.5}$, and its logarithmic transformation, $\ln q = \ln 100 - 0.5(\ln p)$, are shown in figure 17.5.

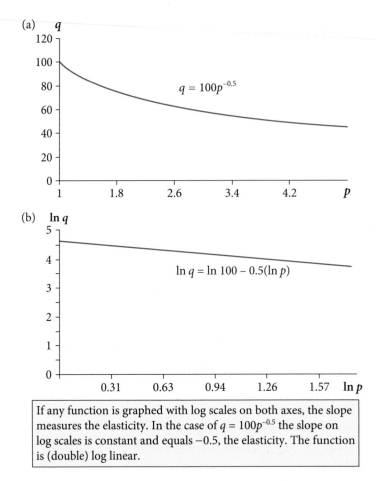

If any function is graphed with log scales on both axes, the slope measures the elasticity. In the case of $q = 100p^{-0.5}$ the slope on log scales is constant and equals -0.5, the elasticity. The function is (double) log linear.

Figure 17.5 The demand function $q = 100p^{-0.5}$ and its logarithmic transformation.

Hint If you found it hard follow the example above, it may help if we define two new variables: $Q = \ln q$ and $P = \ln p$; and a new constant: $C = \ln 100$. When we substitute these into $\ln q = \ln 100 - 0.5(\ln p)$, it becomes

$$Q = C - 0.5P$$

which is more obviously a linear function relating the two variables Q and P, with a constant term of C and a slope of $\frac{dQ}{dP} = -0.5$. But in this derivative, we can replace Q with $\ln q$, and P with $\ln p$. So $\frac{dQ}{dP} = \frac{d(\ln q)}{d(\ln p)} = -0.5$.

17.10 Partial elasticities and logarithmic scales

In the previous section we showed that for any function $y = \mathrm{f}(x)$, $\frac{d(\ln y)}{d(\ln x)} = \frac{x}{y}\frac{dy}{dx}$. This result extends quite straightforwardly to a function of two independent variables, $z = \mathrm{f}(x, y)$. We won't prove it here, but it is quite easy to derive the following rule:

RULE 17.6 Partial elasticities and logarithmic scales

For any function $z = f(x, y)$

$$\frac{\partial(\ln z)}{\partial(\ln x)} = \frac{x}{z}\frac{\partial z}{\partial x} = \text{the partial elasticity of } z \text{ with respect to } x$$

and

$$\frac{\partial(\ln z)}{\partial(\ln y)} = \frac{y}{z}\frac{\partial z}{\partial x} = \text{partial elasticity of } z \text{ with respect to } y$$

In words, this rule means that if we plot the function $z = f(x, y)$ with natural log scales on all three axes, the slope in the x direction of the resulting three-dimensional surface measures the partial elasticity with respect to x, and the slope in the y direction measures the partial elasticity with respect to y. (See section 17.5 for refreshment on partial elasticities.)

EXAMPLE 17.12

Suppose we are given the demand function

$$q = p^{-2}y^{0.5}$$

where $q =$ quantity demanded, $p =$ price, $y =$ consumers' income.

The partial derivative with respect to p is

$$\frac{\partial q}{\partial p} = -2p^{-3}y^{0.5}$$

So the partial elasticity with respect to p is

$$\frac{p}{q}\frac{\partial q}{\partial p} = \frac{p}{q}[-2p^{-3}y^{0.5}] = \frac{1}{q}[-2p^{-2}y^{0.5}] = -2 \quad (\text{since } q = p^{-2}y^{0.5})$$

This, of course, is the own-price elasticity of demand.

Similarly, the partial derivative with respect to y is

$$\frac{\partial q}{\partial y} = 0.5p^{-2}y^{-0.5}$$

So the partial elasticity with respect to y is

$$\frac{y}{q}\frac{\partial q}{\partial y} = \frac{y}{q}[0.5p^{-2}y^{-0.5}] = \frac{1}{q}[0.5p^{-2}y^{0.5}] = 0.5 \quad (\text{since } q = p^{-2}y^{0.5})$$

This is the income elasticity of demand.

To summarize, we have

$$\frac{p}{q}\frac{\partial q}{\partial p} = -2 \quad (\text{own-price elasticity of demand})$$

$$\frac{y}{q}\frac{\partial q}{\partial y} = 0.5 \quad (\text{income elasticity of demand})$$

Note the pattern here: with a function of this (Cobb–Douglas) type, the partial elasticities are given by the exponents of the variables on the right-hand side.

Next, we take natural logs on both sides of the demand function, $q = p^{-2}y^{0.5}$. This gives

$$\ln q = -2[\ln p] + 0.5[\ln y] \quad (\text{the square brackets are not strictly necessary})$$

The partial derivatives are

$$\frac{\partial \ln q}{\partial \ln p} = \frac{\partial}{\partial \ln p} \text{of: } -2[\ln p] + 0.5[\ln y] = -2$$

$$\frac{\partial \ln q}{\partial \ln y} = \frac{\partial}{\partial \ln y} \text{of: } -2[\ln p] + 0.5[\ln y] = 0.5$$

To summarize, we have found that, for the demand function $q = p^{-2}y^{0.5}$,

$$\frac{\partial \ln q}{\partial \ln p} = -2 = \frac{p}{q}\frac{\partial q}{\partial p} \quad (= \text{price elasticity of demand}), \text{ and}$$

$$\frac{\partial \ln q}{\partial \ln y} = 0.5 = \frac{y}{q}\frac{\partial q}{\partial y} \quad (= \text{income elasticity of demand})$$

This means that if we plotted the demand function with natural log scales on all three axes, the slope in the p direction of the resulting surface would measure the price elasticity of demand, and the slope in the y direction would measure the income elasticity of demand. In this example, the slope is constant in both directions: with log scales on all three axes, the surface is a plane. But rule 17.6 holds for any function, whether or not it is a plane.

17.11 The proportionate differential and logs

In this section we show a useful relationship involving the proportionate differential. Recall from rule 17.3 above that the proportionate differential of the function $z = f(x, y)$ is

$$\frac{dz}{z} = \left(\frac{x}{z}\frac{\partial z}{\partial x}\right)\frac{dx}{x} + \left(\frac{y}{z}\frac{\partial z}{\partial y}\right)\frac{dy}{y} \tag{17.8}$$

This formula, derived from the total differential, gives the proportionate differential (that is, $\frac{dz}{z}$) as a function of the proportionate changes in x and y, ($\frac{dx}{x}$ and $\frac{dy}{y}$), and the partial elasticities. It is used frequently in economics because we are often focusing on proportionate rather than absolute changes.

In section 13.8 we showed by means of a numerical example that, if y is any variable that increases by a small amount, Δy, then $\frac{\Delta y}{y}$ approximately equals $\Delta(\ln y)$. That is, the proportionate change in y is approximately equal to the absolute change in the natural log of y. We can write this relationship as

$$\frac{\Delta y}{y} = \Delta(\ln y) \pm \text{a small error} \tag{17.9}$$

The numerical example in section 13.8 also showed (though did not prove) that the error in this approximation can be made as small as we wish by making Δy sufficiently small. Thus there must come a point at which Δy is so small that we are willing to ignore the error and treat $\frac{\Delta y}{y}$ and $\Delta(\ln y)$ as being actually equal to one another. Let us write dy to denote this very small value of Δy. Similarly, we can write $d(\ln y)$ to denote the associated very small change in the log of y. Since by assumption these changes dy and $d(\ln y)$ are so small that we are willing to ignore the error in equation (17.9), we can write the following rule:

RULE 17.7 Proportionate change in a variable and absolute change in its log

If y is any variable, then when dy is sufficiently small,

$$\frac{dy}{y} = d(\ln y)$$

In words, this says that when a variable, y, changes by a small amount, dy, then the proportionate change in y equals the absolute change in the natural log of y. The proof is in appendix W17.2, which you will find of the Online Resource Centre.

Rule 17.7 is true of any variable that changes for whatever reason. Therefore for variables z, x, and y we have

$$\frac{dz}{z} = d(\ln z); \quad \frac{dx}{x} = d(\ln x); \quad \text{and} \quad \frac{dy}{y} = d(\ln y)$$

If we substitute these into the proportionate differential formula (equation (17.8) above), we get

$$d(\ln z) = \left(\frac{x}{z}\frac{\partial z}{\partial x}\right)d(\ln x) + \left(\frac{y}{z}\frac{\partial z}{\partial y}\right)d(\ln y) \qquad (17.10)$$

But we also know from rule 17.6 above that

$$\frac{\partial(\ln z)}{\partial(\ln x)} = \frac{x}{z}\frac{\partial z}{\partial x} \quad \text{and} \quad \frac{\partial(\ln z)}{\partial(\ln y)} = \frac{y}{z}\frac{\partial z}{\partial y}$$

We can substitute these into equation (17.10) and get rule 17.8:

RULE 17.8 The logarithmic differential

For any function $z = f(x, y)$:

$$d(\ln z) = \left(\frac{\partial(\ln z)}{\partial(\ln x)}\right)d(\ln x) + \left(\frac{\partial(\ln z)}{\partial(\ln y)}\right)d(\ln y) \qquad (17.11)$$

Rule 17.8 gives the proportionate differential formula, but now in terms of absolute changes and rates of absolute change of natural logs. The key point is that equations (17.11) and (17.8) are *identically* equal.

In the following example we illustrate the identity of equations (17.11) and (17.8) above for the case of the Cobb–Douglas production function.

EXAMPLE 17.13

Given $Q = AK^{\alpha}L^{\beta}$ where A, α, and β are constants, from example 17.8 above we see that the partial elasticities are

$$\frac{K}{Q}\frac{\partial Q}{\partial K} = \alpha \quad \text{and} \quad \frac{L}{Q}\frac{\partial Q}{\partial L} = \beta \quad \text{(check for yourself)}$$

so the proportionate differential (again from example 17.8) is

$$\frac{dQ}{Q} = \left(\frac{K}{Q}\frac{\partial Q}{\partial K}\right)\frac{dK}{K} + \left(\frac{L}{Q}\frac{L\partial Q}{Q\partial L}\right)\frac{dL}{L} = \alpha\frac{dK}{K} + \beta\frac{dL}{L} \qquad (17.12)$$

We now want to derive the logarithmic equivalent. Given the production function $Q = AK^{\alpha}L^{\beta}$, if we take logs on both sides we get

$$\ln Q = \ln A + \alpha(\ln K) + \beta(\ln L) \qquad (17.13)$$

Hopefully you can see that $\ln Q$ is a linear function of $\ln K$ and $\ln L$, with $\ln A$ a constant. Therefore its partial derivatives are

$$\frac{\partial \ln Q}{\partial \ln K} = \alpha \text{ and } \frac{\partial \ln Q}{\partial \ln L} = \beta \quad (\ln A \text{ is an additive constant, so disappears})$$

and its total differential is, by definition,

$$d\ln Q = \frac{\partial \ln Q}{\partial \ln K}d\ln K + \frac{\partial \ln Q}{\partial \ln L}d\ln L$$

Substituting α and β for the partial derivatives, this becomes

$$d \ln Q = \alpha \, d \ln K + \beta \, d \ln L \qquad (17.14)$$

Now, we want to show that equations (17.12) and (17.14) are identically equal. We can do this using rule 17.7 above, which says that $d \ln y = \frac{dy}{y}$ for any variable y, provided dy is sufficiently small. Using this, we can write $d \ln Q = \frac{dQ}{Q}$, $d \ln K = \frac{dK}{K}$, and $d \ln L = \frac{dL}{L}$. When these are substituted into equation (17.14), it becomes

$$\frac{dQ}{Q} = \alpha\frac{dK}{K} + \beta\frac{dL}{L}$$

which identically equals equation (17.12), as we wished to show.

17.12 Log linearity with several variables

In example 17.11 above, we examined the demand function $q = 100p^{-0.5}$ and saw that when we took logs on both sides we got

$$\ln q = \ln 100 - 0.5(\ln p)$$

which, because it is a linear relationship between $\ln q$ and $\ln p$, is called a (double) log linear function.

In the same way, an equation involving two independent variables can be (triple) log linear, meaning that when it is plotted with log scales on all three axes, the resulting surface is a plane. Any function of the Cobb–Douglas form $z = x^\alpha y^\beta$, where α and β are constants, has this property. We can show this simply by taking logs on both sides (see sections 11.7 and 12.7 for rules of logs), which gives

$$\ln z = \ln(x^\alpha y^\beta) = \ln(x^\alpha) + \ln(y^\beta) = \alpha \ln x + \beta \ln y$$

This log linear property continues to hold as the number of independent variables increases: for example, $z = x^\alpha y^\beta v^\delta$. (Of course, most functions are *not* log linear. The assumption that some functional relationship is log linear is merely a convenient simplification widely used in economics.)

An example of log linearity in economics is the Cobb–Douglas function, much used as both a production function and a utility function. In the production function case we therefore have a result that is sufficiently important to be worth stating as a rule:

RULE 17.9 **Log linearity of the Cobb–Douglas function**

Given the Cobb–Douglas production function

$$Q = AK^\alpha L^\beta$$

If we take logs on both sides we get

$$\ln Q = \ln A + \alpha(\ln K) + \beta(\ln L) \quad \text{(The brackets are not strictly necessary.)}$$

This is a log linear function in three dimensions. If we constructed its three-dimensional graph, with $\ln Q$, $\ln K$, and $\ln L$ on the axes, the resulting surface would be a plane, with an intercept of $\ln A$ on the $\ln Q$-axis and slopes in the $\ln K$ and $\ln L$ directions of α and β respectively.

EXAMPLE 17.14

As another illustration of log linearity in three dimensions, we return to example 17.12 above. There we had the demand function

$$q = p^{-2}y^{0.5}$$

We found in example 17.12 that the partial elasticities were

$$\frac{p}{q}\frac{\partial q}{\partial p} = -2 \quad \text{(price elasticity of demand), and}$$

$$\frac{y}{q}\frac{\partial q}{\partial y} = 0.5 \quad \text{(income elasticity of demand)}$$

So using rule 17.3, the proportionate differential of this demand function is

$$\frac{dq}{q} = (-2)\frac{dp}{p} + (0.5)\frac{dy}{y} \tag{17.15}$$

We now want to derive the logarithmic equivalent. We again take logs on both sides of the demand function, and get

$$\ln q = -2[\ln p] + 0.5[\ln y] \tag{17.16}$$

You can hopefully see that equation (17.16) is a linear relationship between the three variables, $\ln q$, $\ln p$, and $\ln y$. It is another example of a log linear function.

The partial derivatives of equation (17.16) are

$$\frac{\partial \ln q}{\partial \ln p} = -2 \quad \text{and} \quad \frac{\partial \ln q}{\partial \ln y} = 0.5$$

and so the total differential of this function is

$$d(\ln q) = (-2)d(\ln p) + (0.5)d(\ln y) \tag{17.17}$$

Equation (17.17) is the differential of the function: $\ln q = -2[\ln p] + 0.5[\ln y]$: that is, the differential of the demand function after the demand function has been transformed into logarithmic form. Using rule 17.7, we can rewrite equation (17.17) as

$$\frac{dq}{q} = (-2)\frac{dp}{p} + (0.5)\frac{dy}{y}$$

which identically equals equation (17.15), as we wished to show.

Conclusion and generalization

Given any function, if we transform the function by taking natural logs, and then find the differential of the resulting function (such as equation (17.17)), this is equivalent to finding the *proportionate* differential of the original function (such as equation (17.15)).

If you are struggling to grasp all of these results involving logs, hold tight to the basic fact that $d \ln y = \frac{dy}{y}$ for any variable y, provided dy is sufficiently small. In words, this says that, for small changes, the *absolute* change in the log of a variable equals the *proportionate* change in the variable itself. (Check this on your calculator.) All of the more complex relationships that we have examined in this chapter flow from this simple fact.

Chapter summary

In section 17.2 we defined the concept of uniform increasing or decreasing returns to scale as a property of the production function, with constant returns as the boundary case between these two. We showed how uniform returns to scale links with the mathematical concept of homogeneity, as we summarized in table 17.1 (section 17.3). We also noted the possibility that changes in Q might not correspond in any simple way to the changes in K and L, in which case the production function has variable rather than uniform returns to scale.

In section 17.4 we then explored some of the properties of a homogeneous production function. First, the 'parallel tangents' property shows that the perfectly competitive firm's cost-minimizing capital/labour ratio is independent of the scale of production. Second, and most important, we showed that increasing (decreasing) returns means that the perfectly competitive firm's long-run average cost curve is negatively (positively) sloped. Only if there are decreasing returns, so that average and marginal cost increase with output, is there a unique profit-maximizing output. We also examined the 'adding-up problem': if the real rewards of labour and capital equal their marginal productivities (see equation (17.6) and rule 16.2, section 16.7), profits are positive, zero, or negative depending on whether there are respectively decreasing, constant, or increasing returns to scale.

In sections 17.5–17.8 we developed further some ideas from earlier chapters. We introduced partial elasticities, and the proportionate differential, $\frac{dz}{z}$, of the function $z = f(x, y)$, and explained their use in growth accounting.

In sections 17.9–17.12 we showed (though not rigorously) that for any variable y, for sufficiently small changes the absolute change in its log, $d(\ln y)$, equals the proportionate change in y itself, $\frac{dy}{y}$. From this it follows that for any function of two or more independent variables, if we take logs and then differentiate, the resulting partial derivatives will equal the partial elasticities. In particular, the 'log linear' function used frequently in economics has constant partial elasticities.

The Online Resource Centre contains more discussion and examples of the relationship between elasticities and logs, the proportionate differential, and growth accounting.

Progress exercise 17.3

1. The demand function for good 1 is $q_1 = 100 - 3p_1 + 0.5p_2 - 0.75(\frac{p_1}{p_2})^{0.5}$, where p_1 and p_2 are the prices of good 1 and another good, good 2. Find the own-price and cross-price elasticities. (Hint: don't expect very tidy expressions.)

2. (a) Find the partial elasticities of the labour supply and demand functions from exercise 17.2, question 2 above, namely:

$$L^S = A\left(\frac{w}{p}\right)^{\alpha} \text{ and } L^D = B\left(\frac{w}{p}\right)^{-\beta}$$

   (b) Let us write L_w^S to denote the elasticity of labour supply with respect to the money wage, L_p^S to denote the elasticity of labour supply with respect to the price level, and similarly for the labour demand elasticities. Then show that $L_w^S = \alpha$, $L_p^S = -\alpha$, $L_w^D = -\beta$, and $L_p^D = \beta$.

   (c) Are the signs and magnitudes of these partial elasticities as you would expect, given the neoclassical view that the supply and demand for labour should be functions of the real wage?

3. (a) If national product is growing at 2.8% per year, and population is growing at 1.2% per year, calculate the growth rate of national product per person.

   (b) Given the identity $T = tY$, where T = total tax revenue, t = average tax rate, Y = national product. If t is growing at 0.5% per year and Y is growing at 2.8% per year, calculate the growth rate of T.

   (c) If in (b) above population is growing at 1.2% per year, what is the growth rate of tax revenue per person?

4. Given the production function $Q = AK^{\alpha}L^{\beta}$:

   (a) Show that the partial elasticities are α and β. Explain briefly what they measure.

   (b) Show that a 5% increase in input of both K and L will increase output by 5%, more than 5% or less than 5% respectively as $\alpha + \beta = 1$, $\alpha + \beta > 1$, or $\alpha + \beta < 1$.

5. Assume that the aggregate production function for the UK economy is

   $$Q = K^{0.35}L^{0.75}$$

   (a) If capital and labour inputs both grow at 2% per year, what growth rate of output should be expected?

   (b) If due to net investment the capital stock is estimated to be growing by 3% per year (with constant labour supply), how much would this contribute to the annual growth of output? If actual growth of output averaged 2.5%, how might the difference be explained?

   (c) If due to immigration the labour force grew at 1% per year and the capital stock at 3% per year, what growth rate of output should be expected?

   (d) Is there an adding-up problem for this production function? If so, how might it be resolved in the real world?

6. Explain how factor shares (that is, the shares of wages and profits in total income) can be used as proxies for the unobservable partial elasticities of the aggregate production function. What assumptions have to be made to make this procedure valid?

7. Given the labour supply and demand functions from question 2 above, namely

$$L^S = A\left(\frac{w}{p}\right)^{\alpha} \text{ and } L^D = B\left(\frac{w}{p}\right)^{-\beta}$$

For each function, show that it is log linear and that the partial derivatives

$$\frac{\partial(\ln L^S)}{\partial(\ln w)}, \frac{\partial(\ln L^S)}{\partial(\ln p)}, \frac{\partial(\ln L^D)}{\partial(\ln w)}, \text{ and } \frac{\partial(\ln L^D)}{\partial(\ln p)}$$

measure the corresponding partial elasticities.

8. Given the demand function for good 1: $q_1 = p_1^{\alpha} p_2^{\beta} y^{\gamma}$,

where p_2 is the price of another good and y is income:

(a) Show that the function is log linear.

(b) Show that the partial derivatives of the logarithmic transformation

$$\frac{\partial(\ln q_1)}{\partial(\ln p_1)}, \frac{\partial(\ln q_1)}{\partial(\ln p_2)}, \text{ and } \frac{\partial(\ln q_1)}{\partial(\ln y)}$$

measure the corresponding partial elasticities.

(c) What signs would you expect for each of the parameters α, β, and γ?

Checklist

Be sure to test your understanding of this chapter by attempting the progress exercises (answers are at the end of the book). The Online Resource Centre contains further exercises and materials relevant to this chapter www.oxfordtextbooks.co.uk/orc/renshaw3e/.

We have covered a large number of topics in this chapter, all of them rich in relevance to economics. The individual topics were:

✔ **Returns to scale.** The meaning of constant, increasing, and decreasing returns to scale in the production function. How to show, algebraically and graphically, that a production function has constant, increasing or decreasing returns to scale.

✔ **Homogeneity of a function.** Understanding the meaning of homogeneity and how, in the production function, it relates to returns to scale. How to test a function for homogeneity.

✔ **Properties of homogeneous functions.** Understanding the main properties of homogeneous functions and their significance for economic analysis, including the parallel tangents property, Euler's theorem, and the adding-up problem.

✔ **Growth accounting.** Understanding the concept of the proportionate differential and its applications, including the technique of growth accounting.

✔ **Partial elasticities and logarithmic scales.** The concept of partial elasticities. How to evaluate them and their applications in economics, such as the generalized demand function. The relationship between partial elasticities and logarithmic scales.

As all of the above tools and results are used on a daily basis by economists, it is worth pausing to make sure you fully understand the chapter before going on to chapter 18, the first of part 5, where we consider some further areas of maths that are relevant to economics.

Part Five
Some further topics

- Integration
- Matrix algebra
- Difference and differential equations

Chapter 18
Integration

OBJECTIVES

Having completed this chapter you should be able to:

- Recognize a definite integral and understand what it measures.
- Recognize and understand an indefinite integral and evaluate indefinite integrals in simple cases.
- Understand the formula for finding a definite integral and apply it in simple cases.
- Apply integration to derive total cost from marginal cost, and evaluate producers' and consumers' surplus and the present value of a continuous stream of income.

18.1 Introduction

The subject matter of this chapter, integration, can be approached in two distinct ways. One approach is to focus on what is called a **definite integral**. This is a way of measuring, on the graph of a function, the area between the curve and the *x*-axis. The second approach focuses on the **indefinite integral** of a function, which is found by reversing the process of differentiation. Here we begin with the first approach, then move on to the second and explain how the two approaches are related.

Although the applications of integration to economics may not be immediately apparent, they address some important questions, and we look at some of these in the latter part of the chapter. In the first application we work back from information about the firm's marginal cost curve to 'recover' the total cost curve from which it is derived. This uses the property that an indefinite integral is found by 'reverse differentiation', together with the fact that the marginal cost function is the derivative of the total cost function. The second application is consumers' surplus, a measure of consumer satisfaction. To find this we have to measure the area between the inverse market demand curve and the horizontal (quantity) axis. Third, we look at producers' surplus, which is somewhat analogous to consumers' surplus and requires measuring the area under the supply curve as a step towards finding it. Both consumers' and producers' surplus are important in cost–benefit analysis and in analysing the social costs of monopoly. Finally we revisit the question of the present value of a future income or expenditure series. We saw in chapter 12 that the present value of a single future payment, discounted continuously, was found using the natural exponential function $y = ae^{rx}$. Here we are able to show that the present value of a continuous stream of income is given by the area under this curve.

In this chapter we shall develop only the most basic ideas of integration. However, proficiency at integration as a mathematical technique is not of great importance in the study of economics. What is important is that you should grasp the basic principles and above all the relevance of integration to our understanding of economic problems.

18.2 The definite integral

In figure 18.1a we show the graph of some function $y = f(x)$. Suppose we want to measure the shaded area, $aJKb$, where a and b are two points on the x-axis. We will call this shaded area A. We will also assume that the value of x at point a is $x = x_1$ and similarly that at b, $x = x_2$.

Looking now at figure 18.1b, it occurs to us that a crude approximation to the area A is given by the area of rectangle Z: that is, the rectangle $aJLb$. The base of rectangle $aJLb$ (the distance from a to b) is $(x_2 - x_1)$. The height aJ is given by the value of y when $x = x_1$, which is $y_1 = f(x_1)$. So the area of the rectangle is the product of these two: that is,

$$\text{area } Z = f(x_1)(x_2 - x_1)$$

and this is approximately equal to the area that we want to find, A, so we can write

$$A \approx f(x_1)(x_2 - x_1)$$

where '$\approx$' means 'approximately equals'.

The approximation here is very coarse because the error is the wedge-shaped area JKL, which is very large relative to the area we are trying to measure.

It next occurs to us that a closer approximation can be made if we form *two* rectangles Z_1 and Z_2 (see figure 18.1c). The two rectangles are formed by picking a new arbitrary value of x somewhere between a and b. We'll label this point $x = x_2$, and give point b a new label, x_3. The value of y at this new point is $y_2 = f(x_2)$.

The areas of the rectangles are thus

$$Z_1 = f(x_1)(x_2 - x_1)$$
$$Z_2 = f(x_2)(x_3 - x_2)$$

The sum of these two areas is approximately equal to the area A that we are trying to find. So we can write

$$A \approx f(x_1)(x_2 - x_1) + f(x_2)(x_3 - x_2)$$

The error in this case is the sum of the two wedge-shaped areas, JMP and MKR. The error is smaller

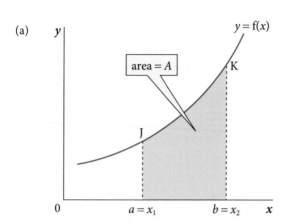

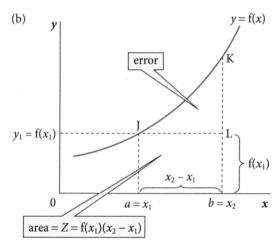

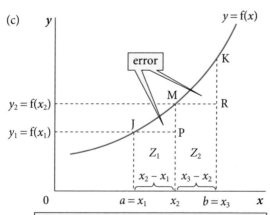

By increasing the number of rectangles Z_1, Z_2 we can reduce the error in our calculation of the area under the curve.

Figure 18.1 Approximating the area under a curve.

than previously because the combined area of these two wedges in figure 18.1c is less than the area of the wedge JKL in figure 18.1b.

The key point here is that we have reduced the error in our approximation by increasing the number of rectangles from one to two. It next occurs to us that we can always increase the accuracy of the approximation further by increasing the number of such rectangles. In general we can create n rectangles by picking $n + 1$ points anywhere along the x-axis between a and b (starting from point a on the left), and where the values of x are $x_1, x_2, x_3, \ldots, x_{n+1}$. Here x_1 is the value of x at point a and x_{n+1} is the value of x at point b.

Following the pattern of figure 18.1c, the area of the first rectangle will be $f(x_1)(x_2 - x_1)$, the area of the second $f(x_2)(x_3 - x_2)$, the area of the third will be $f(x_3)(x_4 - x_3)$ and so on, until we reach the last rectangle on the right, which will have the area $f(x_n)(x_{n+1} - x_n)$.

So with n rectangles, our approximation to the area A is given by the sum of these:

$$A \approx f(x_1)(x_2 - x_1) + f(x_2)(x_3 - x_2) + f(x_3)(x_4 - x_3) + \cdots + f(x_n)(x_{n+1} - x_n) \tag{18.1}$$

We can write this more compactly as

$$A \approx \sum_{r=1}^{n} f(x_r)(x_{r+1} - x_r) \tag{18.2}$$

Here you will hopefully recognize, from chapter 10, the summation sign Σ (the Greek letter 'sigma'), which conventionally means 'the sum of'. The information '$r = 1, \ldots n$' adjacent to the Σ sign tells us that we must form the sum given in equation (18.1) above. Thus the notation tells us that the first term of this sum is obtained by setting $r = 1$, the second term by setting $r = 2$, and so on until we reach the last term where $r = n$. (Check this carefully for yourself.)

As long as n is finite, this is still only an approximation to the shaded area A. This is because, however many rectangles we draw, we will always leave out the wedge-shaped area above each rectangle. These wedges get smaller as we increase the number of rectangles, but they never disappear completely as long as n remains finite, however large a number it may be.

If the function $f(x)$ in figure 18.1 had been negatively sloped, the area of the rectangles would have been an over-estimate rather than an under-estimate of the area under the curve. But the error in each rectangle would still be given by a wedge-shaped area. (Sketch the graph of this case as a check on your understanding.)

However, if we now let the number of rectangles become indefinitely large (that is, we let $n \to \infty$), the sum of their areas will approach closer and closer to the shaded area A. In other words, equation (18.2), which gives the sum of the areas of the rectangles, will approach, as a limiting value, the area A. So we can write

$$A = \text{the limiting value,} \quad \text{as } n \to \infty, \quad \text{of } \sum_{r=1}^{n} f(x_r)(x_{r+1} - x_r)$$

We can write the right-hand side of this equation more compactly as

$$\lim_{n \to \infty} \sum_{r=1}^{n} f(x_r)(x_{r+1} - x_r) \quad \text{therefore}$$

$$A = \lim_{n \to \infty} \sum_{r-1}^{n} f(x_r)(x_{r+1} - x_r) \quad (= \text{area } aJKb \text{ in figure 18.1a})$$

Mathematicians always like to write things in the most compact way possible. Thus a special notation evolved for the right-hand side of this expression. We write it as

$$\lim_{n \to \infty} \sum_{r=1}^{n} f(x_r)(x_{r+1} - x_r) \equiv \int_{a}^{b} f(x)dx \tag{18.3}$$

It is easy to see how this notation evolved. The symbol Σ is replaced by the integral sign $\int$ (which is simply a seventeenth-century 'S', for sum), and the expression $(x_{r+1} - x_r)$, which is the base of each rectangle, quite naturally becomes 'dx'. (Note that as the number of rectangles becomes indefinitely large, so dx, the base of each, becomes indefinitely small.) So we conclude:

RULE 18.1 The definite integral

The area A (= shaded area aJKb in figure 18.1) is given by

$$A = \int_a^b f(x)dx$$

where $\int_a^b f(x)dx$ is as defined in equation (18.3) above. The expression $\int_a^b f(x)dx$ is called the definite integral of the function $y = f(x)$.

Thus the definite integral of any function $y = f(x)$ measures the area between the function and the x-axis between two points a and b. Point a is called the lower limit (in the sense of a boundary, not a limiting value), and b the upper limit, of integration.

The notation is a little off-putting at first. In effect, the $\int_a^b$ is like an opening bracket and the 'dx' is like a closing bracket. What is enclosed between these brackets is the function in question. So, $\int_a^b f(x)dx$ is read as 'the definite integral of the function $f(x)$ between the limits a and b'. Try to associate this expression in your mind with the shaded area in figure 18.1a, for that is what it measures.

18.3 The indefinite integral

Basic ideas

The definite integral $\int_a^b f(x)dx$ that we considered in the previous section is a definite *number* (an area under a curve), which depends for its value on the values of a and b, and on the functional form of $f(x)$.

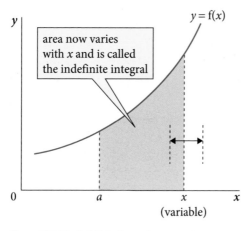

But if the upper limit, b, is not specified but is *variable*, then we have $\int_a^x f(x)dx$, where x is the unknown, or variable, upper limit of integration. Thus the integral now measures an *open-ended* area (see figure 18.2). The value of the integral is therefore no longer a fixed number but is a *variable* number which depends for its value on the value of x, the upper limit of integration. This is called the *indefinite* integral and is written simply as

$$\int f(x)dx$$

Evaluation of the indefinite integral

Figure 18.2 The indefinite integral.

So far in this chapter we have defined two concepts: the definite and the indefinite integral of a function $y = f(x)$. Now we need to know how to evaluate an integral in a specific case. The fundamental theorem of the indefinite integral says that integration (the process of evaluating an indefinite integral) is the reverse of differentiation. We will first work through an example to illustrate how the fundamental theorem works, then derive a general rule.

EXAMPLE 18.1

Suppose we are asked to evaluate

$$\int (2x)\,dx$$

The fundamental theorem tells us that we have to find a function whose derivative is $2x$. What function has the property that its derivative is $2x$? After a moment of thought we see that the answer is the function x^2. So

$$\int (2x)\,dx = x^2$$

and this is true because

$$\frac{d}{dx} \text{of } x^2 = 2x$$

Actually, this answer is not quite complete, but needs minor adjustment. The function x^2 is not the only function whose derivative is $2x$. For example, $x^2 + 10$, $x^2 + 100$, and $x^2 + 26$ all have the property that their derivative is $2x$. This is because they all contain an additive constant that disappears when we differentiate.

Because integration reverses differentiation, when we integrate we have to reintroduce the additive constant that is lost when we differentiate. But we can assign any value we wish to this constant. So the correct answer to the example above is

$$\int (2x)\,dx = x^2 + c$$

where c is an **arbitrary constant**. When we say that c is arbitrary we mean that we can assign any value to it that we wish, without affecting the validity of the result. This is true because whatever value we assign to c, the derivative of $x^2 + c$ is always $2x$.

Generalization of the previous example

Following the pattern of the previous example, to evaluate $\int f(x)\,dx$ (where $f(x)$ is any function of x), we have to find a function, $F(x)$, such that $\frac{d}{dx}$ of $F(x) = f(x)$. Then we have to add an arbitrary constant, c. So the rule is:

RULE 18.2 Fundamental theorem of the indefinite integral

$$\int f(x)\,dx = F(x) + c$$

where $\frac{d}{dx}$ of $F(x) = f(x)$ (and c is an arbitrary constant).

To link with our example above, in that example we had $f(x) = 2x$ and $F(x) = x^2$.

An explanation of *why* the fundamental theorem is true is given in appendix W18.1, to be found at the Online Resource Centre.

18.4 Rules for finding the indefinite integral

Since integration reverses differentiation, the rules of integration are in principle simply the rules of differentiation, reversed. The bad news is that, for many functions, this reversal is either very difficult or impossible. The good news is that in economics we don't have to integrate a function very often; and when we do, it is usually a very simple function.

Let's see how far we can get with reversing the rules of differentiation. We have numbered the rules below to match where possible the numbering of the corresponding rules of differentiation in section 6.7 and also in sections 13.2 and 13.3 for the exponential and logarithmic functions. (If you feel a little rusty on the rules of differentiation, it might be a good idea to revise them now.)

RULE I1 Power rule of integration

The rule says:

$$\int (x^n)\,dx = \frac{x^{n+1}}{n+1} + c$$

(provided $n \neq -1$; see rule (I9) below). Note the additive constant, c. It is added to obey rule 18.2 above.

This rule reverses the power rule of differentiation (see section 6.7, rule D1).

We will illustrate rule I1 with two examples:

EXAMPLE 18.2

We are asked to find $\int (x^3)\,dx$

Here we have $n = 3$, so

$$\frac{x^{n+1}}{n+1} = \frac{x^4}{4}$$

Therefore

$$\int (x^3)\,dx = \frac{x^4}{4} + c$$

which we can also write as $\frac{1}{4}x^4 + c$, or $0.25x^4 + c$ if preferred.

We can check that we have applied the rule correctly by differentiating our answer, which should take us back to the question. Doing this, we get

$$\frac{d}{dx}\left[\frac{x^4}{4} + c\right] = \frac{4x^3}{4} = x^3 \quad \text{(which is what we started with)}$$

Note that the additive constant, c, as usual drops out when we differentiate.

EXAMPLE 18.3

We are asked to find $\int (x^{1/2})\,dx$

Here we have $n = \frac{1}{2}$, so $\dfrac{x^{n+1}}{n+1} = \dfrac{x^{3/2}}{3/2} = \dfrac{2}{3}x^{3/2}$

Therefore

$$\int (x^{1/2})\,dx = \frac{2}{3}x^{3/2} + c$$

Again, we can check that this is correct by differentiating, which gives us

$$\frac{d}{dx}\left[\frac{2}{3}x^{3/2} + c\right] = \frac{2}{3}(3/2)x^{(3/2)-1} = x^{1/2} \quad \text{(which is what we started with)}$$

A special case

Note that rule I1 holds equally when $n = 0$. In that case, the rule says

$$\int (x^0)dx = \frac{x^{0+1}}{0+1} + c = x + c$$

But we know that $x^0 \equiv 1$. So the rule is saying that

$$\int (1)dx \equiv \int (x^0)dx = \frac{x^{0+1}}{0+1} + c = x + c$$

We can check this by differentiation: the derivative of $x + c$ is 1.

In words, rule I1 says that the integral of 1 is x plus an arbitrary constant.

The next rule concerns multiplicative constants:

> **RULE I2 Multiplicative constants**
>
> Stated formally, the rule is
>
> $$\int [Af(x)]dx = A \int [f(x)]dx = A[F(x)] + c$$
>
> where A is any constant, and the derivative of $F(x)$ is $f(x)$.

In words, this rule says that any multiplicative constant, A, reappears when we integrate. Why? We know from the rules of differentiation that a multiplicative constant reappears in the derivative. Since integration reverses differentiation, a multiplicative constant must therefore also reappear when we integrate. Thus rule I2 reverses the equivalent rule of differentiation (see section 6.7, rule D2). The next three examples illustrate this.

EXAMPLE 18.4

We are asked to find: $\int (100x^3)dx$

From rule I2, we know that the multiplicative constant, 100, is going to reappear in the integral. Consequently, we can move it outside of the integral sign. So

$$\int (100x^3)dx \equiv 100 \int (x^3)dx$$

On the right-hand side, the fact that the 100 is now before the integral sign means that it plays no further part in the integration process. This leaves just the x^3 to be integrated. We do this using the power rule (rule I1 above):

$$100 \int (x^3)dx = 100 \left(\frac{x^{n+1}}{n+1} \right) + c = 100 \left(\frac{x^4}{4} \right) + c = 25x^4 + c$$

We can check that our answer is correct by differentiating $25x^4 + c$. The result is $100x^3$, so our integration is correct.

EXAMPLE 18.5

We are asked to find $\int (10x^{0.25})dx$

Applying rules I2 and I1 with $A = 10$, $f(x) = x^{0.5}$ we get

$$\int (10x^{0.25})dx = 10 \int (x^{0.25})dx = 10 \left(\frac{x^{0.25+1}}{0.25+1} \right) + c = 10 \frac{x^{1.25}}{1.25} + c = 8x^{1.25} + c$$

Because the derivative of $8x^{1.25} + c = 10x^{0.25}$, this answer is correct.

A special case

Rule I2 also holds when $f(x) = x^0 \equiv 1$. Applying rule I2, we get

$$\int [Af(x)]dx = \int (A \times 1)dx = \int (A)dx = A\int (1)dx$$

From the special case of rule I1 above, we know that $\int (1)dx = x + c$. So

$$\int (A)dx = \int (A \times 1)dx = A\int (1)dx = Ax + c$$

Because the derivative of $Ax + c$ is A, this answer is correct.

EXAMPLE 18.6

Find $\int [50]dx$

Answer

$$\int [50]dx = \int [50x^0]dx = 50\int [x^0]dx = 50\int [1]dx = 50x + c$$

As the derivative of $50x + c = 50$, this answer is correct.

RULE I3 An additive constant

This has already been covered in rule I1 above: we gain an additive constant when we integrate (because we lose any additive constant when we differentiate).

This reverses the equivalent rule of differentiation (see section 6.7, rule D3).

RULE I4 Sums or differences

This reverses rule D4 of differentiation (section 6.7). Formally stated, the rule is

$$\int [f(x) + g(x)]dx = \int [f(x)]dx + \int [g(x)]dx = F(x) + G(x) + c$$

where $f(x)$ is the derivative of $F(x)$ and $g(x)$ the derivative of $G(x)$.

In words, we can explain rule I4 as follows. We know from rule D4 of differentiation that we find the derivative of a sum (or difference) by simply finding the derivatives of the individual components of that sum (or difference). Thus, for example,

$$\frac{d}{dx} \text{ of } x^2 + x^3 = \frac{d}{dx} \text{ of } x^2 + \frac{d}{dx} \text{ of } x^3 = 2x + 3x^2$$

This rule must also hold in reverse: that is, the integral of the right-hand side of the equation above must equal the left-hand side. So

$$\int [2x + 3x^2]dx = \int [2x]dx + \int [3x^2]dx = x^2 + x^3 + c$$

Note we have added only one arbitrary constant. You may feel that we should add an arbitrary constant when we integrate $2x$, and another one when we integrate $3x^2$. This is formally correct, but we can deal with it by thinking of c as the sum of these two arbitrary constants. The next example illustrates this rule.

EXAMPLE 18.7

Find $\int[5x^3 + 3x^2]dx$

Answer

$$\int[5x^3 + 3x^2]dx = \int[5x^3]dx + \int[3x^2]dx = 5\int[x^3]dx + 3\int[x^2]dx$$

$$= 5\frac{x^4}{4} + 3\frac{x^3}{3} + c = 1.25x^4 + x^3 + c$$

As always, this answer can be checked by differentiation.

RULE I5 **Function of a function rule**

This reverses rule D5 of differentiation (see section 6.7). The rule is very difficult to apply. Here we will consider only the simplest case.

Let's recall the function of a function rule of differentiation. This says

$$\frac{d}{dx} \text{ of } [f(x)]^n = n[f(x)]^{n-1}[f'(x)]$$

Therefore, replacing n with $n + 1$ in the equation above, and dividing both sides by $n + 1$,

$$\frac{d}{dx} \text{ of } \frac{[f(x)]^{n+1}}{n+1} = [f(x)]^n[f'(x)]$$

This rule must also hold in reverse: that is, the integral of the right-hand side of the equation above must equal the left-hand side. So the rule is:

$$\int([f(x)]^n[f'(x)])dx = \frac{[f(x)]^{n+1}}{n+1} + c$$

provided $n \neq -1$ (see rule I9 below).

The next two examples illustrate rule I5.

EXAMPLE 18.8

Find $\int(2x(x^2 + 1)^3)dx$

Since $2x$ is the derivative of $x^2 + 1$, we have the pattern necessary for the function of a function rule to apply. Thus we have

$$f'(x) = 2x \quad \text{and} \quad [f(x)]^n = [x^2 + 1]^3 \quad \text{Therefore}$$

$$\int(2x(x^2 + 1)^3)dx = \frac{[f(x)]^{n+1}}{n+1} + c = \frac{(x^2 + 1)^4}{4} + c$$

Thus this rule may be applied whenever we want to integrate a product in which one factor of the product (in this example, $2x$) is the derivative of the base of the other factor (in this example, $x^2 + 1$).

EXAMPLE 18.9

Find $\int(3x^2 + 3)(x^3 + 3x)^3 \, dx$.

We can use the function of a function rule here because the content of the first pair of brackets, $3x^2 + 3$, is the derivative of the second pair of brackets. Applying rule I5 with $f'(x) = 3x^2 + 3$ and $[f(x)]^n = [x^3 + 3x]^3$ we get

$$\int(3x^2 + 3)(x^3 + 3x)^3dx = \frac{[f(x)]^{n+1}}{n+1} + c = \frac{(x^3 + 3x)^4}{4} + c$$

***RULES 16 AND 17* Product and quotient rule**

There is a single rule of integration which covers both products and quotients, and thus reverses rules D6 and D7 of differentiation (see section 6.7). It is called integration by parts.

However it's quite a difficult rule to apply and anyway doesn't always work. It's not often needed in economics. So we will pass over it.

***RULE 18* The exponential function**

In section 13.2, we looked at the derivative of the exponential function. We found that

$$\frac{d}{dx} \text{ of } e^x = e^x$$

Thus the derivative equals the function itself (see rule 13.1).

Applying this rule in reverse gives rule I8:

$$\int [e^x]dx = e^x + c \quad \text{where } c \text{ is the usual arbitrary constant.}$$

However, this is only the simplest case, where the exponent is simply x. In the general case, the power to which e is raised is some function of x. In this case, we saw in section 13.2 (rule 13.2) that

$$\frac{d}{dx}[e^{f(x)}] = f'(x)e^{f(x)}$$

Applying this rule in reverse, we get

Rule I8a:

$$\int [f'(x)e^{f(x)}]dx = e^{f(x)} + c$$

Note the pattern here. We have a function, $f(x)$, with e raised to the power $f(x)$, and also multiplied by the derivative of $f(x)$: that is, $f'(x)$. The next example illustrates this.

EXAMPLE 18.10

Find $\int [2xe^{x^2}]dx$

Here $2x$ is the derivative of x^2, so we have the correct pattern for applying rule I8a. The integral is therefore

$$\int [2xe^{x^2}]dx = e^{x^2} + c$$

***RULE 19* The natural logarithmic function**

In section 13.3 we looked at the derivative of the natural logarithmic function. We found (rule 13.3) that

$$\frac{d}{dx} \text{ of } \ln x = \frac{1}{x}$$

Applying this rule in reverse gives rule I9:

$$\int \left[\frac{1}{x}\right]dx = \ln x + c$$

Notice that $\frac{1}{x}$ can also be written as x^{-1}. So therefore the rule can also be written as

$$\int [x^{-1}]dx = \ln x + c$$

Note that this is the only exception to the power rule (see rule I1 above). Thus the power rule of integration must not be applied when x is raised to the power -1.

Rule I9 covers only the simplest case, where we are asked to find the integral of $\frac{1}{x}$. In the general case, we may be asked to find the integral of something that involves not just x but some function of x, f(x). In this case, we saw in section 13.3 (rule 13.4) that the derivative is

$$\frac{d}{dx} \text{ of } [\ln f(x)] = \frac{f'(x)}{f(x)}$$

Applying this rule in reverse, we get

Rule I9a:

$$\int \left[\frac{f'(x)}{f(x)} \right] dx = \ln f(x) + c$$

Note the pattern here. We have a quotient, or ratio, in which the numerator, f′(x), is the derivative of the denominator, f(x). The next example illustrates this.

EXAMPLE 18.11

Find $\int \left[\dfrac{3x^2}{x^3 + 1} \right] dx$

Here the numerator is the derivative of the denominator, so we have the correct pattern to apply rule I9a. The integral is therefore

$$\int \left[\frac{3x^2}{x^3 + 1} \right] dx = \ln(x^3 + 1) + c$$

This rule may be applied whenever the numerator is the derivative of the denominator.

Summary of sections 18.1–18.4

In sections 18.1–18.3 we defined two key concepts:

(1) The *definite integral* of the function $y = $ f(x), written as $\int_a^b$ f(x)dx

This measures the area between the x-axis and the curve f(x), between $x = a$ and $x = b$ (see rule 18.1 above).

(2) The *indefinite integral* of the function $y = $ f(x), written as $\int$ f(x)dx

The fundamental theorem of the indefinite integral (rule 18.2 above) states that the indefinite integral is found by reversing the process of differentiation.

In section 18.4 we developed some rules for finding indefinite integrals by applying the fundamental theorem, and saw that we can find the indefinite integrals of some functions, but by no means all. However, this is not really important as most of the functions that we encounter in economic analysis are quite simple to integrate. For the rest, we can usually find a mathematician somewhere to help us. In any case, for your development as an economist, an understanding of how integrals are used in economics is far more important than mere technical proficiency at evaluating them.

Progress exercise 18.1

Find the following indefinite integrals:

(a) $\int x^3\,dx$

(b) $\int x^{-3}\,dx$

(c) $\int -x^{-1.5}\,dx$

(d) $\int x^{0.25}\,dx$

(e) $\int x^{-0.75}\,dx$

(f) $\int 5x^{-1}\,dx$

(g) $\int 2e^x\,dx$

(h) $\int (5x^4 + e^x - 4x^{-0.5})dx$

(i) $\int (2x^{-5} + 9x^2 + x^{0.6})dx$

(j) $\int (10x^{2/3} - 0.5x^{-1})dx$

(k) $\int 100e^{0.1x}\,dx$

(l) $\int 10e^{-0.05x}\,dx$

(m) $\int (16x + 4x^{-1})(2x^2 + \ln x)^3\,dx$

(n) $\int (2x^{0.5} + 6x + 5)^{0.5}(x^{-0.5} + 6)dx$

(o) $\int \dfrac{4x^3 + 10x - 2e^x}{x^4 + 5x^2 - 2e^x}\,dx$

Harder examples (some guesswork needed):

(p) $\int xe^x\,dx$

(q) $\int xe^{2x}\,dx$

18.5 Finding a definite integral

The key rule is:

RULE 18.3 The fundamental theorem of the definite integral

If $\int [f(x)]dx = F(x) + c$

(that is, if F(x) is the indefinite integral of f(x)),

then $\displaystyle\int_a^b [f(x)]dx = [F(x)]_{x=b} - [F(X)]_{x=a}$

where $[F(x)]_{x=b}$ denotes the value of the function F(x) when $x = b$, and similarly for $x = a$.

Like the previous theorem on the indefinite integral, this theorem is a little difficult to grasp at first sight. We give a proof of it in appendix W18.2, which can be downloaded from the Online Resource Centre; and a diagrammatic interpretation is given in the next example.

To see how to apply the theorem, here are some worked examples.

EXAMPLE 18.12

Evaluate the definite integral $\displaystyle\int_2^4 [x^2]dx$

This definite integral measures the shaded area in figure 18.3b. We find this area in two steps:

Step 1. We find the *indefinite* integral. Using the power rule of integration, we get

$$\int [x^2]dx = \frac{x^3}{3} + c \quad \text{where as usual } c \text{ is an arbitrary constant.}$$

As always, we can check that this is correct by differentiating: the derivative of $\frac{x^3}{3} + c$ is x^2, so step 1 is correct.

Step 2. Then we use the result of step 1 to find the *definite* integral, by applying the fundamental theorem (rule 18.3) above with $F(x) = \frac{x^3}{3}$, and $b = 4$ and $a = 2$. This gives

$$\int_2^4 [x^2]dx = \left[\frac{x^3}{3}\right]_{x=4} - \left[\frac{x^3}{3}\right]_{x=2} = \frac{64}{3} - \frac{8}{3} = \frac{56}{3} = 18\frac{2}{3} = \text{shaded area in figure 18.3b.}$$

Figure 18.3 helps us to see what the fundamental theorem means. Rule 18.3 tells us that the shaded *area* in figure 18.3b is equal to the *distance* AB in figure 18.3a, and it is this distance that we have just found to be $18\frac{2}{3}$. (This could be $18\frac{2}{3}$ square inches, or $18\frac{2}{3}$ square metres, depending on the scale of the graph.) Notice that in figure 18.3a the distance AB is independent of the value of c, the arbitrary constant of integration. This is true because

$$\left(\left[\frac{x^3}{3}\right]_{x=4} + c\right) - \left(\left[\frac{x^3}{3}\right]_{x=2} + c\right) \equiv \left[\frac{x^3}{3}\right]_{x=4} - \left[\frac{x^3}{3}\right]_{x=2}$$

That is why c does not appear in rule 18.3.

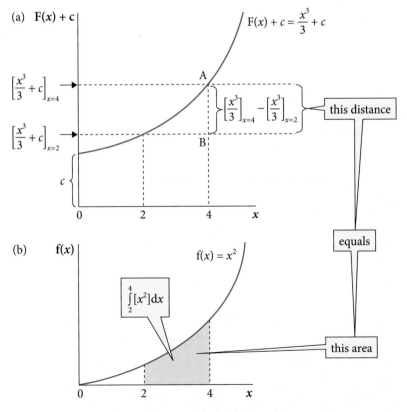

Figure 18.3 Illustration of the fundamental theorem of the definite integral (not to scale).

EXAMPLE 18.13

Evaluate the definite integral $\int_3^5 [2x + 4]dx$

Step 1. We evaluate the *indefinite* integral. Using the power rule, carrying the multiplicative constants through, and remembering that $4 \equiv 4x^0$, this gives

$$\int [2x + 4]dx = 2\frac{x^2}{2} + 4\frac{x^1}{1} + c = x^2 + 4x + c$$

Step 2. Applying the fundamental theorem with $F(x) = x^2 + 4x$, $b = 5$, and $a = 3$:

$$\int_3^5 [2x + 4]dx = [x^2 + 4x]_{x=5} - [x^2 + 4x]_{x=3}$$

$$= [25 + 20] - [9 + 12] = 45 - 21 = 24$$

This integral measures the shaded area in figure 18.4.

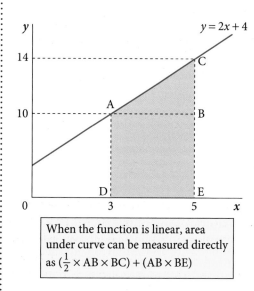

When the function is linear, area under curve can be measured directly as $(\frac{1}{2} \times AB \times BC) + (AB \times BE)$

Figure 18.4 Area under a linear function.

Since in this case we have a linear function, we can directly check that our answer is correct, using elementary geometry. In figure 18.4 the shaded area is composed of the rectangle DABE and the triangle ABC. The rectangle DABE has horizontal side of length 2 and vertical side of length 10, so its area is 20. The triangle ABC has a base of 2 and a height of 4. Using the formula for the area of a triangle (area = half base × height), its area is therefore $0.5 \times 2 \times 4 = 4$.

So the combined area is $20 + 4 = 24$, the same as the answer we just found by integration. Thus our answer is correct.

Note, though, that we are only able to carry out this check because the function we were looking at happened to be linear. But of course the theorem measures the area under curves as well as under linear functions. That is what makes the definite integral such a powerful tool. Next we will look at some of its economic applications.

?

Progress exercise 18.2

Evaluate the following definite integrals:

(a) $\int_0^{50} (100 - 2x)dx$

(b) $\int_{0.5}^{10} x^{-1} dx$

(c) $\int_3^{12} (x^2 - 10x + 24)dx$

(d) $\int_0^{20} (10e^{-0.05x})dx$

(e) $\int_{25}^{50} (5e^{0.1x})dx$

(f) $\int_2^3 (e^{2x} + e^x)dx$

(g) $\int_1^e \left(\frac{1}{x} + 1\right)dx$

(h) $\int_{-1}^{e-2} \left(\frac{1}{x+2}\right)dx$

18.6 Economic applications 1: deriving the total cost function from the marginal cost function

Suppose we are given a marginal cost function $MC = f(q)$ (where q = output). We know that, by definition, MC is obtained by differentiating the total cost (TC) function. We also know that integration reverses differentiation. Therefore, by finding the indefinite integral of the MC function we can get back to the TC function from which the MC function was derived. This is perhaps best understood by means of examples.

EXAMPLE 18.14

Given $MC = 3q + 5$, find the total cost function.

To answer this we find the indefinite integral of $3q + 5$, using the power rule of integration (and recalling that $5 = 5q^0$):

$$\int [3q + 5]\,dq = 3\frac{q^2}{2} + 5q + c = \frac{3}{2}q^2 + 5q + c$$

where c is, of course, the arbitrary constant that reappears when we integrate. Thus the total cost function is

$$TC = \frac{3}{2}q^2 + 5q + c$$

We can check that this is correct by finding the derivative of the TC function. This should be the MC function that we were given:

$$\frac{dTC}{dq} = \frac{d}{dq}\left[\frac{3}{2}q^2 + 5q + c\right] = 3q + 5 \quad \text{(as we were given)}$$

Notice that we cannot by this method find the value of c, the arbitrary constant that reappears when we integrate. Since c can take any value, this means that we cannot *uniquely* find the TC function. In economic terms, $\frac{3}{2}q^2 + 5q$ is total *variable* cost (costs that vary with output), while c is total *fixed* cost (costs that are independent of output) (see section 8.2). Thus from information about MC alone, we can find total variable cost, but we cannot find total fixed cost without additional information.

However, if we are now given the additional information that $TC = 50$ when $q = 4$, we can substitute this information into our TC function, $TC = \frac{3}{2}q^2 + 5q + c$. The result is

$$50 = \frac{3}{2}(4)^2 + 5(4) + c = 44 + c$$

Solving this equation for c gives $c = 6$. So now we know that the full TC function is

$$TC = \frac{3}{2}q^2 + 5q + 6$$

This additional information ($TC = 50$ when $q = 4$), which enabled us to find the arbitrary constant of integration, c, is sometimes called 'the initial condition' (see chapter 20 and also chapter W21, available at the Online Resource Centre).

EXAMPLE 18.15

Given $MC = 3q + 5$, as in the previous example, show that the total variable cost of producing 10 units of output can be measured by the appropriate area under the MC curve.

In this question, 'the appropriate area' is the area under the marginal cost curve between 0 and 10 units of output: that is, $\int_0^{10} [3q + 5]\,dq$ (see figure 18.5). We find this area by following the usual two-step procedure.

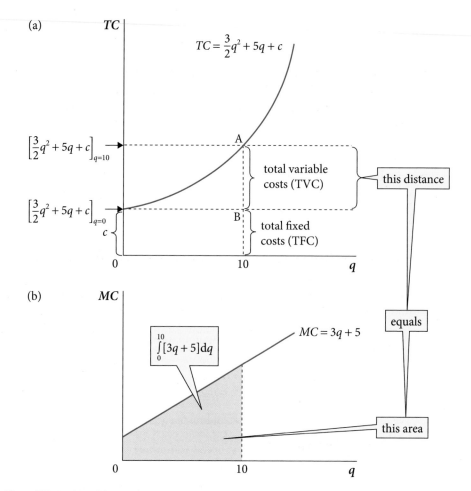

Figure 18.5 Total variable cost as the area under the marginal cost curve (not to scale).

Step 1. We find the indefinite integral of $3q + 5$, which we know from the previous example is

$$\int [3q + 5]\mathrm{d}q = \frac{3}{2}q^2 + 5q + c$$

Step 2. We use this to find the definite integral $\int_0^{10} [3q + 5]\mathrm{d}q$ by applying rule 18.3:

$$\int_0^{10} [3q + 5]\mathrm{d}q = \left[\frac{3}{2}q^2 + 5q\right]_{q=10} - \left[\frac{3}{2}q^2 + 5q\right]_{q=0} = \left[\frac{3}{2}(10)^2 + 5(10)\right] - [0] = 200 \quad (18.4)$$

The left-hand side of equation (18.4), $\int_0^{10} [3q + 5]\mathrm{d}q$, is the area under the MC curve between $q = 0$ and $q = 10$ (see figure 18.5(b)). The right-hand side is $[\frac{3}{2}q^2 + 5q]_{q=10}$, the total variable cost of producing 10 units of output, minus $[\frac{3}{2}q^2 + 5q]_{q=0}$, the total variable cost of producing zero units of output. The latter is, by definition, zero; so the right-hand side is simply the total variable cost of producing 10 units of output (see figure 18.5(a)). Thus, in general, total variable cost may be measured by the corresponding area under the marginal cost curve.

Note that we do not know the level of fixed costs, c, but we don't need to know this as the question asks us only to find total variable cost. In order to draw figure 18.5(a) we assumed an arbitrary level of fixed costs, which gave us the intercept of the TC curve on the vertical axis.

As a variant of this example, we can find the *increase* in total cost that occurs when output increases. Suppose the marginal cost function is $MC = f(q)$. Then if output increases, say from q_0 to q_1 units of output, the definite integral

$$\int_{q_0}^{q_1} [f(q)]dq$$

will give us the increase in variable cost that results. The diagrammatic treatment of this case would be identical to figure 18.5, except that the lower limit of integration would be the output level q_0 rather than 0. Since by definition fixed costs do not change when output increases, the increase in variable cost also gives us the increase in total cost.

18.7 Economic applications 2: deriving total revenue from the marginal revenue function

By analogy with the previous section, if we are given a marginal revenue (MR) function, we can use integration to work backwards and recover the total revenue (TR) function from which MR is derived.

Suppose we are given a marginal revenue function $MR = f(q)$ (where $q =$ output). We know that, by definition, MR is obtained by differentiating the total revenue (TR) function. We also know that integration reverses differentiation. Therefore, by integrating the MR function we can get back to the TR function from which the MR function was derived. Once we have the TR function, it is only a small step to the demand function. Again, this is perhaps best understood by means of an example.

EXAMPLE 18.16

Given $MR = 100 - 2q$, find (a) the total revenue function and (b) the demand function.

Answer

(a) We first find the indefinite integral of $100 - 2q$, using rule I1 of integration (see section 18.4):

$$\int [100 - 2q]dq = 100q - q^2 + c$$

where c is, of course, the arbitrary constant that reappears when we integrate.

In example 18.14 above, we interpreted c as fixed costs. But there can be no constant term in a total revenue function because when $q = 0$, nothing is sold, so there is no revenue. Therefore the TR function must pass through the origin. This additional information, derived from the economics of this case, means that in the above equation the arbitrary constant, c, must equal zero. (This is an example of a common procedure in mathematical economics, in which we use some information derived from economic analysis to move the maths forward.)

With $c = 0$, the total revenue function is

$$TR = 100q - q^2$$

We can check that this is correct by finding the derivative of the TR function. This should be the MR function that we were given:

$$MR \equiv \frac{dTR}{dq} = \frac{d}{dq}[100q - q^2] = 100 - 2q \quad \text{(as we were given)}$$

(b) Now that we have the TR function, we can quickly derive the demand function. By definition, $TR \equiv pq$. So, in general,

$$p \equiv \frac{pq}{q} \equiv \frac{TR}{q}$$

In this example, we have found already that $TR = 100q - q^2$. Substituting this into the previous equation gives

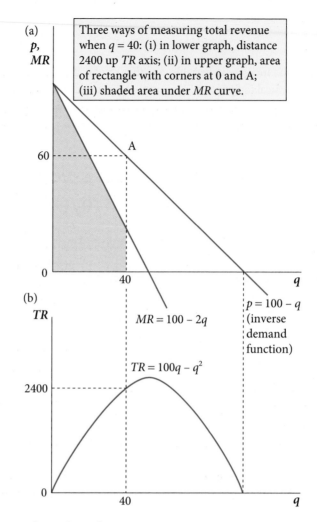

(a)
p,
MR

Three ways of measuring total revenue when $q = 40$: (i) in lower graph, distance 2400 up TR axis; (ii) in upper graph, area of rectangle with corners at 0 and A; (iii) shaded area under MR curve.

60 ---- A

0   40 q

(b)
TR

$MR = 100 - 2q$

$p = 100 - q$
(inverse demand function)

$TR = 100q - q^2$

2400 ----

0   40 q

Figure 18.6 Three ways of measuring total revenue.

$$p \equiv \frac{TR}{q} = \frac{100q - q^2}{q} = 100 - q$$

So the inverse demand function is $p = 100 - q$. The demand function is, of course, the inverse function to this, which is $q = 100 - p$.

A graphical treatment of this example is shown in figure 18.6. The interesting feature is that we now have a total of three ways of measuring total revenue. For example, if $q = 40$, the first way is to read off TR directly from figure 18.6(b) as $TR = 2400$. The second way is by calculating in figure 18.6(a) the area of the rectangle drawn under the demand curve with its corner at A where $q = 40$ and $p = 60$. This area is $pq = 60 \times 40 = 2400$.

The third, and new, way to find TR is by finding the area under the MR curve between $q = 0$ and $q = 40$. This area is shaded in figure 18.6(a). It is given by the definite integral

$$\int_0^{40} [100 - 2q]\mathrm{d}q$$

Applying rule 18.3, with $f(q) = 100 - 2q$ and $F(q) = 100q - q^2$ we get

$$\int_0^{10} [100 - 2q]\mathrm{d}q = [100q - q^2]_{q=40} - [100q - q^2]_{q=0}$$

$$= [100(40) - (40)^2] - 0 = 40(100 - 40) = 2400$$

Thus we see that the area under the MR curve between $q = 0$ and $q = 40$ gives us TR when $q = 40$.

18.8 Economic applications 3: consumers' surplus

Suppose a person's inverse demand function for a good is $p = f(q)$. Suppose the price is p_0 and the quantity purchased q_0. Then, subject to certain economic assumptions that we won't explore here, the area under the demand curve between 0 and q_0 serves as a measure (in units of money) of the total satisfaction derived by that person from the quantity q_0 that she is consuming.

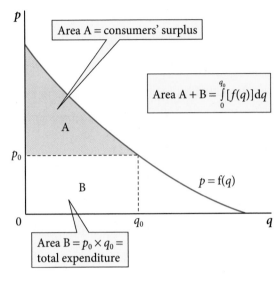

Area A = consumers' surplus

$$\text{Area A} + \text{B} = \int_0^{q_0} [f(q)]\mathrm{d}q$$

A

B

$p = f(q)$

Area B = $p_0 \times q_0$ = total expenditure

Figure 18.7 Consumers' surplus.

Moreover, the same is true of consumers in aggregate, though some additional assumptions are necessary. Therefore let us suppose now that the inverse demand function $p = f(q)$ is the *market* demand function: that is, of all consumers of the good (see figure 18.7). If the price is p_0 and the quantity purchased q_0, then, conditional on certain assumptions, the area under the demand curve between 0 and q_0 serves as a measure (in units of money) of the total satisfaction derived by all consumers from the quantity q_0 that they are consuming. This area is referred to as the consumers' valuation of the quantity q_0.

From section 18.2 above, we know that this area is given by the definite integral:

$$\int_0^{q_0} [f(q)]\mathrm{d}q = \text{area A} + \text{area B in figure 18.7}$$

However, to get a measure of the *net* satisfaction of consumers from consuming the quantity q_0, we must subtract the costs to them of obtaining this quantity. Since they paid price p_0 for quantity q_0, their total expenditure is $p_0 q_0$, which is area B in figure 18.7. So we can measure the *net* satisfaction that consumers enjoy from consuming quantity q_0 as follows:

> **RULE 18.4 Consumers' surplus**
>
> **Consumers' surplus** (CS) measures the net satisfaction consumers enjoy from consuming quantity q_0 when purchased at price p_0. It is found as
>
> $$CS = \int_0^{q_0} [f(q)]\mathrm{d}q - p_0 q_0 = \text{area A} + \text{area B} - \text{area B} = \text{area A in figure 18.7}$$

EXAMPLE 18.17

Find the consumers' surplus when the inverse demand function is $p = 100 - 0.5q$ and the price (p_0) is 20.

Our objective is therefore to find area A in figure 18.7, when $p_0 = 20$ and $f(q) = 100 - 0.5q$. As step 1 we will find area B, which equals $p_0 q_0$. Given $p_0 = 20$, we can find the quantity by substituting $p = 20$ into the inverse demand function, which gives $q_0 = 160$. So total expenditure is $p_0 q_0 = 20 \times 160 = 3200$ (euros, or whatever is the currency unit). This is area B.

Step 2 is to find area $(A + B)$, which is given by

$$\int_0^{160} [100 - 0.5q]\mathrm{d}q \tag{18.5}$$

To evaluate this definite integral, we first find the indefinite integral in the usual way, which is (by the power rule of integration, and remembering that $100 \equiv 100q^0$)

$$\int [100 - 0.5q]dq = 100q - 0.5\frac{q^2}{2} + c = 100q - 0.25q^2 + c$$

(You can check that this is correct by differentiating the right-hand side.)

The definite integral is therefore

$$\int_0^{160} [100 - 0.5q]dq = [100q - 0.25q^2]_{q=160} - [100q - 0.25q^2]_{q=0}$$

$$= [16000 - 6400] - [0] = 9600$$

This is area (A + B).

So, substituting our results into rule 18.4 above gives

Consumers' surplus = area (A + B) − area B = 9600 − 3200 = 6400 (= area A in figure 18.7).

This figure of 6400 is measured in units of money. This is because we have price on one axis and quantity on the other, so any measurement of area is $p \times q$, which is an amount of money.

We described consumers' surplus above as measuring the net satisfaction derived by consumers from buying quantity q_0 at price p_0. One interpretation of this is that it is the maximum amount that consumers would be willing to pay for the right to buy the quantity q_0 at a price of p_0, if the alternative were to be unable to buy any of this product at all.

Thus, if the government regulated the economy by requiring would-be buyers to bid at auction for a licence to buy the product, the 6400 above is the maximum buyers as a group would be willing to pay for a licence to buy the quantity $q = 160$ at a price of $p = 20$. (Of course, the consumers' surplus changes whenever the price and quantity change.)

Note that we have chosen to work with the *market* inverse demand function: that is, the aggregation of the demand functions of individuals for the good. Thus we are analysing consumers' surplus rather than consumer's surplus (the possessive apostrophe is placed after the s when there is more than one consumer). Consumers' surplus is a more questionable concept than consumer's surplus, as the former requires aggregation of satisfaction across individuals, which requires rather strong assumptions. However, the aggregate measure, consumers' surplus, is of much more interest for policymaking, especially in the cost–benefit analysis of important issues such as traffic congestion, rail accidents, and environmental pollution, where we want to measure the aggregate costs and benefits to all consumers.

18.9 Economic applications 4: producers' surplus

Producers' surplus is closely analogous to consumers' surplus, discussed in the previous section. We will explain this through an example.

EXAMPLE 18.18

Suppose the market supply curve (that is, of all sellers combined) for a good is $q = \frac{1}{3}p - \frac{5}{3}$. As we have previously found with the demand function, it is often more convenient to work with the inverse function, giving p as a function of q. In this case, by simple algebra the inverse supply function is $p = 3q + 5$. Then, given certain assumptions that we won't examine here, the area *above* the inverse supply curve between 0 and q_0 serves as a measure of the gain to producers from supplying the quantity q_0. This area is called producers' surplus.

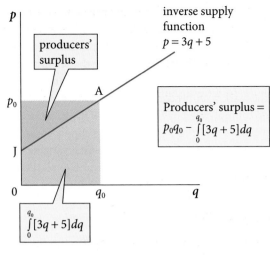

producers' surplus

inverse supply function
$p = 3q + 5$

A

Producers' surplus =
$$p_0 q_0 - \int_0^{q_0} [3q + 5]dq$$

p_0

J

0   q_0 q

$$\int_0^{q_0} [3q + 5]dq$$

Figure 18.8 Producers' surplus.

Thus in figure 18.8, when q_0 units are sold at a price p_0, producers' surplus is measured by the brown triangular area above the inverse supply function between 0 and q_0. We can measure this as follows. First, we take total revenue, $p_0 \times q_0$, which measures the area of the rectangle $0p_0 Aq_0$—that is, the rectangle with one corner at 0 and the opposite corner at A. Second, we evaluate the definite integral $\int_0^{q_0}[3q + 5]dq$, which measures the blue area $0JAq_0$.

Finally, we subtract the second from the first and thereby arrive at the area we wish to measure. Thus:

RULE 18.5 Producers' surplus

Producers' surplus (PS) is the net benefit producers derive from supplying quantity q_0 at price p_0. It is measured as

$$PS = p_0 q_0 - \int_0^{q_0} [3q + 5]dq = \text{area } Jp_0 A \text{ in figure 18.8}$$

In explaining this rule, we chose point A in figure 18.8 randomly, because the rule is valid for *any* quantity and price combination that lies on the supply function. In order to actually calculate the actual producers' surplus we need to know the *equilibrium* quantity and price, which requires introducing the market demand function in to the model. This is illustrated in the next example.

EXAMPLE 18.19

Continuing with the supply curve $p = 3q + 5$ from the previous example, we now assume that the market demand function cuts the supply curve at the point $q_0 = 20$, $p_0 = 65$. Thus we immediately have $p_0 q_0 = 1300$. We can also calculate the definite integral required in rule 18.5 as follows. The indefinite integral is

$$\int [3q + 5]dq = \frac{3}{2}q^2 + 5q + c$$

So the required definite integral is

$$\int_0^{20} [3q + 5]dq = \left[\frac{3}{2}q^2 + 5q\right]_{q=20} - \left[\frac{3}{2}q^2 + 5q\right]_{q=0} = 700$$

Therefore in rule 18.5 we have

$$PS = p_0 q_0 - \int_0^{q_0} [3q + 5]dq = 1300 - 700 = 600$$

A few words on the underlying economics of producers' surplus seem warranted. You may have noticed that the industry (inverse) supply curve in this example, $p = 3q + 5$, is very similar to the MC curve in example 18.15, $MC = 3q + 5$. This is not accidental, because we know that under perfect competition a firm will choose the level of output at which $p = MC$. Given this, we can replace MC with p in the firm's MC curve and thereby transform it into a supply curve for the firm, because it now tells us how much output that firm will produce and sell at any given price (see section 8.18). Finally, provided we are willing to assume that all firms in the industry have the same MC curve, we can transform the supply curve of the firm into a supply curve for all firms in the industry (that is, the market supply curve) by simply taking the amount supplied by one firm at any given price and multiplying it by the number of firms in the industry.

This helps us to understand producers' surplus. Given that the supply curve is also the MC curve, then the area under the supply curve, $\int_0^{q_0}[3q + 5]dq$, is the total variable costs of output q_0 (see example 18.15). Also, $p_0 q_0$ is total revenue. Substituting these two pieces of information into rule 18.5, we have

producers' surplus = total revenue − total variable costs

If we subtract fixed costs from both sides of this equation, it becomes

producers' surplus − fixed costs = profits

Thus *producers' surplus equals profits plus fixed costs.*

This suggests an interesting and useful way of looking at producers' surplus. Suppose the government regulated the economy by requiring would-be sellers to bid at auction for a licence to sell quantity q_0 at price p_0. Then the associated producers' surplus is the maximum that sellers as a group would be willing to pay for the licences, if the alternative were not to sell any of this product at all. This is because, if they sell nothing, they will nevertheless have to meet their fixed costs, by definition. And they will also forgo their profits. So if they sell nothing, they will be worse off by the amount of their fixed costs plus profits forgone. Therefore if they have to pay any amount that is less than this for the right to sell quantity q_0 at price p_0, they will be better off than if they sold nothing. (If there are no fixed costs, producers' surplus simply equals profits.)

Conclusions

Consumer and producer surpluses are a very important tool of economic analysis, especially in the areas of economic policy and cost–benefit analysis. Examples in the UK of policy evaluation using these tools include the congestion charge scheme introduced in February 2003 for vehicles entering central London; the M6 toll road north-east of Birmingham, opened in December 2003; and, most spectacularly, the auction of mobile phone licences in 2000 which netted the government £22 billion.

18.10 Economic applications 5: present value of a continuous stream of income

In section 12.10 we found that the present value (PV) of a single payment of a euros received in x years' time, discounted continuously at $100r\%$ per year, is given by

$$y = \frac{a}{e^{rx}} \quad \text{(which is often written as } y = ae^{-rx}\text{)}$$

Figure 18.9 shows the graph of this function with arbitrary values of the parameters a and r. From this graph we can read off the PV of a *single* payment of a received x years in the future. So, if that payment is received at time x_0, its PV is y_0; if it is received at time x_1, its PV is y_1; and so on. Note that the intercept is at a, since this is the PV of a received now.

It follows directly that if, instead of receiving a single payment of a, we are due to receive a payment of a at time x_0 *and* a payment of a at time x_1, then the combined PV of the two would

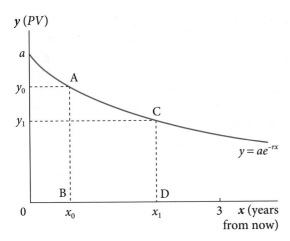

$y = ae^{-rx}$

The length of bar AB ($= y_0$) gives present value of a single payment of a received at future moment in time x_0, when discounting is continuous. The sum of lengths of all such bars (= area under the curve) between 0 and 3, measured by $\int_0^3 [ae^{-rx}]dx$), gives the present value of a continuous income stream of a per year for 3 years.

Figure 18.9 Present value of a continuous income stream.

be $y_0 + y_1$. That is, the combined PV is given by the sum of the two vertical distances, AB + CD.

Extending this further, suppose now that we are due to receive, not a few isolated payments of a but a continuous stream of receipts at a rate of a per year for, say, 3 years. Then the PV of this stream is found by adding together *all* the vertical distances such as AB and CD between 0 and 3 on the x-axis.

At first sight this seems an impossible task, since with a sufficiently sharp pencil we could draw an indefinitely large number of such lines in figure 18.9. However, we can think of each of these vertical lines as being a rectangle with an indefinitely small base. So our task is to find the sum of the areas of these rectangles, and we suddenly remember from section 18.2, rule 18.1 that this sum is, by definition, given by the definite integral:

$$\int_0^3 [ae^{-rx}]dx$$

In figure 18.9 this is the area under the curve between $x = 0$ and $x = 3$. Generalizing, we have a new formula:

⁂ **RULE 18.6 Present value of a continuous stream of receipts**

The PV of a continuous stream of receipts at a rate of a per year for the next x_0 years, discounted at $100r\%$ per year, is given by

$$PV = \int_0^{x_0} [ae^{-rx}]dx$$

This is the area under the curve $y = ae^{-rx}$ between $x = 0$ and $x = x_0$.

This formula can be developed a little further to put it into a more convenient form for doing calculations. To evaluate the PV using rule 18.3, we must first find the indefinite integral, $\int [ae^{-rx}]dx$. This requires a little trick: we multiply and divide the expression in square brackets by $-r$. We then have

$$\int [ae^{-rx}]dx = \int \left[-\frac{a}{r}(-rae^{-rx}) \right]dx$$

On the right-hand side, we can then take the multiplicative constant $(-\frac{a}{r})$ outside of the integral sign, giving

$$\left(-\frac{a}{r} \right)\int [-rae^{-rx}]dx \tag{18.6}$$

This now conforms to rule I8a above, for we have an expression of the form

$$\int [f'(x)e^{f(x)}]dx$$

with $f(x) = -rx$ and $f'(x) = -r$. From rule I8a above, the integral is

$$e^{f(x)} + c$$

Applying this to equation (18.6), we get

$$\left(-\frac{a}{r}\right)\int[-rae^{-rx}]\,dx = \left(-\frac{a}{r}\right)e^{-rx} + c$$

Now that we have the indefinite integral, we can, by using rule 18.3, evaluate the definite integral in rule 18.6 above as

$$\int_0^{x_0}[ae^{-rx}]\,dx = \left[\left(-\frac{a}{r}\right)e^{-rx}\right]_{x=x_0} - \left[\left(-\frac{a}{r}\right)e^{-rx}\right]_{x=0}$$

Since $e^0 = 1$, the right-hand side simplifies to

$$\left(-\frac{a}{r}\right)e^{-rx_0} - \left(-\frac{a}{r}\right) = \frac{a}{r}(1 - e^{-rx_0})$$

And we can drop the zero subscript on x provided it's understood that the x in question is the upper limit of integration (the number of years, in other words). So our final result, stated as a rule, is:

RULE 18.6a The present value of a continuous stream of receipts

The PV of a continuous stream of receipts at a rate of a per year for the next x years, discounted at $100r\%$ per year, is given by

$$PV = \int_0^x[ae^{-rx}]\,dx \equiv \frac{a}{r}(1 - e^{-rx})$$

Rule 18.6a is, of course, simply a variant of rule 18.6.

EXAMPLE 18.20

Let's try out rule 18.6a on an example. Suppose the continuous income stream is at a rate of 100 (euros) per year for 3 years, and the discount rate is 10%. So we have $a = 100$, $r = 0.1$, $x = 3$. Substituting these values into rule 18.6a gives

$$PV = \frac{a}{r}(1 - e^{-rx}) = \frac{100}{0.1}(1 - e^{-(0.1)3}) = 1000(1 - e^{-0.3}) = 1000(1 - 0.74082) = 259.18$$

As always, it's good mental exercise to try to assess whether this answer seems correct. An obvious reference point is that if the discount rate were zero, then 100 per year for 3 years would have a PV of 300. So our answer of 259 is clearly in the right ball-park. Another point of comparison is to calculate the PV of 100 per year received in annual instalments starting 1 year from now. Using rule 10.9 from section 10.10, we get

$$PV = \frac{100}{(1 + 0.1)^1} - \frac{100}{(1 + 0.1)^2} + \frac{100}{(1 + 0.1)^3} = 90.90 + 82.64 + 75.13 = 248.68$$

The PV of these annual instalments is certain to be lower because the series is less front-loaded than the continuous stream; you receive nothing until the end of each year. So our answer of 259.18 seems about right.

Conclusion

You may be thinking that rule 18.6a can have little or no practical application on the grounds that no one ever enjoys a continuous stream of income. But this is not correct. Rock stars enjoy a continuous stream of income from royalties from their recordings, which are being bought 24/7 around the globe. More prosaically, an oil company pumps oil day and night, resulting in a continuous stream of income as the oil is sold. Reflecting this, it is routine for financial analysts to estimate the value of both rock stars and oil companies by calculating the present value of their expected future stream of income. That is why an oil company's shares jump up in value when a new oil discovery is announced, or jump down in value when the company admits that previously announced discoveries have been exaggerated, as Shell confessed in January 2004.

The fact that the share price jumps discontinuously following the arrival of new information, rather than moving gradually up or down over some transitional period, is a key feature of asset markets. The jump occurs because when new information arrives, all rational investors quickly recalculate the present value of the expected profits, using rule 18.6a above. This gives the revised value of one of the company's shares, and the market price immediately jumps to this new value because if the price were any lower there would be a rush to buy, and if it were any higher there would be a rush to sell.

Summary of sections 18.5–18.10

In section 18.5 we looked at the fundamental theorem of the definite integral (rule 18.3), which tells us how to evaluate a definite integral using the indefinite integral (if the latter can be found).

In sections 18.6 and 18.7 we then applied this technique to some key areas of economic analysis. First, we showed how the total cost and total revenue functions could be derived by using the definite integral to find the area under the marginal cost and marginal revenue functions respectively. Then in sections 18.8 and 18.9 we showed how consumers' and producers' surplus, which are important in welfare economics, cost–benefit analysis, and other policy areas, can be measured using the definite integral. Finally, in section 18.10 we showed how the present value of a continuous stream of income, discounted continuously, can be found as the area under the curve $y = ae^{-rx}$. This is the basis of asset pricing.

Progress exercise 18.3

1. A firm's marginal cost function is $MC = 4q + 4$, where q = output. If output is increased from 25 to 50, find the increase in the firm's total cost of production.

2. A firm's marginal revenue function is $MR = 250q^{-0.5}$, where q = output. Find the firm's loss of revenue if it reduces the quantity sold from 400 to 144.

3. Find the consumer's surplus at $q = 8$ for the inverse demand function $p = 100 - q^2$.

4. Show that the present value of a continuous stream of receipts at a rate of a euros per year for the next x years, discounted at $100r\%$ per year, is given by

$$\frac{a}{r}(1 - e^{-rx})$$

5. You are advising an oil company which intends to bid at a public auction for the rights to extract crude oil from beneath a certain sector of the Caspian Sea. The field is expected to yield 500,000 barrels of oil per year for the next 20 years, after which the oil will be exhausted. The oil is expected to sell for $70 per barrel and to cost $50 per barrel to extract.

(a) Assuming a discount rate of 5% per year, what is the maximum bid you would advise the company to make?

(b) Suppose that, due to increased political instability in the region, the company decides that the appropriate discount rate is 10% per year. By how much is the maximum bid decreased? What is the percentage reduction in the maximum bid?

(c) Suppose it now appears likely that the government that is auctioning the rights will renege on the deal after 5 years. What is then the maximum bid, with a discount rate of 10% per year?

Checklist

Be sure to test your understanding of this chapter by attempting the progress exercises (answers are at the end of the book). The Online Resource Centre contains further exercises and materials relevant to this chapter www.oxfordtextbooks.co.uk/orc/renshaw3e/.

In this chapter we have developed the mathematics of integration and shown some of its applications in economics. The topics were:

✔ **The definite integral.** Recognizing a definite integral, $\int_a^b f(x)dx$, and understanding what it measures.

✔ **The indefinite integral.** Recognizing and understanding an indefinite integral, $\int f(x)dx$. Evaluation of indefinite integrals by 'reverse differentiation' in simple cases.

✔ **Evaluation of definite integrals.** Understanding the formula for finding a definite integral and how to apply it in simple cases.

✔ **Economic applications.** Applying integration to derive total variable cost from marginal cost, total revenue from marginal revenue, to evaluate producers' and consumers' surplus and the present value of a continuous stream of income.

There are other economic applications that we do not have space for, such as social cost–benefit analysis, which is used to analyse phenomena such as environmental pollution and climate change, traffic congestion and other crucial policy areas. This chapter therefore deserves careful study.

Chapter 19
Matrix algebra

OBJECTIVES

Having completed this chapter you should be able to:

- Understand the basic concepts and definitions of matrix algebra.
- Express a set of linear equations in matrix notation.
- Evaluate determinants.
- Invert a 3×3 matrix.
- Understand how matrix inversion may be used to solve a set of linear equations.
- Use matrix algebra to find the equilibrium and comparative static properties of simple macroeconomic models.

19.1 Introduction

In section 3.12 we saw that when we try to solve a set of three or more simultaneous equations, the methods of solution that are perfectly adequate for two equations quickly become cumbersome and error-prone. Moreover, the question of whether a solution exists also has to be addressed. Matrix algebra helps overcome these problems and makes it possible to handle large sets of simultaneous linear equations.

First, matrix algebra serves as a notation for writing down such a set of linear equations in a very compact way. Second, it provides a method (determinants) for establishing whether a solution to a set of simultaneous equations exists. Finally, it provides methods—matrix inversion and Cramer's rule—for finding the solution. A limitation of matrix algebra, of course, is that it is capable of handling only linear equations. This appears at first sight to be a major disadvantage. However, in discussing the differential of a function in sections 7.15 and 15.4, we saw that any function could be treated as linear over a sufficiently small range of values. We also saw in chapters 12 and 13 that some exponential functions that are themselves highly non-linear nevertheless become linear when we take logs on both sides of the equation. The question of linear approximation to a non-linear function is also discussed further below in section W21.2, which can be downloaded from the Online Resource Centre. Thus the fact that matrix algebra can handle only linear functions is not such a devastating limitation as might appear.

We conclude the chapter with an illustrative application of matrix algebra to a simple macroeconomic model, showing how the methodology and notation can reveal in a very transparent way much about the relationships between variables and the effects of parameter changes, which would almost certainly be lost if ordinary algebraic methods were used.

19.2 Definitions and notation

A **matrix** is an arrangement of numbers or symbols into rows and columns.

EXAMPLE 19.1

A 3×2 matrix:

$$\begin{bmatrix} 1 & 6 \\ -4 & 9 \\ 11 & 17 \end{bmatrix} \quad \text{a matrix with 3 rows and 2 columns}$$

EXAMPLE 19.2

A 3×3 matrix (called a **square matrix** because the number of rows = number of columns):

$$\begin{bmatrix} 42 & 91 & 78 \\ 11 & -6 & 0 \\ -5 & 99 & 2 \end{bmatrix} \quad \text{a matrix with 3 rows and 3 columns}$$

The numbers 42, 91, and so on are called **elements of the matrix.**

Generalization

A matrix, **A**, is of **order** $m \times n$ if it has m rows and n columns. It may help to remember this as **rc**. So the matrix

$$\mathbf{A} = \begin{bmatrix} a_{11} & a_{12} & a_{13} \\ a_{21} & a_{22} & a_{23} \end{bmatrix}$$

is of order 2×3 because it has 2 rows and 3 columns.

We normally use a **bold** capital letter to write the name of a matrix, such as **A**.

In the matrix **A** the elements, a_{11}, a_{12}, and so on, are identified by their subscripts. The first subscript gives the row, the second the column. Again it may help to remember this as **rc**. So, in general, element a_{ij} is in the ith row and jth column of the matrix **A**.

Note that a matrix is uniquely defined by its elements, so if even one element changes then we have a new matrix. This means that if we have a matrix **A** and another matrix **B**, then we can only write **A** = **B** if *each* element of **A** equals the corresponding element of **B**.

Two special cases of matrices (= plural of matrix)

(1) The **null matrix:** $\begin{bmatrix} 0 & 0 & 0 \\ 0 & 0 & 0 \end{bmatrix}$ (every element is zero)

(2) The **unit**, or **identity, matrix** (which must be square):

$$\begin{bmatrix} 1 & 0 & 0 \\ 0 & 1 & 0 \\ 0 & 0 & 1 \end{bmatrix} \quad \text{(the 'leading' diagonal elements = 1, all the others = zero)}$$

A matrix with only 1 row is called a **row vector**.

A matrix with only 1 column is called a **column vector**.

We normally use a **bold**, lower-case letter to denote a vector, such as **a**.

EXAMPLE 19.3

Row vector of order 1×4: $\mathbf{a} = [a_{11} \quad a_{12} \quad a_{13} \quad a_{14}]$

Column vector of order 3×1: $\mathbf{b} = \begin{bmatrix} b_{11} \\ b_{21} \\ b_{31} \end{bmatrix}$

Having covered basic definitions and notation above, we now explain the rules for manipulation of matrices.

19.3 Transpose of a matrix

When we transpose a matrix, the first row becomes the first column, the second row becomes the second column, and so on.

EXAMPLE 19.4

Given $\mathbf{A} = \begin{bmatrix} a_{11} & a_{12} & a_{13} \\ a_{21} & a_{22} & a_{23} \end{bmatrix}$

the transpose is $\mathbf{A}' = \begin{bmatrix} a_{11} & a_{21} \\ a_{12} & a_{22} \\ a_{13} & a_{23} \end{bmatrix}$

Here the first row of **A** has become the first column of $\mathbf{A}'$, and so on. In general, the ijth element of **A** becomes the jith element of $\mathbf{A}'$. For example a_{23}, which is in row 2, column 3 of **A**, is now in row 3, column 2 of $\mathbf{A}'$. The transpose of matrix **A** can be written as $\mathbf{A}'$ or as $\mathbf{A}^{\mathrm{T}}$.

19.4 Addition/subtraction of two matrices

The rule is that we add/subtract the elements, one by one.

EXAMPLE 19.5

$\mathbf{A} + \mathbf{B} = \mathbf{C}$

means:

$$\begin{bmatrix} a_{11} & a_{12} \\ a_{21} & a_{22} \end{bmatrix} + \begin{bmatrix} b_{11} & b_{12} \\ b_{21} & b_{22} \end{bmatrix} = \begin{bmatrix} a_{11} + b_{11} & a_{12} + b_{12} \\ a_{21} + b_{21} & a_{22} + b_{22} \end{bmatrix}$$

Addition/subtraction of **A** and **B** is impossible unless both matrices are of the *same order* (that is, each must have the same number of rows and the same number of columns—there must be no 'hanging' elements).

SOME FURTHER TOPICS

19.5 Multiplication of two matrices

Suppose we have a matrix **A** and a matrix **B**, both of order 2×2, and we want to multiply them together to create a new matrix, **C**. Then the product

$$\mathbf{A} \times \mathbf{B} = \mathbf{C}$$

is calculated as follows:

$$\begin{bmatrix} a_{11} & a_{12} \\ a_{21} & a_{22} \end{bmatrix} \times \begin{bmatrix} b_{11} & b_{12} \\ b_{21} & b_{22} \end{bmatrix} = \begin{bmatrix} a_{11}b_{11} + a_{12}b_{21} & a_{11}b_{12} + a_{12}b_{22} \\ a_{21}b_{11} + a_{22}b_{21} & a_{21}b_{12} + a_{22}b_{22} \end{bmatrix}$$

In schematic form:

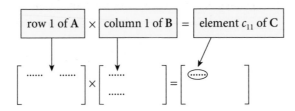

| row 1 of **A** | $\times$ | column 1 of **B** | $=$ | element c_{11} of **C** |

Each element of row 1 of **A** times each element of column 1 of **B**, summed, equals element c_{11} of **C**

The rule is that we construct c_{11}, the top left-hand element of the new matrix, as follows. We take each element in *row* 1 of **A** and multiply it by the corresponding element of *column* 1 of **B**. So a_{11} multiplies b_{11}, and a_{12} multiplies b_{21}. The results are then added and this becomes element c_{11} of the new matrix, **C**. Thus $c_{11} = a_{11}b_{11} + a_{12}b_{21}$ (see diagram).

Then to construct c_{12} we again take each element in *row* 1 of **A** and multiply it now by the corresponding element of *column* 2 of **B**. So a_{11} multiplies b_{12}, and a_{12} multiplies b_{22}. The results are added and this becomes element c_{12} of the new matrix. Thus $c_{12} = a_{11}b_{12} + a_{12}b_{22}$. And so on.

Generalizing, the sum of each of the elements of the *i*th row of **A** times each of the elements of the *j*th column of **B** gives the *ij*th element of **C**.

Thus we again have the pattern: row $\times$ column, or **rc**.

Pre- and post-multiplication

Above we calculated the product $\mathbf{A} \times \mathbf{B}$. Notice that this is *not* the same as the product $\mathbf{B} \times \mathbf{A}$. For, applying the rule above (row $\times$ column), we find that

$$\mathbf{B} \times \mathbf{A} = \mathbf{D}$$

is calculated as

$$\begin{bmatrix} b_{11} & b_{12} \\ b_{21} & b_{22} \end{bmatrix} \times \begin{bmatrix} a_{11} & a_{12} \\ a_{21} & a_{22} \end{bmatrix} = \begin{bmatrix} b_{11}a_{11} + b_{12}a_{21} & b_{11}a_{12} + b_{12}a_{22} \\ b_{21}a_{11} + b_{22}a_{21} & b_{21}a_{12} + b_{22}a_{22} \end{bmatrix}$$

so we can see that the elements of **D** are not the same as the elements of **C**. So, in general, in matrix multiplication **AB** is not equal to **BA**. (This is unlike ordinary multiplication where 3×5 is the same as 5×3.)

To distinguish the two cases, in **AB** we say that **B** is **pre-multiplied** by **A**, while in **BA** we say that **B** is **post-multiplied** by **A**.

As with addition/subtraction, when multiplying matrices there must be no 'hanging' elements. For example, given

$$\mathbf{A} = \begin{bmatrix} a_{11} & a_{12} \\ a_{21} & a_{22} \end{bmatrix} \quad \text{and} \quad \mathbf{B} = \begin{bmatrix} b_{11} & b_{12} & b_{13} \\ b_{21} & b_{22} & b_{23} \\ b_{31} & b_{32} & b_{33} \end{bmatrix}$$

if we try to evaluate the product **AB**, we run into trouble immediately because the number of elements in the first *row* of **A** is less than the number of elements in the first *column* of **B**. So b_{31} is left 'hanging' with no partner element from row 1 of the **A** matrix to marry up with.

Similarly, if we try to evaluate the product **BA**, we have a problem because the number of elements in the first row of **B** is greater than the number of elements in the first column of **A**. So b_{13} is left 'hanging' with no partner element from column 1 of the A matrix to marry up with.

The two matrices **A** and **B** are said to be **non-conformable**. The products **AB** and **BA** do not exist.

Conformability

To avoid this problem of 'hanging' elements, and given the **rc** rule of multiplication, in order for the matrix product **AB** to exist we need the number of elements in any *row* of **A** to be equal to the number of elements in any *column* of **B**. For example, given

$$\mathbf{A} = \begin{bmatrix} a_{11} & a_{12} \\ a_{21} & a_{22} \end{bmatrix} \quad \text{and} \quad \mathbf{B} = \begin{bmatrix} b_{11} & b_{12} & b_{13} \\ b_{21} & b_{22} & b_{23} \end{bmatrix}$$

we see that the rows of **A** and the columns of **B** both have two elements, so the product **AB** exists, and is given by

$$\mathbf{A} \times \mathbf{B} = \mathbf{C}$$

This is

$$\begin{bmatrix} a_{11} & a_{12} \\ a_{21} & a_{22} \end{bmatrix} \times \begin{bmatrix} b_{11} & b_{12} & b_{13} \\ b_{21} & b_{22} & b_{23} \end{bmatrix} = \begin{bmatrix} a_{11}b_{11} + a_{12}b_{21} & a_{11}b_{12} + a_{12}b_{22} & a_{11}b_{13} + a_{12}b_{23} \\ a_{21}b_{11} + a_{22}b_{21} & a_{21}b_{12} + a_{22}b_{22} & a_{21}b_{13} + a_{22}b_{23} \end{bmatrix}$$

Now, here's a tricky point that you will probably need to pause on, and possibly do some scribbling. The rows of **A** have two elements. Each element is in its own column. Therefore **A** must have two columns. In general, the number of elements in any row of a matrix is equal to the number of columns in that matrix.

Similarly, the columns of **B** have two elements. Each element is in its own row. Therefore **B** must have two rows. In general, the number of elements in any column of a matrix is equal to the number of rows in that matrix.

Thus we can say that **AB** exists because the number of *columns* in **A** equals the number of *rows* in **B**.

In diagrammatic form, **AB** exists because

A is of order 2×2 and **B** is of order 2×3

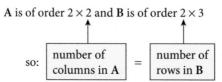

so:

Note that this reverses the **rc** memory aid that we suggested earlier. The rule here is '**AB** exists if number of columns of **A** equals number of rows of **B**', which you might remember as **cr**.

Above we considered the product **AB**. What about the product **BA**? Applying the rule above, we find that **BA** does *not* exist because

B is of order 2×3 and **A** is of order 2×2

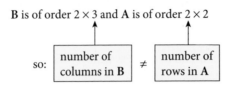

so: ⬚ number of columns in **B** ⬚ $\neq$ ⬚ number of rows in **A** ⬚

Consequently, when we try to calculate **BA**, we get

$$\begin{bmatrix} b_{11} & b_{12} & b_{13} \\ b_{21} & b_{22} & b_{23} \end{bmatrix} \times \begin{bmatrix} a_{11} & a_{12} \\ a_{21} & a_{22} \end{bmatrix}$$

and find straight away that the element b_{13} is hanging because there is no third element in the first column of **A** for it to mate up with.

Generalizing from the analysis above, if we are given two matrices, **A** of order $m \times s$ and **B** of order $s \times n$, then we can say:

(1) The product **AB** exists (because the **cr** test is satisfied: that is, the number of columns, s, in **A** equals the number of rows in **B**). Note also that the product **AB** will be a matrix of order $m \times n$.

(2) The product **BA** does not exist, unless $m = n$.

19.6 Vector multiplication

A special case of matrix multiplication is when **A** is $1 \times n$ (a row vector) and **B** is $n \times 1$ (a column vector). For example, if $n = 3$, then if we pre-multiply **B** by **A**, we get

$$\mathbf{AB} = \begin{bmatrix} a_{11} & a_{12} & a_{13} \end{bmatrix} \begin{bmatrix} b_{11} \\ b_{21} \\ b_{31} \end{bmatrix}$$

$$= a_{11}b_{11} + a_{12}b_{21} + a_{13}b_{31}$$

Note that $a_{11}b_{11} + a_{12}b_{21} + a_{13}b_{31}$ is not a matrix, nor a vector, but is simply an ordinary number. In matrix algebra, an ordinary number is called a **scalar**. So, in general, if a column vector is *pre*-multiplied by a row vector, the result is a scalar.

In contrast, if we post-multiply **B** by **A** we get

$$\mathbf{BA} = \begin{bmatrix} b_{11} \\ b_{21} \\ b_{31} \end{bmatrix} \begin{bmatrix} a_{11} & a_{12} & a_{13} \end{bmatrix}$$

$$= \begin{bmatrix} b_{11}a_{11} & b_{11}a_{12} & b_{11}a_{13} \\ b_{21}a_{11} & b_{21}a_{12} & b_{21}a_{13} \\ b_{31}a_{11} & b_{31}a_{12} & b_{31}a_{13} \end{bmatrix} \quad \text{(that is, a } 3 \times 3 \text{ matrix)}$$

In general, if a column vector is *post*-multiplied by a row vector, the result is a matrix.

19.7 Scalar multiplication

This is when we pre-multiply a matrix, A, by a scalar, k. (As we have already seen, in matrix algebra a constant is known as a scalar.)

The rule is simply that each element of A is multiplied by k. For example, if A is order 2×2,

$$kA = k \begin{bmatrix} a_{11} & a_{12} \\ a_{21} & a_{22} \end{bmatrix} = \begin{bmatrix} ka_{11} & ka_{12} \\ ka_{21} & ka_{22} \end{bmatrix}$$

We can see from this why k is called a scalar. It scales up all the elements of A. For example, if $k = 2$, then all the elements of A are doubled.

19.8 Matrix algebra as a compact notation

We are now ready to see, in a preliminary way at least, how matrix algebra can be used. Suppose we have three simultaneous linear equations:

$$4x + y - 5z = 8$$

$$-2x + 3y + z = 12$$

$$3x - y + 4z = 5$$

If we define a matrix, A, and two column vectors x and b as

$$A = \begin{bmatrix} 4 & 1 & -5 \\ -2 & 3 & 1 \\ 3 & -1 & 4 \end{bmatrix}, \quad x = \begin{bmatrix} x \\ y \\ z \end{bmatrix}, \quad \text{and} \quad b = \begin{bmatrix} 8 \\ 12 \\ 5 \end{bmatrix}$$

then we can write the set of simultaneous equations above as

$$Ax = b$$

We can check that this is correct by calculating the product Ax. Using the rules of matrix multiplication from section 19.5, we get

$$Ax = \begin{bmatrix} 4 & 1 & -5 \\ -2 & 3 & 1 \\ 3 & -1 & 4 \end{bmatrix} \begin{bmatrix} x \\ y \\ z \end{bmatrix} = \begin{bmatrix} 4x + y - 5z \\ -2x + 3y + z \\ 3x - y + 4z \end{bmatrix}$$

That is, we get a 3×1 column vector which equals the column vector b. (The two vectors are equal because each of their elements is equal. For example, $4x + y - 5z = 8$.) Check for yourself that this is correct before moving on.

Thus we see that matrix algebra gives a compact and easily manipulable notation for handling large sets of *linear* simultaneous equations.

? **Progress exercise 19.1**

1. Given $A = \begin{bmatrix} 1 & 2 \\ 3 & 4 \end{bmatrix}$ and $B = \begin{bmatrix} 1 & 3 \\ 4 & 2 \end{bmatrix}$

   Find $2A + 3B$

2. Given $A = \begin{bmatrix} 2 & 1 \\ 1 & 7 \\ 5 & 3 \end{bmatrix}$, $B = \begin{bmatrix} 1 & 3 & 6 \\ -1 & 2 & 4 \end{bmatrix}$

   Find AB and BA

3. Calculate:

   (a) $\begin{bmatrix} 1 & -2 & -3 \end{bmatrix} \begin{bmatrix} 3 \\ 2 \\ 4 \end{bmatrix}$ (b) $\begin{bmatrix} a & b \end{bmatrix} \begin{bmatrix} x \\ y \end{bmatrix}$ (c) $\begin{bmatrix} 3 \\ 2 \\ 1 \end{bmatrix} \begin{bmatrix} 1 & 2 & 3 \end{bmatrix}$

   (d) $\begin{bmatrix} 3 & 2 \\ -1 & -2 \\ 4 & 0 \end{bmatrix} \begin{bmatrix} -3 \\ 0 \end{bmatrix}$ (e) $\begin{bmatrix} 1 & 2 & 3 \end{bmatrix} \begin{bmatrix} 1 & 0 & -1 \\ -1 & -2 & 0 \\ 0 & 1 & 1 \end{bmatrix} \begin{bmatrix} 1 \\ 2 \\ 3 \end{bmatrix}$ (f) $\begin{bmatrix} x & y & z \end{bmatrix} \begin{bmatrix} 1 & 0 & -1 \\ 0 & 2 & 1 \\ -1 & 1 & 3 \end{bmatrix} \begin{bmatrix} x \\ y \\ z \end{bmatrix}$

4. Given $A = \begin{bmatrix} 5 & -3 \\ 2 & 1 \end{bmatrix}$, $B = \begin{bmatrix} 1 & 5 \\ 4 & 0 \end{bmatrix}$, $C = \begin{bmatrix} -1 & 1 \\ 1 & 2 \end{bmatrix}$

   Verify the equations:

   (a) $A(B + C) = AB + AC$ (b) $(AB)C = A(BC)$

   (c) $3(A + B) = 3A + 3B$ (d) $AA^2 = A^2A$

5. A consumer buys 100 apples, 50 bananas, 75 melons, and 25 lemons. The prices are €1.20, €2.00, €1.00, and €0.80 per unit, respectively.

   (a) Use vectors to represent the quantities purchased and their prices, such that the product of the two vectors will give the total cost of the fruit. (Hint: you will need the rule for determining when you can multiply matrices together.)

   (b) Calculate the total cost, using matrix multiplication.

19.9 The determinant of a square matrix

Associated with every square matrix there is a scalar (that is, a single number) called a determinant. We will now explain, by means of examples, how the determinant is defined and calculated.

The determinant of a 2 × 2 matrix

Given $A = \begin{bmatrix} a & b \\ c & d \end{bmatrix}$

The **determinant** of A, written as det. A, or $|A|$, is defined as $ad - cb$.

If $|A| = 0$, the matrix A is said to be **singular**.

EXAMPLE 19.6

Given $\mathbf{A} = \begin{bmatrix} 3 & 4 \\ 2 & -5 \end{bmatrix}$

The determinant of $\mathbf{A}$ is

$|\mathbf{A}| = [3 \times (-5)] - [2 \times 4] = -15 - 8 = -23$

EXAMPLE 19.7

The matrix $\mathbf{A} = \begin{bmatrix} 3 & 2 \\ 6 & 4 \end{bmatrix}$ is singular, because

$|\mathbf{A}| = [3 \times 4] - [6 \times 2] = 12 - 12 = 0$

Notice that in $\mathbf{A}$, each element of row 2 is twice that of row 1. And each element of column 1 is 1.5 times that of column 2. This illustrates a general rule: a matrix is singular if any row (or column) is a multiple of any other row (or column). The multiple includes 1: that is, when a row (or column) equals another row (or column). (Singularity can also occur for slightly more complex reasons.)

The determinant of a 3×3 matrix

Given $\mathbf{A} = \begin{bmatrix} a_{11} & a_{12} & a_{13} \\ a_{21} & a_{22} & a_{23} \\ a_{31} & a_{32} & a_{33} \end{bmatrix}$

We construct the determinant, $|\mathbf{A}|$, as follows.

Step 1. We take the *first* element in row 1, a_{11}. Then we delete the row and column that include a_{11}: that is, the first row and the first column. This leaves us with a 2×2 sub-matrix:

$$\begin{bmatrix} a_{11} & a_{12} & a_{13} \\ a_{21} & a_{22} & a_{23} \\ a_{31} & a_{32} & a_{33} \end{bmatrix}$$

Step 2. Then we find the determinant of this sub-matrix, which is $a_{22}a_{33} - a_{32}a_{23}$. This determinant is called a *minor*, and is written $|\mathbf{M}_{11}|$, where the subscripts tell us that it is the determinant found after deleting row 1 and column 1. Thus

$|\mathbf{M}_{11}| = a_{22}a_{33} - a_{32}a_{23}$

Then we repeat steps 1 and 2, this time taking in step 1 the *second* element in row 1, a_{12}. In this case, when we delete the row and column that include a_{12}, the resulting sub-matrix is

$$\begin{bmatrix} a_{11} & a_{12} & a_{13} \\ a_{21} & a_{22} & a_{23} \\ a_{31} & a_{32} & a_{33} \end{bmatrix}$$

So in step 2 the determinant of the sub-matrix is $a_{21}a_{33} - a_{31}a_{23}$, and the minor is

$|\mathbf{M}_{12}| = a_{21}a_{33} - a_{31}a_{23}$

Third, we again repeat steps 1 and 2, this time taking in step 1 the *third* element in row 1, a_{13}. In this case when we delete the row and column that include a_{13}, the resulting sub-matrix is

$$\begin{bmatrix} a_{11} & a_{12} & a_{13} \\ a_{21} & a_{22} & a_{23} \\ a_{31} & a_{32} & a_{33} \end{bmatrix}$$

So in step 2 the determinant of the sub-matrix is $a_{21}a_{32} - a_{31}a_{22}$, and the minor is

$$|M_{13}| = a_{21}a_{32} - a_{31}a_{22}$$

Finally, we multiply each of the elements of row 1 by its associated minor, and add the results. Before doing this, however, we need to make one final adjustment. The minors are given a + sign if their subscripts add to an even number, and a − sign if their subscripts add to an odd number. So $|M_{11}|$ and $|M_{13}|$ retain their + signs (because $1 + 1 = 2$ and $1 + 3 = 4$ are both even numbers), while $|M_{12}|$ is given a − sign because $1 + 2 = 3$ is odd. The minors, after their signs have been adjusted in this way, are called **signed minors**, or **cofactors**.

With these adjustments to signs, we now have the determinant of **A** as

$$|A| = a_{11}|M_{11}| - a_{12}|M_{12}| + a_{12}|M_{13}|$$

or

$$|A| = +a_{11}(a_{22}a_{33} - a_{32}a_{23}) - a_{12}(a_{21}a_{33} - a_{31}a_{23}) + a_{13}(a_{21}a_{32} - a_{31}a_{22})$$

In the example above, we used the elements from row 1, a_{11}, a_{12}, and a_{13} to derive $|A|$. But this was an arbitrary choice; we could have used any row or column. For example, if we had used column 2, we would have arrived at

$$|A| = -a_{12}|M_{12}| + a_{22}|M_{22}| - a_{32}|M_{32}|$$

Notice how the rule on signing the minors works in this case.

The determinant of a 4 × 4 or higher order matrix

Given the 4 × 4 matrix

$$A = \begin{bmatrix} a_{11} & a_{12} & a_{13} & a_{14} \\ a_{21} & a_{22} & a_{23} & a_{24} \\ a_{31} & a_{32} & a_{33} & a_{34} \\ a_{41} & a_{42} & a_{43} & a_{44} \end{bmatrix}$$

the determinant is found in exactly the same way as in the 3 × 3 case above. We start with a_{11}, then find the minor $|M_{11}|$, which is the determinant of the sub-matrix created by deleting the first row and column:

$$\begin{bmatrix} a_{11} & a_{12} & a_{13} & a_{14} \\ a_{21} & a_{22} & a_{23} & a_{24} \\ a_{31} & a_{32} & a_{33} & a_{34} \\ a_{41} & a_{42} & a_{43} & a_{44} \end{bmatrix}$$

What is new here is that this sub-matrix is a 3 × 3 matrix, so we have to find its determinant by the method of the previous example, unlike the previous example where the sub-matrices were 2 × 2 and consequently we could write down the determinant straight away.

Having found the determinant of this 3 × 3 matrix, we move on to element a_{12} and repeat the step above, which of course means that we must find another 3 × 3 determinant. And so on, until we have worked our way across row 1 (or any other row or column, if we prefer). Having signed all the minors appropriately, our final expression will be

$$|A| = a_{11}|M_{11}| - a_{12}|M_{12}| + a_{13}|M_{13}| - a_{14}|M_{14}|$$

Generalizing from the examples above, we can say that for any $n \times n$ square matrix $\mathbf{A}$, the determinant $|\mathbf{A}|$ is found by taking any row, say row i, and forming the sum

$$|\mathbf{A}| = \pm a_{i1}|\mathbf{M}_{i1}| \pm a_{i2}|\mathbf{M}_{i2}| \pm a_{i3}|\mathbf{M}_{i3}| \pm \ldots \pm a_{in}|\mathbf{M}_{in}|$$

where the minor $|\mathbf{M}_{ij}|$ is the determinant of the sub-matrix obtained by deleting row i and column j of $\mathbf{A}$, and $|\mathbf{M}_{ij}|$ is signed positive (negative) as the sum $i + j$ is even (odd).

Equally, $|\mathbf{A}|$ can be found by taking any column, say column j, and forming the corresponding expression:

$$|\mathbf{A}| = \pm a_{1j}|\mathbf{M}_{1j}| \pm a_{2j}|\mathbf{M}_{2j}| \pm a_{3j}|\mathbf{M}_{3j}| \pm a_{ni}|\mathbf{M}_{ni}|$$

Don't forget that only a square matrix can have a determinant.

19.10 The inverse of a square matrix

In ordinary algebra, we say that $\frac{1}{y}$ is the inverse, or reciprocal, of y because $\frac{1}{y}y = 1$. Since we can write $\frac{1}{y} \equiv y^{-1}$, we can equivalently say that y^{-1} is the inverse of y because $y^{-1}y = 1$.

In the same way, in matrix algebra we say that the square matrix $\mathbf{A}^{-1}$ is the inverse of the square matrix $\mathbf{A}$ if

$$\mathbf{A}^{-1}\mathbf{A} = \mathbf{I}$$

where $\mathbf{I}$ is the unit or identity matrix. (Recall that $\mathbf{I}$ is a square matrix in which every element of the 'leading' diagonal is 1, and every other element is zero.)

So, given any square matrix $\mathbf{A}$, our task is to find another matrix (which we will label $\mathbf{A}^{-1}$) such that $\mathbf{A}^{-1}\mathbf{A} = \mathbf{I}$. The process or technique for finding $\mathbf{A}^{-1}$ is called **matrix inversion**. We will demonstrate this by means of an example.

EXAMPLE 19.8

In section 19.8 above we saw that a set of three simultaneous equations could be written as

$$\mathbf{A}\mathbf{x} = \mathbf{b} \qquad \text{where } \mathbf{A} = \begin{bmatrix} 4 & 1 & -5 \\ -2 & 3 & 1 \\ 3 & -1 & 4 \end{bmatrix}$$

To find the inverse, $\mathbf{A}^{-1}$ of $\mathbf{A}$, we proceed as follows.

Step 1. Choose any row or column of $\mathbf{A}$; let's say row 1. Evaluate all the minors, $|\mathbf{M}_{11}|$, $|\mathbf{M}_{12}|$, and $|\mathbf{M}_{13}|$ of this row and arrange them as row 1 of a new matrix, with their appropriate signs according to whether the sum of the subscripts is even or odd. Repeat for rows 2 and 3 of $\mathbf{A}$, thereby forming rows 2 and 3 of the new matrix. (These signed minors are called cofactors, and the new matrix is called $\mathbf{C}$, the matrix of cofactors.) The result is

$$\mathbf{C} = \begin{bmatrix} \begin{vmatrix} 3 & 1 \\ -1 & 4 \end{vmatrix} & -\begin{vmatrix} -2 & 1 \\ 3 & 4 \end{vmatrix} & \begin{vmatrix} -2 & 3 \\ 3 & -1 \end{vmatrix} \\[2mm] -\begin{vmatrix} 1 & -5 \\ -1 & 4 \end{vmatrix} & \begin{vmatrix} 4 & -5 \\ 3 & 4 \end{vmatrix} & -\begin{vmatrix} 4 & 1 \\ 3 & -1 \end{vmatrix} \\[2mm] \begin{vmatrix} 1 & -5 \\ 3 & 1 \end{vmatrix} & -\begin{vmatrix} 4 & -5 \\ -2 & 1 \end{vmatrix} & \begin{vmatrix} 4 & 1 \\ -2 & 3 \end{vmatrix} \end{bmatrix}$$

Note the signs of the various elements (the cofactors). As explained earlier, each is positive or negative according to whether the sum of its subscripts is even or odd.

Step 2. Evaluate all these 2×2 determinants, giving

$$C = \begin{bmatrix} 13 & 11 & -7 \\ 1 & 31 & 7 \\ 16 & 6 & 14 \end{bmatrix}$$

Step 3. Transpose C, to give C'. (This is also called the adjoint matrix of A, written as adj. A.)

$$C' = \begin{bmatrix} 13 & 1 & 16 \\ 11 & 31 & 6 \\ -7 & 7 & 14 \end{bmatrix}$$

Step 4. Evaluate $|A|$. Using row 1 of A, we get

$$|A| = 4(12 + 1) - 1(-8 - 3) + (-5)(2 - 9) = 98$$

Step 5. Then A^{-1} is given by

$$A^{-1} = \frac{1}{|A|}C' = \frac{1}{98}\begin{bmatrix} 13 & 1 & 16 \\ 11 & 31 & 6 \\ -7 & 7 & 14 \end{bmatrix} = \begin{bmatrix} \frac{13}{98} & \frac{1}{98} & \frac{16}{98} \\ \frac{11}{98} & \frac{31}{98} & \frac{6}{98} \\ \frac{-7}{98} & \frac{7}{98} & \frac{14}{98} \end{bmatrix}$$

Note that if the determinant, $|A|$, equals zero, then the inverse, A^{-1}, does not exist.

We can check whether we have made any mistakes in calculating the inverse matrix A^{-1}. If our calculations above are correct, we should find that $AA^{-1} = I$ (the unit matrix).

To check this, let us assume that $AA^{-1} = D$, where D is some matrix. This implies that

$$\begin{bmatrix} 4 & 1 & -5 \\ -2 & 3 & 1 \\ 3 & -1 & 4 \end{bmatrix}\begin{bmatrix} \frac{13}{98} & \frac{1}{98} & \frac{16}{98} \\ \frac{11}{98} & \frac{31}{98} & \frac{6}{98} \\ \frac{-7}{98} & \frac{7}{98} & \frac{14}{98} \end{bmatrix} = \begin{bmatrix} d_{11} & d_{12} & d_{13} \\ d_{21} & d_{22} & d_{23} \\ d_{31} & d_{32} & d_{33} \end{bmatrix}$$

where d_{11}, d_{12}, and so on are the elements of D.

We can multiply out the left-hand side of this matrix equation. From the first row and column $(r \times c)$ we get

$$4\left(\frac{13}{98}\right) + 1\left(\frac{11}{98}\right) + (-5)\left(\frac{-7}{98}\right) = \frac{52 + 11 + 35}{98} = 1$$

This must equal d_{11}, so we have $d_{11} = 1$.

Similarly, from the first row and second column we get

$$4\left(\frac{1}{98}\right) + 1\left(\frac{31}{98}\right) + (-5)\left(\frac{7}{98}\right) = \frac{4 + 31 + (-35)}{98} = 0$$

This must equal d_{12}, so we have $d_{12} = 0$.

If we continue with these calculations (check this for yourself), we will find that $d_{11} = d_{22} = d_{33} = 1$, while all the other elements of D are zero. So D is the unit matrix, I, and we have

$$\begin{bmatrix} 4 & 1 & -5 \\ -2 & 3 & 1 \\ 3 & -1 & 4 \end{bmatrix}\begin{bmatrix} \frac{13}{98} & \frac{1}{98} & \frac{16}{98} \\ \frac{11}{98} & \frac{31}{98} & \frac{6}{98} \\ \frac{-7}{98} & \frac{7}{98} & \frac{14}{98} \end{bmatrix} = \begin{bmatrix} 1 & 0 & 0 \\ 0 & 1 & 0 \\ 0 & 0 & 1 \end{bmatrix}$$

that is, $AA^{-1} = I$.

In the same way, we could show that $A^{-1}A = I$ also. Both pre-multiplication and post-multiplication of a matrix by its inverse produce the unit matrix.

19.11 Using matrix inversion to solve linear simultaneous equations

We will demonstrate this by continuing with the example in section 19.8.

EXAMPLE 19.9

In section 19.8 above, we looked at the simultaneous equations:

$$4x + y - 5z = 8$$

$$-2x + 3y + z = 12$$

$$3x - y + 4z = 5$$

and saw that we could write these three equations in matrix form as

$$Ax = b \tag{19.1}$$

where

$$A = \begin{bmatrix} 4 & 1 & -5 \\ -2 & 3 & 1 \\ 3 & -1 & 4 \end{bmatrix}, \quad x = \begin{bmatrix} x \\ y \\ z \end{bmatrix}, \quad \text{and} \quad b = \begin{bmatrix} 8 \\ 12 \\ 5 \end{bmatrix}$$

If we find A^{-1} such that $A^{-1}A = I$, we can pre-multiply both sides of equation (19.1) by A^{-1} and thereby get

$$A^{-1}Ax = A^{-1}b \tag{19.2}$$

but by definition, $A^{-1}A = I$, so substituting this on the left-hand side of equation (19.1) gives

$$Ix = A^{-1}b$$

But $Ix \equiv x$, for any column vector x (check this for yourself), so we have

$$x = A^{-1}b \tag{19.3}$$

The left-hand side of (19.3) is a column vector of unknowns, x, y, and z. The right-hand side is a column vector of numerical values, which will give us the solution values of x, y, and z. We will now demonstrate this.

For this example, we found A^{-1} in the previous section, as

$$A^{-1} = \begin{bmatrix} \dfrac{13}{98} & \dfrac{1}{98} & \dfrac{16}{98} \\ \dfrac{11}{98} & \dfrac{31}{98} & \dfrac{6}{98} \\ \dfrac{-7}{98} & \dfrac{7}{98} & \dfrac{14}{98} \end{bmatrix}$$

We also have $x = \begin{bmatrix} x \\ y \\ z \end{bmatrix}$, and $b = \begin{bmatrix} 8 \\ 12 \\ 5 \end{bmatrix}$ from above.

Substituting these into equation (19.3), we have

$$\begin{bmatrix} x \\ y \\ z \end{bmatrix} = \begin{bmatrix} \dfrac{13}{98} & \dfrac{1}{98} & \dfrac{16}{98} \\ \dfrac{11}{98} & \dfrac{31}{98} & \dfrac{6}{98} \\ \dfrac{-7}{98} & \dfrac{7}{98} & \dfrac{14}{98} \end{bmatrix} \begin{bmatrix} 8 \\ 12 \\ 5 \end{bmatrix}$$

Multiplying out the right-hand side, this becomes

$$
\begin{bmatrix} x \\ y \\ z \end{bmatrix} = \begin{bmatrix} (\frac{13}{98})8 + (\frac{1}{98})12 + (\frac{16}{98})5 \\ (\frac{11}{98})8 + (\frac{31}{98})12 + (\frac{6}{98})5 \\ (\frac{-7}{98})8 + (\frac{7}{98})12 + (\frac{14}{98})5 \end{bmatrix} = \begin{bmatrix} 2 \\ 5 \\ 1 \end{bmatrix}
$$

Thus our solution of the simultaneous equations is $x = 2, y = 5, z = 1$.

By substituting these values into the original simultaneous equations in section 19.8 above, we can easily check that these solutions for x, y, and z are indeed correct (check this for yourself).

Progress exercise 19.2

1. Given the matrix $\mathbf{A} = \begin{bmatrix} a & b \\ c & d \end{bmatrix}$, by finding

   $\mathbf{AA}^{-1}$, verify that the inverse of $\mathbf{A}$ is

   $$\mathbf{A}^{-1} = \frac{1}{(ad - bc)} \begin{bmatrix} d & -b \\ -c & a \end{bmatrix}$$

2. Find the inverses (if they exist) of the following matrices:

   (a) $\begin{bmatrix} 1 & 2 \\ 3 & 4 \end{bmatrix}$ (b) $\begin{bmatrix} 4 & -7 \\ 1 & 3 \end{bmatrix}$ (c) $\begin{bmatrix} 6 & 4 \\ 3 & 2 \end{bmatrix}$

3. Using the results of the previous question, solve where possible the following pairs of linear simultaneous equations:

   (a) $x_1 + 2x_2 = 3$ (b) $4x - 7y = 1$ (c) $6x + 4y = 5$
   $3x_1 + 4x_2 = 7$ $x + 3y = 5$ $3x + 2y = 2$

4. Given the matrices

   $$\mathbf{B} = \begin{bmatrix} 2 & 0 & -1 \\ 3 & 2 & -3 \\ -1 & -3 & 5 \end{bmatrix} \quad \mathbf{C} = \begin{bmatrix} 2 & 4 & 1 \\ 4 & 3 & 7 \\ 2 & 1 & 3 \end{bmatrix}$$

   (a) Calculate the determinant of each.

   (b) Calculate det($\mathbf{BC}$) and det($\mathbf{B} + \mathbf{C}$).

5. Find the determinant and all minors and cofactors of the following matrix. Find its inverse.

   $$\begin{bmatrix} 1 & 2 & 0 \\ -5 & 4 & 3 \\ -4 & 1 & 2 \end{bmatrix}$$

19.12 Cramer's rule

Cramer's rule is a little trick which provides us with an alternative route to solving a set of linear simultaneous equations by matrix methods. Let's go back to our example in the previous section, which was

$\mathbf{Ax} = \mathbf{b}$

$$A = \begin{bmatrix} 4 & 1 & -5 \\ -2 & 3 & 1 \\ 3 & -1 & 4 \end{bmatrix}, \quad x = \begin{bmatrix} x \\ y \\ z \end{bmatrix}, \quad \text{and} \quad b = \begin{bmatrix} 8 \\ 12 \\ 5 \end{bmatrix}$$

Cramer's rule says that to find the *first* unknown, x, in the vector x of unknowns, we proceed as follows.

Step 1. Take the matrix A and form a new matrix by replacing the *first* column of A with the column vector b. We can call this new matrix A_1. So A_1 is given by

$$A_1 = \begin{bmatrix} 8 & 1 & -5 \\ 12 & 3 & 1 \\ 5 & -1 & 4 \end{bmatrix}$$

Step 2. Calculate the determinants $|A|$ and $|A_1|$. We already have $|A| = 98$ from example 19.8 above. And, using the first row of A_1, we can calculate $|A_1|$ as

$$\boxed{\text{negative sign because this is minor } M_{12} \text{ and } 1 + 2 = 3 \text{ is odd}}$$

$$|A_1| = 8[(3 \times 4) - (-1 \times 1)] - 1[(12 \times 4) - (5 \times 1)] + (-5)[(12 \times -1) - (5 \times 3)] = 196$$

Step 3. The first unknown, x, in the vector x of unknowns, is then given by

$$x = \frac{|A_1|}{|A|} = \frac{196}{98} = 2$$

To find the *second* unknown, y, in the vector x of unknowns, we follow steps 1–3 above, but in step 1 we form a new matrix, A_2, by replacing the *second* column of A with the column vector b. So A_2 is given by

$$A_2 = \begin{bmatrix} 4 & 8 & -5 \\ -2 & 12 & 1 \\ 3 & 5 & 4 \end{bmatrix}$$

and in step 3 we find that

$$y = \frac{|A_2|}{|A|} = \frac{490}{98} = 5$$

Finally, the third variable, z, is found in the same way by forming a new matrix, A_3, by replacing the *third* column of A with the column vector b. The solution for z is then

$$z = \frac{|A_3|}{|A|} = \frac{98}{98} = 1$$

This solution, $x = 2$, $y = 5$, $z = 1$, is of course the same as we obtained in the previous section by matrix inversion.

Thus a set of simultaneous linear equations may be solved either by the technique of matrix inversion explained in section 19.11, or by using Cramer's rule. The choice between the two methods is to some extent a matter of personal preference, though Cramer's rule has the advantage that less tedious computation is necessary if we are only interested in the solution value of one of the variables. However, economy of effort is not an important consideration as in practice we will usually entrust calculation to a computer program such as Excel®, which can both evaluate determinants and invert matrices.

The Online Resource Centre explains how to use Excel® to evaluate determinants and invert matrices. www.oxfordtextbooks.co.uk/orc/renshaw3e/

19.13 A macroeconomic application

As a simple illustration of the use of matrix algebra in macroeconomics, we can start from the generalized macroeconomic model developed in sections 3.17 and 15.14:

$$Y = C + I \text{ (equilibrium condition)} \tag{19.4}$$

$$C = aY + b \text{ (consumption function, a behavioural relationship)} \tag{19.5}$$

$$I = \bar{I} \text{ (investment, assumed exogenous)} \tag{19.6}$$

We have simplified our earlier model by assuming that $\hat{C}$ (planned consumption) always equals actual consumption, C.

To make the model a little more realistic and interesting we will add a government sector, which levies taxes on all income at a rate t (where obviously $0 < t < 1$). So the government's tax revenue, T, is given by

$$T \equiv tY \tag{19.7}$$

Because of taxes, we must distinguish between income before deduction of taxes, Y, and disposable income, Y_d, defined as income net of taxes. So we have

$$Y_d \equiv Y - T \tag{19.8}$$

and consumption now depends on disposable income, so equation (19.5) must be modified to

$$C = aY_d + b \tag{19.5a}$$

Finally, government spending on goods and services, G, constitutes another component of aggregate spending, so equation (19.4) must be modified to incorporate this, becoming

$$Y = C + I + \bar{G} \tag{19.4a}$$

Note that we have put a bar over G to indicate that we are assuming it to be exogenously determined, like $\bar{I}$. (We do not assume $G \equiv T$ as this would mean that the government's budget was balanced, which is rarely the case as governments typically spend more than their tax revenue, bridging the gap by borrowing. Even more rarely they spend less than their tax revenue, using the resulting surplus to pay off debts inherited from the past. Either way there is no mechanical link between G and T.)

Our model is now complete and consists of the five equations: (19.4a), (19.5a), (19.6), (19.7), and (19.8). However, we can reduce this to three equations by using (19.6) to substitute for I in (19.4a), and (19.8) to substitute for Y_d in (19.5a). Our reduced form is then:

$$Y = C + \bar{I} + \bar{G} \tag{19.9}$$

$$C = a(Y - T) + b \tag{19.10}$$

$$T \equiv tY \tag{19.11}$$

If we were simply interested in finding the solution to this set of simultaneous equations, we wouldn't bother with matrix algebra. The obvious route to the solution would be to use (19.11) to substitute for T in (19.10), then use the resulting equation to substitute for C in (19.9). We would then have an equation with only one unknown, Y. However, the object of this exercise is to demonstrate matrix algebra in action. Before bringing matrix algebra to bear, it is convenient to slightly rearrange the three equations above as:

$$Y - C = \bar{I} + \bar{G} \qquad (19.9a)$$

$$-aY + C + aT = b \qquad (19.10a)$$

$$-tY + T = 0 \qquad (19.11a)$$

The reason for doing this is that it puts all the exogenous variables, $\bar{I}$, $\bar{G}$, and b on the right-hand side. (Strictly speaking, b is a parameter but it does no harm to treat it as an exogenous variable.) We can then write matrix equation (19.9a), (19.10a), and (19.11a) in matrix form as

$$\begin{bmatrix} 1 & -1 & 0 \\ -a & 1 & a \\ -t & 0 & 1 \end{bmatrix} \begin{bmatrix} Y \\ C \\ T \end{bmatrix} = \begin{bmatrix} \bar{I} + \bar{G} \\ b \\ 0 \end{bmatrix} \qquad (19.12)$$

(Check for yourself, by multiplying out this matrix expression, that you get back to equations (19.9a) to (19.11a).) We can write matrix equation (19.12) as

$$\mathbf{Ax} = \mathbf{b}$$

where $\mathbf{A} = \begin{bmatrix} 1 & -1 & 0 \\ -a & 1 & a \\ -t & 0 & 1 \end{bmatrix}$; $\mathbf{x} = \begin{bmatrix} Y \\ C \\ T \end{bmatrix}$; and $\mathbf{b} = \begin{bmatrix} \bar{I} + \bar{G} \\ b \\ 0 \end{bmatrix}$

Note that $\mathbf{A}$ is a matrix of parameters (the marginal propensity to consume, a, and the tax rate, t); $\mathbf{x}$ is a vector of unknowns (Y, C, and T); and $\mathbf{b}$ is a vector of exogenous variables ($\bar{I}$, b, and $\bar{G}$). This is a very common set-up in economic modelling.

As in section 19.11 above, we can solve this set of equations by finding $\mathbf{A}^{-1}$, for then we can write

$$\mathbf{x} = \mathbf{A}^{-1}\mathbf{b}$$

To find $\mathbf{A}^{-1}$ we follow the procedure explained in section 19.10 above. The matrix of cofactors is

$$\mathbf{C} = \begin{bmatrix} \begin{vmatrix} 1 & a \\ 0 & 1 \end{vmatrix} & -\begin{vmatrix} -a & a \\ -t & 1 \end{vmatrix} & \begin{vmatrix} -a & 1 \\ -t & 0 \end{vmatrix} \\ -\begin{vmatrix} -1 & 0 \\ 0 & 1 \end{vmatrix} & \begin{vmatrix} 1 & 0 \\ -t & 1 \end{vmatrix} & -\begin{vmatrix} 1 & -1 \\ -t & 0 \end{vmatrix} \\ \begin{vmatrix} -1 & 0 \\ 1 & a \end{vmatrix} & -\begin{vmatrix} 1 & 0 \\ -a & a \end{vmatrix} & \begin{vmatrix} 1 & -1 \\ -a & 1 \end{vmatrix} \end{bmatrix}$$

When we multiply out all the 2×2 determinants, this becomes

$$\mathbf{C} = \begin{bmatrix} 1 & a(1-t) & t \\ 1 & 1 & t \\ -a & -a & 1-a \end{bmatrix} \quad \text{which, transposed, is} \quad \mathbf{C}' = \begin{bmatrix} 1 & 1 & -a \\ a(1-t) & 1 & -a \\ t & t & 1-a \end{bmatrix}$$

Next we find the determinant of $\mathbf{A}$, $|\mathbf{A}|$. Using row 1 of $\mathbf{A}$ we get

$$|\mathbf{A}| = 1\begin{vmatrix} 1 & a \\ 0 & 1 \end{vmatrix} - (-1)\begin{vmatrix} -a & a \\ -t & 1 \end{vmatrix} + 0\begin{vmatrix} -a & 1 \\ -t & 0 \end{vmatrix}$$

$$= 1 - a + at = 1 - a(1-t)$$

Finally, therefore, we have

$$\mathbf{A}^{-1} = \frac{1}{|\mathbf{A}|}\mathbf{C}' = \frac{1}{1-a(1-t)}\begin{bmatrix} 1 & 1 & -a \\ a(1-t) & 1 & -a \\ t & t & 1-a \end{bmatrix}$$

so the solution to our set of simultaneous equations is

$\mathbf{x} = \mathbf{A}^{-1}\mathbf{b}$, that is

$$\begin{bmatrix} Y \\ C \\ T \end{bmatrix} = \frac{1}{1-a(1-t)} \begin{bmatrix} 1 & 1 & -a \\ a(1-t) & 1 & -a \\ t & t & 1-a \end{bmatrix} \begin{bmatrix} \bar{I} + \bar{G} \\ b \\ 0 \end{bmatrix} \qquad (19.13)$$

The attraction of matrix equation (19.13) is, first, that we can quickly extract the solution value of any of the unknowns. For example, by multiplying out the first row and column on the right-hand side, we get the solution for Y as

$$Y = \frac{1}{1-a(1-t)}(\bar{I} + \bar{G} + b)$$

Second, in equation (19.13) we have our solution in a completely general and transparent form, so if an exogenous variable ($\bar{I}$, $\bar{G}$, or b), or a parameter (a or t) changes in value, we can immediately calculate the effect on the unknowns. Finally, and perhaps most importantly, this method of solution can easily accommodate an increase in the number of equations and unknowns. This is vital because real-world macroeconomic models used by government economists and others for analysis and forecasting may well consist of several hundred equations.

19.14 Conclusions

In order to use matrix algebra we have had to make a heavy investment in learning new concepts and manipulative rules. Using matrix algebra may seem an unnecessarily complicated way of solving a set of simultaneous linear equations compared with the simple algebraic techniques that we explored in chapter 3. For only three or so simultaneous equations, this is a fair comment. But for larger sets of equations, such as we encounter in some areas of economics, matrix methods are invaluable, not only because these methods make it easier to find the solution, but also because they help us understand the nature of the solution and to conduct comparative static exercises, for example on the effects of changes in the values of exogenous variables or parameters. In theoretical work, understanding the nature of the solution is usually far more important than merely getting the right numerical answer to a problem. This was illustrated by the macroeconomic application above.

Chapter summary

In sections 19.1–19.7 we introduced the basic concepts, definitions, and manipulative rules of matrix algebra including addition, subtraction, and multiplication of matrices and vectors. In sections 19.8–19.10 we saw that matrix algebra gives us a compact notation for handling possibly large sets of simultaneous linear equations, and showed how to find the determinant and the inverse of a square matrix. In sections 19.11–19.12 we showed how to solve a set of simultaneous linear equations by matrix inversion or by Cramer's rule, and finally in section 19.13 illustrated this by solving a simple macroeconomic model.

Progress exercise 19.3

1. A factory can produce three products, Widgets, Votsits, and Hummets using parts A and B. To produce 1 Widget, it needs 2 units of A and 6 units of B. To produce 1 Votsit, it needs 3 units of A and 1 unit of B. To produce 1 Hummet, 1 unit of A and 2 units of B are required.

(a) Construct a parts requirements matrix *using parts as column headings*.

(b) An order is received for 6 Widgets, 2 Votsits, and 7 Hummets. Would you use a row or a column vector to represent this order?

(c) Use matrix multiplication to compute the total number of parts required to fulfil the order in (b).

(d) Given that the cost of 1 unit of A is €2 and the cost of 1 unit of B is €3, express the costs as a vector and hence find the total cost of the order in (b) *by matrix multiplication.*

2. The equilibrium conditions for two related markets (whisky and brandy) are given by:

$$18P_b - P_w = 87$$

$$-2P_b + 36P_w = 98$$

where P_b and P_w are the prices of brandy and whisky. Express this system of equations in matrix form, and hence find the equilibrium prices.

3. (a) Verify that the inverse of

$$\begin{bmatrix} 1 & 2 & 0 \\ -5 & 4 & 3 \\ -4 & 1 & 2 \end{bmatrix} \quad \text{is} \quad \begin{bmatrix} 5 & -4 & 6 \\ -2 & 2 & -3 \\ 11 & -9 & 14 \end{bmatrix}$$

(b) Hence solve the system of equations:

$$x + 2y = 12$$

$$-5x + 4y + 3z = 13$$

$$-4x + y + 2z = -1$$

4. Use Cramer's rule to solve the system of equations in question 3.

5. Verify that the inverse of

$$\begin{bmatrix} 1 & 2 & 3 \\ -4 & 1 & 6 \\ 2 & 7 & 5 \end{bmatrix} \quad \text{is} \quad -\frac{1}{63}\begin{bmatrix} -37 & 11 & 9 \\ 32 & -1 & -18 \\ -30 & -3 & 9 \end{bmatrix}$$

6. Solve the following system of linear equations: (a) using the inverse matrix from question 5, and (b) using Cramer's rule.

$$x_1 + 2x_2 + 3x_3 = 9$$

$$-4x_1 + x_2 + 6x_3 = -9$$

$$2x_1 + 7x_2 + 5x_3 = 13$$

7. A certain macroeconomic model is defined as follows:

national product/income: $Y = C + I + G^*$

(where government expenditure $G^* > 0$)

consumption: $C = \alpha Y + \beta$

(where $0 < \alpha < 1, \beta > 0$)

investment: $I = \gamma r + \delta$

(where $\gamma < 0$, $\delta > 0$, r = interest rate)

money supply: $M_s^* = k_1 Y + k_2 r$

(where $k_1 > 0$, $k_2 < 0$, $M_s^* > 0$)

An asterisk (*) denotes an exogenous variable.

The parameters are α, β, γ, δ, k_1, and k_2.

(a) Write this system in matrix form $\mathbf{Ax} = \mathbf{b}$, where

$$\mathbf{x} = [Y \quad C \quad I \quad r]^T \text{ and } \mathbf{b} = [G^* \quad \beta \quad \gamma \quad M_s^*]^T$$

(where the superscript T denotes transposition).

(b) Use Cramer's rule to show that

$$r = \frac{M_s^*(1 - \alpha) - k_1(\beta + \delta + G^*)}{k_2(1 - \alpha) + \gamma k_1}$$

(c) Using (b), find an expression for the effect on the interest rate of a rise in government expenditure, and determine whether this effect is positive or negative. (Hint: find the relevant partial derivative.)

Checklist

Be sure to test your understanding of this chapter by attempting the progress exercises (answers are at the end of the book). The Online Resource Centre contains further exercises and materials relevant to this chapter www.oxfordtextbooks.co.uk/orc/renshaw3e/.

In this chapter we have introduced the mathematical technique of matrix algebra. The points you should understand before continuing are:

✔ **Concepts and basic operations.** Understanding the basic concepts and operations: definition of a matrix and a vector; addition/subtraction, multiplication/division of matrices. How to express a set of linear equations in matrix notation.

✔ **Determinants.** Evaluating the determinant of a square matrix.

✔ **Matrix inversion.** Purpose and technique of inverting a 3×3 matrix.

✔ **Solving equations.** Using matrix inversion to solve a set of linear equations.

As with integration in the previous chapter, there are economic applications of matrix algebra that we do not have space for, such as analysing the capital and labour intensities of different industries, and 'environmental audit' in which the contribution of an industry to total pollution is measured. The techniques of this chapter are therefore worth careful study.

Chapter 20
Difference and differential equations

OBJECTIVES

Having completed this chapter you should be able to:

- Recognize and solve a first order linear difference equation with constant coefficients.
- Analyse the solution qualitatively.
- Understand how a first order difference equation may arise in an economics context such as achieving market equilibrium.
- Recognize and solve a first order linear differential equation with constant coefficients.
- Analyse the solution qualitatively.
- Use a differential equation model to analyse dynamic stability of a market.

20.1 Introduction

In part 3 of this book we examined various economic variables that were growing (or in some cases declining) through time, such as money deposited in a bank, GDP, and the price level. We developed two formulae, $y = a(1 + r)^x$ and $y = ae^{rx}$, the first describing a variable growing discretely, in annual jumps; and the second describing a variable growing continuously. In both cases we assumed that the growth rate was constant.

We now want to broaden out this analysis to consider more generally how to analyse economic variables that are varying through time. We may encounter such variables in two ways: first, through observed real-world data for a number of years (GDP, industrial production, unemployment, and so on), in which case the data we obtain are called time-series data (see section 1.13); second, in theoretical economic analysis where we are examining how variables such as the price or output of some product may be expected to vary through time in the course of adjusting to equilibrium, or readjusting following some disturbance. An example is the change in export and import values that follows a change in exchange rate between domestic and foreign currencies, or the change in quantities of beer or cigarettes consumed when their tax rates are increased. This latter analysis is called *economic dynamics* and is the focus of this chapter.

As we saw in part 3, when analysing a variable that is changing through time, we often have the choice of whether to treat the variable as varying discretely or continuously. We will consider both cases in this chapter. In sections 20.2–20.5 we consider models in which variables change discretely with time, for example in annual jumps. Then in sections 20.6–20.9 we consider variables that change continuously with time.

20.2 Difference equations

Before going any further we have to introduce two slight variations on our customary notation in order to fall in line with conventional practice in this area of economics. First, we denote time by t rather than x, the symbol used in part 3 of this book. Second, we will write y_t to denote the value of the variable y in time period t. This time period could have any length, and we shall arbitrarily assume here that time is measured in years. Thus y_0 is the value of y in year 0, y_1 is the value of y in year 1, and so on.

Now we will begin our analysis with an example. Suppose in analysing some problem we come to the view that some economic variable, y, increases through time in annual jumps at a constant rate of, say, 10% per year from an initial level of y_0 in year 0. This means that we will have the following sequence for y:

in year 1:　$y_1 = y_0 + (0.1)y_0 = (1.1)y_0$

in year 2:　$y_2 = y_1 + (0.1)y_1 = (1.1)y_1$

in year 3:　$y_3 = y_2 + (0.1)y_2 = (1.1)y_2$

and in general:

in year t:　$y_t = y_{t-1} + (0.1)y_{t-1} = (1.1)y_{t-1}$

This last expression:

$$y_t = (1.1)y_{t-1} \tag{20.1}$$

is called a **difference equation** because it expresses the value of y in any period, t, as a function of its own value in the previous period, $t - 1$. Because the difference is only one time period, it is called a **first order difference equation**. Because neither y_t nor y_{t-1} is raised to any power other than 1, it is called a **linear difference equation**. Because y_t and y_{t-1} are both multiplied by constants (1 and 1.1 respectively), it is called a first order linear difference equation with *constant coefficients*. Because there is no additive constant on either side of the equation, it is said to be **homogeneous**. Thus equation (20.1) is a first order, linear, homogeneous difference equation with constant coefficients. (Note that 'homogeneous' here has a different meaning from that in chapter 17.)

Given the information in equation (20.1), it would be good if we could find the equation that gives the general relationship between y and t, as this would enable us to find y_t for any value of t. We can find this equation by using equation (20.1) to write out the sequence of values for y as follows:

$y_1 = (1.1)y_0$

$y_2 = (1.1)y_1 = (1.1)(1.1)y_0 = (1.1)^2 y_0$　　(after substituting for y_1 from the line above)

$y_3 = (1.1)y_2 = (1.1)(1.1)^2 y_0 = (1.1)^3 y_0$　　(after substituting for y_2 from the line above)

Hopefully you can see the pattern developing here. We have

$y_1 = (1.1)^1 y_0$　　$y_2 = (1.1)^2 y_0$　　$y_3 = (1.1)^3 y_0$

So, in any unspecified period, t, we will have

$y_t = (1.1)^t y_0$

which is usually written as

$$y_t = y_0(1.1)^t \tag{20.2}$$

Equation (20.2) is the *solution* to the difference equation in equation (20.1); that is, $y_t = (1.1)y_{t-1}$. It is the solution in the sense that we can use it to quickly compute the value of y_t for any given values of t and y_0 instead of having to write out a possibly lengthy sequence.

Let's just check that equation (20.2) is indeed the solution to the difference equation in equation (20.1). Recall that a solution to an equation is anything that, when substituted into that equation, causes it to collapse into an identity. For example, $x = 2$ is the solution to the equation $x + 3 = 5$ because $2 + 3 = 5$ is an identity. So we expect that when we substitute (20.2) into (20.1), the latter equation will collapse into an identity.

From equation (20.2), we have immediately that $y_t = y_0(1.1)^t$, and also that $y_{t-1} = y_0(1.1)^{t-1}$. Substituting these into equation (20.1), we have

$$y_0(1.1)^t = (1.1)y_0(1.1)^{t-1}$$

which is an identity, since the right-hand side equals $y_0(1.1)^t$.

Notice that equation (20.2) gives the solution whatever the initial value, y_0, may be. In that sense, y_0 is an *arbitrary* multiplicative constant, and we normally replace it with the symbol A to remind ourselves of this. We therefore write equation (20.2) as

$$y_t = A(1.1)^t \quad \text{where } A \text{ is arbitrary} \tag{20.3}$$

Equation (20.3) is called the **general solution** to equation (20.1) because it encompasses all possible particular solutions that arise when a specific value is assigned to A.

It is a characteristic of a first order difference equation that its general solution contains one arbitrary constant. The intuition is that a difference equation tells us only about *changes* in the variable from one period to the next, but nothing about the *level* of the variable. The arbitrary constant A allows the level of y_t to take any value without violating the difference equation.

The initial condition

However, suppose we somehow acquire the information that the initial value of the variable, in year $t = 0$, is $y_0 = 100$. This given value for y_0 is called the **initial condition**. We can then immediately substitute this information into equation (20.2), which then becomes

$$y_t = 100(1.1)^t \tag{20.4}$$

Equation (20.4) is a **unique solution** to the given difference equation (equation 20.1). It is a unique solution because it is valid only for $y_0 = 100$; that is, it satisfies the initial condition, because when $t = 0$ equation (20.4) becomes $y_0 = 100(1.1)^0 = 100$.

Generalization

We can generalize from the example above by substituting an unspecified parameter, b, in place of 1.1 in equations (20.1), (20.2), and (20.3). We can then conclude as follows:

RULE 20.1 Linear first order homogeneous difference equations

The linear first order homogeneous difference equation

$$y_t = by_{t-1} \quad \text{(where } b \text{ is a parameter)} \tag{20.1a}$$

has the general solution

$$y_t = A(b)^t \quad \text{(where } A \text{ is arbitrary)} \tag{20.3a}$$

If the initial value of y, y_0, is known, the equation $y_t = by_{t-1}$ has the unique solution

$$y_t = y_0(b)^t \tag{20.4a}$$

(To be precise, the unique solution can be found if *any* specific value for y_t, not just y_0, is known. For example, if we know that $b = 5$ and $y_2 = 50$, we can substitute this information into our general solution, $y_t = A(b)^t$, and get $y_2 = A(5)^2 = 50$, from which $A = 2$.)

Non-homogeneous equations

A non-homogeneous difference equation has the form

$$y_t = by_{t-1} + c \qquad (20.5)$$

where b and c are constants. This differs from equation (20.1a) in having the additive constant c (which, of course, may be either positive or negative). Introducing this constant complicates the method of solution, which is now found in three steps.

Step 1

We drop the additive constant, c, which leaves us with

$$y_t = by_{t-1} \quad \text{(that is, the same as equation (20.1a) above)}$$

We then solve this difference equation. From equation (20.3a) above, we know that its solution is

$$y_t = A(b)^t \quad \text{(where } A \text{ is arbitrary)}$$

This solution is called the **complementary function** (CF).

Step 2

We find *any* solution to the given equation $y_t = by_{t-1} + c$. This turns out to be easier than we might have expected. As noted above, a solution to an equation is any value for the unknown that causes the equation to collapse into an identity. So any values for y_t and y_{t-1} that turn equation (20.5) into an identity will do. A possibility is $y_t = Z$, where Z is some constant. Since Z by assumption is a constant, it is independent of t, and therefore if $y_t = Z$, then $y_{t-1} = Z$ also. Substituting these values into equation (20.5), we get

$$Z = bZ + c$$

where b and c are given constants. This equation is satisfied if

$$Z = \frac{c}{1-b}$$

So $Z = \frac{c}{1-b}$ is a solution, because it satisfies equation (20.5). This solution is called the **particular solution** (PS). (Check for yourself that this is indeed a solution by substituting $y_t = Z = \frac{c}{1-b}$ and $y_{t-1} = Z = \frac{c}{1-b}$ into equation (20.5).)

Step 3

The *general solution* (GS) to the given equation is then given by the sum of the CF (from step 1) and the PS (from step 2). So we have

$$GS = CF + PS; \text{ that is, the GS is given by}$$

$$y_t = A(b)^t + \frac{c}{1-b} \quad \text{(where } A \text{ is arbitrary)} \qquad (20.6)$$

Since we haven't given any explanation of why this is the solution, much less a proof, let's at least check that it *is* the solution. If it is a solution, it turns the difference equation that we started with into an identity. From equation (20.6), for time periods t and $t-1$, we have

$$y_t = A(b)^t + \frac{c}{1-b} \quad \text{and} \quad y_{t-1} = A(b)^{t-1} + \frac{c}{1-b}$$

If we substitute these values into the given difference equation (equation 20.5), we get

$$A(b)^t + \frac{c}{1-b} = b\left[A(b)^{t-1} + \frac{c}{1-b} \right] + c$$

The right-hand side of this equation, after multiplying out the square brackets and simplifying, equals the left-hand side. So we have an identity, which validates our statement that equation (20.6) is a solution to equation (20.5). Moreover, it is the *general* solution because it contains an arbitrary constant, A.

The initial condition

As in the homogeneous case considered earlier, if we are now told, say, that $y_0 = 100$, we can substitute this information into equation (20.6) and thereby solve for the arbitrary constant, A. We then have a unique solution.

With $y_0 = 100$, and $t = 0$, equation (20.6) becomes

$$100 = A(b)^0 + \frac{c}{1-b} = A + \frac{c}{1-b} \quad \text{(since } (b)^0 = 1\text{)}$$

A slight rearrangement of this equation gives

$$A = 100 - \frac{c}{1-b}$$

Generalizing this, for any given value of y_0 we can find A as

$$A = y_0 - \frac{c}{1-b}$$

When this is used to replace A in equation (20.6), that equation becomes

$$y_t = A(b)^t + \frac{c}{1-b} \quad \text{where } A = y_0 - \frac{c}{1-b} \tag{20.7}$$

This is a unique solution, giving y_t as a function of its initial value, y_0, and the parameters b and c.

Before moving on, a minor point should be mentioned. The general solution (equation 20.6) fails if $b = 1$. In that case equation (20.5) becomes $y_t = y_{t-1} + c$. We can find the solution to this by writing out a series of values and noting the pattern that emerges, as we did with the homogeneous case earlier. This is left to you as an exercise; you should quickly find that the pattern is $y_t = y_0 + ct$. Since this is true whatever the value of y_0, we can replace y_0 with an arbitrary constant, A. Thus the general solution is $y_t = A + ct$. When y_0 is specified as an initial condition, the particular solution is $y_t = y_0 + ct$.

20.3 Qualitative analysis

We said earlier that one reason for wanting to find the solution to a difference equation is that we can use it to quickly compute the value of y_t for any given values of t and y_0. More importantly, the solution enables us to quickly see the **time path** of the variable in question: that is, how the variable will increase (or decrease) as time passes. It turns out that there are four distinct types of time path, which we will illustrate with four examples.

EXAMPLE 20.1

Suppose we are given the difference equation

$$y_t = 2y_{t-1} + 1 \quad \text{with initial condition } y_0 = 9 \tag{20.8}$$

Comparing this with our general form:

$$y_t = by_{t-1} + c \tag{20.5}$$

and its solution:

$$y_t = A(b)^t + \frac{c}{1-b} \quad \text{where } A = y_0 - \frac{c}{1-b} \tag{20.7}$$

we see that $b = 2$ and $c = 1$, from which we can quickly calculate that $\frac{c}{1-b} = -1$ and $A = y_0 - \frac{c}{1-b} = 10$. Thus the unique solution to equation (20.8) is

$$y_t = 10(2)^t - 1 \tag{20.9}$$

From equation (20.9) we can calculate the values of y_t for $t = 0$ to $t = 8$ shown in table 20.1 and plotted as a graph in figure 20.1. The graph is a step function because, by assumption, y varies in annual jumps.

The path of y through time is dramatic. After only 8 years, y has grown from 9 to more than 2500. This behaviour is known as **monotonic divergence**. This type of behaviour occurs whenever b in equation (20.7) is greater than 1, for then $(b)^t$ gets bigger and bigger as t increases. In this example, the monotonic divergence is positive (y increases without limit as time passes) because A in equation (20.7) is positive. If the values of y_0, a, and b had been such as to make A negative, the monotonic divergence would have been negative (y decreasing without limit). (Monotonic divergence is also known as monotonic growth or decline.)

Table 20.1 Values of $y_t = 10(2)^t - 1$ for $t = 0, 1, \ldots, 8$.

t	0	1	2	3	4	5	6	7	8
y	9	19	39	79	159	319	639	1279	2559

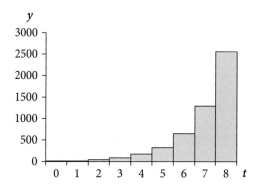

Figure 20.1 Graph of $y_t = 10(2)^t - 1$, showing monotonic divergence in y_t.

EXAMPLE 20.2

Suppose we now change the parameter b in equation (20.8) above from $b = 2$ to $b = 0.5$, with all other parameters unchanged. Our difference equation becomes

$$y_t = 0.5y_{t-1} + 1 \tag{20.10}$$

and its unique solution, found in the same way as in example 20.1, is

$$y_t = 7(0.5)^t + 2 \tag{20.11}$$

From this solution, values of y_t for $t = 0$ to $t = 8$ are shown in table 20.2, and these values are plotted as a graph in figure 20.2. From the graph we see that the path of y through time is one of convergence towards a limiting value. This behaviour is described as **monotonic convergence**. From equation (20.11) we can see that convergence occurs because, as t increases, $(0.5)^t$ gets smaller and smaller, and the term $7(0.5)^t$ eventually becomes

Table 20.2 Values of $y_t = 7(0.5)^t + 2$.

t	0	1	2	3	4	5	6	7	8
y	9	5.5	3.75	2.875	2.437	2.218	2.109	2.054	2.027

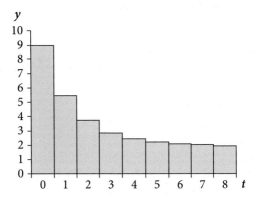

Figure 20.2 Graph of $y_t = 7(0.5)^t + 2$, showing monotonic convergence in y_t.

negligible. In equation (20.11) we are then left with only $y_t = 2$ as the limiting value, to which y_t approaches closer and closer as time passes.

Since any positive fraction becomes smaller and smaller as it is raised to higher and higher powers, we can see that whenever, in equation (20.7), b is a positive fraction, monotonic convergence occurs. If A in equation (20.7) is positive, as in this example, the convergence is from above; that is, y is always greater than its limiting value. If A is negative, convergence is from below.

EXAMPLE 20.3

For our third example we change the parameter b in equation (20.8) above to $b = -0.5$, again with all other parameters unchanged. Our difference equation becomes

$$y_t = -0.5y_{t-1} + 1 \tag{20.12}$$

and its unique solution, found in the same way as in example 20.1, is

$$y_t = 8\tfrac{1}{3}(-0.5)^t + \tfrac{2}{3} \tag{20.13}$$

From this solution, values of y_t for $t = 0$ to $t = 8$ are shown in table 20.3, and these values are plotted as a graph in figure 20.3. In this case, we see from the graph that y_t again approaches a limiting value, but the change in y_t from one period to the next alternates between positive and negative values on the way. This behaviour is described as **oscillatory convergence**. Looking at equation (20.13), we can see that this sign alternation occurs because $(-0.5)^t$ is positive or negative as t is even or odd. At the same time, in *absolute* value $(-0.5)^t$ becomes smaller and smaller as t increases (compare tables 20.2 and 20.3). Therefore y_t approaches the constant term $\tfrac{2}{3}$ in equation (20.13), but oscillates between slightly more and slightly less than $\tfrac{2}{3}$. Oscillatory convergence (also known as cyclical convergence) occurs whenever the parameter b in equation (20.7) is a negative fraction.

Table 20.3 Values of $y_t = 8\tfrac{1}{3}(-0.5)^t + \tfrac{2}{3}$.

t	0	1	2	3	4	5	6	7	8
y	9.000	−3.500	2.750	−0.375	1.187	0.406	0.797	0.601	0.699

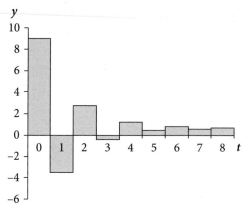

Figure 20.3 Graph of $y_t = 8\frac{1}{3}(-0.5)^t + \frac{2}{3}$, showing oscillatory convergence in y_t.

EXAMPLE 20.4

For our final example we give b in equation (20.8) the value $b = -2$, with all other parameters unchanged. Then our difference equation becomes

$$y_t = -2y_{t-1} + 1 \tag{20.14}$$

and its unique solution, found in the same way as above, is

$$y_t = 8\frac{2}{3}(-2)^t + \frac{1}{3} \tag{20.15}$$

From this solution, values of y_t for $t = 0$ to $t = 8$ are shown in table 20.4, and these values are plotted as a graph in figure 20.4. In this case, because the parameter $b = -2$ is greater than 1 in *absolute* value, then as in example 20.1 above it gets larger and larger in absolute value as it is raised to higher and higher powers. But because $b = -2$ is negative, it oscillates in sign according to whether it is raised to an odd or even power (compare tables 20.1 and 20.4). So as time passes, the change in y oscillates between positive and negative values, while getting larger and larger in absolute value. This behaviour is described as **oscillatory divergence** (also sometimes known as explosive oscillation or cyclical divergence). It occurs whenever b in equation (20.7) is less than -1.

Table 20.4 Values of $y_t = 8\frac{2}{3}(-2)^t + \frac{1}{3}$.

t	0	1	2	3	4	5	6	7	8
y	9	−17	35	−69	139	−277	555	−1109	2219

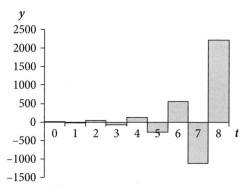

Figure 20.4 Graph of $y_t = 8\frac{2}{3}(-2)^t + \frac{1}{3}$, showing oscillatory divergence in y_t.

We can collect our findings on the solution to non-homogeneous difference equations, and their qualitative behaviour, in the form of a rule:

RULE 20.2 Linear first order non-homogeneous difference equations

The linear first order non-homogeneous difference equation

$$y_t = by_{t-1} + c \quad \text{(see equation (20.5))}$$

has the general solution

$$y_t = A(b)^t + \frac{c}{1-b} \quad \text{where } A = y_0 - \frac{c}{1-b} \quad \text{(see equation (20.7))}$$

There are four distinct types of time path for y, depending on the parameter b:

(1) Monotonic divergence   if $b > 1$ (see example 20.1)

(2) Monotonic convergence   if $0 < b < 1$ (see example 20.2)

(3) Oscillatory convergence   if $-1 < b < 0$ (see example 20.3)

(4) Oscillatory divergence   if $b < -1$ (see example 20.4)

The influence of the other parameters, $\frac{c}{1-b}$ and $A = y_0 - \frac{c}{1-b}$, on the solution is more straightforward. If the path is convergent, then $\frac{c}{1-b}$ is the long-run equilibrium value, to which y approaches closer and closer as time passes. The parameter A is determined by y_0 for given values of b and c. Since A is a multiplicative constant, its absolute magnitude affects only the scale on the vertical axis when we graph the time path of y. When monotonic divergence occurs, the sign of A determines whether y goes off towards $+\infty$ or $-\infty$. When monotonic convergence occurs, the sign of A determines whether y converges from above or below.

Progress exercise 20.1

1. Solve the following homogeneous first order difference equations and comment on the time path of y.

(a) $y_t = 2y_{t-1}$ (b) $y_{t+1} = 0.5y_t$ (c) $y_{t+4} = -0.3y_{t+3}$ (d) $y_t = -2y_{t-1}$

2. Solve the following first order difference equations and comment on the time path of y.

(a) $y_t = 2y_{t-1} - 1$ (b) $y_{t+1} = 0.5y_t + 2$ (c) $y_{t+4} = -0.3y_{t+3} - 3$ (d) $y_t = -2y_{t-1} + 100$

3. Solve the following first order difference equations, using the initial conditions, and comment on the time path of y.

(a) $y_t = 3y_{t-1} + 3; y_0 = 0$ (b) $y_{t+1} = 0.3y_t - 1; y_0 = 1$

(c) $y_{t+4} = -0.4y_{t+3} + c; y_0 = -4$ (d) $y_t = -4y_{t-1} + 100; y_0 = 20$

4. Solve the following first order difference equations and comment on the time path of y.

(a) $y_t = 3y_{t-1} + 3; y_3 = 0$ (b) $y_{t+1} = 0.3y_t - 1; y_1 = 1$

(c) $y_{t+4} = -0.4y_{t+3} + c; y_2 = -4$ (d) $y_t = -y_{t-1} + 100; y_1 = 20$

20.4 The cobweb model of supply and demand

In sections 3.14 and 4.13 we considered some simple supply and demand models of the market for a good. There we implicitly assumed that the quantities supplied and demanded were functions of the *current* price, and that adjustment of both quantity and price to their equilibrium values was instantaneous. Therefore we concluded that there was no need to consider the time dimension of the adjustment process at all.

In this section we will develop a simple but instructive model of supply and demand which explicitly considers the adjustments of price and quantity as processes taking place through time. This type of economic analysis is called **dynamic economics**.

We assume that *within* each time period, price adjusts instantaneously to equate supply and demand. The link between one time period and the next, which is the defining characteristic of dynamic economics, is provided in the following way. Suppose the good in question takes time to produce, and also can't be easily or cheaply stored. Then in deciding how much to produce now, producers have to forecast what the price will be at the future date when the product is sold. We will assume that the production process takes exactly one period, so the goods produced in period $t-1$ will be sold in period t at the price, p_t, effective in period t. Since producers cannot know this price when they make their production decisions in period $t-1$, they must forecast this future price. We will suppose that their forecast is very naïve: they simply assume that the price will not change between the current period, $t-1$, and the next period, t. Therefore they decide how much to produce in period $t-1$ by reference to price p_{t-1}, and these goods are then sold in period t at whatever the price in period t turns out to be. Given these assumptions, we can write the supply function as

$$q_t^S = f(p_{t-1})$$

that is, the quantity supplied in period t is a function of the price in period $t-1$. This is called a **lagged supply function**, as the quantity supplied responds to price only after a lag (of one time period in this case). We can make this idea more specific by thinking, say, of an agricultural product where the time period of our analysis has the same length as the growing season, and the crop cannot be stored.

There is no reason to imagine any comparable lags on the demand side, so we assume

$$q_t^D = g(p_t)$$

that is, the quantity demanded in any period is, in the standard way, a function of the price in that period.

As we only know how to solve *linear* difference equations, to make progress we are forced to assume that the supply and demand functions are both linear:

$$q_t^S = hp_{t-1} - j \quad \text{and} \quad q_t^D = -mp_t + n$$

where h, j, m, and n are positive constants. As usual the equilibrium condition is that the market 'clears': that is, in any period supply and demand are equal. Thus we have

$$q_t^D = q_t^S$$

Combining these three equations in the usual way (see section 3.14) gives us

$$-mp_t + n = hp_{t-1} - j$$

which, after a little rearrangement, becomes

$$p_t = \left(-\frac{h}{m}\right)p_{t-1} + \frac{j+n}{m} \quad \text{(the brackets are not strictly necessary)} \tag{20.16}$$

This conforms to the pattern of equation (20.5) above:

$$y_t = by_{t-1} + c \quad \text{with} \quad b = \left(-\frac{h}{m}\right) \quad \text{and} \quad c = \frac{j+n}{m}$$

The general solution to equation (20.5) was

$$y_t = A(b)^t + \frac{c}{1-b} \quad \text{(where A is arbitrary)}$$

Therefore the solution to equation (20.16) follows this pattern, with $b = (-\frac{h}{m})$ and $c = \frac{j+n}{m}$:

$$p_t = A\left(-\frac{h}{m}\right)^t + \frac{\frac{j+n}{m}}{1-(-\frac{h}{m})} \quad \text{which tidies up to give}$$

$$p_t = A\left(-\frac{h}{m}\right)^t + \frac{j+n}{m+h} \tag{20.17}$$

In this equation, A is arbitrary but if we are given the initial price, p_0 we can solve for A following the pattern of equation (20.7). This gives

$$A = \left[p_0 - \frac{j+n}{(m+h)}\right] \tag{20.18}$$

You are probably thinking that we have plodded through a lot of heavy algebra without much to show for it at the end. This is understandable, but hopefully you will feel it was worthwhile after we have worked through the two following examples. These examples show that we have developed here a tool of economic analysis that throws useful light on some important real-world phenomena.

EXAMPLE 20.5

In this example we assume that the parameters of the supply and demand functions are $h = 5, j = 24, m = 8$, and $n = 80$. So the equations are

$$q_t^S = 5p_{t-1} - 24 \quad \text{and} \quad q_t^D = -8p_t + 80$$

Equation (20.16) above therefore becomes

$$p_t = \left(-\frac{h}{m}\right)p_{t-1} + \frac{j+n}{m} = \left(-\frac{5}{8}\right)p_{t-1} + 13 \tag{20.19}$$

and the general solution to this difference equation, from equation (20.17) above, is

$$p_t = A\left(-\frac{h}{m}\right)^t + \frac{j+n}{m+h} = A\left(-\frac{5}{8}\right)^t + 8 \tag{20.20}$$

Assume now that we know that the initial price is $p_0 = 9$. Then we can solve for A from equation (20.18):

$$A = \left[p_0 - \frac{j+n}{(m+h)}\right] = [9 - 8] = 1 \tag{20.21}$$

So the unique solution is

$$p_t = 1\left(-\frac{5}{8}\right)^t + 8 \tag{20.22}$$

The key point now is that by looking at this equation we can see what the path of p through time will be. This case is like example 20.3 in section 20.3 above. Because $-\frac{5}{8}$ is negative, $(-\frac{5}{8})^t$ alternates between positive and negative values as t is even or odd. Because $\frac{-5}{8}$ is a fraction, it becomes smaller and smaller in absolute value as t increases. Thus in equation (20.22), as time passes p_t approaches a limiting value of $p_t = 8$, and we observe oscillatory convergence of the price towards this limiting value. The values of p for the first 8 time periods are given in table 20.5. It is clear that p approaches its limiting value of 8 quite quickly.

Table 20.5 also gives the path of q through time, found by substituting our solution for p (equation 20.22) into either the demand function or the supply function. To show how this works out, the demand function is $q_t^D = -8p_t + 80$ and if we substitute for p_t from equation (20.22) we get

Table 20.5 Values of $p_t = 1(-\frac{5}{8})^t + 8$.

t	0	1	2	3	4	5	6	7	8
p_t	9.00	7.38	8.39	7.76	8.15	7.90	8.06	7.96	8.02
q_t		21.00	12.88	17.95	14.78	16.76	15.52	16.30	15.81

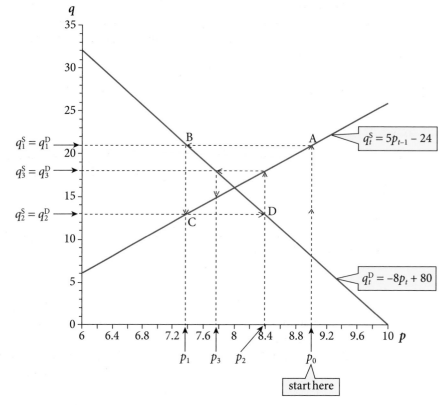

Figure 20.5 Cobweb diagram for $p_t = 1(-\frac{5}{8})^t + 8$, showing oscillatory convergence of p_t.

$$q_t^D = -8\left[\left(-\frac{5}{8}\right)^t + 8\right] + 80 = -8\left(-\frac{5}{8}\right)^t + 16$$

This was the equation used to calculate the values of q in table 20.5.

We can also solve for q_t by substituting equation (20.22) into the supply function $q_t^S = 5p_{t-1} - 24$, but in this case we must remember to lag p by one period. Then we get

$$q_t^S = 5\left[\left(-\frac{5}{8}\right)^{t-1} + 8\right] - 24 = 5\left(-\frac{5}{8}\right)^{t-1} + 16 = 5\left(-\frac{5}{8}\right)^t\left(-\frac{8}{5}\right) + 16 = -8\left(-\frac{5}{8}\right)^t + 16$$

In figure 20.5 we show the paths of p and q in an interesting and informative way. With q on the vertical and p on the horizontal axis, we first graph the supply and demand functions $q_t^S = 5p_{t-1} - 24$ and $q_t^D = -8p_t + 80$. We can then trace the paths of price and quantity through time on the graph, starting from the given initial price, $p_0 = 9$. Given that the supply function has a one-period lag, from $p_0 = 9$ we can read off the supply in period 1 at point A on the supply curve. From table 20.5 we know that this is $q_1^s = 21$. In order for the market to clear, we need $q_1^D = 21$ also, which means that buyers must be at point B on their demand curve with a price of $p_1 = 7.375$. Thus the general pattern is that in each period, supply is pre-determined by the price in the previous period, and the current period's price must adjust to make demand equal to this pre-determined supply.

Given the price $p_1 = 7.375$, we can read off the supply in period 2 at point C, which is $q_1^S = 12.875$. In order for demand to equal 12.875 also, buyers must be at D on their demand curve, which requires a price of $p_2 = 8.39$. This process continues from one period to the next as we follow the arrows in figure 20.5. Because the path is convergent, the arrows get closer and closer to the intersection of the supply and demand curves, this intersection being the static equilibrium of the type we examined in chapters 3 and 4. From our solution for p (equation 20.22) we can see that, as time passes, the price approaches closer and closer to its static equilibrium value, $p = 8$. From our equations above for the path of quantity, $q_t^S = q_t^D = -8(-\frac{5}{8})^t + 16$, we can also see that the static equilibrium quantity is $q = 16$. Of course, these values for p and q are never actually reached. It is because figure 20.5 resembles a spider's web that the model we are using here, with a lagged supply function, is known as the **cobweb model** of market dynamics.

EXAMPLE 20.6

This example is identical to example 20.5 except for one change: we give the supply function a slope of 10 instead of 5. So the supply and demand functions are now

$$q_t^S = 10p_{t-1} - 24 \quad \text{and} \quad q_t^D = -8p_t + 80$$

We also assume now an initial price of $p_0 = 7$ instead of $p_0 = 9$, but this is purely for convenience in calculations and changes nothing of significance. Following the steps of example 20.5, we find the unique solution as

$$p_t = 1.222\left(-\frac{10}{8}\right)^t + 5.778 \tag{20.23}$$

Looking at this solution, we can see what will happen to p as time passes. This case is like example 20.4 in section 20.3 above. Because $-\frac{10}{8}$ is negative, $(-\frac{10}{8})^t$ alternates between positive and negative values as t is even or odd, resulting in oscillations. Because $-\frac{10}{8}$ is greater than 1 in absolute value, $(-\frac{10}{8})^t$ becomes larger and larger in absolute value as t increases, so these oscillations are divergent.

This is confirmed when we use equation (20.23) to calculate the time paths of p and q, as explained in the previous example. These values for the first six time periods are shown in table 20.6. It is clear from the tables that we have divergent oscillations in p and q. This would become more apparent if we continued for more than six time periods, but there seems little point in this, as in period 6 quantity, according to the model, becomes negative (see the last column of table 20.6). It is difficult to know how to interpret negative supply in terms of the behaviour of economic agents, but we can safely presume that the model ceases to be valid at this point.

It's interesting to look at the cobweb diagram in this example (see figure 20.6). Starting from $p_0 = 7$ we can read off the supply in period 1 at point A on the supply curve. From table 20.6 we see that this supply is $q_1^S = 46$. To clear the market we require $q_1^D = 46$ also, so buyers must be at point B on their demand curve, which requires a price of $p_1 = 4.25$. This price in period 1 then leads us to the supply in period 2, and so on. The key point is that in this example, as we follow the arrows, the oscillations of both p and q become larger and larger until in period 6 the price of the previous period, p_5, which determines supply, is so low that supply is negative, as noted above.

Table 20.6 Values of $p_t = 1.222(-\frac{10}{8})^t + 5.778$.

t	0	1	2	3	4	5	6
p_t	7.00	4.25	7.69	3.39	8.76	2.05	10.44
q_t		46.00	18.50	52.88	9.91	63.62	−3.52

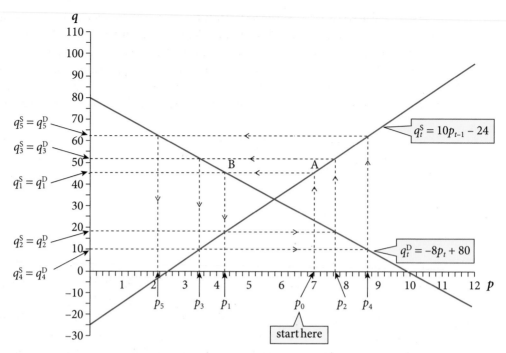

Figure 20.6 Cobweb diagram for $p_t = 1.222(-\frac{10}{8})^t + 5.778$, showing oscillatory divergence in p_t.

20.5 Conclusions on the cobweb model

Examples 20.5 and 20.6 both have the supply and demand functions

$$q_t^S = hp_{t-1} - j \quad \text{and} \quad q_t^D = -mp_t + n \tag{20.24}$$

and both have as their solution

$$p_t = A\left(-\frac{h}{m}\right)^t + \frac{j+n}{m+h} \tag{20.17}$$

In both examples the time path of price (and therefore quantity) is oscillatory. This oscillation occurs because $-\frac{h}{m}$ is negative, and therefore $(-\frac{h}{m})^t$ is alternately positive or negative as t is even or odd. This leads to an important economic conclusion. From equations (20.24) we see that h is the slope of the supply function and $-m$ is the slope of the demand function. As supply curves normally slope upwards, we expect that $h > 0$, and similarly as demand curves normally slope downwards, we expect $-m < 0$ (or equivalently, $m > 0$). Thus our expectation is that in normal markets $-\frac{h}{m}$ will be negative. This will produce oscillations in this model of market equilibrium which incorporates a lag in supply.

The only difference between the two examples is in the value of h, the slope of the supply curve, which is $h = 5$ in example 20.5 and $h = 10$ in example 20.6.

In example 20.5 we have $-\frac{h}{m} = -\frac{5}{8}$, which is a fraction. In the solution (equation 20.17) this means that $(-\frac{h}{m})^t$ becomes smaller and smaller in absolute value as t increases, resulting in convergence to an equilibrium value (accompanied by oscillation because $-\frac{h}{m}$ is negative, as just discussed). Consequently the cobweb in figure 20.5 is convergent.

Thus convergence requires $-\frac{h}{m}$ to be a fraction. For $-\frac{h}{m}$ to be a fraction requires $h < m$. Here h is the slope of the supply function, while m is the absolute slope of the demand function (that is, ignoring its sign). So we conclude that convergence requires the slope of the supply curve to be less than the absolute slope of the demand curve.

In example 20.6 we increased the slope of the supply curve from 5 to 10. Now h, the slope of the supply curve, is greater than m, the absolute slope of the demand curve. Thus we have

$-\frac{h}{m} = -\frac{10}{8}$, which is greater than 1 in absolute value. In the solution this now means that $(-\frac{h}{m})^t$ becomes larger and larger in absolute value as t increases, resulting in an explosion in p (again accompanied by oscillation because $-\frac{h}{m}$ is negative). The cobweb in figure 20.6 is divergent.

Which of these two outcomes—convergent oscillation or explosive oscillation—is the more likely? We cannot answer this, since we have no basis for saying whether in any particular market the slope of the supply curve is likely to be greater or less than the absolute slope of the demand curve.

Finally, for the sake of completeness we should also mention the boundary case that separates convergent and explosive oscillation. This boundary case occurs when $h = m$; that is, when the absolute slopes of the supply and demand curves are equal. When $h = m$, equation (20.17) becomes

$$p_t = A(-1)^t + \frac{j+n}{m+h}$$

Since $(-1)^t$ equals -1 when t is odd and equals $+1$ when t is even, we see that p_t will oscillate between $A + (j+n)/(m+h)$ and $-A + (j+n)/(m+h)$ as t switches between even and odd. Thus we have a regular oscillation; that is, one that neither converges nor diverges.

If you have found all the algebra in this section a little overwhelming, take heart. It's actually quite easy to confirm diagrammatically our conclusion that convergent or explosive oscillation occurs according to whether the slope of the supply curve is less or greater (respectively) than the absolute slope of the demand curve. If you simply use a ruler and pencil (no algebra is necessary!) to draw a supply and demand diagram, then pick an arbitrary initial price and trace out the resulting cobweb, you will quickly find after a few examples that the only thing that matters for convergence or divergence is the slope of the supply curve relative to the absolute slope of the demand curve. Note though that our conclusions concerning relative slopes assume that p rather than q is on the horizontal axis; that is, following the convention of mathematics rather than economics. When we talk of slope, we therefore mean $\frac{dq}{dp}$, not $\frac{dp}{dq}$.

Progress exercise 20.2

1. The following set of equations describes the behaviour of buyers and sellers in the market for a particular commodity:

$$Q_t^D = 120 - 0.5p_t$$
$$Q_t^S = -30 + 0.3P_t$$
$$P_t = P_{t-1} - \alpha(Q_{t-1}^S - Q_{t-1}^D)$$

where Q_t^D is quantity demanded, Q_t^S is quantity supplied, P_t is price, and α is a positive parameter.

   (a) Solve for the long-run equilibrium price.

   (b) Solve the first order difference equation in the price and find the solution if P_0 is 200.

   (c) For what values of α will the path of price through time exhibit

      (i) no oscillations,

      (ii) damped (convergent) oscillations, or

      (iii) explosive (divergent) oscillations?

2. Consider the following macroeconomic model:

$$C_t = 70 + \tfrac{2}{3}Y_{t-1} \quad \text{(consumption equation, with } C_t = \text{planned and actual consumption)}$$

$$Y_t = C_t + I_t \quad \text{(income/expenditure identity)}$$

(a) Suppose that I_t is kept constant at the value I_0. Obtain a first order difference equation for Y_t and solve it. Show that the system will reach a long-term equilibrium in which Y_t and C_t are also constant and find their values when $I_0 = 20$.

(b) Suppose that I_t has been fixed at 20 for many years, and then suddenly rises to a new fixed value of 30. Find the new equilibrium value for Y_t. How many years will it take for Y_t to complete two-thirds of the rise to its new equilibrium value?

3. In the Harrod–Domar macroeconomic growth model, it is assumed that savings S_t in period t are some proportion, s, of income, Y_t and investment I_t in period t is some proportion v of the change in income since the last period, where $0 < s < 1$ and (usually) $v > 1$. Thus we have

$$S_t = sY_t$$
$$I_t = v(Y_t - Y_{t-1})$$

For equilibrium, we require $S_t = I_t$.

(a) Obtain a first order difference equation for Y_t and find its solution given that $Y = Y_0$ when $t = 0$.

(b) According to this model, what happens in the long run?

20.6 Differential equations

We now consider situations in which the variable that we are interested in, y, varies continuously through time rather than discretely in annual jumps. Thus we now have $y = f(t)$ as a smooth and continuous function rather than the step functions analysed so far in this chapter. Consequently, the derivative $\frac{dy}{dt}$ exists, and the nature of our problem is that we have an equation involving both y and $\frac{dy}{dt}$ and wish to find the original function $y = f(t)$ which gave rise to this derivative. This immediately brings to mind the integration that we did in chapter 18, and indeed integration is precisely the technique we use to solve **differential equations**.

Suppose for example we are given

$$\frac{dy}{dt} = 3y \tag{20.25}$$

This is called a **first order differential equation** because it involves only the first derivative and not higher derivatives such as $\frac{d^2y}{dt^2}$. It is a **linear differential equation** because neither y nor its derivative is raised to any power other than 1. It has *constant coefficients* because neither y nor its derivative is multiplied by anything except a constant, and it is **homogeneous** because there is no additive constant. So equation (20.25) is a first order, linear, homogeneous differential equation with constant coefficients.

How are we to solve equation (20.25)? An immediate problem is that the right-hand side contains the dependent variable, y, rather than the independent variable, t, that we are accustomed to seeing there. We can deal with this to some extent, as follows. If we assume that the solution is something of the form $y = f(t)$, we can substitute this into equation (20.25) and get

$$\frac{dy}{dt} = 3f(t) \tag{20.25a}$$

But this doesn't take us very far as we don't know what $f(t)$ is because it's the solution. However, you may already have spotted something familiar about equation (20.25a). It says that the function $f(t)$ and its derivative $\frac{dy}{dt}$ are *linearly* related. In chapters 12 and 13 we met a function with exactly that property—the exponential function

$y = Ae^{rt}$ (where A and r are constants). For then the derivative is

$$\frac{dy}{dt} = rAe^{rt} = ry$$

which is precisely equation (20.25) with $r = 3$. So we can conclude that the solution to equation (20.25) is

$y = Ae^{3t}$ (where A is an arbitrary constant) (20.26)

This is the *general solution* to equation (20.25).

Although it was only by luck that we managed to solve equation (20.25), it is a general rule that the solution to any linear first order differential equation must be a natural exponential function with a linear exponent because this is the only function possessing the property that the function and its derivative are related by a multiplicative constant.

The initial condition

If we are given an initial condition, such as $y = 10$ when $t = 0$, we can substitute this into our general solution (20.26) and get $10 = Ae^0$, from which $A = 10$ (since $e^0 = 1$). Substituting $A = 10$ into (20.26) we then have

$y = 10e^{3t}$

as a *unique* solution that satisfies both equation (20.25) and the initial condition.

We can state our conclusions up to this point as a rule:

RULE 20.3 Linear first order homogeneous differential equations

The linear first order homogeneous differential equation

$$\frac{dy}{dt} = by$$ (where b is a parameter) (20.27)

has the *general solution*

$y = Ae^{bt}$ (where A is arbitrary) (20.28)

If the initial value of y, when $t = 0$, is known, we can find A. For example, if we know that $y = y(0)$ when $t = 0$, then on substitution into (20.28) we have

$y(0) = Ae^0 = A$ (since $e^0 = 1$)

We can then substitute $y(0)$ in place of A in (20.28) and obtain the *unique* solution, which satisfies both (20.27) and the initial condition

$y = [y(0)]e^{bt}$

(To be precise, the unique solution can be found if we know *any* specific value for y, not just its value when $t = 0$. For example, if we know that $y = y(2)$ when $t = 2$, we can substitute this information into (20.28) and get $y(2) = Ae^{2b}$, from which $A = \frac{y(2)}{e^{2b}}$.)

The non-homogeneous case

This is the same as equation (20.27), but with an additive constant. Thus the general form is

$$\frac{dy}{dt} = by + c$$ (where b and c are constants) (20.29)

As we found with the difference equations in section 20.2, this additive constant dictates a more formal approach to the solution, which is now in three steps.

Step 1

We drop the additive constant, leaving us with $\frac{dy}{dt} = by$. We know from equation (20.27) that this has the general solution $y = Ae^{bt}$, where A is arbitrary. This solution is called the *complementary function* (CF).

Step 2

We find any solution to the given equation. Lacking any other ideas, we naïvely try the solution $y = Z$, where Z is some constant. If this is a solution, then it implies $\frac{dy}{dt} = 0$ (since the derivative of a constant is zero). To test whether we have found a solution, we substitute $y = Z$ and $\frac{dy}{dt} = 0$ into equation (20.29). If $y = Z$ is a solution, equation (20.29) should become an identity. On substitution, we get

$$0 = bZ + c \quad \text{which is an identity if} \quad Z = -\frac{c}{b}$$

So we have $Z = -\frac{c}{b}$ as a *particular solution* (PS) to the given equation.

Step 3

The *general solution* (GS) to equation (20.29) is then given by the sum of the CF and the PS: that is, GS = CF + PS. Thus the general solution is

$$y = Ae^{bt} - \frac{c}{b} \quad \text{where } A \text{ is arbitrary} \tag{20.30}$$

Let's check that this is correct by substituting into equation (20.29) to see whether we get an identity. Differentiating equation (20.30) gives

$$\frac{dy}{dt} = bAe^{bt} \tag{20.31}$$

Substituting equations (20.30) and (20.31) into equation (20.29), we get

$$bAe^{bt} = b\left(Ae^{bt} - \frac{c}{b} \right) + c$$

Since the right-hand side simplifies to bAe^{bt}, this is an identity.

The initial condition

As with a homogeneous equation, if we are given an initial condition, say $y = 50$ when $t = 0$, we can substitute this into the general solution (20.30) and get $50 = Ae^0 - \frac{c}{b} = A - \frac{c}{b}$ (since $e^0 = 1$). Substituting $A = 50 + \frac{c}{b}$ into (20.30) we obtain

$$y = \left(50 + \frac{c}{b} \right)e^{bt} - \frac{c}{b}$$

as a *unique* solution that satisfies both equation (20.29) and the initial condition.

RULE 20.4 Linear first order non-homogeneous differential equations

The linear first order non-homogeneous differential equation

$$\frac{dy}{dt} = by + c \quad \text{(where } b \text{ and } c \text{ are constants)} \tag{20.29}$$

has the *general* solution

$$y = Ae^{bt} - \frac{c}{b} \quad \text{where } A \text{ is arbitrary} \tag{20.30}$$

As with a homogeneous equation, if the initial value of y, when $t = 0$, is known, we can find A. For example, if we know that $y = y(0)$ when $t = 0$, then on substitution into (20.30) we find $A = y(0) + \frac{c}{b}$ (see rule 20.3 for details). So (20.29) then has the *unique* solution

$$y = \left[y(0) + \frac{c}{b} \right] e^{bt} - \frac{c}{b} \tag{20.32}$$

A *unique* solution satisfies both equation (20.29) and the initial condition.

20.7 Qualitative analysis

The main reason for wanting to solve a differential equation is in order to analyse the path of y through time. In the case of $\frac{dy}{dt} = by + c$ we can do this analysis now that we know that the solution is

$$y = Ae^{bt} - \frac{c}{b} \quad \text{where} \quad A = y(0) + \frac{c}{b}$$

There are four possible paths of y through time, depending on whether b is positive or negative, and whether A is positive or negative:

Case 1. Assume $b > 0$. Then e^{bt} has an initial value of 1 (when $t = 0$) and increases towards $+\infty$ as t increases. Then there are two sub-cases:

1(a) Assume $A > 0$, which is true if $y(0) > -\frac{c}{b}$. (Recall that $y(0)$ is the value of y when $t = 0$, and in figure 20.7 is the vertical intercept of y.) With A positive, this means that Ae^{bt} also increases towards $+\infty$ as t increases, and so therefore does $y = Ae^{bt} - \frac{c}{b}$. The $-\frac{c}{b}$ term becomes relatively insignificant. The path of y in this case is shown in figure 20.7(a). We can label this path 'monotonic increasing'.

1(b) Assume $A < 0$, which is true if $y(0) < -\frac{c}{b}$. With A negative, this means that Ae^{bt} decreases towards $-\infty$ as t increases, and so therefore does $y = Ae^{bt} - \frac{c}{b}$. Again, the constant term, $-\frac{c}{b}$, becomes relatively insignificant. The path of y in this case is shown in figure 20.7(b). We can label this path 'monotonic decreasing'.

Case 2. Assume $b < 0$. Then e^{bt} has an initial value of 1 (when $t = 0$) and decreases continuously as t increases, and approaches a limiting value of zero. Then there are two sub-cases:

2(a) Assume $A > 0$, which is true if $y(0) > -\frac{c}{b}$. With A positive, this means that Ae^{bt} has an initial value of A (when $t = 0$), which by assumption is positive, and then decreases continuously as t increases, approaching a limiting value of zero. Therefore $y = Ae^{bt} - \frac{c}{b}$ approaches $-\frac{c}{b}$. The path of y in this case is shown in figure 20.7(c). We can label this path 'convergent from above'.

2(b) Assume $A < 0$, which is true if $y(0) < -\frac{c}{b}$. Then Ae^{bt} has an initial value of A (when $t = 0$), which by assumption is negative, and increases continuously as t increases, approaching

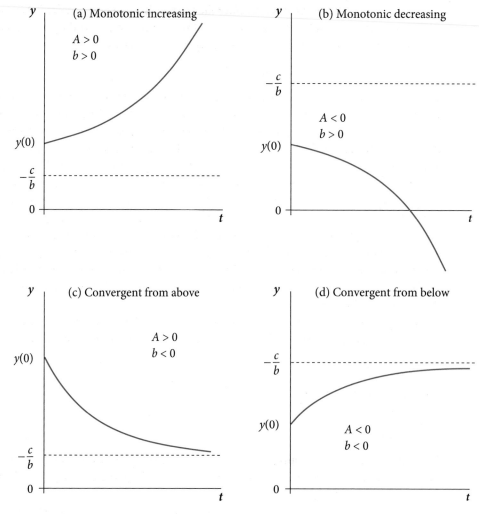

Figure 20.7 Possible paths of $y = Ae^{bt} - \frac{c}{b}$ through time.

a limiting value of zero. Therefore $y = Ae^{bt} - \frac{c}{b}$ approaches the constant term, $-\frac{c}{b}$ (see figure 20.7(d)). We can label this path 'convergent from below'.

Thus when y is described by a first order linear differential equation, we see only two possible time paths for y: either explosive growth or decline, when $b > 0$; or convergence towards a limiting value, when $b < 0$. This is rather unexciting when compared with the four different types of path that resulted from the first order linear difference equation, where there was the possibility not only of growth or convergence, but also of oscillatory behaviour in both cases. (Note that in drawing figure 20.7 we have assumed that both $y(0)$ and $-\frac{c}{b}$ are positive, but these are not necessary assumptions as it is only their relative size that matters.)

20.8 Dynamic stability of a market

In this section we examine a simple but very important economic application of a first order differential equation. In sections 3.14 and 4.13 we analysed market equilibrium as the solution to a set of simultaneous equations, without giving much thought to the question of whether equilibrium would necessarily be achieved, and if so, how. Addressing these questions systematically is a topic in dynamic economics, and is similar to the cobweb model considered in section 20.4. However, whereas the cobweb model arises from the rather special assumption of a lagged supply function, here we consider the broader issue of how equilibrium is achieved when supply and demand are both functions of the current price. Accordingly, we assume that the supply and demand functions are both linear and are given by

$$q^S = hp - j \quad \text{and} \quad q^D = -mp + n \tag{20.33}$$

where h, j, m, and n are parameters. At this stage, we don't want to make any assumptions about the signs or magnitude of these parameters, as to do so would restrict the possible outcomes of our enquiry into the dynamics of the model. However, we might expect, as the 'normal' case, that the supply function is positively sloped, so $h > 0$; and that there is some minimum price at which supply falls to zero, so $j > 0$. Similarly in the 'normal' case we expect the demand function to be negatively sloped, with a positive intercept on the q-axis, implying $m > 0$ and $n > 0$.

Note that the variables p and q are assumed to be continuous functions of time, so the equations (20.33) describe quantities supplied and demanded, and the associated price, which exist at a *moment* of time, rather than *during a period* of time as was the case when we were working with difference equations earlier in this chapter. Consequently it now does not make sense to use subscripts to denote time periods. Nevertheless this is no longer a static or timeless market model of the kind we examined in chapters 3 and 4, because the clock is ticking all the time and as it ticks, quantities and prices change, as we shall see.

Given the supply and demand functions, we can find the equilibrium values, $p = p^*$ and $q = q^*$, by imposing the market-clearing condition that $q^S = q^D$. This implies that

$$hp - j = -mp + n$$

Solving this equation for p gives

$$p^* = \frac{n + j}{m + h} \tag{20.34}$$

From this we can find q^* by substituting p^* into either the supply function or the demand function, in the usual way. (We will assume that the values of the parameters n, j, m, and h are such as to make both p^* and q^* positive.)

However, whereas in chapters 3 and 4 we implicitly assumed that the market cleared instantaneously, we now assume that price adjustment is not instantaneous. Instead, starting from some arbitrary initial price, we assume that the actual market price p adjusts gradually over a period of time. During this period the market is in disequilibrium. To model the behaviour of p we make an assumption that is frequently used over a wide range of economic analysis. We assume that the process of price adjustment through time is governed by the equation

$$\frac{dp}{dt} = k(q^D - q^S) \quad \text{(where } k \text{ is a positive parameter)} \tag{20.35}$$

This is known as a **price adjustment equation**.

Let us think carefully about what this equation is saying. Suppose that, at the currently ruling price, there is excess demand (that is, $q^D > q^S$). Since the parameter k is assumed positive, this means that $\frac{dp}{dt}$ is positive: that is, the price rises through time. Conversely, if there is excess supply ($q^D < q^S$), then $\frac{dp}{dt}$ is negative and the price falls through time. But equation (20.35) does more than merely tell us whether the price is rising or falling. It also says that when the excess demand (supply) is large, then the price rises (falls) more rapidly than when the excess demand (supply) is small.

Equation (20.35) seems a sensible reflection of the likely behaviour of individual market participants. For a given value of k, if at the current price demand is much greater than supply, there are many frustrated buyers, all of whom by definition are willing to pay a higher price. Consequently, we would expect the market price to rise rapidly. Conversely, when demand is only slightly greater than supply, there are only a few frustrated buyers, and for the same value of k we would expect the market price to rise only slowly. The same reasoning explains the speed with which price falls when there is excess supply. (Whether k itself is large or small is a further interesting question that we won't pursue here.)

If we now substitute the supply and demand functions (equations 20.33) into the price adjustment equation (equation 20.35), it becomes

$$\frac{dp}{dt} = k[(-mp + n) - (hp - j)]$$

which can be rearranged as

$$\frac{dp}{dt} = -k(h + m)p + k(n + j) \tag{20.36}$$

Comparing this with equation (20.29) above, we see that this is a first order linear differential equation of the form

$$\frac{dp}{dt} = bp + c \quad \text{where} \quad b = -k(h + m) \quad \text{and} \quad c = k(n + j)$$

From equation (20.30) above, we therefore know that its solution is

$$p = Ae^{bt} - \frac{c}{b} \text{ where } A = p(0) + \frac{c}{b} \tag{20.37}$$

Since $b = -k(h + m)$ and $c = k(n + j)$, we have $\frac{c}{b} = -\frac{n+j}{h+m} = -p^*$ (see equation (20.34)). Substituting these values for b, c, and $\frac{c}{b}$ in equation (20.37), we get

$$p = (p(0) - p^*)e^{-k(h+m)t} + p^* \tag{20.37a}$$

This is the solution to our dynamic supply and demand model. It gives us the path of p through time.

In analysing equation (20.37a), there are four cases to be considered, depending on the signs of $-k(h + m)$ and $(p(0) - p^*)$:

Case 1. In case 1 we assume $h + m < 0$. Since we have assumed $k > 0$ (see equation 20.35), this implies that $-k(h + m)$ is positive, meaning that $e^{-k(h+m)t}$ will increase monotonically through time. There are then two sub-cases:

1(a) Assume $(p(0) - p^*) > 0$. This means that the initial price $p(0)$ is greater than the equilibrium price, p^*, and therefore that the actual price needs to fall through time to achieve equilibrium. However, since $e^{-k(h+m)t}$ increases through time, the assumption that $(p(0) - p^*)$ is positive means that their product, $(p(0) - p^*)e^{-k(h+m)t}$, will increase too. Therefore in equation (20.37a) we see that this means that p *increases* through time, which is a movement *away* from equilibrium, since we know that the actual price needs to *fall* through time to achieve equilibrium. This case is shown as case 1(a) in figure 20.8.

1(b) Assume that $(p(0) - p^*)$ is negative. This means that the initial price $p(0)$ is less than the equilibrium price, p^*, and therefore that the actual price needs to rise through time to achieve equilibrium. However, since $e^{-k(h+m)t}$ increases through time, the assumption that $(p(0) - p^*)$ is negative means that their product, $(p(0) - p^*)e^{-k(h+m)t}$, will decrease through time, a movement *away* from equilibrium. This is case 1(b) in figure 20.8.

Thus in case 1 the market is **dynamically unstable** because in case 1(a) the price rises when it needs to fall to achieve equilibrium, and in case 1(b) the price falls when it needs to rise to achieve equilibrium. The instability occurs because $h + m$ is negative.

Case 2. In case 2 we assume $h + m > 0$. Since we have $k > 0$, this implies that $-k(h + m)$ is negative, meaning that $e^{-k(h+m)t}$ will decrease through time and approaches zero. There are then two sub-cases:

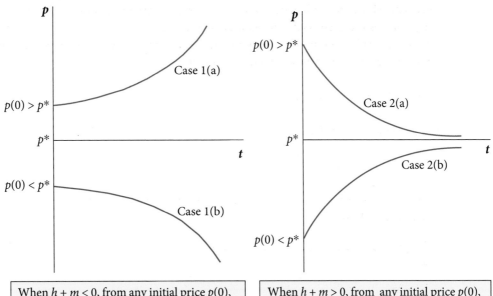

When $h + m < 0$, from any initial price $p(0)$, p diverges increasingly through time from its equilibrium value p^*.

When $h + m > 0$, from any initial price $p(0)$, p converges through time towards its equilibrium value p^*.

Figure 20.8 Cases 1(a) and 1(b): a dynamically unstable market; cases 2(a) and 2(b): a dynamically stable market.

2(a) Assume $(p(0) - p^*) > 0$. This means that the initial price $p(0)$ is greater than the equilibrium price, p^*, and therefore that the actual price needs to fall through time to achieve equilibrium. Since $e^{-k(h+m)t}$ decreases through time, the assumption that $(p(0) - p^*)$ is positive means that their product, $(p(0) - p^*)e^{-k(h+m)t}$, will decrease and approach zero. Therefore in equation (20.37a) we see that this means that p *decreases* through time, which is a movement *towards* equilibrium, since we know that the actual price needs to *fall* through time to achieve equilibrium. As $(p(0) - p^*)e^{-k(h+m)t}$ approaches zero, so p approaches p^*. This case is shown as case 2(a) in figure 20.8.

2(b) Assume that $(p(0) - p^*) < 0$. This means that the initial price $p(0)$ is less than the equilibrium price, p^*, and therefore that the actual price needs to rise through time to achieve equilibrium. Since $e^{-k(h+m)t}$ decreases through time and approaches zero, the assumption that $(p(0) - p^*)$ is negative means that their product, $(p(0) - p^*)e^{-k(h+m)t}$, will become decreasingly negative; that is, will increase through time and approach zero, a movement *towards* equilibrium. As $(p(0) - p^*)e^{-k(h+m)t}$ approaches zero, so p approaches p^*. This is case 2(b) in figure 20.8.

Thus in case 2 the market is **dynamically stable** because in case 2(a) the price is initially too high and falls towards equilibrium, and in case 2(b) the initial price is too low and rises towards equilibrium. This stability is present because $h + m$ is positive.

20.9 Conclusions on market stability

To summarize our results from the model of the previous section, whatever the initial price may be, this market is:

- dynamically unstable if $h + m$ is negative (figure 20.8, cases 1(a) and 1(b))
- dynamically stable if $h + m$ is positive (figure 20.8, cases 2(a) and 2(b)).

Can we say anything about the likelihood of $h + m$ being either positive or negative? First, h is the slope of the supply curve, which we can reasonably assume is positive. Second, $-m$ is the slope of the demand curve, which we can reasonably assume is negative, in which case m is positive. With h and m both positive, $h + m$ is positive, and the market is stable. Note that the assumptions required to reach this conclusion are not very strong: we assume only that the supply function slopes upwards, and that the demand function slopes downwards. We also assumed $k > 0$, meaning that the price rises (falls) in response to excess demand (supply). If real-world markets conform to these rather undemanding assumptions, they should be dynamically stable.

However, a positively sloped supply curve and a negatively sloped demand curve are *sufficient* conditions for stability, but not *necessary* conditions (see section 2.13). For even in an abnormal case where the demand curve is positively sloped, so that $-m$ is positive and therefore m is negative, we can still have stability provided $h + m$ remains positive, which will be true provided h is larger than m in absolute value. In other words, if both the supply curve and the demand curve are upward sloping, the market will remain stable as long as the supply curve is steeper in slope than the demand curve (see figure 20.9a). A second abnormal case arises if the

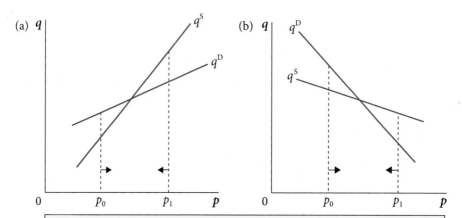

In case (a) the demand curve slopes upward less steeply than the supply curve. In case (b) the supply curve slopes downward less steeply than the demand curve. In both cases the market is **stable** because if the initial price is $p_0(p_1)$ there is excess demand (supply), hence p rises (falls) towards its equilibrium value. (Arrows indicate direction of price change.)

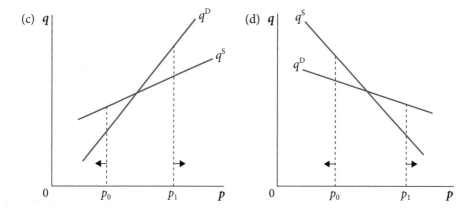

In case (c) the demand curve slopes upward more steeply than the supply curve. In case (d) the supply curve slopes downward more steeply than the demand curve. In both cases the market is **unstable** because if the initial price is $p_0(p_1)$ there is excess supply (demand), hence p falls (rises) away from its equilibrium value. (Arrows indicate direction of price change.)

Figure 20.9 Stability conditions with an upward sloping demand curve or downward sloping supply curve.

supply curve is negatively sloped ($h < 0$). In this case, provided h is smaller than m in absolute value, $h + m$ remains positive and we have stability (see figure 20.9b). The unstable counterparts to these two cases are shown in figures 20.9c and 20.9d.

Note though that, as with the difference equation case considered in section 20.4, we are here obeying the mathematics convention, with quantity as the dependent variable and measured on the vertical axis, so the slopes of the supply and demand curves are $\frac{dq}{dp}$, not $\frac{dp}{dq}$.

Conclusions

In this chapter we have covered only the simplest type of difference and differential equations: the first order, linear variety. Despite this simplicity, we have seen that the technique permits us to analyse some complex and important issues in economic dynamics. A larger menu of modelling possibilities opens up when we study second order equations; these are considered in chapter W21, section 21.4, which you can read or download from the Online Resource Centre.

Chapter summary

In sections 20.1–20.3 we introduced the idea of a difference equation, a relationship in which the dependent variable changes through time in discrete jumps in response to discrete jumps in the independent variable. We saw how to solve first order linear difference equations (see rules 20.1 and 20.2), and saw that there were four possible time paths for the dependent variable (see figures 20.1–20.4).

In sections 20.4–20.5 we analysed the 'cobweb' model of supply and demand and found the time path of price by solving a difference equation. Price was certain to oscillate around its equilibrium value, but the oscillations could be either convergent or divergent (or neither, as a boundary case).

In sections 20.6–20.7 we examined linear first order differential equation models, in which time now varies continuously rather than in jumps and the path of the dependent variable is given by a derivative. The solutions were stated (rules 20.3 and 20.4) and possible time paths were found to be either smoothly convergent or smoothly divergent from the equilibrium value. We applied this technique in sections 20.8–20.9 to analyse the stability of a market in which price varies continuously in response to excess demand or supply. On reasonable assumptions about the slopes of the supply and demand functions, we found that the market was stable, in that price would converge to its equilibrium value (at which supply and demand were equal) from any arbitrary initial value.

Progress exercise 20.3

1. Solve the following first order differential equations with the given initial conditions:

   (a) $\frac{dy}{dt} = -4y + 12$; $y(0) = 2$ (b) $\frac{dy}{dt} = -10y + 15$; $y(0) = 0$ (c) $\frac{dy}{dy} = -y + 4$; $y(0) = 0$

   (d) $\frac{dy}{dt} = 5y$; $y(0) = 2$ (e) $\frac{dy}{dt} = 7y + 7$; $y(0) = 7$

2. In the market for houses the supply and demand functions respectively are

   $$q^S = 0.25p + 10; \text{ and } q^D = 0.5p + 5$$

   When there is excess demand, price adjusts according to the equation

   $$\frac{dp}{dt} = 2(q^D - q^S)$$

   (a) Find the equilibrium price, $p^\star$ (that is, the price at which there is no excess demand or supply).

   (b) Formulate and solve the first order differential equation giving p as a function of time, t. Is the housing market dynamically stable or unstable?

   (c) If the initial price is $p = 21$, what will be the value of p when $t = 12$?

   (d) Illustrate with a sketch graph showing $p^\star$, the initial price $p(0)$ and the path of the actual price.

   (e) Suggest some possible explanations, in terms of behaviour of buyers and sellers, which might explain your results.

3. Suppose in the model of the previous question that both the supply and demand functions are negatively sloped. What would be the condition for dynamic stability in this case?

Checklist

Be sure to test your understanding of this chapter by attempting the progress exercises (answers are at the end of the book). The Online Resource Centre contains further exercises and materials relevant to this chapter www.oxfordtextbooks.co.uk/orc/renshaw3e/.

In this chapter we have covered rather a lot of ground rather quickly, but hopefully this has not left you completely breathless. Although we have considered only the simplest types of difference and differential equations, we have seen that they have some interesting applications in economics, addressing such fundamental questions as the stability of the market economy. The topics were:

✔ **Difference equations.** Recognizing and solving a first order linear difference equation with constant coefficients, $y_t = by_{t-1} + c$. Analysing the solution qualitatively to determine whether behaviour is convergent or divergent, oscillatory or non-oscillatory.

✔ **Economic application.** Understanding how a first order difference equation may arise in economics and solving the 'cobweb' model of supply and demand.

✔ **Differential equations.** Recognizing and solving a first order linear differential equation with constant coefficients, $\frac{dy}{dt} = by + c$. Analysing the solution qualitatively to determine whether behaviour is convergent or divergent.

✔ **Economic application.** Using a differential equation model to analyse dynamic stability of a market.

In the final chapter of this book we introduce some more advanced maths as an extension of work in earlier chapters and also as a guide to future directions for the study of mathematical economics. Due to space constraints we have been unable to include this chapter in the third edition of the book. It is to be found as chapter W21 at the Online Resource Centre.

Answers to progress exercises

The answers given here are supplemented by further material at the book's Online Resource Centre, **www.oxfordtextbooks.co.uk/orc/renshaw3e/**.

Progress exercise 1.1
(a) 20; (b) 53; (c) 4; (d) −8; (e) −14; (f) −2

Progress exercise 1.2
(a) −21; (b) 2; (c) 18; (d) −2; (e) −10; (f) −3

Progress exercise 1.3
(a) $6(3+1-6)$ (b) $10(15+22)$

(c) $3(-4-1+2)$ or $-3(4+1-2)$

(d) 135   (e) −42   (f) 3

Progress exercise 1.4
(a) $\frac{1}{5}$;   (b) $\frac{1}{11}$;   (c) $-\frac{2}{13}$;   (d) $\frac{3}{11}$

Progress exercise 1.5
(a) $\frac{9}{10}$;   (b) 0;   (c) $2\frac{3}{4}$;   (d) $6\frac{9}{16}$;   (e) $\frac{1}{6}$;   (f) $1\frac{11}{16}$

Progress exercise 1.6
1. (a) $\frac{1}{5}$;   (b) $\frac{2}{3}$;   (c) $\frac{3}{32}$;   (d) $7\frac{1}{8}$;   (e) $10\frac{1}{2}$;   (f) $1\frac{7}{20}$

2. (a) $\frac{1}{4}(1+\frac{3}{2})$ (b) won't factorize   (c) $\frac{3}{8}(\frac{1}{2}+1)$

   (d) $\frac{1}{3}(\frac{5}{4}-1)$ (e) $\frac{1}{4}(\frac{11}{4}+\frac{5}{3})$ (f) $\frac{1}{15}(3+5)$

Progress exercise 1.7
1. (a) 1.35; (b) 127.5; (c) 0.032; (d) 0.0465; (e) 0.325; (f) 3; (g) 0.016; (h) 0.05

2. (a) $\frac{9}{25}$; (b) $\frac{7}{8}$; (c) $3\frac{51}{1000}$

3. (a) 0.06; (b) 0.375; (c) 0.15; (d) 0.666 (recurring)

Progress exercise 1.8
1. (a) 20%; (b) 22.22 (recurring); (c) 25; (d) 125; (e) 15; (f) 60; (g) 4; (h) 3.5

2. (a) 428.4; (b) 508,042.5; (c) 0.0383

3. Suppose the price is 60 (euros, pounds, or whatever). Then you get 3 for 120, equivalent to a price of 40. So the effective price reduction is from 60 to 40, a reduction of one-third or $33\frac{1}{3}$%.

4. $\frac{14.894}{100} \times 100 = 14.894$%.

5. A's fuel consumption is $\frac{10-5}{5} \times 100 = 100$% greater than B's. If I switch to B, I will use 5 litres where I would previously have used 10, a change of $\frac{5-10}{5} \times 100 = -50$%. (That is, a reduction, or negative change, of 50%.)

Progress exercise 1.9
1. (a) 81; (b) 32; (c) 6 and −6; (d) 3; (e) 125; (f) 3; (g) 20; (h) $\frac{1}{8}$

Progress exercise 2.1
1. (a) $-6a$ (b) $-4a-2b+6$ (c) $8a+3b$

   (d) $8x+y-10z$ (e) ab (f) $x(6a-4b)+b(2-c)$

2. (a) $12ab$; 120   (b) $-\frac{1}{c}$ or $\frac{-1}{c}$; $\frac{1}{4}$;

   (c) $30abc$; −1200   (d) $\frac{ab}{2c}$; −1.25

Progress exercise 2.2
1. (a) $c(b-a)$ (b) $a-5(b-c)$ (c) $3y$

   (d) −2   (e) −10   (f) $2a(4-b)$

2. (a) $4(x+2y)$ (b) $5(x-2y)$ (c) $-3x(1+3y)$

   (d) $-a(x+2y)$ (e) $-5b(a-3)$ (f) $4y(a-3x)$

   (g) $3x(a-3y)$ (h) $5y(3xz+z-2x)$

Progress exercise 2.3
1. (a) $\frac{3b-6a}{ab}$ (b) $\frac{ab-cb}{bd}$ (c) $\frac{5b-6a}{3ab}$

   (d) $\frac{a(b-1)}{b}$ (e) $\frac{20bd-6ac}{15ad}$ (f) $\frac{5a(c+d)+3c(a+b)}{(a+b)(c+d)}$

   (g) $\frac{b(c+d)-2ac}{a(c+d)}$ (h) $\frac{d+e-d(b+c)}{(b+c)(b+e)}$

2. (a) $\frac{ad}{bc}$; (b) $\frac{abc}{abc} \equiv 1$; (c) ad; (d) $\frac{ab}{d}$; (e) $\frac{2a}{c}$; (f) $\frac{x+2}{2(y+1)}$

Progress exercise 2.4
1. (a) a^2; $-a^3$; a^4; $-a^5$ (b) $4a^2$; $-8a^3$; $16a^4$; $-32a^5$

   (c) $\frac{c^2}{9}$; $-\frac{c^3}{27}$; $\frac{c^4}{81}$; $-\frac{c^5}{243}$ (d) $-2t$

   (e) $32x^7$ (f) $-\frac{3x}{y}$ or $\frac{-3x}{y}$

2. (a) x^2y^2 (b) $xy^{1/3}$ (c) $10x^{1/2}$

   (d) $\frac{x^2}{y^2}$ (e) $\frac{1}{xy^4}$ (f) xy

   (g) $\frac{1}{x^{5/6}y^{5/6}}$ or $x^{-5/6}y^{-5/6}$ (h) $\frac{y}{x^2}$

Progress exercise 3.1
1. (a) Total cost is: $9500 + (500 \times 6.22) + (200 \times 2.45) = 4550$ pence = 45.50 pounds.

   (b) I will consume 50 units less at the day rate, saving $50 \times \frac{6.22}{100}$ pounds; and 50 units more at the night rate, costing $50 \times \frac{2.45}{100}$. So the change in my total cost of units, in pounds, is $50 \times \frac{2.45}{100} - 50 \times \frac{6.22}{100} = 50\left(\frac{2.45}{100} - \frac{6.22}{100}\right) = 50\left(\frac{2.45-6.22}{100}\right) = -1.885$. Negative change = saving.

2. (a) (i) Monthly cost is $C_1 = 1.75J$, where J is the number of journeys. (ii) Monthly cost is $C_2 = 28$. (Both measured in pounds.) The total cost is the same when $C_1 = C_2$, which is true when $1.75J = 28$. Solving this equation gives $J = 28/1.75 = 16$. So if I make more than 16 journeys per month, it is cheaper to buy a season ticket.

3. Assuming I make more than 3 minutes of call each day, the average total daily cost is $(3 \times 35) + 12(D-3)$, where D is the average number of minutes of calls per day. (If the idea of an average confuses you, just assume I make *exactly* D minutes of calls per day.) So my annual expenditure will be $365[(3 \times 35) + 12(D-3)]$. If $D = 15$, my annual bill in cents will be $365[(3 \times 35) + 12(15-3)] = 90885$ cents, or 908.85 euros.

4. (a) $x = \frac{c-b}{a}$ (b) $x = \frac{c}{d} - \frac{a+b}{dk}$

    (c) $x = \pm\sqrt{\frac{3a(c-b)}{81}}$ or $x = \pm\frac{\sqrt{3a(c-b)}}{9}$ (d) $x = \frac{a-c(b-d)}{1-c}$

    (e) $x = \pm\sqrt{\frac{ad-c}{b}}$ (f) $x = \frac{bc}{b-a}$

    (g) $x = \frac{c-b}{a}$ (h) $x = cd - b$

5. (a) $x = \frac{9}{15} = \frac{3}{5}$; (b) $x = 6$; (c) $x = 18$; (d) $x = -\frac{6}{25}$

Progress exercise 3.2

1. (a) $(-4, 4)$ (b) $(0.8, 4.8)$ (c) $(-2.8, 0)$

    (d) $(0, 2.4)$ (e) $(-7.6, -4.8)$ (f) $(7.2, -4.4)$

    (g) $(-2.8, -1.6)$ (h) $(3.2, -6)$

2. See Online Resource Centre.

3. See Online Resource Centre.

4. For graphical solutions, see Online Resource Centre. Algebraic solutions:

    (a) $x = \frac{2}{3}$; (b) $x = -2$; (c) $x = -10$; (d) $x = 3$; (e) $x = \frac{5}{3}$

5.

	slope	y-intercept	x-intercept
(a)	2	5	$-\frac{5}{2}$
(b)	$\frac{1}{3}$	12	-36
(c)	$\frac{1}{2}$	20	-40
(d)	0.25	0.5	-2
(e)	$\frac{1}{5}$	5	-25

Progress exercise 3.3

1. For sketch graphs, see Online Resource Centre. Algebraic solutions:

    (a) $x = 7, y = 9$ (b) $x = -2, y = -5.5$

    (c) $x = y = 1$ (d) $x = 2, y = 0$ (e) $x = 4, y = -1$

Progress exercise 3.4

1. (a) For the graphs, see Online Resource Centre. Algebraic solutions: (i) The equilibrium conditions are $p^D = p^S$ and $q^D = q^S$. Therefore we can drop the superscripts and write simply $q = -0.75p + 10$; $p = 2p - 1$. Solving this pair of simultaneous equations, the solution is $p = p^D = p^S = 4$, $q = q^D = q^S = 7$. (ii) Using the same method as in (i), $p = 11, q = 18$.

    (b) For the graph, see Online Resource Centre. Using the same method as (a)(i), the solution is $p = 5, q = -1$. Since the quantity cannot be negative, we must conclude that the equations are inconsistent. Thus no equilibrium in this market is possible.

2. Using same method as (1), $q = 7, p = 9$.

3. (a) Using same method as (1) above, $p = 10, q = 50$.

    (b) By elementary algebraic operations we obtain the inverse supply and demand functions as: $p^S = \frac{1}{15}q^S + \frac{100}{15}$; $p^D = -\frac{1}{5}q^D + 20$. Solving using the method of (1) gives the same solution as in (a).

    (c) See Online Resource Centre.

4. (a) Using the same method as (2) above, the solution is $p = 10, q = 20$.

    (b) Let p^{S+} denote the supply price including tax, and let t denote the tax rate. Then we have: $p^{S+} = p^S + tp^S = 1.5p^S$. The equilibrium conditions are now $p^D = p^{S+}$ and $q^D = q^S$. Using the same method as in (a) above, the solution is: $q = 7.5, p^D = p^{S+} = 13.125, p^S = 8.75$.

    (c) See Online Resource Centre.

5. (a) Combining the five equations we obtain $Y = \frac{650}{1 - 0.8} = 3250$, from which $C = \hat{C} = 0.8(3250) + 100 = 2700$. For the graphical treatment see Online Resource Centre.

    (b) Using the same method as in (a), new $Y = 4000$, new $C = 3300$. $\Delta Y = 1750, \Delta C = 600$. Investment multiplier is $\frac{\Delta Y}{\Delta I} = \frac{750}{150} = 5$.

    (c) The general solution to this model is $Y = \frac{100 + I}{1 - c}$, where c is the MPC. Here we need $Y = 3250$ with $I = 700$, so we have $3250 = \frac{100 + 700}{1 - c}$, with solution $c = 0.754$ (to 3 dp).

6. (a) $Y = 5000$, $C = 4350$. For the graphical treatment, see Online Resource Centre.

    (b) $Y = 3333, C = 2933$.

    (c) Required $MPC = 0.9$, an increase of 0.05.

Progress exercise 4.1

1. For answer, see figure 4.4 in book.

2. (a) $x^2 + 5x + 6$ (b) $x^2 + x - 20$ (c) $2x^2 + \frac{3}{2}x + \frac{1}{4}$

    (d) $x^2 - 8x + 16$ (e) $x^2 - 25$

Progress exercise 4.2

1. (a) $x = 1$ or 2 (b) $x = -4$ or -1 (c) $x = -5$ or 1

    (d) $x = -0.5$ or -0.25 (e) $x = 4$ or -4 (f) $x = -3$ or -1

    (g) $x = 1.5$ or -2 (h) $x = -4$ or -4 (a perfect square)

    (i) $x = -1$ or 2;   (j) no real roots $(b^2 < 4ac)$

    (h) and (j) are different because (h) is a perfect square $(b^2 = 4ac)$ and (j) has no real roots $(b^2 < 4ac)$. All the others have $b^2 > 4ac$, hence two distinct roots (solutions).

Progress exercise 4.3

1. For graphs, see Online Resource Centre. Factors are:

    (a) $(x + 3)(x - 2)$ (b) $(x - \frac{5}{2})(x + 1)$

    (c) $(x - 3)(x + 1)$ (d) $(x + 1)(x + 1)$

    (e) No real roots; cannot be factorized.

2. Using the factors found in (1) above, the exact solutions are:

    (a) 2, –3; (b) 2.5, –1; (c) –1, 3; (d) –1, –1

    Note: the graphs sketched in (1) above should cut the x-axis at approximately the above solution values.

    (e) No (real) solutions because $b^2 = 0$ and $4ac = 32$, so $b^2 < 4ac$. The graph in 1(e) never cuts the x-axis.

    Case (d) is different because $b^2 = 64$ and $4ac = 64$, so $b^2 = 4ac$. This is a perfect square. Case (e) is different because there are no real roots, as explained immediately above.

3. For sketch graphs see Online Resource Centre.

    (a) Using example 4.5, we have $y = a(x + A)(x + B) \equiv ax^2 + bx + c$, with $A = -2, B = 3$. So we have $y = a(x - 2)(x + 3) = a(x^2 + x - 6) = ax^2 + ax - 6a \equiv ax^2 + bx + c$. We are also given $y = -12$ when $x = 0$, which implies that $y = -6a = -12$ and therefore $a = 2$. So the quadratic function we are seeking is $y = ax^2 + ax - 6a = 2x^2 + 2x - 12$.

    (b) Same method as (a). We have $y = a(x + A)(x + B) \equiv ax^2 + bx + c$ with $a = -2, -A = -4, -B = -2$. Therefore $y = -2(x + 4)(x + 2) = -2x^2 - 12x - 16$.

4. Same method as (3). For sketch graphs see Online Resource Centre; solution is $b = -10, c = -12, y = -2x^2 - 10x - 12$.

5. Same method as (3). For sketch graphs see Online Resource Centre; the solution is $b = -3, c = 6, x_0 = 1, y = -3x^2 - 3x + 6$.

Progress exercise 4.4

1. For graphs see Online Resource Centre.

   (a) Solutions are $x=-3, y=-3$; or $x=4, y=18$. Solutions are at the intersection of the two graphs.

   (b) $x=2, y=6$; or $x=-3, y=9$.

2. Impose the equilibrium condition $q^S = q^D$, then solve as simultaneous equations, giving $p=6, q=16$. For graphs see Online Resource Centre.

3. Impose the equilibrium condition $p^S = p^D$, then solve as simultaneous equations, giving $q=10, p=70$. For graphs see Online Resource Centre.

For progress exercises 5.1–5.3, see Online Resource Centre.

Progress exercise 5.4

1. Given $0 < x < 2$. After multiplying by 2 and adding 1, this becomes $1 < 2x + 1 < 5$. Therefore $0 < x < 2$ implies that $y = 2x + 1$ must be greater than 1 but less than 5.

2. Given $\frac{1}{x} + 1 > -5$. Subtracting 1 gives $\frac{1}{x} > -6$. Two cases: (i) Assume $x > 0$. Multiplying both sides of $\frac{1}{x} > -6$ by x gives: $1 > -6x$ (inequality not reversed, since $x > 0$). Dividing this by -6, we get: $-\frac{1}{6} < x$ (inequality reversed because we divided by a negative number). This is the same as $x > -\frac{1}{6}$. This condition is satisfied when $x > 0$ or $-\frac{1}{6} < x \le 0$. But the latter case is ruled out due to our assumption $x > 0$. Therefore the given condition, $\frac{1}{x} + 1 > -5$, is satisfied when x is positive. (ii) Assume $x < 0$. Then multiplying both sides of $\frac{1}{x} > -6$ by x gives: $1 < -6x$ (inequality reversed). Dividing this by -6, we get: $-\frac{1}{6} > x$ (inequality again reversed). This is the same as $x < -\frac{1}{6}$. So when x is negative, the condition $\frac{1}{x} + 1 > -5$ is satisfied when $x < -\frac{1}{6}$. Combining (i) and (ii), the given inequality is satisfied when $x > 0$ or $x < -\frac{1}{6}$.

   This solution is confirmed by figure S1, where we see that $y = \frac{1}{x} + 1$ is a rectangular hyperbola, while $y = -5$ is a horizontal line 5 units below the x-axis. From the graph, we see that the graph of $y = \frac{1}{x} + 1$ lies above the graph of $y = -5$ in two cases: (a) when $x > 0$; (b) when x is negative and less than x_0. We have found that $x_0 = -\frac{1}{6}$.

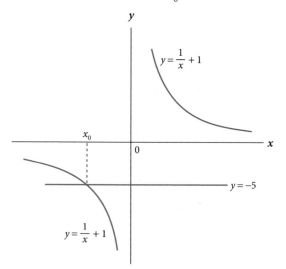

Figure S1 Sketch graph for question 2 of exercise 5.4.

3. Method similar to (2). Given $\frac{1}{x} + 1 > 2x$, two cases: (i) Assume $x > 0$, then multiply both sides by x, giving $1 + x > 2x^2$ (since $x > 0$, the inequality is not reversed). This rearranges as $2x^2 - x - 1 < 0$. Factorizing the left side,

$(2x+1)(x-1) < 0$. This is true when $(2x+1)$ and $(x-1)$ have opposite signs. Two case: (a) $(2x+1) < 0$ and $(x-1) > 0$. The first requires $x < -\frac{1}{2}$ and the second requires $x > 1$. These cannot be true simultaneously, so this case is impossible; (b) $(2x+1) > 0$ and $(x-1) < 0$. The first requires $x > -\frac{1}{2}$ and the second $x < 1$. Given our assumption that $x > 0$, this case therefore requires $0 < x < 1$. (ii) Assume $x < 0$, then proceed as in (i), giving $(2x+1)(x-1) > 0$. This is true when $(2x+1)$ and $(x-1)$ have the same signs. Two case: (a) $(2x+1) > 0$ and $(x-1) > 0$. The first requires $x > -\frac{1}{2}$ and the second requires $x > 1$. Given the assumption $x < 0$, this is impossible. (b) $(2x+1) < 0$ and $(x-1) < 0$. The first requires $x < -\frac{1}{2}$ and the second $x < 1$. Given our assumption that $x < 0$, this case therefore requires $x < -\frac{1}{2}$.

Collecting results, the given inequality was satisfied under (i) when $0 < x < 1$ and under (ii) when $x < -\frac{1}{2}$. This solution is confirmed by figure S2, where we see that the graph of $y = \frac{1}{x} + 1$ lies above graph of $y = 2x$ when x is positive but less than 1, and when x is negative and less than $-\frac{1}{2}$. (Finally, note that we have considered $x > 0$ and $x < 0$, but not $x = 0$. In that case, $\frac{1}{x}$ is undefined, and the question whether the given inequality is satisfied becomes literally meaningless.)

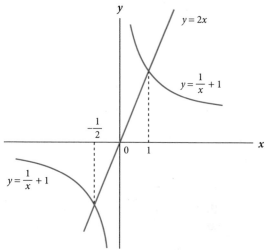

Figure S2 Sketch graph for question 3 of exercise 5.4.

4. (Here $\Rightarrow$ means 'implies'.) Given $(x-1)^2 < 2 - 2x$. Multiply out the left-hand side $\Rightarrow x^2 - 2x + 1 < 2 - 2x \Rightarrow x^2 - 1 < 0$. Factorize the left-hand side $\Rightarrow (x+1)(x-1) < 0$. The left-hand side is negative if $x + 1$ and $x - 1$ have opposite signs. Now, $x + 1$ and $x - 1$ are both positive if $x > 1$, so we require $x < 1$ for them to have opposite signs. However, they are both negative if $x < -1$, so we also require $x > -1$ for them to have opposite signs. Combining these, the given inequality is satisfied if $-1 < x < 1$. This solution is confirmed by a graph of $y = x^2 - 1$ (not drawn) which lies below the x-axis if $-1 < x < 1$.

5. We can see immediately that $y = (x - 1)^2 = 0$ when $x = 1$. When $x \ne 1$, and given $0 < x < 2$, there are two cases: (i) When $0 < x < 1$. Subtracting 1 gives $-1 < x - 1 < 0$. Multiply by $x - 1$ (which reverses the inequality, since $x - 1 < 0$). This gives $-(x-1) > (x-1)^2 > 0$, which rearranges as: $0 < (x-1)^2 < 1-x$, where $1 - x < 1$ since x is positive. Therefore $0 < (x-1)^2 < 1$ when $0 < x < 1$. (ii) When $1 < x < 2$. Subtracting 1 gives $0 < x - 1 < 1$. Multiply by $x - 1$ (which does not reverse the inequality, since $x - 1 > 0$). This gives $0 < (x-1)^2 < x - 1$, where $x - 1 < 1$ since x is less than 2. So again $0 < (x-1)^2 < 1$. Overall conclusion: when $0 < x < 2$, $0 \le (x-1)^2 < 1$ (note

the solution includes $0 = (x - 1)^2$, when $x = 1$. The answer may be confirmed by the graph of $y = (x - 1)^2 = x^2 - 2x + 1$ (not drawn).

6. (a) (i) $3x + 4y \leq 120$; (ii) $y \leq 30 - \frac{3}{4}x$.

 (b) Opportunity cost of x in terms of y is $\frac{P_x}{P_y} = \frac{3}{4}$. This means that in order to increase her purchases of x by 1 unit, the consumer must reduce her purchases of y by $\frac{3}{4}$ of a unit.

 (c) (i) 30 (the maximum quantity of y that her money income can buy).
 (ii) 40 (the maximum quantity of x that her money income can buy).

 (d) No, because the total expenditure is then $112\ (< 120)$.

 (e) 40% of 120 is 48, so she buys $48/3 = 16$ units of x. Similarly she buys $72/4 = 18$ units of y.

Progress exercise 6.1

1. $\frac{\Delta y}{\Delta x} =$: (a) 2; (b) $\frac{1}{2}$; (c) 50; (d) -3; (e) 0; (f) $-\frac{1}{2}$

2. (a) For unspecified values x_1 and x_0, question 1(a) above becomes:
$$\frac{\Delta y}{\Delta x} = \frac{(2x_1 + 4) - 2(x_0 + 4)}{x_1 - x_0} = \frac{2(x_1 - x_0)}{x_1 - x_0} = 2$$

 Similarly, 1(b) becomes:
$$\frac{\Delta y}{\Delta x} = \frac{(\frac{1}{2}x_1 - 2) - (\frac{1}{2}x_0 - 2)}{x_1 - x_0} = \frac{\frac{1}{2}(x_1 - x_0)}{x_1 - x_0} = \frac{1}{2}$$

 And 1(c) becomes:
$$\frac{\Delta y}{\Delta z} = \frac{50x_1 - 50x_0}{x_1 - x_0} = 50$$

 (b) If $y = ax + b$, then with unspecified values x_1 and x_0 we get:
$$\frac{\Delta y}{\Delta x} = \frac{(ax_1 + b) - (ax_0 + b)}{x_1 - x_0} = \frac{a(x_1 - x_0)}{x_1 - x_0} = a$$

Progress exercise 6.2

$\frac{dy}{dx} =$: (a) $10x$; (b) $4x^3$; (c) $36x^2$
 (d) $0.5x^{-0.5}$ (e) $3x^2 + 2x$ (f) $12x^3 + 20x$
 (g) $12x^2 - 16$ (h) $30x^2 + 10x - 9 - 5x^{-2}$
 (i) $\frac{3}{5}x^2 + x + 5$ (j) $\frac{1}{2}x^{-1/2} + 2x^{-3}$
 (k) $\frac{1}{2}x^{-1/2} + 2x^{-3}$ (l) $4x^{-3} - 0.5x^{-0.5} + 1$
 (m) $-x^{-2} + 0.25x^{-1.25}$ (n) $1 - 2x^{-3}$ $\left(\text{since } \frac{1}{x^2} = x^{-2}\right)$
 (o) $-10x^{-4} + 0.09x^{-1.3} + \frac{1}{4}x^{-1/2}$ (p) a
 (q) $2ax + b$ (r) $\frac{dq}{dp} = 6p^2 + 10p$
 (s) $\frac{dq}{dp} = -\alpha A p^{-\alpha - 1}$ (t) $\frac{dz}{dt} = 1.5t^2 + 4$

Progress exercise 6.3

$\frac{dy}{dx} =$: (a) $4(1 + 2x)$ (b) $-x(1 - x^2)^{-1/2}$
 (c) $5(x^2 - 2x^3)^4(2x - 6x^2)$ (d) $0.5(x^2 - x)^{-0.5}(2x - 1)$
 (e) $12x - 1$ (f) $(4x^3 - 3)(10x^4 + 1) + (2x^5 + x)(8x)$
 (g) $\frac{4x}{(1 - x^2)^2}$ (h) $\frac{x(x^2 + 3)}{(1 - x^2)^2(x^2 + 1)^{1/2}}$

 (i) Using the function of a function rule, $\frac{dx}{dy} = 0.5(y + 1)^{-0.5}$
 By the inverse function rule, $\frac{dy}{dx} = \frac{1}{\frac{dx}{dy}} = \frac{1}{0.5}(y + 1)^{0.5}$
 $= 2(y + 1)^{0.5} = 2x$ (since $x = (y + 1)^{0.5}$)

 (j) $\frac{1}{(1 - x)^2}$.

Progress exercise 7.1

(a) $\frac{dy}{dx} = -2$. As this is negative for all x, the function is always negatively sloped (sufficient condition for function to be classified as strictly decreasing).

(b) $\frac{dy}{dx} = 3x^2 + 1$. This is positive for all x, sufficient condition for function to be classified as strictly increasing.

(c) Function is strictly decreasing when $\frac{dy}{dx} = -6x + 4 < 0$, which requires $x > \frac{2}{3}$.

(d) $\frac{dy}{dx} = 3x^2 + 12x + 12 = 3(x + 2)^2$. When $x > -2$, $\frac{dy}{dx} > 0$ and the function is strictly increasing. (Note, $\frac{dy}{dx} > 0$ and the function is strictly increasing also when $x < -2$.)

(e) $\frac{dy}{dx} = 6x^2 + 42x + 60 = 6(x + 5)(x + 2)$. So $\frac{dy}{dx} > 0$ when $(x + 5)$ and $(x + 2)$ have the same sign (that is, both positive or both negative). They are both positive when $x > -2$ and both negative when $x < -5$.

Progress exercise 7.2

For graphs and extended answers see Online Resource Centre.

(a) Min at $x = 3$

(b) Max at $x = 5$

(c) Min at $x = \frac{3}{2}$

(d) Min at $x = 5$; max at $x = 1$.

(e) Max at $x = 9$; min at $x = 1$.

Progress exercise 7.3

For graphs and extended answers see Online Resource Centre.

(a) Stationary point which is a point of inflection at $x = -2$.

(b) Minimum at $x = 5$; maximum at $x = 2$.

(c) Stationary point which is a point of inflection at $x = \frac{1}{2}$.

(d) Maximum at $x = -5$; minimum at $x = -1$.

(e) Stationary point which is a pt of inflection at $x = 2$.

Progress exercise 7.4

For graphs and extended answers see Online Resource Centre.

(a) Minimum at $x = 5$; maximum at $x = 1$; point of inflection at $x = 3$.

(b) Maximum at $x = 9$, min at $x = 1$, point of inflection at $x = 5$.

(c) Max at $x = 6$, min at $x = 12$, point of inflection at $x = 9$.

(d) Maximum at $x = -\frac{1}{2}$; minimum at $x = \frac{1}{2}$; point of inflection at $x = 0$.

(e) Minimum at $x = 10$; maximum at $x = 1$; point of inflection at $x = 5\frac{1}{2}$.

Progress exercise 7.5

For graphs see Online Resource Centre.

1. (a) $\frac{dy}{dx} = 2x - 6 > 0$ when $x \geq 4$. Also $\frac{d^2y}{dx^2} = 2 > 0$ when $x \geq 4$. So the slope is positive and increasing (that is, convex from below) when $x \geq 4$.

 (b) $\frac{dy}{dx} = -2x + 10 = -2(x - 5) > 0$ when $x \geq 4$. Also $\frac{d^2y}{dx^2} = -2 < 0$ when $x \leq 4$.
 So when $x \leq 4$, the slope is positive but decreasing; that is, concave.

 (c) $\frac{dy}{dx} = 4x - 6 = 2(2x - 3) < 0$ when $x < 1.5$. Also $\frac{d^2y}{dx^2} = 4 > 0$ when $x < 1.5$.
 So when $x < 1.5$, the slope is negative and increasing; that is, convex.

(d) $\frac{dy}{dx} = 2x + 4 = 2(x + 2) > 0$ when $x > -2$. Also $\frac{d^2y}{dx^2} = 2 > 0$ when $x > -2$. So when $x > -2$, the slope is positive and increasing; that is, convex.

(e) $\frac{dy}{dx} = -3x^2 + 30x - 27 = -3(x - 9)(x - 1)$. Also $\frac{d^2y}{dx^2} = -6x + 30 = -6(x - 5)$. So when $x > 9$, $(x - 9)$ and $(x - 1)$ are both positive so $\frac{dy}{dx} < 0$; and $\frac{d^2y}{dx^2} < 0$. So the slope is negative and decreasing; that is, concave. When $x < 1$, $(x - 9)$ and $(x - 1)$ are both negative so $\frac{dy}{dx} < 0$; but $\frac{d^2y}{dx^2} > 0$. So the slope is negative and increasing; that is, convex.

2. (a) Point of inflection: $\frac{dy}{dx} = 3x^2 = 0$ when $x = 0$, so $x = 0$ is a SP. Also $\frac{d^2y}{dx^2} = 6x = 0$ when $x = 0$, and $\frac{d^3y}{dx^3} = 6 \neq 0$ when $x = 0$. So $x = 0$ is a SP (since $\frac{dy}{dx} = 0$) and is also a point of inflection since $\frac{d^2y}{dx^2} = 0$ and $\frac{d^3y}{dx^3} \neq 0$.

(b) Convexity/concavity: when $x < 0$, $\frac{dy}{dx} = 3x^2 > 0$ and $\frac{d^2y}{dx^2} = 6x < 0$, therefore concave. When $x > 0$, $\frac{dy}{dx} > 0$ and $\frac{d^2y}{dx^2} > 0$, therefore convex.

3. When $x = 1$, $y = 13$; when $x = 1.1$, $y = 13.641$. So the true change in y is $\Delta y = 13.641 - 13 = 0.641$. Using the differential formula, $dy = \frac{dy}{dx} \cdot dx$, where $\frac{dy}{dx} = 3x^2 + 2x + 1$. When $x = 1$, $\frac{dy}{dx} = 3x^2 + 2x + 1 = 3 + 2 + 1 = 6$. So when $dx = 0.1$, $dy = \frac{dy}{dx} dx = 6(0.1) = 0.6$. So the absolute error, which we can define as the actual change, minus the estimated change, is $0.641 - 0.6 = 0.041$. As a percentage of the actual change, this error is $\frac{0.041}{0.641} \times 100 = 6.40\%$ to two decimal places.

Progress exercise 8.1

1. (a) See Online Resource Centre.

(b) $MC \equiv \frac{dTC}{dq} = 0.5$. $AC \equiv \frac{TC}{dq} = \frac{0.5q + 2}{q} = 0.5 + \frac{2}{q}$. For graphs, see Online Resource Centre.

(c) AC includes fixed costs per unit of output ($= \frac{2}{q}$ in this example); MC does not.

(d) If TC is linear, say $TC = aq + b$, we have $MC = a$ and $AC = a + \frac{b}{q}$. So $MC = AC$ only if $b = 0$; that is, no fixed costs.

2. (a) See Online Resource Centre.

(b) $MC = 6q + 5$. $AC = 3q + 5 + \frac{48}{q}$.

(c) Stationary point of AC when $\frac{dAC}{dq} = 3 - \frac{48}{q^2} = 0$. Solving gives $q = 4$ (ignoring negative solution). $\frac{d^2AC}{dq^2} = \frac{96}{q^3} > 0$ when $q = 4$, so this is a minimum. When $q = 4$, $MC = AC = 29$. Alternatively, set $MC = AC$, giving $6q + 5 = 3q + 5 + \frac{48}{q}$. Solving this equation gives $q = 4$.

(d) See Online Resource Centre.

3. (a) Set $\frac{dAC}{dq} = 4q - 2 - 24q^{-2} = 0$. Solution is $q = 2$. And $\frac{d^2AC}{dq^2} = 4 + \frac{48}{q^3} > 0$ when $q = 2$. So $q = 2$, a minimum of AC.

(b) When $q = 2$, $MC = \frac{dTC}{dq} = 6q^2 - 4q + 5 = 21$, and $AC = 2q^2 - 2q + \frac{5}{q} = 21$. Alternatively, set $MC = AC$ and solve this equation, giving $q = 4$.

(c) $\frac{dMC}{dq} = 6q^2 - 4q + 5 = 0$ when $q = \frac{1}{3}$, and $\frac{d^2MC}{dq^2} = 12$ which is > 0 for all q, including $q = \frac{1}{3}$. So $q = \frac{1}{3}$ is a minimum of MC.

(d), (e) See Online Resource Centre.

4. (a) $\frac{dAC}{dq} = 12q - 20 - 144q^{-2}$. When $q = 3$, $\frac{dAC}{dq} = 0$ and $\frac{d^2AC}{dq^2} = 12 + \frac{288}{q^3}$ is positive. So $q = 3$ is a minimum of AC.

(b) From (a), there is a minimum of AC at $q = 3$. When $q = 3$, $AC = 67$ and $MC = 67$. So $MC = AC$. Alternatively, set $MC = AC$ and solve this equation, giving $q = 3$.

(c) $\frac{dMC}{dq} = 36q - 40 = 0$ when $q = \frac{40}{36} = 1\frac{1}{9}$, and $\frac{d^2MC}{dq^2} = 36 > 0$. So $q = 1\frac{1}{9}$ is a minimum of MC.

(d), (e) See Online Resource Centre.

Progress exercise 8.2

1. (a) $TR \equiv pq = (-\frac{1}{8}q + 250)q = -\frac{1}{8}q^2 + 250q$; and $MR \equiv \frac{dTR}{dq} = -\frac{1}{4}q + 250$.

(b) Set $MR \equiv \frac{dTR}{dq} = -\frac{1}{4}q + 250 = 0$ with solution $q = 1000$, and $\frac{d^2TR}{dq^2} = \frac{dMR}{dq} = -\frac{1}{4} < 0$. So $q = 1000$ is a maximum of TR.

(c), (d) See Online Resource Centre.

2. (a) Inverse demand function is $p = -0.4q + 48$. So $TR \equiv pq = -0.4q^2 + 48q$, and $MR \equiv \frac{dTR}{dq} = -0.8q + 48$.

(b) Setting $MR = 0$, the solution is $q = 60$; and $\frac{d^2TR}{dq^2} = \frac{dMR}{dq} = -0.8 < 0$. So $q = 60$ is a maximum of TR. When $q = 60$, $p = -0.4q + 48 = 24$.

(c), (d) See Online Resource Centre.

3. (a) $q = 300 - 4p^2$ (demand function). Because TR is, by convention, defined as a function of q, in order to find TR we first have to find the inverse demand function (as in questions 1 and 2 above). From $q = 300 - 4p^2$ we can get the inverse demand function as: $p = (75 - \frac{1}{4}q)^{1/2}$. Therefore, $TR \equiv pq = (75 - \frac{1}{4}q)^{1/2} \cdot q$ (a function of q, as required). The derivative of this function then gives us MR as: $MR \equiv \frac{dTR}{dq} = (75 - \frac{1}{4}q)^{1/2} - \frac{1}{8}q(75 - \frac{1}{4}q)^{-1/2}$ (using both the function of a function and product rules).

(b) TR maximum requires $MR = (75 - \frac{1}{4}q)^{1/2} - \frac{1}{8}q(75 - \frac{1}{4}q)^{-1/2} = 0$. But this equation is not easy to solve so, following the hint in the question, we find TR in terms of p as: $TR \equiv pq = p(300 - 4p^2) = 300p - 4p^3$ (using the demand function to eliminate q).
Set $\frac{dTR}{dp} = 300 - 12p^2 = 0$; the solution is $p = 5$, and $\frac{d^2TR}{dp^2} = -24p < 0$. So $p = 5$ gives maximum TR. From the demand function with $p = 5$, $q = 300 - 4(5)^2 = 200$.

(c) See Online Resource Centre.

4. (a) Given $p = \frac{16}{q+2} - 2$. The p intercept is where $q = 0 \Rightarrow p = \frac{16}{0+2} - 2 = 6$. The q intercept is where $p = 0 \Rightarrow p = \frac{16}{q+2} - 2 = 0 \Rightarrow q = 6$.

(b) $TR \equiv pq = \frac{16q}{q+2} - 2q$. So $MR \equiv \frac{dTR}{dq} = \frac{(q+2)16 - 16q}{(q+2)^2} - 2$.

(c) Maximum TR when $MR = \frac{32 - 2(q+2)^2}{(q+2)^2} = 0$. This is true when $32 - 2(q + 2)^2 = 0$. This rearranges as $-2q^2 - 8q + 24 = 0$. This quadratic equation has roots $q = 2$ or -6 (but we discard the negative solution). When $q = 2$, $p = \frac{16}{2+2} - 2 = 2$. Maximum $TR \equiv pq = 4$.

(d) See Online Resource Centre.

5. Inverse demand function is $p = 5$. Therefore $TR \equiv pq = 5q$ and $MR \equiv \frac{dTR}{dq} = 5$. (This is the inverse demand function for the product of a perfectly competitive firm.) For graphs see Online Resource Centre.

6. (a) See Online Resource Centre.

(b) See appendix 8.2 to chapter 8, especially rule 8.4.

Progress exercise 8.3

1. (a) Profit function is $\Pi = 67q - (q^2 - 3q + 500) = -q^2 + 70q - 500$. For maximum Π we set $\frac{d\Pi}{dq} = -2q + 70 = 0$, with solution $q = 35$. Since $\frac{d^2\Pi}{dq^2} = -2 < 0$, $q = 35$ is a maximum. Max $\Pi = -(35)^2 + 70(35) - 500 = 725$.

    (b) $AC = q - 3 + \frac{500}{q}$. Minimum AC is when $\frac{dAC}{dq} = 1 - 500q^{-2} = 0 \Rightarrow q = 22.36$. Since $\frac{d^2AC}{dq^2} = \frac{1000}{q^3} > 0$ when $q = 22.36$, this is a minimum. The most profitable output ($q = 35$) > output at which AC is minimized ($q = 22.36$). In general there is no reason to suppose the two will coincide as profit depends on TR and TC. AC in itself has no relevance.

    (c), (d) See Online Resource Centre.

2. (a) In question 1, $MC \equiv \frac{dTR}{dq} = 2q - 3$. We also have $TR = 67q$ so $MR \equiv \frac{dTR}{dq} = 67$. We know Π maximum requires $MC = MR$ (see section 8.16). So Π maximum requires $2q - 3 = 67 \Rightarrow q = 35$. (We assume the second order condition for a maximum is satisfied.) If p rises to 68, the $MC = MR$ condition becomes $2q - 3 = 68 \Rightarrow q = 35\frac{1}{2}$. If p rises to 69 this becomes $2q - 3 = 69 \Rightarrow q = 36$, and so on (see the graph on Online Resource Centre).

    (b) If the price has some unspecified value p^*, the $MC = MR$ condition becomes $2q - 3 = p^*$. This can be rearranged as $q = \frac{1}{2}p^* - \frac{3}{2}$. This is the firm's supply curve, as it tells us what quantity the firm will choose to supply, at any given market price p^*, in order to maximize profit. (Note: this assumes perfect competition.)

    (c) AC is minimized when $q = 22.36$ (from 1(b) above). The firm produces output $q = \frac{1}{2}p^* - \frac{3}{2}$ (from 2(b) above). So the firm will produce output $q = 22.36$ only if $22.36 = \frac{1}{2}p^* - \frac{3}{2}$, from which $p^* = 41.72$. Thus when $q = 22.36$, $p^* = AC = 41.72$. For the graph, see Online Resource Centre.

3. (a) From 1(a) above, $\Pi = -q^2 + 70q - 500$ where $500 = $ fixed costs. To find Π max, we solve the equation $\frac{d\Pi}{dq} = -2q + 70 = 0$. An increase in fixed costs from 500 to 1000 (or any other level) has no effect on $\frac{d\Pi}{dq}$ and therefore does not change the profit maximizing value of q. However, fixed costs do appear in our expression for Π above, so an increase in fixed costs fom 500 to 1000 will reduce profits by 500.

    (b) If fixed costs rise to 1250, $q = 35$ remains the profit maximizing output, but maximum profit is now -25 (= minimum loss).

4. (a) $MC = \frac{dTC}{dq} = 2q - 3$. Also, $MR = -2q + 105$. Setting $MC = MR$ (for max Π) and solving for q gives $q = 27$ (and $p = -q + 105 = 78$). The second order condition for a maximum of Π, $\frac{dMC}{dq} > \frac{dMR}{dq}$, is satisfied since $\frac{dMC}{dq} = 2$ and $\frac{dMR}{dq} = -2$ (see equation 8.8 in section 8.20). When $q = 27$, $\Pi = TR - TC = -q^2 + 105q - (q^2 - 3q + 500) = 958$.

    (b) From 1(b), the minimum AC is at $q = 22.36$ which is less than $q = 27$ which maximizes profit.

    (c), (d) See Online Resource Centre.

5. (a) New TR is $TR = -2q^2 + 159q$. So new MR is $MR = -4q + 159$. TC is unchanged. Setting $MR = MC$ gives $-4q + 159 = 2q - 3$, with solution $q = 27$ (no change). From the new inverse demand function, $p = -2(27) + 159 = 105$ (compared with old price, 78). New Π is $\Pi = -2q^2 + 159q - (q^2 - 3q + 500) = -3q^2 + 162q - 500 = 1687$ when $q = 27$ (compared with old profits, 958).

    (b) New MR curve cuts the unchanged MC curve at the same point as the old MR curve, hence the output is unchanged. But the new inverse demand function and the new MR function have higher intercept and steeper slope. Consequently, for any given MR, p is now higher. In turn, this means that profits are higher since TR is higher but TC is unchanged.

    (c) See Online Resource Centre.

6. (a) $TC = 20q + 500$. So $MC = 20$. Given $p = -3q + 146$, we have $TR \equiv pq = (-3q + 146)q = -3q^2 + 146q$. So $MR = -6q + 146$. For maximum profit we set $MC = MR \Rightarrow 20 = -6q + 146 \Rightarrow q = 21$. We then have $\Pi = TR - TC = -3q^2 + 146q - (20q + 500) = 823$ when $q = 21$.

    (b) No, because $AC = 20 + \frac{500}{q}$ which falls continuously as q increases. (That is, the AC curve has no minimum.)

    (c), (d), (e) See Online Resource Centre.

7. (a) As in (6) above, $MC = 20$. But now $p = 25$, so $TR = 25q$, from which $MR = 25$. Thus $MR > MC$ at all output levels. Therefore the firm can always increase its profits by increasing its output, hence would wish to increase output without limit. There is no equilibrium.

    (b) Break-even output is where $TR = TC$ and profits are therefore zero. We have $TR = 25q$ and $TC = 20q + 500$. Setting $TR = TC$ and solving the resulting equation gives $q = 100$. (Note, dividing both sides of $TR = TC$ by q gives $AR = AC$, so break-even can also be found by setting $AR = AC$.)

    (c) If $p = 15$, then $MR = p = 15$. And $MC = 20$ (unchanged). Again there is no equilibrium, but this time because $MC > MR$ at all output levels. Therefore the firm can always increase its profits (= reduce its losses) by reducing output, hence would reduce output to zero. (This is sometimes known as a 'corner solution'; see chapter W21.)

    (d) No, because MC is the only cost measure that is relevant in choosing the most profitable output, and MC is independent of fixed costs. (See also the answer to (3), above.) A change in fixed costs changes profits but does not change the profit maximizing output.

    (e) For algebra, see above. For figures see Online Resource Centre.

8. (a) We have $MC = 6q^2 - 96q + 700$.

    Given $p = 532$ we have $TR = 532q$, so $MR = 532$.

    (i) Setting $MR = MC \Rightarrow 532 = 6q^2 - 96q + 700 \Rightarrow 6q^2 - 96q + 168 = 0$. The roots of this quadratic equation are $q = 2$ and $q = 14$ (that is, two SPs). To determine which of these is a maximum of Π we must examine the second order condition. The second order condition for a maximum is $\frac{dMC}{dq} > \frac{dMR}{dq}$ and for a minimum $\frac{dMC}{dq} < \frac{dMR}{dq}$ (see equation 8.8 in section 8.20). Here, $\frac{dMC}{dq} = 12q - 96$ and $\frac{dMR}{dq} = 0$. When $q = 2$, $\frac{dMC}{dq} = 24 - 96 < 0$ so $\frac{dMC}{dq} < \frac{dMR}{dq}$. Therefore $q = 2$, a minimum of Π. When $q = 14$, $\frac{dMC}{dq} = 12(14) - 96 > 0$ so $\frac{dMC}{dq} > \frac{dMR}{dq}$. Therefore $q = 14$, a maximum of Π.

    (ii) $\Pi = TR - TC = -2q^3 + 48q^2 + 168q - 100$. Therefore $\frac{d\Pi}{dq} = -6q^2 + 96q + 168 = 0$ for SP. The roots of this quadratic equation are $q = 2$ and $q = 14$. For maximum profits we need $\frac{d^2\Pi}{dq^2} < 0$. Here $\frac{d^2\Pi}{dq^2} = -12q + 96$. This is positive when $q = 2$, so this is a minimum of Π; and negative when $q = 14$ so this is a maximum of Π. At $q = 2$, $\Pi = -260$ (loss). At $q = 14$, $\Pi = 1468$.

    (b) See Online Resource Centre.

(c) When $q = 2$, MC is falling while MR is constant. Therefore the second order condition for a *minimum* of Π, $\frac{dMC}{dq} < \frac{dMR}{dq}$, is satisfied, and any change in output (increase or decrease) will increase profits. When $q = 14$, MC is now rising so $\frac{dMC}{dq} > \frac{dMR}{dq}$. Thus the second order condition for a *maximum* of Π is satisfied, and any change in output reduces profits.

(d) A firm would only choose to produce at an output at which MC was decreasing if MR at that output was decreasing even faster, for only then would the second order condition for maximum profit be satisfied. Graphically, both the MC and MR curves would be negatively sloped, with MR steeper.

9. (a) (i) We have $MC = 12q^2 - 370q + 4500$. Given $q = -\frac{1}{5}p + 600$ (demand function) $\Rightarrow p = -5q + 3000$ (inverse demand function). Therefore $TR = -5q^2 + 3000q$ and $MR = -10q + 3000$. Setting $MR = MC \Rightarrow 12q^2 - 360q + 1500 = 0$. Solving this quadratic equation gives $q = 5$ or 25. Second order condition. Here $\frac{dMC}{dq} = 24q - 370$, and $\frac{dMR}{dq} = -10$. When $q = 5$, $\frac{dMC}{dq} < \frac{dMR}{dq}$ so minimum of Π. When $q = 25$, $\frac{dMC}{dq} > \frac{dMR}{dq}$ so maximum of Π. Profit levels: we have $\Pi = TR - TC = -4q^3 + 180q^2 - 1500q + 500$. So when $q = 5$, $\Pi = -4000$; when $q = 25$, $\Pi = 12000$.

(ii) From (i) above, $\Pi = -4q^3 + 180q^2 - 1500q - 500$. So for maximum profit we want $\frac{d\Pi}{dq} = -12q^2 + 360q - 1500 = 0 \Rightarrow -12(q^2 + 30q - 125) = 0$. From (i) above we know that the solution to this quadratic equation is $q = 5$ or 25. Second order condition: $\frac{d^2\Pi}{dq^2} = -12(2q + 30) = 24(15 - q)$. This is positive at $q = 5$ so is a minimum; and negative at $q = 25$ so is a maximum. Profits are obviously the same as in (i).

(b) See Online Resource Centre.

(c) As in question 8(c) above.

10. We know that when AC is at its minimum, $MC = AC$. We also know that when profit is at its maximum, $MC = MR$. For both conditions to be satisfied simultaneously therefore requires $MR = MC = AC$. (In both the monopoly case and the perfect competition case, we will assume that the necessary second order conditions for a minimum of AC and a maximum of Π are satisfied.)

(a) Monopoly. For a monopolist, the MR curve is downward sloping, so $MR = MC = AC$ implies that the MR curve cuts the AC curve at the latter's minimum.

(b) Perfect competition. For a perfectly competitive firm, $MR = p = $ a constant (independent of output) so the MR curve is a horizontal straight line. Then if $MR = MC = AC$ this implies that the horizontal MR curve is tangent to the minimum point on the AC curve.

For a graphical treatment, see Online Resource Centre.

Progress exercise 9.1

1. (a) (i) Arc elasticity $\equiv \dfrac{\frac{\Delta q}{\Delta p}}{\frac{q_0}{p_0}} = \dfrac{2}{\frac{5}{5}} = 2$; (ii) $\dfrac{2}{1.9} = 1.052632.$

(b) (i) Point elasticity $\equiv \dfrac{\frac{dq}{dp}}{\frac{q}{p}} = \dfrac{2}{\frac{5}{5}} = 2$; (ii) $\dfrac{2}{1.9} = 1.052632.$

(c) Point elasticity $\equiv \dfrac{\frac{dq}{dp}}{\frac{q}{p}} = \dfrac{2}{\frac{2p-5}{p}} = \dfrac{2p}{2p-5}.$

As p increases without limit, the difference between $2p$ and $2p - 5$ becomes insignificant, and their ratio approaches 1.

(d) For the graph see Online Resource Centre.

2. (a) Arc elasticity $= \dfrac{\frac{\Delta q}{\Delta p}}{\frac{q_0}{p_0}} = \dfrac{7.1}{5} = 1.42;$

Point elasticity $= \dfrac{\frac{dq}{dp}}{\frac{q}{p}} = \dfrac{7}{5} = 1.4.$

The two elasticity measures are very similar because, in the demand function, the coefficient of p^2 (0.1) is small relative to the coefficient of p (5), so the function is almost linear for small changes in p.

(b) See Online Resource Centre.

3. (a) For $p_0 = 10$, $p_1 = 11$, arc elasticity is $\frac{4.37}{3.7} = 1.181$ to three decimal places.

Point elasticity at $p = 10$ is $\dfrac{\frac{dq}{dp}}{\frac{q}{p}} = \dfrac{-0.06p + 5}{3.7} = \dfrac{4.4}{3.7} = 1.189$ to 3 dp.

As in (2), the elasticity measures are close because, for small changes in p, the demand function is almost linear.

(b) See Online Resource Centre.

Progress exercise 9.2

1. (a) $E_A^D = $ arc elasticity of demand $= \dfrac{\frac{\Delta q}{\Delta p}}{\frac{q_0}{p_0}} = \dfrac{\Delta q}{\Delta p} \cdot \dfrac{p_0}{q_0}.$

   (i) $p_0 = 16$, $p_1 = 17 \Rightarrow q_0 = 128$, $q_1 = 126$

   $E_A^D = \dfrac{-2}{\frac{128}{16}} = -2 \cdot \dfrac{16}{128} = -0.25.$

   (ii) $p_0 = 48$, $p_1 = 49 \Rightarrow q_0 = 64$, $q_1 = 62$

   $E_A^D = \dfrac{-2}{\frac{64}{48}} = -2 \cdot \dfrac{48}{64} = -1.5.$

(b) $E^D = $ point elasticity of demand $= \dfrac{\frac{dq}{dp}}{\frac{q}{p}} = \dfrac{dq}{dp} \cdot \dfrac{p_0}{q_0}.$

   (i) $p_0 = 16$, $q_0 = 128$, $\dfrac{dq}{dp} = -2 \Rightarrow E^D = \dfrac{-2}{\frac{128}{16}} = -0.25.$

   (ii) $p_0 = 48$, $q_0 = 64$, $\dfrac{dq}{dp} = -2 \Rightarrow E^D = \dfrac{-2}{\frac{64}{48}} = -1.5.$

(c) See Online Resource Centre.

(d) In both E_A^D and E^D, $\frac{q_0}{p_0}$ appears in the denominator. When p_0 is low (and consequently q_0 is high), then $\frac{q_0}{p_0}$ is large and therefore E_A^D and E^D are small (in absolute value). When p_0 is high, the reverse is true. So even though the slope of the demand function in this example is constant, the elasticity increases in absolute value (that is, demand becomes more elastic) as the price increases. With q on the vertical axis, $\frac{q}{p}$ is the slope of a ray from the origin to the demand curve. This slope decreases as p increases.

2. (a) (i) $p_0 = 5$, $p_1 = 6$; $q_0 = 684.5$, $q_1 = 639$

   $\Rightarrow E_A^D = \dfrac{\frac{-45.5}{1}}{\frac{684.5}{5}} = -0.3324$ (to 4 dp).

   (ii) $p_0 = 12$, $p_1 = 13$; $q_0 = 198$, $q_1 = 96.5$

   $\Rightarrow E_A^D = \dfrac{\frac{-101.5}{1}}{\frac{198}{12}} = -6.1515.$

(b) (i) $p_0 = 5$, $q_0 = 684.5$, $\frac{dq}{dp} = -8p - 1.5 = -41.5$ when

$p = 5 \Rightarrow E^D = \frac{5}{664.5}(-41.5) = -0.3123$ (to 4 dp).

(ii) $p_0 = 12$, $q_0 = 198$, $\frac{dq}{dp} = -97.5$ when $p = 12$

$\Rightarrow E^D = \frac{12}{198}(-97.5) = -5.9091$.

(c) See Online Resource Centre.

(d) In this example, as p increases the demand curve becomes steeper, so both $\frac{dq}{dp}$ and $\frac{\Delta q}{\Delta p}$ increase in absolute value (though negative, of course). At the same time, the slope of the ray from the origin $(= \frac{q}{p})$ decreases. Thus both arc and point elasticities increase in absolute value because the numerator is increasing and the denominator is decreasing. Demand becomes more elastic. (Specifically, the absolute value of the point elasticity is 0.3123 when $p = 5$, but 5.9091 when $p = 12$.) Note that the demand elasticity increases, as p increases, more rapidly than in the case of the linear demand function of question (1).

3. (a) Demand function is $q = \frac{50}{p+2}$.

(b) Using the quotient rule, $\frac{dq}{dp} = \frac{(p+2)(0) - 50(1)}{(p+2)^2} = \frac{-50}{(p+2)^2}$.

So $E^D = \frac{p}{q}\frac{dq}{dp} = \frac{p}{\frac{50}{p+2}} \cdot \frac{-50}{(p+2)^2} = \frac{-50p}{(p+2)^2} \cdot \frac{p+2}{50} = -\frac{p}{p+2}$.

Because $p + 2 > p$ (assuming $p > 0$), it follows that $\frac{p}{p+2}$ is positive but less than 1, that is:

$0 < \frac{p}{p+2} < 1$. Multiplying by -1 we get

$0 > -\frac{p}{p+2} > -1 \Rightarrow 0 > E^D > -1$; that is, inelastic demand at any price.

(c) See Online Resource Centre.

4. (a) Demand function is $q = \frac{50}{p-2} = 50(p-2)^{-1}$.

(b) Using the function of a function rule, $\frac{dq}{dp} = -50(p-2)^{-2}$.

So $E^D \equiv \frac{p}{q}\frac{dq}{dp} = \frac{p}{50(p-2)^{-1}} \cdot \frac{-50}{(p-2)^2} = -\frac{p}{p-2}$.

Assuming $p > 0$, $\frac{p}{p-2} > 1$. Multiplying by -1,

$-\frac{p}{p-2} < -1 \Rightarrow E^D < -1$; that is, elastic demand at any p.

(c) See Online Resource Centre.

5. (a) Demand function $q = \frac{50}{p} = 50p^{-1}$ so $\frac{dq}{dp} = -50p^{-2}$.

(b) $E^D = \frac{p}{q}\frac{dq}{dp} = \frac{p}{50p^{-1}} - 50p^{-2} = \frac{-50p^{-1}}{50p^{-1}} = -1$ (at any price).

(c) See Online Resource Centre.

Progress exercise 9.3

1. Given $q = -2p + 160$ the inverse function is $p = -\frac{1}{2}q + 80$.

(a) $TR \equiv pq = -\frac{1}{2}q^2 + 80q$; $MR \equiv \frac{dTR}{dq} = -q + 80$.

(b) TR is a maximum when $MR = -q + 80 = 0 \Rightarrow q = 80$. Also $\frac{d^2TR}{dq^2} = -1 < 0$ when $q = 80$ so the second order condition for a maximum is satisfied. Thus $q^* = 80$ and $p^* = -\frac{1}{2}q^* + 80 = 40$.

(c) Given $q = -2p + 160$, $\frac{dq}{dp} = -2$,

so $E^D \equiv \frac{p}{q}\frac{dq}{dp} = \frac{p}{-2p+160}(-2) = \frac{p}{p-80}$.

So when $p = p^* = 40$, $E^D = \frac{40}{40-80} = -1$.

(d) If $p > 40$, then we can suppose that $p = 40 + h$, where h is some unknown but positive number. (Here h is called a 'slack variable', which converts the inequality $p > 40$ into the equation $p = 40 + h$.) Then we have

$E^D = \frac{p}{p-80} = \frac{40+h}{40+h-80} = \frac{40+h}{h-40} = -\frac{40+h}{40-h}$.

Since $40 + h > 40 - h$ (because h is positive), we have $\frac{40+h}{40-h} > 1$. Multiplying by -1 gives: $E^D = -\frac{40+h}{40-h} < -1$ so demand is elastic when $p > 40$.

(e) If $p < 40$ we can repeat all the steps above, but this time h is negative, so $40 + h < 40 - h$ and therefore $\frac{40+h}{40-h} < 1$. Multiplying by -1 gives

$E^D = -\frac{40+h}{40-h} > -1$ so demand is inelastic when $p < 40$.

(f) For graphs, see Online Resource Centre. In words, when demand is elastic, this means that following a price reduction the gain in total revenue from the increase in q is greater than the loss of revenue from the fall in p, hence the total revenue rises. But if the total revenue rises when q rises, then by definition the marginal revenue is positive. So when demand is elastic, marginal revenue is positive. Similarly when demand is inelastic, marginal revenue is negative.

2. (a) Given $q = -4p^2 - 1.5p + 792$, we have $TR \equiv pq = -4p^3 - 1.5p^2 + 792p$ (as a function of p).

(b) Set $MR = -12p^2 - 3p + 792 = 0$ for maximum TR. Solving this quadratic gives $p = -8.25$ or 8. We have $\frac{d^2TR}{dp^2} = -24p - 3 < 0$ when $p = 8$ so the second order condition for the maximum is satisfied. When $p = 8$, $q = 524$, and $TR = pq = 4192$.

(c) From (a), $\frac{dq}{dp} = -8p - 1.5$, so

$E^D = \frac{p}{q}\frac{dq}{dp} = \frac{p}{-4p^2 - 1.5p + 792} \cdot (-8p - 1.5)$

$= \frac{-8p^2 - 1.5p}{-4p^2 - 1.5p + 792}$.

(d) From (c), $E^D = -1 \Rightarrow -8p^2 - 1.5p = 4p^2 + 1.5p - 792 \Rightarrow -12p^2 - 3p + 792 = 0$. But from (b) above, we know that $MR = -12p^2 - 3p + 792 = 0$ (and TR is at its maximum) when $p = 8$. So $E^D = -1$ when TR is at its maximum.

(e) Elastic demand $\Rightarrow E^D < -1$. From (d), $E^D < -1 \Rightarrow -8p^2 - 1.5p < 4p^2 + 1.5p - 792 \Rightarrow -12p^2 - 3p + 792 = 0$. As we know from (d) that $-12p^2 - 3p + 792 = 0$ when $p = 8$, and the signs of p^2 and p are both negative, it follows that $-12p^2 - 3p + 792 < 0$ when $p > 8$. In the same way, we can show that $E^D > -1$ (inelastic demand) when $p < 8$.

3. (a) Given $p = \frac{50}{q} - 2$, $TR = pq = 50 - 2q$; $MR = -2$. See Online Resource Centre for graphs.

(b) From (a), $MR = -2$. Since MR always negative, this implies that the TR function is always negatively sloped. This also follows from the fact that demand is inelastic at any price, as we found in exercise 9.2(3). Inelastic demand means that a price reduction, and associated quantity increase, reduces TR.

4. (a) In this case we have $TR = pq = 50 + 2q$. $MR = 2$. See Online Resource Centre for graphs. Since MR always positive, this implies that the TR function is always positively sloped. This also follows from the fact that demand is elastic at any price, as we found in exercise 9.2(4). Elastic demand means that a price reduction, and associated quantity increase, increases TR.

(b) It seems very unlikely that any demand curve would have constant elasticity at *every* price, as in these two example. But it is not unreasonable to suppose that a demand curve might have constant elasticity over a range of values of p, and this assumption is frequently made to simplify problems in economic analysis, both theoretical and applied.

Progress exercise 9.4

1. Given $TC = 9q^2 + 2q + 8100$.

(a) $MC \equiv \frac{dTC}{dq} = 18q + 2$: $AC \equiv \frac{TC}{q} = 9q + 2 + \frac{8100}{q}$.

(b) AC is a minimum where $\frac{dAC}{dq} = 9 - \frac{8100}{q^2} = 0 \Rightarrow 9q^2 = 8100 \Rightarrow q = 30$. Since $\frac{d^2AC}{dq^2} = \frac{16200}{q^3} > 0$ when $q = 30$, this is a minimum.

$MC = AC \Rightarrow 18q + 2 = 9q + 2 + \frac{8100}{q} \Rightarrow 9q^2 = 8100 \Rightarrow q = 30$.

(c) $MC < AC \Rightarrow 18q + 2 < 9q + 2 + \frac{8100}{q} \Rightarrow 9q^2 < 8100 \Rightarrow q < 30$. Also, AC is falling when $\frac{dAC}{dq} = 9 - \frac{8100}{q^2} < 0 \Rightarrow 9q^2 < 8100 \Rightarrow q < 30$. So when $q < 30$, $MC < AC$ and AC is falling. Similarly we can show that when $q > 30$, $MC > AC$ and AC is rising.

(d) $E_{TC} \equiv \frac{\frac{dTC}{dq}}{\frac{TC}{q}} \equiv \frac{MC}{AC}$.

(e) $E_{TC} < 1 \Rightarrow \frac{MC}{AC} < 1 \Rightarrow MC < AC$. But, from (c) $MC < AC \Rightarrow AC$ is falling. So $E_{TC} < 1 \Rightarrow AC$ is falling. Similarly, $E_{TC} > 1 \Rightarrow AC$ is rising.

(f) See Online Resource Centre.

2. (a) $MC \equiv \frac{dTC}{dq} = 2aq + b; AC \equiv \frac{TC}{q} = aq + b + \frac{c}{q}$.

(b) AC is a minimum where $\frac{dAC}{dq} = a - \frac{c}{q^2} = 0 \Rightarrow q = \sqrt{\frac{c}{a}}$. Also, $\frac{d^2AC}{dq^2} = \frac{2c}{q^3}$. We assume $q > 0$. Assume c (fixed costs) is also positive, so then $\frac{d^2AC}{dq^2} > 0$ so the AC curve has a minimum when $q > 0$.

$MC = AC \Rightarrow 2aq + b = aq + b + \frac{c}{q} \Rightarrow aq = \frac{c}{q} \Rightarrow q = \sqrt{\frac{c}{a}}$. But from above, we know that the minimum of AC is where $q = \sqrt{\frac{c}{a}}$. So $MC = AC$ at minimum AC.

(c) $MC < AC \Rightarrow 2aq + b < aq + b + \frac{c}{q} \Rightarrow aq < \frac{c}{q} \Rightarrow q^2 < \frac{c}{a}$. (Note, we here multiplied both sides of the inequality by q and divided by a. As both are positive, this did not reverse the inequality.)

AC is falling when $\frac{dAC}{dq} = a - \frac{c}{q^2} < 0 \Rightarrow a < \frac{c}{q^2} \Rightarrow q^2 < \frac{c}{a}$. So we have shown that when AC is falling, $MC < AC$. In the same way, we can show that when AC is rising, $MC > AC$.

(d) $E_{TC} \equiv \frac{\frac{dTC}{dq}}{\frac{TC}{q}} \equiv \frac{MC}{AC}$.

(e) $E_{TC} < 1 \Rightarrow \frac{MC}{AC} < 1 \Rightarrow MC < AC$. But, from (c) $MC < AC \Rightarrow AC$ falling. So $E_{TC} < 1 \Rightarrow AC$ falling. Similarly, $E_{TC} > 1 \Rightarrow AC$ rising.

(f) The figure is identical to that in part (f) of the previous question (see Online Resource Centre), except that we cannot calibrate the axes. Recall the assumptions we made in proving the results above, that in the general quadratic TC function, $TC = aq^2 + bq + c$, both a and c were positive.

3. Given $C = 0.9Y + 100$.

(a) $MPC \equiv \frac{dC}{dY} = 0.9$: $APC \equiv \frac{C}{Y} = 0.9 + \frac{100}{Y}$.

(b) $MPC < APC \Rightarrow 0.9 < 0.9 + \frac{100}{Y} \Rightarrow 0 < \frac{100}{Y}$ which is true if $Y > 0$ (as we assume is the case).

Since $\frac{100}{Y}$ decreases as Y increases, APC approaches MPC as Y increases without limit.

(c) Elasticity by definition equals $\frac{Y}{C}\frac{dC}{dY} \equiv \frac{\frac{dC}{dY}}{\frac{C}{Y}} \equiv \frac{MPC}{APC}$.

Given $C = 0.9Y + 100$, elasticity $= \frac{0.9}{\frac{0.9 \cdot 100}{Y}} = \frac{0.9Y}{0.9Y + 100}$.

As $0.9Y < 0.9Y + 100$, and $Y > 0$, elasticity is less than 1.

(d) See figure S3.

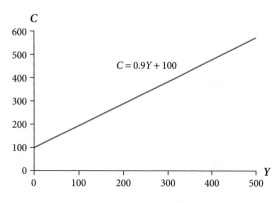

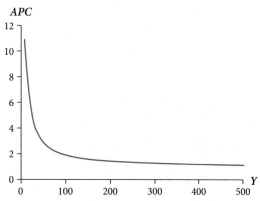

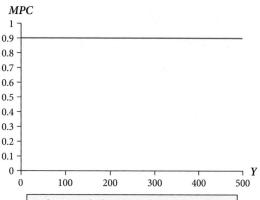

In this example the APC is always greater than the MPC due to the additive constant in the consumption function. However, the APC falls continuously as Y increases, and approaches the MPC as Y approaches infinity. Note that $C > Y$ when $Y < 1000$, implying the consumer is able to borrow or draw on past savings.

Figure S3 Graphs for question 3(d) of exercise 9.4.

4. (a) Given $C = aY + b$

$MPC \equiv \frac{dC}{dY} = a$: $APC \equiv \frac{C}{Y} = a + \frac{b}{Y}$.

So $MPC > APC \Rightarrow a > a + \frac{b}{Y} \Rightarrow 0 > \frac{b}{Y}$

which is true if $b < 0$ (since we assume $Y > 0$). The consumption function would then have a negative intercept on the Y-axis. This implies there is some positive level of income at which consumption falls to zero, which seems very unlikely.

(b) $MPC = APC \Rightarrow a = a + \frac{b}{Y} \Rightarrow b = 0$. This consumption function passes through the origin.

Progress exercise 10.1

1. Using rule 10.1, $\sum_{50} = \frac{50}{2}\{2 + (50 - 1)1\} = 1275$.

2. Using rule 10.2, 11th term is $AR^{n-1} = 25(2)^{10} = 25,600$.

3. Using rule 10.2, $\sum_{10} = \frac{A(R^n - 1)}{R - 1} = \frac{1(2^{10} - 1)}{2 - 1} = 2^{10} - 1 = 1023$.

4. Using rule 10.2, $\sum_{10} = \frac{\frac{1}{2}\left(\left(\frac{1}{2}\right)^{10} - 1\right)}{\frac{1}{2} - 1} = 0.999023$ to 6 dp.

Using rule 10.2a, $\sum_{\infty} = \frac{A}{1 - R} = \frac{\frac{1}{2}}{1 - \frac{1}{2}} = 1$.

5. (a)

Month	C	I	Y = C + I	ΔY
1	75	25	100	
2	75	35	110	10
3	82.5	35	117.5	7.5
4	88.125	35	123.125	5.625
5	92.344	35	127.344	4.219
6	95.508	35	130.508	3.164
7	97.881	35	132.881	2.373

(b) From col. 5 of the table above, we see that the increases in Y, ΔY, form the series: $10, 10 \times 0.75, (10 \times 0.75)0.75,$ ... This series is a GP with first term 10, common ratio 0.75. The increase after 1 year is the cumulative sum of these monthly increases. Using rule 10.2, this sum is

$$\sum_{12} = \frac{10(1 - (0.75)^{12})}{1 - 0.75} = 38.733.$$

(c) As the common ratio lies between 0 and 1, we can apply rule 10.2a. The sum to infinity is thus

$$\sum_{\infty} = \frac{A}{1 - R} = \frac{10}{1 - 0.75} = 40.$$

(d) Although the rise in Y never quite reaches 40, (c) above gives us a close approximation to the increase in Y after 1 year, 2 years, and so on; and is easy to calculate.

Progress exercise 10.2

1. Using rule 10.3, GDP after 25 years will be: $y = 100(1 + 0.025)^{25} = 185.39$.

2. Using rule 10.3, we have $538 = 411(1 + r)^8 \Rightarrow \left(\frac{538}{411}\right)^{1/8} = 1 + r = 1.03423$.

So $r = 1.03423 - 1 = 0.03423 = 3.423\%$.

3. Using rule 10.3, we have $115 = 100(1 + r)^1 \Rightarrow r = \frac{115}{100} - 1 = 0.15 = 15\%$. Current height is 115, so after 5 more years height will be $y = 115(1 + 0.15)^5 = 115(2.0114) = 231.3$.

4. (a) $y = a(1 + r)^x = 100(1 + 0.06)^5 = 133.82$.

(b) $y = a\left(1 + \frac{r}{n}\right)^{nx} = 100\left(1 + \frac{0.06}{2}\right)^{2 \times 5} = 134.39$.

(c) $y = a\left(1 + \frac{r}{n}\right)^{nx} = 100\left(1 + \frac{0.06}{12}\right)^{12 \times 5} = 134.89$.

5. Bank A: after 1 year, 1 euro becomes $y = 1(1 + 0.05)^1 = 1.05$ euros.

So effective annual rate = nominal rate = 5%.

Bank B: after 1 year, 1 euro becomes $y = 1\left(1 + \frac{0.049}{12}\right)^{12} = (1.00408)^{12} = 1.0501$.

So effective annual rate = 5.01%, while nominal annual rate = 4.9%.

6. $y = a(1 + r)^x$ with $a = 100,000$; r unknown; $x = 5$; $y = 20,000$ $\Rightarrow 20,000 = 100,000(1 + r)^5 \Rightarrow \left(\frac{20,000}{100,000}\right)^{1/5} = 1 + r = 0.7248$.

So $r = 0.7248 - 1 = -0.2752 = -27.52\%$.

Progress exercise 10.3

1. (a) Using rule 10.5, $PV = y = \frac{20,000}{(1 + 0.05)^{10}} = \frac{20,000}{1.629} = 12,277.5$.

(b) I must invest 12,277.5 because, from (a) above, $(12,277.5)(1 + 0.05)^{10} = 20,000$.

(c) $x = 12,277.5$, because from (b) above, 12,277.5 now plus 10 years' interest will amount to 20,000 in 10 years' time.

2. (a) Using rule 10.7,

$$PV = \frac{10,000}{1.05} + \frac{10,000}{(1.05)^2} + \frac{10,000}{(1.05)^3} + \frac{10,000}{(1.05)^4} + \frac{10,000}{(1.05)^5}$$
$$+ \frac{10,000}{(1.05)^6} + \frac{10,000}{(1.05)^7} + \frac{10,000}{(1.05)^8} + \frac{10,000}{(1.05)^9} + \frac{10,000}{(1.05)^{10}}$$
$$= 77,217.35.$$

(b) $x = 77,217.35$; that is, the answer to (a) above. This is because, if I can borrow or lend freely at 5% per year, I can exchange in the market place a lump sum of x now for the pension rights, and vice versa.

3. PV of series A: $\frac{50}{1.1} + \frac{100}{(1.1)^5} + \frac{150}{(1.1)^{10}} = 165.38$.

PV of series B: $\frac{40}{1.1} + \frac{100}{(1.1)^5} + \frac{165}{(1.1)^{10}} = 162.07$.

Although series A has lower total receipts (200 vs. 205), their PV is higher because they occur earlier than B's; that is, they are more 'front loaded' (the term used in the business world).

4. (a) (i) Discounted at 5%, A has the higher PV (see table).

(ii) At 10%, B has the higher PV.

	Present values		
	Series A (nuclear)	Series B (coal)	Higher PV
Discounted at 5% per year	233.68	222.05	A
Discounted at 10% per year	197.47	200.43	B

(b) If series A and B are costs (not profit or revenue), then it makes sense to choose the project with the *lower PV*. The point of this question is to show that, on this criterion, which project will be chosen may depend on the discount rate. Specifically, when the discount rate is high (10%), the heavy costs of decommissioning the nuclear power plant in year 5 are more heavily discounted, so nuclear 'wins' over coal. At a low discount rate (5%) the reverse is true; coal 'wins'.

Progress exercise 10.4

1. Using rule 10.9 to find the first repayment of principal,

$$P_1 = 5000\left[\frac{0.05}{(1.05)^5 - 1}\right] = 5000[0.18098] = 904.88.$$

The first year's interest is $rk = 0.05(5000) = 250$, so the first year payment is $P_1 + rk = 1154.88$, and this also equals the total payment in every year.

2. Using rule 10.9, $P_1 = 200\left[\dfrac{0.015}{(1.015)^{24} - 1}\right] = 200[0.03492]$

$$= 6.98.$$

The first month's interest is $rk = 0.15(200) = 3$, so the monthly payments are 9.98.

3. (a) Using rule 10.9, $P_1 = 100{,}000\left[\dfrac{0.06}{(1.06)^{25} - 1}\right]$

$$= 100{,}000\left[\frac{0.06}{3.2919}\right] = 1822.66.$$

The first year's interest is $rk = 0.06(100{,}000) = 6000$, so the annual payments are $1822.66 + 6000 = 7822.66$.

(b) The capital repayments are given by the series:

Year	1	2	3	4	5
Repayment of capital	P_1	$P_1(1+r)$	$P_1(1+r)^2$	$P_1(1+r)^3$	$P_1(1+r)^4$

So total repayments are the sum of these; that is,

$$P_1 + P_1(1+r) + P_1(1+r)^2 + P_1(1+r)^3 + P_1(1+r)^4$$
$$= P_1[1 + (1+r) + (1+r)^2 + (1+r)^3 + (1+r)^4] = 10{,}274.5$$

when $P_1 = 1822.66$ and $r = 0.06$

(c) From (a) above, the annual payments are 7822.66 so the total payments over 25 years are $7822.66 \times 25 = 195{,}566.5$. But this includes repayment of the loan of 100,000, so the total interest is $195{,}566.5 - 1000{,}000 = 95{,}566.5$.

(d) As stated in the question, at the end of each year you are charged interest on the amount owed at the beginning of that year. But if you pay monthly instalments, you are repaying some of this capital during the year. So at the end of the year you are charged interest on money that you have already repaid (except for the December instalment).

Progress exercise 11.1

1. Without using a calculator we can say:

(a) 1, because $10^1 = 10$; (b) 2, because $10^2 = 100$.

(c) If $x = \log 2$, then by definition $10^x = 2$, so x must be quite small, because even $10^{0.5} = \sqrt{10}$, which is more than 3 since $3^2 = 9$. So x must be less than 0.5, and is probably about 0.3.

(d) If $x = \log 2$, then $-x = \log\left(\frac{1}{2}\right)$; see below.

(e) If $10^x = 2$, then by definition $x = \log 2$. And if $10^x = 2$, $10^{-x} = \frac{1}{10^x} = \frac{1}{2}$. Then, by definition, $-x = \log\left(\frac{1}{2}\right)$. So $\log 2 = -\log\left(\frac{1}{2}\right)$ (and in general $\log A = -\log\frac{1}{A}$).

2. (a) See figure 11.2(a).

(b) In figure 11.2(a) the xs are the logs of the ys. So log 500 lies between 2 and 3 and looks to be nearer to 3 than 2. From (1)(e) above, $\log\left(\frac{1}{500}\right) = -\log 500$, so $\log\left(\frac{1}{500}\right)$ is between -2 and -3 and is nearer to -3 than to -2.

(c) See (1)(e) above.

(d) (i) Because $10^1 = 10$, log 10 = 1. Similarly, because $10^0 = 1$, log 1 = 0. Therefore the log of any number between 10 and 1 must lie between 1 and 0. And if log 1 = 0, the log of any number which is less than 1 must be less than 0; that is, negative.

(ii) Also, since $y = 10^x$ is positive for all x, and the xs are the logs of the ys, it follows that when y is negative there is no corresponding x, hence negative numbers have no logs.

3. (a) $\log y = 0.379$.

(b) $y_0 = \dfrac{1}{y_1}$ (see (1)(e) above).

4. (a) See figure 11.2(b).

(b) Since the xs are the logs of the ys, we can see that log 50 lies between 1 and 2 and is probably nearer to 2 than to 1. Also, $\log\frac{1}{50} = -\log 50$ so lies between -1 and -2.

(c) This graph is identical to that of $y = 10^x$ except that x is now on the vertical axis. If the labels on the two variables are interchanged, we then have the graph of $y = \log x$, which is the inverse of $x = 10^y$. (Confusing, isn't it?)

Progress exercise 11.2

1. From a calculator, log 2 = 0.3010. Without a calculator:

(a) $\log 4 = \log(2 \times 2) = \log 2 + \log 2 = 0.6020.$

(b) $\log(0.5) = \log\left(\frac{1}{2}\right) = -\log 2 = -0.3010.$

(c) $\log\sqrt{0.5} = \log((0.5)^{1/2}) = \frac{1}{2}\log(0.5) = -1505.$

(d) $\log\left(\frac{1}{8}\right) = \log 1 - \log 8 = 0 - \log(2^3) = -3\log 2 = -0.9030.$

2. Assume growth in annual jumps, $y = a(1 + r)^x$, with $y = 250{,}000$, $a = 50{,}000$, $r = 0.15$, x unknown. Thus $250{,}000 = 50{,}000(1.15)^x \Rightarrow 5 = (1.15)^x \Rightarrow \log 5 = x\log(1.15) \Rightarrow x = \frac{\log 5}{\log 1.15} = \frac{0.6990}{0.0607} = 11.52$ (years).

3. Assume growth in annual jumps, $y = a(1 + r)^x$ with $y = \frac{1}{2}a$, $r = -0.05$, x unknown. Thus: $\frac{1}{2}a = a(0.95)^x \Rightarrow \log\left(\frac{1}{2}\right) = \log[(0.95)^x] = x\log(0.95) \Rightarrow x = \frac{\log\left(\frac{1}{2}\right)}{\log 0.95} = \frac{-0.3010}{-0.0223} = 13.5$ (years).

4. (a) See figure 11.6.

(b) Same as figure 11.6 but with all numbers on x-axis multiplied by 4.

(c) Same as figure 11.6 but with all numbers on x-axis divided by 2.

Comparing $y_1 = 10^{0.25x}$ with $y_0 = 10^x$, $y_1 = 10^{0.25x} = (10^x)^{0.25} = y_0^{0.25}$.

Comparing $y_2 = 10^{2x}$ with $y_0 = 10^x$, $y_2 = (10^x)^2 = y_0^2$.

They all have the same intercept because $10^x = 10^{0.25x} = 10^{2x} = 1$ when $x = 0$.

5. Relation between these graphs and the figures in book:

(a) See figure 11.6.

(b) Take the graph of $y = 10^{0.5x}$ (figure 11.5), then increase the scale on the y-axis by a factor of 50. (For example, when $x = 2$, we now have $y = 500$ instead of $y = 10$.)

(c) Take the graph of $y = 10^x$ (figure 11.2a); increase the scale on the y-axis by a factor of 25, then shift the curve upwards by 5 units. (For example, when $x = 2$, we now have $y = 25 \times 100 + 5$ instead of $y = 100$.)

(d) Since $10^{-0.25x} = (10^{-x})^{0.25}$, we take the graph of $y = 10^{-x}$ (figure 11.6), then for every x value we raise the corresponding y value to the power 0.25. This gives the graph of $y = 10^{-0.25x}$. We then scale up all y values by a factor of 3 to get the graph of $y = 3(10^{-0.25x})$. (For example, when $x = -16$, $y = 10^{-x} = 10^{16}$ (a very large number); $y = 10^{-0.25x} = 10^4 = 10000$; $y = 3(10^{-0.25x}) = 30000$.)

Progress exercise 12.1

1. (a) Without using a calculator:

(1) If $x = \ln 30$, then by definition $e^x = 30$. Since e is (very roughly) 3, and $3^3 = 27$, x must be a little more than 3; say 3.5.

(2) $\ln 900 = \ln(30 \times 30) = \ln[(30)^2] = 2\ln 30$. From (i), $2\ln 30$ must be about $2 \times 3.5 = 7$.

(3) If $x = \ln 10$, $e^x = 10$ and since e equals roughly 3 and $3^2 = 9$, x must be a little more than 2, say 2.3.

(4) $\ln(0.1) = \ln\left(\frac{1}{10}\right) = \ln 1 - \ln 10 = 0 - 2.3 = -2.3$ (using (3) above). So the answer to (4) is the same as the answer to (3) but with sign reversed.

(5) If $x = \ln 2.71828$, then by definition $e^x = 2.71828$. But we know that $e = 2.71828$ (to 5 dp) so we have $e^x = e$. So x must equal 1.

(6) Since $e = 2.71828$ (to 5 dp), $e^3 = (2.71828)^3$. Since $3^3 = 27$, $(2.71828)^3$ must be a little less than 27; say 23.

(7) $e^1 = e = 2.71828$ approx.

(8) If $x = \ln 1$, then $e^x = 1$ by definition. So $x = 0$.

(9) If $x = \ln e$, $e^x = e$. So $x = 1$.

(10) $e^0 = 1$.

(b) (1) $e^{3.401} = 30$ (keystrokes SHIFT or INV or 2nd F (depending on the make of calculator), then ln, then 3.401. There will be rounding error.);

(2) $e^{6.802} = 900$;

(3) $e^{2.303} = 10$;

(4) $e^{-2.303} = 0.1$;

(5) $e^1 = 2.78128$ (keystrokes SHIFT/ln/1);

(6) $e^3 = 20.086$;

(7) $\ln 2.71828 = 1$;

(8) $e^0 = 1$ (keystrokes SHIFT/ln/0);

(9) $e^1 = 2.78128$;

(10) $\ln 1 = 0$.

2. (a) See figure 12.2.

(b) In figure 12.2 the xs are the natural logs of the ys, so for example, $3 = \ln 20.081$. Therefore $\ln 20$ must be very slightly less than 3, say 2.95. From question 1(a)(4) above we know that $\ln\left(\frac{1}{x}\right) = -\ln x$, so $\ln 0.05 = \ln\left(\frac{1}{20}\right) = -\ln 20 = -2.95$. Or, from figure 12.2 we can read off $\ln(0.05)$ as -3 (approx.).

3. (a) See figure 12.8.

(b) If $x = \ln(2.5)$, then $e^x = 2.5$ by definition. Since $e^1 = 2.718$ approx., which is more than 2.5, x must be a bit less than 1; say 0.85. This is confirmed by figure 12.9. (Calculator gives 0.916.) If $x = \ln(0.4)$, then $e^x = 0.4$. Since $e^0 = 1$, which is a lot more than 0.4, x must be quite a lot less than 0; say -0.85. (Calculator gives -0.916.) Note that $0.4 = \frac{1}{2.5}$, and therefore $\ln 0.4 = -\ln 2.5$.

(c) If $y = \ln x$, then by definition the inverse function is $x = e^y$. If we interchange the labels on the axes of the graph of $y = e^x$ (figure 12.2) it becomes the graph $x = e^y$, which is the inverse of $y = \ln x$.

4. (a) Apart from a change of scale on the x-axis (all values divided by 10), this graph is identical to that of $y = e^{0.5x}$ (see figure 12.4).

(b) See figure 12.5.

(c) Same as $y = e^x$ but with all labels on x-axes divided by 2 and all labels on y-axes multiplied by 100.

(d) See figure 12.8.

Progress exercise 12.2

1. (a) Annual jumps so $y = a(1 + r)^x$ with $a = 100$, $r = 0.03$. After 1 year, $y = 100(1.03)^1 = 103$. After 25 years $y = 100(1.03)^{25} = 209.4$ (using ^ key on calculator, or y^x key on some calculators). Graph is identical in shape to figure 12.9(b) but with slightly smaller vertical steps (3% instead of 4% growth per year).

(b) Continuous growth so $y = ae^{rx}$ with $a = 100$, $r = 0.03$. After 1 year $y = 100e^{0.03} = 103.045$ (using ^ key). After 25 years $y = 100e^{0.03(25)} = 100e^{0.75} = 211.7$. Note the compounding effect; y grows by 111.7%, compared with 75% (25 years at 3% per year) if growth is simple rather than compound growth. The graph is the same as figure 12.9(a) but turns up slightly more slowly as growth is only 3% in this case.

2. Continuous growth so $y = ae^{rx}$ with $a = 1$ (say), so $y = 2$ (doubling from 1 to 2), $r = 0.03$, x unknown. So $2 = e^{0.03x}$. Taking logs both sides:

$\ln 2 = \ln(e^{0.03x}) = 0.03x \ln e = 0.03x$ (since $\ln e = 1$)

so $x = \dfrac{\ln 2}{0.03} = \dfrac{0.6931}{0.03} = 23.105$ (years).

If $r = 0.23$ with x unknown, we have $2 = e^{r23} \Rightarrow \ln 2 = 23r$

so $r = \dfrac{\ln 2}{23} = \dfrac{0.6931}{23} = 0.03$.

Comparing the two problems we see that in both, $rx = 0.6931$. So when a variable doubles, the product of the growth rate, r, and the number of years, x, is always 0.6931. (The rule of 69.)

Progress exercise 12.3

1. (a) $y = ae^{rx} = 100e^{0.05x}$. The graph is like $y = e^{0.5x}$ in figure 12.4 with labels on x-axis multiplied by 10, as x now needs to be 10 times as large for a given y; and labels on y-axis multiplied by 100. (Note that the y intercept then becomes $y = 100$ instead of $y = 1$.)

(b) $y = 100e^{-0.05x}$. The graph is like figure 12.5 but with labels on the x-axis multiplied by 20, and on the y-axis by 100.

(c) In (a) $y = 100e^{0.05x}$, so taking natural logs on both sides gives

$\ln y = \ln 100 + \ln(e^{0.05x}) = \ln 100 + 0.05x \ln e$
$= \ln 100 + 0.05x$ (since $\ln e = 1$).

Sketch graph: see figure S4

In (b), following the same steps we get: $\ln y = \ln 100 - 0.05x$.

Graph: as figure S4, but with negative slope of -0.05

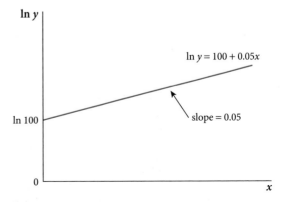

Figure S4 Sketch graph for question 1(c) of exercise 12.3.

2. $y = 100e^{0.05x}$. The graph is like figure 12.2 in the book, but with labels on the x-axis multiplied by 20, and on the y-axis by 100. Thus, for example, after 20 years we have $y = 100e^{0.05x} = 100e^1 = 271.828$.

After 30 years $(x = 30)$ $y = 100e^{1.5} = 100(4.482) = 448.2$ (using the 'shift' and 'ln' keys).

Absolute growth: $\Delta y = 448.2 - 100 = 348.2$.

Proportionate growth is $\dfrac{\Delta y}{y} = \dfrac{348.2}{100} = 3.482$.

Percentage growth is $\dfrac{\Delta y}{y} \times 100 = 348.2\%$.

3. $y = 100e^{-0.05x}$. The graph is like figure 12.5, but with labels on the x-axis multiplied by 20, and on the y-axis by 100.

After 30 years, $y = 100e^{-0.05(30)} = 100e^{-1.5} = 100(0.223) = 22.3$.

Absolute growth: $\Delta y = 22.3 - 100 = -77.7$.

Proportionate growth $= \dfrac{\Delta y}{y} = -\dfrac{77.7}{100} = -0.777$.

Percentage growth $= -0.777 \times 100 = -77.7\%$ (that is, the variable has decreased by 77.7% and thus is only 22.3% of its initial value). Note that when a variable is declining (negative growth) the compounding effect *reduces* the decline. In this case, if growth were simple rather than compound, a variable decrease by 5% a year for 30 years would decline by $-5 \times 30 = -150\%$, so its level would fall from 100 to -50 (assuming of course that the variable can take negative values). With compounding, it falls from 100 to 22.3.

4. (a) Graph: see figure S5.

  (b) The graph in figure S5 becomes steeper as x increases, so the *absolute* growth of y is increasing, but it is impossible to say with confidence what is happening to proportionate or percentage growth.

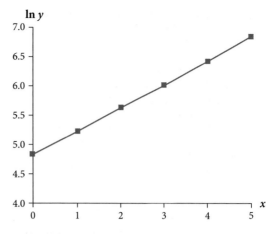

Figure S6 Graph for question 4(c) of exercise 12.3.

  (e) The above inference is confirmed, as we have:
  $$\frac{187.5}{125} = \frac{281.25}{187.5} = \frac{421.88}{281.25} = \frac{632.81}{421.88} = \frac{949.22}{632.81} = 1.5.$$
  So growth is 50% in every year.

  (f) With a log scale on the vertical axis, the slope of the graph gives us the proportionate growth rate of the variable in question.

5. (a) Graph: see figure S7.

  (b) The graph in figure S7 becomes steeper as x increases, but as in the previous question we can't be sure what is happening to the proportionate growth rate.

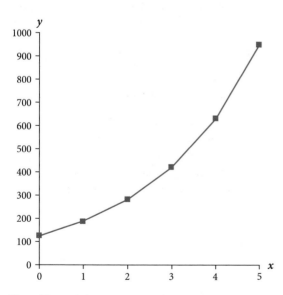

Figure S5 Graph for question 4(a) of exercise 12.3.

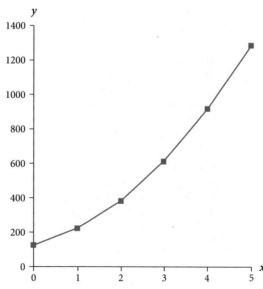

Figure S7 Graph for question 5(a) of exercise 12.3.

  (c) Graph: see figure S6.

  (d) The graph in figure S6 is clearly linear, which tells us that the absolute growth of log y is constant. Because the *absolute* growth of the log of any variable equals the *proportionate* growth of the variable itself, this tells us that the growth rate of y is, in fact, constant (neither increasing nor declining).

  (c) Graph: see figure S8.

  (d) In the graph (figure S8) we see that the slope of the graph is decreasing, so we can infer that the proportionate growth rate of y is decreasing.

  (e) This is confirmed when we check the actual growth rates, which are 80%, 70%, 60%, 50%, and 40% in years 1, 2, 3, 4, and 5, respectively.

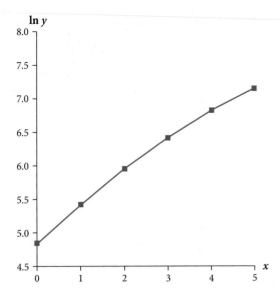

Figure S8 Graph for question 5(c) of exercise 12.3.

6. Let G = level of government expenditure, with initial level G_o. Let Y = GDP with initial level Y_o.

(a) Assuming continuous growth, government expenditure at any moment in time is given by: $G = G_o e^{0.03x}$ and similarly $Y = Y_o e^{0.02x}$ where x = time.

Therefore $\dfrac{G}{Y} = \dfrac{G_o e^{0.03x}}{Y_o e^{0.02x}} = \dfrac{G_o}{Y_o} e^{(0.03-0.02)x} = \dfrac{G_o}{Y_o} e^{0.01x}$.

(Remark: This tells us that the ratio of government expenditure to GDP is growing at 1% per year.)

We are also given $\dfrac{G_o}{Y_o} = 0.4$, so the above equation becomes: $\dfrac{G}{Y} = 0.4 e^{0.01x}$.

Setting $\dfrac{G}{Y} = 0.5$ in the above equation, we get: $0.5 = 0.4 e^{0.01x}$. (Solve for x to get the number of years until $G/Y = 0.5$.) Solution:

$0.5 = 0.4 e^{0.01x} \Rightarrow \dfrac{0.5}{0.4} = e^{0.01x} \Rightarrow 1.25 = e^{0.01x}$.

Taking logs on both sides:

$\ln 1.25 = 0.01x$ (since $\ln e = 1$)

$\Rightarrow x = \dfrac{\ln 1.25}{0.01} = \dfrac{0.02231}{0.01} = 22.31$ (years).

(Remark: Note a tempting error. If $\dfrac{G}{Y}$ is now 40% and is growing at 1% per year, it is tempting to infer that in 10 years' time $\dfrac{G}{Y}$ will be 50%. In fact it takes 22 years, because 1% of 40% is 0.4, not 1, so after 1 year G/Y increases from 40% to 40.4%, not to 41%; and so on in subsequent years.)

(b) Growth in annual jumps.

$G = G_o(1 + 0.03)^x; \ Y = Y_o(1 + 0.02)^x; \ \dfrac{G_o}{Y_o} = 0.4$

$\Rightarrow \dfrac{G}{Y} = \dfrac{G_o(1.03)^x}{Y_o(1.02)^x} = (0.4)\dfrac{(1.03)^x}{(1.02)^x}$. As before, set this

equal to 0.5 $\Rightarrow (0.4)\dfrac{(1.03)^x}{(1.02)^x} = 0.5 \Rightarrow \log[(1.03)^x] - \log[(1.02)^x] = \log 1.25$

$x[\log(1.03)] - \log(1.02)] = \log 1.25$

$x = \dfrac{\log 1.25}{\log(1.03) - \log(1.02)} = 22.87$.

7. (a) Let Y_o^s and Y_o^d be the GDP of Stagnatia and Dynamica in 1995.

We have $\dfrac{Y_o^s}{Y_o^d} = 2$ (given). If Stagnatia's GDP is growing continuously at an annual rate of α, its GDP in year x is given by $Y^S = Y_o^S e^{\alpha x}$ where x = years since 1995. Similarly Dynamica's GDP in year x is given by $Y^D = Y_o^D e^{\beta x}$ where β is Dynamica's growth rate.

So the ratio of GDPs in year x is $\dfrac{Y^S}{Y^D} = \dfrac{Y_o^S e^{\alpha x}}{Y_o^D e^{\beta x}} = 2e^{(\alpha-\beta)x}$.

We also know that in 2005 ($x = 10$), the ratio of GDPs was 1.5. So with $x = 10$, we have

$\dfrac{Y^S}{Y^D} = 2e^{(\alpha-\beta)10} = 1.5 \Rightarrow e^{(\alpha-\beta)10} = 0.75$. Taking logs on both sides:

$10(\alpha - \beta) = \ln 0.75$ (since $\ln e = 1$)

$\alpha - \beta = \dfrac{\ln 0.75}{10} = = -0.02877 = -2.877\%$.

So $\beta - \alpha = 2.877\%$; that is, Dynamica's growth rate is 2.877% faster than Stagnatia's.

(b) From (a) we have $\dfrac{Y^S}{Y^D} = 2e^{(\alpha-\beta)x} = 2e^{(-0.02877)x}$.

If we set $Y^S = Y^D$, we have $1 = 2e^{(-0.02877)x}$

$\Rightarrow \ln\left(\dfrac{1}{2}\right) = \ln[e^{-(0.02877)x}] = (-0.02877)x$.

So $x = \dfrac{\ln\left(\dfrac{1}{2}\right)}{-0.02877} = 24.0927$.

So $Y^S = Y^D$ 24 years after $Y^S = 2Y^D$, which was 1990. So Y^D will catch up with S Y^S in 2014.

Progress exercise 13.1

1. (a) $15e^{0.4x}(0.4) = 6e^{0.4x}$ (b) $-5e^{-0.05x}$ (c) $\dfrac{2}{x}$

(d) $-\dfrac{0.5}{x}$ (e) $\dfrac{1}{x^2} \cdot 2x = \dfrac{2}{x}$ (f) $\dfrac{-0.5}{x^{-0.5}} = -0.5x^{0.5}$

(g) $e^{-0.5x^2}(-x) = -xe^{-0.5x^2}$ (h) $e^{-1/2x}\left(-\dfrac{1}{2}\right) = -\dfrac{1}{2}e^{-1/2x}$

For graphs see Online Resource Centre.

2. (a) $\dfrac{2010}{1990} = \left(\dfrac{114.5 - 71.5}{71.5}\right)100 = 60.14\%$

(b) Increase in price level between 1990 and 1991 is calculated as:

$\dfrac{1991}{1990} = \left(\dfrac{76.8 - 71.5}{71.5}\right)100 = 7.41\%$. In later years, calculated in the same way:

1992	4.30
1993	2.50
1994	2.07
1995	2.63
1996	2.44
1997	1.82
1998	1.56
1999	1.32
2000	0.87

(c) (i) 1990–97. Method (a): $y = a(1 + r)^x \Rightarrow 89.7 = 71.5(1 + r)^7 \Rightarrow \log\left(\dfrac{89.7}{71.5}\right) = 7\log(1 + r) \Rightarrow r = 3.29\%$.

Method (b): $y = a(1 + r)^x \Rightarrow 89.7 = 71.5(1 + r)^7 \Rightarrow \left(\dfrac{89.7}{71.5}\right)^{1/7} = 1 + r \Rightarrow r = 3.29\%$ (Note, this method uses the $\wedge$ key on the calculator but does not use logs.)

(ii) 1997–2010. Using either method (a) or (b) above: 1.90%.

Progress exercise 13.2

1. (a) For any function $y = f(x)$, the instantaneous rate of proportionate growth is, by definition, $\frac{1}{y}\frac{dy}{dx}$.

    If $y = ae^{rx}$, $\frac{dy}{dx} = ae^{rx}(r) = rae^{rx} = ry$ (since $y = ae^{rx}$).

    Divide both sides by $y \Rightarrow \frac{1}{y}\frac{dy}{dx} = r$.

    (b) See figure S9. In both cases the instantaneous rate of growth at time x_0 is given by the slope of the tangent RS $\left(= \frac{dy}{dx}\right)$ divided by the distance JM $(= y_0)$. Thus

    $$\frac{\text{slope of RS}}{\text{distance JM}} = \frac{\frac{dy}{dx}}{y_0} = \frac{1}{y_0}\frac{dy}{dx}.$$

(i)

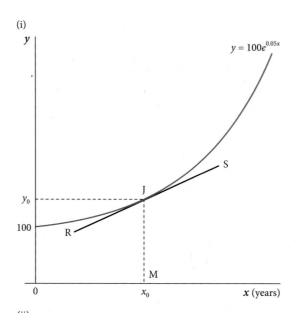

(ii)

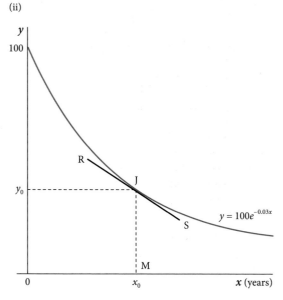

Figure S9 Sketch graphs for question 1(b)(i) and (ii) of exercise 13.2.

2. (a) Given any function $y = f(x)$, take logs on both sides $\Rightarrow$ $\ln y = \ln f(x)$. Differentiate $\Rightarrow \frac{d(\ln y)}{dx} = \frac{1}{f(x)}\cdot f'(x) \equiv \frac{1}{y}\frac{dy}{dx}$.

    Here $\frac{d(\ln y)}{dx}$ is the slope of a graph with $\ln y$ on the vertical and x on the horizontal axis. Thus the slope

of this graph measures $\frac{1}{y}\frac{dy}{dx}$, the instantaneous rate of proportionate growth. In the case where $y = f(x) = ae^{rx}$, taking logs gives $\ln y = \ln a + rx$, so the derivative is:

$\frac{d(\ln y)}{dx} = r = \frac{1}{y}\frac{dy}{dx}$. Thus the instantaneous rate of growth $= r$ (a constant) and the graph of $\ln y = \ln a + rx$ is a straight line with slope r.

   (b) See figure S10.

(a)

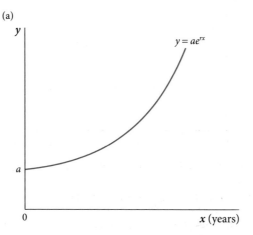

(b)

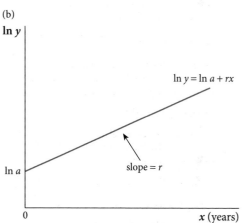

Figure S10 Sketch graphs for question 2(b) of exercise 13.2.

3. (a) See figure S11(a). From the graph there appears to be a rather sharp bend in the curve in about 1997, suggesting that the rate of increase of the CPI (the inflation rate) was fairly constant during 1985–97, then increased rather sharply in about 1997, and thereafter was fairly constant again. (Note we are discussing here not the inflation rate, but changes in the inflation rate. Innumerate politicians and journalists will often confuse 'inflation has fallen' (meaning that the price level is rising, but more slowly) with 'prices have fallen' (meaning that the price level has fallen). This confusion, between the level of a variable and its rate of change, is common.)

   (b) When we re-plot the graph with a log scale on the vertical axis (see the red line in figure S11(b)), the slope now gives the inflation rate. The blue broken line connecting the first and last data points gives an indication of the trend of inflation (though there are other, better ways of measuring trend). We see that the inflation rate (the slope of the pink line) fell below trend in 1992–98, then rose above trend in 1998–2002, and thereafter fell

(a)

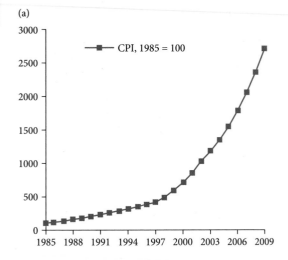

(c) Given $\ln q = \ln 16 - 0.5 \ln p$, we have

$$\frac{d(\ln q)}{d(\ln p)} = -0.5 = \frac{p}{q}\frac{dq}{dp} \quad \text{(from (a) above)}.$$

The left-hand side is the slope of the 'log-log' function (that is, with log scales on both axes), and the right-hand side is the elasticity. Thus the slope of the 'log-log' function measures the elasticity of the function itself (in this case, a demand function). (The elasticity happens to be constant in this example, but that is incidental. The relationship holds for any function, whether or not the elasticity is constant.)

(d) See figure S12. The presentational advantage is that we can immediately infer the elasticity by inspecting the slope of the 'log-log' graph.

(b)

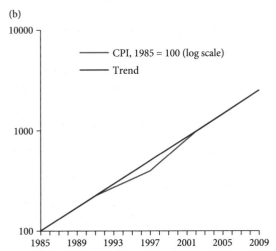

Figure S11 Graphs for question 3 of exercise 13.2.

(a)

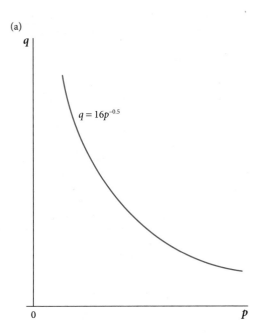

again to its trend rate. This is very different from the impression we gained from figure S11(a). It is clear that the log scale graph gives more, and more reliable, information.

(c) Because the slope with a semi-log scale measures the growth rate of the variable; for the maths see the answer to question (2) above. With a natural scale (figure S11(a)) the slope gives only the absolute rate of change, when it is the proportionate rate of change that we are interested in.

4. (a) $\dfrac{dq}{dp} = -0.5(16p^{-1.5})$

So $E^D \equiv \dfrac{p}{q}\dfrac{dq}{dp} = \dfrac{p}{q}[-0.5(16p^{-1.5})] = -0.5\left[\dfrac{16p^{-1.5}(p)}{q}\right]$

$= -0.5\left[\dfrac{16p^{-0.5}}{q}\right] = -0.5 \quad \text{(since } q = 16p^{-1.5}\text{)}.$

(b) Given $q = 16p^{-1.5}$, taking logs: $\ln q = \ln 16 - 0.5\ln p$.

This is a linear relationship between the two variables $\ln q$ and $\ln p$, with an intercept of $\ln 16$ and slope -0.5. It is 'log-log linear' because it is linear when log scales are taken on both axes.

(b)

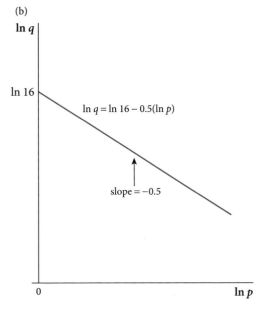

Figure S12 Sketch graphs for question 4(d) of exercise 13.2.

Progress exercise 14.1

1. (a) Along an iso-x section, x is constant at, say, $x = x_0$. Then we have $z = (120 - 3x_0) - 0.5y$. Since $(120 - 3x_0)$ is constant, this is a linear relationship between z and y, with an intercept of $(120 - 3x_0)$ on the z-axis (when $y = 0$), and a slope of -0.5. Whatever the value of x_0, the slope is always -0.5, so we can infer that all iso-x sections are linear with slope -0.5.

   Similarly an iso-y section has the equation $z = (120 - 0.5y_0) - 3x$, a linear relationship between z and x with an intercept on the z-axis (when $x = 0$) of $120 - 0.5y_0$ and a slope of -3.

   An iso-z section has the equation $z_0 = 120 - 3x - 0.5y$, which rearranges to $y = (240 - 2z_0) - 6x$, which again is linear with intercept $240 - 2z_0$ on the y-axis (when $x = 0$) and slope -6.

   Since we have found that the slope of the surface is linear in the xz, yz, and xy planes, the surface must be flat in all directions (that is, a plane).

   (b) By definition, at any point on the z-axis, x and y are both zero. So we can find the z intercept of the function by setting $x = 0$ and $y = 0$. This gives $z = 120$. Similarly at the x intercept, $z = y = 0$, giving $0 = 120 - 3x$, from which $x = 40$. Finally at the y intercept, $z = x = 0$, giving $0 = 120 - 0.5y$, from which $y = 240$. For sketch graph, see figure S13.

2. (a) With $y = y_0 =$ some constant, we have $z = x^2 + 3y_0$ which is a quadratic function relating z and x, with a z intercept of $3y_0$. With $x = x_0 =$ some constant, we have $z = (x_0)^2 + 3y$ which is a linear function relating z and y, with a z intercept of $(x_0)^2$. The surface is therefore U-shaped in the xz plane, and linear in the yz plane.

   (b) In general the z intercept is where $x = y = 0$. In this case, $z = 0$ when $x = y = 0$. So the z intercept is at $z = 0$. Similarly, in this example the x and y intercepts are at $x = 0$ and $y = 0$, respectively. Thus the surface passes through the origin. See figure S14.

3. (a) Along an iso-x section we have $x = x_0 =$ constant, so the equation of any iso-x section is $z = x_0 y$. This is a linear relationship between z and y with a slope given by x_0. Since x_0 is constant, the surface is linear in the yz plane

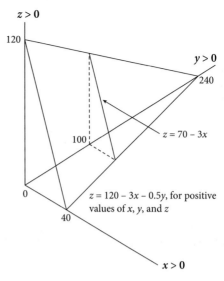

Figure S13 Sketch graph for question 1(b) of exercise 14.1.

with a slope that varies with the fixed value of x. Similarly the equation of any iso-y section is $z = y_0 x$. This is a linear relationship between z and x with a slope given by y_0. The surface is thus also linear in the xz plane with a slope that varies with the fixed value of y.

Recall that, in two dimensions, at any point on the y-axis, $x = 0$. Similarly, in three dimensions, at any point on the y-axis, both $x = 0$ and $z = 0$ (see section 14.2 of the book). In the case of this function, when $x = 0$, $z = 0$. Therefore the surface passes through the y-axis since when $x = 0$, $z = 0$ whatever the value of y. Similarly the surface passes through the x-axis since when $y = 0$, $z = 0$ whatever the value of x.

(b) From (a) we know that iso-x and iso-y sections are linear. A typical iso-z section has equation $z_0 = xy$ or $y = \frac{z_0}{x}$. This is a rectangular hyperbola. Therefore we can deduce that the surface has the shape shown in figure S15, which shows two iso-x sections, two iso-y sections, and two iso-z sections through the surface.

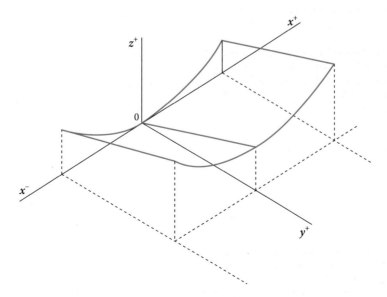

Figure S14 Sketch graph for question 2(b) of exercise 14.1.

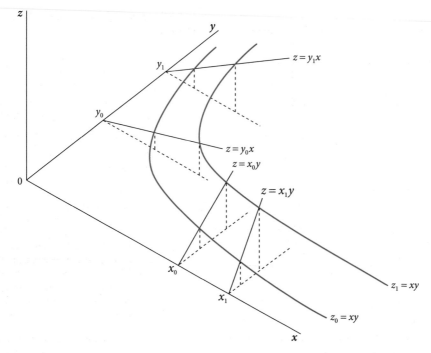

Figure S15 Sketch graph for question 3(b) of exercise 14.1.

Progress exercise 14.2

$\dfrac{\partial z}{\partial x}$:	$\dfrac{\partial z}{\partial y}$:
(a) $2x+2y$	$2x+2y$
(b) $9x^2+4xy$	$2x^2+2y+1$
(c) $x^2-x^{-2}y^{-1}$	$0.5y^{-0.5}-y^{-2}x^{-1}$
(d) $\frac{3}{2}x^2(x^3+y^2)^{-0.5}$	$y(x^3+y^2)^{-0.5}$
(e) $\dfrac{(x-y)(3x^2)-(x^3+y^2)}{(x-y)^2}$	$\dfrac{(x-y)(2y)+(x^3+y^2)}{(x-y)^2}$
(f) $(x^2+y)+(x-y^2)(2x)$	$(x^2+y)(-2y)+(x-y^2)$
(g) $200e^{2x+3y}$	$300e^{2x+3y}$
(h) $\dfrac{3x^2}{x^3+y^2}-\dfrac{1}{x}$	$\dfrac{2y}{x^3+y^2}-\dfrac{1}{y}$

Remarks: In (d), use function of a function rule; in (e), quotient rule; in (f), product rule.

Progress exercise 14.3

1. (a) and (b)

Given $z=x^3+5xy+4y^2$:

$\dfrac{\partial z}{\partial x}=3x^2+5y$. Measures slope in the x direction.

$\dfrac{\partial^2 z}{\partial x^2}=6x$. Measures how the slope in the x direction changes as x increases with y constant.

$\dfrac{\partial^2 z}{\partial y\partial x}=5$. Measures how the slope in the x direction changes as y increases with x constant.

$\dfrac{\partial z}{\partial y}=5x+8y$. Measures the slope in the y direction.

$\dfrac{\partial^2 z}{\partial y^2}=8$. Measures how the slope in the y direction changes as y increases with x constant.

$\dfrac{\partial^2 z}{\partial x\partial y}=5$. Measures how the slope in the y direction changes as x increases with y constant.

2.

	$\dfrac{\partial z}{\partial x}$	$\dfrac{\partial^2 z}{\partial x^2}$	$\dfrac{\partial z}{\partial y}$	$\dfrac{\partial^2 z}{\partial x\partial y}$
(a)	$2x+2$	2	$-3+2y$	0
(b)	$2x+y$	2	$x+2y$	1
(c)	$3x^2+6x-2y$ $-y^2+2xy$	$6x+6+2y$	$-2x-2xy$ $+6y+x^2$	$-2-2y+2x$
(d)	$15+6x-2y$	6	$-2x-4y+12$	-2
(e)	$4x^{-0.6}y^{1.5}$	$-2.4x^{-1.6}y^{1.5}$	$15x^{0.4}y^{0.5}$	$6x^{-0.6}y^{0.5}$
(f)	$\alpha x^{\alpha-1}y^{\beta}$	$(\alpha-1)\alpha x^{\alpha-2}y^{\beta}$	$\beta x^{\alpha}y^{\beta-1}$	$\alpha\beta x^{\alpha-1}y^{\beta-1}$
(g)	$\dfrac{0.25}{x}$	$\dfrac{-0.25}{x^2}$	$\dfrac{0.5}{y}$	0
(h)	$\dfrac{\alpha}{x}$	$\dfrac{-\alpha}{x^2}$	$\dfrac{\beta}{y}$	0
(i)	$\dfrac{2x}{x^2+3y}$	$\dfrac{(x^2+3y)(2)-2x(2x)}{(x^2+3y)^2}$	$\dfrac{3}{x^2+3y}$	$\dfrac{-6x}{(x^2+3y)^2}$
(j)	$2e^{2x+3y}$	$4e^{2x+3y}$	$3e^{2x+3y}$	$6e^{2x+3y}$

3.

	$\dfrac{\partial z}{\partial x}$	$\dfrac{\partial^2 z}{\partial x^2}$	$\dfrac{\partial z}{\partial y}$
(a)	$(2x-3y)^{-0.5}$	$-(2x-3y)^{-1.5}$	$-\dfrac{3}{2}(2x-3y)^{-0.5}$
(b)	$\dfrac{6x}{4y^2}$	$\dfrac{6}{4y^2}$	$\dfrac{-3x^2}{2y^3}$
(c)	$\dfrac{-2xy(1+y)}{(x^2-y)^2}$	$\dfrac{(8x^2y-2y(x^2-y))(1+y)}{(x^2-y)^3}$	$\dfrac{2y(x^2-y)+x^2+y^2}{(x^2-y)^2}$
(d)	$\dfrac{1}{(1-y^3)^{0.5}}$	0	$\dfrac{3xy^2}{2(1-y^3)^{1.5}}$
(e)	$2xe^{3y}$	$2e^{3y}$	$3x^2e^{3y}$
(f)	$\dfrac{2x}{x^2+1}$	$\dfrac{2x(x-1)^2}{(x^2+1)^2}$	$\dfrac{-2y}{y^2-1}$
(g)	$\dfrac{0.5}{x}$	$\dfrac{-0.5}{x^2}$	$\dfrac{0.25}{y}$

4. Since $\dfrac{\partial z}{\partial x}$ and $\dfrac{\partial^2 z}{\partial x^2}$ are both positive, we know that the surface is positively sloped and convex in the x direction. Since $\dfrac{\partial z}{\partial y} > 0$ but $\dfrac{\partial^2 z}{\partial y^2} < 0$, we know that the function is positively sloped and concave in the y direction (see figure S16). Since $\dfrac{\partial^2 z}{\partial y \partial x} < 0$, the slope in the x direction decreases as y increases, and therefore in figure S16, the slope in the x direction at B is less positive than at P.

5. Since $\dfrac{\partial z}{\partial x}$ is negative and $\dfrac{\partial^2 z}{\partial x^2}$ is positive, we know that the surface is negatively sloped and convex in the x direction. Since $\dfrac{\partial z}{\partial y}$ and $\dfrac{\partial^2 z}{\partial y^2}$ are both negative, we know that the function is negatively sloped and concave in the y direction (see figure S17). Since $\dfrac{\partial^2 z}{\partial y \partial x} < 0$, the slope in the x direction decreases as y increases, and therefore in figure S17, the

slope in the x direction at B is less positive (more negative) than at Q.

Progress exercise 14.4

1. (a) (i) If $Q = 100$, $1 = K^{0.25}L^{0.75}$. Therefore $1^4 = (K^{0.25})^4(L^{0.75})^4 = KL^3 \Rightarrow K = \dfrac{1}{L^3}$.

   (ii) On isoquant $K = \dfrac{1}{L^3}$, $\dfrac{dK}{dL} = -3L^{-4} = -\dfrac{3}{L^4}$ which is negative since $L^4 > 0$. So isoquant is negatively sloped. Also $\dfrac{d^2K}{dL^2} = 12L^{-5} = \dfrac{12}{L^5}$ which is positive since L is positive and so therefore is L^5. So the isoquant is convex.

   (iii) Graph is very similar to a rectangular hyperbola, asymptotic to the K and L axes.

   (iv) Negative slope reflects the assumption embodied in this production function, that marginal products of L and K are always positive. Therefore if one input is increased, the other must be reduced in order to hold output constant. The positive second derivative (which tells us that the isoquant is convex) indicates diminishing absolute (that is, ignoring sign) MRS between K and L.

   (v) Yes, because for any output Q_0, the isoquant is $K = \dfrac{\left(\dfrac{Q_0}{100}\right)^4}{L^3}$ and the signs of the first and second derivatives are the same as above.

   (b) (i)&(ii) $MPL \equiv \dfrac{\partial Q}{\partial L} = 75K^{0.25}L^{-0.25}$ which is positive since K and L are positive and a positive number raised to any power is positive (see section 2.12 of the book). The same is true of $MPK \equiv \dfrac{\partial Q}{\partial K} = 25K^{-0.75}L^{0.75}$.

   (iii) No, it's quite possible that MPL could become negative when L is sufficiently high, for a given K. In effect, there are so many workers with so little capital that they get in one another's way, hence total output would increase if some of them stayed at home. And similarly for a sufficiently large K,

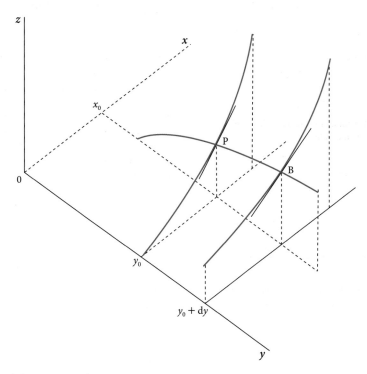

Figure S16 Sketch graph for question 4 of exercise 14.3.

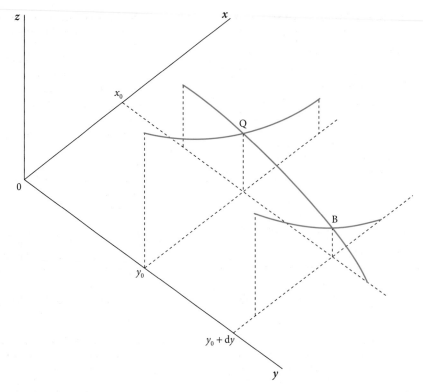

Figure S17 Sketch graph for question 5 of exercise 14.3.

with a given L – too few workers with too much machinery.

(c) From (b), $MPL \equiv \frac{\partial Q}{\partial L} = 75K^{0.25}L^{-0.25}$. So $\frac{\partial}{\partial L}(MPL) \equiv \frac{\partial^2 Q}{\partial L^2} = (-0.25)75K^{0.25}L^{-1.25}$ which is negative since K and L are positive and a positive number raised to any power is positive. For the same reason, $\frac{\partial}{\partial K}(MPK) \equiv \frac{\partial^2 Q}{\partial K^2} = (-0.75)(25)K^{-1.75}L^{0.75}$ is also negative.

(d) See figures 14.17 and 14.18 in the book. The slopes are positive because MPL and MPK are positive, and concave because $\frac{dMPL}{dL}$ and $\frac{dMPK}{dK}$ are negative.

(e) (i) By reasoning similar to (b)(iii) above, we can say that as L increases the fixed quantity of K has to be divided between more and more workers, causing capital per worker and hence output per worker to fall. This offsets the increase in output from additional workers, so the marginal product (which of course is measured *net*) falls. And similarly with increases in K with L constant. In the case of the Cobb–Douglas production function this is true at all levels of output but this is not true for all production functions.

(ii) It is tempting to suppose that if DMP is present so that MPL is falling, then APL must be falling too. However in figure 14.21 we see that, between P and N, MPL is falling while APL is rising. The correct condition for APL to be falling is that $MPL < APL$.

(iii) There are two methods of showing APL is always falling for this production function: first method: $APL \equiv \frac{Q}{L} = 100K^{0.25}L^{-0.25}$, so $\frac{\partial APL}{\partial L} = (-0.25)100K^{0.25}L^{-1.25}$ which is always negative. Second method: from (b), $MPL \equiv \frac{\partial Q}{\partial L} = 75K^{0.25}L^{-0.25}$ and $APL \equiv \frac{Q}{L} = 100K^{0.25}L^{-0.25}$. Since $75 < 100$, $MPL < APL$ so APL is always falling. And similarly

for APK. Their graphs are similar to APL in figure 14.19b.

(f) (i) From 1(b) above, $MPL \equiv \frac{\partial Q}{\partial L} = 75K^{0.25}L^{-0.25}$, so $\frac{\partial}{\partial K}(MPL) = \frac{\partial^2 Q}{\partial K \partial L} = (0.25)75K^{-0.75}L^{-0.25}$ which is positive since K and L are positive. Thus a (small) increase in the capital input increases the MPL. Similarly $\frac{\partial}{\partial L}(MPK) = \frac{\partial^2 Q}{\partial L \partial K} = (0.75)25K^{-0.75}L^{-0.25} > 0$.

(ii) It seems plausible that an increase in K should not only increase the output of existing workers (that is, increase Q) but also increase the amount of extra output that one more worker produces (that is, increase MPL). But it is not a logical necessity.

(g) (i) Since $(160,000)^{0.25} = 20$, $Q = 2000L^{0.75}$. The graph has the same shape as those in figure 14.19a.

(ii) The slope is positive indicating MPL is positive; and the slope decreases as L increases, indicating a 'law' of diminishing MPL.

(iii) Since $(810,000)^{0.25} = 30$, $Q = 3000L^{0.75}$. Since $3000L^{0.75} > 2000L^{0.75}$, this graph lies entirely above the graph of $Q = 2000L^{0.75}$, indicating that for any given L, Q is higher because K input is higher.

(iv) Slope of $Q = 3000L^{0.75}$ is: $MPL = \frac{\partial Q}{\partial L} = 2250L^{-0.25}$. For any given L, this is greater than the slope of $Q = 2000L^{0.75}$, which is $\frac{\partial Q}{\partial L} = 1500L^{-0.25}$. This difference reflects the fact that $\frac{\partial Q}{\partial K \partial L} > 0$.

(h) (i) $Q = 800,000K^{0.25}$. This has the same shape as the quasi-short-run production function in (g) above, except for scale. This is because the relationships between Q and L, and between Q and K, in the production function $Q = 100K^{0.25}L^{0.75}$ are identical except for the difference in their exponents. The shape of $Q = 800,000K^{0.25}$ reflects our earlier findings that $\frac{\partial Q}{\partial K}$ was always positive and $\frac{\partial^2 Q}{\partial K^2}$ always negative.

2. (a) $MPK \equiv \frac{\partial Q}{\partial K} = \alpha AK^{\alpha-1}L^{\beta}$. This is positive since α, A, K, and L are all positive. Also $\frac{\partial}{\partial K}(MPK) \equiv \frac{\partial^2 Q}{\partial K^2} = (\alpha - 1)\alpha AK^{\alpha-2}L^{\beta}$. This is positive if $\alpha > 1$. Similarly, $MPL \equiv \frac{\partial Q}{\partial L} = \beta AK^{\alpha}L^{\beta-1}$ is positive, and $\frac{\partial}{\partial L}(MPL) \equiv \frac{\partial^2 Q}{\partial L^2} = (\beta - 1)\beta AK^{\alpha}L^{\beta-2}$ is positive if $\beta > 1$.

    (b) If $\alpha > 1$ this means that MPK increases as K increases with L constant; and if $\beta > 1$, MPL increases as L increases with K constant. Although each could be true over a certain range of values for L and K, it is hard to imagine how this could be true for all L and K. Thus we normally assume $\alpha < 1$ and $\beta < 1$ in the Cobb–Douglas production function.

3. (a) (i) Indifference curve for $U = 27$ is: $27 = X^{0.75}Y^{1.5}$ (an implicit function). Raise both sides to power $\frac{2}{3}$ gives: $9 = X^{0.5}Y \Rightarrow Y = \frac{9}{X^{0.5}}$ (an explicit function).

        (ii) From (i), $Y = 9X^{-0.5}$, so $\frac{dY}{dX} = -4.5X^{-1.5} = -\frac{4.5}{X^{1.5}}$ This is negative because $X^{1.5}$ is positive, so the indifference curve is downward sloping. We also have $\frac{d^2Y}{dX^2} = (-1.5)(-4.5)X^{-2.5}$ which is positive, so the slope increases as X increases (that is, becomes flatter) so the indifference curve is convex from below.

        (iii) Negative slope of the indifference curve reflects the assumption that the marginal utilities are always positive (non-satiation). Therefore if consumption of one good is increased, the other must be reduced in order to hold utility constant. The positive second derivative (which tells us that the indifference curve is convex) indicates diminishing absolute MRS between X and Y.

        (iv) Shape of the indifference curve $Y = \frac{9}{X^{0.5}}$ is similar to that in figures 14.23 and 14.24, and is similar to a rectangular hyperbola. It is asymptotic to both axes, because $Y \to 0$ as $X \to \infty$, and there is a discontinuity at $X = 0$.

    (b) (i) Indifference curve is $Y = \frac{4}{X^{0.5}}$. Slope is $\frac{dY}{dX} = -2X^{-1.5} = -\frac{2}{X^{1.5}}$ which is negative. We also have $\frac{d^2Y}{dX^2} = 3X^{-2.5}$ which is positive, so the indifference curve is convex. Economic interpretation, and sketch, same answers as 3(a)(iii) and (iv).

        (ii) The level of utility is 27 at every point on the indifference curve in (a) and 8 at every point on the indifference curve in (b), but these numerical values have no significance. We can only *rank* utility levels; that is, utility level 27 is higher than, and preferred to, level 8, because $27 > 8$.

        (iii) Given the utility function $U = X^{0.75}Y^{1.5}$, the indifference curve for $U = U_0$ has the equation $U_0 = X^{0.75}Y^{1.5}$ (implicit function) or $Y = \left(\frac{U_0}{X^{0.75}}\right)^{2/3}$ (explicit function). Thus all indifference curves have the same general shape (asymptotic to both axes).

    (c) (i), (ii) $MU_X \equiv \frac{\partial U}{\partial X} = 0.75X^{-0.25}Y^{1.5}$, which is positive since X and Y are positive. $MU_Y \equiv \frac{\partial U}{\partial Y} = 1.5X^{0.75}Y^{0.5}$ is positive for the same reason.

        (iii) No, because an individual can become satiated with a good, and increasing consumption beyond this level will by definition add nothing to her welfare.

    (d) (i) $\frac{\partial^2 U}{\partial X^2} = (-0.25)0.75X^{-1.25}Y^{1.5}$ which is negative, so consumption of good X is subject to diminishing marginal utility at all levels of consumption. But $\frac{\partial^2 U}{\partial Y^2} = (0.5)(1.5)X^{0.75}Y^{0.5}$ which is positive, so consumption of good Y is subject to increasing marginal utility at all levels of consumption ('appetite grows with eating').

        (ii) As the units in which U is measured are arbitrary, no significance can be attached to the absolute values of the second derivatives. For example the consumer could choose a new arbitrary unit, V, to measure her utility. Then we might have $V = U^2$, so then $V = X^{1.5}Y^3$, and $\frac{\partial^2 V}{\partial X^2}$ and $\frac{\partial^2 V}{\partial Y^2}$ are then both positive (check for yourself). Or, she could choose another unit, W, for measuring her utility, such that $W = U^{1/3}$. Then $W = X^{0.25}Y^{0.5}$ and $\frac{\partial^2 W}{\partial X^2}$ and $\frac{\partial^2 W}{\partial Y^2}$ are both negative.

    (e) From (c) and (d) above, we have $\frac{\partial U}{\partial X} > 0$ and $\frac{\partial^2 U}{\partial X^2} < 0$ (diminishing MU). So the curve giving U as function of X with Y constant is concave (see figure S18(a)). But for Y, from (c) and (d) above, we have $\frac{\partial U}{\partial Y} > 0$ and $\frac{\partial^2 U}{\partial Y^2} > 0$ (increasing MU). So the curve giving U as function of X with Y constant is convex (see figure S18(b)).

    (f) (i) Given $U = X^{0.75}Y^{1.5}$ with $\frac{\partial U}{\partial X} = 0.75X^{-0.25}Y^{1.5}$ (see (c) above). Then $\frac{\partial^2 U}{\partial Y\partial X} = (1.5)(0.75)X^{-0.25}Y^{0.5}$ which is positive. Similarly $\frac{\partial U}{\partial Y} = 1.5X^{0.75}Y^{0.5}$ (from (c) above), so $\frac{\partial^2 U}{\partial X\partial Y} = (0.75)1.5X^{-0.25}Y^{0.5}$ which

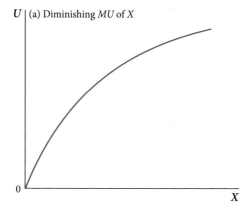

U | (a) Diminishing MU of X

0 — X

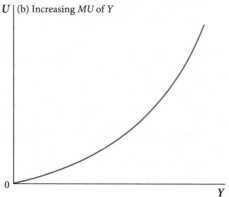

U | (b) Increasing MU of Y

0 — Y

Figure S18 Sketch graphs for question 3(e) of exercise 14.4.

ANSWERS TO PROGRESS EXERCISES

is positive. So for this utility function a small increase in consumption of one good increases the *MU* of the other.

(ii) No significance, for reasons given in (d)(ii) above. See also Exercise 15.4, question 5, and its answer.

4. (a) An 'extreme' bundle (lots of X, not much Y) is $X = 256$, $Y = 9$. Then $U = (256)^{0.75}(9)^{1.5} = 64 \times 27 = 1728$. Another 'extreme' bundle (lots of Y, not much X) is $X = 81$, $Y = 16$. Then $U = (81)^{0.75}(16)^{1.5} = 27 \times 64 = 1728$. Thus these two bundles lie on the same indifference curve, $U = 1728$.

The average or 'middling' bundle is $X = \frac{256 + 81}{2}$ $= 168.5$, $Y = \frac{9 + 16}{2} = 12.5$. Then $U = (168.5)^{0.75}(12.5)^{1.5}$ $= 2067$. So the 'middling' bundle gives higher utility than the extreme bundles.

(b) See figure S19.

(c) Suppose that in figure S20 the extreme bundles are point R with coordinates (X_1, Y_1) and point S with coordinates (X_2, Y_2). We can connect these two points by means of a straight line with equation $Y = a - bX$ where a and b are parameters. Thus we have $Y_1 = a - bX_1$ and $Y_2 = a - bX_2$.

Now consider the average bundle, consisting of $\frac{X_1 + X_2}{2}$ units of X and $\frac{Y_1 + Y_2}{2}$ units of Y (point A in figure S20). This point also lies on the straight line $Y = a - bX$.

Proof: If the point $\left(\frac{X_1 + X_2}{2}, \frac{Y_1 + Y_2}{2}\right)$ lies on $Y = a - bX$, then by definition $\frac{Y_1 + Y_2}{2} = a - b\left[\frac{X_1 + X_2}{2}\right]$. After substituting $Y_1 = a - bX_1$ and $Y_2 = a - bX_2$ the left-hand side becomes $\frac{(a - bX_1) + (a - bX_2)}{2} = \frac{2a - b(X_1 + X_2)}{2} = a - \frac{b(X_1 + X_2)}{2} =$ the right-hand side (an identity).

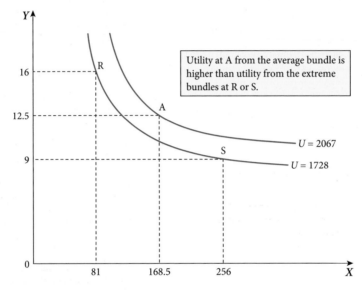

Figure S19 Sketch graph for question 4(b) of Exercise 14.4.

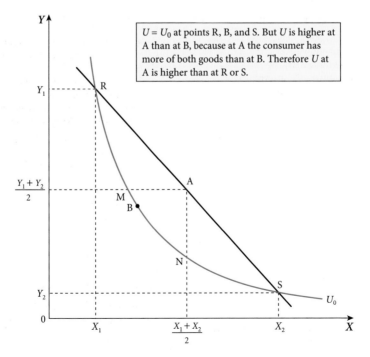

Figure S20 Graph for question 4(c) of exercise 14.4.

Now since by assumption the indifference curve is convex and passes through (X_1, Y_1) and (X_1, Y_1), the utility at A in figure S20 must be higher than at any point in the section of the indifference curve between M and N, such as B, because more of both goods is consumed at A than at B and both goods have positive marginal utility due to the assumption of non-satiation. However, utility *anywhere* on the indifference curve is the same as it is at B. Combining these two facts, utility at A is higher than utility anywhere on the indifference curve, including of course points R and S, as we wished to prove.

5. (a) Repeat of question 3 for utility function $U = XY$:

Repeat of 3(a):

(i) Given $U = XY$, the equation of the indifference curve for some fixed level of utility, U^*, is $U^* = XY$, which rearranges as $Y = \frac{U^*}{X}$ or $Y = U^*X^{-1}$.

(ii) $\frac{dY}{dX} = -U^*X^{-2} = -\frac{U^*}{X^2} < 0$, so the indifference curve is negatively sloped.

$\frac{d^2Y}{dX^2} = 2U^*X^{-3} = \frac{2U^*}{X^3} > 0$, so the indifference curve is convex from below.

(iii) Answer is the same as answer to question 3(a)(iii) above.

(iv) The graph of any indifference curve has the shape of a rectangular hyperbola.

Repeat of 3(b):

(i) Already answered in (a)(i) above.

(ii) and (iii) Already discussed in answer to 3(b) above.

Repeat of 3(c):

(i) (ii) Given $U = XY$, $MU_X \equiv \frac{\partial U}{\partial X} = Y$ and $MU_Y \equiv \frac{\partial U}{\partial Y} = X$. These are both positive assuming quantities X and Y of both goods are positive.

(iii) See answer to 3(c)(iii) above.

Repeat of 3(d):

(i) From (c) above, we have $\frac{\partial U}{\partial X} = Y$ so $\frac{\partial^2 U}{\partial X^2} = 0$. So the marginal utility of X is determined by the fixed value of Y, and does not vary as X varies. Similarly $\frac{\partial^2 U}{\partial Y^2} = 0$, so the marginal utility of Y is determined by the fixed value of X, and does not vary as Y varies. See figure S21.

(ii) No economic significance; see answer to 3(d) above.

Repeat of 3(e):

Figure S21 shows how U varies with X, when Y is fixed at 5 and 10.

Repeat of 3(f):

From (d) above, $\frac{\partial U}{\partial X} = Y$, so $\frac{\partial^2 U}{\partial Y \partial X} = 1$. Similarly $\frac{\partial^2 U}{\partial X \partial Y} = 1$. No economic significance.

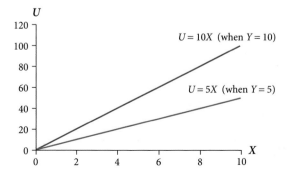

Figure S21 Graphs for question 3(e) of exercise 14.4.

(b) Repeat of question (4) for the utility function $U = XY$:

Repeat of 4(a): Suppose our 'extreme' bundles are $X_0 = 1$, $Y_0 = 99$ (giving $U_0 = 1 \times 99 = 99$) and $X_1 = 99$, $Y_1 = 1$ (giving $U_1 = 99 \times 1 = 99$). Then the average of these two bundles is $X_2 = \frac{1 + 99}{2} = 50$ and $Y_2 = \frac{99 + 1}{2} = 50$, so the utility of this 'middling' bundle is $U_2 = 50 \times 50 = 2500$. Thus the 'middling' bundle gives higher utility that the extreme bundles, as we wished to show.

Repeat of 4(b): Sketch would be in essence the same as figure S19 for question 4(b) above.

Repeat of 4(c): See answer to 4(c) and figure S20.

(c) This utility function, although very simple, has one of the key characteristics of a plausible utility function: downward sloping and convex indifference curves. However, the fact that marginal utilities are constant is somewhat restrictive. (Note that this utility function illustrates the point made in the book that diminishing marginal utility is not a necessary condition for convexity of indifference curves.)

Progress exercise 15.1

1. The partial derivatives are: $\frac{\partial z}{\partial x} = 60 - 12x - 4y$ and $\frac{\partial z}{\partial y} = 34 - 4x - 6y$. Setting these equal to zero and solving simultaneously gives a stationary point at $y = 3$, $x = 4$. The second derivatives are: $\frac{\partial^2 z}{\partial x^2} = -12$; $\frac{\partial^2 z}{\partial y^2} = -6$; and $\frac{\partial^2 z}{\partial x \partial y} = \frac{\partial^2 z}{\partial y \partial x} = -4$. Since at $x = 4$, $y = 3$, $\frac{\partial^2 z}{\partial x^2}$ and $\frac{\partial^2 z}{\partial y^2}$ are both negative, and $\frac{\partial^2 z}{\partial x^2} \cdot \frac{\partial^2 z}{\partial y^2} > \frac{\partial^2 z}{\partial x \partial y} \cdot \frac{\partial^2 z}{\partial y \partial x}$, the SP is a maximum.

2. The simultaneous equations are: $\frac{\partial z}{\partial x} = 8x - y - 3x^2 = 0$, and $\frac{\partial z}{\partial y} = -x + 2y = 0$, with solutions $x = y = 0$ and $x = \frac{5}{2}$, $y = \frac{5}{4}$. The second derivatives are: $\frac{\partial^2 z}{\partial x^2} = 8 - 6x$, $\frac{\partial^2 z}{\partial y^2} = 2$, and $\frac{\partial^2 z}{\partial x \partial y} = \frac{\partial^2 z}{\partial y \partial x} = -1$. At $x = y = 0$, $\frac{\partial^2 z}{\partial x^2}$ and $\frac{\partial^2 z}{\partial y^2}$ are both positive and $\frac{\partial^2 z}{\partial x^2} \cdot \frac{\partial^2 z}{\partial y^2} = (8)(2) = 16$, while $\frac{\partial^2 z}{\partial x \partial y} \cdot \frac{\partial^2 z}{\partial y \partial x} = (-1)(-1) = 1$, so $\frac{\partial^2 z}{\partial x^2} \cdot \frac{\partial^2 z}{\partial y^2} > \frac{\partial^2 z}{\partial x \partial y} \cdot \frac{\partial^2 z}{\partial y \partial x}$ and this point is therefore a minimum. At $x = \frac{5}{2}$, $y = \frac{5}{4}$, $\frac{\partial^2 z}{\partial x^2} = 8 - 6(\frac{5}{2}) = -7$ and $\frac{\partial^2 z}{\partial y^2} = 2$ (opposite signs). And $\frac{\partial^2 z}{\partial x^2} \cdot \frac{\partial^2 z}{\partial y^2} < \frac{\partial^2 z}{\partial x \partial y} \cdot \frac{\partial^2 z}{\partial y \partial x}$. So a saddle point.

3. Using the same method as (1) and (2) above: (a) minimum at $x = 5$, $y = 6$; (b) maximum at $x = 10$, $y = 4$; (c) minimum at $X = 6$, $Y = 2$; (d) $f_x = 3x^2 - 3 = 0 \Rightarrow x = 1$ or -1; $f_y = 3y^2 - 12 = 0 \Rightarrow y = 2$ or -2; $f_{xx} = 6x$; $f_{yy} = 6y$; $f_{yx} = 0$; $f_{xy} = 0$. This example is unusual in that f_x and therefore f_{xx} are functions only of x; and f_y and therefore f_{yy} are functions only of y. So we have four cases in which f_x and f_y are both zero:

(i) $x = 1$, $y = 2$. Here $f_{xx} > 0$, $f_{yy} > 0$, $f_{xx} \cdot f_{yy} > f_{yx} \cdot f_{xy} \Rightarrow$ minimum;

(ii) $x = 1$, $y = -2$. Here $f_{xx} > 0$, $f_{yy} < 0$, $f_{xx} \cdot f_{yy} < f_{yx} \cdot f_{xy} \Rightarrow$ saddle point;

(iii) $x = -1$, $y = 2$. Here $f_{xx} < 0$, $f_{yy} > 0$, $f_{xx} \cdot f_{yy} < f_{yx} \cdot f_{xy} \Rightarrow$ saddle point;

(iv) $x = -1$, $y = -2$. Here $f_{xx} < 0$, $f_{yy} < 0$, $f_{xx} \cdot f_{yy} > f_{yx} \cdot f_{xy} \Rightarrow$ maximum.

Progress exercise 15.2

1. (a) $dz = (2x + y)dx + (x + 2y)dy$

(b) $dz = (0.25x^{-0.75}y^{0.5})dx + (0.5x^{0.25}y^{-0.5})dy$

(c) $dz = (\alpha x^{\alpha-1}y^{\beta})dx + (\beta x^{\alpha}y^{\beta-1})dy$

(d) $\frac{\partial z}{\partial x} = 0.5(x^2 + y^3)^{-0.5}(2x) = x(x^2 + y^3)^{-0.5}$;

$\frac{\partial z}{\partial y} = 0.5(x^2 + y^2)^{-0.5}(3y^2) = \frac{3}{2}y^2(x^2 + y^2)^{-0.5}$;

So $dz = [x(x^2 + y^3)^{-0.5}]dx + [\frac{3}{2}y^2(x^2 + y^3)^{-0.5}]dy$.

2. (a) If the lawn has sides x and y, the area A is: $A = xy$ with $\frac{\partial A}{\partial x} = y$ and $\frac{\partial A}{\partial y} = x$. So the total differential is: $dA = y\,dx + x\,dy$. Here $x = 10$, $y = 5$ and $dx = dy = 1$, so: $dA = 5(1) + 10(1) = 15$. So, if I use the total differential to make the calculation, I conclude that I must buy 15 square metres of turf.

The true value is $A_1 - A_0$ where A_0 = initial area, A_1 = new area. So $A_0 = 10 \times 5 = 50$, $A_1 = 11 \times 6 = 66$ so $A_1 - A_0 = 16$ square metres. So using the total differential I would buy 15 square metres when I actually need 16.

(b) The diagram is identical to figure 15.12 with $x_0 = 10$, $y_0 = 5$ and $dx = dy = 1$. Error = area C = 1 metre $\times$ 1 metre = 1 square metre.

3. Total differential is: $dz = 2x\,dx + 2y\,dy$. Here $x = 5$, $y = 10$ and $dx = dy = 0.1$, so: $dz = 10(0.1) + 20(0.1) = 3$. The true change is $z_1 - z_0 = (x_1^2 + y_1^2) - (x_0^2 + y_0^2)$ where $x_0 = 5$, $y_0 = 10$, $x_1 = 5.1$, $y_1 = 10.1$. So $z_1 - z_0 = (5.1)^2 + (10.1)^2 - (5^2 + 10^2) = 3.02$. So the error is 0.02 which, as a percentage of the true value, is $\frac{0.02}{3.02}(100) = 0.662\%$.

Progress exercise 15.3

1. (a) From $z = x^3 + y^2$, the total differential is $dz = 3x^2\,dx + 2y\,dy$. Divide through by $dx \Rightarrow \frac{dz}{dx} = 3x^2 + 2y\frac{dy}{dx}$. Given $y = x^2$ so $\frac{dy}{dx} = 2x$. Using this to substitute $2x$ in place of $\frac{dy}{dx}$, we get: $\frac{dz}{dx} = 3x^2 + 2y\frac{dy}{dx} = 3x^2 + 2y(2x) = 3x^2 + 4xy$. Finally, using $y = x^2$ to substitute for y, this becomes: $\frac{dz}{dx} = 3x^2 + 4x^3$.

(b) $dz = (6x + 4y^3)dx + 12xy^2\,dy$. Divide through by $dx \Rightarrow \frac{dz}{dx} = 6x + 4y^3 + 12xy^2\frac{dy}{dx}$. Given: $y = (x + 1)^2 \Rightarrow \frac{dy}{dx} = 2(x + 1)$. So $\frac{dz}{dx} = 6x + 4y^3 + 24xy^2(x + 1)$. Using $y = (x + 1)^2$ to substitute for y, this becomes: $\frac{dz}{dx} = 6x + 4(x + 1)^6 + 24x(x + 1)^5$.

(c) $dz = u\,dv + v\,du$. Divide through by $dx \Rightarrow \frac{dz}{dx} = u\frac{dv}{dx} + v\frac{du}{dx}$. Given $u + 3x + 2$, $\frac{du}{dx} = 3$. Given $v = 3x^2$, $\frac{dv}{dx} = 6x$. So: $\frac{dz}{dx} = u(6x) + v(3) = (3x + 2)(6x) + 3x^2(3) = 18x^2 + 12x + 9x^2 = 27x^2 + 12x = 3x(9x + 4)$.

2. (a) Given: $TR = pq$, the total differential is: $dTR = p\,dq + q\,dp$. Divide through by $dq \Rightarrow \frac{dTR}{dq} = p + q\frac{dp}{dq}$. Given $p = 50e^{-0.5q}$, $\frac{dp}{dq} = -25e^{-0.5q}$. So $\frac{dTR}{dq} = 50e^{-0.5q} - q25e^{-0.5q} = (2 - q)25e^{-0.5q}$.

(b) By direct substitution, $TR = pq = q(50e^{-0.5q})$. Since we now have TR as a function of q alone, we can simply differentiate, using the product rule, and find $\frac{dTR}{dq}$ as: $\frac{dTR}{dq} = q(-25e^{-0.5q}) + (50e^{-0.5q})(1) = (2 - q)25e^{-0.5q}$.

(c) From $MR \equiv \frac{dTR}{dq} = p + q\frac{dp}{dq}$ we see that MR has two components: (i) the change in revenue from selling 1 more unit, which obviously equals p; (ii) the change in revenue from the price reduction necessary to sell 1 more unit, which equals $q\frac{dp}{dq}$ (= the price reduction $\times$ initial quantity sold). This is negative since $\frac{dp}{dq} < 0$.

(d) $\frac{dTR}{dq} \equiv$ marginal revenue.

3. (a) Let $z = 2x - 3y$. Then $dz = 2\,dx - 3\,dy$ but $z = 0$ so $dz = 0 \Rightarrow 2\,dx - 3\,dy = 0$. This rearranges to give: $\frac{dy}{dx} = \frac{2}{3}$.

For (b)–(d) use the same method as (a), giving:

(b) $(2x + y)dx + (y + 2x)dy = 0$, from which $\frac{dy}{dx} = -\frac{2x + y}{x + 2y}$.

(c) $(\alpha x^{\alpha-1}y^\beta)dx + (\beta x^\alpha y^{\beta-1})dy = 0$, so $\frac{dy}{dx} = -\frac{\alpha x^{\alpha-1}y^\beta}{\beta x^\alpha y^{\beta-1}} = -\frac{\alpha}{\beta}\frac{y}{x}$.

(d) $0.5(x^2 + 3y)^{-0.5}(2x)dx + 0.5(x^2 + 3y)^{-0.5}(3)dy = 0$, so $\frac{dy}{dx} = -\frac{2}{3}x$.

4. (a) Given $z = x^3 + y^2$ with differential: $dz = 3x^2\,dx + 2y\,dy$. But along any iso-z section, z is constant by definition so $dz = 0$. So: $3x^2\,dx + 2y\,dy = 0$, from which: $\frac{dy}{dx} = -\frac{3x^2}{2y} =$ slope of any iso-z section.

(b) Same method as (a): $dz = (6x + 4y^3)dx + 12xy^2\,dy = 0$, from which: $\frac{dy}{dx} = -\frac{6x + 4y^3}{12xy^2}$.

(c) Same method as (a): $dz = u\,dv + v\,du = 0$, from which: $\frac{dv}{du} = -\frac{v}{u}$.

Progress exercise 15.4

1. (a) $\frac{\partial Q}{\partial L} \equiv MPL = 0.5K^{0.5}L^{-0.5}$; and $\frac{\partial Q}{\partial K} \equiv MPK = 0.5K^{-0.5}L^{0.5}$. Both are always positive because K and L are positive and a positive number raised to any power is positive. Graphs of MPL and MPK have the same shape as figures 14.19(a) (sic).

(b) (i) $Q = 10$ isoquant is: $10 = K^{0.5}L^{0.5}$ (an implicit function). To obtain an explicit function, raise both sides to power 2, giving: $100 = KL \Rightarrow K = \frac{100}{L}$ or $L = \frac{100}{K}$.

(ii) Using the same method as (i): $K = \frac{10,000}{L}$ or $L = \frac{10,000}{K}$.

(c) Total differential is $dQ \equiv \frac{\partial Q}{\partial K}dK + \frac{\partial Q}{\partial L}dL$. Along any isoquant $dQ = 0$ so $\frac{\partial Q}{\partial K}dK = -\frac{\partial Q}{\partial L}dL \Rightarrow \frac{dK}{dL} = -\frac{\frac{\partial Q}{\partial L}}{\frac{\partial Q}{\partial K}} = -\frac{0.5K^{0.5}L^{-0.5}}{0.5K^{-0.5}L^{0.5}} = -K^{0.5-(-0.5)}L^{-0.5-0.5} = -\frac{K}{L}$.

The slope is negative provided K and L are positive. By definition, $\frac{\partial Q}{\partial L} = MPL$ and $\frac{\partial Q}{\partial K} = MPK$.

(d) The graphs are very similar to figure 15.14.

(e) To stay on the same isoquant when L changes by a small amount dL, we require the differential $dQ = MPL\,dL + MPK\,dK = 0$. We can rearrange this as $dK = -\frac{MPL}{MPK}dL$, which gives us the change in K, dK, necessary to offset dL and leave Q unchanged. For a given dL, dK is greater, the larger the MPL (for if MPL is large, dL has a large effect on output); and the smaller is MPK (for if MPK is small, dK must be large to achieve the necessary change in output).

2. (a) $MPL \equiv \frac{\partial Q}{\partial L} = 4K - 2L$. Assuming K and L both positive, $MPL > 0$ when $\frac{K}{L} > \frac{1}{2}$. Similarly, $MPK \equiv \frac{\partial Q}{\partial K} = 4L - 2K > 0$ when $\frac{K}{L} < 2$. So both MPL and MPK are positive when both of these inequalities are satisfied; that is, when $\frac{1}{2} < \frac{K}{L} < 2$.

The graph of MPL is a straight line with slope -2 and intercept on MPL (vertical) axis of $4K$, where K is a given constant. And similarly for the graph of MPK.

(b) From 1(c) above, we know that the slope of any isoquant is

$\frac{dK}{dL} = -\frac{MPL}{MPK} = -\frac{4K - 2L}{4L - 2K}$ in this example.

Any isoquant is negatively sloped if MPK and MPL are both positive. From (a) this is true when $\frac{1}{2} < \frac{K}{L} < 2$.

(c) See figure S22, which shows a typical isoquant, for $Q = 108$. The graph confirms (b) above, that the marginal products are not always positive. If either marginal product is negative, $\frac{dK}{dL} = -\frac{MPL}{MPK}$ is positive and the isoquant is positively sloped. The isoquant is horizontal when $\frac{dK}{dL} = -\frac{4K - 2L}{4L - 2K} = 0$, which is true when the numerator equals zero. With $Q = 108$, this requires

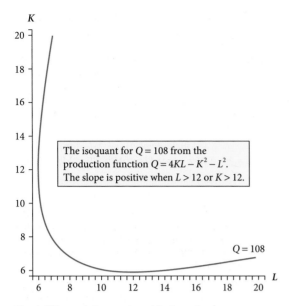

Figure S22 Graph for question 2(c) of exercise 15.4.

$K = 6$, $L = 12$. The isoquant is vertical when the denominator equals zero, which, with $Q = 108$, requires $K = 12$, $L = 6$.

(d) If we fix K at, say, $K = 10$, the short run production function is $Q = 40L - 100 - L^2$ (see figure S23). The intercepts on the L-axis are where $40L - 100 - L^2 = 0$. This quadratic has approximate solutions $L = 2.68$ or 37.3. Q is maximized when $MPL = 40 - 2L = 0$; that is, when $Q = 20$.

Generalizing from this, if we fix K at K_0, the short run production function is $Q = 4K_0L - K_0^2 - L^2$. The slope equals $\frac{\partial Q}{\partial L} = MPL = 4K_0 - 2L$. The maximum of Q is thus where $4K_0 - 2L = 0$. If L is increased beyond $L = 2K_0$, MPL becomes negative and total output falls.

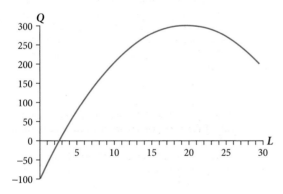

Figure S23 Graph for question 2(d) of exercise 15.4.

3. (a) Differential is $dU = \frac{\partial U}{\partial X}dX + \frac{\partial U}{\partial Y}dY$. Along any isoquant,

$$dU = 0 \text{ so } \frac{dY}{dX} = -\frac{\frac{\partial U}{\partial X}}{\frac{\partial U}{\partial Y}} = -\frac{MU_x}{MU_y} = -\frac{0.75X^{-0.25}Y^{1.5}}{1.5X^{0.75}Y^{0.5}} = -\frac{1}{2}\frac{Y}{X}.$$

This gives the slope of any indifference curve at any point.

(b) In exercise 14.4, question 3(a)(ii), we found that when $U = 27$, $Y = \frac{9}{X^{0.5}}$. Substituting $Y = \frac{9}{X^{0.5}}$ into our expression in (a) above for $\frac{dY}{dX}$, we obtain $\frac{dY}{dX} = -\frac{1}{2}\frac{Y}{X} = -\frac{4.5}{X^{1.5}}$, which equals the value of $\frac{dY}{dX}$ that we found in exercise 14.4, question 3(a)(ii).

4. (a) For this utility function, by implicit differentiation, the slope of any indifference curve is:

$$\frac{dY}{dX} = -\frac{\frac{\partial U}{\partial X}}{\frac{\partial U}{\partial Y}} = -\frac{MU_x}{MU_y} = -\frac{0.5(X+1)^{-0.5}(Y+2)^{0.5}}{0.5(X+1)^{0.5}(Y+2)^{-0.5}}$$

$$= -\frac{Y+2}{X+1}.$$

(b) Equation of indifference curve for $U = \bar{U}$ is $\bar{U} = (X+1)^{0.5}(Y+2)^{0.5}$. After squaring both sides and a little rearrangement, this becomes: $Y = \frac{\bar{U}^2}{X+1} - 2$. The first derivative is: $\frac{dY}{dX} = -\bar{U}^2(X+1)^{-2} = \frac{-\bar{U}^2}{(X+1)^2}$, which is negative (provided $X \neq -1$, which we rule out) for all U including $U = 6$. The second derivative is: $\frac{d^2Y}{dX^2} = \bar{U}^2(X+1)^{-3} = \frac{\bar{U}^2}{(X+1)^3}$ which is positive for all U including $U = 6$. So all indifference curves are downward sloping and convex.

(c) Given $U = (X+a)^{\alpha}(Y+b)^{\beta}$, from (a) above:

$$\frac{dY}{dX} = -\frac{MU_x}{MU_y} = -\frac{\alpha(X+a)^{\alpha-1}(Y+b)^{\beta}}{\beta(X+a)^{\alpha}(Y+b)^{\beta-1}} = -\frac{\alpha}{\beta}\frac{Y+b}{X+a}.$$

5. Since Larry's utility function is the same as that in 4(c) above, we know that the slope of one of Larry's indifference curves is: $\frac{dY}{dX} = -\frac{\alpha}{\beta}\frac{Y+b}{X+a}$ (equation 1). Given Milly's utility function $U_M = [(X+a)^{\alpha}(Y+b)^{\beta}]^2 = (X+a)^{2\alpha}(Y+b)^{2\beta}$, by implicit differentiation the slope of one of Milly's indifference curves is:

$$\frac{dY}{dX} = -\frac{2\alpha(X+a)^{2\alpha-1}(Y+b)^{2\beta}}{2\beta(X+a)^{2\alpha}(Y+b)^{2\beta-1}} = -\frac{\alpha}{\beta}\frac{Y+b}{X+a} \quad \text{(equation 2)}.$$

Comparing equations (1) and (2) we see that for any given values of X and Y, Larry's and Milly's indifference curves have the same slope at *any* point. This means that Milly's indifference curves must be identical to Larry's; that is, they have identical tastes or preferences. Only the utility levels differ; the utility that Milly derives from any given bundle of X and Y equals the square of Larry's utility level from an identical bundle. But this is of no significance, since the units in which utility is measured are arbitrary.

Progress exercise 16.1

1. (i) (a) Direct substitution method: $z = 4x^2 - 200x + 3200$ $\Rightarrow \frac{dz}{dx} = 8x - 200 = 0 \Rightarrow x = 25; \frac{d^2z}{dx^2} = 8 > 0$. So the minimum of z is when $x = 25$, $y = 40 - x = 15$. (Note, substitution of the constraint into the objective function ensures the constraint is satisfied.)

(b) Equating slopes method: Slope of an iso-z section, by implicit differentiation, is $-\frac{\frac{\partial z}{\partial x}}{\frac{\partial z}{\partial y}} = -\frac{2x-y}{-x+4y}$. Slope of constraint $y = 40 - x$ is $\frac{dy}{dx} = -1$. Equating these: $-\frac{2x-y}{-x+4y} = -1 \Rightarrow x = \frac{5}{3}y$. Substituting into constraint, $y = 40 - \frac{5}{3}y \Rightarrow y = 15 \Rightarrow x = \frac{5}{3}y = 25$, as before.

(c) The shape of $z = x^2 - xy + 2y^2$ is, roughly speaking, a cone with its vertex at the origin. So the iso-z sections are concave when viewed from the origin. The tangency of the iso-z contour to the constraint is therefore like the tangency at W in figure 16.6 in the book. So $x = 25$, $y = 15$ is a constrained *minimum* of z. That is, it is the lowest value of z with the constraint satisfied. A non-rigorous test of this is to look at another point on the constraint, say $y = 16$, $x = 24$, which gives $z = 704(> 700)$. This is a point like J or T in figure 16.6.

(ii) (a) Using the method of (i)(a) above, $y = 20$, $x = 36$, giving $z = 9744$.

   (b) Using the method of (i)(b) above, the same answer is obtained.

   (c) Shape is similar to (i) above. The constrained SP seems to be a minimum, since at a nearby point on the constraint, such as $x = 35$ and $y = 21$, $z = 9751$ (> 9744).

(iii) (a) Using the method of (i) above, $y = 8$, $x = 15$, giving $z = 245$.

   (b) Using the method of (i)(b) above, the same answer is obtained.

   (c) The function $z = (x - 1)^2 + (y - 1)^2$ has an *unconstrained* minimum at $x = y = 1$, with $z = 0$. Its shape is roughly conical with its vertex at $x = y = 1$. The constrained SP seems to be a minimum, since when $x = 16$ and $y = 6$, $z = 275$ (> 245).

Progress exercise 16.2

1. (i) Lagrangean expression is: $V = x^2 - xy + 2y^2 + \lambda(40 - x - y)$, so taking partial derivatives and setting them equal to zero gives:

$$V_x = 2x - y - \lambda = 0 \tag{1}$$
$$V_y = -x + 4y - \lambda = 0 \tag{2}$$
$$V_\lambda = 40 - x - y = 0 \tag{3}$$

From (1), $2x - y = \lambda$. From (2), $-x + 4y = \lambda$. Therefore $2x - y = -x + 4y \Rightarrow x = \frac{5}{3}y$. Therefore, in (3) $40 - \frac{5}{3}y - y = 0 \Rightarrow y = 15$, so $x = \frac{5}{3}y = 25$. So $z = (25)^2 - 25(15) + 4(15)^2 = 700$. Substituting $x = 25$, $y = 15$ into, say, equation (1) we get $2(25) - 15 - \lambda = 0 \Rightarrow \lambda = 35$.

If we re-solve equations (1) to (3) with a constraint of $y = 41 - x$ instead of $y = 40 - x$, we get $y = 15\frac{3}{8}$, $x = 25\frac{5}{8}$, $z = 735.4375$. So the increase in z is 35.4375, which is very close to the value of λ, 35.

(ii) Lagrangean expression is: $V = 4x^2 + 3xy + 6y^2 + \lambda(56 - x - y)$, so:

$$V_x = 4x + 3y - \lambda = 0 \tag{1}$$
$$V_y = 3x + 12y - \lambda = 0 \tag{2}$$
$$V_\lambda = 56 - x - y = 0 \tag{3}$$

Solving these simultaneously gives the same solution as in exercise 16.1(ii).

(iii) Lagrangean expression is: $V = (x - 1)^2 + (y - 1)^2 + \lambda(38 - 2x - y)$, so:

$$V_x = 2(x - 1) - 2\lambda = 0 \tag{1}$$
$$V_y = 2(y - 1) - \lambda = 0 \tag{2}$$
$$V_\lambda = 38 - 2x - y = 0 \tag{3}$$

Solving these simultaneously gives the same solution as in exercise 16.1(iii).

Progress exercise 16.3

1. (a) The problem is to minimize $TC = wL + rK = L + 4r$, subject to $Q = K^{0.5}L^{0.5} = 100$. Lagrangean equation is therefore $V = L + 4K + \lambda(K^{0.5}L^{0.5} - 100)$. The partial derivatives are:

$$V_L = 1 + 0.5\lambda K^{0.5}L^{-0.5} = 0 \tag{1}$$
$$V_K = 4 + 0.5\lambda K^{-0.5}L^{0.5} = 0 \tag{2}$$
$$V_\lambda = K^{0.5}L^{0.5} - 100 = 0 \tag{3}$$

From (1), after multiplying both sides by 4: $2\lambda K^{0.5}L^{-0.5} = -4$.

From (2): $0.5\lambda K^{-0.5}L^{0.5} = -4$. Combining these, $2\lambda K^{0.5}L^{-0.5} = 0.5\lambda K^{-0.5}L^{0.5}$.

Divide both sides by the right-hand side, $\Rightarrow \frac{2\lambda K^{0.5}L^{-0.5}}{0.5\lambda K^{-0.5}L^{0.5}} = 1 \Rightarrow 4KL^{-1} = 1$.

So $L = 4K$. Substituting this into equation (3):

$K^{0.5}(4K)^{0.5} = 100 \Rightarrow 2K = 100 \Rightarrow K = 50$; $L = 4K = 200$.

Minimized TC is then $TC = L + 4K = 200 + 4(50) = 400$.

(b) Since $MPL \equiv \frac{\partial Q}{\partial L} = 0.5K^{0.5}L^{-0.5}$ and $MPK \equiv \frac{\partial Q}{\partial K} = 0.5K^{-0.5}L^{0.5}$, in equations (1) and (2) we have: $w + \lambda MPL = 0 \Rightarrow MPL = -\frac{1}{\lambda}w$; and $r + \lambda MPK = 0 \Rightarrow MPK = -\frac{1}{\lambda}r$.

Dividing the first equation by the second gives: $\frac{MPL}{MPK} = \frac{w}{r}$, as the question requires us to show.

(c) If we rework (a) above with $Q = 200$, equations (1) and (2) are unchanged so we get $L = 4K$ as before. Equation (3) is now: $K^{0.5}L^{0.5} - 200 = 0$; and when we substitute $L = 4K$ into this we get: $(4K)^{0.5}K^{0.5} = 200 \Rightarrow 2K = 200 \Rightarrow K = 100$, $L = 4K = 400$.

As the quantity of both inputs has exactly doubled, and input prices are unchanged, it is obvious that TC is also doubled. And as output has also doubled, it is clear that the average cost, $\frac{TC}{Q}$, is unchanged. (Specifically, when $Q = 100$, $AC = \frac{TC}{Q} = \frac{L + 4K}{Q} = \frac{400}{100} = 4$; when $Q = 200$, $AC = \frac{800}{200} = 4$.) So the average cost is constant, at least for $Q = 100$ and $Q = 200$.

(d) In this particular example, we have from (b) above $\frac{MPL}{MPK} = \frac{0.5K^{0.5}L^{-0.5}}{0.5K^{-0.5}L^{0.5}} = K^{0.5-(-0.5)}L^{-0.5-0.5} = \frac{K}{L}$. We also know that for cost minimization we require $\frac{MPL}{MPK} = \frac{w}{r}$. Combining these, $\frac{K}{L} = \frac{w}{r} \Rightarrow rK = wL$. Thus the cost of machines, rK, equals the wage bill, wL, so each will account for 50% of total cost. In this example, this is true for all values of w and r.

2. (a) Lagrangean equation is: $V = 3.2L + 7.5K + \lambda(K^{0.75}L^{0.5} - 40)$. So:

$$V_L = 3.2 + 0.5\lambda K^{0.75}L^{-0.5} = 0 \tag{1}$$
$$V_K = 7.5 + 0.75\lambda K^{-0.25}L^{0.5} = 0 \tag{2}$$
$$V_\lambda = K^{0.75}L^{0.5} - 40 = 0 \tag{3}$$

From (1) and (2), we can get $L = \frac{25}{16}K$. Substituting this into (3) gives: $K^{0.75}\left(\frac{25}{16}K\right)^{0.5} = 40 \Rightarrow K^{1.25} = \left(\frac{16}{25}\right)^{0.5} 40 = 32 \Rightarrow K = (32)^{1/1.25} = 16$. So $L = \frac{25}{16}K = 25$. Minimum total cost is then $TC = 3.2L + 7.5K = 200$, as required.

(b) From equations (1) and (2) above, we can get: $\frac{0.5K^{0.75}L^{-0.5}}{0.75K^{-0.25}L^{0.5}} = \frac{3.2}{7.5}$. Here the left-hand side is $\frac{MPL}{MPK} =$ (absolute) slope of any isoquant. The right-hand side is $\frac{w}{r} =$ (absolute) slope of any isocost line. So cost minimization requires these slopes be equal (that is, a tangency between linear isocost line and convex isoquant).

(c) In (a) above we found $L = \frac{25}{16}K$, using only equations (1) and (2). As the level of output affects only equation (3) above, cost minimization requires $L = \frac{25}{16}K$ whatever the output (unless w or r changes).

3. (a) & (b) Problem is to maximize $Q = K^{0.4}L^{0.5}$ subject to the constraint $3L + 4K = 270$. So Lagrangean equation is: $V = K^{0.4}L^{0.5} + \lambda(3L + 4K - 270)$. So:

$$V_L = 0.5K^{0.4}L^{-0.5} + 3\lambda = 0 \tag{1}$$
$$V_K = 0.4K^{-0.6}L^{0.5} + 4\lambda = 0 \tag{2}$$
$$V_\lambda = 3L + 4K - 270 = 0 \tag{3}$$

Dividing (1) by (2) $\Rightarrow \frac{0.5K^{0.4}L^{-0.5}}{0.4K^{-0.6}L^{0.5}} = \frac{-3\lambda}{-4\lambda} = \frac{3}{4} \Rightarrow K^{0.4-(-0.6)}L^{-0.5-0.5} = \frac{3}{4} \cdot \frac{4}{5} \Rightarrow \frac{K}{L} = \frac{3}{5} \Rightarrow K = \frac{3}{5}L$.

Substituting this into equation (3) $\Rightarrow 3L + 4(\frac{3}{5}L) = 270$ $\Rightarrow L = 50$. So $K = \frac{3}{5}L = 30$. Therefore maximum output is $Q = (30)^{0.4}(50)^{0.5} = 3.898(7.071) = 27.56$.

(c) They are methodologically identical. In order to achieve maximum output from a given budget, it is a logical necessity that each and every unit of output be produced at minimum cost.

Progress exercise 16.4

1. (a) Problem is to maximize $\Pi = PQ - (wL + rK)$ subject to the constraint $K^{0.4}L^{0.5} - Q = 0$. Lagrangean equation is: $V = PQ - (wL + rK) + \lambda(K^{0.4}L^{0.5} - Q) = (20)Q - 5L - 4K + \lambda(K^{0.4}L^{0.5} - Q)$. Partial derivatives are:

$$V_Q = 20 - \lambda = 0 \tag{1}$$
$$V_L = -5 + \lambda 0.5 K^{0.4}L^{-0.5} = 0 \tag{2}$$
$$V_K = -4 + \lambda 0.4 K^{-0.6}L^{0.5} = 0 \tag{3}$$
$$V_\lambda = K^{0.4}L^{0.5} - Q = 0 \tag{4}$$

From (1), $\lambda = 20$. Substitute this into (2) and (3). Then, from (2): $10K^{0.4}L^{-0.5} = 5 \Rightarrow 2K^{0.4} = L^{0.5} \Rightarrow L = 4K^{0.8}$. Substitute this into (3) (with $\lambda = 20$) $\Rightarrow -4 + 8K^{-0.6}(4K^{0.8})^{0.5} = 0 \Rightarrow 16K^{-0.6}K^{0.4} = 4 \Rightarrow K^{-0.2} = \frac{1}{4} \Rightarrow K^{0.2} = 4 \Rightarrow K = 4^5 = 1024$. So $L = 4K^{0.8} = 4(256) = 1024$.

Substituting $K = L = 1024$ into (4), we get: $(1024)^{0.4}(1024)^{0.5} = Q = 512$.

So profits are $\Pi = PQ - (wL + rK) = 20(512) - 5(1024) - 4(1024) = 1024$.

(b) If P rose from 20 to 22, re-solving gives: $K = L = 2655.992$; $Q = 1206.863$; $\Pi = 2647.058$. Thus the 10% rise in P (from 20 to 22) induces an increase in output of 135% (from 512 to 1206.863) so (arc) supply elasticity is $\frac{135}{10} = 13.5$; highly elastic.

(c) From equation (2) in (1)(a) above, we have: $\lambda 0.5 K^{0.4}L^{-0.5} = 5$ and similarly from equation (3): $\lambda 0.4 K^{-0.6}L^{0.5} = 4$. Dividing the first equation by the second: $\frac{0.5}{0.4} \cdot \frac{K^{0.4}L^{-0.5}}{K^{-0.6}L^{0.5}} = \frac{5}{4}$, which simplifies to $K = L$.

Diagrammatically, if we draw the isoquants with (as usual) K on the vertical and L on the horizontal axis, then draw a line from the origin with slope of 1 (or 45 degrees), then where this line cuts the isoquants gives the cost-minimizing combination of K and L; that is, $K = L$.

(d) We have: $TC = wL + rK = 5L + 4K$. From (c) above we have $K = L$ when costs are minimized. Combining these, $TC = wL + rL = (w + r)L = (w + r)K$. Also, with $K = L$ the production function can be written as $Q = K^{0.4}K^{0.5} = K^{0.9}$ from which $K = Q^{1/0.9} = Q^{1.111}$. Substituting this into the TC function gives: $TC = (w + r)Q^{1.111} = 9Q^{1.111}$. The graph of this becomes steeper as Q increases, because the exponent of Q is greater than 1. We can confirm this by finding marginal cost as: $MC \equiv \frac{dTC}{dQ} = (1.111)9Q^{0.111}$. Here we see that MC increases as Q increases; more rigorously, if we take the second derivative $\frac{d^2TC}{dQ^2}$ (the slope of the MC curve) we find it is positive. Also the average cost, $AC \equiv \frac{TC}{Q} = \frac{9Q^{1.111}}{Q} = 9Q^{0.111}$, increases as Q increases. Note that $MC > AC$ at every level of output, because $MC = (1.111)9Q^{0.111} > AC = 9Q^{0.111}$. This is expected, because we know that when AC is rising, we must have $MC > AC$.

The marginal cost curve is upward sloping; whereas the marginal revenue is constant (and equal to price) irrespective of output, so is a horizontal line. The intersection of the upward sloping MC curve and the horizontal MR curve gives the most profitable output.

Progress exercise 16.5

1. (a) The problem is to maximize $U = X^2Y^3$ subject to the budget constraint $B = P_X X + P_Y Y$; in this case, $120 = 4X + 3Y$. The Lagrangean equation is: $V = X^2Y^3 + \lambda[4X + 3Y - 120]$, so:

$$V_X = 2XY^3 + 4\lambda = 0 \tag{1}$$
$$V_Y = 3Y^2X^2 + 3\lambda = 0 \tag{2}$$
$$V_\lambda = 4X + 3Y - 120 = 0 \tag{3}$$

Dividing equation (1) by equation (2), after slight rearrangement $\Rightarrow \frac{2XY^3}{3Y^2X^2} = \frac{-4\lambda}{-3\lambda} = \frac{4}{3}$. This simplifies to: $Y = 2X$. Substituting this into equation (3) we get: $4X + 3(2X) = 120 \Rightarrow X = 12$ so $Y = 2X = 24$.

(b) The graph is identical to figure 16.14, with $X_0 = 12$, $Y_0 = 24$, $B_0 = 120 = 4X + 3Y$, and the maximized level of utility, $U_0 = (12)^2(24)^3$.

(c) (i) $X = 12$, so $P_X X = 4(12) = 48$. Similarly, $Y = 24$, so $P_Y Y = 3(24) = 72$.

(ii) $\frac{P_X X}{B} = \frac{48}{120} = 40\%$; $\frac{P_Y Y}{B} = \frac{72}{120} = 60\%$.

(d) If $P_X = 6$, equation (1) becomes $2XY^3 + 6\lambda = 0$. Following the same steps as in (a) we get: $\frac{Y}{X} = \frac{6}{3} \cdot \frac{3}{2} = 3$, so $Y = 3X$. Thus in equation (3): $6X + 3(3X) = 120 \Rightarrow X = 8$, so $Y = 3X = 24$. Thus consumption of X falls from 12 to 8, but consumption of Y is unchanged. (Unchanged consumption of Y following a change in P_X is not generally to be expected, but is a feature of the Cobb–Douglas utility function.)

(e) New expenditures are $P_X X = 6 \times 8 = 48$ and $P_Y Y = 3 \times 24 = 72$. Expenditure shares are unchanged at $\frac{48}{120} = 40\%$ and $\frac{72}{120} = 60\%$.

(f) Following an increase in the price of X, the quantity purchased falls so as to leave the total expenditure, $P_X X$, unchanged. This is a property of a demand function with a constant own-price elasticity equal to -1.

2. (a) The Lagrangean is now: $V = X^\alpha Y^\beta + \lambda(P_X X + P_Y Y - B)$. So:

$$V_X = \alpha X^{\alpha-1}Y^\beta + \lambda P_X = 0 \tag{1}$$
$$V_Y = \beta X^\alpha Y^{\beta-1} + \lambda P_Y = 0 \tag{2}$$
$$V_\lambda = P_X X + P_Y Y - B = 0 \tag{3}$$

Dividing (1) by (2), preceded by a little rearrangement, gives:

$$\frac{\alpha X^{\alpha-1}Y^\beta}{\beta X^\alpha Y^{\beta-1}} = \frac{P_X}{P_Y} \tag{4}$$

The left-hand side of this equation is the ratio of the marginal utility of X to the marginal utility of Y; that is, the consumer's marginal rate of substitution between X and Y. This must equal the ratio of prices (the right-hand side) as a *necessary* condition for utility maximization. (It is not however a *sufficient* condition for utility maximization, since it is also necessary that equation (3) be satisfied, and also that the relevant second order conditions are satisfied – which we did not consider in the book.)

(b) The relevant diagram is again figure 16.14.

(c) Equation (4) simplifies to: $\frac{\alpha}{\beta}\frac{Y}{X} = \frac{P_X}{P_Y} \Rightarrow \frac{Y}{X} = \frac{\beta}{\alpha}\frac{P_X}{P_Y}$. (5)

So the ratio of Y consumption to X consumption depends only on the price ratio $\frac{P_X}{P_Y}$ (given the parameter values α and β). The ratio is independent of the budget, B. See figure S24.

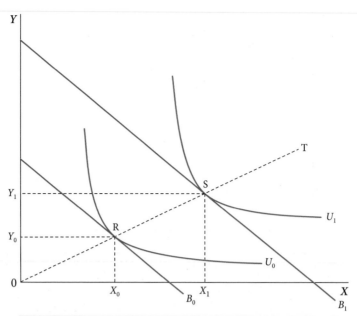

When the budget increases from B_0 to B_1 the equilibrium shifts from R to S, but the proportion in which the two goods are consumed is unchanged. This is because R and S lie on a ray from the origin, so $\frac{Y_0}{X_0} = \frac{Y_1}{X_1}$. The slope of the ray 0T gives the proportion in which the two goods are consumed, and is given by $\frac{Y}{X} = \frac{\beta}{\alpha}\frac{P_X}{P_Y}$ (equation (5) in the answer).

Figure S24 Graph for question 2(c) of exercise 16.5.

(d) From equation (5) above, $Y = \frac{\beta}{\alpha}\frac{P_X}{P_Y}X$. Substitute this into equation (3): $\Rightarrow B = P_X X + P_Y\left(\frac{\beta}{\alpha}\frac{P_X}{P_Y}\right)X = P_X X\left(1 + \frac{\beta}{\alpha}\right)$
$= P_X X\left(\frac{\alpha+\beta}{\alpha}\right) \Rightarrow X = \frac{\alpha}{\alpha+\beta}\frac{B}{P_X}$. Similarly $Y = \frac{\beta}{\alpha+\beta}\frac{B}{P_Y}$.

(e) From the two equations immediately above we can easily get: $\frac{P_X X}{B} = \frac{\alpha}{\alpha+\beta}$ and $\frac{P_Y Y}{B} = \frac{\beta}{\alpha+\beta}$. Here $\frac{P_X X}{B}$ gives the proportion of the budget spent on X, and similarly $\frac{P_Y Y}{B}$ gives the proportion of the budget spent on Y.

(f) From (d) we have the demand function for X as: $X = \frac{\alpha}{\alpha+\beta}\frac{B}{P_X}$. The graph of this relationship between X and P_X (with α, β, and B as parameters) is a rectangular hyperbola. Total expenditure on X is $P_X X = \frac{\alpha}{\alpha+\beta}B$. The right-hand side of this is constant, unless B changes (or the consumer's tastes, reflected in α and β, change). So if P_X doubles, X will halve and vice versa. The same is true of the demand function for Y, which is: $Y = \frac{\beta}{\alpha+\beta}\frac{B}{P_Y}$.

(g) Again, using (e) above, we have the expenditure shares as: $\frac{P_X X}{B} = \frac{\alpha}{\alpha+\beta}$ and $\frac{P_Y Y}{B} = \frac{\beta}{\alpha+\beta}$. Since α and β are parameters, if B doubles or halves then $P_X X$ and $P_Y Y$ must also double or halve to continue to satisfy these equations. Since P_X and P_Y are constants, this necessitates that X and Y, the quantities, must double or halve.

(h) If B, P_X and P_Y increase to $2B$, $2P_X$ and $2P_Y$ then the right-hand side of the equation $\frac{\alpha X^{\alpha-1}Y^\beta}{\beta X^\alpha Y^{\beta-1}} = \frac{P_X}{P_Y}$ obtained in (a) above becomes $\frac{2P_X}{2P_Y}$ and since the 2s cancel this means that the equation is left unchanged. Equation (3) becomes $V_\lambda = 2P_X X + 2P_Y Y - 2B = 2(P_X X + P_Y Y - B) = 0$ which has the same solution as the original equation (3). So doubling B, P_X and P_Y does not change the equilibrium values of X and Y. This is not surprising because the consumer's real income, and the real price of each good are unchanged. (In the real world, a doubling of prices and the budget might affect demand for goods because the consumer's wealth might be changed.)

Progress exercise 17.1

1. (a) See figure S25.
   (b) See figure S26.
      Neither production function can be homogeneous.

2. (a) $Q_0 = 3L_0 K_0 + L_0^2$; $Q_1 = 3(\lambda L_0)(\lambda K_0) + (\lambda L_0)^2 = \lambda^2(3L_0 K_0 + L_0^2) = \lambda^2 Q_0$. The function is homogeneous of degree 2.
   (b) $Q_0 = (3L_0 K_0 + L_0^2)^{1/2}$; $Q_1 = [3(\lambda L_0)(\lambda K_0) + (\lambda L_0)^2]^{1/2} = [\lambda^2(3L_0 K_0 + L_0^2)]^{1/2} = \lambda(3L_0 K_0 + L_0^2)^{1/2} = \lambda Q_0$. The function is homogeneous of degree 1.
   (c) $Q_0 = 4K_0^2 + 0.5L_0 K_0 + 3L_0^2$; $Q_1 = 4(\lambda K_0)^2 + 0.5(\lambda L_0)(\lambda K_0) + 3(\lambda L_0)_0^2 = \lambda^2(4K_0^2 + 0.5L_0 K_0 + 3L_0^2) = \lambda^2 Q_0$. The function is homogeneous of degree 2.
   (d) $Q_0 = 3K_0^2 + 10L_0 K_0^{0.5} + 5L_0^2$; $Q_1 = 3(\lambda K_0)^2 + 10(\lambda L_0)(\lambda K_0)^{0.5} + 5(\lambda L_0)^2 = \lambda^2 3K_0^2 + \lambda^{1.5}10L_0 K_0^{0.5} + \lambda^2 5L_0^2$. Since λ raised to some power is not a common factor (as it is in (a)–(c) above), the function is not homogeneous.

3. (a) If initially $Q_0 = 100K_0^{0.3}L_0^{0.7}$, then K increases to λK_0 and L increases to λL_0, and the new value of Q is: $Q_1 = 100(\lambda K_0)^{0.3}(\lambda L_0)^{0.7} = 100\lambda^{0.3}K_0^{0.3}\lambda^{0.7}L_0^{0.7} = \lambda 100K_0^{0.3}L_0^{0.3} = \lambda Q_0$. So the proportionate increase in Q equals the proportionate increase in K and L. This production function is therefore homogeneous of degree 1 and (equivalently) there are constant returns to scale.
   (b) Initial value: $Q_0 = 50K_0^{0.6}L_0^{0.5}$. New value:
      $Q_1 = 50(\lambda K_0)^{0.6}(\lambda L_0)^{0.5} = 50\lambda^{0.6}K_0^{0.6}\lambda^{0.5}L_0^{0.5}$
      $= \lambda^{1.1}50K_0^{0.6}L_0^{0.5} = \lambda^{1.1}Q_0$.

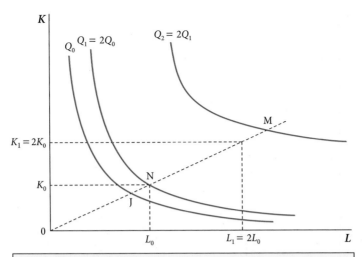

Because J, N, and M lie on the same ray from the origin, they all have the same ratio of K to L; they differ only in the scale of output. The movement from J to N doubles output, but less than doubles inputs (increasing returns). The movement from N to M doubles output, but more than doubles inputs (decreasing returns).

Figure S25 Graph for question 1(a) of exercise 17.1.

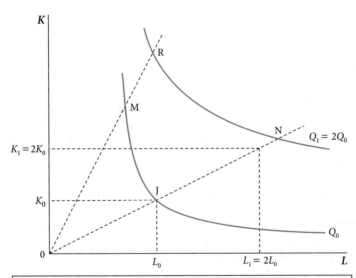

At M, K is large relative to L (capital intensive production). In moving from M to R, Q doubles but, because R is quite close to M, we see that K and L are less than doubled (increasing returns). At J, K is small relative to L (labour intensive production). In moving from J to N, Q doubles but K and L are more than doubled (decreasing returns).

Figure S26 Graph for question 1(b) of exercise 17.1.

This function is homogeneous of degree 1.1. If $\lambda = 2$, say, then L and K both double but Q increases by the factor $2^{1.1}$ which is greater than 2. So there are increasing returns to scale.

(c) $Q_0 = 25K_0^{0.3}L_0^{0.5}$;

$Q_1 = 25(\lambda K_0)^{0.3}(\lambda L_0)^{0.5} = \lambda^{0.8}25K_0^{0.3}L_0^{0.5} = \lambda^{0.8}Q_0$.

This function is homogeneous of degree 0.8. If $\lambda = 2$ this means that L and K double but Q increases by the factor $2^{0.8}$ which is less than 2. So there are decreasing returns to scale.

(d) $Q_0 = AK_0^{\alpha}L_0^{\beta}$;

$Q_1 = A(\lambda K_0)^{\alpha}(\lambda L_0)^{\beta} = \lambda^{\alpha+\beta}AK_0^{\alpha}L_0^{\beta} = \lambda^{\alpha+\beta}Q_0$.

The function is homogeneous of degree $\alpha + \beta$. If $\alpha + \beta > 1$ there are increasing returns to scale; if $\alpha + \beta = 1$ there are constant returns to scale; and if $\alpha + \beta < 1$ there are decreasing returns to scale.

(e) Same as (d) but with $\beta = 1 - \alpha$. So $Q_1 = \lambda^{\alpha+(1-\alpha)}Q_0 = \lambda Q_0$. Whatever the value of α, the function is homogeneous of degree 1 and there are constant returns to scale.

4. $Q_0 = A[bK_0^{\alpha} + (1-b)L_0^{\alpha}]^{1/\alpha};$

$Q_1 = A[b(\lambda K_0)^{\alpha} + (1-b)(\lambda L_0)^{\alpha}]^{1/\alpha} = A[\lambda^{\alpha}bK_0^{\alpha}$
$+ \lambda^{\alpha}(1-b)L_0^{\alpha}]^{1/\alpha} = A[\lambda^{\alpha}(bK_0^{\alpha} + (1-\alpha)L_0^{\alpha})]^{1/\alpha}$
$= A[\lambda(bK_0^{\alpha} + (1-\alpha)L_0^{\alpha})^{1/\alpha}] = \lambda Q_0.$

This production function is homogeneous of degree 1 and thus has constant returns to scale.

Progress exercise 17.2

1. (a) $Q_0 = AK_0^{\alpha}L_0^{\beta}; Q_1 = A(\lambda K_0)^{\alpha}(\lambda L_0)^{\beta} = \lambda^{\alpha+\beta}AK_0^{\alpha}L_0^{\beta} = \lambda^{\alpha+\beta}Q_0.$ So the function is homogeneous of degree $\alpha + \beta$. If $\alpha + \beta = 1$ there are constant returns to scale. If $\alpha + \beta > 1$ there are increasing returns to scale; if $\alpha + \beta < 1$ there are decreasing returns to scale.

    (b) We know from chapter 16 that cost minimization requires $\frac{MPL}{MPK} = \frac{w}{r}$. In this case, $MPL \equiv \frac{\partial Q}{\partial L} = \beta AK^{\alpha}L^{\beta-1}$ and $MPK \equiv \frac{\partial Q}{\partial K} = \alpha AK^{\alpha-1}L^{\beta}$. So: $\frac{MPL}{MPK} = \frac{\beta AK^{\alpha}L^{\beta-1}}{\alpha AK^{\alpha-1}L^{\beta}} = \frac{\beta}{\alpha} \cdot \frac{K}{L}$. Combining the two equations, cost minimization requires $\frac{\beta}{\alpha}\frac{K}{L} = \frac{w}{r}$. Since the left-hand side contains only the *ratio* of K to L this expression is independent of the absolute values of K and L and therefore independent of output.

    (c) Your diagram should be like figure 17.3 in the book. There, points J and M have the same value of $\frac{w}{r}$, because the isocost line aa′ is parallel to the isocost line bb′. They also have the same value of $\frac{K}{L}$, because they lie on the same ray from the origin. Thus the condition $\frac{\beta}{\alpha}\frac{K}{L} = \frac{w}{r}$ is satisfied at both J and M, and indeed at any point on the ray 0R. (The slope of the ray 0R is $K = \frac{\alpha}{\beta}\frac{w}{r}L$.) So the answer is that, as output varies, K and L vary in such a way as to maintain $\frac{K}{L}$ constant.

    (d) With w and r constant, and therefore $\frac{w}{r}$ constant, we know from (c) above that the firm will not change $\frac{K}{L}$ when it increases its output.

    Case (1): If $\alpha + \beta > 1$ we have increasing returns to scale so if K and L are doubled, Q will more than double. Since w and r are constant, this means that total cost $(= wL + rK)$ will exactly double. Since output has more than doubled, this means that average cost $\left(= \frac{wL+rK}{Q}\right)$ has fallen. As this conclusion is not dependent on the precise level of output, it must hold at all output levels, and therefore average cost falls continuously as output increases. This means that marginal cost must also fall continuously. This is true because, if AC is falling, we must have $MC < AC$ and to maintain this inequality MC must fall as AC falls. The graphical illustration is figure 17.3, with $Q_1 > 2Q_0$ because we are assuming increasing returns to scale.

    Case (2): If $\alpha + \beta = 1$ we have constant returns to scale, so if K and L are doubled, Q will also double. Since w and r are constant, this means that total cost $(= wL + rK)$ will exactly double. This means that $AC = \frac{wL+rK}{Q}$ will be unchanged. So AC is constant at all levels of output, and therefore $MC = AC$ (because we know that when $MC < AC$, AC is falling, and when $MC > AC$, AC is rising). The graphical illustration is figure 17.3, with $Q_1 = 2Q_0$.

    Case (3) If $\alpha + \beta < 1$ we have decreasing returns to scale and when K and L double, Q must less than double. Since $TC = wL + rK$ doubles but Q less than doubles, $AC = \frac{wL+rK}{Q}$ must increase. This is true at all levels of output, and with AC increasing we must have $MC > AC$ and MC increasing too. The graphical illustration is figure 17.3, with $Q_1 < 2Q_0$.

(e) The answer requires you to reproduce example 16.7, chapter 16.

(f) Figures 16.11, 16.12, and 16.13 give the answer here. We know that the most profitable output is where $MC = MR$ and $\frac{d^2MC}{dQ^2} > \frac{d^2MR}{dQ^2}$ (the second order condition for a maximum). For a perfectly competitive firm, $MR = P =$ a constant, so the MR curve is horizontal. There are then three cases: first, *increasing* returns to scale. In (d) and (e) above we found that with increasing returns to scale, AC and MC fall as output increases so the MC curve is negatively sloped (see figure 16.12(b)). Then the MC curve cuts the MR curve at output level Q_1, so that $MC = MR$ at this output. So the first order condition for maximum profit is satisfied. However, because the slope of the MC curve is negative while the slope of the MR curve is zero, the second order condition for a maximum of profit is not fulfilled. In fact this is a *minimum* of profit. If the firm increases its output beyond this level, it will find that $MR > MC$. And as MC is falling, the firm will wish to increase output without limit.

Second, if there are *decreasing* returns to scale, the MC curve is upward sloping and at some level of output must cut the MR curve (figure 16.11). The second order conditions are satisfied here because the MC curve is upward sloping so there is a unique equilibrium at output Q_0.

Finally, if there are constant returns to scale, we found above that $AC = MC = $ constant (figure 16.13). With MC and MR now both horizontal, there are three possibilities. One is that $MC > MR$ at all output levels, in which case the firm will shut down as no output is profitable. The second is that $MC < MR$ at all output levels, in which case the firm will wish to expand output without limit. The third possibility is that $MC = MR$ at all output levels, in which case one level of output is as good as any other, judged by the criterion $MC = MR$. It is then impossible for the firm to decide on a unique output.

Thus we conclude that decreasing returns to scale are necessary for a unique and determinate solution to the firm's profit maximizing problem with perfect competition. However, this conclusion is based on the assumption that the ruling market price is independent of output. This is true for an individual firm, but not for all firms in the industry taken together.

2. Suppose initially $w = w_0$ and $P = P_0$, then they increase to λw_0 and λP_0. Initially we have $L_0^S = A\left(\frac{w_0}{P_0}\right)^{\alpha}$, then $L_1^S = A\left(\frac{\lambda w_0}{\lambda P_0}\right)^{\alpha} = A\left(\frac{w_0}{P_0}\right)^{\alpha} = L_0^S$. Thus the labour supply function is homogeneous of degree zero in w and P, meaning that if w and P both change in the same proportion, labour supply is unchanged. By repeating the steps above we find that the labour demand function is also homogeneous of degree zero. Both supply and demand for labour change only if w changes relative to P. The absolute values of w and P are, in themselves, of no significance.

Progress exercise 17.3

1. Own-price elasticity: $\frac{p_1}{q_1}\frac{\partial q_1}{\partial p_1} = \frac{p_1}{q_1}\left[-3 - \frac{0.75}{2}\left(\frac{p_1}{p_2}\right)^{-1/2}\left(\frac{1}{p_2}\right)\right]$

$= \frac{p_1}{q_1}\left[-3 - \frac{0.75}{2}(p_1p_2)^{-1/2}\right] = \frac{1}{q_1}\left[-3p_1 - \frac{3}{8}\left(\frac{p_1}{p_2}\right)^{1/2}\right].$

Cross-price elasticity: $\frac{p_2}{q_1}\frac{\partial q_1}{\partial p_2} = \frac{p_2}{q_1}\left[0.5 - \frac{0.75}{2}\left(\frac{p_1}{p_2}\right)^{-1/2}\left(-\frac{p_1}{p_2^2}\right)\right]$

$= \frac{1}{q_1}\left(0.5p_2 + \frac{3}{8}\left(\frac{p_1}{p_2}\right)^{0.5}\right).$

2. (a)&(b) Write $L^S = A\left(\frac{w}{P}\right)^\alpha = Aw^\alpha P^{-\alpha}$. Then money wage elasticity of labour supply is: $\frac{w}{L^S}\frac{\partial L^S}{\partial w} = \frac{w}{L^S}(\alpha Aw^{\alpha-1}P^{-\alpha}) = \frac{\alpha Aw^\alpha P^{-\alpha}}{L^S} = \alpha$ (since $L^S = Aw^\alpha P^{-\alpha}$). Price elasticity of labour supply is: $\frac{P}{L^S}\frac{\partial L^S}{\partial P} = \frac{P}{L^S}(-\alpha Aw^\alpha P^{-\alpha-1}) = -\alpha\frac{Aw^\alpha P^{-\alpha}}{L^S} = -\alpha$. Similarly the wage and price elasticities of labour demand are $-\beta$ and β, respectively (note reversal of signs compared to labour supply).

(c) We find that the money wage and price elasticities of labour supply are α and $-\alpha$; that is, equal in absolute magnitude but with opposite signs. This means that a 1% increase in the money wage has the same effect on labour supply as a 1% fall in the price level. The effect is the same because both have the same effect on the real wage and it is, by assumption, the real wage that influences labour supply. Exactly the same is true of labour demand, where the money wage and price elasticities are $-\beta$ and β. Their signs are reversed, compared with labour supply, because a rise in real wage $\left(\frac{w}{P}\right)$ is assumed to increase labour supply but reduce labour demand.

3. (a) Let Y = national product, N = population. National product per person is $y = \frac{Y}{N}$, so the proportionate differential, $\frac{dy}{y}$, is (see example 17.10): $\frac{dy}{y} = \frac{dY}{Y} - \frac{dN}{N} = 0.028 - 0.012 = 0.016 = 1.6\%$ per year.

(b) $T = tY$ so the proportionate differential, $\frac{dT}{T}$, is (see example 17.9) $\frac{dT}{T} = \frac{dt}{t} + \frac{dY}{Y} = 0.005 + 0.028 = 0.033 = 3.3\%$ per year.

(c) Tax revenue per person is $V = \frac{T}{N}$, so the proportionate differential is: $\frac{dV}{V} = \frac{dT}{T} - \frac{dN}{N} = 0.033 - 0.012 = 0.021 = 2.1\%$ per year.

Remark: using the differential always involves an error. For example, in (b) the error is $dt \times dY$ (see chapter 15, example 15.2). As the time period of the analysis increases, dt and dY become larger and the error increases. In general, using the differential to answer questions relating to growth rates therefore becomes increasingly inaccurate as the time interval being considered lengthens.

4. (a) Partial elasticity with respect to K is:
$\frac{K}{Q}\frac{\partial Q}{\partial K} = \frac{K}{Q}(\alpha AK^{\alpha-1}L^\beta) = \frac{\alpha AK^\alpha L^\beta}{Q} = \alpha$
(since $Q = AK^\alpha L^\beta$).
Similarly, partial elasticity with respect to L is: $\frac{L}{Q}\frac{\partial Q}{\partial L} = \beta$.

The partial elasticity with respect to K measures the instantaneous rate of proportionate change in Q with respect to a proportionate change in K. After multiplying by 100 it measures *approximately* the percentage increase in Q that results from a small percentage increase in K, with L constant. Similarly the partial elasticity with respect to L measures the effect on Q of a small percentage increase in labour input, with K constant. (The error increases as the percentage change in K or L increases.)

(b) If $Q = AK^\alpha L^\beta$, the proportionate differential is: $\frac{dQ}{Q} = \left(\frac{K}{Q}\frac{\partial Q}{\partial K}\right)\frac{dK}{K} + \left(\frac{L}{Q}\frac{\partial Q}{\partial L}\right)\frac{dL}{L} = \alpha\frac{dK}{K} + \beta\frac{dL}{L}$. If $\frac{dK}{K} = \frac{dL}{L}$ (that is, both inputs are increased in the same proportion) we have: $\frac{dQ}{Q} = (\alpha+\beta)\frac{dL}{L} = (\alpha+\beta)\frac{dK}{K}$. If $\frac{dL}{L} = \frac{dK}{K} = 5\% = 0.05$, $\frac{dQ}{Q} = (\alpha+\beta)(0.05)$. So $\frac{dQ}{Q}$ is greater, equal to, or less than 0.05 according as $(\alpha+\beta)$ is greater, equal to, or less than 1. (Note that the error in this calculation using the differential is not insignificant when the proportionate change is as large as 5%, as in this example.)

5. (a) $\frac{dQ}{Q} = 0.35\frac{dK}{K} + 0.75\frac{dL}{L}$. If $\frac{dK}{K} = \frac{dL}{L} = 2\% = 0.02$, then $\frac{dQ}{Q} = (0.35 + 0.75)0.02 = 0.022 = 2.2\%$. (Note: increasing returns to scale so output grows faster than input.)

(b) In this case $\frac{dK}{K} = 3\% = 0.03$, and $\frac{dL}{L} = 0$, so $\frac{dQ}{Q} = 0.35(0.03) = 0.0105 = 1.05\%$. If actual growth of output were faster, this could be due to technical advances making production more efficient, so that Q becomes larger for any given K and L.

(c) Here $\frac{dK}{K} = 0.03$ and $\frac{dL}{L} = 0.01$, so $\frac{dQ}{Q} = (0.35)0.03 + (0.75)0.01 = 0.018 = 1.8\%$. That is, immigration of 1% per year adds 0.75% to the growth rate.

(d) There is an adding up problem because this production function has increasing returns to scale. This means, according to Euler's theorem, that: $Q < K\frac{\partial Q}{\partial K} + L\frac{\partial Q}{\partial L}$ (check this for yourself). We know that under perfect competition the real rewards of K and L will equal their marginal products; that is: $\frac{r}{P} = \frac{\partial Q}{\partial K}$ and $\frac{w}{P} = \frac{\partial Q}{\partial L}$. Combining these equations, we have: $Q < \frac{rK}{P} + \frac{wL}{P} \Rightarrow PQ < rK + wL$. Thus total revenue, PQ, is less than the reward of capital (owners of machines), rK, plus the reward of labour, wL. Therefore there is simply not enough sales revenue to reward factors according to their marginal products. The same problem occurs in reverse when there are decreasing returns to scale. It is not clear how this can be resolved; however, the problem disappears if we are convinced, or are willing to assume, that returns to scale are constant.

6. The partial elasticities of the production function $Q = f(K, L)$ are, by definition, $\frac{K}{Q}\frac{\partial Q}{\partial K}$ and $\frac{L}{Q}\frac{\partial Q}{\partial L}$. If the real rewards of K and L equal their marginal products (as discussed in question 5 above), then the partial elasticities may be written as $\frac{rK}{PQ}$ and $\frac{wL}{PQ}$, respectively. Here rK is the income of owners of capital, so $\frac{rK}{PQ}$ is the income of capital as a proportion of total income; that is, the share of capital in total income. Similarly $\frac{wL}{PQ}$ is the share of labour in total income. So factor shares will equal the respective partial elasticities, if the factors' real rewards equal their marginal products. However, Euler's theorem says that, if there are increasing returns to scale, $K\frac{\partial Q}{\partial K} + L\frac{\partial Q}{\partial L} > Q \Rightarrow \frac{K}{Q}\frac{\partial Q}{\partial K} + \frac{L}{Q}\frac{\partial Q}{\partial L} > 1$ that is, the sum of the partial elasticities exceeds 1. But in that case, it is logically impossible to have: $\frac{K}{Q}\frac{\partial Q}{\partial K} = \frac{rK}{PQ}$ and $\frac{L}{Q}\frac{\partial Q}{\partial L} = \frac{wL}{PQ}$.

This impossibility is because the right-hand sides of these two equations are the shares of capital and labour in total income, and these shares must, from their definition, add up to 1. And if the right-hand sides add up to 1, so must the left-hand sides, but we know from Euler's theorem that this is not the case when there are increasing returns.

The same problem arises if there are decreasing returns, for then Euler's theorem says that: $\frac{K}{Q}\frac{\partial Q}{\partial K} + \frac{L}{Q}\frac{\partial Q}{\partial L} < 1$. Thus we can use factor shares as proxies for partial elasticities only when the sum of the partial elasticities equals 1, which is true only with constant returns to scale.

7. Given: $L^S = A\left(\frac{w}{P}\right)^\alpha$. Taking logs: $\ln L^S = \ln A + \alpha(\ln w - \ln P)$; so the partial derivatives are: $\frac{\partial \ln L^S}{\partial \ln w} = \alpha$ and $\frac{\partial \ln L^S}{\partial \ln P} = -\alpha$. In question 2 we showed that the partial elasticities of $L^S = A\left(\frac{w}{P}\right)^\alpha$ were: $\frac{w}{L^S}\frac{\partial L^S}{\partial w} = \alpha$ and $\frac{P}{L^S}\frac{\partial L^S}{\partial P} = -\alpha$. So this example confirms the general truth that with log scales on both axes the slopes $\frac{\partial \ln L^S}{\partial \ln w}$ and $\frac{\partial \ln L^S}{\partial \ln P}$ equal the elasticities $\frac{w}{L^S}\frac{\partial L^S}{\partial w}$ and $\frac{P}{L^S}\frac{\partial L^S}{\partial P}$. The same is true of the labour demand function $L^D = B\left(\frac{w}{P}\right)^{-\beta}$.

8. (a) Taking logs: $\ln q_1 = \alpha(\ln p_1) + \beta(\ln p_2) + \gamma(\ln y)$. Since α, β, and γ are constants (parameters) we can see that this defines a linear relationship (a four-dimensional hyper-plane) between the four variables: $\ln q_1$, $\ln p_1$, $\ln p_2$, and $\ln y$.

(b) The partial derivatives of the logarithmic function are

$$\frac{\partial \ln q_1}{\partial \ln p_1} = \alpha; \frac{\partial \ln q_1}{\partial \ln p_2} = \beta; \text{ and } \frac{\partial \ln q_1}{\partial \ln y} = \gamma.$$

From the demand function $q_1 = p_1^\alpha p_2^\beta y^\gamma$, the partial elasticities are: Own price elasticity: $\frac{p_1}{q_1}\frac{\partial q_1}{\partial p_1} = \frac{p_1}{q_1}(\alpha p_1^{\alpha-1}p_2^\beta y^\gamma) = \frac{\alpha p_1^\alpha p_2^\beta y^\gamma}{q_1} = \alpha$ (since $q_1 = p_1^\alpha p_2^\beta y^\gamma$). Similarly, cross-price elasticity is: $\frac{p_2}{q_1}\frac{\partial q_1}{\partial p_2} = \beta$; and income elasticity is: $\frac{y}{q_1}\frac{\partial q_1}{\partial y} = \gamma$. So the partial derivatives of the logarithmic transformation measure the partial elasticities of the original function.

(c) We have: α = own price elasticity; normally negative; β = cross-price elasticity; either positive or negative depending on whether the two goods are substitutes or complements; γ= income elasticity; if positive, good 1 is said to be a 'normal' good; if negative, an 'inferior' good.

Progress exercise 18.1

(a) $\frac{1}{4}x^4 + c$ (b) $-0.5x^{-2} + c$ (c) $2x^{-0.5} + c$

(d) $0.8x^{1.25} + c$ (e) $4x^{0.25} + c$ (f) $5\ln x + c$

(g) $2e^x + c$ (h) $x^5 + e^x - 8x^{0.5} + c$

(i) $-0.5x^{-4} + 3x^3 + 0.625x^{1.6} + c$ (j) $6x^{5/3} - 0.5\ln x + c$

(k) $1000e^{0.1x} + c$ (l) $-200e^{-0.05x} + c$

(m) $(2x^2 + \ln x)^4 + c$ (n) $\frac{2}{3}(x^{0.5} + 6x + 5)^{3/2} + c$

(o) $\ln(x^4 + 5x^2 - 2e^x) + c$ (p) $xe^x - e^x + c$

(q) $0.5xe^{2x} - 0.25e^{2x} + c$

Progress exercise 18.2

(1) 2500 (2) 2.9957 (3) 108 (4) 126.4241 (5) 6811.5333 (6) 187.112 (7) e (8) 1

Progress exercise 18.3

1. To answer this, we find the definite integral of $4 + 4q$ from 25 to 50. Answer is 3850.

2. To answer this, we find the definite integral of MR from 144 to 400, which equals 4000. So answer is -4000; that is, a loss.

3. To answer this, we first find the consumer's valuation, which is the area under the inverse demand curve between $q = 0$ and $q = 8$. This is

$$\int_0^8 (100 - q^2)dq = 629\frac{1}{3}.$$

Then we subtract what the consumer has to pay for 8 units. When $q = 8$, then from the inverse demand function $p = 100 - q^2$ we have $p = 36$. So the consumer pays $8 \times 36 = 288$ for 8 units. Her consumer surplus is therefore $629\frac{1}{3} - 288 = 341\frac{1}{3}$.

4. The PV is the integral: $PV = \int_0^x ae^{-rx}dx$. With some manipulation (see rules 18.6 and 18.6a) this becomes: $PV = \frac{a}{r}(1 - e^{-rx})$.

5. In general, the present value (PV) of the expected profit (see rules 18.6 and 18.6a), discounted continuously, is

$$PV = \int_0^x ae^{-rx}dx = \frac{a}{r}(1 - e^{-rx}) \text{ million dollars}$$

where a = expected profit per year, x = number of years, r = discount rate.

(a) Here $a = 20 \times 0.5 = 10$ (million dollars per year); $r = 0.05$, $x = 20$. The PV of expected profits is then $\frac{a}{r}(1 - e^{-rx}) = \frac{10}{0.05}(1 - e^{-0.05(20)}) = 126.42$ million dollars. The Maximum it is worth paying for the right to extract is therefore the PV, minus 1 dollar.

(b) Repeat (a) with $r = 0.1$, giving $PV = 86.47$. So the maximum bid is reduced by $126.42 - 86.47 = 39.93$. Percentage reduction is $100(39.93/126.42) = 31.6$.

(c) Repeat (b) with $x = 5$, giving $PV = 39.35$.

Progress exercise 19.1

1. $2A + 3B = 2\begin{bmatrix} 1 & 2 \\ 3 & 4 \end{bmatrix} + 3\begin{bmatrix} 1 & 3 \\ 4 & 2 \end{bmatrix} = \begin{bmatrix} 2 & 4 \\ 6 & 8 \end{bmatrix} + \begin{bmatrix} 3 & 9 \\ 12 & 6 \end{bmatrix}$

$= \begin{bmatrix} 5 & 13 \\ 18 & 14 \end{bmatrix}$

2. $AB = \begin{bmatrix} 2 & 1 \\ 1 & 7 \\ 5 & 3 \end{bmatrix}\begin{bmatrix} 1 & 3 & 6 \\ -1 & 2 & 4 \end{bmatrix} = \begin{bmatrix} 1 & 8 & 16 \\ -6 & 17 & 34 \\ 2 & 21 & 42 \end{bmatrix}$

$BA = \begin{bmatrix} 1 & 3 & 6 \\ -1 & 2 & 4 \end{bmatrix}\begin{bmatrix} 2 & 1 \\ 1 & 7 \\ 5 & 3 \end{bmatrix} = \begin{bmatrix} 35 & 40 \\ 20 & 25 \end{bmatrix}$ (Note that $AB \neq BA$.)

3. (a) -13 (a scalar; neither a matrix nor a vector);

(b) $ax + by$ (a scalar);

(c) The matrix: $\begin{bmatrix} 3 & 6 & 9 \\ 2 & 4 & 6 \\ 1 & 2 & 3 \end{bmatrix}$;

(d) The column vector: $\begin{bmatrix} -9 \\ 3 \\ -12 \end{bmatrix}$;

(e) 3 (scalar);

(f) $x^2 + 2y^2 + 3z^2 - 2xz + 2yz$ (scalar).

4. You should calculate each side of each equality separately. Each side should be equal to:

(a) $\begin{bmatrix} -15 & 24 \\ 5 & 14 \end{bmatrix}$ (b) $\begin{bmatrix} 32 & 43 \\ 4 & 26 \end{bmatrix}$

(c) $\begin{bmatrix} 18 & 6 \\ 18 & 3 \end{bmatrix}$ (d) $\begin{bmatrix} 59 & -75 \\ 50 & -41 \end{bmatrix}$

5. (a) Price vector (row): [1.20 2.00 1.00 0.80].

Quantities (column): $\begin{bmatrix} 100 \\ 50 \\ 75 \\ 25 \end{bmatrix}$

(b) Total cost = $[1.20 \quad 2.00 \quad 1.00 \quad 0.80]\begin{bmatrix} 100 \\ 50 \\ 75 \\ 25 \end{bmatrix} = 315$.

Progress exercise 19.2

1. By direct multiplication, $AA^{-1} = \begin{bmatrix} a & b \\ c & d \end{bmatrix} \cdot \frac{1}{(ad - bc)}\begin{bmatrix} d & -b \\ -c & a \end{bmatrix}$

$= \frac{1}{(ad - bc)}\begin{bmatrix} ad - bc & 0 \\ 0 & ad - bc \end{bmatrix} = \begin{bmatrix} 1 & 0 \\ 0 & 1 \end{bmatrix} = I$

(the identity matrix)

2. Inverses are: (a) $-\frac{1}{2}\begin{bmatrix} 4 & -2 \\ -3 & 1 \end{bmatrix}$; (b) $\frac{1}{19}\begin{bmatrix} 3 & 7 \\ -1 & 4 \end{bmatrix}$

(c) no inverse (because determinant = 0).

3. (a) In matrix form, the pair of simultaneous equations can be written as:

$$\begin{bmatrix} 1 & 2 \\ 3 & 4 \end{bmatrix}\begin{bmatrix} x_1 \\ x_2 \end{bmatrix}=\begin{bmatrix} 3 \\ 7 \end{bmatrix}$$

Therefore we can use the inverse matrix from 2(a) above, to solve as:

$$\begin{bmatrix} x_1 \\ x_2 \end{bmatrix}=-\frac{1}{2}\begin{bmatrix} 4 & -2 \\ -3 & 1 \end{bmatrix}\begin{bmatrix} 3 \\ 7 \end{bmatrix}=-\frac{1}{2}\begin{bmatrix} -2 \\ -2 \end{bmatrix}=\begin{bmatrix} 1 \\ 1 \end{bmatrix}$$

(b) $\begin{bmatrix} x \\ y \end{bmatrix}=\frac{1}{19}\begin{bmatrix} 3 & 7 \\ -1 & 4 \end{bmatrix}\begin{bmatrix} 1 \\ 5 \end{bmatrix}=\frac{1}{19}\begin{bmatrix} 38 \\ 19 \end{bmatrix}=\begin{bmatrix} 2 \\ 1 \end{bmatrix}$

(using same method as (a))

(c) From 2(c) we know that the matrix $\begin{bmatrix} 6 & 4 \\ 3 & 2 \end{bmatrix}$ has no inverse, therefore the given pair of simultaneous equations has no solution.

4. (a) det B = 9; det C = 10

(b) $B+C=\begin{bmatrix} 4 & 4 & 0 \\ 7 & 5 & 4 \\ 1 & -2 & 8 \end{bmatrix};\quad \det(B+C)=-16$

$BC=\begin{bmatrix} 2 & 7 & -1 \\ 8 & 15 & 8 \\ -4 & -8 & -7 \end{bmatrix};\quad \det BC=90$

5. $\det(A)=1$. Minors M_{ij} are:

$\begin{array}{lll} M_{11}=5 & M_{12}=2 & M_{13}=11 \\ M_{21}=4 & M_{22}=2 & M_{23}=9 \\ M_{31}=6 & M_{32}=3 & M_{33}=14 \end{array}$

and cofactors A_{ij} are:

$\begin{array}{lll} A_{11}=5 & A_{12}=-2 & A_{13}=11 \\ A_{21}=-4 & A_{22}=2 & A_{23}=-9 \\ A_{31}=6 & A_{32}=-3 & A_{33}=14 \end{array}$

so $A^{-1}=\frac{1}{|A|}\begin{bmatrix} A_{11} & A_{21} & A_{31} \\ A_{12} & A_{22} & A_{32} \\ A_{13} & A_{23} & A_{33} \end{bmatrix}=\begin{bmatrix} 5 & -4 & 6 \\ -2 & 2 & -3 \\ 11 & -9 & 14 \end{bmatrix}$

Progress exercise 19.3

1. (a) Parts requirement matrix:

$\begin{array}{c}\\ \text{Widgets} \\ \text{Votsits} \\ \text{Hummets} \end{array}\begin{array}{c} A \quad B \\ \begin{bmatrix} 2 & 6 \\ 3 & 1 \\ 1 & 2 \end{bmatrix} \end{array}$

(b) We need to use a row vector to represent the quantities (in order for us to use matrix multiplication):

(c) Quantities: [6 2 7]

Total number of parts required $=\begin{bmatrix} 6 & 2 & 7 \end{bmatrix}\begin{bmatrix} 2 & 6 \\ 3 & 1 \\ 1 & 2 \end{bmatrix}=$

[25 52] (that is, 25 of part A and 52 of B)

(d) Cost vector: $\begin{bmatrix} \text{unit cost of }A \\ \text{unit cost of }B \end{bmatrix}=\begin{bmatrix} 2 \\ 3 \end{bmatrix}$:

Total Cost $=\begin{bmatrix} 25 & 52 \end{bmatrix}\begin{bmatrix} 2 \\ 3 \end{bmatrix}=206$

2. In matrix form, the equations are: $\begin{bmatrix} 18 & -1 \\ -2 & 36 \end{bmatrix}\begin{bmatrix} P_b \\ P_p \end{bmatrix}=\begin{bmatrix} 87 \\ 98 \end{bmatrix}$

Solving for prices:

$\begin{bmatrix} P_b \\ P_p \end{bmatrix}=\begin{bmatrix} 18 & -1 \\ -2 & 36 \end{bmatrix}^{-1}\begin{bmatrix} 87 \\ 98 \end{bmatrix}=\frac{1}{646}\begin{bmatrix} 36 & 1 \\ 2 & 18 \end{bmatrix}\begin{bmatrix} 87 \\ 98 \end{bmatrix}$

$=\frac{1}{646}\begin{bmatrix} 3230 \\ 1938 \end{bmatrix}=\begin{bmatrix} 5 \\ 3 \end{bmatrix}$

3. (a) $\begin{bmatrix} 1 & 2 & 0 \\ -5 & 4 & 3 \\ -4 & 1 & 2 \end{bmatrix}\begin{bmatrix} 5 & -4 & 6 \\ -2 & 2 & -3 \\ 11 & -9 & 14 \end{bmatrix}=\begin{bmatrix} 1 & 0 & 0 \\ 0 & 1 & 0 \\ 0 & 0 & 1 \end{bmatrix}$

so they are inverses.

(b) In matrix form, the equations are: $\begin{bmatrix} 1 & 2 & 0 \\ -5 & 4 & 3 \\ -4 & 1 & 2 \end{bmatrix}\begin{bmatrix} x \\ y \\ z \end{bmatrix}=\begin{bmatrix} 12 \\ 13 \\ -1 \end{bmatrix}$

so, using (a), the solution is:

$\begin{bmatrix} x \\ y \\ z \end{bmatrix}=\begin{bmatrix} 5 & -4 & 6 \\ -2 & 2 & -3 \\ 11 & -9 & 14 \end{bmatrix}\begin{bmatrix} 12 \\ 13 \\ -1 \end{bmatrix}=\begin{bmatrix} 2 \\ 5 \\ 1 \end{bmatrix}$

4. $\det(A)=1$, $\det(A_1)=2$, $\det(A_2)=5$, $\det(A_3)=1$, so

$x=\dfrac{\det A_1}{\det A}=2;\quad y=\dfrac{\det A_2}{\det A}=5;\quad z=\dfrac{\det A_3}{\det A}=1.$

5. Verify by the same method as in 3(a) above.

6. (a) Using the same method as 3(b) above, the solution is:

$\begin{bmatrix} x_1 \\ x_2 \\ x_3 \end{bmatrix}=-\dfrac{1}{63}\begin{bmatrix} -37 & 11 & 9 \\ 32 & -1 & -18 \\ -30 & -3 & 9 \end{bmatrix}\begin{bmatrix} 9 \\ -9 \\ 13 \end{bmatrix}=\begin{bmatrix} 5 \\ -1 \\ 2 \end{bmatrix}$

(b) $x_1=\dfrac{\det A_1}{\det A}=\dfrac{-315}{-63}=5;\ x_2=\dfrac{\det A_2}{\det A}=\dfrac{63}{-63}=-1;$

$x_3=\dfrac{\det A_3}{\det A}=\dfrac{-126}{-63}=2$

7. (a) If $A=\begin{bmatrix} 1 & -1 & -1 & 0 \\ -\alpha & 1 & 0 & 0 \\ 0 & 0 & 1 & -\gamma \\ k_1 & 0 & 0 & k_2 \end{bmatrix}$; $x=\begin{bmatrix} Y \\ C \\ I \\ r \end{bmatrix}$; $b=\begin{bmatrix} G^{\star} \\ \beta \\ \delta \\ M_S^{\star} \end{bmatrix}$, then

$Ax=b.$

We can check this by multiplying out. For example, the first row of A times the column vector x, set equal to the first element of b, gives:

$Y-C-I=G^{\star}$ or $Y=C+I+G^{\star}$ (the first equation of the model).

(b) $\det A=\begin{vmatrix} 1 & -1 & -1 & 0 \\ -\alpha & 1 & 0 & 0 \\ 0 & 0 & 1 & -\gamma \\ k_1 & 0 & 0 & k_2 \end{vmatrix}$

$=-k_1\begin{vmatrix} -1 & -1 & 0 \\ 1 & 0 & 0 \\ 0 & 1 & -\gamma \end{vmatrix}+k_2\begin{vmatrix} 1 & -1 & -1 \\ -\alpha & 1 & 0 \\ 0 & 0 & 1 \end{vmatrix}$

$=\gamma k_1+k_2(1-\alpha)$

(using the fourth row of A). Because r is the fourth element in the column vector of unknowns, we need to find $\det A_4$, the determinant of the sub-matrix that results from deleting the fourth column of A and replacing it with the vector of unknowns. We get:

$\det A_4=\begin{vmatrix} 1 & -1 & -1 & G^{\star} \\ -\alpha & 1 & 0 & \beta \\ 0 & 0 & 1 & \delta \\ k_1 & 0 & 0 & M_S^{\star} \end{vmatrix}$

$=-k_1(\beta+\delta+G^{\star})+M_S^{\star}(1-\alpha).$

Therefore $r=\det A_4/\det A=\dfrac{-k_1(\beta+\delta+G^{\star})+M_S^{\star}(1-\alpha)}{\gamma k_1+k_2(1-\alpha)}$, the required result.

(c) From (b) we have $r=\dfrac{-k_1(\beta+\delta+G^{\star})+M_S^{\star}(1-\alpha)}{\gamma k_1+k_2(1-\alpha)}.$

The partial derivative with respect to $G^{\star}$ is

$\dfrac{\partial r}{\partial G^{\star}}=\dfrac{-k_1}{\gamma k_1+k_2(1-\alpha)}.$

Since we are told that both γ and k_2 are negative, the denominator is negative and therefore the expression is positive, so an increase in government spending causes the interest rate to rise.

Progress exercise 20.1

1. (a) $y_t = A2^t$, non-convergent.

    (b) $y_t = A(0.5)^t$, convergent.

    (c) $y_t = A(-0.3)^t$, oscillating and convergent.

    (d) $y_t = A(-2)^t$, oscillating and non-convergent.

2. (a) $y_t = A2^t + 1$, non-convergent.

    (b) $y_t = A(0.5)^t + 4$, convergent to limit of 4.

    (c) $y_t = A(-0.3)^t - 3/1.3$, oscillating and convergent to $-3/1.3$.

    (d) $y_t = A(-2)^t + 100/3$, oscillating and non-convergent.

3. (a) $y_t = (3/2)3^t + (-3/2)$, non-convergent and increasing.

    (b) $y_t = (1 + 1/0.7)(0.3)^t - 1/.7$, convergent and decreasing to $-1/.7$.

    (c) $y_t = -(4 + c/1.4)(-0.4)^t + c/1.4$, oscillating and convergent to $c/1.4$.

    (d) $y_t = 20$ and constant since begins at equilibrium (but would be non-convergent if disturbed from this equilibrium).

4. (a) $y_t = (1/6)3^t + (-3/2)$, non-convergent and increasing.

    (b) $y_t = (1/0.3)(1 + 1/.7)(0.3)^t - 1/.7$, convergent and decreasing to $-1/.7$.

    (c) $y_t = -(1/0.16)(4 + c/1.4)(-0.4)^t + c/1.4$, oscillating and convergent to $c/1.4$.

    (d) $y_t = 20$ and constant since is at equilibrium in period 1 (but would be non-convergent if disturbed from this equilibrium).

Progress exercise 20.2

1. (a) At equilibrium, $Q_t^D = Q_t^S$, and $P_{t-1} = P_t$, so that $120 - 0.5P = -30 + 0.3P$, from which $P = 187.5$. (We can drop the time subscript on P when referring to the long run equilibrium value.)

    (b) By substitution, the difference equation is $P_t = (1 - 0.8\alpha)P_{t-1} + 150\alpha$.

    General solution: $P_t = A(1 - 0.8\alpha)^t + 187.5$.

    With $P_0 = 200$, solution is $P_t = 12.5(1 - 0.8\alpha)^t + 187.5$.

    (c) The behaviour of the solution depends on the value of $a = (1 - 0.8\alpha)$.
    (i) There are no oscillations when $a \geq 0$, i.e. when $0 < \alpha \leq 1.25$.
    (ii) The solution has damped oscillations when $-1 < a < 0$, i.e. when $1.25 < \alpha < 2.5$.
    (iii) The oscillations are explosive when $a < -1$, i.e. when $\alpha > 2.5$. When $\alpha = 2.5$ ($a = -1$), the solution has constant oscillations.

2. (a) $C_t = Y_t - I_0$, so $Y_t = \left(\frac{2}{3}\right)Y_{t-1} + (70 + I_0)$. So the solution is $Y_t = A\left(\frac{2}{3}\right)^t + 210 + 3I_0$. As t increases indefinitely, $\left(\frac{2}{3}\right)^t \Rightarrow 0$ so long run equilibrium is $Y_t = 210 + 3I_0$. Similarly the solution for C_t is $C_t = A\left(\frac{2}{3}\right)^t + 210 + 2I_0 \Rightarrow 210 + 2I_0$ as t increases indefinitely. When $I_0 = 20$, $Y_t \Rightarrow 270$, $C_t \Rightarrow 250$.

    (b) If I_t has been 20 for many years, Y_t must be close to its long run value $Y_t = 210 + 3I_0 = 270$ (from (a) above). Then we can restart the clock with $Y_0 = 270$ and the new level of I, 30. So the new solution for Y_t is $Y_t = A\left(\frac{2}{3}\right)^t + 210 + 3I_0$ (from (a)) with $I_0 = 30$; that is, $Y_t = A\left(\frac{2}{3}\right)^t + 300$. So the new long run equilibrium solution is $Y_t = 300$.

    To find how many years it will take for Y to rise from 270 to 290, we must solve for A. We can do this using the initial condition $Y_0 = 270$. Then in the solution we have $270 = A\left(\frac{2}{3}\right)^0 + 300$, from which $A = -30$. So the solution is now $Y_t = (-30)\left(\frac{2}{3}\right)^t + 300$. Setting $Y_t = 290$ gives $290 = (-30)\left(\frac{2}{3}\right)^t + 300$. Solving this for t gives $\left(\frac{2}{3}\right)^t = \frac{1}{3}$, Using logs, the solution is $t = 2.7$ years (approx).

3. (a) The difference equation is $Y_t = \frac{v}{v-s}Y_{t-1}$, which has solution $Y_t = Y_0\left(\frac{v}{v-s}\right)^t$.

    (b) If $v > s$, $\frac{v}{v-s} > 1$, and therefore Y_t is explosive but non-oscillating. Income expands indefinitely. (Economic analysis suggests this case is likely.) If $v < s$, $\frac{v}{v-s} < 0$, and Y_t is oscillating. Then, two sub-cases: (i) $-1 < \frac{v}{v-s} < 0$, which is true when $v < s$ and $2v < s$. Then, convergent oscillation in Y_t. (ii) $\frac{v}{v-s} < -1$, which is true when $v < s$ but $2v > s$. Then, explosive oscillation.

Progress exercise 20.3

1. (a) $y(t) = 3 - e^{-4t}$ (b) $y(t) = 1.5(1 - e^{-10t})$
    (c) $y(t) = 4(1 - e^{-t})$ (d) $y(t) = 2e^{5t}$
    (e) $y(t) = -1 + 8e^{7t}$

2. (a) $\frac{dp}{dt} = 2(q^D - q^S) = 2(0.25p - 5)$, so equilibrium price (where demand and supply are equal and price does not change) is 20.

    (b) Solution is: $p = Ae^{0.5t} + 20$, where $A = p(0) - 20$. Since $e^{0.5t}$ increases without limit as t increases, price is divergent rather than convergent. If $A > 0$, p increases without limit, while if $A < 0$, p decreases without limit. (We need to know $p(0)$ in order to determine the value of A.)

    (c) When $p(0) = 21$, $A = 1$ and $p(12) = 423.4$.

    (d) If $p_0 > 20$, then $A = p(0) - 20 > 0$, and the path of p is like figure 20.8 case 1(a). If $p_0 < 20$, then $A = p(0) - 20 < 0$, and the path of p is like figure 20.8 case 1(b). Thus the market is unstable, because if the price is initially above its equilibrium value of 20, then the price rises, while if the price is initially below its equilibrium value of 20, then the price falls.

    (e) The market is unstable (the price is divergent) because the slope of the demand function > the slope of the supply function (see figure 20.9(c)). This could be because buyers interpret a high current price as indicating that the future price will be even higher.

3. The demand curve would need to be more negatively sloped than the supply curve for the path to be stable (see figure 20.9(b)).

Answers to chapter 1 self-test

1. (a) 29

    (b) $\frac{2}{3}$

    (c) 35

    (d) 5

    (e) $\frac{1}{4}$

    (f) $2\frac{1}{2}$

    (g) $\frac{9}{100}$

    (h) 1

    (i) $\frac{15}{16}$

2. (a) $3(\frac{1}{8} - \frac{3}{11})$

    (b) $\frac{5}{16}(1 - \frac{3}{2})$

    (c) $\frac{5}{9}(1 - \frac{5}{2})$

3. (a) 417.79; (b) 418

4. (a) $\frac{1}{4}$; (b) $\frac{1}{8}$

5   (a) 0.375; (b) 0.067

6. (a) 37.5; (b) 17.5

7. (a) 216; (b) 5; (c) 2

8. (a)

Year	GDP (index 1990 = 100)
1990	100
1991	98.63
1992	98.83
1993	101.13
1994	105.60
1995	108.60

    (b) (i) 2.84; (ii) 8.60

9. (a) 22 727; (b) 17 727

Glossary

The number in each entry gives the section in which the term is first used. Cross-references within the glossary are in *italics*.

Absolute value 1.16 the value of an expression when its sign is treated as positive. Thus the absolute value of -4 is $+4$.

Absolute, proportionate and percentage change 9.2 if a variable increases from 2000 to 2100, the absolute change is $2100 - 2000 = 100$ units. The proportionate change is $\frac{2100 - 2000}{2000} = 0.05$. The percentage change is the proportionate change $\times 100 = 5\%$.

Ad valorem tax 3.16 tax whose amount depends on the value of the transaction; for example, value added tax.

Annuity method 10.13 a method of calculating the equal annual payments for a given number of years (or some other time unit) that will repay a capital sum together with the interest payments on the amount outstanding.

Arbitrary constant 18.3 the indefinite integral of a function includes an arbitrary constant; that is, a constant to which any value may be assigned. This restores the additive constant which is lost on differentiation.

Arc elasticity 9.3 for a function $y = f(x)$, $\frac{x}{y}\frac{\Delta y}{\Delta x}$, measures the *rate of proportionate change* of y per unit of proportionate change in x, when x increases by a discrete amount Δx.

Arc partial elasticity 17.5 for a *function* $z = f(x, y)$, arc partial elasticity with respect to x is $\frac{x}{z}\frac{\Delta z}{\Delta x}$, measures the *rate of proportionate change* of z per unit of proportionate change in x, when x increases by a discrete amount Δx with y held constant. Analogously, arc partial elasticity with respect to y is $\frac{y}{z}\frac{\Delta z}{\Delta y}$.

Arguments (of a function) W21.2 where a function is described as a function of a number of variables, each of these variables is alternatively called an argument of the function.

Asymptote 5.6 for a function $y = f(x)$, a straight line which a function approaches closer and closer but never quite meets; a limiting value which the function approaches.

Auxiliary (characteristic) equation W21.4 the auxiliary or characteristic equation of a difference equation is the polynomial equation that is solved by the roots of the difference equation. It thus defines the dynamic properties of the difference equation.

Base 1.14 in an expression of the general form a^b, a is the base and b is the power or exponent; we say that a is 'raised to the power b'.

Base year 1.13 when a time series such as annual value of a country's exports is expressed in index number form, the export value of a chosen year is set equal to 1 (or, more commonly, 100); this year is called the base year.

Behavioural relationship 3.15 identifies key influences on individuals' behaviour; for example, price as an influence on quantity demanded.

Break-even point 8.15 a level of a firm's output at which total cost and total revenue are equal and profits are therefore zero.

Budget constraint 5.2 same as *budget line*.

Budget line 5.2 with goods X and Y on the axes, a downward sloping line giving all the combinations of goods the consumer can buy with a given budget B, given the prices p_x and p_y. The equation of the budget line is $p_x X + p_y Y = B$. Generalizes to any number of goods.

Calculus see *differentiation*.

Cancelling 1.6 eliminating a multiplicative factor common to both numerator and denominator; for example, $\frac{6}{10}$ becomes $\frac{3}{5}$ after cancelling the common factor, 2.

Capital intensity 14.8 for any production function, capital intensity may be defined as $\frac{K}{L}$ (capital per unit of labour), or as $\frac{K}{Q}$ (capital per unit of output).

Capitalization 10.10 finding the *present discounted value* of a series of future receipts or payments.

Chord 6.4 a straight line joining two points on a curve.

Cobweb model 20.4 a dynamic supply and demand model in which the path of price through time resembles, when graphed, a spider's web.

Coefficient 3.3 a multiplicative *parameter*; for example, in $y = 3x$, 3 is the coefficient of x.

Cofactor 1.5 in elementary algebra, 5 and 3 are cofactors of 15 because $5 \times 3 = 15$. In matrix algebra 19.9, a *minor* $|M_{ij}|$ which has been multiplied by -1 if $i + j$ is an odd number.

Column vector 19.2 a column vector is a matrix with 1 column; a row vector is a matrix with 1 row.

Common denominator 1.7 the fractions $\frac{1}{2}$ and $\frac{1}{3}$ have a common denominator, 6, because 6 is divisible by 2 and 3 without remainder. Similarly $\frac{a}{b}$ and $\frac{c}{d}$ have a common denominator bd.

Common factor 1.5 ab and ac have a common multiplicative term, a, called a common factor.

Common logarithm 11.3 if x is the common logarithm of y, then, by definition, $10^x = y$.

Comparative statics 3.16 methodology that compares an initial *equilibrium* in an *economic model* with the new equilibrium that follows a shift in an exogenous variable or parameter.

Complementary function 20.2 in a *difference or differential equation* that is not homogeneous, the complementary function is the complete solution to the reduced-form equation, this latter being the equation obtained by dropping the constant term from the given equation.

Complex numbers W21.4 a number composed of a real part and an imaginary part; the latter is a multiple of i, the square root of -1.

Consumers' surplus 18.8 the maximum amount consumers in aggregate would willingly pay for a given quantity of a good, rather than go without it entirely, minus the amount they actually have to pay.

Consumer's surplus the same as *consumers' surplus* but referring now to an individual consumer.

Consumption function 3.17 in a Keynesian macroeconomic model, *behavioural relationship* explaining how expenditure on consumption by a person (or the personal sector in aggregate) is determined. See also *savings function*.

Continuous variable 3.6 variable that can take any value; contrasts with *discrete variable*, which can vary only in jumps.

Contour see *section*.

Convex or concave function 7.13 a *function* $y = f(x)$ is convex (concave) from below if the second derivative is positive (negative).

Coordinates 3.6 in two-dimensional space where variables x and y are measured along the axes, the unique value of x and the unique value of y associated with any point are called the coordinates of that point; in n-dimensional space the coordinates of any point are the unique values of each of the variables $x_1, \ldots, x_n$ associated with that point.

Cross partial derivative 14.6 for a *function* $z = f(x, y)$, the cross partial derivative $\frac{\partial^2 z}{\partial y \partial x}$ measures how the slope in the x direction is affected by a small increase in y; and similarly for $\frac{\partial^2 z}{\partial y \partial y}$.

Decreasing marginal rate of substitution of a convex *isoquant* 14.8, characteristic that successive increases in labour input require smaller and smaller decreases in capital input to maintain constant output; of a convex *indifference curve* 14.12, characteristic that successive increases in consumption of good X require smaller and smaller decreases in consumption of good Y to maintain constant utility.

Definite integral 18.1 for a function $y = f(x)$, the area bounded by the curve and the x-axis, between two values of x, $x = a$ and $x = b$.

Demand function 3.14 *behavioural relationship* between quantity of a good demanded by consumers and independent variables that influence quantity demanded, notably price.

Demand price 3.15 maximum price at which a given quantity of a certain good will be demanded.

Denominator 1.6 in any fraction or ratio $\frac{a}{b}$, a is called the numerator and b the denominator.

Dependent variable 3.5 in a function of the form $y = f(x)$, y is considered to be the dependent variable because its value is determined as soon as a value for x is substituted into the equation, whereas the converse is not always true. Causation running from x to y may or may not be implied. See also *independent variable*.

Derivative 6.5 for a function $y = f(x)$, the derivative $\frac{dy}{dx}$ measures the rate of change of y (the slope of a curve) at any point.

Determinant 19.9 a *scalar* associated with any square *matrix*.

Difference equation 20.2 a relationship between current and earlier values of a variable that *varies discretely* through time.

Difference quotient 6.2 for a function $y = f(x)$, $\frac{\Delta y}{\Delta x}$; measures the rate of change associated with a discrete change in x, Δx.

Differential 7.15 for a function $y = f(x)$, the differential is given by $dy = f'(x)dx$, where $f'(x)$, is the *derivative* of $f(x)$ and dx and dy are small changes in x and y respectively. See also *total differential*.

Differential equation 20.6 a relationship between the current value of a *variable* and its *derivative*(s), when the variable varies continuously through time.

Differentiation 6.1 application of technique for finding the *derivative* of a function.

Diminishing marginal productivity 14.8 of a production function, characteristic that successive increases in one input (with others held constant) yield smaller and smaller increases in output.

Diminishing marginal utility 14.12 of an individual's utility function, characteristic that successive increases in consumption of one good (with others held constant) yield smaller and smaller increases in utility or satisfaction.

Discontinuity 5.6 for a *function* $y = f(x)$, a jump or gap where there is no finite value of y corresponding to a specific value of x.

Discrete growth 10.6 growth in a *variable* that proceeds in discrete jumps, rather than smoothly and continuously.

Discrete variable see *continuous variable*.

Dynamic economics see *economic dynamics*.

Dynamic instability when the *solution* to a *difference* or *differential equation* shows that the *dependent variable* diverges increasingly from its equilibrium value as time passes.

Dynamic model see *economic dynamics*.

Dynamic stability 20.8 when the *solution* to a *difference* or *differential equation* shows that the dependent variable approaches closer and closer to its equilibrium value as time passes.

Economic dynamics 20.4 economic relationship(s) in which values of variables vary in a specified way through time.

Economic model 3.15 aims to identify the key features of an economic process and to draw conclusions.

Elasticity 9.4 a *function* $y = f(x)$ is elastic if $\frac{x}{y}\frac{dy}{dx} > 1$, inelastic if $\frac{x}{y}\frac{dy}{dx} < 1$, and of unit elasticity if $\frac{x}{y}\frac{dy}{dx} = 1$.

Elasticity of substitution W21.3 the proportionate response of a ratio of two input quantity variables to a proportionate change in their marginal rate of substitution. When the marginal rate of substitution is equal to a ratio of prices, the elasticity of substitution measures the responsiveness of the cost-minimizing mix of input quantities to changes in their relative prices.

Element of a matrix 19.2 an entry in a specific row and column of a *matrix*.

Elementary operations 3.2 addition, subtraction, multiplication, division, raising to any power, or taking the logarithm.

Endogenous variable 3.17 a *variable* whose value is determined by solving the system of equations in which it is embedded; see also *exogenous variable*.

Envelope W21.3 the upper (or lower) extreme of a set of curves.

Equation 3.1 any mathematical expression containing an 'equals' sign.

Equilibrium condition 3.14 an equation which, if satisfied, implies no tendency for any *endogenous variable* to change through time.

Excess demand 3.14 quantity demanded minus quantity supplied at any given price.

Excess supply 3.14 quantity supplied minus quantity demanded at any given price.

Exogenous variable 3.17 a *variable* whose value is determined independently of the system of equations in which it is embedded; see also *endogenous variable*.

Expanding 1.4 the operation of multiplication performed on a product; for example, the expansion of $(x+2)(x+3)$ is $x^2 + 5x + 6$.

Expansion path 17.4 with K and L on the axes, shows how quantities of K and L chosen by the firm vary as output is increased, with w and r constant.

Explicit function 3.7 a function in which the *dependent variable* is identified by being isolated on one side of the equation; not an *implicit function*.

Explosive growth or decline see *monotonic divergence*.

Explosive oscillation see *oscillatory divergence*.

Exponent see *base*.

Factor shares 17.8 shares in total revenue received by the various factors of production such as labour and capital.

Factorization 1.5 separating out a *common factor*; for example, in $ab + ac$, a is a common factor, and after factorization $ab + ac$ becomes $a(b + c)$.

First order condition 7.8 for a *function* $y = f(x)$, the first order condition for a stationary value is that the first derivative $f'(x)$ should equal zero at the point in question. If the function has two or more independent variables, then first order conditions require that all partial derivatives of the function should be zero.

First order difference equation 20.2 a *difference equation* in which the current value of the variable is related to its value one period earlier.

First order differential equation 20.6 a differential equation in which the current value of a variable is related to its first derivative.

First order partial derivative 14.4 for a *function* $z = f(x, y)$, the first order partial derivatives $\frac{\partial z}{\partial x}$ and $\frac{\partial z}{\partial y}$ measure the slope in the x and y directions respectively. Generalizes to many variables.

Fixed costs 4.14 a component of a firm's total costs which, over a relevant time interval, remains the same whatever the level of output may be.

Fraction 1.3 any number (except 0) divided by any other number (except 0) is called a fraction or ratio.

Function 3.5 an *equation* specifying a relationship between two or more variables.

Future compounded value 10.4 the value at some future date of a given capital sum invested now, together with compounded interest.

General solution 20.2 the general solution to a difference or differential equation of order n is any expression that contains n arbitrary constants and which, when substituted into the given equation, turns it into an identity.

Hessian matrix W21.2 a matrix where the element in the ith row and jth column is the second order derivative of a function, the first derivative being with respect to the ith variable in the function, and the second derivative being then taken with respect to the jth variable in the function.

Homogeneous difference equation 20.2 or differential equation 20.6 an equation that contains no additive constant.

Identity 3.1 an equation that is satisfied whatever value(s) is (are) assigned to the variables that appear in it; denoted by the symbol '≡'.

Implicit differentiation 15.7 technique for finding the *derivative* of an *implicit function*. If the function has more than two variables, the technique finds the partial derivatives and is then called implicit partial differentiation.

Implicit function 3.7 a *function* of the form $f(x, y) = 0$ which defines a relationship of mutual dependency between two or more variables; not an *explicit function*.

Inconsistent equations 3.11 two or more *equations* such that there is no combination of values of the *endogenous variables* that can satisfy all of the equations simultaneously.

Increment 6.2 increase.

Indefinite integral 18.3 for a *function* $y = f(x)$, the indefinite integral is $F(x) + c$, where $f(x)$ is the *derivative* of $F(x)$ and c is an arbitrary constant.

Independent equations 3.11 given two equations, if one can be obtained from the other by means of *elementary operations*, then the two are not independent. Given three equations, if one can be obtained by combining the other two by means of elementary operations, then the three are not independent; and so on.

Independent variable 3.5 in a function of the form $y = f(x)$, x is considered to be the independent variable; see also *dependent variable*.

Index confusingly, index may mean an *index number*, but also occasionally means a *power* or *exponent*.

Index number series 1.13 a set of related numbers in which one number is assigned the value 1 (or, more commonly, 100) and all other numbers are measured relative to this.

Indifference curve 14.12 an iso-*U* section of the utility function $U = f(X, Y)$; thus the equation of an indifference curve is $U_0 = f(X, Y)$, where U_0 is the constant level of utility.

Initial condition 20.2 in the context of a difference or *differential equation*, an initial condition is any given value of the *dependent variable* during a specific time period (in the case of a difference equation) or at a specified moment in time (in the case of a differential equation).

Instantaneous rate of growth 13.6 for a *function* $y = f(x)$, where x denotes time, instantaneous rate of growth is $\frac{1}{y}\frac{dy}{dx}$; measures at a point in time the rate

of *proportionate change* of y per unit of absolute change in x. See also *rate of growth*.

Inverse function 3.5 given a *function* $y = f(x)$ in which y is the *dependent variable*, the function inverse to this makes x the dependent variable and is often written as $x = f^{-1}(y)$.

Investment multiplier 3.17 in a Keynesian macroeconomic model, the investment multiplier gives the change in aggregate income resulting from a small change in investment.

Iso-*x*, iso-*y* and iso-*z* sections 14.3 for a *function* $z = f(x, y)$, an iso-*x* section is a set of points at which x is constant; thus the equation of an iso-*x* section is $z = f(x_0, y)$, where x_0 is the constant value of x. Similarly, the equations of an iso-*y* and iso-*z* section are $z = f(x, y_0)$ and $z_0 = f(x, y)$ respectively. See also *section*.

Isocost line 16.6 with inputs K and L on the axes, a downward-sloping line giving all the combinations of inputs that have the same total cost, TC_0, given the input prices w and r. The equation of the isocost line is $wL + rK = TC_0$. Generalizes to any number of inputs.

Isoquant 14.8 an iso-*Q* section of the production function $Q = f(K, L)$; thus the equation of an isoquant is $Q_0 = f(K, L)$, where Q_0 is the constant level of output.

Iterative methods 5.3 techniques for solving an equation by 'trial and error'.

Lagged supply function 20.4 a *supply function* in which the quantity supplied in any period varies, not with the price in that period, but with the price of an earlier period.

Limit, limiting value 5.6 for a *function* $y = f(x)$, a value that y approaches closer and closer, but does not reach, as x becomes larger and larger in absolute value.

Linear approximation 7.15 for a *function* $y = f(x)$, the *differential* gives the change in y that would result from a small change in x, if $f(x)$ were linear in the neighbourhood of the point in question; this is an approximation to the actual change and is called a linear approximation. For a function $z = f(x, y)$, the *total differential* is similarly a linear approximation.

Linear difference equation 20.2 a *difference equation* in which neither the current nor lagged values of the dependent variable are raised to any power other than 1.

Linear differential equation 20.6 a *differential equation* in which neither the dependent variable nor any of its derivatives are raised to any power other than 1.

Log linear 13.9 a *function* $y = f(x)$ is (semi) log linear if its graph is a straight line when plotted with log y on the vertical axis and x on the horizontal axis.

The function is log-log, or double log, linear if its graph is a straight line when plotted with log y on the vertical axis and log x on the horizontal axis.

Marginal products 14.8 for the production function $Q = f(K, L)$, the marginal products of inputs K and L are measured by the partial derivatives f_K and f_L respectively.

Marginal productivity same as *marginal product*.

Marginal productivity conditions 16.7 conditions necessary for profit maximization: *marginal product* of each input should equal its real marginal cost.

Marginal revenue product 16.7 *marginal product* of an input multiplied by the price of the product. For a perfectly competitive firm, measures (approximately) the additional revenue from a small increase in the input in question. Same as *value of the marginal product*.

Marginal utilities 14.12 for the utility function $U = f(X, Y)$, the marginal utilities of goods X and Y are measured by the *partial derivatives* $\frac{\partial U}{\partial X}$ and $\frac{\partial U}{\partial Y}$ respectively.

Market value 10.9 value of an asset, good or service if sold on the open market.

Matrix 19.2 an arrangement of numbers or symbols into rows and columns.

Matrix inversion 19.10 the inverse matrix of the matrix $\mathbf{A}$, written $\mathbf{A}^{-1}$, has the property that $\mathbf{A}\mathbf{A}^{-1} \equiv \mathbf{A}^{-1}\mathbf{A} = \mathbf{I}$, where $\mathbf{I}$ is the unit matrix. The inverse may not exist.

Minor 19.9 an element a_{ij} of a matrix $\mathbf{A}$ has an associated minor $|\mathbf{M}_{ij}|$ which is the determinant of the sub-matrix $\mathbf{M}_{ij}$ formed by deleting row i and column j of $\mathbf{A}$.

Monopoly 4.14 a market in which there is only one producer of the product in question.

Monotonic convergence 20.3 when, as time passes, the *dependent variable* in a *difference* or *differential equation* approaches closer and closer, without oscillation, to its long-run equilibrium value.

Monotonic divergence 20.3 when, as time passes, the *dependent variable* in a *difference* or *differential equation* departs increasingly, without oscillation, from its long-run equilibrium value.

Multiplying out see *expanding*.

Natural logarithm 12.6 if x is the natural logarithm of y, then, by definition, $e^x = y$, where $e \equiv \lim_{n \to \infty} (1 + \frac{1}{n})^n$.

Non-conformability 19.5 two matrices $\mathbf{A}$ and $\mathbf{B}$ are said to be non-conformable if the products $\mathbf{AB}$ and $\mathbf{BA}$ do not exist.

Non-satiation 14.12 characteristic of an individual's utility function, that *marginal utility* is always positive.

Null matrix 19.2 a matrix in which every element is zero.

Numerator see *denominator*.

Objective function 16.2 a *function*, say $z = f(x, y)$, for which we seek the maximum or minimum value when the values of x and y are in some way constrained.

Opportunity cost 5.12 of a given quantity of a good X is the cost of the most desired alternative that the buyer must forgo in order to obtain that quantity of X.

Optimization 7.3 process of finding *stationary points* of a function.

Order 19.2 a matrix is of order $m \times n$ if it has m rows and n columns.

Origin 3.5 the zero point on a graph from which all distances along the horizontal and vertical axes are measured.

Oscillatory convergence 20.3 when, as time passes, the *dependent variable* in a *difference* or *differential equation* approaches closer and closer to its long-run equilibrium value but is alternately above or below it.

Oscillatory divergence 20.3 when, as time passes, the dependent variable in a difference or differential equation diverges further and further from its long-run equilibrium value and is alternately above or below it.

Parabola 4.8 the graph of $y = ax^2 + bx + c$ is a parabola that is symmetrical around an axis parallel with the y-axis.

Parameter 3.3 in a general functional form such as the linear function $y = ax + b$ the *variables* are x and y and the parameters are a and b. A specific case of the general form is obtained by assigning fixed values to the parameters; for example, if $a = 3$ and $b = 4$, we obtain $y = 3x + 4$.

Partial differential 15.4 the *total differential* when x or y alone changes.

Particular solution same as *unique solution*.

Percentage 1.12 when we say that 10 is 50 per cent of 20, we are expressing 10 as a percentage of 20.

Perfect competition 8.12 a market in which there are many small producers of a homogeneous product, with the consequence that no individual producer can affect the ruling market price but each producer can sell as much or as little as it wishes at that price.

Perfect square 4.7 an expression of the form $(x + a)^2$, which measures the area of a square with sides $x + a$.

Perpetual bond 10.12 a bond with no fixed redemption date (that is, date at which it will be redeemed (repurchased) by the person or organization that issued it).

Plane 14.2 in three-dimensional space, a set of points with one coordinate in common. For example, points in the $0zx$ plane have a y coordinate of $y = 0$.

Point elasticity 9.8 for a *function* $y = \mathrm{f}(x)$, point elasticity is $\frac{x}{y}\frac{dy}{dx}$; measures at any point the *rate of proportionate change* of y per unit of proportionate change in x. Generalizes to many variables.

Point of inflection 7.10 for the *function* $y = \mathrm{f}(x, y)$, a point at which the slope reaches a maximum or minimum value. If the maximum or minimum value of the slope is zero, the point of inflection is also a *stationary point*.

Point partial elasticity 17.5 for a *function* $z = \mathrm{f}(x, y)$, point partial elasticity with respect to x is $\frac{x}{z}\frac{\partial z}{\partial x}$; measures at any point the *rate of proportionate change* of z per unit of proportionate change in x, with y held constant. Analogously, point partial elasticity with respect to y is $\frac{y}{z}\frac{\partial z}{\partial y}$.

Polynomial function 5.3 with a, b, c, and d constants, $y = ax^2 + bx + c$ is a polynomial of degree 2; $y = ax^3 + bx^2 + cx + d$ is a polynomial of degree 3; and so on.

Post-multiplication see *pre-multiplication*.

Power see *base*.

Pre-multiplication 19.5 if we write AB, the matrix B is pre-multiplied by the matrix A; if we write BA, the matrix B is post-multiplied by the matrix A.

Preferences 14.12 an individual's subjective ranking of various bundles of goods or services according to their desirability.

Present discounted value 10.8 the capital sum required now to generate a given future series of receipts or payments, given the interest rate (discount rate). Same as *present value*.

Present value same as *present discounted value*.

Price adjustment equation 20.8 specifies the process of price adjustment through time.

Principal 10.4 a capital sum borrowed or lent.

Producers' surplus 18.9 the amount of money producers in aggregate actually receive for a given quantity of a good, minus the minimum amount they would willingly accept rather than sell nothing at all.

Product 1.3 the result of multiplying one number by another; the product of 3 and 5 is 15.

Product wage 16.7 $\frac{w}{p}$, where w = money wage and p = price of the product produced by the firm in question. Measures the cost to the firm, in units of output, of hiring one more unit of labour.

Programming W21.2 mathematical programming is the set of constrained optimization techniques that includes evaluating a finite number of possible candidates for an optimum and comparing these according to some algorithm.

Proportion 1.11 when we divide 10 by 20 we are expressing 10 as a proportion of 20; the result is $^1/_2$.

Quadratic expression 4.2 an expression of the form $ax^2 + bx + c$, where a, b, and c are constants.

Quasi-concavity W21.3 if a *function* is increasing in its variables, then quasi-concavity implies that its contour lines are convex to the origin.

Quotient 1.3 when one number is divided by another, the result is called the quotient. Same as *ratio*.

Rate of change 6.2 for a *function* $y = \mathrm{f}(x)$, the change in y per unit of change in x.

Rate of growth 13.5 for a *function* $y = \mathrm{f}(x)$, the proportionate change in y per unit of *absolute* change in x. See also *instantaneous rate of growth*.

Rate of proportionate change 9.5 same as *elasticity*.

Ratio 1.3 same as *quotient*; 1.11 same as *fraction* or *proportion*; the ratio of 10 to 20 is $^1/_2$.

Ray 8.3 a straight line from the origin in a graph.

Real cost, or real reward, of capital 16.7 analogue of *real wage* with w replaced by r, the money cost of hiring one more unit of capital.

Real income 5.12 the good or bundle of goods that a given money income can buy.

Real wage 16.7 real wage $= \frac{w}{p}$ where w = money wage. If p = price of the product produced by a firm, then real wage has the same meaning as *product wage*. If p = index (average) of prices generally, $\frac{w}{p}$ measures cost in units of general purchasing power of hiring one more unit of labour.

Reciprocals 1.8 a and $\frac{1}{a}$ are reciprocals because $a \times \frac{1}{a} \equiv 1$ for any a (except $a = 0$).

Reduced form 3.15 if two simultaneous equations are reduced to one by substituting from one into the other, the resulting equation is called a reduced form; similarly, if three equations are reduced to two in the same way; and so on; see also *substitution*. See also *complementary function* for another meaning.

Relation 3.5 a broader term than *function* as the former includes cases where there is more than one value of the dependent variable associated with a given value of the independent variable. For example, $y = \pm\sqrt{x}$ is a relation.

Roots 4.4 the solutions to a quadratic equation, a cubic equation, or higher degree polynomial equation of the form $a_n x^n + a_{n-1} x^{n-1} + \cdots + a_0 x^0 = 0$.

Rounding 1.9 for simplicity, using an approximation in place of a number; for example, 1.85 becomes 2 when rounded to the nearest whole number (integer).

Row vector see *column vector*.

Saddle point 15.3 for the function $z = \mathrm{b}(x, y)$, a *stationary point* such that some small variations in x and y take us uphill, and others downhill.

Savings function 3.17 in a Keynesian macroeconomic model, *behavioural relationship* explaining how savings by a person (or the personal sector in aggregate) is determined. The sum of savings and consumption expenditure necessarily equals income. See also *consumption function*.

Scalar 19.6 in matrix algebra, a scalar is a number; that is, it is neither a matrix nor a vector.

Second order condition 7.8 for a *function* $y = f(x)$, the second order condition for a maximum or minimum is that the second derivative $f''(x)$ should be negative or positive, respectively, at the point in question. If the function has many variables, additional second order conditions are required relating to changing more than one variable at once; these are stated in terms of determinants of a sequence of sub-matrices of the *Hessian matrix*.

Second order partial derivatives 14.6 for a *function* $z = f(x, y)$ the direct second order partial derivative $\frac{\partial^2 z}{\partial x^2}$ measures how the slope in the x direction is affected by a small increase in x; and similarly for $\frac{\partial^2 z}{\partial y^2}$. See also *cross partial derivative*.

Section 14.3 for a *function* $z = f(x, y)$, any set of points lying on the surface and chosen according to some rule; for example, points with the same value of x. Also known as a *contour*. See also **iso-x, iso-y, and iso-z sections**.

Short-run production function 14.8 for the production function $Q = f(K, L)$, the *function* obtained by setting the capital input, K, equal to some constant, reflecting the assumption that by its nature most types of capital input cannot quickly be varied.

Singular matrix 19.9 a square *matrix* of which the determinant is zero.

Sinking fund 10.13 method of repaying a loan together with interest due in which all periodic instalments except the last are equal.

Slack variable W21.2 the addition of a non-negative slack variable to a non-strict inequality can transform this into an *equation*.

Solution 3.1 for one *equation* containing one *variable* or *unknown*, the solution is any value or values for that variable which, when substituted into the given equation, turn the equation into an *identity*. For n independent simultaneous equations containing n variables, a solution is any set of values for the variables which, when substituted into the given equations, turn them into *identities*.

Specific tax 3.16 tax whose amount depends on the quantity sold.

Square matrix 19.2 a *matrix* with an equal number of rows and columns.

Square root 1.14 $b > 0$ is a square root of a if $b \times b = a$.

Stationary point 7.4 for a *function* $y = f(x)$, a point at which the tangent is horizontal and therefore $\frac{dy}{dx} = 0$. A stationary point may be a maximum, minimum, or *point of inflection*.

Step function 3.6 a function of the form $y = f(x)$ in which x is a *discrete variable* and consequently y varies in jumps rather than smoothly; the graph of this relationship resembles a flight of stairs.

Strictly decreasing function 7.2 for a function $y = f(x)$, an increase in x always results in a decrease in y; the graph is downward sloping from left to right.

Strictly increasing function 7.2 for a *function* $y = f(x)$ an increase in x always results in an increase in y; the graph is upward sloping from left to right.

Subscript, superscript 3.3 if we write, for example, x_1, then the 1 is called a subscript and is often used to refer to a specific value of the variable x. A superscript such as x^2 may be used in the same way. A superscript should not be confused with a power or exponent.

Substitution 3.9 if we know that, for example, $y = x^2$ and also that $x = 3z$, then, by using the second equation to substitute for x in the first, we can obtain $y = (3z)^2$.

Supply function 3.14 *behavioural relationship* between quantity of a good supplied by producers and independent variables that influence quantity supplied, notably price.

Supply price 3.15 minimum price at which a given quantity of a certain good will be supplied.

Symmetrical function 14.3 a function such as $z = x^2 + 2xy + y^2$, in which interchanging x and y leaves the function unchanged.

Table of values 3.6 for any function, shows values of the dependent variable associated with the value(s) assigned to the independent variable(s); these values may then be used to plot the graph of the function.

Tangent 6.4 a straight line that touches a curve at one and only one point.

Taylor's approximations W21.2 the polynomial approximation found from Taylor's theorem, disregarding the remainder or error term.

Taylor's theorem W21.2 states the remainder (or error) in approximating any function by a polynomial function of order n, for any n.

Time path 20.3 in a *difference* or *differential equation*, the successive values of the dependent variable as time passes.

Time series 1.13 any series of observations of a variable taken at successive intervals, such as monthly coal production.

Time-series data see *time series*.

Total derivative 15.5 the *derivative* $\frac{dy}{dx}$ in a case where, for example, $z = f(x, y)$ and $y = f(x)$, so that z depends solely on x.

Total differential 15.4 for a *function* $z = f(x, y)$, the total differential dz is defined as $dz = \frac{\partial z}{\partial x}dx + \frac{\partial z}{\partial y}dy$. It measures approximately the change in z, dz, that follows a small change in x, dx, and a simultaneous small change in y, dy. See also *differential*.

Transitive (consumer preferences) 14.12 if there are two goods X and Y, and a consumer prefers the bundle (X_1, Y_1) to the bundle (X_2, Y_2), and also prefers the bundle (X_2, Y_2) to the bundle (X_3, Y_3); then, if her preferences are transitive, she will also prefer the bundle (X_1, Y_1) to the bundle (X_3, Y_3).

Unique solution 20.2 (also known as a particular solution) to a *difference* or *differential equation* of order n is any expression which contains fewer than n arbitrary constants and which, when substituted into the given equation, turns it into an identity. See also *general solution*.

Unit or identity matrix 19.2 a *matrix* in which every element in the leading diagonal is 1 and all other elements are zero.

Unknown 3.3 a *variable* when embedded in an *equation*.

Value of marginal product same as *marginal revenue product*.

Variable 3.3 an unspecified number, usually denoted by a letter from the end of the alphabet such as x.

Variable costs 8.2 elements of a firm's costs that vary with output (in contrast to *fixed costs*).

Index